CRC Handbook of Materials Science

VOLUME III

NONMETALLIC MATERIALS AND APPLICATIONS

Editor
Charles T. Lynch, Ph.D.
Senior Scientists for Environmental Effects
Metals Behavior Branch, USAF
Air Force Materials Laboratory
Wright-Patterson Air Force Base, Ohio

CRC

CRC Press, Inc.
Boca Raton, Florida

Library of Congress Cataloging in Publication Data
Lynch, Charles T.
CRC Handbook of materials science.

Vol. 2 includes bibliography and index.
CONTENTS:
-v. 2. Metals, composites, and refractory materials.
1. Materials–Handbooks, manuals, etc.
TA403.4L94 620.1'12 73-90240
ISBN 0-8493-0233-1 (Vol. III)

This book represents information obtained from authentic and highly regarded sources. Reprinted material is quoted with permission, and sources are indicated. A wide variety of references are listed. Every reasonable effort has been made to give reliable data and information, but the author and the publisher cannot assume responsibility for the validity of all materials or for the consequences of their use.

Direct all inquiries to CRC Press, Inc., 2000 Corporate Blvd., N.W., Boca Raton, Florida, 33431.

Second Printing, 1983
Third Printing, 1984

International Standard Book Number 0-8493-0233-1 (Volume III)
International Standard Book Number 0-8493-0230-7 (Complete Set)
Former International Standard Book Number 0-87819-233-6 (Volume III)
Former International Standard Book Number 0-87819-234-4 (Complete Set)

Library of Congress Card Number 73-90240
Printed in the United States

PREFACE

It has been the goal of the *CRC Handbook of Materials Science* to provide a current and readily accessible guide to the physical properties of solid state and structural materials. Interdisciplinary in approach and content, it covers the broadest variety of types of materials consistent with a reasonable size for the volumes, including materials of present commercial importance plus new biomedical, composite, and laser materials. This volume, Nonmetallic Materials and Applications, is the third of the three-volume *Handbook*; General Properties is the first, and Metals, Composites, and Refractory Materials is the second.

During the approximately four years that it has taken to formulate and compile this *Handbook*, the importance of materials science has taken on a new dimension. The term "materials limited" has come into new prominence, enlarged from the narrower consideration of technical performance of given materials in given conditions of stress, environment, and so on, to encompass the availability of materials in commerce at a reasonable price. Our highly industrialized society, with its immense per capita consumption of raw materials, today finds itself facing the long-prophesized shortages of materials in many diverse areas of our economy. Those future shortages have become today's problems. As we find ourselves "materials limited" with respect to availability and price, with a growing concern for where our raw materials come from and how supplies may be manipulated to our national disadvantage, increased economic utilization of all our resources, and particularly our materials resources, becomes an American necessity. With this changing background the purpose for this type of compilation has broadened beyond a collection of data on physical properties to one of concern for comparative properties and alternative employment of materials. Therefore, at this time it seems particularly appropriate to enter this new addition to the CRC Handbook Series.

Most of the information presented in this *Handbook* is in tabular format for easy reference and comparability of various properties. In some cases it has seemed more advisable to retain written sections, but these have been kept to a minimum. The importance of having critically evaluated property data available on materials to solve modern problems is well understood. In this *Handbook* we seek to bridge the gap between uncritical data collections carrying all the published information for a single material class and general reference works with only limited property and classification data on materials. On the basis of advice from many and varied sources, numerous limitations and omissions have been necessary to retain a reasonable size. This reference is particularly aimed at the nonexperts, or those who are experts in one field but seek information on materials in another field. The expert normally has his own specific original sources available to guide him in his own area of expertise. He often needs assistance, however, to get started on something new. There is also considerable general information of interest to almost all scientists, engineers, and many administrators in the field of materials and materials applications. Comments and suggestions, and the calling to our attention of typographical errors, will be welcomed and are encouraged.

My sincere thanks is extended to all who have advised on the formulation, content, and coverage of this *Handbook*. I am grateful to many colleagues in industry, academic circles, and government for countless suggestions and specific contributions, and am particularly indebted to the Advisory Board and Contributors who have put so much of their time, effort, and talent into this compilation. Special appreciation is extended to the editorial staff of CRC Press, to Karen A. Gajewski, the Administrative Editor, and to Gerald A. Becker, Director of Editorial Operations.

I want to pay special tribute to my wife, Betty Ann, for her magnificant patience, encouragment, and assistance, and to our children, Karen, Ted Jr., Richard, and Thomas, for giving their Dad some space, quiet, and assistance in the compilation of a considerable amount of data.

Charles T. Lynch
Fairborn, Ohio
April 1975

THE EDITOR

Charles T. Lynch, Ph.D., is Senior Scientist for Environmental Effects in the Metals Behavior Branch of the Air Force Materials Laboratory, Wright-Patterson Air Force Base, Ohio.

Dr. Lynch graduated from the George Washington University in 1955 with a B.S. degree in chemistry. He received his M.S. and Ph.D. degrees in analytical chemistry in 1957 and 1960, respectively, from the University of Illinois, Urbana.

Dr. Lynch served in the Air Force for several years before joining the Air Force Materials Laboratory as a civilian employee in 1962. Prior to his current position, he served as a research engineer, group leader for ceramic research, and Chief of the Advanced Metallurgical Studies Branch.

Dr. Lynch is a member of the American Chemical Society, American Ceramic Society, American Association for the Advancement of Science, Ohio Academy of Science, New York Academy of Science, Sigma Xi-RESA, and the Metallurgical Society of the AIME. He holds 13 patents and has published more than 60 research papers, over 70 national and international presentations, and one book, *Metal Matrix Composites,* written with J. P. Kershaw and published by The Chemical Rubber Company (now CRC Press) in 1972.

CONTRIBUTORS

C. Howard Adams
SPI Research Associate
National Bureau of Standards
Washington, D.C. 20234

Allen M. Alper
Director of Research and Engineering
Chemical and Metallurgical Division
GTE Sylvania, Incorporated
Towanda, Pennsylvania 18848

Ray E. Bolz
Vice President and Dean of the Faculty
Worcester Polytechnic Institute
Worcester, Massachusetts 01609

Allen Brodsky
Radiation Physicist
Mercy Hospital
Pittsburgh, Pennsylvania 15219

D. F. Bunch
Atomics International
Canoga Park, California 91304

Donald E. Campbell
Senior Research Associate – Chemistry
Research and Development Division, Technical Staffs Services Laboratories
Corning Glass Works
Sullivan Park
Corning, New York 14830

William B. Cottrell
Director, Nuclear Safety Program
Oak Ridge National Laboratory
Oak Ridge, Tennessee 37830

Joseph E. Davison
Assistant Professor of Materials Engineering
University of Dayton
300 College Park
Dayton, Ohio 45409

Edward B. Fernsler (retired)
114 Willoughby Avenue
Huntington, West Virginia 25705

Francis S. Galasso
Chief, Materials Science
United Aircraft Research Laboratories
East Hartford, Connecticut 06108

Henry E. Hagy
Senior Research Associate – Physics
Research and Development Division, Technical Staffs Services Laboratories
Corning Glass Works
Sullivan Park
Corning, New York 14830

C. R. Hammond
Emhart Corporation
P.O. Box 1620
Hartford, Connecticut 06102

Michael Hoch
Professor, Department of Materials Science and Metallurgical Engineering
University of Cincinnati
Clifton Avenue
Cincinnati, Ohio 45221

Bernard Jaffe
Vernitron Piezoelectric Division
232 Forbes Road
Beford, Ohio 44146

Richard N. Kleiner
Section Head, Ceramics Department
Precision Materials Group
Chemical and Metallurgical Division
GTE Sylvania, Incorporated
Towanda, Pennsylvania 18848

George W. Latimer, Jr.
Group Leader, Analytical Methods
Mead Johnson Company
Evansville, Indiana 47721

Robert I. Leininger
Project Director, Biomaterials
Biological, Ecological, and Medical Sciences Department
Battelle/Columbus Laboratories
505 King Avenue
Columbus, Ohio 43201

Robert S. Marvin
Office of Standard Reference Data
National Bureau of Standards
U.S. Department of Commerce
Washington, D.C. 20234

Eugene F. Murphy
Director, Research Center for Prosthetics
U.S. Veterans Administration
252 Seventh Avenue
New York, New York 10001

A. Pigeaud
Research Associate
Department of Metallurgy and Material Science
University of Cincinnati
Clifton Avenue
Cincinnati, Ohio 45221

B. W. Roberts
Director, Superconductive Materials Data Center
General Electric Corporate Research and Development
Box 8
Schenectady, New York 12301

Gail D. Schmidt
Chief, Radioactive Materials Branch
Division of Radioactive Materials and Nuclear Medicine
Bureau of Radiological Health
U.S. Public Health Service
Rockville, Maryland 20852

James E. Selle
Senior Research Specialist
Mound Laboratory
Monsanto Research Corporation
Miamisburg, Ohio 45342

Gertrude B. Sherwood
Office of Standard Reference Data
National Bureau of Standards
U.S. Department of Commerce
Washington, D.C. 20234

Ward F. Simmons
Associate Director, Defense Metals Information Center
Battelle/Columbus Laboratories
505 King Avenue
Columbus, Ohio 43201

George L. Tuve
2625 Exeter Road
Cleveland Heights, Ohio 44118

A. Bennett Wilson, Jr.
Executive Director, Committee on Prosthetics Research and Development
National Research Council
Washington, D.C. 20037

TABLE OF CONTENTS

VOLUME III

Section 1

Polymers

Table 1–1
ABS RESINS – MOLDED, EXTRUDED

	ASTM No.	Type: Medium impact	High impact	Very high impact	Low temperature impact	Heat resistant
Physical properties						
Specific gravity	D792	1.05–1.07	1.02–1.04	1.01–1.06	1.02–1.04	1.06–1.08
Thermal conductivity, Btu/hr/ft^2/°F/ft	C177	0.08–0.18	0.12–0.16	0.01–0.14	0.08–0.14	0.12–0.20
Coefficient of thermal expansion, 10^6/°F	D696	3.2–4.8	5.5–6.0	5.0–6.0	5.0–6.0	3.0–4.0
Specific heat, Btu/lb/°F	–	0.36–0.38	0.36–0.38	0.36–0.38	0.35–0.38	0.37–0.39
Water absorption (24 hr), %	D570	0.2–0.4	0.2–0.45	0.2–0.45	0.2–0.45	0.2–0.4
Flammability, ipm	D635	1.0–1.6	1.3–1.5	1.3–1.5	1.0–1.5	1.3–2.0
Heat distortion temperature (264 psi), °F	D648	185–223	180–215	180–218	185–224	220–245
Mechanical properties						
Modulus of elasticity in tension, 10^5 psi	D638	3.3–4.0	2.6–3.1	2.0–3.1	2.0–3.1	3.5–4.2
Tensile strength, 1,000 psi	D638	6.3–8.0	5.0–6.0	4.5–6.0	4–6	7.0–8.0
Elongation (in 2 in.), %	D638	5–20	5–50	20–50	30–200	20
Hardness (Rockwell)	D785	R108–115	R95–105	R85–105	R75–95	R107–116
Impact strength (Izod), ft-lb/in. notch	D256	2.0–4.0	3.0–5.0	5.0–7.0	6–10	2.0–4.0
Impact strength (–40 F), ft-lb/in.	D256	0.8–1.0	1.5–2.0	2.0–5.0	2.5–3.5	0.8–1.5
Modulus of elasticity in flex, 10^5, psi[a]	D790	3.5–4.0	2.5–3.2	2.0–3.2	2.0–3.2	3.5–4.2
Flexural strength, 1,000 psi[a]	D790	9.9–11.8	7.5–9.5	6.0–9.8	5–8	11.0–12.0
Compressive strength, 1,000 psi	D695	0.5–11.0	7.0–9.0	–	–	9.3–11.0
Electrical properties						
Volume resistivity, ohm-cm	D257	2.7×10^{16}	$1\text{-}4 \times 10^{16}$	$1\text{-}4 \times 10^{16}$	$1\text{-}4 \times 10^{16}$	$1\text{-}5 \times 10^{16}$
Dielectric strength (short time), V/mil	D149	385	350–440	300–375	300–415	360–400
Dielectric constant						
60 cycles	D150	2.8–3.2	2.8–3.2	2.8–3.5	2.5–3.5	2.7–3.5
10^6 cycles	D150	2.75–3.0	2.7–3.0	2.4–3.0	2.4–3.0	2.8–3.2
Dissipation factor						
60 cycles	D150	0.003–0.006	0.005–0.007	0.005–0.010	0.005–0.01	0.030–0.040
10^6 cycles	D150	0.008–0.009	0.007–0.015	0.008–0.016	0.008–0.016	0.005–0.015

Table 1–1 (continued)
ABS RESINS – MOLDED, EXTRUDED

	ASTM No.	Type: Medium impact	High impact	Very high impact	Low temperature impact	Heat resistant
Fabricating properties						
Bulk factor	D1182	1.5–2.0	1.5–2.0	1.5–2.0	1.5–3.0	1.5–2.0
Compression molding						
Pressure, 1,000 psi	–	1	1	1	1	1
Temperature, °F	–	325–375	325–375	325–375	325–375	375–450
Injection molding						
Pressure, 1,000 psi	–	6–30	6–30	6–30	6–30	6–30
Temperature, °F	–	425–500	375–500	0.400–525	350–500	475–550
Mold shrinkage, in./in.	D955	0.004–0.006	0.004–0.007	0.006–0.007	0.005–0.008	0.003–0.007
Chemical resistance	Highly resistant to aqueous acids, alkalis, salts. Resistant to concentrated phosphoric and hydrochloric acids, alcohols, and animal, vegetable, and mineral oils. Disintegrated by concentrated sulfuric or nitric acids. Soluble in esters, ketones, ethylene dichloride.					
Uses	Pipe, appliance housings, housewares, lawn and garden equipment, chrome-plated parts, highway safety devices, extruded profiles, shoe heels, fume hoods and ducts, toys, office equipment; also available as formable sheet for such uses as cases, luggage, refrigerator linings.					

[a]For a bar 2½ × ½ × ¼ in.

From *1973 Materials Selector,* Reinhold Publishing, Stamford, Conn., 1972, 242. With permission.

Table 1–2
ACRYLICS – CAST, MOLDED, EXTRUDED

	ASTM No.	Type: Cast resin sheets, rods — General purpose type I[a]	Type: Cast resin sheets, rods — General purpose type II[a]	Type: Moldings — Grades 5, 6, 8[b]	Type: Moldings — High impact grade
Physical properties					
Specific gravity	D792	1.17–1.19	1.18–1.20	1.18–1.19	1.12–1.16
Thermal conductivity Btu/ft^2/°F/ft	c	0.12	0.12	0.12	0.12
Coefficient of thermal expansion, 10^{-6}/°F	D696	4.5	4.5	3–4	4–6
Specific heat, Btu/lb/°F	–	0.35	0.35	0.35	0.34
Refractive index	D542	1.485–1.500	1.485–1.495	1.489–1.493	–
Transmittance (luminous, 0.125 in.), %	D791	91–92	91–92	>92	–
Haze, %	D672	1–2	1–2	< 3	–
Water absorption (24 hr), %	D570	0.3–0.4	0.2–0.4	0.3–0.4	0.2–0.3
Flammability (0.125 in.), ipm	D635	0.5–2.2	0.5–1.8	0.9–1.2	0.8–1.2
Mechanical properties					
Modulus of elasticity in tension, 10^5 psi	D638	3.5–4.5	4.0–5.0	3.5–5.0	2.3–3.3
Tensile strength, 1,000 psi	D638	6–9	8–10	9.5–10.5	5.5–8.0
Elongation (in 2 in.), %	D638	2–7	2–7	3–5	>25
Hardness (Rockwell)	D785	M80–90	M96–102	M80–103	L60–94
Impact strength (Izod notched), ft-lb/in.	D256	0.4	0.4	0.2–0.4	0.8–2.3
Modulus of elasticity in flexure, 10^5 psi	D790	3.5–4.5	4.0–5.0	3.5–5.0	2.8–3.6
Flexural strength, 1,000 psi	D790	12–14	15–17	15–16	8.7–12.0
Compressive yield strength (0.1% offset), 1,000 psi	D695	12–14	14–18	14.5–17	7.3–12.0
Electrical properties					
Volume resistivity, ohm-cm	D257	$>10^{15}$	$>10^{15}$	$>10^{14}$	2.0×10^{16}
Dielectric strength (short time), V/mil	D149	450–530	450–500	400	400–500
Dielectric constant					
60 cycles	D150	3.5–4.5	3.5–4.5	3.5–3.9	3.5–3.9
10^6 cycles	D150	2.7–3.2	2.7–3.2	2.7–2.9	2.5–3.0
Dissipation factor					
60 cycles	D150	0.05–0.06	0.05–0.06	0.04–0.06	0.03–0.04
10^6 cycles	D150	0.02–0.03	0.02–0.03	0.02–0.03	0.01–0.02
Arc resistance, sec.		No track	No track	No track	No track

Table 1–2 (continued)
ACRYLICS – CAST, MOLDED, EXTRUDED

	ASTM No.	Type: Cast resin sheets, rods — General purpose type I[a]	Type: Cast resin sheets, rods — General purpose type II[a]	Type: Moldings — Grades 5, 6, 8[b]	Type: Moldings — High impact grade
Fabricating properties					
Bulk factor		–	–	1.8–2.2	–
Injection molding					
Pressure, 1,000 psi		–	–	10–20	10–20
Temperature, °F		–	–	320–500	400–490
Hot forming temperature, °F		250–320	280–340	240–350	–
Extruding temperature, °F		–	–	350–450	–
Heat resistance					
Maximum recommended service temperature, °F		140–160	180–200	155–190	–
Heat distortion temperature, °F		150–180	190–225	166–250[d]	169–205
Chemical resistance		Resists weak alkalis, acids, and aliphatic hydrocarbons. Attacked by esters, ketones, aromatic hydrocarbons, chlorinated hydrocarbons, and concentrated acids.			
Uses		Transparent aircraft enclosures, radio and television parts, lighting, drafting equipment, signs.		Decorative and functional automotive parts, reflectors, protective goggle lenses, radio and television parts, household appliance parts.	Shoe heels, control knobs, business machine and piano keys, pump parts, sprinkler heads, tool handles.

[a]ASTM D702.
[b]Range includes typical values for Grades 5, 6, and 8, and may be superior to minimum or maximum requirements for these grades as detailed in ASTM D788.
[c]Cenco-Fitch.
[d]D788 specified values for Grades 5, 6, and 8: 149 F, 162 F, 183 F, respectively.

From *1973 Materials Selector,* Reinhold Publishing, Stamford, Conn., 1972, 243. With permission.

Table 1–3
ALKYDS AND THERMOSET CARBONATE

	ASTM No.	Material				
			Alkyds – molded			
		Allyl diglycol carbonate	Putty (encapsulating)	Rope (general purpose)	Granular (high speed molding)	Glass reinforced (heavy duty parts)
Physical properties						
Specific gravity	D792	1.32	2.05–2.15	2.20–2.22	2.21–2.24	2.02–2.10
Thermal conductivity, Btu/hr/ft^2/°F/in.	–	1.45	0.35–0.60	0.35–0.60	0.35–0.60	0.20–0.30
Coefficient of thermal expansion/°F	D696	6×10^{-5}	1.3×10^{-5}	1.3×10^{-5}	1.3×10^{-5}	1.3×10^{-5}
Specific heat, Btu/lb/°F	–	0.3	–	–	–	–
Water absorption (24 hr), %	D570	0.20	0.10–0.15	0.05–0.08	0.08–0.12	0.007–0.10
Flammability, ipm	D635	0.35	Nonburning	Self extinguishing	Self extinguishing	Nonburning
Transparency (visible light), %	–	89–92	Opaque	Opaque	Opaque	Opaque
Refractive index, n_D	D542	1.50	–	–	–	–
Mechanical properties						
Tensile strength, 1,000 psi	D638	5–6	4–5	7–8	3–4	5–9
Tensile modulus, 10^5 psi	D638	3.0	20–27	19–20	24–29	20–25
Elongation, %	D638	–	–	–	–	–
Impact strength, (Izod notched), ft-lb/in.	D256	0.2–0.4	0.25–0.35	2.2	0.30–0.35	8–12
Flexural strength, 1,000 psi	D790	–	8–11	19–20	7–10	12–17
Modulus of elasticity in flexure 10^5 psi	D790	2.5–3.3	–	22–27	22–27	22–28
Compressive strength, 1,000 psi	D690	22.5	20–25	28	16–20	24–30
Hardness (Barcol)	D785	M95–M100	60–70	70–75	60–70	70–80
Electrical properties						
Volume resistivity, ohm-cm	D257	4×10^{14}	10^{14}	10^{14}	1×10^{14}–1×10^{15}	10^{14}
Dielectric strength (step by step), V/mil	D149	290	300–350	290	300–350	300–350
Dielectric constant	D150					
60 cycles		4.4	5.4–5.9	7.4	5.7–6.3	5.2–6.0
10^6 cycles		3.5–3.8	4.5–4.7	6.8	4.8–5.1	4.5–5.0
Dissipation factor	D150					
60 cycles		0.03–0.04	0.030–0.045	0.019	0.030–0.040	0.02–0.03
10^6 cycles		0.1–0.2	0.016–0.020	0.023	0.017–0.020	0.015–0.022
Arc resistance, sec	D495	185	180	180	180	180

Table 1–3 (continued)
ALKYDS AND THERMOSET CARBONATE

	ASTM No.	Material				
			Alkyds – molded			
		Allyl diglycol carbonate	Putty (encapsulating)	Rope (general purpose)	Granular (high speed molding)	Glass reinforced (heavy duty parts)
Fabricating properties						
Bulk factor	–		1.1–1.2	>1	1.95–2.15	9–11
Compression molding	–					
Pressure, 1,000 psi		(Only available as stock sheet for machining: 48 × 72 × 1 in. maximum)	0.8–1.0	0.6–1.0	0.8–1.0	2–3
Temperature, °F			270–330	270–300	270–330	270–330
Mold shrinkage, in./in.			0.001–0.004	0.004–0.007	0.001–0.004	
Transfer molding	–					
Pressure, 1,000 psi			0.5–1	1–2	1–2	1–2
Temperature, °F			270–330	270–330	270–330	290–330
Mold shrinkage, in./in.			0.004–0.008	0.001–0.004	0.004–0.007	0.002–0.006
Heat resistance						
Maximum recommended service temperature, °F	–	212	250	300	300	300
Deflection temperature (264 psi), °F	D648	–	350–400	>400	350–400	>400
Chemical resistance		Resists nearly all solvents including acetone, benzene, and gasoline, and practically all chemicals except highly oxidizing acids.	Resistant to weak acids; attacked by alkalis; practically unattacked by organic liquids such as alcohols, hydrocarbons, and fatty acids.			

Table 1–3 (continued)
ALKYDS AND THERMOSET CARBONATE

	ASTM No.	Material				
			Alkyds – molded			
		Allyl diglycol carbonate	Putty (encapsulating)	Rope (general purpose)	Granular (high speed molding)	Glass reinforced (heavy duty parts)
Uses		Aircraft windows, lenses, marine glazing, vending machine windows, slides, watch crystals, safety windows.	Encapsulation of resistors, coils, and small electronic parts.	Molding of tube bases and sockets, connectors, tuning devices, electrical instrument parts, switches, and relays. Parts for transformers, motor controllers, and automotive ignition systems.		Heavy duty circuit breaker and switchgear, standoff insulators electrical motor brush holders, and end plates.

From *1973 Materials Selector,* Reinhold Publishing, Stamford, Conn., 1972, 244. With permission.

Table 1–4
CELLULOSE ACETATE – MOLDED, EXTRUDED

	ASTM No.	ASTM grade[a] H6-1	H4-1	H2-1	MH-1, MH-2	MS-1, MS-2	S2-1
Physical properties							
Specific gravity	D792		1.29–1.31	1.25–1.31	1.24–1.31	1.23–1.30	1.22–1.30
Thermal conductivity, Btu/hr/ft^2/°F/ft	C177	0.10–0.19	0.10–0.19	0.10–0.19	0.10–0.19	0.10–0.19	0.10–0.19
Coefficient of thermal expansion, 10^{-5}/°F	D696	4.4–9.0	4.4–9.0	4.4–9.0	4.4–9.0	4.4–9.0	4.4–9.0
Refractive index	D542	1.46–1.50	1.46–1.50	1.46–1.50	1.46–1.50	1.46–1.50	1.46–1.50
Specific heat, Btu/lb/°F	–	0.3–0.42	0.3–0.42	0.3–0.42	0.3–0.42	0.3–0.42	0.3–0.42
Luminous transmittance, %	D791	75–90	75–90	80–90	80–90	80–90	80–95
Haze, %	D672	2–15	2–15	2–10	2–10	2–10	2–8
Water absorption (24 hr), %	D570		1.7–2.7	1.7–2.7	1.8–4.0	2.1–4.0	2.3–4.0
Flammability, ipm[b]	D635	0.5–2.0	0.5–2.0	0.5–2.0	0.5–2.0	0.5–2.0	0.5–2.0
Mechanical properties							
Tensile strength at fracture, 1,000 psi	D638	–	7–8	5.8–7.2	4.8–6.3	3.9–5.3	3.0–4.4
Hardness (Rockwell R)	D785	–	103–120	89–112	74–104	54–96	49–88
Impact strength (Izod), ft-lb/in, of notch	D256	–	1.1–3.1	1.5–3.9	2.5–4.9	2.9–6.5	4.0–6.8
Modulus of elasticity in flexure, 10^5 psi	D747	–	2.0–2.55	1.50–2.35	1.50–2.15	1.25–1.90	1.05–1.65
Flexural strength at yield, 1,000 psi	D790	–	8.1–11.15	6.0–10.0	4.4–8.65	3.8–7.1	3.5–5.7
Compressive strength at yield, 1,000 psi	D695	–	6.5–10.6	4.3–9.6	4.4–8.4	3.2–7.2	3.15–6.1
Electrical properties							
Volume resistivity, ohm-cm	D257	10^{10}–10^{13}	10^{10}–10^{13}	10^{10}–10^{13}	10^{10}–10^{13}	10^{10}–10^{13}	10^{10}–10^{13}
Dielectric strength (short time), V/mil	D149	250–600	250–600	250–600	250–600	250–600	250–600
Dielectric constant							
60 cycles	D150	3.5–7.5	3.5–7.5	3.5–7.5	3.5–7.5	3.5–7.5	3.5–7.5
10^6 cycles	D150	3.2–7.0	3.2–7.0	3.2–7.0	3.2–7.0	3.2–7.0	3.2–7.0
Dissipation factor							
60 cycles	D150		0.01–0.06	0.01–0.06	0.01–0.06	0.01–0.06	0.01–0.06
10^6 cycles	D150		0.01–0.10	0.01–0.10	0.01–0.10	0.01–0.10	0.01–0.10
Fabricating properties							
Bulk factor		2.0–2.6	2.0–2.6	2.0–2.6	2.0–2.6	2.0–2.6	2.0–2.6
Compression molding							
Pressure, psi		500–5,000	500–5,000	500–5,000	500–5,000	500–5,000	500–5,000
Temperature, °F		390–475	375–450	350–425	325–400	300–370	290–330
Injection molding							
Pressure, psi		8–32	8--32	8–32	8–32	8–32	8–32

Table 1–4 (continued)
CELLULOSE ACETATE – MOLDED, EXTRUDED

ASTM No.	ASTM grade[a] H6-1	H4-1	H2-1	MH-1, MH-2	MS-1, MS-2	S2-1
Fabricating properties (cont.)						
Temperature, °F	420–490	410–480	390–460	370–440	350–420	335–395
Molding shrinkage, in./in.	0.003–0.006	0.003–0.006	0.003–0.006	0.003–0.006	0.003–0.006	0.003–0.006
Extrusion temperature, °F	420–450	405–455	390–420	370–400	350–385	335–365
Heat resistance						
Heat deflection temperature, °F						
66 psi	–	172–203	145–188	145–170	136–153	132–141
264 psi	–	145–188	120–172	128–155	123–141	117–129
Chemical resistance	Unattacked by water, salt water solutions, white gasoline, oleic acid, 5% acetic acid, and dilute sulfuric acid. Decomposed by 30% sulfuric, 10% nitric, and 10% hydrochloric acids, sodium hydroxide, and 10% ammonium hydroxide. Dissolved by acetone and ethyl acetate.					
Uses	Film, tape, blister packaging, appliance housings, optical parts, tool handles, brush handles, toys and novelties, toothbrushes, buttons, tags.					

[a]According to ASTM D706-63.
[b]Self-extinguishing compositions are available.

From *1973 Materials Selector,* Reinhold Publishing, Stamford, Conn., 1972, 245. With permission.

Table 1–5
CELLULOSE ACETATE BUTYRATE AND CELLULOSE ACETATE PROPIONATE – MOLDED, EXTRUDED

	ASTM No.	ASTM grade[a]					
		Cellulose acetate butyrate			Cellulose acetate propionate		
		H4	MH	S2	1	3	6
Physical properties							
Specific gravity	D792	1.22	1.18–1.20	1.15–1.18	1.22	1.20–1.21	1.19
Thermal conductivity, Btu/hr/ft^2/°F/ft	C177	0.10–0.19	0.10–0.19	0.10–0.19	0.10–0.19	0.10–0.19	0.10–0.19
Coefficient of thermal expansion, 10^{-5}/°F	D696	(6–9) × 10^{-5}	(6–9) × 10^{-5}	(6–9) × 10^{-5}	(6–9) × 10^{-5}	(6–9) × 10^{-5}	(6–9) × 10^{-5}
Refractive index	D543	1.46–1.49	1.46–1.49	1.46–1.49	1.46–1.49	1.46–1.49	1.46–1.49
Specific heat, Btu/lb/°F		0.3–0.4	0.3–0.4	0.3–0.4	0.3–0.4	0.3–0.4	0.3–0.4
Luminous transmittance, %	D791	75–92	80–92	85–95	80–92	80–92	80–92
Haze, %	D672	2–5	2–5	2–5	2–5	2–5	2–5
Water absorption (24 hr), %	D570	2.0	1.3–1.6	0.9–1.3	1.6–2.0	1.3–1.8	1.6
Flammability, ipm	D635	0.5–1.5	0.5–1.5	0.5–1.5	0.5–1.5	0.5–1.5	0.5–1.5
Mechanical properties							
Tensile strength at fracture, 1,000 psi	D638	6.9	5.0–6.0	3.0–4.0	5.9–6.5	5.1–5.9	4.0
Hardness (Rockwell R)	D785	114	80–100	23–42	100–109	92–96	57
Impact strength (Izod), ft-lb/in. of notch	D256	3.0	4.4–6.9	7.5–10.0	1.7–2.7	3.5–5.6	9.4
Modulus of elasticity in flexures, 10^5 psi	D747	1.80	1.20–1.40	0.70–0.90	1.7–1.8	1.45–1.55	1.1
Flexural strength at yeild, 1,000 psi	D790	9.0	5.6–6.7	2.5–3.95	6.8–7.9	5.6–6.2	–
Compressive strength at yield, 1,000 psi	D695	8.8	5.3–7.1	2.6–4.3	6.2–7.3	4.9–5.8	–
Electrical properties							
Volume resistivity, ohm-cm	D257	10^{11}–10^{14}	10^{11}–10^{14}	10^{11}–10^{14}	10^{11}–10^{14}	10^{11}–10^{14}	10^{11}–10^{14}
Dielectric strength (short time), V/mil	D149	250–400	250–400	250–400	300–450	300–450	300–450
Dielectric constant							
60 cycles	D150	3.5–6.4	3.5–6.4	3.5–6.4	3.7–4.0	3.7–4.0	3.7–4.0
10^6 cycles	D150	3.2–6.2	3.2–6.2	3.2–6.2	3.4–3.7	3.4–3.7	3.7–3.4
Dissipation factor							
60 cycles	D150	0.01–0.04	0.01–0.04	0.01–0.04	0.01–0.04	0.01–0.04	0.01–0.04
10^6 cycles	D150	0.02–0.05	0.02–0.05	0.02–0.05	0.02–0.05	0.02–0.05	0.02–0.05

Table 1–5 (continued)
CELLULOSE ACETATE BUTYRATE AND CELLULOSE ACETATE PROPIONATE – MOLDED, EXTRUDED

	ASTM No.	ASTM grade[a] Cellulose acetate butyrate H4	MH	S2	Cellulose acetate propionate 1	3	6
Fabricating properties							
Bulk factor		2.0–2.4	2.0–2.4	2.0–2.4	2.0–2.4	2.0–2.4	2.0–2.4
Compression molding							
Pressure, psi		500–5,000	500–5,000	500–5,000	500–5,000	500–5,000	500–5,000
Temperature, °F		335–390	305–340	265–305	335–390	325–360	305–340
Injection molding							
Pressure, 1,000 psi		8–32	8–32	8–32	8–32	8–32	8–32
Temperature, °F		400–480	375–440	335–395	400–475	380–450	350–420
Molding, shrinkage, in./in.		0.003–0.006	0.003–0.006	0.003–0.006	0.003–0.006	0.003–0.006	0.003–0.006
Extrusion temperature, °F		420–440	400–420	380–400	410–440	400–430	390–420
Heat resistance							
Heat deflection temperature, °F							
66 psi		222	171–184	136–147	191–201	169–187	163
264 psi		196	146–160	118–130	163–173	141–157	129
Chemical resistance		Unaffected by 3% sulfuric, 5% acetic, 10% hydrochloric and oleic acids; discolored by 10% nitric acid. Unaffected by 1% sodium hydroxide and 2% sodium carbonate; slightly softened by 10% sodium hydroxide and discolored by 10% ammonium hydroxide; unaffected by white gasoline, but swollen or dissolved by ethyl alcohol, acetone, ethyl acetate, ethylene dichloride, carbon tetrachloride, and toluene. Unaffected by water, salt water, and 3% hydrogen peroxide.					
Uses		Vacuum-formed outdoor signs and molded letters, blister packaging, TV and radio knobs, handles, pipe, pens, optical parts, containers.			Telephones, steering wheels, blister packaging, toothbrushes, pens, knobs, containers, optical parts.		

[a]According to ASTM D707-63 and D1562-60, respectively.

From *1973 Materials Selector,* Reinhold Publishing, Stamford, Conn., 1972, 246. With permission.

Table 1–6
CHLORINATED POLYETHER, CHLORINATED POLYVINYL CHLORIDE, POLYCARBONATES

	ASTM No.	Material: Chlorinated polyether	Chlorinated polyvinyl chloride	Polycarbonate	Polycarbonate (40% glass fiber reinforced)
Physical properties					
Specific gravity	D792	1.4	1.54	1.20	1.51
Thermal conductivity, Btu/hr/ft²/°F/in.	–	0.91	0.95	0.11	0.13
Coefficient of thermal expansion, 10^{-5}/°F	D696	6.6	4.4	3.75	1.0–1.1
Specific heat, Btu/lb/°F	–	–	0.3	0.30	–
Water absorption (24 hr), %	D570	0.01	0.11	0.15	0.08
Flammability, ipm	D635	Self extinguishing	Nonburning	Self extinguishing	Self extinguishing
Transparency (visible light), %	–	Opaque	Opaque	75–85	Translucent
Refractive index, η_D	D542	–	–	1.586	–
Mechanical Properties					
Tensile strength, 1,000 psi	D638	6	7.3	9.5	18
Yield strength, 1,000 psi	D638	5.9	–	8.5	–
Tensile modulus 10^5 psi	D638	1.5	3.7	3.45	17
Elongation, %	D638	130 (in 2 in.)	–	110	0–5
Yield elongation, %	D638	15	–	5	–
Impact strength, (Izod notched), ft-lb/in.	D256	0.4 (D758)	6.3	16	–
Unnotched	–	>33	–	>60	16
Flexural strength, 1,000 psi	D790	5 (0.1% offset)	14.5	13.5	27
Modulus of elasticity in flexure, 10^5 psi	D790	1.3 (0.1% offset)	3.85	3.4	12
Fatigue strength (Krause), 1,000 psi	–	–	–	1	7.5
Compressive strength, 1,000 psi	D690	9.0	–	12.5	18.5
Hardness (Rockwell)	D785	R100	R118	M70	M97
Coefficient of static friction (against self)	–	–	–	0.52	–
Abrasion resistance (Taber, CS-17 wheel), mg/1,000 cycles	D104	–	–	10	40

Table 1–6 (continued)
CHLORINATED POLYETHER, CHLORINATED POLYVINYL CHLORIDE, POLYCARBONATES

	ASTM No.	Material: Chlorinated polyether	Chlorinated polyvinyl chloride	Polycarbonate	Polycarbonate (40% glass fiber reinforced)
Electrical properties					
Volume resistivity, ohm-cm	D257	1.5×10^{16}	$1 \times 10^{15}-2 \times 10^{16}$	2.1×10^{16}	1.4×10^{15}
Dielectric strength (short time), V/mil	D149	400	1,250–1,550	400	475
Dielectric constant	D150				
60 cycles		3.10	3.08	3.17	3.80
10^6 cycles		2.92	3.2–3.6	2.96	3.58
Dissipation factor	D150				
60 cycles		0.011	0.0189–0.0208	0.0009	0.006
10^6 cycles		0.011	0.020	0.010	0.007
Arc resistance, sec	D495	–	–	120 (tungsten electrode)	120 (tungsten electrode)
Fabricating properties					
Bulk factor	–	–	–	1.7	–
Injection molding	–				
Pressure, 1,000 psi		10–20	–	15–20	15–20
Temperature, °F		440–465	–	525–625	575–650
Mold shrinkage, in./in.		0.004–0.008	0.007	0.005–0.007	0.002
Extrusion temperature, °F	–	360–450	–	475–580	–
Heat resistance					
Maximum recommended service temperature, °F	–	250–275	210	250	250
Deflection temperature (264 psi), °F	D648	210	212	270	295
Chemical resistance		Excellent resistanct to both inorganic and organic chemicals to 250°F. Resistant to all inorganic acids except fuming nitric and fuming sulfuric.	Generally resistant to alkalis and weak acids; moderate to poor resistance to strong acids. Not resistant to ketones and esters; aromatic hydrocarbons produce swelling.	Insoluble in aliphatic hydrocarbons, ether, and alcohols; partially soluble in aromatic hydrocarbons; soluble in chlorinated hydrocarbons. High stability to water and mineral and organic acids.	More stable to solvents that tend to act as stress cracking agents than is base resin alone. Reinforced material generally displays chemical resistance normally associated with polycarbonates.

Table 1–6 (continued)
CHLORINATED POLYETHER, CHLORINATED POLYVINYL CHLORIDE, POLYCARBONATES

		Material			
	ASTM No.	Chlorinated polyether	Chlorinated polyvinyl chloride	Polycarbonate	Polycarbonate (40% glass fiber reinforced)
Uses		Valves, pump parts, water meter parts, tank linings, pipe, sheet, and coatings for high temperature corrosive environments.	Valves, pump parts, pipe for high temperature corrosive environments.	Electrical parts, housings, structural parts, electronic components, safety helmets, street light globes, portable tool housings.	Military parts, module cases, pump impellers, weapons components, aircraft parts, automotive parts, portable tool housings.

From *1973 Materials Selector,* Reinhold Publishing, Stamford, Conn., 1972, 247. With permission.

Table 1–7
DIALLYL PHTHALATES – MOLDED

	ASTM No.	Type			
		Orlon filled	Dacron filled	Asbestos filled	Glass fiber filled
Physical properties					
Specific gravity	D792	1.31–1.35	1.40–1.65	1.50–1.96	1.55–1.85
Coefficient of thermal expansion/°F	D696	5.0×10^{-5}	5.2×10^{-5}	4.0×10^{-5}	$2.2.\text{-}2.6 \times 10^{-5}$
Water absorption (122 F, 48 hr), %	–	0.2–0.5	0.2–0.5	0.4–0.7	0.2–0.4
Flammability (ignition time), sec	–	68	84–90[e]	70[e]	70–400[e]
Mechanical properties					
Modulus of elasticity in tension, psi[a]	D638	6×10^5	–	12×10^5	–
Tensile strength, psi	D638	4,500–6,000	4,600–5,500	4,000–6,500	5,500–9,500
Hardness (Rockwell)	D785	M108	–	M107	M108
Impact strength (Izod notched), ft-lb/in.	D256	0.5–1.2	1.7–4.5	0.30–0.50	0.5–15.0
Flexural strength, 1,000 psi	D790	7.5–10.5	9–11.5	8–10	10–18
Compressive strength, 1,000 psi	D695	20–25	20–30	18–25	25
Electrical properties					
Dielectric strength, V/mil					
Short time (dry)	D149	400	376–390	350–450	350–430
Short time (wet[b])	D149	375	360–391	300–400	300–420
Step by step (dry)	D149	350	350–374	300–400	300–420
Step by step (wet[b])	D149	325	350–361	250–350	275–420
Dielectric breakdown, kV					
Short time (dry)	–	65–75	65	55–80	63–70
Short time (wet[b])	–	60–65	60	55	45–65
Step by step (dry)	–	55–60	60	38–70	55–65
Step by step (wet[b])	–	46–60	55	39–60	45–65
Dissipation factor[c]					
Dry	D150	0.023, 0.015	0.008, 0.015	0.05, 0.03	0.01, 0.015
Wet[d]	D150	0.045, 0.040	0.009, 0.017	0.154, 0.050	0.012, 0.020
Dielectric constant[e]					
Dry	D150	3.9, 3.3	3.8, 3.6	5.2, 4.5	4.5, 4.2
Wet[d]	D150	4.1, 3.4	3.9, 3.7	6.5, 4.8	4.6, 4.4
Volume resistivity, megohm-cm[d]	D257	60,000–6,000,000	100–25,000	100–5,000	10,000–50,000
Surface resistivity, megohm[d]	D257	25,000–2,500,000	500–25,000	100–5,000	10,000–100,000
Arc resistance, sec	D495	85–115	105-125	125–140	125–135

Table 1–7 (continued)
DIALLYL PHTHALATES – MOLDED

	ASTM No.	Type			
		Orlon filled	**Dacron filled**	**Asbestos filled**	**Glass fiber filled**
Fabricating properties					
Bulk factor	D392	3.5–5.2	3.5–5.2	1.9–2.4	1.9–6.0
Compression molding					
Pressure, psi	–	500–2,000	500–2,000	500–2,000	500–2,000
Temperature, °F	–	270–290	270–290	270–320	270–320
Mold shrinkage, in./in.	–	0.009–0.011	0.009–0.010	0.004–0.008	0.000–0.005
Transfer molding					
Pressure, psi	–	1,000–5,000	1,000–5,000	1,000–5,000	1,000–5,000
Temperature, °F	–	270–290	270–290	270–310	270–310
Mold shrinkage, in./in.	–	0.012–0.015	0.010	0.005–0.009	0.001–0.005
Post-mold shrinkage[f], in./in.	–	0.001	0.0006	0.0005	0.0002
Heat resistance					
Maximum recommended service temperature, °F	–	300	300–370	350–450	400–450
Heat distortion temperature, °F	D648	240–266	270–290	300–350	350–500
Chemical resistance		Unaffected by weak acids and alkalis and organic solvents; slightly affected by strong acids and alkalis			
Uses		Molding compounds – connectors, potentiometers, plugboards, housings, appliance fixtures, resistors, insulators, etc. Prepregs – radomes, aircraft leading edges, housings, nose cones, air ducts, etc. Laminates – decorative sheets for surfacing real and grain-printed wood and fabrics, etc.			

[a]Conditioned 48 hr at 122°F.
[b]Tested after 48-hr immersion in water at 122°F.
[c]Values given for frequencies of 1 kc and 1 mc, in that order.
[d]Conditioned 30 days at 100% RH and 158°F.
[e]Flame-resistant type is available.
[f]480 hr, 257°F.

From *1973 Materials Selector,* Reinhold Publishing, Stamford, Conn., 1972, 248. With permission.

Table 1–8

FLUOROCARBONS – MOLDED, EXTRUDED

	ASTM No.	Type: Polytrifluoro chloroethylene (PTFCE)	Polytetrafluoro-ethylene (PTFE)	Ceramic reinforced (PTFE[a])	Fluorinated ethylene propylene (FEP)	Polyvinylidene-fluoride (PVF_2)
Physical properties						
Specific gravity	D792	2.10–2.15	2.1–2.3	2.2–2.4	2.14–2.17	1.77
Thermal conductivity, Btu/hr/ft^2/°F	b	0.145	0.14	–	0.12	0.14
Coefficient of thermal expansion/°F	D696	3.88×10^{-5}	5.5×10^{-5}	$1.7–2.0 \times 10^{-5}$ d	$8.3–10.5 \times 10^{-5}$	8.5×10^{-5}
Refractive index	D542	1.43	1.35	–	1.34	1.42
Specific heat, Btu/lb/°F		0.22	0.25	–	0.28	0.33
Transmittance (luminous), %	D791	80–92	–	–	–	–
Water absorption (24 hr), %	D570	0.00	0.01	>0.2	0.01	0.03
Flammability		Noninflammable	Noninflammable	Noninflammable	Noninflammable	Self extinguishing
Mechanical properties						
Modulus of elasticity in compression, psi	D638	1.8×10^5	$0.70–0.90 \times 10^5$	$1.5–2.0 \times 10^5$	$0.6–0.8 \times 10^5$	$1.7–2 \times 10^5$
Modulus of elasticity in tension, psi	D638	$1.9–3.0 \times 10^5$	$0.38–0.65 \times 10^5$	$1.5–2.0 \times 10^5$	$0.5–0.7 \times 10^5$	$1.7–2 \times 10^5$
Tensile strength, 1,000 psi	D638	4.6–5.7	2.5–6.5	0.75–2.5	2.5–3.5	7.2–8.6
Elongation (in 2 in.), %	D638	125–175	250–350	10–200	250–330	200–300
Hardness (Rockwell)	D785	R110–115	52D	R35–55	58D	R110
Abrasion resistance, g/cycle	c	0.0080	–	–	–	0.0006–0.0012
Impact (Izod notched), ft-lb/in.	D256	3.50–3.62	2.5–4.0	–	No break	3.8
Modulus of elasticity in flexure, psi	D747	$2.0–2.5 \times 10^5$	0.6×10^5	4.64×10^5	0.8×10^5	–
Flexural strength (0.1% offset), 1,000 psi	D790	3.5	–	–	–	2
Compressive strength (0.1% offset), 1,000 psi	D695	2.0	0.7–1.8	1.4–1.8	1.6	12.8–14.2
Electrical properties						
Volume resistivity, ohm-cm						
(short time), V/mil	D257	10^{18}	$>10^{18}$	10^{15}	$>2 \times 10^{18}$	5×10^{14}
Dielectric strength	D149	530–600	400–500	300–400	500–600	260
Dielectric constant						
60 cycles	D150	2.6–2.7	2.1	2.9–3.6	2.1	10.0
10^6 cycles	D150	2.30–2.37	2.1	2.9–3.6	2.1	7.5
Dissipation factor						
60 cycles	D150	0.02	0.0002	0.0005–0.0015	0.0003	0.050
10^6 cycles	D150	0.007–0.010	0.0002	0.0005–0.0015	0.0003	0.184
Arc resistance, sec		>360	>200	–	>165	50–60

Table 1–8 (continued)
FLUOROCARBONS – MOLDED, EXTRUDED

		Type				
	ASTM No.	Polytrifluoro chloroethylene (PTFCE)	Polytetrafluoro-ethylene (PTFE)	Ceramic reinforced (PTFE[a])	Fluorinated ethylene propylene (FEP)	Polyvinylidene-fluoride (PVF_2)
Fabricating properties						
Injection molding	5–30	–	–	5–20	15–20	
Pressure, 1,000 psi		420–620	–	–	625–760	400–500
Compression molding						
Pressure, 1,000 psi		0.1–15	–	–	1–2	2
Temperature, °F		445–525	–	–	600–750	390–460
Bulk factor		2.5	–	–	–	2
Mold shrinkage, in./in.		0.005–0.010	–	–	0.03–0.06	0.015–0.030
Heat resistance						
Maximum recommended service temperature, °F		380	550	450–500	400	340
Heat distortion temperature, °F						
66 psi	D648	196–291	–	350–480	–	300
264 psi	D648	151–178	–	170–220	–	232
Chemical resistance		Impervious to corrosive chemicals; highly resitant to most organic solvents.	Inert to most chemicals and solvents with exception of alkali metals. Halogenated solvents at high temperatures and pressure have some effect.			Resistant to most acids and bases except fuming sulfuric.
Uses		Chemical pipes, gaskets, pump parts, electrical cables, tank linings, connectors, coil forms, connector inserts, valve diaphragms, insulation.	Chemical pipes, valves and valve liners, gaskets and packings, pump bearings and impellers, electrical equipment, anti-adhesive coatings.	Bearings, bushings, sliding and wear surfaces, electrical insulators, gaskets, packings, washers, valve seats operating in corrosive conditions.	Molded electronic and instrument components, valve linings, laminates, corrosion resistant and nonadhesive coatings.	Seals; chemical pipe and fittings, gaskets; electrical jackets and primary insulation; finishes.

[a] A proprietary material consisting of polytetrafluoroethylene and special constituents designed to improve TFE's mechanical and thermal properties while retaining its electrical and chemical characteristics.

365

Table 1–8 (continued)
FLUOROCARBONS – MOLDED, EXTRUDED

[b] Cenco-Fitch.
[c] Federal Spec L-P–406A No. 1092.1
[d] From 73 to 500°F.

From *1973 Materials Selector,* Reinhold Publishing, Stamford, Conn., 1972, 249. With permission.

Table 1–9
EPOXIES – CAST, MOLDED, REINFORCED

		Material					
		Standard epoxies (diglycidyl ethers of bisphenol A)					
	ASTM No.	Cast rigid[a]	Cast flexible[b]	Molded[c]	General purpose glass cloth laminate[d]	High strength laminate[e]	Filament wound composite[f]
Physical properties							
Specific gravity	D792	1.15	1.14–1.18	1.80–2.0	1.8	1.84	2.18–2.17
Thermal conductivity, Btu/hr/ft^2/°F/ft	D325	0.1–0.3	–	0.1–0.5	–	2.35	–
Coefficient of thermal expansion, 10^{-5}/°F	D696	3.3	3–5	1–2	$3.3–4.8 \times 10^{-6}$	$3.3–4.8 \times 10^{-6}$	2–6
Specific heat, Btu/lb/°F	–	0.4–0.5	–	–	–	0.21	0.24
Water absorption (24 hr), %	D570	0.1–0.2	0.4–0.1	0.3–0.8	0.05–0.07	0.05	0.05–0.07
Flammability, ipm	D635 or D757	0.3–0.34	–	Self extinguishing	Slow burn to Self extinguishing	Self extinguishing	Self extinguishing
Transparency (visible light), %	–	90	85	Opaque	Opaque	Opaque	–
Refractive index, n_D	D542	1.61	1.61	–	–	–	–
Mechanical properties							
Tensile strength, 1,000 psi	D638	9.5–11.5	1.4–7.6	8–11	50–58	160	230–240 (hoop)
Tensile modulus, 10^5 psi	D638	4.5	0.5–2.5	–	33–36	57–58	72–64
Elongation, %	D638	4.4	1.5–60	–	–	–	–
Impact strength (Izod notched), ft-lb/in.	D256	0.2–0.5	0.3–0.2	0.4–0.5	12–15	60–61 (edgewise)	–

Table 1–9 (continued)
EPOXIES – CAST, MOLDED, REINFORCED

	ASTM No.	Material: Standard epoxies (diglycidyl ethers of bisphenol A)					
		Cast rigid[a]	Cast flexible[b]	Molded[c]	General purpose glass cloth laminate[d]	High strength laminate[e]	Filament wound composite[f]
Mechanical properties (cont.)							
Flexural strength 1,000 psi	D790	14–18	1.2–12.7	19–22	80–90	165–177	180–170
Modulus of elasticity in flexure, 10^5 psi	D790	4.5–5.4	0.36–3.9	15–25	36–39	53–55	69–75
Compressive strength, 1,000 psi	D695	16.5–24	–	34–38	50–60	80–90 (edgewise)	–
Hardness (Rockwell)	D785	106M	50–100M	75–80 (Barcol)	115–117M	70–72 (Barcol)	98–120M
Electrical properties							
Volume resitivity, ohm-cm	D257	6.1×10^{15}	9.1×10^{8} 6.7×10^{9}	$1–5 \times 10^{15}$	–	$6.6 \times 10^{7}–10^{9}$	–
Dielectric strength (step by step), V/mil	D149	>400	400–410	360–400	450–550	650–750	–
Dielectric constant	D150						
60 cycles		4.02	4.43–4.79	4.4–5.4	5.3–5.4	–	–
10^6 cycles		3.42	2.78–3.52	4.1–4.6	4.7–4.8	4.8–5.2	–
Dissipation factor	D150						
60 cycles		0.0074	0.0048–0.0380	0.011–0.018	0.004–0.006	–	–
10^6 cycles		0.032	0.0369–0.0622	0.013–0.020	0.024–0.026	0.010–0.017	–
Arc resistance, sec	D495	100	75–98	135–190	130–180	–	–
Fabricating properties							
Cure time, hr	–	16	24	–	2	10 min	2
Cure temperature, °F	–	77	77	–	250	330	175
Postcure time, hr	–	3	–	–	5	16	3
Postcure temperature, °F	–	212	–	–	320	280	300
Shrinkage, %	–	3	–	–	–	–	–
Bulk factor	–	–	–	2.1–2.3	–	–	–
Transfer molding	–						
Pressure, 1,000 psi		–	–	0.100–2.0	–	–	–
Temperature, °F		–	–	290–350	–	–	–
Mold shrinkage, %		–	–	0.002–0.003	–	–	–

Table 1–9 (continued)
EPOXIES – CAST, MOLDED, REINFORCED

	ASTM No	Material: Standard epoxies (diglycidyl ethers of bisphenol A) — Cast rigid[a]	Cast flexible[b]	Molded[c]	General purpose glass cloth laminate[d]	High strength laminate[e]	Filament wound composite[f]
Heat resistance							
Maximum recommended service temperature, °F	–	175–190	100–125	<400	250–350	250–350	250–350
Heat deflection (264 psi), °F	D648	230	90–155	340–400	–	–	280–295
Chemical resistance		Highly resistant to water and strong alkaline environments; less resistant to sulfuric and acetic acids, and oxidizing agents.					
Uses		Potting and encapsulation of electronic components, precision castings, tools and dies, patching compounds.		Electrical moldings, such as condensers, switch plates, connector plugs, resistor bobbins and wirewound resistors, molded coils, relay assemblies.	High strength parts such as laminated tools for metal forming, aircraft structural parts, pipe, leaf or coil springs, high strength electrical or chemical resistant parts.		Rocket motor cases, chemical tanks, pipe, pressure bottles, high strength tubing, shotgun barrels, missile bodies.

[a]13 phr of TETA curing agent.
[b]30 to 80 phr of flexible curing agent.
[c]Mineral-glass reinforced.
[d]23 phr aromatic amine curing agent, 12 plies E-181 glass cloth with Volan® A finish.
[e]36% resin, 64% unidirectional nonwoven glass fiber reinforcement.
[f]NOL rings made with 12 end E-HTS glass; 15 phr metaphenylenediamine curing agent.

From *1973 Materials Selector,* Reinhold Publishing, Stamford, Conn., 1972, 250. With permission.

Table 1–10
EPOXIES – MOLDED, EXTRUDED

	ASTM No.	Type				
		Epoxy novolacs			High performance resins (cycloaliphatic diepoxides)	
		Cast, rigid[a]	Molded[b]	Glass cloth laminate[c]	Cast, rigid[d]	Glass cloth laminate[e]
Physical properties						
Specific gravity	D792	1.24	1.7	1.97	1.22	1.97
Thermal conductivity, Btu/hr/ft²/°F/ft	–	–	–	–	–	–
Coefficient of thermal expansion/°F × 10^{-6}	D696	–	1.7–2.2	–	1.6–3.0	–
Specific heat	–	–	–	–	–	–
Water absorption (24 hr), %	D570	–	0.11–0.2	0.04–0.06	0.1–0.7[f]	–
Flammability, ipm	D635	Self extinguishing	Self extinguishing	Self extinguishing	–	–
Transparency (visible light), %	–	–	Opaque	Opaque	–	Opaque
Mechanical properties						
Tensile strength, 1,000 psi	D638	8–12	5.2–5.3	50–52	9.6–12.0	59.2
Tensile modulus, 10^5 psi	D638	4–5	–	32–33	4.8–5.0	27.5
Elongation, %	D638	2–5	–	–	2.2–4.8	–
Impact strength (Izod notched), ft-lb/in.	D256	0.5	0.3–0.5	13–17	–	–
Flexural strength 1,000 psi	D790	11–16	10–12	70–72	12–13	84–89
Modulus of elasticity in flexure, 10^5 psi	D790	4–5	–	28–31	4.4–4.8	32–35
Compressive strength, 1,000 psi	D695	17–19	22–26	67–71	30–50	48–57
Hardness (Rockwell)	D785	107–112[g]	94–96D[h]	75–80	–	–
Electrical properties						
Volume resistivity, ohm-cm	D257	2.10 × 10^{14}	1.4–5.5 × 10^{14}	–	>10^{16}	–
Dielectric strength (step by step), V/mil	D149	–	280–400	–	444 (short time)	–
Dielectric constant	D150					
60 cycles		3.96–4.02	4.7–5.7	–	3.34–3.39	4.41–4.43
10^6 cycles		3.53–3.58	4.3–4.8	5.1[g]	–	–
Dissipation factor	D150					
60 cycles		0.0055–0.0074	0.0071–0.025	–	0.001–0.007	–
10^6 cycles		0.029–0.028	–	0.015[g]	–	–
Arc resistance, sec	D495	–	180–185	–	120	–

Table 1–10 (continued)
EPOXIES – MOLDED, EXTRUDED

	ASTM No.	Type				
		Epoxy novolacs			High performance resins (cycloaliphatic diepoxides)	
		Cast, rigid[a]	Molded[b]	Glass cloth laminate[c]	Cast, rigid[d]	Glass cloth laminate[e]
Fabricating properties						
Cure temperature, °F	–	195	–	380–400	250	250
Cure time, hr	–	2	–	20 min	1	0.5
Postcure temperature, °F	–	16	–	420	250	320
Postcure time, hr	–	390	–	24	3	7
Transfer molding	D1895					
Bulk factor		–	1.9–2.2	–	–	–
Pressure, psi		–	50–500	–	–	–
Temperature, °F		–	250–350	–	–	–
Mold shrinkage, %		–	0.003–0.006	–	–	–
Heat resistance						
Maximum recommended service temperature, °F	–	450	450–500	450–500	450–500	450–500
Heat deflection temperature (264 psi), °F	D648	300–400	300–425	–	300–525	–
Chemical resistance		Resistant to water and strong alkalis; more resistant to sulfuric and acetic acid and oxidizing agents than standard epoxy systems.			Outstanding weather resistance compared to other epoxy systems. Highly resistant to water, strong alkaline environments; less resistant to sulfuric and acetic acids, oxidizing agents.	
Uses		Impregnation and potting requiring high heat resistance; adhesives.	Electrical and electronic encapsulation designed for high temperature.	High temperature tooling, structural laminates, ablatives.	Encapsulation, impregnation, and potting requiring outstanding arc and tracking resistance.	Electrical laminates requiring outstanding weather resistance.

Table 1–10 (continued)
EPOXIES – MOLDED, EXTRUDED

[a]28 phr methylene dianilene; cure: 16 hr at 130°F, 2 hr at 257°F, 2 hr at 347°F.
[b]Mineral-filled proprietary compounds.
[c]12 plies glass cloth with Volan® A finish, 26 to 27% resin; cure: 20 min at 383 to 400°F and contact pressure, plus 24 min at 419°F postcure.
[d]12 phr of hexahydrophthalic and anhydride, 12 phr sodium alcoholate accelerator; cure: 24 phr at 250°F and postcure of 3 phr at 400°F.
[e]100 phr resin, 85 parts anhydride curing agent, 181 Volan glass cloth.
[f]1 hr at 212°F.
[g]1 mc.
[h]Durometer.

From *1973 Materials Selector,* Reinhold Publishing, Stamford, Conn., 1972, 251. With permission.

Table 1–11
MELAMINES – MOLDED

	ASTM No.	Filler and type			
		Unfilled	Cellulose electrical	Glass fiber	Alpha cellulose and mineral
Physical properties					
Specific gravity	D792	1.48	1.43–1.50	1.9–2.0	1.49
Thermal conductivity, Btu/hr/ft²/°F/ft		–	0.17–0.20	0.28	–
Coefficient of thermal expansion/°F	D696	–	1.11–2.78 × 10^{-5}	0.82 × 10^{-5}	–
Transmittance (luminous), %		Good	Opaque	–	–
Water absorption (24 hr), %	D570	0.2–0.5	0.27–0.80	0.09–0.60	0.5
Flammability		Self extinguishing	Self extinguishing	Self extinguishing	Self extinguishing
Mechanical properties					
Modulus of elasticity in tension, psi	D638	–	10–11 × 10^5	–	–
Tensile strength, 1,000 psi	D638	–	5–9	5–10	5
Elongation (in 2 in.), %	D638	–	0.6	–	–
Hardness (Rockwell)	D785	E110	M115–125	–	–
Impact strength (Izod notched), ft-lb/in.	D256	–	0.27–0.36	0.5–12.0	0.30
Modulus of elasticity in flexure, psi	D790	10–13 × 10^5	1.0–1.3 × 10^6	24 × 10^5	–
Compressive strength, 1,000 psi	D695	40–45	25–35	20–32	–
Flexural strength, 1,000 psi	D790	9.5–14	6–15	10–24	8

Table 1–11 (continued)
MELAMINES – MOLDED

	ASTM No.	Filler and type			
		Unfilled	Cellulose electrical	Glass fiber	Alpha cellulose and mineral
Electrical properties					
Volume resistivity, ohm-cm	D257	–	10^{12}–10^{13}	1–7 × 10^{11}	10^{12}
Dielectric strength (short time), V/mil	D149	–	350–400	250–300	375
Dielectric constant					
60 cycles	D150	7.9–11.0	6.2–7.7	7.0–11.1	–
10^6 cycles	D150	6.3–7.3	5.2–6.0	6.9–7.9	6.4
Dissipation factor					
60 cycles		0.048–0.162	0.026–0.192	0.14–0.23	–
10^6 cycles		0.031–0.040	0.032–0.12	0.013–0.03	0.031
Arc resistance, sec	D495	100–145	70–135	180–186	125
Fabricating properties					
Bulk factor	D392	2.0	2.2–2.6	5–7	2.4
Compression molding					
Pressure, 1,000 psi		2–5	1.5–3	2–8	2–5
Temperature, °F		300–340	290–360	280–340	280–350
Transfer molding					
Pressure, 1,000 psi		–	6–20	8–20	2–10
Temperature, °F		–	300–330	290–310	285–350
Mold shrinkage, in./in.		0.011–0.012	0.006–0.008	0.001–0.004	0.006–0.007
Heat resistance					
Maximum recommended service temperature, °F		210	250–280	300–400	275–325
Heat distortion temperature (264 psi), °F	D648	293–298	265	400	300
Chemical resistance	Resistant to weak acids, weak alkalis, organic solvents, greases, and oils. Attacked by strong acids and strong alkalis.				

Table 1–11 (continued)
MELAMINES – MOLDED

	ASTM No.	Filler and type			
		Unfilled	Cellulose electrical	Glass fiber	Alpha cellulose and mineral
Uses		Pearlescent buttons, moldings, ornamental applications.	General mechanical and electrical applications, particularly at elevated temperatures, Applications requiring improved holding power for metallic inserts such as electrical and electronic parts.	Applications requiring high shock resistance, good electrical properties, and high resistance to burning. Switchgear, terminal strips, stand-off insulators, coil forms.	Primarily electrical applications requiring low after-shrinkage, good dimensional stability, and excellent molding characteristics.

From *1973 Materials Selector,* Reinhold Publishing, Stamford, Conn., 1972, 252. With permission.

Table 1–12
NYLONS – MOLDED, EXTRUDED

		Material						
		Type 6						
	ASTM No.	General purpose[a]	Glass fiber (30%) reinforced[a]	Cast	Flexible[a] copolymers	Type 8[b]	Type 11	Type 12
Physical properties								
Specific gravity	D792	1.14	1.37	1.15	1.12–1.14	1.09	1.04	1.01
Thermal conductivity, Btu/hr/ft²/°F/in.	–	1.2	1.2–1.7	1.2–1.7	–	–	1.5	1.7
Coefficient of thermal expansion, 10^{-5}/°F	D696	4.8	1.2	4.4	–	–	5.5	7.2
Specific heat, Btu/lb/°F	–	0.4	–	0.4	–	0.4	0.58	0.28
Refractive, index, n_D	D542	–	–	–	–	–	–	–
Water absorption (24 hr), %	D570	1.7–1.8	1.3	0.6	0.8–1.4	9.5	0.4	0.25
Flammability, ipm	D635	Self extinguishing	Slow burn	Self extinguishing	Slow burn, 0.6	Self extinguishing	Self extinguishing	c
Mechanical properties								
Tensile strength (2 in./min)[d]	D638							
Ultimate		9.5–12.5	21–23	12.8	7.5–10.0	–	–	7.1–8.5
Yield		8.5–12.5	–	12.8	7.5–10.0	3.9	8.5	5.5–6.5
Elongation (2 in./min), %	D638							
Ultimate		30–220	2–4	20	200–320	400	100–120	120–350
Yield		–	–	5	–	–	–	5.8
Modulus of elasticity in tension, 10^5 psi	D638	–	10–12	5.4	–	0.3	1.78–1.85	1.7–2.1
Flexural strength, 1,000 psi	D790	Unbreakable	26–34	16.5	3.4–16.4	–	–	–
Modulus of elasticity in flexure, 10^5 psi	D790	1.4–3.7	10–12	5.05	0.92–3.2	0.4	1.51	–
Impact strength (Izod notched), ft-lb/in.	D256	0.8–1.2	3–2.3	1.2	1.5–19	>16	3.3–3.6	1.2–4.2
Compressive strength (1% offset)[d]	D695	9.7	19–20	14	–	–	–	–
Hardness (Rockwell)	D785	R118–R120	R121	R116	R72–R119	–	R100–R108	R106
Coefficient of dynamic friction	–	–	–	0.32[c]	–	–	–	–
Abrasion resistance (Taber, CS-17), mg/1,000 cycles	D1044	5	–	2.7	–	–	–	–
Electrical properties								
Volume resistivity, ohm-cm	D257	4.5×10^{13}	$2.8–10^{14}$ – 1.5×10^{15}	2.6×10^{14}	–	1.5×10^{11}	2×10^{13}	10^{14}–10^{15}
Dielectric strength (short time), V/mil	D149	385	400–450	380	440	340	425	840
Dielectric constant	D150							
60 cycles		4.0–5.3	4.6–5.6	4.0	3.2–4.0	9.3	3.3 (10^3 cps)	3.6 (10^3 cps)
10^6 cycles		3.6–3.8	3.9–5.4	3.3	3.0–3.6	4.0	–	–

Table 1–12 (continued)
NYLONS – MOLDED, EXTRUDED

		Material						
		Type 6						
	ASTM No.	General purpose[a]	Glass fiber (30%) reinforced[a]	Cast	Flexible[a] copolymers	Type 8[b]	Type 11	Type 12
Electrical properties (cont.)								
Dissipation factor	D150							
60 cycles		0.06–0.014	0.022–0.008	0.015	0.007–0.010	0.19	0.03	0.04 (10^3 cps)
10^6 cycles		0.03–0.04	0.019–0.015	0.05	0.010–0.015	0.08	0.02	–
Arc resistance, sec	D495	–	92–81	–	–	–	–	–
Fabricating properties								
Bulk factor	D1895	–	2.23	–	–	–	2.2	1.9
Injection molding	–							
Pressure, 1,000 psi		10–20	10–20	–	14–22	8–12	4–10,	8.5–14.2
Temperature, °F		440–550	500–570	–	420–550	300–400	390–520	375–460
Shrinkage, in./in.		0.12–0.20	0.003	–	0.015–0.025	0.008–0.025	0.010–0.021	0.003–0.016
Extrusion temperature	–	450–600	500–600	–	420–575	300–380	350–550	375–460
Heat resistance								
Maximum recommended service temperature, °F	–	250–300[f]	250–300	250–300	175–200	–	212–250	175–230
Deflection temp	D648							
66 psi		360	425–428	420	260–350	129	302	–
264 psi		155–160	420–419	410	115–130	–	131	–
Chemical resistance		Resists esters, ketones, alkalis, weak acids, alcohols, and common solvents. Not resistant to concentrated mineral acids.		Resists most organic chemicals such as alcohols, ketones, hydrocarbons, and chlor. solvents. Attacked by strong acids, phenols, strong oxidizing agents.	Resists esters, ketones, alkalis, weak acids, alcohols, and common solvents. Not resistant to concentrated mineral acids.	Excellent to resistance aqueous alkalis, aliphatic and aromatic hydrocarbons, ether and mineral oils; poor to good resistance to dilute mineral acids, alcohols, aromatic acids.	Resists alkalis, petroleum products, and common organic solvents. Not resistant to phenols and concentrated acids and oxidants.	Resists alkalis, petroleum products, and common organic solvents. Not resistant to phenols and concentrated acids and oxidants.

Table 1–12 (continued)
NYLONS – MOLDED, EXTRUDED

		Material						
		Type 6						
	ASTM No.	General purpose[a]	Glass fiber (30%) reinforced[a]	Cast	Flexible[a] copolymers	Type 8[b]	Type 11	Type 12
Uses		Bearings, gears, bushings, coil forms, brush backs, rod, tubing, tape.	General purpose Type 6 parts requiring greater stiffness and dimensional stability.	Bearings, wearplates, bushings, gears, rollers, stock shapes.	Parts requiring high impact strength or flexibility.	Molded parts requiring flexibility and chemical resistance	Electrical insulation and other nylon uses where low moisture absorption is needed.	Filament, rod, tubing, sheet, moldings requiring dimensional stability and low moisture absorption.

[a]Dry, as-molded properties.
[b]Non-cross-linked; can be cross-linked.
[c]Dynamic, no lubrication, nylon to steel.
[d]1,000 psi.
[e]0.4, self extinguishing to slow burning.
[f]Heat stabilized.

From *1973 Materials Selector,* Reinhold Publishing, Stamford, Conn., 1972, 253. With permission.

Table 1–13
NYLONS – MOLDED AND EXTRUDED

	ASTM No.	Type: 6/6 Nylon, General purpose molding[a]	6/6 Nylon, Glass fiber reinforced[b]	6/6 Nylon, Glass fiber molybdenum disulfide filled[c]	6/6 Nylon, General purpose extrusion[a]	6/10 Nylon, General purpose[a]	6/10 Nylon, Glass fiber (30%) reinforced[c]
Physical properties							
Specific gravity	D792	1.13–1.15, –	1.37, 1.47	1.37–1.41	1.13, 1.15	1.07–1.09, –	1.30
Thermal conductivity, Btu/hr/ft²/°F/in.	–	1.7, –	1.5, 3.3	–	1.7, –	1.5	3.5
Coefficient of thermal expansion, 10^{-5}/°F	D696	4.5, –	2.1, 1.4	1.75	–	5	2.5
Specific heat, Btu/lb/°F	–	0.3–0.5	–	–	0.3–0.5	0.3–0.5	–
Refractive, index, n_D	D542	Translucent	Opaque	Opaque	Opaque	Opaque	Opaque
Water absorption (24 hr), %	D570	1.5, –	0.9, 0.8	0.5–0.7	1.5	0.4	0.2
Flammability, ipm	D635	Self extinguishing	Slow burn	Slow burn	Self extinguishing	Self extinguishing	Slow burn
Coefficient of static frict (against self)	–	0.04–0.13, –	–	–	–	–	–
Mechanical properties							
Tensile strength, 1,000 psi	D638						
Ultimate		11.8, 11.2	25, 30	19–22	1.26, 8.6	8.5, 7.1	19
Yield		11.8, 8.5	–	–	12.6, 8.6	8.5, 7.1	–
Elongation, %	D638						
Ultimate		60, 300	1.8, 2.2	3	90, 240	85, 220	1.9
Yield		5, 25	–	–	5, 30	5, 30	–
Modulus of elasticity in tension, 10^5 psi	D638	4.75, 3.85	14, 20	–	–	2.8–3.0, –	–
Flexural strength, 1,000 psi	D790	Unbreakable	26, 35	26–28	–	8	23
Modulus of elasticity in flexure, 10^5 psi	D790	410, 175	10, 18	11–13	4.1, 1.75	2.8, 1.6	8.5
Impact strength (Izod notched), ft-lb/in.	D638	1.0, 2.0	2.5, 3.4[e]	–	1.3, –	0.6, 1.6	3.4
Compressive strength (1%), 1,000 psi	D695	4.9, –	20, 24	–	4.9 (1%), –	3.0 (1%), –	18
Hardness (Rockwell)	D785	R118, R108	E60, E80	M95–100	R118–108	R111	E40–50
Abrasion resistance (Taber CS-17), mg/1,000 cycles	D1044	3–5, 6–8	–	–	–, 3–5	–	–
Electrical properties							
Volume resistivity, ohm-cm	D257	10^{14}–10^{15}	5.5×10^{15} 2.6×10^{15}	–	10^{15}	10^{15}	–
Dielectric strength (short time), V/mil	D149	385	400, 480	300–400	–	470	
Dielectric constant	D150						
60 cycles		4.0, –	4.0, 4.4	–	–	3.9, –	
10^6 cycles		3.6, –	3.5, 4.1	–	–	3.5, –	

Table 1–13 (continued)
NYLONS – MOLDED AND EXTRUDED

	ASTM No.	6/6 Nylon: General purpose molding[a]	6/6 Nylon: Glass fiber reinforced[b]	6/6 Nylon: Glass fiber molybdenum disulfide filled[c]	6/6 Nylon: General purpose extrusion[a]	6/10 Nylon: General purpose[a]	6/10 Nylon: Glass fiber (30%) reinforced[c]
Electrical properties (cont.)							
Dissipation factor	D150						
60 cycles		0.014, 0.04	0.018, 0.009	–	–	0.04, –	–
10^6 cycles		0.04, –	0.017, 0.018	–	–	–	–
Arc resistance, sec	D495	120	148, 100	135	120	120	–
Heat resistance							
Maximum recommended service temperature, °F	–	250–300[d]	250–300[d]	250–300[d]	250–300[d]	255–300[d]	250–300[d]
Deflection temperature, °F	D648						
66 psi		470	507, 509	–	470	300	430
264 psi		220	495, 500	–	220	135	420
Fabricating properties							
Bulk factor	D392	2.1	2.23, 2.30	1.7	–	2.2	–
Injection molding	–						
Pressure, 1,000 psi		10–20	–	–	–	10–20	–
Temperature, °F		520–650	–	–	–	450–600	–
Shrinkage, in./in.		0.015	–	–	–	0.015	0.0035 0.0045
Extrusion temperature	–	–	–	–	530–570	460–500	–
Chemical resistance		Inert to most organic chemicals such as esters, ketones, alcohols, and hydrocarbons. Resist alkalis and salt solutions, but attacked by phenols, formic acid, strong mineral acids, and strong oxidizing agents.					
Uses		Bearings, gears, bushings, coil forms, brush backs, rod, tubing, tape.		Mechanical parts where lubrication is undesirable or difficult.	Tubing. rod, pipe, sheeting, laminations.	Jacketing for wire and cable, special molded parts.	

Table 1–13 (continued)
NYLONS – MOLDED AND EXTRUDED

[a]Where two values are given, the first is for dry, as-molded material, and the second for moisture equilibrium in air; single value pertains to dry material.
[b]First value is for 30% glass fiber, the second for 40%. All values are at moisture equilibrium.
[c]30% glass fiber.
[d]Heat stabilized for maximum heat resistance.
[e]¼ in.

From *1973 Materials Selector,* Reinhold Publishing, Stamford, Conn., 1972, 254. With permission.

Table 1–14
PHENOLICS – MOLDED

		Type and filler			
	ASTM No.	General–woodflour and flock	Shock–paper, flock or pulp	High shock–chopped fabric or cord	Very high shock–glass fiber
Physical properties					
Specific gravity	D792	1.32–1.46	1.34–1.46	1.36–1.43	1.75–1.90
Thermal conductivity, Btu/hr/ft²/°F/ft	C177	0.097–0.3	0.1–0.16	0.097–0.170	0.20
Coefficient of thermal expansion, 10^{-5}/°F	D696	1.66–2.50	1.6–2.3	1.60–2.22	0.88
Specific heat, Btu/lb/°F		0.35–0.40	–	0.30–0.35	0.28–0.32
Water absorption (24 hr), %	D570	0.3–0.8	0.4–1.5	0.4–1.75	0.1–1.0
Flammability	D635	Self extinguishing	Self extinguishing	Self extinguishing	Self extinguishing
Mechanical properties					
Modulus of elasticity in tension, 10^5 psi	D638	8–13	8–12	9–14	30–33
Tensile strength, 1,000 psi	D638				
	D651	5.0–8.5	5.0–8.5	5–9	5–10
Elongation (in 2 in.), %	D638	0.4–0.8	–	0.37–0.57	0.2
Hardness (Rockwell)	D785	E85–100	E85–95	E80–90	E50–70
Impact strength (Izod notched), ft-lb/in.	D256	0.24–0.50	0.4–1.0	0.6–8.0	10–33
Modulus of elasticity in flexure, 10^5 psi	D790	8–12	8–12	9–13	30–33
Flexural strength, 1,000 psi	D790	8.5–12	8.0–11.5	8–15	10–45
Compressive strength, 1,000 psi	D695	22–36	24–35	15–30	17–30

Table 1–14 (continued)
PHENOLICS – MOLDED

	ASTM No.	Type and filler: General–woodflour and flock	Shock–paper, flock or pulp	High shock–chopped fabric or cord	Very high shock–glass fiber
Electrical properties					
Volume resistivity, ohm-cm	D257	10^9-10^{13}	$1-50 \times 10^{11}$	$>10^{10}$	$7-10 \times 10^{12}$
Dielectric strength (short time), V/mil	D149	200–425	250–350	200–350	200–370
Dielectric constant					
60 cycles	D150	5.0–9.0	5.6–11.0	6.5–15.0	7.1–7.2
10^6 cycles	D150	4.0–7.0	4.5–7.0	4.5–7.0	4.6–6.6
Dissipation factor					
60 cycles	D150	0.05–0.30	0.08–0.35	0.08–0.45	0.02–0.03
10^6 cycles	D150	0.03–0.07	0.03–0.07	0.03–0.09	0.02
Arc resistance, sec	D495	5–60	5–60	5–60	60
Fabricating properties					
Bulk factor		2.1–4.4	2.3–5.7	3.0–18.0	5–7
Compression molding					
Pressure, 1,000 psi		1.5–5.0	2–5	2–6.5	1–5
Temperature, °F		290–380	290–380	280–380	280–350
Transfer molding					
Pressure, 1,000 psi		2–10	2–10	2–12	4–8
Temperature, °F		275–340	275–340	275–340	280–350
Injection molding					
Pressure, 1,000 psi		4–8	6–9	6–12	–
Temperature, °F		320–340	–	–	–
Mold shrinkage, in./in.		0.005–0.008	0.004–0.009	0.002–0.009	0.0
Heat resistance					
Maximum recommended service temperature, °F		300–350	300	250–300	350–450
Deflection temperature, °F	D648	260–360	290–340	250–340	600
Chemical resistance		Severely attacked by strong acids and strong alkalis. Effects of dilute acids, alkalis, and organic solvents vary with the reagent. Chemical resistance varies with the particular formulation and not all materials of a type are equally resistant.			
Uses		Mechanical applications include pulleys, wheels, motor housings, handles. Electrical uses include coil forms, ignition parts, condenser housings, fuse blocks, instrument panels. Thermal applications include handles, appliance connector plugs. Chemical uses include photographic development tanks, rayon spinning buckets and parts, milking machine cups. Decorative uses include radio and television cabinets, handles, knobs, buttons.			

From *1973 Materials Selector,* Reinhold Publishing, Stamford, Conn., 1972, 255. With permission.

Table 1–15
PHENOLICS – MOLDED

	ASTM No.	Type and filler: Arc resistant–mineral	Rubber phenolic–woodflour or flock	Rubber phenolic–chopped fabric	Rubber phenolic–asbestos
Physical properties					
Specific gravity	D792	1.6–3.0	1.24–1.35	1.30–1.35	1.60–1.65
Thermal conductivity, Btu/hr/ft^2/°F/ft	C177	0.24–0.34	0.12	0.05	0.04
Coefficient of thermal expansion, 10^{-5}/°F	D696	–	0.83–2.20	1.7	2.2
Specific heat, Btu/lb/°F		0.28–0.32	0.33	–	–
Water absorption (24 hr), %	D570	0.2	0.5–2.0	0.5–2.0	0.10–0.50
Flammability	D635	Self extinguishing	Self extinguishing	Self extinguishing	Self extinguishing
Mechanical properties					
Modulus of elasticity in tension, 10^5 psi	D638	10–30	4–6	3.5–6	5–9
Tensile strength, 1,000 psi	D638				
	D651	6	4.5–9	3.5	4
Elongation (in 2 in.), %	D638	–	0.75–2.25	–	–
Hardness (Rockwell)	D785	E80–90	M40–90	M57	M50
Impact strength (Izod notched), ft-lb/in.	D256	0.32	0.34–1.0	2.0–2.3	0.3–0.4
Modulus of elasticity in flexure, 10^5 psi	D790	10–30	4–6	3.5	5.0
Flexural strength, 1,000 psi	D790	10	7–12	7	7
Compressive strength, 1,000 psi	D695	20	12–20	10–15	10–20
Electrical properties					
Volume resistivity, ohm-cm	D257	6×10^{12}	10^8–10^{11}	10^{11}	10^{11}
Dielectric strength (short time), V/mil	D149	380	250–375	250	350
Dielectric constant					
60 cycles	D150	7.4	9–16	15	15
10^6 cycles	D150	5.0	5	5	5
Dissipation factor					
60 cycles	D150	0.13–0.16	0.15–0.60	0.5	0.15
10^6 cycles	D150	0.10	0.1–0.2	0.09	0.13
Arc resistance, sec	D495	180	7–20	10–20	5–20
Fabricating properties					
Bulk factor		2.4	2.5–4.0	4.6–8.0	2.5
Compression molding					
Pressure, 1,000 psi		2–5	2–6	2–6	2–6
Temperature, °F		285–350	300–360	300–350	300–350

Table 1–15 (continued)
PHENOLICS – MOLDED

	ASTM No.	Type and filler: Arc resistant–mineral	Rubber phenolic–woodflour or flock	Rubber phenolic–chopped fabric	Rubber phenolic–asbestos
Fabricating properties (cont.)					
Transfer molding					
Pressure, 1,000 psi		2–10	1–12	2–12	2–12
Temperature, °F		285–350	300–350	300–350	300–350
Injection molding					
Pressure, 1,000 psi		2–8	–	–	–
Temperature, °F		–	–	–	–
Mold shrinkage, in./in.		0.004	0.005–0.010	0.003–0.006	0.005–0.008
Heat resistance					
Maximum recommended service temperature, °F		400	212–300	212–225	225–360
Deflection temperature, °F	D648	335	220–270	220–280	250–300
Chemical resistance		Severely attacked by strong acids and strong alkalis. Effects of dilute acids, alkalis, and organic solvents vary with the reagent. Chemical resistance varies with the particular formulation and not all materials of a type are equally resistant.			
Uses		Mechanical applications include pulleys, wheels, motor housings, handles. Electrical uses include coil forms, ignition parts, condenser housings, fuse blocks, instrument panels. Thermal applications include handles, appliance connector plugs. Chemical uses include photographic development tanks, rayon spinning buckets and parts, milking machine cups. Decorative uses include radio and television cabinets, handles, knobs, buttons.			

From *1973 Materials Selector*, Reinhold Publishing, Stamford, Conn., 1972, 256. With permission.

Table 1–16
ABS-POLYCARBONATE ALLOY, PVC-ACRYLIC ALLOY, POLYIMIDES

		Material					
					Polymides		
	ASTM No.	ABS-polycarbonate	PVC-acrylic[a] sheet	PVC-acrylic injection molded	Unreinforced		Glass reinforced
Physical properties							
Specific gravity	D792	1.14	1.35	1.30	14.3	1.47	1.90
Thermal conductivity, Btu/hr/ft²/°F/in.	Cenco Fitch	2.46 (per ft)	1.01	0.98	6.78[b]	3.8[b]	3.59[b]
Coefficient of thermal expansion, 10^{-5}/°F	D696	6.12	3.5	–	2.5	3.0	0.8
Specific heat, Btu/lb/°F	–	–	0.293	–	0.31[b,c]	0.28[b,c]	0.27[b,c]
Refractive index, n_D	D542	Opaque	Opaque	Opaque	Opaque	Opaque	Opaque
Water absorption (24 hr), %	D570	0.21	0.06	0.13	0.47	0.68	0.2
Flammability, ipm	D635	0.90	Nonburning	Nonburning	IBM Class A	IBM Class A	UL SE-0
Coefficient of friction	–	0.2 (to itself)	–	–	–	–	–
Mechanical properties							
Tensile strength, 1,000 psi	D638						
Ultimate		8.2	–	–	–	–	–
Yield		8.2	6.5	5.5	7.5	5	28
Elongation, %	D638						
Ultimate		110	>100	150	<1	1.2	<1
Yield		–	–	–	–	–	–
Modulus of elasticity in tension, 10^5 psi	D638	3.7	3.35	2.75	7.5	5.4	45
Flexural strength, 1,000 psi	D790	14.3	10.7	8.7	11	6.6	56
Modulus of elasticity in flexure, 10^5 psi	D790	4.0	4.0	3.0	7.0	5.0	38.4
Impact strength (Izod notched), ft-lb/in.	D638	10	15	15	0.5[d]	0.5[d]	17[d]
Compressive strength, 1,000 psi	D695	11.1–11.8	8.4	6.2	27.4	18.4	42
Hardness (Rockwell)	D785	R118	R105	R104	99	95	114E
Abrasion resistance (Taber, CS-10 wheel), mg/1,000 cycles, g loss	D1044	–	0.073	0.0058	0.080[a]	0.004[e]	20
Electrical properties							
Volume resistivity, ohm-cm	D257	2.2×10^{16}	$1-5 \times 10^{13}$	5×10^{15}	f	4×10^{15}	9.2×10^{15}
Dielectric strength (short time), V/mil	D149	500	>429	400	f	310	300
Dielectric constant	D150						
60 cycles		2.74	3.86	4.0	f	4.12	4.84
10^6 cycles		2.69	3.44	3.4	f	3.96	4.74
Dissipation factor	D150						
60 cycles		0.0026	0.076	0.037	f	0.003	0.0034
10^6 cycles		0.0059	0.094	0.031	f	0.011	0.0055
Arc resistance, sec	D495	96	80	25	f	152	50–180

Table 1–16 (continued)
ABS-POLYCARBONATE ALLOY, PVC-ACRYLIC ALLOY, POLYIMIDES

		Material					
					Polymides		
	ASTM No.	ABS-polycarbonate	PVC-acrylic[a] sheet	PVC-acrylic injection molded	Unreinforced		Glass reinforced
Heat resistance							
Maximum recommended service temperature, °F	–	240	160	–	500	500	500
Deflection temperature, °F	D648						
66 psi		261	177	180	–	–	–
264 psi		246	160	171	582	615	660
Fabricating properties							
Bulk factor	–	–	–	–	1.9	2.0	8.3
Injection molding	–						
Pressure, 1,000 psi		10–20	–	–	2–6	2–6	2–6
Temperature, °F		490–530	–	340–380	440–480	480–520	340–480
Shrinkage, in./in.		0.005	–	0.007	0.008	0.008	0.002
Chemical resistance		Not affected by weak acids or weak bases; attacked by oxidizing acids and strong bases. Will dissolve and swell in organic hydrocarbons.	Meets ASTM D1784 standard for Type 1 grade 2 rigid PVC. Excellent resistance to acids, bases, and oils.	Excellent resistance to acids, bases, and oils.	Resists polar and nonpolar organic solvents, dilute acids, and bases.		

Table 1–16 (continued)
ABS-POLYCARBONATE ALLOY, PVC-ACRYLIC ALLOY, POLYIMIDES

		Material				
					Polymides	
	ASTM No.	ABS-polycarbonate	PVC-acrylic[a] sheet	PVC-acrylic injection molded	Unreinforced	Glass reinforced
Uses		Machine housings, safety helmets, tote boxes and trays, luggage, carrying cases, sports equipment.	Machine housings, safety helmets, tote boxes and trays, corrosion-resistant ducts, seating, luggage.	Appliances, power tool housings, safety helmets, sporting goods, corner guards.	Bushings, valve seats and high-temperature mechanical parts.	High temperature uses such as jet engine components.

[a]Extruded sheet.
[b]G.E. test.
[c]cal/g/°C.
[d]ASTM D256.
[e]in./1,000 hr dry, 10,000 PV.
[f]This grade is an electrical conductor.

From *1973 Materials Selector,* Reinhold Publishing, Stamford, Conn., 1972, 257. With permission.

Table 1–17
POLYACETALS

	ASTM No.	Homopolymer[a] Standard	Homopolymer[a] 20% glass reinforced	Homopolymer[a] 22% TFE reinforced	Copolymer[b] Standard	Copolymer[b] 25% glass reinforced	Copolymer[b] High flow
Physical properties							
Specific gravity	D792	1.425	1.56	1.54	1.410	1.61	1.410
Thermal conductivity, Btu/hr/ft²/°F/ft	–	0.13	–	–	0.16	–	1.6
Coefficient of thermal expansion, 10^{-5}/°F	D696	4.5	2.0–4.5	4.5	4.7	2.2–4.7	4.7
Specific heat, Btu/lb/°F	–	0.35	–	–	0.35	–	0.35
Refractive index, n_D	D542	Opaque	Opaque	Opaque	Opaque	Opaque	Opaque
Water absorption (24 hr), %	D570	0.25	0.25	0.20	0.22	0.29	0.22
Flammability, ipm	D635	1.1	0.8	0.8	1.1	1.0	1.1
Mechanical properties							
Tensile strength, 1,000 psi	D638						
Ultimate	–	10.0	8.5	6.9	8.8	18.5	8.8
Yield	–	10.0	–	–	8.8	18.5	8.8
Elongation, %	D638			12			
Ultimate	–	25	7	–	60–75	3	40
Yield	–	12	–	–	12	3	12
Modulus of elasticity in tension, 10^5 psi	D638	5.2	–	–	4.1	12.5	4.13
Flexural strength, 1,000 psi	D790	14.1	–	–	13	28	13
Modulus of elasticity in flexure, 10^5 psi	D790	4.1	8.8	4.0	3.75	11	3.75
Impact strength (Izod, notched), ft-lb/in.	D638	1.4	0.8	0.7	1.3	1.8	1.0
Compressive strength (1%), 1,000 psi	D695	5.2	5.2	4.5	4.5	–	4.5
Hardness (Rockwell)	D785	M94	M90	M78	M80	M79	M80
Coefficient of static friction (against steel)	–	0.1–0.3	0.1–0.3	0.05–0.15	0.15	0.15	0.15
Abrasion resistance (Taber, CS-17), mg/ 1,000 cycles	D1044	14–20	33	9	14	40	14
Electrical properties							
Volume resistivity, ohm-cm	D257	1×10^{15}	5×10^{14}	–	1×10^{14}	1.2×10^{14}	1.0×10^{14}
Dielectric strength (short time), V/mil	D149	500	500	–	500	580	500
Dielectric constant	D150						
60 cycles	–	3.7	4.0	–	3.7 at 100	3.9 at 100	3.7 at 100
10^6 cycles	–	3.7	4.0	–	3.7	3.9	3.7
Dissipation factor	D150						
60 cycles	–	0.0048	0.0047	–	0.001 at 100	0.003 at 100	0.001 at 100
10^6 cycles	–	0.0048	0.0036	–	0.006	0.006	0.006
Arc resistance, sec	D495	129[c]	188	–	240	136	240

Table 1–17 (continued)
POLYACETALS

		Material					
		Homopolymer[a]			Copolymer[b]		
	ASTM No.	Standard	20% glass reinforced	22% TFE reinforced	Standard	25% glass reinforced	High flow
Heat resistance							
Maximum recommended service temperature, °F	–	195	195	195	220	220	220
Deflection temperature, °F	D648						
66 psi	–	338	345	329	316	331	316
264 psi	–	255	3.5	212	230	325	230
Fabricating properties							
Bulk factor	–	1.78	–	–	1.78	–	1.78
Injection molding	–						
Pressure, 1,000 psi	–	15–25	15–25	15–25	15–20	15–20	15–20
Temperature, °F	–	380–420	380–420	380–410	360–440	375–440	375–440
Shrinkage, in./in.	–	0.025	0.01–0.025	0.025	0.020	0.018–0.004	0.018
Extrusion temperature, °F	–	370–400	370–400	370–400	360–440	360–440	360–440
Chemical resistance		Excellent resistance to most organic solvents, including aliphatic and aromatic hydrocarbons. Not recommended for use with strong acids and alkalis.	Same as standard homopolymer.	Same as standard homopoly mer.	Excellent resistance to strong alkalis. Most organic solvents including alcohols, ketones, esters, aliphatic and aromatic hydrocarbons, and glycols do not seriously alter properties. Not recommended for use in strong mineral acids and oxidizing reagents.		

Table 1–17 (continued)
POLYACETALS

	ASTM No.	Material					
		Homopolymer[a]			Copolymer[b]		
		Standard	20% glass reinforced	22% TFE reinforced	Standard	25% glass reinforced	High flow
Uses		Appliance parts, gears, bushings, aerosol bottles, auto, plumbing, textile, consumer uses.	Same as homopolymer. Where high stiffness and dimensional stability are required.	Same as homopolymer. Where low friction and high resistance to wear are required.	Appliance parts, gears, bushings, aerosol bottles, various automotive, plumbing, textile machinery, and consumer products.		

[a]Delrin® is most common trade name.
[b]Celcon® is most common trade name.
[c]15-mil specimen.

From *1973 Materials Selector*, Reinhold Publishing, Stamford, Conn., 1972, 258. With permission.

Table 1–18
POLYESTER – THERMOPLASTIC

	ASTM No.	Type: Injection moldings: Polybutylene terephthalates: General purpose grade	Glass-reinforced grades	Glass reinforced self extinguishing	Polytetramethylene terephthalate: General purpose grade	Glass-reinforced grade	Asbestos-filled grade
Physical properties							
Specific gravity	D792	1.31	1.52	1.58	1.31	1.45	1.46
Specific heat, Btu/lb/°F	–	–	–	–	0.36–0.55	–	–
Thermal conductivity, Btu/hr/ft²/°F/in.	C177	1.1	1.3	1.3	–	–	–
Coefficient of thermal expansion, 10^{-5}/in./in./°F	D696	5.3	2.7–3.3	3.5	4.9–13.0[a]	–	–
Volume resistivity, 10^{16} ohm-cm	D257	4	3.2–3.3	3.4	2×10^{15}	–	3×10^{14}
Dielectric strength (short time), V/mil	D149	590	560–750	750	420–540	–	580
Dielectric constant	D150	3.1–3.3	3.7–4.2	3.7–3.8	3.16	–	3.5–4.2
Dissipation factor (to 10^3 Hz)	D150	0.002	0.002–0.003	0.002	0.023	–	0.015
Arc resistance, sec	D495	190	130	80	125	–	108
Water absorption (24 hr), %	D570	0.08	0.06–0.07	0.07	0.09	0.07	0.1
Flammability	D635	Slow burn	Slow burn	Self extinguishing	Slow burn	Slow burn	–
	UL94[b]	Slow burn	Slow burn	Self extinguishing	–	–	–
Mechanical Properties							
Tensile strength, 10^3 psi	D638	8	17–17.3	17	8.2	14	12
212°F			7			–	–
302°F			5.5			–	–
Elongation, %	D638	300	1–5	5	250	<5	<5
Flexural strength, 10^3 psi	D790	12.8	22–24	23	12	19	19
Flexural modulus, 10^6 psi	D790	0.34	1.2–1.5	1.2	3.3	8.7	9.0
212°F		–	0.63	–	–	–	–
302°F		–	0.53	–	–	–	–
Compressive strength, 10^3 psi	D695	13	16–18	18	–	–	–
Shear strength 10^3 psi	D732	7.7	8.9	9	–	–	–
Impact strength, (Izod), ft-lb/in.	D256						
Notched		1.2	1.3–2.2	1.8	1	1	0.5
Unnotched		–	4.2–15	15	–	7	6.0
Hardness (Rockwell)	D785	R117	R118–M90	R119	R117	R117–M85	M85
Heat deflection temperature, °F	D648						
66 psi		310	420	420	302	–	–
264 psi		130	415–416	415	122	380	330

Table 1–18 (continued)
POLYESTER – THERMOPLASTIC

		Type					
		Injection moldings					
		Polybutylene terephthalates			Polytetramethylene terephthalate		
	ASTM No.	General purpose grade	Glass-reinforced grades	Glass reinforced self extinguishing	General purpose grade	Glass-reinforced grade	Asbestos-filled grade
Mechanical properties (cont.)							
Endurance limit (10^7 cycles), 10^3 psi	D671	2.85	5	5	–	–	–
Creep, %	D674	1.1 (1,000 psi)	0.44 (4,000 psi)	0.44 (4,000 psi)	–	–	–
Abrasion resistance,[c] mg/10^3 cycles	D1044	6.5	9–50	11	–	–	–
Coefficient of friction	D1894						
Self		0.17	0.16	0.16	–	–	–
Steel		0.13	0.14	0.14	–	–	–
Fabricating properties							
Injection molding							
Pressure, psi		–	–	–	800–1,000	800–1,000	–
Temperature, °F		100–175	150–200	150–200	100–150	150	100
Shrinkage, in./in.		0.020	0.002–0.003	0.002–0.003	0.020	0.003	0.006
Extrusion temperature,[d] °F		–	–	–	520	–	–
Machining							
Sawing		Standard equipment; speeds 2,500–3,000 ft/min; moderate feeds.					
Drilling		Standard equipment; 1,000–3,000 rpm.					
Milling		Standard equipment; 2,700–3,500 rpm; feed 20–36 ipr.					
Finishing		File, sand, grind, ream, tap; avoid overheating.					
Decorating		Hot stamp, silk screen, vacuum metallize, paint.					
Fastening		Standard screws; inserts; ultrasonic, epoxy bond.					
Chemical resistance		Resistant to aliphatic hydrocarbons, gasoline, carbon tetrachloride, perchloroethylene, oils, fats, alcohols, glycols, ethers, high molecular weight esters and ketones, dilute acids and bases, detergents, most aqueous salt solutions, at ambient temperatures. Attacked by strong acids and bases. Resistant to potable water at ambient temperatures. Prolonged use in water above 150°F not recommended. Swollen by ethylene dichloride, low molecular weight ketones, and substituted aromatic compounds.					

Table 1–18 (continued)
POLYESTER – THERMOPLASTIC

		Type					
		Injection moldings					
		Polybutylene terephthalates			Polytetramethylene terephthalate		
	ASTM No.	General purpose grade	Glass-reinforced grades	Glass reinforced self extinguishing	General purpose grade	Glass-reinforced grade	Asbestos-filled grade
Uses		Gears, bearings, valves, pump parts, fittings, rollers, cams, bushings, electronic parts, textile machinery parts, tape cassettes, fasteners.					

[a]For 32 to 302°F range.
[b]Underwriters' Laboratories specification.
[c]Taber abrasion; CS-17 wheel, 1,000-g load.

Note: Properties shown are at approximately room temperature unless otherwise noted.

From *1973 Materials Selector,* Reinhold Publishing, Stamford, Conn., 1972, 259. With permission.

Table 1–19
POLYESTERS – THERMOSETS

		Type				
		Cast polyester		Reinforced polyester moldings		
	ASTM No.	Rigid	Flexible	High strength (glass fibers)	Heat and chemical resistant (asbestos)	Sheet molding compounds general purpose
Physical properties						
Specific gravity	D792	1.12–1.46	1.06–1.25	1.8–2.0	1.5–1.75	1.65–1.80
Thermal conductivity, Btu/hr/ft² /°F/in.	–	0.10–0.12	–	1.32–1.68	–	–
Coefficient of thermal expansion/°F	D696	$3.9–5.6 \times 10^{-5}$	–	$13–19 \times 10^{-6}$	–	–
Specific heat, Btu/lb/°F	–	0.30–0.55	–	0.25–0.35	–	0.20–0.25
Water absorption (24 hr), %	D570	0.20–0.60	0.12–2.5	0.5–0.75	0.25–0.50	0.15–0.25
Flammability, ipm	D635	0.87 to self extinguishing	Slow burn to self extinguishing	Self extinguishing	Self extinguishing	Self extinguishing
Transparency (visible light), %	–	–	–	Opaque	Opaque	Opaque
Refractive index, n_D	D542	1.53–1.58	1.50–1.57	–	–	–
Mechanical properties						
Tensile strength, 1,000 psi	D638	4–10	1–8	5–10	4–6	15–17
Tensile modulus, 10^5 psi	D638	1.5–6.5	0.001–0.10	16–20	12–15	15–20
Elongation, %	D638	1.7–2.6	25–300	0.3–0.5	–	–
Impact strength (Izod notched) ft–lb/in.	D256	0.18–0.40	4.0	1–10	0.45–1.0	5–15
Flexural strength, 1,000 psi	D790	14–18	4–16	6–26	10–13	26–32
Modulus of elasticity in flexure, 10^5 psi	D790	1–9	0.001–0.39	15–25	–	15–18
Compressive strength, 1,000 psi	D690	12–37	1–17	20–26	20–25	22–36
Hardness (Barcol)	D785	35–50	6–40	60–80	40–70	45–60
Electrical properties						
Volume resistivity, ohm-cm	D257	10^{13}	10^{12}	1×10^{12} – 1×10^{13}	1×10^{12} – 1×10^{13}	6.4×10^{15} – 2.2×10^{16}
Dielectric strength (step by step), V/mil	D149	300–400	300–400	200–400	350	400–440
Dielectric constant	D150					
60 cycles		2.8–4.4	3.18–7.0	–	–	4.62–5.0
10^6 cycles		2.8–4.4	3.7–6.1	–	–	4.55–4.75
Dissipation factor	D150					
60 cycles		0.003–0.04	0.01–0.18	–	–	0.0087–0.04
10^6 cycles		0.006–0.04	0.02–0.06	–	–	0.0086–0.022
Arc resistance, sec	D495	115–135	125–145	130–170	–	130–180

Table 1–19 (continued)
POLYESTERS – THERMOSETS

	ASTM No.	Type: Cast polyester		Type: Reinforced polyester modligns		
		Rigid	Flexible	High strength (glass fibers)	Heat and chemical resistant (asbestos)	Sheet molding compounds general purpose
Fabricating properties						
Bulk factor	–	–	–	3–4	–	–
Matched die molding	–					
Pressure, 1,000 psi		–	–	0.3–2.5	0.4–0.6	0.5
Temperature, °F		–	–	250–350	250–320	265–340
Mold shrinkage, %		–	–	0.006–0.007	–	0.0014–0.0016
Transfer molding	–					
Pressure, 1,000 psi		–	–	1–5	–	–
Temperature, °F		–	–	250–350	–	–
Mold shrinkage, %		–	–	0.006–0.007	–	–
Heat resistance						
Maximum recommended service temperature, °F	–	250–300	150–250	250–400	300	300
Heat deflection temperature (264 psi), °F	D648	120–400	–	400	375–400	375–400
Chemical resistance		Slightly to heavily attacked by strong acids; attacked by strong alkalis, ketones, and solvents.		Good to excellent resistance to weak acids, organic solvents, and weak alkalis; good resistance to strong acids; poor to fair resistance to strong alkalis.	Good to excellent resistance to weak acids, organic solvents, and weak alkalis; good resistance to strong acids; fair to good resistance to strong alkalis.	Good to excellent resistance to weak acids, organic solvents, and weak alkalis; good resistance to strong acids; poor to fair resistance to strong alkalis.

Table 1–19 (continued)
POLYESTERS – THERMOSETS

	ASTM No.	Type: Cast polyester — Rigid	Type: Cast polyester — Flexible	Type: Reinforced polyester moldings — High strength (glass fibers)	Type: Reinforced polyester moldings — Heat and chemical resistant (asbestos)	Type: Reinforced polyester moldings — Sheet molding compounds general purpose
Uses		Electrical components, buttons, decorative architectural uses.	Flooring, tooling encapsulants, buttons, and shields.	Chairs, housings, covers, trays, molded panels, bezels, motor shrouds, electrical parts, fans, helmets.		

From *1973 Materials Selector,* Reinhold Publishing, Stamford, Conn., 1972, 260. With permission.

Table 1–20
PHENYLENE OXIDES, POLYSULFONES, POLYARYLSULFONE

	ASTM No.	Material: Phenylene oxides (Noryl) SE-100	SE-1	Glass fiber reinforced[a]	Polysulfones: Standard	Glass fiber reinforced[b]	Polyarylsulfone
Physical properties							
Specific gravity	D792	1.10	1.06	1.21, 1.27	1.24	1.41, 1.55	1.36
Thermal conductivity, Btu/hr/ft²/°F/in.	C177	1.10	1.5	1.15, 1.1	1.8	–	1.1
Coefficient of thermal expansion, 10^{-5}/°F	D696	3.8	3.3	2.0, 1.4	3.1	1.6, 1.2	2.6
Specific heat, Btu/lb/°F	–	–	–	–	0.24	–	–
Refractive index, n_D	D542	Opaque	Opaque	Opaque	1.63	Opaque	–
Water absorption (24 hr), %	D570	0.07	0.07	0.06	0.22	0.22, 0.18	1.8
Flammability, ipm	D635	Self extinguishing	Self extinguishing	Self extinguishing	Self extinguishing	Self extinguishing	Self extinguishing
Mechanical properties							
Tensile strength, 1,000 psi	D638						
Ultimate		–	–	–	–	–	13
Yield		7.8	9.6	14.5, 17.0	10.2	17, 19	8
Elongation, %	–						
Ultimate		50	60	4–6	50–100	–	15–20
Yield		–	–	–	5.6	2, 1.6	13
Modulus of elasticity in tension, 10^5 psi	D638	3.8	3.55	9.25, 13.3	3.6	10.9, 14.9	3.7
Flexural strength, 1,000 psi	D790	12.8	13.5	20.5, 22	15.4	25, 28	17.2
Modulus of elasticity in flexure, 10^5 psi	D790	3.6	3.6	7.4, 10.4	3.9	12, 15.5	4.0
Impact strength (Izod notched), ft-lb/in.	D638	5.0	5.0	2.3	1.3	1.8, 2.0	5.0
Compressive strength, 1,000 psi	D695	12	16.4	17.6, 17.9	13.9	–	17.8
Hardness (Rockwell)	D785	R115	R119	L106, L108	R120	M84	M110
Coefficient of static friction (against self)	–	–	–	–	0.67	–	0.1–0.3
Abrasion resistance (Taber, CS-17), mg/1,000 cycles	–	100	20	35	20	–	40
Electrical properties							
Volume resistivity ohm-cm	D257	10^{17}	10^{17}	10^{17}	5×10^{16}	10^{17}	3.2×10^{16}
Dielectric strength (short time), V/mil	D149	400 (1/8 in.)	500 (1/8 in.)	1,020 (1/32 in.)	425	480	350
Dielectric constant	D150						
60 cycles		2.65	2.69	2.93	3.06	3.55	3.94
10^6 cycles		2.64	2.68	2.92	3.03	3.41	3.7
Dissipation factor	D150						
60 cycles		0.0007	0.0007	0.0009	0.0008	0.0019	0.003
10^6 cycles		0.0024	0.0024	0.0015	0.0034	0.0049	0.012
Arc resistance, sec	D495	75	75	120	122	114	67

Table 1–20 (continued)
PHENYLENE OXIDES, POLYSULFONES, POLYARYLSULFONE

	ASTM No.	Material					
		Phenylene oxides (Noryl)			Polysulfones		
		SE-100	SE-1	Glass fiber reinforced[a]	Standard	Glass fiber reinforced[b]	Polyarylsulfone
Heat resistance							
Maximum recommended service temperature, °F		–	212	–	340	350	500
Deflection temperature, °F	D648						
66 psi		230	279	293, 317	358	389	–
264 psi		212	265	282, 310	345	365	525
Fabricating properties							
Bulk factor	D1895	–	–	–	1.8	–	2
Injection molding	–						
Pressure, 1,000 psi		12–18	12–18	15–20	15–25	15–25	15–40
Temperature, °F		450–600	450–600	525–600	625–750	625–750	700–800
Shrinkage, in./in.		0.005–0.007	0.005–0.007	0.001–0.002	0.007	0.003	0.008
Extrusion temperature	–	450–600	450–600	–	600–750	–	650–800
Chemical resistance		Excellent resistance to aqueous media such as detergents and weak and strong acids and bases even at elevated temperatures. Many halogenated and aromatic hydrocarbons will soften or partially dissolve these materials			Resistant to inorganic acids, alkalis, and aliphatic hydrocarbons; partially soluble or swells in ketones and aromatic hydrocarbons, soluble in chlorinated hydrocarbons.		Resistant to aqueous acids and bases, fuels, oils, fluorinated solvents. Dissolves in highly polar solvents such as DMF, DMAC, NMP.

Table 1–20 (continued)
PHENYLENE OXIDES, POLYSULFONES, POLYARYLSULFONE

	ASTM No.	Material					
		Phenylene oxides (Noryl)			Polysulfones		
		SE-100	SE-1	Glass fiber reinforced[a]	Standard	Glass fiber reinforced[b]	Polyarylsulfone
Uses		Automotive dashboards and electrical connectors; appliance and business machine housings; cabinets, consoles, and covers; coffee brewers and dispensers; pump and plumbing parts; valves; tape cartridge platforms; coil assemblies; bus bar insulators; switch housings; terminal blocks; tuner bars; light fixtures.			Coil bobbins, switches, terminal blocks, battery cases, connectors, circuit carriers, sockets, tube bases, range hardware, coffee maker parts, sight glasses, auto parts, lamp bezels, aircraft ducts, housing and side wall panels, meter housings and components, projector transparencies.		Molded parts requiring strength and toughness at high temperatures. Extrusions, coatings, filled compositions for bearing use.

[a]Where two values are given, the first applies to 20% glass fiber, the second to 30%; otherwise, the same value applies to both.
[b]Where two values are given, the first applies to 30% glass fiber, the second to 40%; otherwise, the same value applies to both.

From *1973 Materials Selector,* Reinhold Publishing, Stamford, Conn., 1972, 261. With permission.

Table 1–21
POLYPROPYLENE, POLYPHENYLENE SULFIDE

	ASTM No.	Material: Polypropylene, General purpose	Polypropylene, High impact	Polypropylene, Asbestos filled	Polypropylene, Glass reinforced	Polypropylene, Flame retardant	Polyphenylene sulfide, Standard	Polyphenylene sulfide, 40% glass reinforced
Physical properties								
Specific gravity	D792	0.900–0.910	0.900–0.910	1.11–1.36	1.04–1.22	1.2	1.34	1.64
Thermal conductivity, Btu/hr/ft²/°F/in.	–	1.21–1.36	1.72	–	–	–	2.0	–
Coefficient of thermal expansion, 10^{-5}/°F	D696	3.8–5.8	4.0–5.9	2–3	1.6–2.4	–	3.0	–
Specific heat, Btu/lb/°F	–	0.45	0.45–0.48	–	–	–	0.26	–
Refractive index, n_D	D542	Translucent-opaque	Translucent-opaque	Opaque	Opaque	Opaque	Opaque	Opaque
Water absorption (24 hr), %	D570	<0.01–0.03	<0.01–0.02	0.02–0.04	0.02–0.05	0.02–0.03	–	–
Flammability, ipm	D635	0.7–1	1	1	1	Self extinguishing	Non-burning	Non-burning
Mechanical properties								
Tensile strength, 1,000 psi	D638, C							
Maximum		4.8–5.5	–	–	–	–	–	–
Yield		4.8–5.2	2.8–4.3	3.3–8.2	6–10	3.6–4.2	11	21
Elongation, %	–							
Break		30–>200	30–>200	3–20	2–4	3–15	3	3–9
Yield		9–15	7–13	5	–	–	–	–
Modulus of elasticity in tension, 10^5 psi	D638, B	1.6–2.2	1.3	–	812	1.5–2.4	4.8	11.2
Flexural yield	D790, B	6–7	4.1	7.5–9	8–11	–	20	37
Modulus of elasticity in flexure, 10^5 psi	D790, B	1.7–2.5	1.0–2.0	3.4–6.5	4–8.2	1.9–6.1	6.0	22.0
Impact strength (Izod notched), ft-lb/in.	D256	0.4–2.2	1.5–12	0.5–1.5	0.5–2	2.2	–	–
Compressive yield strength, 1,000 psi	D695	5.5–6.5	4.4	7.0	6.5–7	–	–	–
Hardness (Rockwell)	D785	R80–R100	R28–95	R90–R110	R90–R115	R60–R105	R124	R123
Electrical properties								
Volume resistivity, ohm-cm	D257	$>10^{17}$	10^{17}	1.5×10^{15}	1.7×10^{16}	$4 \times 10^{16}-10^{17}$	–	–
Dielectric strength (short time), V/mil	D149	650(125 mil)	450–650	450	317–475	485–700	595	490
Dielectric constant	D150							
60 cycles		2.20–2.28	2.20–2.28	2.75	2.3–2.5	2.46–2.79	–	–
10^6 cycles		2.23–2.24	2.23–2.27	2.6–3.17	2–2.25	2.45–2.70	3.22	3.88

Table 1–21 (continued)
POLYPROPYLENE, POLYPHENYLENE SULFIDE

	ASTM No.	Material: Polypropylene, General purpose	Polypropylene, High impact	Polypropylene, Asbestos filled	Polypropylene, Glass reinforced	Polypropylene, Flame retardant	Polyphenylene sulfide, Standard	Polyphenylene sulfide, 40% glass reinforced
Electrical properties (cont.)								
Dissipation factor	D150							
60 cycles		0.0005–0.0007–	<0.0016	0.007	0.002	0.0007–0.017	–	–
10^6 cycles		0.0002–0.0003	0.0002–0.0003	0.002	0.003	0.0006–0.003	0.0007	0.0041
Arc resistance, sec	D495	125–136	123–140	121–125	73–77	15–40	–	–
Heat resistance								
Maximum recommended service temperature, °F	–	230	–	250	250	205	500	500
Deflection temperature, °F	D648							
66 psi		205–230	190–235	270–290	275–310	245–280	–	–
264 psi		135–140	120–140	170–220	250–300	155	278	425
Fabricating properties								
Bulk factor	–	1.8–2.2	1.8–2.2	1.8–2.2	–	1.31	–	–
Injection molding	–							
Pressure, 1,000 psi		10–20	10–20	10–20	15–20	10–20	10–15	10–15
Temperature, °F		400–550	400–550	400–550	450–575	400–550	600–700	600–700
Shrinkage, in./in.		0.010–0.025	0.010–0.025	0.003–0.008	0.001–0.008	0.010–0.020	0.008	0.004
Extrusion temperature	–	380–430	380–430	–	430–575	380–430	600–700	600–700
Chemical resistance		Resistant to most acids, alkalis, and saline solutions, even at higher temperatures; resistant to higher aliphatic solvents and polar substances. Above 175°F soluble in such aromatic substances as toluene and xylene, and chlorinated hydrocarbons.					Excellent resistance to organic solvents below 375°F. Unaffected by strong alkalis or aqueous inorganic salt solutions.	

Table 1–21 (continued)
POLYPROPYLENE, POLYPHENYLENE SULFIDE

		Material						
		Polypropylene					Polyphenylene sulfide	
	ASTM No.	General purpose	High impact	Asbestos filled	Glass reinforced	Flame retardant	Standard	40% glass reinforced
Uses		Hospital ware, housewares, appliances, radio and TV housings, film fibers.	Luggage seating packaging, housings, automotive parts, containers, wire coating.	Housings, automobile fan shrouds, covers.	Housings, shrouds, cases, panels, and mechanical parts.	Electrical uses to meet UL requirements, housings and shields.	Corrosion-resistant pump components, valves, and pipe.	Pump vanes, valve parts, gaskets, fuel cells, and auto parts requiring chemical resistance at higher temperatures.

From *1973 Materials Selector,* Reinhold Publishing, Stamford, Conn., 1972, 262. With permission.

Table 1–22
POLYETHYLENES – MOLDED, EXTRUDED

		Type								
		Type 1–lower density (0.910–0.925)			Type II–medium density (0.926–0.940)		Type III–higher density (0.941–0.965)			
	ASTM No.	Melt index 0.3–3.6	Melt index 6–26	Melt index 200	Melt index 20	Melt index 1.0–1.9	Melt index 0.2–0.9	Melt index 0.1–12.0	Melt index 1.5–15	High molecular weight
Physical properties										
Specific gravity	D792	0.910–0.925	0.918–0.925	0.910	0.930	0.930–0.940	0.96	0.950–0.955	0.96	0.94
Thermal conductivity, Btu/hr/ft^2/°F/ft	C177	0.19	0.19	0.19	0.19	0.19	0.19	0.19	0.19	0.19
Coefficient of thermal expansion, 10^{-5}/°F	D696	8.9–11.0	8.9–11.0	11.0	8.3–16.7	8.3–16.7	8.3–16.7	8.3–16.7	8.3–16.7	–
Refractive index	D542	1.51	1.51	1.51	1.51	1.51	1.54	1.54	1.54	–
Specific heat, Btu/lb/°F		0.53–0.55	0.53–0.55	0.53–0.55	0.53–0.55	0.53–0.55	0.46–0.55	0.46–0.55	0.46–0.55	–
Water absorption (24 hr), %	D570	<0.01	<0.01	<0.01	<0.01	<0.01	<0.01	<0.01	<0.01	<0.01
Flammability, ipm	D635	1.0	1.0	1.0	1.0	1.0	1.0	1.0	1.0	1.0
Mechanical properties										
Modulus of elasticity in tension, 10^5 psi	D638	0.21–0.27	0.20–0.24	–	–	–	–	–	–	1.0
Tensile strength, 1,000 psi	D412	1.4–2.5	1.4–2.0	0.9–1.1	2.0	2.3–2.4	4.4	2.9–4.0	4.4	5.4
Elongation, %	D412	500–725	125–675	80–100	200	200–425	700–1,000	50–1,000	100–700	400
Hardness (Shore)	D785	C73, D50–52	C73, D47–53	D45	D55	D55–D56	D68–70	D60–70	D68–70	60–65
Impact strength (Izod), ft-lb in. notch	D256	–	–	–	–	–	4.0–14	0.4–6.0	1.2–2.5	>20
Brittleness temperature, °F		<-94	<-4	<14	<-148	<-148	-106–-180	<-76–<-170	-100–-180	<-100
Modulus of elasticity in flexure, 10^3 psi	D747	13–27	12–30	10	35–50	35–50	130–150	90–125	150	75
Shear strength, 1,000 psi		1.6–1.85	1.4–1.7	1	–	–				
Electrical properties										
Volume resistivity, ohm-cm	D257	10^{17}–10^{19}	10^{17}–10^{19}	10^{17}–10^{19}	$>10^{15}$	$>10^{15}$	$>10^{15}$	$>10^{15}$	$>10^{15}$	$>10^{15}$

Table 1–22 (continued)
POLYETHYLENES – MOLDED, EXTRUDED

	ASTM No.	Type 1 – lower density (0.910–0.925)			Type II–medium density (0.926–0.940)		Type III–higher density (0.941–0.965)			
		Melt index 0.3–3.6	Melt index 6–26	Melt index 200	Melt index 20	Melt index 1.0–1.9	Melt index 0.2–0.9	Melt index 0.1–12.0	Melt index 1.5–15	High molecular weight
Electrical properties (cont.)										
Dielectric strength (short time), V/mil	D149	480	480	480	480	480	480	480	480	480
Dielectric constant	D150	2.3	2.3	2.3	2.3	2.3	2.3	2.3	2.3	2.3
Dissipation factor	D150	<0.0005	<0.0005	<0.0005	<0.0005	<0.0005	<0.0005	<0.0005	<0.0005	<0.0005
Fabricating properties										
Bulk factor		1.6–2.2	1.6–2.2	1.6–2.2	1.6–2.2	1.6–2.2	1.6–2.2	1.6–2.2	1.6–2.2	0.35–0.45[a]
Injection molding										
Pressure, 1,000 psi		5–22	5–15	2–10	10–15	10–15	10–15	10–15	10–15	10–20[b]
Temperature, °F		275–650	275–650	250–350	300–500	300–500	330–530	330–530	330–530	450–500[b]
Mold shrinkage, in./in.		0.02–0.05	0.01–0.04	0.01–0.02	0.02–0.05	0.02–0.05	0.02–0.05	0.02–0.05	0.02–0.05	0.03–0.05[b]
Heat resistance										
Vicat softening point, °F		176–201	176–201	–	215–230	220–235	258–266	240–255	250–260	–
Chemical resistance		Excellent resistance to acids and alkalis at normal temperature, except oxidizing acids such as nitric, chlorosulfonic, and fuming sulfuric. Below 122°F insoluble in organic solvents; at higher temperatures soluble to varying degrees in hydrocarbons and halogenated hydrocarbons, but insoluble in more polar liquids. Generally, a higher melt index material has greater solubility.					Same basic chemical resistance as Types I and II, but better resistance to some specific chemicals.			
Uses		Injection moldings; kitchen utilityware, toys, process tank liners, closures, packages, sealing rings, battery parts. Blow moldings: squeeze bottles for packaging, containers for drugs. Film: wrapping materials for food, clothes, other items. Wire and cable: high frequency insulation, jacketing. Pipe: chemical handling, irrigation systems, natural gas transmission.					Refrigerator parts, packaging, structural housing panels, pipe, defroster and heater ducts, sterilizable housewares and hospital equipment, hoops, battery parts, blow molded containers and parts, film wrapping materials, wire and cable insulations, and chemical resistant pipe.			

[a]Powder.
[b]Flow molding.

From *1973 Materials Selector,* Reinhold Publishing, Stamford, Conn., 1972, 264. With permission.

Table 1–23
OLEFIN COPOLYMERS – MOLDED

		Type					
	ASTM No.	EEA (ethylene ethyl acrylate)	EVA (ethylene vinyl acetate)	Ethylene-butene	Propylene-ethylene	Ionomer	Polyallomer
Physical properties							
Specific gravity	D792	0.93	0.94	0.95	0.91	0.94	0.898–0.904
Tensile impact, ft-lb/in.^{2}	D1822	500	690	–	150	400	–
Tensile strength, 100 psi	D638	2.0	3.6	3.5	4	4	30–43
Izod impact strength (notched), ft-lb/in.	D256	–	–	0.4	1.1	9–14	1.5
Hardness, Shore D	D785	35	36	65	–	60	–
Elongation (in 1 in.), %	D638	650	650	20	–	450	300–400
Flexural modulus, psi	D790	–	–	165	140	–	$0.7–1.3 \times 10^5$
Electrical properties							
Volume resistivity, 10^{15} ohm-cm	D257	2.4	0.15	–	–	10	$>10^{16}$
Dielectric strength (short time), V/mil	D149	550	525	–	–	1,000	500–650
Dielectric constant 60 cps	D150	2.8	3.16	–	–	2.4	2.3
Dissipation factor, 60 cps	D150	0.001	0.003	–	–	0.003	>0.0005
Thermal properties							
Heat deflection temperature (66 psi) °F	D648	–	–	–	104	105	122–133[a]
Brittleness temperature, °F	D746	–155	–148	–35	–	–160	–
Vicat softening point	D1525	147	147	243	–	162	–
Chemical resistance		Resists most weak mineral acids, alkalis; attacked by chlorinated hydrocarbons, straight-chain paraffinic solvents, benzene.		Generally satisfactory to chemicals; good resistance to stress corrosion.		Resistant to organic attacked by oxidizing acids.	Resistant to strong alkalis, weak acids, organic solvents; attacked slowly by acid reagents.

Table 1–23 (continued)
OLEFIN COPOLYMERS – MOLDED

	ASTM No.	Type: EEA (ethylene ethyl acrylate)	EVA (ethylene vinyl acetate)	Ethylene-butene	Propylene-ethylene	Ionomer	Polyallomer
Uses		Tubing, seals, bushings, dampering pads, rug backing, electric insulation, floor mats, other flexible items.		Packaging, appliances, furniture, wire and cable insulation, rigid parts.		Skin packaging; coated substrates; clear, flexible parts.	Molded parts requiring toughness and good hinge properties.

[a]For 264 psi.

From *1973 Materials Selector,* Reinhold Publishing, Stamford, Conn., 1972, 264. With permission.

Table 1–24
POLYSTYRENES – MOLDED

	ASTM No.	Material: Polystyrenes: General purpose	Polystyrenes: Medium impact	Polystyrenes: High impact	Polystyrenes: Glass fiber (30%) reinforced	Styrene acrylonitrile (SAN)	Glass fiber (30%) reinforced SAN
Physical properties							
Specific gravity	D792	1.04	1.04–1.07	1.04–1.07	1.29	1.04–1.07	1.35
Thermal conductivity, Btu/hr/ft²/°F/ft	–	0.058–0.090	0.024–0.090	0.024–0.090	0.117	–	–
Coefficient of thermal expansion, 10^{-5}/°F	D696	3.3–4.8	3.3–4.7	2.2–5.6	1.8	3.6–3.7	1.6
Specific heat, Btu/lb/°F	–	0.30–0.35	0.30–0.35	0.30–0.35	0.256	0.33	–
Refractive index, n_D	D542	1.60	Opaque	Opaque	Opaque	1.565–1.569	Opaque
Water absorption (24 hr), %	D570	0.30–0.2	0.03–0.09	0.05–0.22	0.07	0.20–0.35	0.15
Flammability, ipm	D635	1.0–1.5	0.5–2.0	0.5–1.5	–	0.8	–
Mechanical properties							
Tensile strength, 1,000 psi	D638						
Ultimate		5.0–10	6.0	3.3–5.1	14	9.5–12.0	18
Yield		5.0–10	6.0	2.8–5.3	14	–	18
Elongation, %	D638						
Ultimate		1.0–2.3	3.0–40	–	1.1	0.5–3.7	1.4
Yield		1.0–2.3	1.2–3.0	1.5–2.0	1.1	–	1.4
Modulus of elasticity in tension, 10^5 psi	D638	4.6–5.0	3.9–4.7	1.50–3.80	12.1	4.0–5.0	17.5
Flexural strength, 1,000 psi	D790	10–15	–	–	17	–	22
Modulus of elasticity in flexure, 10^5 psi	D790	4–5	3.5–5.0	2.3–4.0	12	–	14.5
Impact strength (Izod notched), ft-lb/in.	D638	0.2–0.4	0.5–0.7	0.8–1.8	2.5	0.30–0.45	3.0
Compressive strength, 1,000 psi	D695	11.5–16.0	4–9	4–9	19	–	2.3
Hardness (Rockwell)	D785	M72	M47–65	M3–43	M85–95	M80–85	M90–100
Abrasion resistance (Taber), mg/1,000 cycles	–	–	–	–	164	–	–
Electrical properties							
Volume resistivity, ohm-cm	D257	$>10^{16}$	$>10^{16}$	$>10^{16}$	3.6×10^{16}	$>10^{16}$	4.4×10^{16}
Dielectric strength (short time), V/mil	D149	>500	>425	300–650	396	400–500	515
Dielectric constant	D150						
60 cycles		2.45–2.65	2.45–4.75	2.45–4.75	3.1	2.6–3.4	3.5
10^6 cycles		2.45–2.65	2.4–3.8	2.5–4.0	3.0	2.6–3.02	3.4

Table 1–24 (continued)
POLYSTYRENES – MOLDED

	ASTM No.	Material					
		Polystyrenes					
		General purpose	Medium impact	High impact	Glass fiber (30%) reinforced	Styrene acrylonitrile (SAN)	Glass fiber (30%) reinforced SAN
Electrical properties (cont.)							
Dissipation factor	D150						
60 cycles		0.0001–0.0003	0.0004–0.002	0.0004–0.002	0.005	>0.006	0.005
10^6 cycles		0.0001–0.0005	0.0004–0.002	0.0004–0.002	0.002	0.007–0.010	0.009
Arc resistance, sec	D495	60–135	20–135	20–100	28	100–150	65
Heat resistance							
Maximum recommended service temperature, °F	–	160–205	125–165	125–165	190–200	175–190	–
Deflection temperature, °F	D648						
66 psi		–	–	–	230	–	230
264 psi		220 maximum	210 maximum	210 maximum	220	210–220	220
Fabricating properties							
Bulk factor	–	1.6–2.3	1.6–2.3	1.6–2.3	–	–	–
Injection molding	–						
Pressure, 1,000 psi		10–24	10–24	10–24	0.5–1.2 (line)	10–24	0.5–1.2 (line)
Temperature, °F		325–650	300–600	300–450	450–625	375–550	430–550
Shrinkage, in./in.		0.002–0.008	0.002–0.008	0.002–0.008	0.001–0.003	0.003–0.007	0.0005–0.002
Extrusion temperature (Vicat soft)	–	194–224	187–216	190–220	–	–	–

Table 1–24 (continued)
POLYSTYRENES – MOLDED

		Material					
		Polystyrenes					
	ASTM No.	General purpose	Medium impact	High impact	Glass fiber (30%) reinforced	Styrene acrylonitrile (SAN)	Glass fiber (30%) reinforced SAN
Chemical resistance		Resists alkalis, salts, low alcohols, glycols, and water. Fair resistance to mineral chemicals and vegetable oils. Not resistant to aromatic and chlorinated hydrocarbons.			No effect by weak acids, strong acids; attacked by oxidizing acids; no effect by weak alkalis; attacked slowly by strong alkalis; soluble in aromatic and chlorinated hydrocarbons.	Resistant to alkalis and acids, animal and vegetable oils, soaps, detergents, and household chemicals.	No effect by weak acids, alkalis, strong acids; attacked by oxidizing acids, strong alkalis; soluble in ketones, esters, some chlorinated hydrocarbons.
Uses		Thin parts, long flow parts, toys, appliances, containers, film, monofilaments, and housewares.	Radio cabinets, toys, containers, packaging, and closures.	Containers, cups, lids, large thin wall parts, auto parts, TV cabinets, trays, and appliance housings.	Auto dashboard skeletons, camera housings and frames, tape reels, fan blades.	Kitchenware, tumblers, broom bristles, ice buckets, closures, film, containers, lenses, battery cases.	Camera housings and frames, auto bezels, electrical components, handles, auto panels.

From *1973 Materials Selector,* Reinhold Publishing, Stamford, Conn., 1972, 265. With permission.

Table 1–25
POLYVINYL CHLORIDE AND COPOLYMERS – MOLDED, EXTRUDED

	ASTM No.	Type: Polyvinyl chloride, polyvinyl chloride acetate – Nonrigid–general	Nonrigid–electrical	Rigid–normal impact	Vinylidene chloride[a]
Physical properties					
Specific gravity	D792	1.20–1.55	1.16–1.40	1.32–1.44	1.68–1.75
Thermal conductivity, Btu/hr/ft^2/°F/ft	D325	0.07–0.10	0.07–0.10	0.07–0.10	0.053
Coefficient of thermal expansion, 10^{-5}/°F	D696	–	–	2.8–3.3	8.78
Refractive index	D542	–	–	–	1.60–1.63
Specific heat, Btu/lb/°F	D570	–	–	–	0.32
Water absorption (24 hr), %	D635	0.2–1.0	0.40–0.75	0.03–0.40	>0.1
Flammability		Self extinguishing	Self extinguishing	Self extinguishing	Self extinguishing
Mechanical properties					
Modulus of elasticity in tension, 10^5 psi	D412	0.004–0.03	0.01–0.03	3.5–4.0[b]	0.7–2.0
Tensile strength, 1,000 psi	D412	1–3.5	2–3.2	5.5–8	4–8, 15–40
Elongation (in 2 in.), %	D638	200–450	220–360	1–10	15–25, 20–30
Hardness (Rockwell)	D785	–	–	R110–120	M50–65
Hardness (Shore)	D676	A50–100	A78–100	D70–85	>A95
Impact strength (Izod notched), ft-lb/in.	D256	Variable	Variable	0.5–10	2–8, 0.053
Modulus of elasticity in flexure, psi	D790	–	–	$3.8–5.4 \times 10^5$	–
100% modulus, psi		600–2,800	600–2,800	–	–
Flexural strength, 1,000 psi	D790	–	–	11–16	15–17, flexible
Compressive strength, 1,000 psi	D695	–	–	11–12	–
Compressive yield strength, 1,000 psi	D695	–	–	10–11	75–85
Cold flexural temperature, °F	D1043	-70–0	-7–+20	–	–
Cold bend temperature, °F		-40–-4	-49–-4	–	–
Electrical properties					
Volume resistivity, ohm-cm	D257	$1–700 \times 10^{12}$	$4–300 \times 10^{11}$	10^{14}–$>10^{16}$	10^{14}–10^{16}
Dielectric strength, (short time), V/mil	D149	–	24–500	725–1,400	–
Dielectric constant (60 cycles)	D150	5.5–9.1	6.0–8.0	2.3–3.7	3–5
Dissipation factor (60 cycles)	D150	0.05–0.15	0.08–0.11	0.020–0.03	0.03–0.15
Loss factor (60 cycles)	D150	–	1.0–1.2	0.030–0.072	–

Table 1–25 (continued)
POLYVINYL CHLORIDE AND COPOLYMERS – MOLDED, EXTRUDED

	ASTM No.	Type: Polyvinyl chloride, polyvinyl chloride acetate – Nonrigid–general	Nonrigid–electrical	Rigid–normal impact	Vinylidene chloride[a]
Fabricating properties					
Bulk factor		2.4–2.6	2.4–2.6	2.0–2.4	–
Compression molding					
Pressure, 1,000 psi		0.5–2	0.5–2	>1	0.5–5
Temperature, °F		285–350	285–350	275–400[c]	250–350
Injection molding					
Pressure, 1,000 psi		7–15	12–20	>20	10–30
Temperature, °F		320–350	325–375	300–375	300–400
Mold shrinkage, in./in.		0.02–0.05	0.02–0.06	0.001–0.004	0.008–0.012
Extrusion temperature, °F		325–400	350–385	360–420[d]	<375
Heat resistance					
Maximum recommended service temperature, °F		150–220	140–220	150–165	170–212
Heat distortion temperature, °F					
66 psi	D648	–	–	170–185	190–210
264 psi	D648	–	–	140–170	130–150
Softening point, °F		–	–	–	240–280
Chemical resistance		Generally resistant to alkalis and weak acids. Moderately to not resistant to strong acids. Not resistant to ketones and esters; aromatic hydrocarbons produce swelling.			Excellent to all acids and most common alkalis.[e]

Table 1–25 (continued)
POLYVINYL CHLORIDE AND COPOLYMERS – MOLDED, EXTRUDED

	ASTM No.	Type			
		Polyvinyl chloride, polyvinyl chloride acetate			Vinylidene chloride[a]
		Nonrigid–general	Nonrigid–electrical	Rigid–normal impact	
Uses		Parts made by molding, high speed extrusion, calendering. Blown extruded film. Vacuum cleaner parts, handlebar grips, doll parts, hair curlers, safety goggle cups, grommets, toy tires, garden hose, and protective garments.	Parts made by calendering, extrusion. Insulation and jacketing for communication and low tension power wire and cable, building wiring, appliance and machine tool cords, and switchboard cable.	Parts made by calendering, laminating, molding, extrusion. Fume hoods and ducts, storage tanks, chemical piping, plating tanks, phonograph records. Sheets and shapes for decorative panels, other building uses.	Extrusions: gasket rods, valve seats, flexible chemical tubing and pipe, tape for wrapping joints, chemical conveyor belts. Moldings: spray-gun handles, acid dippers, parts for rayon producing equipment.

[a]Where two values or ranges are given, they represent unoriented and oriented forms, respectively.
[b]Modulus of elasticity in compression.
[c]Barrel temperature.
[d]Stock temperature.
[e]Unaffected by aliphatic and aromatic hydrocarbons, alcohols, esters, etc.

From *1973 Materials Selector,* Reinhold Publishing, Stamford, Conn., 1972, 266. With permission.

Table 1–26
SILICONES – MOLDED, LAMINATED

	ASTM No.	Material		
		Fibrous (glass) reinforced silicones	Granular (silica) reinforced silicones	Woven glass fabric/ silicone laminate
Physical properties				
Specific gravity	D792	1.88	1.86–2.00	1.75–1.8
Thermal conductivity, Btu/hr/ft^2/°F/ft	–	0.18	0.25–0.5	0.075–0.125
Coefficient of thermal expansion, 10^{-5}/°F	D696	3.17–3.23	2.5–5.0	–
Specific heat, Btu/lb/°F	–	–	–	0.246
Water absorption (24 hr), %	D570	0.1–0.15	0.08–0.1	0.03–0.05
Flammability, ipm	D635	Nonburning	Nonburning	0.120
Transparency (visible light), %	–	Opaque	Opaque	Opaque
Refractive index, n_D	D542	–	–	–
Mechanical properties				
Tensile strength, 1,000 psi	D651	6.5	4–6	30–35
Tensile modulus, 10^5 psi	D651	–	–	28
Elongation, %	D651	<3	<3	–
Impact strength (Izod notched), ft-lb/in.	D256	10	0.34	10–25
Flexural strength, 1,000 psi	D790	16–19	6–10	33–47
Modulus of elasticity in flexure, 10^5 psi	D790	25	14–17	26–32
Compressive strength, 1,000 psi	D690	10–12.5	10.6–17.0	15–24
Hardness (Rockwell)	D785	M87	M71–95	75 (Barcol)
Abrasion resistance (Taber)	–	–	–	–
Electrical properties				
Volume resistivity, ohm-cm (dry)	D257	9×10^{14}	5×10^{14}	$2–5 \times 10^{14}$
Dielectric strength (short time), V/mil	D149	280 (in oil)	380 (in oil)	725
Dielectric constant	D150			
60 cycles		4.34	4.1–4.5	3.9–4.2
10^6 cycles		4.28	3.4–4.3	3.8–3.97
Dissipation factor	D150			
60 cycles		0.01	0.002–0.004	0.020
10^6 cycles		0.004	0.001–0.004	0.002
Arc resistance, sec	D495	240	250–310	225–250

Table 1–26 (continued)
SILICONES – MOLDED, LAMINATED

	ASTM No.	Material: Fibrous (glass) reinforced silicones	Granular (silica) reinforced silicones	Woven glass fabric/ silicone laminate
Fabricating properties				
Bulk factor	–	5–7	2.0	–
Compression molding	–			Laminating
Pressure, 1,000 psi		1–3	1–3	0.9–1.0
Temperature, °F		350	300–350	350
Shrinkage, %		0.006–0.010	0.006–0.010	–
Transfer molding	–			
Pressure, 1,000 psi		0.05–10	0.150–5	–
Temperature, °F		350	310–350	–
Shrinkage, %		0.0005	0.004–0.006	–
Heat resistance				
Maximum recommended service temperature, °F	–	>500	>500	450–500
Heat deflection temperature (264 psi), °F	D648	>900	520–>900	>900
Chemical resistance		Resistant to aviation gasoline, lubricating oil, and sulfuric and hydrochloric acids. Slightly softened and pitted by sodium hydroxide, except some mineral-filled materials. Should be tested if resistance to ketones, toluene, ethylenes, etc., is required.		Satisfactory resistance to aviation gas, lube oils, 40% sulfuric acid, 5% hydrochloric acid, and Freon 114®. Slightly attacked by 5% hydrochloric acid. Severely attacked by many organic solvents.

Table 1–26 (continued)
SILICONES – MOLDED, LAMINATED

	ASTM No.	Material		
		Fibrous (glass) reinforced silicones	Granular (silica) reinforced silicones	Woven glass fabric/ silicone laminate
Uses		Connector plugs, and other structural electronic parts requiring heat resistance.	Electronic component encapsulation such as transistors, diodes, resistors, and capacitors.	Special high-temperature structural or electrical parts such as aircraft radomes and ductwork, thermal and arc barriers, covers and cases for high-frequency equipment.

From *1973 Materials Selector,* Reinhold Publishing, Stamford, Conn., 1972, 267. With permission.

Table 1–27
UREAS – MOLDED

	ASTM No.	Type: Alpha-cellulose filled (ASTM Type 1)	Cellulose filled (ASTM Type 2)	Woodflour filled
Physical properties				
Specific gravity	D792	1.45–1.55	1.52	1.45–1.49
Thermal conductivity, Btu/hr/ft²/°F/ft	C177	0.17–0.244	–	–
Coefficient of thermal expansion/°F	D696	1.22–1.50 × 10^{-5}	–	–
Transmittance (luminous), %	–	21.8	0	0
Water absorption (24 hr), %	D570	0.4–0.8	2.0	–
Flammability	D635	Self extinguishing	Self extinguishing	Self extinguishing
Mechanical properties				
Modulus of elasticity in tension, psi	D638	13–16 × 10^5	–	11–14 × 10^5
Tensile strength, 1,000 psi	D638	5–10	–	–
Elongation (in 2 in.), %	D638	1.0	–	–
Hardness (Rockwell)	D785	E94–97, M116–120	–	M116–120
Impact strength (Izod notched), ft-lb/in.	D256	0.20–0.35	0.20–0.275	0.25–0.35
Flexural strength, 1,000 psi	D790	8–18	7.5–13	7.5–12.0
Compressive strength, 1,000 psi	D695	25–38	–	25–35
Shear strength, 1,000 psi	–	11–12	–	–
Electrical properties				
Volume resistivity, ohm-cm	D257	0.5–5 × 10^{11}	5–8 × 10^{10}	–
Dielectric strength (short time), V/mil	D149	300–400	340–370	300–400
Dielectric constant				
60 cycles	D150	7.0–9.5	7.2–7.3	7.0–9.5
10^6 cycles	D150	6.4–6.9	6.4–6.5	6.4–6.9
Dissipation factor				
60 cycles	D150	0.035–0.043	0.042–0.044	0.035–0.040
10^6 cycles	D150	0.028–0.032	0.027–0.029	0.028–0.032
Loss factor				
60 cycles	–	0.24–0.38	0.30–0.32	0.24–0.38
10^6 cycles	–	0.18–0.22	0.17–0.19	0.18–0.22
Arc resistance, sec	D495	110–130	85–110	80–110

Table 1–27 (continued)
UREAS – MOLDED

	ASTM No.	Type: Alpha-cellulose filled (ASTM Type 1)	Type: Cellulose filled (ASTM Type 2)	Type: Woodflour filled
Fabricating properties				
Bulk factor	–	2.4–3.0	2.5	2.2–2.5
Compression molding				
Pressure, 1,000 psi	–	2–8	2–5	2–8
Temperature, °F	–	275–325	275–320	275–325
Mold shrinkage, in./in.	–	0.006–0.014	0.007–0.008	0.006–0.014
Heat resistance				
Heat distortion temperature, °F	D648	266–280	–	270–280
Chemical resistance		High resistance to organic solvents, oils, and greases. Poor resistance to acids and alkalis, depending on concentration.		
Uses		Housings for radios, business machines, food equipment; toilet seats, household electrical switches and plugs, buttons, cosmetic containers, and closures.	Low cost items; available only in dark color; especially suited for electric switch plates, wiring devices, and electrical parts requiring high arc resistance.	

From *1973 Materials Selector,* Reinhold Publishing, Stamford, Conn., 1972, 268. With permission.

Table 1–28A
PLASTIC FILMS

	ASTM No.	Material: Nylon 6	Nylon 66	Nylon 610	Nylon 12	Polycarbonate	Polyester[a]	Polystyrene (oriented)
General characteristics								
Method of production	–	Extrusion	Extrusion	Extrusion	Extrusion	Extrusion casting	Extrusion	Extrusion
Forms available	–	Sheet, rolls, tapes	–	–	Sheet, rolls, tapes	Sheet, rolls	Sheet, rolls	Sheet, rolls
Clarity	–	Transparent	Transparent	Transparent, translucent	Transparent, translucent	Transparent, translucent	Transparent, opaque	Transparent, translucent, opaque
Minimum thickness, in.	–	0.0005	–	–	0.0005	0.0005	0.00015	0.001
Maximum width, in.	–	120	–	–	120	54	55	40
Area factor, 1,000 $in.^2$/lb/mil	–	24.5	24.3	25	27.3	23.1	20	26.1
Physical and mechanical properties								
Specific gravity	D792	1.12	1.14	1.11	1.01	1.20	1.39	1.05–1.07
Tensile strength, 1,000 psi	D882	9–13	12	10	7–9	8	17–18	7–12
Elongation, %	D882	>400	>250	>250	120–350	85–105	70–130	3–10
Burst strength (Mullen), psi	D774	–	–	–	–	25–35	45	30–60
Tear strength (Elmendorf), gm/mil	D689	50	50	70	–	10–16	18	2–8
Fold endurance	D643	Excellent	Excellent	Excellent	Excellent	250–400	Excellent	–
Heat sealing range, °F	–	400–450	490–540	420–470	350–400	400–430	490	220–300
Water absorption (24 hr), %	D570	8.0	1.5	0.4	0.25	0.35	Nil	0.04–0.06
Water vapor permeability, g/100 $in.^2$/24 hr/mil	E96	18.0[b]	–	–	3.6–4.6	8.0	1.8	6.2
Gas permeability, cm^3/100 $in.^2$/24 hr/mil	–							
Oxygen		2.6	2.5	4.5	51–89	142	5.7	213
Nitrogen		–	0.7	–	13–18	28	0.9	42
Carbon dioxide		–	11	–	152–330	680	17.5	926
Electrical properties								
Dielectric strength (77°F, 60 cps), V/mil	D149	480	385	470	–	3,200	7,500	400–600
Dielectric constant (77°F, 60 cps)	D150	4.8	4.0	3.6	4.2	3.09	3.25	2.4–2.7
Dissipation factor (77°F, 60 cps)	D150	0.014–0.040	–	–	0.009	0.0003	0.0021	0.005
Surface resistivity, ohm	D257	–	–	–	6×10^{12}	1.4×10^{12}	$>1 \times 10^{17}$	–

Table 1–28A (continued)
PLASTIC FILMS

	ASTM No.	Material: Nylon 6	Nylon 66	Nylon 610	Nylon 12	Polycarbonate	Polyester[a]	Polystyrene (oriented)
Chemical resistance								
Strong acids	D543 or D1239	Poor	Poor	Poor	Poor to good	Good	Excellent	Good
Strong alkalis	–	Excellent	Excellent	Excellent	Excellent	Poor	Excellent	Excellent
Greases and oils	–	Excellent	Excellent	Excellent	Excellent	Good	Excellent	Good
Solvents	–							
Ketone and ester		Excellent	Excellent	Excellent	Excellent	Fair	Excellent	Poor
Chlorinated		–	–	–	Poor	Poor	Excellent	Good
Hydrocarbon		Excellent	Excellent	Excellent	Excellent	Fair	Excellent	Good
Environment properties								
Maximum cont. service temperature, °F	–	380	300	300	230	270–280	250	160–180
Minimum service temperature, °F	–	<–100	<–100	<–100	<–100	<–212	–80	–
Resistance to sunlight	–	Fair	Fair	Fair	Fair	Fair	Fair	Fair
Dimensional change, %	–	Nil	Nil	Nil	Nil	Nil	Nil	Nil
Storage stability	–	Excellent	Excellent	Excellent	Excellent	Excellent	Excellent	Good
Rate of burning	–	Self extinguishing	Self extinguishing	Self extinguishing	Self extinguishing to slow burn	Slow	Self extinguishing	Slow

[a] Polyethylene terephthalate.
[b] Procedure E.

From *1973 Materials Selector,* Reinhold Publishing, Stamford, Conn., 1972, 269. With permission.

Table 1–28B
PLASTIC FILMS

	ASTM No.	Type: Cellophane	Fluorocarbon				Polypropylene	
		Coated	CTFE	FEP	PVF	PVF_2	Set cast	Biaxially oriented
General properties								
Method of production		Extrusion	Extrusion	Extrusion	Extrusion	Extrusion	Extrusion, calendering	Extrusion
Forms available		Sheets, rolls	Rolls	Rolls	Rolls	Sheets, rolls[a]	Sheets, rolls, tapes	Sheets, rolls
Clarity		Transparent	Transparent	Transparent	Transparent	Transparent	Transparent	Transparent
Minimum thickness, in.		0.0009	0.0005	0.0005	0.0005	0.00036	0.00075	0.0005
Maximum width, in.		46	48	48	48	40[a]	60	72
Area factor, 1,000 in.^{2}/lb/mil		12–25	13.5	13	7	12–14	31	31
Physical properties								
Specific gravity	D792	1.40–1.55	2.1	2.15	1.5	1.79–1.80	0.90	0.90
Tensile strength, 1,000 psi	D882	7–16	5–8	2.5–3	7–18	24–36	4–10	18–32
Elongation, %	D882	15–50	50–150	300–400	115–250	150–200	>400	40–80
Burst strength (Mullen), psi	D774	45–70	23–31	10–15	19–70	–	–	–
Tear strength (Elmendorf), g/mil	D689	2–15	10–26	100–150	12–100	–	20–100	5–10
Fold endurance	D643	–	Good	4,000	–	–	Excellent	Excellent
Heat seal range, °F		200–350	370–500	600–700	–	360–500	320	–
Water absorption (24 hr), %	D570	High	Negligible	0.01	0.05	0.03	Negligible	Negligible
Water vapor permeability, g/100 in.^{2}/24 hr/mil	E96	0.2–1.0	0.025	0.40	–	0.31	0.4–1.0	0.25–0.5
Gas permeability, cm^3/100 in.^{2}/24 hr/mil	D1434-66							
Oxygen		Low	7–12	750	3.2	0.2	150	150
Nitrogen		Low	2.5	320	0.3	0.017	–	–
Carbon dioxide		Low	16–40	1,670	11.1	0.75	–	–

Table 1–28B (continued)
PLASTIC FILMS

	ASTM No.	Type						
		Cellophane	Fluorocarbon				Polypropylene	
		Coated	CTFE	FEP	PVF	PVF[a]	Set cast	Biaxially oriented
Chemical resistance								
Strong acids	D543 or D1239	–	Excellent	Excellent	Excellent	Excellent	Excellent	Excellent
Strong alkalis		–	Excellent	Excellent	Excellent	Excellent	Excellent	Excellent
Greases and oils		–	Excellent	Excellent	Excellent	Excellent	Good	Good
Solvents								
Ketone and ester		–	Good	Excellent	Excellent	Good	–	–
Chlorinated		–	Good	Excellent	Excellent	Good	–	–
Hydrocarbon		–	Excellent	Excellent	Excellent	Excellent	Good	–
Permanence								
Maximum cont. service temperature, °F		300–375	300–390	400	225	300	285	–
Minimum service temperature, °F		0	–320	–400	–100	–76	–	<–40
Resistance to sunlight		Good	Excellent	Excellent	Excellent	Excellent	Fair	Fair
Dimensional change, %		2–5	Nil	Nil	Nil	Nil	Nil	–
Storage stability		Good	Excellent	Excellent	Excellent	Excellent	Excellent	Excellent
Flammability (rate of burning)		Fast	Nil	Nil	Slow	Nil	Slow	Slow

[a]Biaxially oriented rolls.

From *1973 Materials Selector,* Rienhold Publishing, Stamford, Conn., 1972, 270. With permission.

Table 1–28C
PLASTIC FILMS

	ASTM No.	Material						
		Polyethylene			Polyvinyl chloride (including copolymers)			
		Type I	Type II	Type III	Rigid	Nonrigid	Rubber hydrochloride	Polyimide
General characteristics								
Method of production	–	Extrusion, calendering	Extrusion, calendering	Extrusion	Casting, calendering, extrusion	Casting, calendering, extrusion	Casting	Casting
Forms available	–	Sheets, rolls, tapes, tubes	Sheets, rolls tapes	Sheets, rolls, tapes	Sheets, rolls, tapes	Sheets, rolls, tapes, tubes	Sheets, rolls, tapes	Sheets, rolls, tapes
Clarity	–	Transparent, translucent, opaque	Transparent, translucent, opaque	Transparent, translucent	Transparent, translucent, opaque	Transparent, translucent, opaque	Transparent, translucent, opaque	Transparent
Minimum thickness, in.	–	0.00075	0.00075	0.00075	0.001	0.005	0.0004	0.0005
Maximum width, in.	–	72	60	60	54	104	60	–
Area factor 1,000 in.^{2}/lb/mil	–	30	30	29	19.5–22.5	20–23	24	19.4
Physical and mechanical properties								
Specific gravity	D792	0.92	0.935–0.938	0.940–0.945	1.36–1.50	1.15–1.50	1.12–1.15	1.42
Tensile strength, 1,000 psi	D882	1.6–3.0	2.5–3.5	3.5–8.0	6.5–8.5	1–5	5–6	24–25
Elongation, %	D882	300–800	>200	50–400	5–200	50–500	350–500	65–70
Burst strength (Mullen), psi	D774	10–15	–	–	–	9–20	–	75
Tear strength (Elmendorf), g/mil	D689	100–125	93–97	10–350	20–150	30–1,400	1,000–1,500	–
Fold endurance	D643	Good	Good	Good	Poor	Good	–	Good
Heat sealing range, °F	–	400–450	250–375	250–375	260–400	200–400	225–350	Not possible
Water absorption (24 hr), %	D570	0.01	Negligible	Negligible	Negligible	Negligible	Negligible	2.9
Water vapor permeability, g/100 in.^{2}/24 hr/mil	E96	1.5	0.5–0.7	0.3–0.4	0.5 (0.005 in.)	0.7 (0.005 in.)	0.5–15.5	5.4
Gas permeability, cm^3/100 in.^{2}/24 hr/mil	–							
Oxygen		5.50 (1 mil)	280 (1 mil)	200 (1 mil)	3 (0.005 in.)	–	2–405	25
Nitrogen		300 (1 mil)	–	42 (1 mil)	–	–	–	6
Carbon dioxide		2,500 (1 mil)	990 (1 mil)	580 (1 mil)	8 (0.005 in.)	–	36–2,616	45
Electrical properties								
Dielectric strength (77°F, 60 cps), V/mil	D149	450	450	500	250–1,300	250–1,300	–	7,000
Dielectric constant (77°F, 60 cps)	D150	2.3	2.3	2.3	3.0–8.0	3.0–8.0	–	3.5 (1,000 cps)
Dissipation factor (77°F, 60 cps)	D150	0.0005	0.0005	0.0005	0.009–0.16	0.009–0.16	–	0.003 (1,000 cps)
Surface resistivity, ohm	D257	$>10^{16}$	$>10^{16}$	$>10^{16}$	–	–	–	$>10^{16}$

Table 1–28C (continued)
PLASTIC FILMS

	ASTM No.	Material						
		Polyethylene			Polyvinyl chloride (including copolymers)		Rubber hydrochloride	Polyimide
		Type I	Type II	Type III	Rigid	Nonrigid		
Chemical resistance								
Strong acids	D543 or D1239	Excellent	Excellent	Excellent	Excellent	Excellent	Good	Excellent
Strong alkalis	–	Excellent	Excellent	Excellent	Excellent	Excellent	Good	Poor
Greases and oils	–	Fair	Fair	Fair	Good	Fair	Excellent	Excellent
Solvents	–							
Ketone and ester		Good	Good	Good	Poor	Poor	Fair	Excellent
Chlorinated		Fair	Fair	Fair	Fair	Fair	Fair	Excellent
Hydocarbon		Fair	Fair	Fair	Excellent	Good	Excellent	Excellent
Environmental properties								
Maximum cont. service temperature, °F	–	180	230	250	160–180	140–160	205	550
Minimum service temperature, °F	–	–70	<–100	<–100	–30	–50	–20	–450
Resistance to sunlight	–	Fair	Fair	Fair	Good	Good	Fair	Excellent
Dimensional change, %	–	Nil	Nil	Nil	Nil	Nil	Slight	Nil
Storage stability	–	Excellent	Excellent	Excellent	Excellent	Excellent	Good	Excellent
Rate of burning	–	Slow	Slow	Slow	Self extinguishing	Self extinguishing	Self extinguishing	Self extinguishing

From *1973 Materials Selector,* Reinhold Publishing, Stamford, Conn., 1972, 271. With permission.

Table 1–29
PLASTIC FOAMS – RIGID

Table A

	ASTM No.	ABS	Cellulose acetate	Epoxy	Syntactic epoxy[a]		Phenolic	Poly-ethylene	Poly-propylene
Density, lb/ft^3		31	6–8	5–8	36	42	2–4	34	5
Thermal conductivity, Btu/hr/ft^2/°F/in.	177	0.58	0.31–0.32	0.24–0.28	4.56	–	0.20–0.22	0.92	0.27
Coefficient of thermal expansion, 10^{-5}/°F	D696	9.7	2.5	2.3	4.5	–	0.5	4.18	–
Water absorption, % vol	C272	0.6	13–17	–	–	1.5	<3	0.22	–
Flammability, ipm	D1692	1.04	4.9	–	–	–	SE	–	1.6
Dielectric constant at 10^6 cps	–	1.59	1.10–1.12	2.0	1.55	–	–	1.48	–
Dissipation factor at 10^6 cps	–	0.007	0.003	0.005	0.01	–	–	0.0003	–
Maximum recommended service temperature, °F	–	200	350	400–500	300	300	270	195	230
Tensile strength, psi	D1623	1,400	170	50–200	3,300	4,600	20–55	1,000	170
Ultimate tensile elongation, %	D1623	–	–	–	–	–	–	–	–
Modulus of elasticity in tension, 1,000 psi	D1623	2.4	–	–	–	610	–	–	–
Compressive strength, psi (10%)	D1621	–	125–150	60–90	9,600	13,400	20–90	800	55
Modulus of elasticity in compression, 1,000 psi	D1621	–	5.5–13	2.1–6.5	373	480	–	–	1.2
Flexural strength, psi	D790	2.4	150	200–800	3,800	6,000	25–65	1,900	230
Modulus of elasticity in flexure, 1,000 psi	D790	9	5.5	2.5–6	–	–	–	88	9.6
Shear strength, psi	C273	–	140	–	3,800	4,400	15–30	–	–
Modulus of elasticity in shear, 1,000 psi	C273	–	–	–	–	–	0.4–0.75	–	–
Hardness (Shore D)	–	60	–	–	80–85	–	–	–	–

[a]Glass-microsphere-filled epoxy.

Table 1–29 (continued)
PLASTIC FOAMS – RIGID

Table B

	ASTM. No.	Polyvinyl chloride	Polystyrene (expanded)		Urea	Urethane		
Density, lb/ft³		3	2	6	0.8–1.2	2–3	4–7	18–25
Thermal conductivity, Btu/hr/ft²/°F/in.	177	0.15–0.20	0.20–0.28	0.20–0.25	0.18–0.21	0.11–0.23	0.15–0.28	0.29–0.52
Coefficient of thermal expansion, 10^{-5}/°F	D696	2	2.7–4	2.7–4	–	3–4	4	4
Water absorption, % vol	C272	0.1	<0.1	<0.1	–	3–4	1.5–2	0.2
Flammability, ipm	D1692	Self extinguishing-nonburning	2–8	2–8	Self extinguishing	Nonburning	Nonburning	Nonburning
Dielectric constant at 10^6 cps	–	–	1.02–1.24	–	–	–	–	–
Dissipation factor at 10^6 cps	–	–	<0.0005	–	–	–	–	–
Maximum recommended service temperature, °F	–	180	175	175	120	200–250	250–300	300–400
Tensile strength, psi	D1623	100–200	50–55	120	–	20–70	90–250	700–1,300
Ultimate tensile elongation, %	D1623	5–20	5	2	–	–	–	–
Modulus of elasticity in tension, 1,000 psi	D1623	3–4	740	6,100	–	–	–	–
Compressive strength, psi (10%)	D1621	70–100	25–30	100–150	5	20–50	65–275	1,200–2,000
Modulus of elasticity in compression, 1,000 psi	D1621	3–4	0.55–2	3–6	–	0.3–0.6	1.5–4.5	10–40
Flexural strength, psi	D790	120–160	55–75	200–300	17	60–100	200–350	700–2,000
Modulus of elasticity in flexure, 1,000 psi	D790	3–4	1.3–3.8	5–15	0.7	0.8–0.9	0.8–15	12–100
Shear strength, psi	C273	60–80	35	150	–	20–30	60–130	7,600
Modulus of elasticity in shear, 1,000 psi	C273	2–2.5	1.15–1.6	3	–	0.17–0.21	0.5–1.5	3–9

From *1973 Materials Selector,* Reinhold Publishing, Stamford, Conn., 1972, 272. With permission.

Table 1–30
PLASTICS AND RUBBER FOAMS – FLEXIBLE

	Type						
	Polyethylene (cellular)			Silicone (foam in place)	Urethane (prefoamed		Vinyl (open cell)
	Density, lb/ft³						
	2.2	33[a]	41[b]	10	1.5–2.0	4.0–5.0	4 and up
Maximum recommended service temperature, °F	160	–	–	500	–	–	–
Electrical resistivity, microhm-cm	1.5×10^{17}	–	–	3.8×10^{18}	–	–	–
Dielectric strength (short time), V/mil	55	220	220	50	–	–	–
Dielectric constant, 10^6 cycles	1.05[c]	1.5	1.84	1.17[d]	–	–	–
Dissipation factor, 10^6 cycles	0.0002[c]	0.0004	0.0006	0.001[d]	–	–	–
Tensile strength, psi	20–30	600	2,100	5	12–20	18–20	10–200
Elongation, %	60	300	350	40	150–250	100–125	75–300
Tear strength, lb/in.	6	–	–	–	2.0–3.0	1.0–2.0	–
Compression set (22 hr at 158°F), %	15	–	–	4	4–8	2–9	15
Compression–deflection, psi							
Stress for 25% deflection	8.5[e]	–	–	1	0.35	0.9–2.2	3–500
Stress for 50% deflection	17.5	–	–	3	0.45	–	–

[a]Based on low-density polymer; typical expanded insulation (about 46% gas by volume).
[b]Based on high-density polymers; typical expanded insulation (about 30% gas by volume).
[c]At 10^9 cps.
[d]10^5 cycles.
[e]Recovery at same temperature as compression.

From *1973 Materials Selector,* Reinhold Publishing, Stamford, Conn., 1972, 272. With permission.

Section 2

Electronic Materials

SUPERCONDUCTIVITY

These tables on superconductivity include superconductive properties of chemical elements (Table 2–5), thin films (Table 2–6), a selected list of compounds and alloys (Table 2–7), and high-magnetic-field superconductors (Table 2–8).

The historically first observed and most distinctive property of a superconductive body is the near-total loss of resistance at a critical temperature (T_c) that is characteristic of each material. Figure 1A illustrates schematically two types of possible transitions. The sharp vertical discontinuity in resistance is indicative of that found for a single crystal of a very pure element or one of a few well annealed alloy compositions. The broad transition, illustrated by broken lines, suggests the transition shape seen for materials that are not homogeneous and contain unusual strain distributions. Careful testing of the resistivity limits for superconductors shows that it is less than 4×10^{-23} ohm-cm, while the lowest resistivity observed in metals is of the order of 10^{-13} ohm-cm. If one compares the resistivity of a superconductive body to that of copper at room temperature, the superconductive body is at least 10^{17} times less resistive.

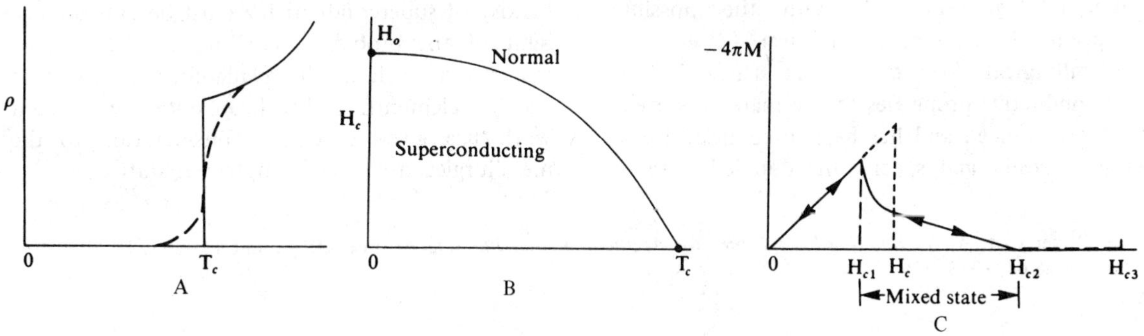

FIGURE 1.1. Physical properties of superconductors. A. Resistivity versus temperature for a pure and perfect lattice (solid line); impure and/or imperfect lattice (broken line). B. Magnetic-field temperature dependence for Type 1 or "soft" superconductors. C. Schematic magnetization curve for Type II or "hard" superconductors.

The temperature interval ΔT_c, over which the transition between the normal and superconductive states takes place, may be of the order of as little as 2×10^{-5} K or as much as several K in width, depending on the material state. The narrow transition width was attained in 99.9999% pure gallium single crystals.

A Type I superconductor below T_c, as exemplified by a pure metal, exhibits perfect diamagnetism and excludes a magnetic field up to some critical field H_c, whereupon it reverts to the normal state as shown in the H-T diagram of Figure 1B.

The difference in entropy near absolute zero between the superconductive and normal states relates directly to the electronic specific heat, γ: $(S_s - S_n)_{T\to 0} = -\gamma T$.

The magnetization of a typical high-field superconductor is shown in Figure 1C. The discovery of the large current-carrying capability of Nb_3Sn and other similar alloys has led to an extensive study of the physical properties of these alloys. In brief, a high-field superconductor, or Type II superconductor, passes from the perfect diamagnetic state at low magnetic fields to a mixed state and finally to a sheathed state before attaining the normal resistive state of the metal. The magnetic-field values separating the four stages are given as H_{c1}, H_{c2}, and H_{c3}. The superconductive state below H_{c1} is perfectly diamagnetic, identical to the state of most pure metals of the "soft" or Type I superconductor. Between H_{c1} and H_{c2} a "mixed superconductive state" is found in which fluxons (a minimal unit of magnetic flux) create lines of normal superconductor in a superconductive matrix. The volume of the normal state is proportional to $-4\pi M$ in the "mixed state" region. Thus, at H_{c2} the fluxon density has become so great as to drive the interior volume of the superconductive body completely normal. Between H_{c2} and H_{c3} the superconductor has a sheath of current-carrying superconductive material at the body surface, and above H_{c3} the normal state exists. With several types of careful measurement, it is possible to determine H_{c1}, H_{c2}, and H_{c3}. Table 2–8 contains some of the available data on high-field superconductive materials.

High-field superconductive phenomena are also related to specimen dimension and configuration. For example, the Type I superconductor, Hg, has entirely different magnetization behavior in high-magnetic fields when contained in the very fine sets of filamentary tunnels found in an unprocessed Vycor glass. The great majority of superconductive materials are Type II. The elements in very pure form and a very few precisely stoichiometric and well-annealed compounds are Type I, with the possible exceptions of vanadium and niobium.

Metallurgical Aspects – The sensitivity of superconductive properties to the material state is most pronounced and has been used in a reverse sense to study and specify the detailed state of alloys. The mechanical state, the homogeneity, and the presence of impurity atoms and other electron-scattering centers are all capable of controlling the critical temperature and the current-carrying capabilities in high-magnetic fields. Well-annealed specimens tend to show sharper transitions than those that are strained or inhomogeneous. This sensitivity to mechanical state underlines a general problem in the tabulation of properties for superconductive materials. The occasional divergent values of the critical temperature and of the critical fields quoted for a Type II superconductor may lie in the variation in sample preparation. Critical temperatures of materials studied early in the history of superconductivity must be evaluated in light of the probable metallurgical state of the material as well as the availability of less pure starting elements. It has been noted that recent work has given extended consideration to the metallurgical aspects of sample preparation.

From Roberts, B. W., in *Handbook of Tables for Applied Engineering Science,* Bolz, R. E. and Tuve, G. L., Eds., CRC Press, Cleveland, 1973, 229.

REFERENCES

References to the data presented in this section, to additional entries of superconductive materials, and to those materials specifically tested and found non-superconductive to some low temperature may be found in the following publications:

Roberts, B. W., Superconductive materials and some of their properties, in *Progress in Cryogenics,* Vol. 4, Mendelssohn, K., Ed., Academic Press, New York, 1964, 160.

Roberts, B. W., Properties of Selected Superconductive Materials, National Bureau of Standards Technical Notes 482, 724, and 825, U.S. Government Printing Office, Washington, D. C., 1969, 1972, and 1974.

Table 2–1
ELECTRICAL CONDUCTORS FOR GENERAL WIRING

A. Allowable Ampacities of Insulated Copper Conductors

Not More than Three Conductors in Raceway or Cable or Direct Burial (Based on Ambient Temperature of 30°C/86°F)

Size	Temperature rating of conductor[a]							
AWG; MCM	60°C (140°F)	75°C (167°F)	85°C (185°F)	90°C (194°F)	110°C (230°F)	125°C (257°F)	200°C (392°F)	250°C (482°F)
	Types RUW (14-2), T, TW, UF	Types RH, RHW, RUH (14-2), THW, THWN, XHHW, USE	Types V, MI	Types TA, TBS, SA, AVB, SIS, FEP, FEPB, RHH, THHN, XHHW	Types AVA, AVL	Types AI (14-8), AIA	Types A (14-8), AA, FEP, FEPB	Type TFE (nickel or nickel-coated copper only)
18	–	–	–	21	–	–	–	–
16	–	–	22	22	–	–	–	–
14	15	15	25	25[b]	30	30	30	40
12	20	20	30	30[b]	35	40	40	55
10	30	30	40	40[b]	45	50	55	75
8	40	45	50	50	60	65	70	95
6	55	65	70	70	80	85	95	120
4	70	85	90	90	105	115	120	145
3	80	100	105	105	120	130	145	170
2	95	115	120	120	135	145	165	195
1	110	130	140	140	160	170	190	220
1/0	125	150	155	155	190	200	225	250
2/0	145	175	185	185	215	230	250	280
3/0	165	200	210	210	245	265	285	315
4/0	195	230	235	235	275	310	340	370
250	215	255	270	270	315	335	–	–
300	240	285	300	300	345	380	–	–
350	260	310	325	325	390	420	–	–
400	280	335	360	360	420	450	–	–
500	320	380	405	405	470	500	–	–

Table 2–1 (continued)
ELECTRICAL CONDUCTORS FOR GENERAL WIRING

A. Allowable Ampacities of Insulated Copper Conductors (continued)

Not More than Three Conductors in Raceway or Cable or Direct Burial (Based on Ambient Temperature of 30°C/86°F)

Size	Temperature rating of conductor[a]							
AWG; MCM	60°C (140°F)	75°C (167°F)	85°C (185°F)	90°C (194°F)	110°C (230°F)	125°C (257°F)	200°C (392°F)	250°C (482°F)
	Types RUW (14-2), T, TW, UF	Types RH, RHW, RUH (14-2), THW, THWN, XHHW, USE	Types V, MI	Types TA, TBS, SA, AVB, SIS, FEP, FEPB, RHH, THHN, XHHW	Types AVA, AVL	Types AI (14-8), AIA	Types A (14-8), AA, FEP, FEPB	Type TFE (nickel or nickel-coated copper only)
600	355	420	455	455	525	545	–	–
700	385	460	490	490	560	600	–	–
750	400	475	500	500	580	620	–	–
800	410	490	515	515	600	640	–	–
900	435	520	555	555	–	–	–	–
1,000	455	545	585	585	680	730	–	–
1,250	495	590	645	645	–	–	–	–
1,500	520	625	700	700	785	–	–	–
1,750	545	650	735	735	–	–	–	–
2,000	560	665	775	775	840	–	–	–

[a]See Section E of this table.
[b]The ampacities for Types FEP, FEPB, RHH, THHN, and XHHW conductors for Sizes 14, 12, and 10 in this table are the same as designated for 75°C conductors.

Table 2–1 (continued)
ELECTRICAL CONDUCTORS FOR GENERAL WIRING

B. Allowable Ampacities of Insulated Copper Conductors

Single Conductor in Free Air (Based on Ambient Temperature of 30°C/86°F)

Size	Temperature rating of conductor[a]								
AWG; MCM	60°C (140°F)	75°C (167°F)	85°C (185°F)	90°C (194°F)	110°C (230°F)	125°C (257°F)	200°C (392°F)	250°C (482°F)	
	Types RUW (14-2), T, TW	Types RH, RHW, RUH (14-2), THW, THWN, XHHW	Types V, MI	Types TA, TBS, SA, AVB, SIS, FEP, FEPB, RHH THHN, XHHW	Types AVA, AVL	Types AI (14-8), AIA	Types A (14-8), AA, FEP, FEPB	Type TFE (nickel or nickel-coated copper only)	Bare and covered conductors
18	–	–	–	25	–	–	–	–	–
16	–	–	27	27	–	–	–	–	–
14	20	20	30	30[b]	40	40	45	60	30
12	25	25	40	40[b]	50	50	55	80	40
10	40	40	55	55[b]	65	70	75	110	55
8	55	65	70	70	85	90	100	145	70
6	80	95	100	100	120	125	135	210	100
4	105	125	135	135	160	170	180	285	130
3	120	145	155	155	180	195	210	335	150
2	140	170	180	180	210	225	240	390	175
1	165	195	210	210	245	265	280	450	205
1/0	195	230	245	245	285	305	325	545	235
2/0	225	265	285	285	330	355	370	605	275
3/0	260	310	330	330	385	410	430	725	320
4/0	300	360	385	385	445	475	510	850	370
250	340	405	425	425	495	530	–	–	410
300	375	445	480	480	555	590	–	–	460
350	420	505	530	530	610	655	–	–	510
400	455	545	575	575	665	710	–	–	555
500	515	620	660	660	765	815	–	–	630

Table 2–1 (continued)
ELECTRICAL CONDUCTORS FOR GENERAL WIRING

B. Allowable Ampacities of Insulated Copper Conductors (continued)

Single Conductor in Free Air (Based on Ambient Temperature of 30°C/86°F)

Size	Temperature rating of conductor[a]								
AWG; MCM	60°C (140°F)	75°C (167°F)	85°C (185°F)	90°C (194°F)	110°C (230°F)	125°C (257°F)	200°C (392°F)	250°C (482°F)	
	Types RUW (14-2), T, TW	Types RH, RHW, RUH (14-2), THW, THWN, XHHW	Types V, MI	Types TA, TBS, SA, AVB, SIS, FEP, FEPB, RHH THHN, XHHW	Types AVA, AVL	Types AI (14-8), AIA	Types A (14-8), AA, FEP, FEPB	Type TFE (nickel or nickel-coated copper only)	Bare and covered conductors
600	575	690	740	740	855	910	–	–	710
700	630	755	815	815	940	1,005	–	–	780
750	655	785	845	845	980	1,045	–	–	810
800	680	815	880	880	1,020	1,085	–	–	845
900	730	870	940	940	–	–	–	–	905
1,000	780	935	1,000	1,000	1,165	1,240	–	–	965
1,250	890	1,065	1,130	1,130	–	–	–	–	–
1,500	980	1,175	1,260	1,260	1,450	–	–	–	1,215
1,750	1,070	1,280	1,370	1,370	–	–	–	–	–
2,000	1,155	1,385	1,470	1,470	1,715	–	–	–	1,405

[a]See Section E of this table.
[b]The ampacities for Types FEP, FEPB, RHH, THHN, and XHHW conductors for Sizes 14, 12, and 10 in this table are the same as designated for 75°C conductors.

Table 2–1 (continued)

ELECTRICAL CONDUCTORS FOR GENERAL WIRING

C. Allowable Ampacities of Insulated Aluminum and Copper-clad Aluminum Conductors

Not More than Three Conductors in Raceway or Cable or Direct Burial (Based on Ambient Temperature of 30° C/86° F)

Size	Temperature rating of conductor[a]						
AWG; MCM	60° C (140° F)	75° C (167° F)	85° C (185° F)	90° C (194° F)	110° C (230° F)	125° C (257° F)	200° C (392° F)
	Types RUW (12-2), T, TW, UF	Types RH, RHW, RUH (12-2), THW, THWN, XHHW, USE	Types V, MI	Types TA, TBS, SA, AVB, SIS, RHH, THHN, XHHW	Types AVA, AVL	Types AI (12-8), AIA	Types A (12-8), AA
12	15	15	25	25b	25	30	30
10	25	25	30	30b	35	40	45
8	30	40	40	40	45	50	55
6	40	50	55	55	60	65	75
4	55	65	70	70	80	90	95
3	65	75	80	80	95	100	115
2	75	90	95	95	105	115	130
1	85	100	110	110	125	135	150
1/0	100	120	125	125	150	160	180
2/0	115	135	145	145	170	180	200
3/0	130	155	165	165	195	210	225
4/0	155	180	185	185	215	245	270
250	170	205	215	215	250	270	–
300	190	230	240	240	275	305	–
350	210	250	260	260	310	335	–
400	225	270	290	290	335	360	–
500	260	310	330	330	380	405	–
600	285	340	370	370	425	440	–
700	310	375	395	395	455	485	–
750	320	385	405	405	470	500	–
800	330	395	415	415	485	520	–
900	355	425	455	455	–	–	–
1,000	375	445	480	480	560	600	–

Table 2–1 (continued)
ELECTRICAL CONDUCTORS FOR GENERAL WIRING

C. Allowable Ampacities of Insulated Aluminum and Copper-clad Aluminum Conductors (continued)

Not More than Three Conductors in Raceway or Cable or Direct Burial (Based on Ambient Temperature of 30°C/86°F)

Size	Temperature rating of conductor[a]						
AWG; MCM	60°C (140°F)	75°C (167°F)	85°C (185°F)	90°C (194°F)	110°C (230°F)	125°C (257°F)	200°C (392°F)
	Types RUW (12-2), T, TW, UF	Types RH, RHW, RUH (12-2), THW, THWN, XHHW, USE	Types V, MI	Types TA, TBS, SA, AVB, SIS, RHH, THHN, XHHW	Types AVA, AVL	Types AI (12-8), AIA	Types A (12-8), AA
1,250	405	485	530	530	–	–	–
1,500	435	520	580	580	650	–	–
1,750	455	545	615	615	–	–	–
2,000	470	560	650	650	705	–	–

[a]See Section E of this table.
[b]The ampacities for Types RHH, THHN, and XHHW conductors for Sizes 12 and 10 in this table are the same as designated for 75°C conductors.

Table 2–1 (continued)
ELECTRICAL CONDUCTORS FOR GENERAL WIRING

D. Allowable Ampacities of Insulated Aluminum and Copper-clad Aluminum Conductors

Single Conductor in Free Air (Based on Ambient Temperature of 30°C/86°F)

Size	Temperature rating of conductor[a]							
AWG; MCM	60°C (140°F)	75°C (167°F)	85°C (185°F)	90°C (194°F)	110°C (230 °F)	125°C (257°F)	200°C (392°F)	
	Types RUW (12-2), T, TW	Types RH, RHW, RUH (12-2), THW, THWN, XHHW	Types V, MI	Types TA, TBS, SA, AVB, SIS, RHH, THHN, XHHW	Types AVA, AVL	Types AI (12-8), AIA	Types A (12-8), AA	Bare and covered conductors
12	20	20	30	30[b]	40	40	45	30
10	30	30	45	45[b]	50	55	60	45
8	45	55	55	55	65	70	80	55
6	60	75	80	80	95	100	105	80
4	80	100	105	105	125	135	140	100
3	95	115	120	120	140	150	165	115
2	110	135	140	140	165	175	185	135
1	130	155	165	165	190	205	220	160
1/0	150	180	190	190	220	240	255	185
2/0	175	210	220	220	255	275	290	215
3/0	200	240	255	255	300	320	335	250
4/0	230	280	300	300	345	370	400	290
250	265	315	330	330	385	415	–	320
300	290	350	375	375	435	460	–	360
350	330	395	415	415	475	510	–	400
400	355	425	450	450	520	555	–	435
500	405	485	515	515	595	635	–	490
600	455	545	585	585	675	720	–	560
700	500	595	645	645	745	795	–	615
750	515	620	670	670	775	825	–	640
800	535	645	695	695	805	855	–	670
900	580	700	750	750	–	–	–	725
1,000	625	750	800	800	930	990	–	770

Table 2–1 (continued)
ELECTRICAL CONDUCTORS FOR GENERAL WIRING

D. Allowable Ampacities of Insulated Aluminum and Copper-clad Aluminum Conductors (continued)

Single Conductor in Free Air (Based on Ambient Temperature of 30°C/86°F)

Size	Temperature rating of conductor[a]							
AWG; MCM	60°C (140°F)	75°C (167°F)	85° (185°F)	90°C (194°F)	110°C (230°F)	125°C (257°F)	200°C (392°F)	
	Types RUW (12-2), T, TW	Types RH, RHW, RUH (12-2), THW, THWN, XHHW	Types V, MI	Types TA, TBS, SA, AVB, SIS, RHH, THHN, XHHW	Types AVA, AVL	Types AI (12-8), AIA	Types A (12-8), AA	Bare and covered conductors
1,250	710	855	905	905	–	–	–	–
1,500	795	950	1,020	1,020	1,175	–	–	985
1,750	875	1,050	1,125	1,125	–	–	–	–
2,000	960	1,150	1,220	1,220	1,423	–	–	1,165

[a]See Section E of this table.
[b]The ampacities for Types RHH, THHN, and XHHW conductors for Sizes 12 and 10 in this table are the same as designated for 75°C conductors.

Table 2–1 (continued)
ELECTRICAL CONDUCTORS FOR GENERAL WIRING

E. Correction Factors

Ambient Temperatures Over 30°C/86°F

Celsius	Fahrenheit	60°C (140°F)	75°C (167°F)	85°C (185°F)	90°C (194°F)	110°C (230°F)	125°C (257°F)	200°C (392°F)	250°C (482°F)
40	104	0.82	0.88	0.90	0.91	0.94	0.95	–	–
45	113	0.71	0.82	0.85	0.87	0.90	0.92	–	–
50	122	0.58	0.75	0.80	0.82	0.87	0.89	–	–
55	131	0.41	0.67	0.74	0.76	0.83	0.86	–	–
60	140	–	0.58	0.67	0.71	0.79	0.83	0.91	0.95
70	158	–	0.35	0.52	0.58	0.71	0.76	0.87	0.91
75	167	–	–	0.43	0.50	0.66	0.72	0.86	0.89
80	176	–	–	0.30	0.41	0.61	0.69	0.84	0.87
90	194	–	–	–	–	0.50	0.61	0.80	0.83
100	212	–	–	–	–	–	0.51	0.77	0.80
120	248	–	–	–	–	–	–	0.69	0.72
140	284	–	–	–	–	–	–	0.59	0.59
160	320	–	–	–	–	–	–	–	0.54
180	356	–	–	–	–	–	–	–	0.50
200	392	–	–	–	–	–	–	–	0.43
225	437	–	–	–	–	–	–	–	0.30

Note: Use of conductors with higher operating temperatures – where the room temperature is within 10°C of the maximum allowable operating temperature of the insulation, it is desirable to use an insulation with a higher maximum allowable operating temperature, although insulation can be used in a room temperature approaching its maximum allowable operating temperature limit if the current is reduced in accordance with the correction factors for different room temperatures as shown in this correction factor table.

Table 2–1 (continued)
ELECTRICAL CONDUCTORS FOR GENERAL WIRING

Key to Conductor Insulation and Maximum Operating Temperature

Trade name	Type letter	Maximum operating temperature	Insulation	AWG or MCM	Thickness of insulation, mils	Outer covering
Heat-resistant rubber	RH	75°C, 167°F	Heat-resistant rubber	14–12[a]	30	Moisture-resistant flame-retardant, nonmetallic covering[b]
				10	45	
				8-2	60	
				1-4/0	80	
				213-500	95	
Heat-resistant rubber	RHH	90°C, 194°F		501-1,000	110	
				1,001-2,000	125	
Moisture- and heat-resistant rubber	RHW	75°C, 167°F	Moisture- and heat-resistant rubber	14-10	45	Moisture-resistant, flame-retardant, nonmetallic covering[b]
				8-2	60	
				1-4/0	80	
				213-500	95	
				501-1,000	110	
				1,001-2,000	125	
Heat-resistant latex rubber	RUH	75°C, 167°F	90% unmilled, grainless rubber	14-10	18	Moisture-resistant, flame-retardant, nonmetallic covering
				8-2	25	
Moisture-resistant latex rubber	RUW	60°C, 140°F	90% unmilled, grainless rubber	14-10	18	Moisture-resistant, flame-retardant nonmetallic covering
				8-2	25	
Thermoplastic	T	60°C, 140°F	Flame-retardant, thermoplastic compound	14–10	30	None
				8	45	
				6–2	60	
				1-4/0	80	
				213-500	95	
				501-1,000	110	
				1,001-2,000	125	

Table 2–1 (continued)
ELECTRICAL CONDUCTORS FOR GENERAL WIRING

Key to Conductor Insulation and Maximum Operating Temperature (continued)

Trade name	Type letter	Maximum operating temperature	Insulation	AWG or MCM	Thickness of insulation, mils	Outer covering
Moisture-resistant thermoplastic	TW	60°C, 140°F	Flame-retardant, moisture-resistant thermoplastic	14-10	30	None
				8	45	
				6-2	60	
				1-4/0	80	
				213-500	95	
				501-1,000	110	
				1,001-2,000	125	
Heat-resistant thermoplastic	THHN	90°C, 194°F	Flame-retardant, heat-resistant thermoplastic	14-12	15	Nylon jacket
				10	20	
				8-6	30	
				4-2	40	
				1-4/0	50	
				250-500	60	
				501-1,000	70	
Moisture- and heat-resistant thermoplastic	THW	75°C, 167°F; 90°C, 194°F	Flame-retardant, moisture- and heat-resistant thermoplastic	14-10	45	None
				8-2	60	
				1-4/0	80	
				213-500	95	
				501-1,000	110	
				1,001-2,000	125	
Moisture- and heat-resistant thermoplastic	THWN	75°C, 167°F	Flame-retardant, moisture- and heat resistant thermoplastic	14-12	15	Nylon jacket
				10	20	
				8-6	30	
				4-2	40	
				1-4/0	50	
				250-500	60	
				501-1,000	70	
Moisture- and heat-resistant cross-linked synthetic polymer	XHHW	90°C, 194°F; 75°C, 167°F	Flame-retardant cross-linked synthetic polymer	14-10	30	None
				8-2	45	
				1-4/0	55	
				213-500	65	
				501-1,000	80	
				1,001-2,000	95	

Table 2–1 (continued)
ELECTRICAL CONDUCTORS FOR GENERAL WIRING

Key to Conductor Insulation and Maximum Operating Temperature (continued)

Trade name	Type letter	Maximum operating temperature	Insulation	AWG or MCM	Thickness of insulation, mils		Outer covering
					(A)	(B)	
Moisture-, heat-, and oil-resistant thermoplastic	MTW	60°C, 140°F; 90°C, 194°F	Flame-retardant, moisture-, heat- and oil-resistant thermoplastic	22-12	30	15	(A) None, (B) nylon jacket
				10	30	20	
				8	45	30	
				6	60	30	
				4-2	60	40	
				1-4/0	80	50	
				213-500	95	60	
				501-1,000	110	70	
Extruded polytetra-fluoroethylene	TFE	250°C, 482°F	Extruded polytetrafluoroethylene	14-10		20	None
				8-2		30	
				1-4/0		45	
					Thermoplastic	Asbestos	
Thermoplastic and asbestos	TA	90°C, 194°F	Thermoplastic and asbestos	14-8	20	20	Flame-retardant, nonmetallic covering
				6-2	30	25	
				1-4/0	40	30	
Thermoplastic and fibrous outer braid	TBS	90°C, 194°F	Thermoplastic	14-10		30	Flame-retardant, nonmetallic covering
				8		45	
				6-2		60	
				1-4/0		80	
Synthetic heat resistant	SIS	90°C, 194°F	Heat-resistant rubber	14-10		30	None
				8		45	
				6-2		60	
				1-4/0		80	
Mineral insulation (metal sheathed)	MI	85°C, 185°F; 250°C, 482°F	Magnesium oxide	16-10		36	Copper
				9-4		50	
				3-250		55	

Table 2–1 (continued)
ELECTRICAL CONDUCTORS FOR GENERAL WIRING

Key to Conductor Insulation and Maximum Operating Temperature (continued)

Trade name	Type letter	Maximum operating temperature	Insulation	AWG or MCM	Thickness of insulation, mils	Outer covering
Underground feeder and branch circuit cable single conductor	UF	60°C, 140°F; 75°C, 167°F	Moisture-resistant	14-10	60[c]	Integral with insulation
			Moisture and heat resistant	8-2	80[c]	
				1-4/0	95[c]	
Underground service entrance cable-single conductor	USE	75°C, 167°F	Heat and moisture resistant	12-10	45	Moisture-resistant nonmetallic covering
				8-2	60	
				1-4/0	80	
				213-500	95	
				501-1,000	110	
				1,001-2,000	125	
Silicone-asbestos	SA	90°C, 194°F; 125°C, 257°F	Silicone rubber	14-10	45	Asbestos or glass
				8-2	60	
				1-4/0	80	
				213-500	95	
				501-1,000	110	
				1,001-2,000	125	
Fluorinated ethylene propylene	FEP or FEPB	90°C, 194°F; 200°C, 392°F	Fluorinated ethylene propylene	14-10	20	None
				8-2	30	
				14-8	14	Glass braid
				6-2	14	Asbestos braid
Varnished cambric	V	85°C, 185°F	Varnished cambric	14-8	45	Nonmetallic covering or lead sheath
				6-2	60	
				1-4/0	80	
				213-500	95	
				500-1,000	110	
				1,001-2,000	125	

Table 2–1 (continued)
ELECTRICAL CONDUCTORS FOR GENERAL WIRING

Key to Conductor Insulation and Maximum Operating Temperature (continued)

Trade name	Type letter	Maximum operating temperature	Insulation	AWG or MGM	Thickness of insulation, mils				Outer covering
Asbestos and varnished cambric	AVA	110°C, 230°F	Impregnated asbestos and varnished cambric		1st asbestos	Varnished cambric asbestos	AVA 2nd asbestos	AVL 2nd	AVA-asbestos braid or glass
				14-8 (solid only)	–	30	20	25	
				14-8	10	30	15	25	
				6-2	15	30	20	25	
				1-4/0	20	30	30	30	
				213-500	25	40	40	40	
	AVL	110°C, 230°F		501-1,000	30	40	40	40	AVL-lead sheath
				1,001-2,000	30	50	50	50	
							Varnished cambric	Asbestos	
Asbestos and varnished cambric	AVB	90°C, 194°F		18-8			30	20	Flame-retardant cotton braid (switchboard wiring)
				6-2			40	30	
				1-4/0			40	40	
						Asbestos	Varnished cambric	2nd Asbestos	
			Impregnated asbestos and varnished cambric	14-8		10	30	15	Flame-retardant cotton braid
				6-2		15	30	20	
				1-4/0		20	30	30	
				213-500		25	40	40	
				501-1,000		30	40	40	
				1,001-2,000		30	50	50	
Asbestos	A	200°C, 392°F	Asbestos	14				30	Without asbestos braid
				12-8				40	

Table 2–1 (continued)
ELECTRICAL CONDUCTORS FOR GENERAL WIRING

Key to Conductor Insulation and Maximum Operating Temperature (continued)

Trade name	Type letter	Maximum operating temperature	Insulation	AWG or MCM	Thickness of insulation, mils		Outer covering
Asbestos	AA	200°C, 392°F	Asbestos	14		30	With asbestos braid or glass
				12-8		30	
				6-2		40	
				1-4/0		60	
Asbestos	AI	125°C, 257°F	Impregnated asbestos	14		30	Without asbestos braid
				12-8		40	
					Sol.	Str.	
Asbestos	AIA	125°C, 257°F	Impregnated asbestos	14	30	30	With asbestos braid or glass
				12-8	30	40	
				6-2	40	60	
				1-4/0	60	75	
				213-500		90	
				501-1,000		105	
Paper		85°C, 185°F	Paper				Lead sheath

[a]For 14–12 sizes RHH shall be 45 mils thickness insulation.
[b]Outer covering shall not be required over rubber insulations that have been specifically approved for the purpose.
[c]Includes integral jacket.

Note: For insulated aluminum and copper-clad aluminum conductors, the minimum size shall be No. 12.

Table 2–2
AMPERE CAPACITIES – HEAVY COPPER WIRE AND CABLE

Copper size, AWG (or MCM)	AIEE–IPCEA power cable ampacities[a] 90°C conductor, 40°C ambient air, and 20°C ambient earth					National Electrical Code[b] for buildings 75°C conductor, 30°C ambient	
	Single conductor in air, one cable per support	Underground, direct burial, single-conductor cable	Underground, direct burial, 3-conductor cable	Underground duct; one 3-conductor cable	Underground duct bank; six 3-conductor cables	In free air	In conduit, one to three conductors
8	83	108	83	59	46	65	45
6	109	139	106	78	60	95	65
4	145	180	137	102	77	125	85
2	192	231	178	133	98	170	115
1	223	261	201	154	112	195	130
0	258	297	229	177	127	230	150
00	298	337	260	202	144	265	175
000	345	384	297	231	163	310	200
0000	400	434	335	264	185	360	230
(250)	445	472	367	292	202	405	255
(350)	552	569	442	354	242	505	310
(500)	695	690	531	429	289	620	380
(750)	898	847	648	529	348	785	475
(1,000)	1,076	980	729	599	390	935	545
(1,250)	1,228	1,083				1,065	590
(1,500)	1,367	1,176				1,175	625
(1,750)	1,493	1,257				1,280	650
(2,000)	1,606	1,325				1,385	665

[a]From Power Cable Ampacities, AIEE–IPCEA Publications S-135-1 and P-46-426. Ampere capacities in these five columns are for unshielded conductors for potentials to 3,000 volts, 100% load factor. At 75% load factor (max 1 hr/max 24 hr) the capacities underground are almost 15% higher. For higher conductor temperatures or lower ambient temperatures, the ampacities will be higher. For effects of temperature, shielding, voltage, or other cable arrangements, see the above source. At high frequencies the ampacities will be lower than above.
[b]Maximum capacities depend on insulation.

From Bolz, R. E. and Tuve, G. L., Eds., *Handbook of Tables for Applied Engineering Science,* 2nd ed., CRC Press, Cleveland, 1973, 226.

Table 2–3
STANDARD FLAT CONDUCTOR CABLE

Compact Cable for Aerospace, Communication, Instrument, and Automotive Applications

Flat conductor cable is made up of rolled flat conductors (equivalent round wire sizes 10 to 30 AWG), laminated between thin, flexible plastic-insulating films. Standard sizes cover the range from 2 to 57 conductors as shown in this table. Various insulating materials are used; the standard thickness of the insulating layers is 2 mils, resulting in a very compact cable assembly. Connector plugs and receptacles are available. In addition to the standard cables shown in this table, there are high-temperature, high-density, and power cables of similar design.

All dimensions in the table are in inches; to obtain the equivalents in mm, multiply the tabular dimensions by 25.4.

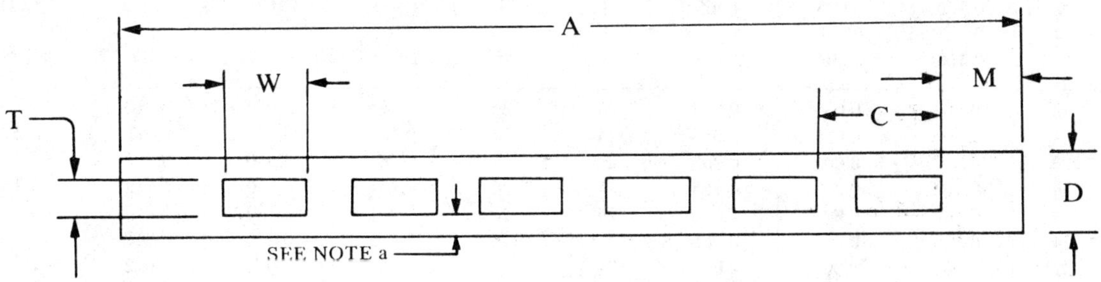

Conductor				*Nearest AWG wire size*	*Nominal conductor resistance, ohm/1000 ft*[b]	*Cable*			
No.	*Spacing "C" ±0.005*	*Size*				*Width "A" ±0.005*	*Margin "M" ±0.008*	*"D"*	*Weight, g/ft*
		T ±0.0004	*W ±0.0002*						
4	0.050	0.003	0.025	30	111	0.275	0.050	0.010	1.2
7	0.050	0.003	0.025	30	111	0.425	0.050	0.010	1.9
17	0.050	0.003	0.025	30	111	0.925	0.050	0.010	4.2
27	0.050	0.003	0.025	30	111	1.425	0.050	0.010	6.5
37	0.050	0.003	0.025	30	111	1.925	0.050	0.010	8.8
47	0.050	0.003	0.025	30	111	2.425	0.050	0.010	11.2
57	0.050	0.003	0.025	30	111	2.925	0.050	0.010	13.5
3	0.075	0.004	0.025	29	83	0.290	0.057	0.011	1.3
6	0.075	0.004	0.025	29	83	0.515	0.057	0.011	2.4
12	0.075	0.004	0.025	29	83	0.965	0.057	0.011	4.5
18	0.075	0.004	0.025	29	83	1.415	0.057	0.011	6.6
25	0.075	0.004	0.025	29	83	1.940	0.057	0.011	9.1
31	0.075	0.004	0.025	29	83	2.465	0.057	0.011	11.5
38	0.075	0.004	0.025	29	83	2.915	0.057	0.011	13.6
3	0.075	0.003	0.040	28	69	0.290	0.050	0.010	1.3
6	0.075	0.003	0.040	28	69	0.515	0.050	0.010	2.4
12	0.075	0.003	0.040	28	69	0.965	0.050	0.010	4.5
18	0.075	0.003	0.040	28	69	1.415	0.050	0.010	6.7
25	0.075	0.003	0.040	28	69	1.940	0.050	0.010	9.2
32	0.075	0.003	0.040	28	69	2.465	0.050	0.010	11.7
38	0.075	0.003	0.040	28	69	2.915	0.050	0.010	13.8

Table 2–3 (continued)
STANDARD FLAT CONDUCTOR CABLE

Compact Cable for Aerospace, Communication, Instrument, and Automotive Applications (continued)

Conductor				Nearest AWG wire size	Nominal conductor resistance, ohm/1000 ft[b]	Cable			
No.	Spacing "C" ±0.005	Size T ±0.0004	Size W ±0.0002			Width "A" ±0.005	Margin "M" ±0.008	"D"	Weight, g/ft
2	0.100	0.004	0.065	25	32	0.265	0.050	0.011	1.5
4	0.100	0.004	0.065	25	32	0.465	0.050	0.011	2.7
9	0.100	0.004	0.065	25	32	0.965	0.050	0.011	5.9
14	0.100	0.004	0.065	25	32	1.465	0.050	0.011	9.1
19	0.100	0.004	0.065	25	32	1.965	0.050	0.011	12.3
24	0.100	0.004	0.065	25	32	2.465	0.050	0.011	15.5
29	0.100	0.004	0.065	25	32	2.965	0.050	0.011	18.6
2	0.100	0.005	0.065	24	26	0.265	0.050	0.012	1.7
4	0.100	0.005	0.065	24	26	0.465	0.050	0.012	3.2
9	0.100	0.005	0.065	24	26	0.965	0.050	0.012	6.9
14	0.100	0.005	0.065	24	26	1.465	0.050	0.012	10.7
19	0.100	0.005	0.065	24	26	1.965	0.050	0.012	14.4
24	0.100	0.005	0.065	24	26	2.465	0.050	0.012	18.2
29	0.100	0.004	0.065	24	26	2.965	0.050	0.012	21.9
3	0.150	0.004	0.115	22	18	0.555	0.070	0.011	3.5
6	0.150	0.004	0.115	22	18	1.005	0.070	0.011	6.6
9	0.150	0.004	0.115	22	18	1.445	0.070	0.011	9.7
12	0.150	0.004	0.115	22	18	1.905	0.070	0.011	12.8
16	0.150	0.004	0.115	22	18	2.505	0.070	0.011	17.0
19	0.150	0.004	0.115	22	18	2.955	0.070	0.011	20.2
3	0.150	0.005	0.115	21	14	0.555	0.070	0.012	4.1
6	0.150	0.005	0.115	21	14	1.005	0.070	0.012	7.7
9	0.150	0.005	0.115	21	14	1.445	0.070	0.012	11.5
12	0.150	0.005	0.115	21	14	1.905	0.070	0.012	15.1
16	0.150	0.005	0.115	21	14	2.505	0.070	0.012	20.1
19	0.150	0.005	0.115	21	14	2.955	0.070	0.012	23.8

[a]The thickness of the insulation on each side of the conductors shall be uniform to within 0.001 in.
[b]1000 ft = 304.8 m.

From Flat Conductor Cable Technology, NASA Report Sp-5043, 1968.

Table 2–4
ELECTRICAL WIRE – FLEXIBLE CORDS AND FIXTURE WIRING

At Room Temperature, 30°C (86°F)

Description	*National Electrical Code letter classification examples*	*Maximum amperes for AWG wire sizes*				
		18	*16*	*14*	*12*	*10*
Flexible cord, various services, rubber insulation; also rubber fixture wire	C, K, PO, PD, PW, RF	5	7	15	20	25
Flexible fixture cord for high-temperature location; also heat-resistant fixture wire	CFC, AFC, CFPO, AFPO, CF, AF, SF, TF	6	8	17	23	28
Flexible cord, various services, silicone rubber, or thermoplastic	SO, SJ, SP, ST, SJT, SPT	7	10	15	20	25
Rubber and asbestos or neoprene heater cord	AFS, HC, HPD, HSJ, HS, HPN	10	15	20	30	35
Asbestos: flame- and moisture-resistant	AVPO, AVPD	17	22	28	36	47

Notes: Some of the above types are not furnished in the larger wire sizes.
Tinsel cord, size 27 AWG, is furnished with rubber (TP, TS) or thermoplastic (TPT, TST) insulation and is limited to 0.5 amperes.
Cords shall not be smaller than required for the rated current of the total connected equipment.

From Bolz, R. E. and Tuve, G. L., Eds., *Handbook of Tables for Applied Engineering Science,* 2nd ed., CRC Press, Cleveland, 1973, 228.

Table 2–5
SELECTED PROPERTIES OF THE SUPERCONDUCTIVE ELEMENTS

Element	T_c(K)	H_o(oersteds)[a]	θ_D(K)	γ(mJmole^{-1} deg·K^2)
Al	1.175	104.93	420	1.35
Be	0.026			0.21
Cd	0.518, 0.52	29.6	209	0.688
Ga	1.0833	59.3	325	0.60
Ga (β)	5.90, 6.2	560		
Ga (γ)	7.62	950		
Ga (δ)	7.85	815		
Hg (α)	4.154	411	87, 71.9	1.81
Hg (β)	3.949	339	93	1.37
In	3.405	281.53	109	1.672
Ir	0.14, 0.11	19	425	3.27
La (α)	4.88	808, 798	142	10.0, 11.3
La (β)	6.00	1096	139	11.3
Mo	0.916	90, 98	460	1.83
Nb	9.25	1970	277, 238	7.80
Os	0.655	65	500	2.35
Pa	1.4			
Pb	7.23	803	96.3	3.0
Re	1.697	188, 211	415	2.35
Ru	0.493	66	580	3.0
Sb	2.6–2.7[b]	HF		
Sn	3.721	305	195	1.78
Ta	4.47	831	258	6.15
Tc	7.73, 7.78	1410	411	4.84, 6.28
Th	1.39	159.1	165	4.31
Ti	0.39	56, 100	429, 412	3.32
Tl	2.332, 2.39	181	78.5	1.47
V	5.43, 5.31	1100, 1400	382	9.82
W	0.0154	1.15	550	0.90
Zn	0.875	55	319.7	0.633
Zr	0.53	47	290	2.78
Zr (ω)	0.65			

[a]To convert oersteds to ampere/meters, multiply by 79.57.
[b]Metastable.

From Roberts, B. W., Properties of Selected Superconductive Materials – 1974 Supplement, NBS Technical Note 825, National Bureau of Standards, U.S. Government Printing Office, Washington, D.C., 1974, 10.

Table 2–6
THIN FILMS OF SUPERCONDUCTIVE ELEMENTS

A. Thin Films Condensed at Various Temperatures

Element	T_c (K)	Element	T_c(K)
Al	1.18–~5.7	Pb	~2–7.7
Be	~.03, ~9.6, 6.5–10.6,[a] 10.2[b]	Re	~7
Bi	~2–~5, 6.11, 6.154, 6.173	Sn	3.6, 3.84–6.0
Cd	0.53–0.91	Ta	<1.7–4.25, 3.16–4.8
Ga	6.4–6.8, 7.4–8.4, 8.56	Ti	1.3
In	3.43–4.5, 3.68–4.17[c]	Tl	2.64
La	5.0–6.74	V	5.14–6.02
Mo	3.3–3.8, 4–6.7	W	<1.0–4.1
Nb	6.2–10.1	Zn	0.77–1.48

[a]With KCl.
[b]With Zn etioporphyrin.
[c]In glass pores.

B. Data for Elements Studied Under Pressure

Element	T_c (K)	Pressure[a]	Element	T_c (K)	Pressure[a]
As	0.31–0.5	220–140 kbar	P	4.7	>100 kbar
	0.2–0.25	~140–100 kbar		5.8	170 kbar
Ba II	~1.3	55 kbar	Pb II	3.55, 3.6	160 kbar
Ba III	3.05	85–88 kbar	Sb	3.55	85 kbar
	~5.2	>140 kbar		3.52	93 kbar
Bi II	3.916	25 katm		3.53	100 kbar
	3.90	25.2 katm		3.40	~150 kbar
	3.86	26.8 katm	Se II	6.75, 6.95	~130 kbar
Bi III	6.55	~37 kbar	Si	6.7, 7.1	120 kbar
	7.25	27–28.4 katm	Sn II	5.2	125 kbar
Bi IV	7.0	43, 43–62 kbar		4.85	160 kbar
Bi V	8.3, 8.55	81 kbar	Sn III	5.30	113 kbar
Bi VI	8.55	90, 92–101 kbar	Te II	2.05	43 kbar
Ce	1.7	50 kbar		3.4	50 kbar
Cs	~1.5	> ~125 kbar	Te III	4.28	70 kbar
Ga II	6.24, 6.38	⩾35 katm	Te IV	4.25	84 kbar
Ga II′	7.5	⩾35 katm (P → O)	Tl, cub.	1.45	35 kbar
Ge	4.85–5.4	~120 kbar	Tl, hex.	1.95	35 kbar
	5.35	115 kbar	U	2.3	10 kbar
La	~5.5–11.93	0–~140 kbar	Y	~1.2, ~2.7	120–170 kbar

[a]To convert katm to N/m^2, multiply by 1.013×10^8; to convert kbar to N/m^2, multiply by 1×10^8.

From Roberts, B. W., Properties of Selected Superconductive Materials, 1974 Supplement, National Bureau of Standard Technical Note 825, U.S. Government Printing Office, Washington, D.C., 1974, 11.

Table 2–7
SELECTED SUPERCONDUCTIVE COMPOUNDS AND ALLOYS

All compositions are denoted on an atomic basis, i.e., AB, AB_2, or AB_3 for compounds, unless noted. Solid solutions or odd compositions may be denoted as A_xB_{1-x} or A_zB. A series of three or more alloys is indicated as A_zB_{1-x} or by actual indication of the atomic fraction range, such as $A_{0-0.6}B_{1-0.4}$. The critical temperature of such a series of alloys is denoted by a range of values or possibly the maximum value.

The selection of the critical temperature from a transition in the effective permeability, or the change in resistance, or possibly the incremental changes in frequency observed by certain techniques is not often obvious from the literature. Most authors choose the mid-point of such curves as the probable critical temperature of the idealized material, while others will choose the highest temperature at which a deviation from the normal-state property is observed. In view of the previous discussion concerning the variability of the superconductive properties as a function of purity and other metallurigical aspects, it is recommended that appropriate literature be checked to determine the most probable critical temperature or critical field of a given alloy.

A very limited amount of data on critical fields, H_o, is available for these compounds and alloys; these values are given at the end of the table.

Symbols:

n = number of normal carriers per cubic centimeter for semiconductor superconductors.

Substance	T_c, *K*	*Crystal structure type*[a]
$Ag_xAl_yZn_{1-x-y}$	0.5–0.845	
$Ag_7BF_4O_8$	0.15	Cubic
$AgBi_2$	3.0–2.78	
$Ag_7F_{0.25}N_{0.75}O_{10.25}$	0.85–0.90	
Ag_7FO_8	0.3	Cubic
Ag_2F	0.066	
$Ag_{0.8-0.3}Ga_{0.2-0.7}$	6.5–8	
Ag_4Ge	0.85	Hex., c.p.
$Ag_{0.438}Hg_{0.562}$	0.64	$D8_2$
$AgIn_2$	~2.4	C16
$Ag_{0.1}In_{0.9}Te$ (n = 1.40×10^{22})	1.20–1.89	B1
$Ag_{0.2}In_{0.8}Te$ (n = 1.07×10^{22})	0.77–1.00	B1
AgLa (9.5 kbar)	1.2	B2
Ag_7NO_{11}	1.04	Cubic
Ag_xPb_{1-x}	7.2 max.	
Ag_xSn_{1-x} (film)	2.0–3.8	
Ag_xSn_{1-x}	1.5–3.7	
$AgTe_3$	2.6	Cubic
$AgTh_2$	2.26	C16
$Ag_{0.03}Tl_{0.97}$	2.67	
$Ag_{0.94}Tl_{0.06}$	2.32	
Ag_xZn_{1-x}	0.5–0.845	
Al (film)	1.3–2.31	
Al (1 to 21 katm)	1.170–0.687	A1
$AlAu_4$	0.4–0.7	Like A13
Al_2CMo_3	10.0	A13
Al_2CMo_3	9.8–10.2	A13 + trace 2nd phase
Al_2CaSi	5.8	
$Al_{0.131}Cr_{0.088}V_{0.781}$	1.46	Cubic
$AlGe_2$	1.75	
$Al_{0.5}Ge_{0.5}Nb$	12.6	A15
$Al_{\sim 0.8}Ge_{\sim 0.2}Nb_3$	20.7	A15
$AlLa_3$	5.57	DO_{19}
Al_2La	3.23	C15
Al_3Mg_2	0.84	Cubic, f.c.
$AlMo_3$	0.58	A15
$AlMo_6Pd$	2.1	
AlN	1.55	B4
Al_2NNb_3	1.3	A13
$AlNb_3$	18.0	A15
Al_xNb_{1-x}	<4.2–13.5	$D8_b$
Al_xNb_{1-x}	12–17.5	A15
$Al_{0.27}Nb_{0.73-0.48}V_{0-0.25}$	14.5–17.5	A15
$AlNb_xV_{1-x}$	<4.2–13.5	
AlOs	0.39	B2
Al_3Os	5.90	
AlPb (films)	1.2–7	
Al_2Pt	0.48–0.55	C1
Al_5Re_{24}	3.35	A12
Al_3Th	0.75	DO_{19}
$Al_xTi_yV_{1-x-y}$	2.05–3.62	Cubic
$Al_{0.108}V_{0.892}$	1.82	Cubic
Al_xZn_{1-x}	0.5–0.845	
$AlZr_3$	0.73	$L1_2$
AsBiPb	9.0	
AsBiPbSb	9.0	
$As_{0.33}InTe_{0.67}$ (n = 1.24×10^{22})	0.85–1.15	B1
$As_{0.5}InTe_{0.5}$ (n = 0.97×10^{22})	0.44–0.62	B1
$As_{0.50}Ni_{0.06}Pd_{0.44}$	1.39	C2
AsPb	8.4	
$AsPd_2$ (low-temperature phase)	0.60	Hexagonal
$AsPd_2$ (high-temp. phase)	1.70	C22

[a]See key at end of table.

Table 2–7 (continued)
SELECTED SUPERCONDUCTIVE COMPOUNDS AND ALLOYS

Substance	T_c, *K*	*Crystal structure type*[a]	*Substance*	T_c, *K*	*Crystal structure type*[a]
$AsPd_5$	0.46	Complex	BW_2	3.1	C16
AsRh	0.58	B31	B_6Y	6.5–7.1	
$AsRh_{1.4-1.6}$	<0.03–0.56	Hexagonal	$B_{12}Y$	4.7	
AsSn	4.10		BZr	3.4	Cubic
AsSn			$B_{12}Zr$	5.82	
(n = 2.14×10^{22})	3.41–3.65	B1	$BaBi_3$	5.69	Tetragonal
$As_{\sim 2}Sn_{\sim 3}$	3.5–3.6,		$Ba_xO_3Sr_{1-x}Ti$		
	1.21–1.17		(n = 4.2–11 $\times 10^{19}$)	<0.1–0.55	
As_3Sn_4			$Ba_{0.13}O_3W$	1.9	Tetragonal
(n = 0.56×10^{22})	1.16–1.19	Rhombohedral	$Ba_{0.14}O_3W$	<1.25–2.2	Hexagonal
Au_5Ba	0.4–0.7	$D2_d$	$BaRh_2$	6.0	C15
AuBe	2.64	B20	$Be_{22}Mo$	2.51	Cubic,
Au_2Bi	1.80	C15			like $Be_{22}Re$
Au_5Ca	0.34–0.38	$C15_b$	$Be_8Nb_5Zr_2$	5.2	
AuGa	1.2	B31	$Be_{0.98-0.92}Re_{0.02-0.08}$		
$Au_{0.40-0.92}Ge_{0.60-0.08}$	<0.32–1.63	Complex	(quenched)	9.5–9.75	Cubic
AuIn	0.4–0.6	Complex	$Be_{0.957}Re_{0.043}$	9.62	Cubic,
AuLu	<0.35	B2			like $Be_{22}Re$
$AuNb_3$	11.5	A15	BeTc	5.21	Cubic
$AuNb_3$	1.2	A2	$Be_{22}W$	4.12	Cubic,
$Au_{0-0.3}Nb_{1-0.7}$	1.1–11.0				like $Be_{22}Re$
$Au_{0.02-0.98}Nb_3Rh_{0.98-0.02}$	2.53–10.9	A15	$Be_{13}W$	4.1	Tetragonal
$AuNb_{3(1-x)}V_{3x}$	1.5–11.0	A15	Bi_3Ca	2.0	
$AuPb_2$	3.15		$Bi_{0.5}Cd_{0.13}Pb_{0.25}Sn_{0.12}$		
$AuPb_2$ (film)	4.3		(weight fractions)	8.2	
$AuPb_3$	4.40		BiCo	0.42–0.49	
$AuPb_3$ (film)	4.25		Bi_2Cs	4.75	C15
Au_2Pb	1.18, 6–7	C15	Bi_xCu_{1-x}		
$AuSb_2$	0.58	C2	(electrodeposited)	2.2	
AuSn	1.25	$B8_1$	BiCu	1.33–1.40	
Au_xSn_{1-x} (film)	2.0–3.8		$Bi_{0.019}In_{0.981}$	3.86	
Au_5Sn	0.7–1.1	A3	$Bi_{0.05}In_{0.95}$	4.65	α-phase
Au_3Te_5	1.62	Cubic	$Bi_{0.10}In_{0.90}$	5.05	α-phase
$AuTh_2$	3.08	C16	$Bi_{0.15-0.30}In_{0.85-0.70}$	5.3–5.4	α- and β-phases
AuTl	1.92		$Bi_{0.34-0.48}In_{0.66-0.52}$	4.0–4.1	
AuV_3	0.74	A15	Bi_3In_5	4.1	
Au_xZn_{1-x}	0.50–0.845		$BiIn_2$	5.65	β-phase
$AuZn_3$	1.21	Cubic	Bi_2Ir	1.7–2.3	
Au_xZr_y	1.7–2.8	A3	Bi_2Ir (quenched)	3.0–3.96	
$AuZr_3$	0.92	A15	BiK	3.6	
$BCMo_2$	5.4	Orthorhombic	Bi_2K	3.58	C15
$B_{0.03}C_{0.51}Mo_{0.47}$	12.5		BiLi	2.47	$L1_0$, α-phase
$BCMo_2$	5.3–7.0	Orthorhombic	$Bi_{4-9}Mg$	0.7–∼1.0	
BHf	3.1	Cubic	Bi_3Mo	3–3.7	
B_6La	5.7		BiNa	2.25	$L1_0$
$B_{12}Lu$	0.48		$BiNb_3$ (high pressure		
BMo	0.5 (extrap-		and temperature)	3.05	A15
	olated)		BiNi	4.25	$B8_1$
BMo_2	4.74	C16	Bi_3Ni	4.06	Orthorhombic
BNb	8.25	B_f	$Bi_{1-0}Pb_{0-1}$	7.26–9.14	
BRe_2	2.80, 4.6		$Bi_{1-0}Pb_{0-1}$ (film)	7.25–8.67	
$B_{0.3}Ru_{0.7}$	2.58	$D10_2$	$Bi_{0.05-0.40}Pb_{0.95-0.60}$	7.35–8.4	Hexagonal,
$B_{12}Sc$	0.39				c.p., to
BTa	4.0	B_f			ε-phase
B_6Th	0.74		BiPbSb	8.9	

[a]See key at end of table.

Table 2–7 (continued)
SELECTED SUPERCONDUCTIVE COMPOUNDS AND ALLOYS

Substance	T_c, K	Crystal structure type[a]
$Bi_{0.5}Pb_{0.31}Sn_{0.19}$ (weight fractions)	8.5	
$Bi_{0.5}Pb_{0.25}Sn_{0.25}$	8.5	
$BiPd_2$	4.0	
$Bi_{0.4}Pd_{0.6}$	3.7–4	Hexagonal, ordered
BiPd	3.7	Orthorhombic
Bi_2Pd	1.70	Monoclinic, α-phase
Bi_2Pd	4.25	Tetragonal, β-phase
BiPdSe	1.0	C2
BiPdTe	1.2	C2
BiPt	1.21	$B8_1$
BiPtSe	1.45	C2
BiPtTe	1.15	C2
Bi_2Pt	0.155	Hexagonal
Bi_2Rb	4.25	C15
$BiRe_2$	1.9–2.2	
BiRh	2.06	$B8_1$
Bi_3Rh	3.2	Orthorhombic, like NiB_3
Bi_4Rh	2.7	Hexagonal
Bi_3Sn	3.6–3.8	
BiSn	3.8	
Bi_xSn_y	3.85–4.18	
Bi_3Sr	5.62	$L1_2$
Bi_3Te	0.75–1.0	
Bi_5Tl_3	6.4	
$Bi_{0.26}Tl_{0.74}$	4.4	Cubic, disordered
$Bi_{0.26}Tl_{0.74}$	4.15	$L1_2$, ordered?
Bi_2Y_3	2.25	
Bi_3Zn	0.8–0.9	
$Bi_{0.3}Zr_{0.7}$	1.51	
$BiZr_3$	2.4–2.8	
CCs_x	0.020–0.135	Hexagonal
C_8K (gold)	0.55	
$CGaMo_2$	3.7–4.1	Hexagonal, H-phase
$CHf_{0.5}Mo_{0.5}$	3.4	B1
$CHf_{0.3}Mo_{0.7}$	5.5	B1
$CHf_{0.25}Mo_{0.75}$	6.6	B1
$CHf_{0.7}Nb_{0.3}$	6.1	B1
$CHf_{0.6}Nb_{0.4}$	4.5	B1
$CHf_{0.5}Nb_{0.5}$	4.8	B1
$CHf_{0.4}Nb_{0.6}$	5.6	B1
$CHf_{0.25}Nb_{0.75}$	7.0	B1
$CHf_{0.2}Nb_{0.8}$	7.8	B1
$CHf_{0.9-0.1}Ta_{0.1-0.9}$	5.0–9.0	B1
Ck (excess K)	0.55	Hexagonal
C_8K	0.39	Hexagonal
$C_{0.40-0.44}Mo_{0.60-0.56}$	9–13	
CMo	6.5, 9.26	
CMo_2	12.2	Orthorhombic

Substance	T_c, K	Crystal structure type[a]
$C_{0.44}Mo_{0.56}$	1.3	B1
$C_{0.5}Mo_xNb_{1-x}$	10.8–12.5	B1
$C_{0.6}Mo_{4.8}Si_3$	7.6	$D8_8$
$CMo_{0.2}Ta_{0.8}$	7.5	B1
$CMo_{0.5}Ta_{0.5}$	7.7	B1
$CMo_{0.75}Ta_{0.25}$	8.5	B1
$CMo_{0.8}Ta_{0.2}$	8.7	B1
$CMo_{0.85}Ta_{0.15}$	8.9	B1
CMo_xTi_{1-x}	10.2 max.	B1
$CMo_{0.83}Ti_{0.17}$	10.2	B1
CMo_xV_{1-x}	2.9–9.3	B1
CMo_xZr_{1-x}	3.8–9.5	B1
$C_{0.1-0.9}N_{0.9-0.1}Nb$	8.5–17.9	
$C_{0-0.38}N_{1-0.62}Ta$	10.0–11.3	
CNb (whiskers)	7.5–10.5	
$C_{0.984}Nb$	9.8	B1
CNb (extrapolated)	~14	
$C_{0.7-1.0}Nb_{0.3-0}$	6–11	B1
CNb_2	9.1	
CNb_xTa_{1-x}	8.2–13.9	
CNb_xTi_{1-x}	<4.2–8.8	B1
$CNb_{0.6-0.9}W_{0.4-0.1}$	12.5–11.6	B1
$CNb_{0.1-0.9}Zr_{0.9-0.1}$	4.2–8.4	B1
CRb_x (gold)	0.023–0.151	Hexagonal
$CRe_{0.01-0.08}W$	1.3–5.0	
$CRe_{0.06}W$	5.0	
CTa	~11 (extrapolated)	
$C_{0.987}Ta$	9.7	
$C_{0.848-0.987}Ta$	2.04–9.7	
CTa (film)	5.09	B1
CTa_2	3.26	L'_3
$CTa_{0.4}Ti_{0.6}$	4.8	B1
$CTa_{1-0.4}W_{0-0.6}$	8.5–10.5	B1
$CTa_{0.2-0.9}Zr_{0.8-0.1}$	4.6–8.3	B1
CTc (excess C)	3.85	Cubic
$CTi_{0.5-0.7}W_{0.5-0.3}$	6.7–2.1	B1
CW	1.0	
CW_2	2.74	L'_3
CW_2	5.2	Cubic, f.c.
$CaIr_2$	6.15	C15
$Ca_xO_3Sr_{1-x}Ti$ ($n = 3.7–11.0 \times 10^{19}$)	<0.1–0.55	
$Ca_{0.1}O_3W$	1.4–3.4	Hexagonal
CaPb	7.0	
$CaRh_2$	6.40	C15
$Cd_{0.3-0.5}Hg_{0.7-0.5}$	1.70–1.92	
CdHg	1.77, 2.15	Tetragonal
$Cd_{0.0075-0.05}In_{1-x}$	3.24–3.36	Tetragonal
$Cd_{0.97}Pb_{0.03}$	4.2	
CdSn	3.65	
$Cd_{0.17}Tl_{0.83}$	2.3	
$Cd_{0.18}Tl_{0.82}$	2.54	
$CeCo_2$	0.84	C15
$CeCo_{1.67}Ni_{0.33}$	0.46	C15

[a]See key at end of table

Table 2–7 (continued)
SELECTED SUPERCONDUCTIVE COMPOUNDS AND ALLOYS

Substance	T_c, *K*	*Crystal structure type* [a]	*Substance*	T_c, *K*	*Crystal structure type* [a]
$CeCo_{1.67}Rh_{0.33}$	0.47	C15	CuSSe	1.5–2.0	C18
$Ce_xGd_{1-x}Ru_2$	3.2–5.2	C15	$CuSe_2$	2.3–2.43	C18
$CeIr_3$	3.34		CuSeTe	1.6–2.0	C18
$CeIr_5$	1.82		Cu_xSn_{1-x}	3.2–3.7	
$Ce_{0.005}La_{0.995}$	4.6		Cu_xSn_{1-x} (film)		
Ce_xLa_{1-x}	1.3–6.3		(made at 10°K)	3.6–7	
$Ce_xPr_{1-x}Ru_2$	1.4–5.3	C15	Cu_xSn_{1-x} (film)		
Ce_xPt_{1-x}	0.7–1.55		(made at 300°K)	2.8–3.7	
$CeRu_2$	6.0	C15	$CuTe_2$	<1.25–1.3	C18
$Co_xFe_{1-x}Si_2$	1.4 max.	C1	$CuTh_2$	3.49	C16
$CoHf_2$	0.56	$E9_3$	$Cu_{0-0.027}V$	3.9–5.3	A2
$CoLa_3$	4.28		Cu_xZn_{1-x}	0.5–0.845	
$CoLu_3$	~0.35		Er_xLa_{1-x}	1.4–6.3	
$Co_{0-0.01}Mo_{0.8}Re_{0.2}$	2–10		$Fe_{0-0.04}Mo_{0.8}Re_{0.2}$	1–10	
$Co_{0.02-0.10}Nb_3Rh_{0.98-0.90}$	2.28–1.90	A15	$Fe_{0.05}Ni_{0.05}Zr_{0.90}$	~3.9	
$Co_xNi_{1-x}Si_2$	1.4 max.	C1	Fe_3Th_7	1.86	D10
$Co_{0.5}Rh_{0.5}Si_2$	2.5		Fe_xTi_{1-x}	3.2 max.	Fe in α-Ti
$Co_xRh_{1-x}Si_2$	3.65 max.		Fe_xTi_{1-x}	3.7 max.	Fe in β-Ti
$Co_{\sim 0.3}Sc_{\sim 0.7}$	~0.35		$Fe_xTi_{0.6}V_{1-x}$	6.8 max.	
$CoSi_2$	1.40, 1.22	C1	FeU_6	3.86	$D2_c$
Co_3Th_7	1.83	$D10_2$	$Fe_{0.1}Zr_{0.9}$	1.0	A3
Co_xTi_{1-x}	2.8 max.	Co in α-Ti	$Ga_{0.5}Ge_{0.5}Nb_3$	7.3	A15
Co_xTi_{1-x}	3.8 max.	Co in β-Ti	$GaLa_3$	5.84	
$CoTi_2$	3.44	$E9_3$	Ga_2Mo	9.5	
CoTi	0.71	A2	$GaMo_3$	0.76	A15
CoU	1.7	B2, distorted	Ga_4Mo	9.8	
CoU_6	2.29	$D2_c$	GaN (black)	5.85	B4
$Co_{0.28}Y_{0.72}$	0.34		$GaNb_3$	14.5	A15
CoY_3	<0.34		$Ga_xNb_3Sn_{1-x}$	14–18.37	A15
$CoZr_2$	6.3	C16	$Ga_{0.7}Pt_{0.3}$	2.9	C1
$Co_{0.1}Zr_{0.9}$	3.9	A3	GaPt	1.74	B20
$Cr_{0.6}Ir_{0.4}$	0.4	Hexagonal, c.p.	GaSb (120 kbar, 77°K,		
$Cr_{0.65}Ir_{0.35}$	0.59	Hexagonal, c.p.	annealed)	4.24	A5
$Cr_{0.7}Ir_{0.3}$	0.76	Hexagonal, c.p.	GaSb (unannealed)	~5.9	
$Cr_{0.72}Ir_{0.28}$	0.83		$Ga_{0.1}Sn_{1-0}$ (quenched)	3.47–4.18	
Cr_3Ir	0.45	A15	$Ga_{0-1}Sn_{1-0}$ (annealed)	2.6–3.85	
$Cr_{0-0.1}Nb_{1-0.9}$	4.6–9.2	A2	Ga_5V_2	3.55	Tetragonal,
$Cr_{0.80}Os_{0.20}$	2.5	Cubic			Mn_2Hg_5 type
Cr_xRe_{1-x}	1.2–5.2		GaV_3	16.8	A15
$Cr_{0.40}Re_{0.60}$	2.15	$D8_b$	$GaV_{2.1-3.5}$	6.3–14.45	A15
$Cr_{0.8-0.6}Rh_{0.2-0.4}$	0.5–1.10	A3	$GaV_{4.5}$	9.15	
Cr_3Ru (annealed)	3.3	A15	Ga_3Zr	1.38	
Cr_2Ru	2.02	$D8_b$	Gd_xLa_{1-x}	<1.0–5.5	
$Cr_{0.1-0.5}Ru_{0.9-0.5}$	0.34–1.65	A3	$Gd_xOs_2Y_{1-x}$	1.4–4.7	
Cr_xTi_{1-x}	3.6 max.	Cr in α-Ti	$Gd_xRu_2Th_{1-x}$	3.6 max.	C15
Cr_xTi_{1-x}	4.2 max.	Cr in β-Ti	GeIr	4.7	B31
$Cr_{0.1}Ti_{0.3}V_{0.6}$	5.6		Ge_2La	1.49, 2.2	Orthorhombic,
$Cr_{0.0175}U_{0.9825}$	0.75	β-phase			distorted
$Cs_{0.32}O_3W$	1.12	Hexagonal			$ThSi_2$-type
$Cu_{0.15}In_{0.85}$ (film)	3.75		$GeMo_3$	1.43	A15
$Cu_{0.04-0.08}In_{1-x}$	4.4		$GeNb_2$	1.9	
CuLa	5.85		$GeNb_3$ (quenched)	6–17	A15
Cu_xPb_{1-x}	5.7–7.7		$Ge_{0.29}Nb_{0.71}$	6	A15
CuS	1.62	B18	$Ge_xNb_3Sn_{1-x}$	17.6–18.0	A15
CuS_2	1.48–1.53	C18	$Ge_{0.5}Nb_3Sn_{0.5}$	11.3	

[a]See key at end of table.

Table 2–7 (continued)
SELECTED SUPERCONDUCTIVE COMPOUNDS AND ALLOYS

Substance	T_c, *K*	*Crystal structure type* [a]
GePt	0.40	B31
Ge_3Rh_5	2.12	Orthorhombic, related to $InNi_2$
Ge_2Sc	1.3	
Ge_3Te_4 ($n = 1.06 \times 10^{22}$)	1.55–1.80	Rhombohedral
Ge_xTe_{1-x} ($n = 8.5 - 64 \times 10^{20}$)	0.07–0.41	Bl
GeV_3	6.01	A15
Ge_2Y	3.80	C_c
$Ge_{1.62}Y$	2.4	
$H_{0.33}Nb_{0.67}$	7.28	Cubic, b.c.
$H_{0.1}Nb_{0.9}$	7.38	Cubic, b.c.
$H_{0.05}Nb_{0.95}$	7.83	Cubic, b.c.
$H_{0.12}Ta_{0.88}$	2.81	Cubic, b.c.
$H_{0.08}Ta_{0.92}$	3.26	Cubic, b.c.
$H_{0.04}Ta_{0.96}$	3.62	Cubic, b.c.
$HfN_{0.989}$	6.6	Bl
$Hf_{0-0.5}Nb_{1-0.5}$	8.3–9.5	A2
$Hf_{0.75}Nb_{0.25}$	>4.2	
$HfOs_2$	2.69	C14
$HfRe_2$	4.80	C14
$Hf_{0.14}Re_{0.86}$	5.86	A12
$Hf_{0.99-0.96}Rh_{0.01-0.04}$	0.85–1.51	
$Hf_{0-0.55}Ta_{1-0.45}$	4.4–6.5	A2
HfV_2	8.9–9.6	C15
Hg_xIn_{1-x}	3.14–4.55	
HgIn	3.81	
Hg_2K	1.20	Orthorhombic
Hg_3K	3.18	
Hg_4K	3.27	
Hg_8K	3.42	
Hg_3Li	1.7	Hexagonal
Hg_2Na	1.62	Hexagonal
Hg_4Na	3.05	
Hg_xPb_{1-x}	4.14–7.26	
HgSn	4.2	
Hg_xTl_{1-x}	2.30–4.109	
Hg_5Tl_2	3.86	
Ho_xLa_{1-x}	1.3–6.3	
$InLa_3$	9.83, 10.4	Ll_2
$InLa_3$ (0–35, kbar)	9.75–10.55	
$In_{1-0.86}Mg_{0-0.14}$	3.395–3.363	
$InNb_3$ (high pressure and temp.)	4–8, 9.2	A15
$In_{0-0.3}Nb_3Sn_{1-0.7}$	18.0–18.19	A15
$In_{0.5}Nb_3Zr_{0.5}$	6.4	
$In_{0.11}O_3W$	<1.25–2.8	Hexagonal
$In_{0.95-0.85}Pb_{0.05-0.15}$	3.6–5.05	
$In_{0.98-0.91}Pb_{0.02-0.09}$	3.45–4.2	
InPb	6.65	
InPd	0.7	B2
InSb (quenched from 170 kbar into liquid N_2)	4.8	Like A5

Substance	T_c, *K*	*Crystal structure type* [a]
InSb	2.1	
$(InSb)_{0.95-0.10}Sn_{0.05-0.90}$ (various heat treatments)	3.8–5.1	
$(InSb)_{0-0.07}Sn_{1-0.93}$	3.67–3.74	
In_3Sn	~5.5	
In_xSn_{1-x}	3.4–7.3	
$In_{0.82-1}Te$ ($n = 0.83–1.71 \times 10^{22}$)	1.02–3.45	Bl
$In_{1.000}Te_{1.002}$	3.5–3.7	Bl
In_3Te_4 ($n = 0.47 \times 10^{22}$)	1.15–1.25	Rhombohedral
In_xTl_{1-x}	2.7–3.374	
$In_{0.8}Tl_{0.2}$	3.223	
$In_{0.62}Tl_{0.38}$	2.760	
$In_{0.78-0.69}Tl_{0.22-0.31}$	3.18–3.32	Tetragonal
$In_{0.69-0.62}Tl_{0.31-0.38}$	2.98–3.3	Cubic, f.c.
Ir_2La	0.48	C15
Ir_3La	2.32	$Dl0_2$
Ir_3La_7	2.24	$Dl0_2$
Ir_5La	2.13	
Ir_2Lu	2.47	C15
Ir_3Lu	2.89	C15
IrMo	<1.0	A3
$IrMo_3$	8.8	A15
$IrMo_3$	6.8	$D8_b$
$IrNb_3$	1.9	A15
$Ir_{0.4}Nb_{0.6}$	9.8	$D8_b$
$Ir_{0.37}Nb_{0.63}$	2.32	$D8_b$
IrNb	7.9	$D8_b$
$Ir_{0.02}Nb_3Rh_{0.98}$	2.43	A15
$Ir_{0.05}Nb_3Rh_{0.95}$	2.38	A15
$Ir_{0.287}O_{0.14}Ti_{0.573}$	5.5	$E9_3$
$Ir_{0.265}O_{0.035}Ti_{0.65}$	2.30	$E9_3$
Ir_xOs_{1-x}	0.3–0.98 (max.)–0.6	
IrOsY	2.6	C15
$Ir_{1.5}Os_{0.5}$	2.4	C14
Ir_2Sc	2.07	C15
$Ir_{2.5}Sc$	2.46	C15
$IrSn_2$	0.65–0.78	Cl
Ir_2Sr	5.70	C15
$Ir_{0.5}Te_{0.5}$	~3	
$IrTe_3$	1.18	C2
IrTh	<0.37	B_f
Ir_2Th	6.50	C15
Ir_3Th	4.71	
Ir_3Th_7	1.52	$Dl0_2$
Ir_5Th	3.93	$D2_d$
$IrTi_3$	5.40	A15
IrV_2	1.39	A15
IrW_3	3.82	
$Ir_{0.28}W_{0.72}$	4.49	
Ir_2Y	2.18, 1.38	C15
$Ir_{0.69}Y_{0.31}$	1.98, 1.44	C15
$Ir_{0.70}Y_{0.30}$	2.16	C15

[a] See key at end of table.

Table 2–7 (continued)
SELECTED SUPERCONDUCTIVE COMPOUNDS AND ALLOYS

Substance	T_c, *K*	*Crystal structure type*[a]
Ir_2Y	1.09	C15
Ir_2Y_3	1.61	
Ir_xY_{1-x}	0.3–3.7	
Ir_2Zr	4.10	C15
$Ir_{0.1}Zr_{0.9}$	5.5	A3
$K_{0.27-0.31}O_3W$	0.50	Hexagonal
$K_{0.40-0.57}O_3W$	1.5	Tetragonal
$La_{0.55}Lu_{0.45}$	2.2	Hexagonal, La type
$La_{0.8}Lu_{0.2}$	3.4	Hexagonal, La Type
$LaMg_2$	1.05	C15
LaN	1.35	
$LaOs_2$	6.5	C15
$LaPt_2$	0.46	C15
$La_{0.28}Pt_{0.72}$	0.54	C15
$LaRh_3$	2.60	
$LaRh_5$	1.62	
La_7Rh_3	2.58	$D10_2$
$LaRu_2$	1.63	C15
La_3S_4	6.5	$D7_3$
La_3Se_4	8.6	$D7_3$
$LaSi_2$	2.3	C_c
La_xY_{1-x}	1.7–5.4	
LaZn	1.04	B2
LiPb	7.2	
$LuOs_2$	3.49	C14
$Lu_{0.275}Rh_{0.725}$	1.27	C15
$LuRh_5$	0.49	
$LuRu_2$	0.86	C14
$Mg_{\sim 0.47}Tl_{\sim 0.53}$	2.75	B2
Mg_2Nb	5.6	
Mn_xTi_{1-x}	2.3 max.	Mn in α-Ti
Mn_xTi_{1-x}	1.1–3.0	Mn in β-Ti
MnU_6	2.32	$D2_c$
MoN	12	Hexagonal
Mo_2N	5.0	Cubic, f.c.
Mo_xNb_{1-x}	0.016–9.2	
Mo_3Os	7.2	A15
$M_{0.62}Os_{0.38}$	5.65	$D8_b$
Mo_3P	5.31	DO_e
$Mo_{0.5}Pd_{0.5}$	3.52	A3
Mo_3Re	10.0	
Mo_xRe_{1-x}	1.2–12.2	
$MoRe_3$	9.25, 9.89	A12
$Mo_{0.42}Re_{0.58}$	6.35	$D8_b$
$Mo_{0.52}Re_{0.48}$	11.1	
$Mo_{0.57}Re_{0.43}$	14.0	
$Mo_{\sim 0.60}Re_{0.395}$	10.6	
MoRh	1.97	A3
Mo_xRh_{1-x}	1.5–8.2	Cubic, b.c.
MoRu	9.5–10.5	A3
$Mo_{0.61}Ru_{0.39}$	7.18	$D8_b$
$Mo_{0.2}Ru_{0.8}$	1.66	A3
Mo_3Sb_4	2.1	

Substance	T_c, *K*	*Crystal structure type*[a]
Mo_3Si	1.30	A15
$MoSi_{0.7}$	1.34	
Mo_xSiV_{3-x}	4.54–16.0	A15
Mo_xTc_{1-x}	10.8–15.8	
$Mo_{0.16}Ti_{0.84}$	4.18, 4.25	
$Mo_{0.913}Ti_{0.087}$	2.95	
$Mo_{0.04}Ti_{0.96}$	2.0	Cubic
$Mo_{0.025}Ti_{0.975}$	1.8	
Mo_xU_{1-x}	0.7–2.1	
Mo_xV_{1-x}	0–~5.3	
Mo_2Zr	4.27–4.75	C15
NNb (whiskers)	10–14.5	
NNb (diffusion wires)	16.10	
NNb (film)	6–9	B1
$N_{0.988}Nb$	14.9	B1
$N_{0.824-0.988}Nb$	14.4–15.3	B1
$N_{0.70-0.795}Nb$	11.3–12.9	Cubic and tetragonal
NNb_xO_y	13.5–17.0	B1
NNb_xO_y	6.0–11	
$N_{100-42\ w/o}Nb_{0-58\ w/o}Ti$[b]	15–16.8	
$N_{100-75\ w/o}Nb_{0-25\ w/o}Zr$[b]	12.5–16.35	
NNb_xZr_{1-x}	9.8–13.8	B1
$N_{0.93}Nb_{0.85}Zr_{0.15}$	13.8	B1
$N_xO_yTi_z$	2.9–5.6	Cubic
$N_xO_yV_z$	5.8–8.2	Cubic
$N_{0.34}Re$	4–5	Cubic, f.c.
NTa	12–14 (extrapolated)	B1
NTa (film)	4.84	B1
$N_{0.6-0.987}Ti$	<1.17–5.8	B1
$N_{0.82-0.99}V$	2.9–7.9	B1
NZr	9.8	B1
$N_{0.906-0.984}Zr$	3.0–9.5	B1
$Na_{0.28-0.35}O_3W$	0.56	Tetragonal
$Na_{0.28}Pb_{0.72}$	7.2	
NbO	1.25	
$NbOs_2$	2.52	A12
Nb_3Os	1.05	A15
$Nb_{0.6}Os_{0.4}$	1.89, 1.78	$D8_b$
$Nb_3Os_{0.02-0.10}Rh_{0.98-0.90}$	2.42–2.30	A15
$Nb_{0.6}Pd_{0.4}$	1.60	$D8_f$ plus cubic
$Nb_3Pd_{0.02-0.10}Rh_{0.98-0.90}$	2.49–2.55	A15
$Nb_{0.62}Pt_{0.38}$	4.21	$D8_b$
Nb_3Pt	10.9	A15
Nb_5Pt_3	3.73	$D8_b$
$Nb_3Pt_{0.02-0.98}Rh_{0.98-0.02}$	2.52–9.6	A15
$Nb_{0.38-0.18}Re_{0.62-0.82}$	2.43–9.70	A12
Nb_3Rh	2.64	A15
$Nb_{0.60}Rh_{0.40}$	4.21	$D8_b$ plus other
$Nb_3Rh_{0.98-0.90}Ru_{0.02-0.10}$	2.42–2.44	A15
Nb_xRu_{1-x}	1.2–4.8	
NbS_2	6.1–6.3	Hexagonal, $NbSe_2$ type

[a]See key at end of table.
[b]*w/o* denotes weight percent.

Table 2–7 (continued)
SELECTED SUPERCONDUCTIVE COMPOUNDS AND ALLOYS

Substance	T_c, K	*Crystal structure type*[a]	*Substance*	T_c, K	*Crystal structure type*[a]
NbS_2	5.0–5.5	Hexagonal, three-layer type	Os_2Zr	3.0	C14
			Os_xZr_{1-x}	1.50–5.6	
			PPb	7.8	
$Nb_3Sb_{0-0.7}Sn_{1-0.3}$	6.8–18	A15	$PPd_{3.0-3.2}$	<0.35–0.7	DO_{11}
$NbSe_2$	5.15–5.62	Hexagonal, NbS_2 type	P_3Pd_7 (high temperature)	1.0	Rhombohedral
			P_3Pd_7 (low temp.)	0.70	Complex
$Nb_{1-1.05}Se_2$	2.2–7.0	Hexagonal, NbS_2 type	PRh	1.22	
			PRh_2	1.3	Cl
Nb_3Si	1.5	Ll_2	PW_3	2.26	DO
Nb_3SiSnV_3	4.0		Pb_2Pd	2.95	C16
Nb_3Sn	18.05	A15	Pb_4Pt	2.80	Related to C16
$Nb_{0.8}Sn_{0.2}$	18.18, 18.5	A15	Pb_2Rh	2.66	C16
Nb_xSn_{1-x} (film)	2.6–18.5		PbSb	6.6	
$NbSn_2$	2.60	Orthorhombic	PbTe (plus 0.1 *w/o* Pb)[b]	5.19	
Nb_3Sn_2	16.6	Tetragonal	PbTe (plus 0.1 *w/o* Tl)[b]	5.24–5.27	
$NbSnTa_2$	10.8	A15	$PbTl_{0.27}$	6.43	
Nb_2SnTa	16.4	A15	$PbTl_{0.17}$	6.73	
$Nb_{2.5}SnTa_{0.5}$	17.6	A15	$PbTl_{0.12}$	6.88	
$Nb_{2.75}SnTa_{0.25}$	17.8	A15	$PbTl_{0.075}$	6.98	
$Nb_{3x}SnTa_{3(1-x)}$	6.0–18.0		$PbTl_{0.04}$	7.06	
NbSnTaV	6.2	A15	$Pb_{1-0.26}Tl_{0-0.74}$	7.20–3.68	
$Nb_2SnTa_{0.5}V_{0.5}$	12.2	A15	$PbTl_2$	3.75–4.1	
$NbSnV_2$	5.5	A15	Pb_3Zr_5	4.60	$D8_8$
Nb_2SnV	9.8	A15	$PbZr_3$	0.76	A15
$Nb_{2.5}SnV_{0.5}$	14.2	A15	$Pd_{0.9}Pt_{0.1}Te_2$	1.65	C6
Nb_xTa_{1-x}	4.4–9.2	A2	$Pd_{0.05}Ru_{0.05}Zr_{0.9}$	~9	
$NbTc_3$	10.5	A12	$Pd_{2.2}S$ (quenched)	1.63	Cubic
Nb_xTi_{1-x}	0.6–9.8		$PdSb_2$	1.25	C2
$Nb_{0.6}Ti_{0.4}$	9.8		PdSb	1.50	$B8_1$
Nb_xU_{1-x}	1.95 max.		PdSbSe	1.0	C2
$Nb_{0.88}V_{0.12}$	5.7	A2	PdSbTe	1.2	C2
$Nb_{0.75}Zr_{0.25}$	10.8		Pd_4Se	0.42	Tetragonal
$Nb_{0.66}Zr_{0.33}$	10.8		$Pd_{6-7}Se$	0.66	Like Pd_4Te
$Ni_{0.3}Th_{0.7}$	1.98	$D10_2$	$Pd_{2.8}Se$	2.3	
$NiZr_2$	1.52		Pd_xSe_{1-x}	2.5 max.	
$Ni_{0.1}Zr_{0.9}$	1.5	A3	PdSi	0.93	B31
$O_3Rb_{0.27-0.29}W$	1.98	Hexagonal	PdSn	0.41	B31
O_3SrTi			$PdSn_2$	3.34	
(n = 1.7–12.0 × 10^{19})	0.12–0.37		Pd_2Sn	0.41	C37
O_3SrTi			Pd_3Sn_2	0.47–0.64	$B8_2$
(n = 10^{18}–10^{21})	0.05–0.47		PdTe	2.3, 3.85	$B8_1$
O_3SrTi			$PdTe_{1.02-1.08}$	2.56–1.88	$B8_1$
(n = ~10^{20})	0.47		$PdTe_2$	1.69	C6
OTi	0.58		$PdTe_{2.1}$	1.89	C6
$O_3Sr_{0.08}W$	2–4	Hexagonal	$PdTe_{2.3}$	1.85	C6
$O_3Tl_{0.30}W$	2.0–2.14	Hexagonal	$Pd_{1.1}Te$	4.07	$B8_1$
OV_3Zr_3	7.5	$E9_3$	$PdTh_2$	0.85	C16
OW_3 (film)	3.35, 1.1	A15	$Pd_{0.1}Zr_{0.9}$	7.5	A3
OsReY	2.0	C14	PtSb	2.1	$B8_1$
Os_2Sc	4.6	C14	PtSi	0.88	B31
OsTa	1.95	A12	PtSn	0.37	$B8_1$
Os_3Th_7	1.51	$D10_2$	PtTe	0.59	Orthorhombic
Os_xW_{1-x}	0.9–4.1		PtTh	0.44	B_f
OsW_3	~3		Pt_3Th_7	0.98	$D10_2$
Os_2Y	4.7	C14	Pt_5Th	3.13	

[a]See key at end of table.
[b]*w/o* denotes weight percent.

Table 2–7 (continued)
SELECTED SUPERCONDUCTIVE COMPOUNDS AND ALLOYS

Substance	T_c, K	*Crystal structure type*[a]
$PtTi_3$	0.58	A15
$Pt_{0.02}U_{0.98}$	0.87	β-phase
$PtV_{2.5}$	1.36	A15
PtV_3	2.87–3.20	A15
$PtV_{3.5}$	1.26	A15
$Pt_{0.5}W_{0.5}$	1.45	A1
Pt_xW_{1-x}	0.4–2.7	
Pt_2Y_3	0.90	
Pt_2Y	1.57, 1.70	C15
Pt_3Y_7	0.82	$D10_2$
PtZr	3.0	A3
$Re_{0.64}Ta_{0.36}$	1.46	A12
$Re_{24}Ti_5$	6.60	A12
Re_xTi_{1-x}	6.6 max.	
$Re_{0.76}V_{0.24}$	4.52	$D8_b$
$Re_{0.92}V_{0.08}$	6.8	A3
$Re_{0.6}W_{0.4}$	6.0	
$Re_{0.5}W_{0.5}$	5.12	$D8_b$
Re_2Y	1.83	C14
Re_2Zr	5.9	C14
Re_6Zr	7.40	A12
$Rh_{17}S_{15}$	5.8	Cubic
$Rh_{\sim 0.24}Sc_{\sim 0.76}$	0.88, 0.92	
Rh_xSe_{1-x}	6.0 max.	
Rh_2Sr	6.2	C15
$Rh_{0.4}Ta_{0.6}$	2.35	$D8_b$
$RhTe_2$	1.51	C2
$Rh_{0.67}Te_{0.33}$	0.49	
Rh_xTe_{1-x}	1.51 max.	
RhTh	0.36	B_f
Rh_3Th_7	2.15	$D10_2$
Rh_5Th	1.07	
Rh_xTi_{1-x}	2.25–3.95	
$Rh_{0.02}U_{0.98}$	0.96	
RhV_3	0.38	A15
RhW	~3.4	A3
RhY_3	0.65	
Rh_2Y_3	1.48	
Rh_3Y	1.07	C15
Rh_5Y	0.56	
$RhZr_2$	10.8	C16
$Rh_{0.005}Zr$ (annealed)	5.8	
$Rh_{0-0.45}Zr_{1-0.55}$	2.1–10.8	
$Rh_{0.1}Zr_{0.9}$	9.0	Hexagonal, c.p.
Ru_2Sc	1.67	C14
Ru_2Th	3.56	C15
RuTi	1.07	B2
$Ru_{0.05}Ti_{0.95}$	2.5	
$Ru_{0.1}Ti_{0.9}$	3.5	
$Ru_xTi_{0.6}V_y$	6.6 max.	
$Ru_{0.45}V_{0.55}$	4.0	B2
RuW	7.5	A3

Substance	T_c, K	*Crystal structure type*[a]
Ru_2Y	1.52	C14
Ru_2Zr	1.84	C14
$Ru_{0.1}Zr_{0.9}$	5.7	A3
SbSn	1.30–1.42, 1.42–2.37	Bl or distorted Bl
$SbTi_3$	5.8	A15
Sb_2Tl_7	5.2	
$Sb_{0.01-0.03}V_{0.99-0.97}$	3.76–2.63	A2
SbV_3	0.80	A15
Si_2Th	3.2	C_c, α-phase
Si_2Th	2.4	C32, β-phase
SiV_3	17.1	A15
$Si_{0.9}V_3Al_{0.1}$	14.05	A15
$Si_{0.9}V_3B_{0.1}$	15.8	A15
$Si_{0.9}V_3C_{0.1}$	16.4	A15
$SiV_{2.7}Cr_{0.3}$	11.3	A15
$Si_{0.9}V_3Ge_{0.1}$	14.0	A15
$SiV_{2.7}Mo_{0.3}$	11.7	A15
$SiV_{2.7}Nb_{0.3}$	12.8	A15
$SiV_{2.7}Ru_{0.3}$	2.9	A15
$SiV_{2.7}Ti_{0.3}$	10.9	A15
$SiV_{2.7}Zr_{0.3}$	13.2	A15
Si_2W_3	2.8, 2.84	
$Sn_{0.174-0.104}Ta_{0.826-0.896}$	6.5–<4.2	A15
$SnTa_3$	8.35	A15, highly ordered
$SnTa_3$	6.2	A15, partially ordered
$SnTaV_2$	2.8	A15
$SnTa_2V$	3.7	A15
Sn_xTe_{1-x} ($n = 10.5–20 \times 10^{20}$)	0.07–0.22	Bl
Sn_xTl_{1-x}	2.37–5.2	
SnV_3	3.8	A15
$Sn_{0.02-0.057}V_{0.98-0.943}$	2.87–~1.6	A2
$Ta_{0.025}Ti_{0.975}$	1.3	Hexagonal
$Ta_{0.05}Ti_{0.95}$	2.9	Hexagonal
$Ta_{0.05-0.75}V_{0.095-0.25}$	4.30–2.65	A2
$Ta_{0.8-1}W_{0.2-0}$	1.2–4.4	A2
$Tc_{0.1-0.4}W_{0.9-0.6}$	1.25–7.18	Cubic
$Tc_{0.50}W_{0.50}$	7.52	α plus σ
$Tc_{0.60}W_{0.40}$	7.88	σ plus α
Tc_6Zr	9.7	A12
$Th_{0-0.55}Y_{1-0.45}$	1.2–1.8	
$Ti_{0.70}V_{0.30}$	6.14	Cubic
Ti_xV_{1-x}	0.2–7.5	
$Ti_{0.5}Zr_{0.5}$ (annealed)	1.23	
$Ti_{0.5}Zr_{0.5}$ (quenched)	2.0	
V_2Zr	8.80	C15
$V_{0.26}Zr_{0.74}$	≈5.9	
W_2Zr	2.16	C15

[a]See key at end of table

Table 2–7 (continued)
SELECTED SUPERCONDUCTIVE COMPOUNDS AND ALLOYS

Critical Field Data

Substance	H_o, *oersteds*	*Substance*	H_o, *oersteds*
Ag_2F	2.5	InSb	1,100
Ag_7NO_{11}	57	In_xTl_{1-x}	252–284
Al_2CMo_3	1,700	$In_{0.8}Tl_{0.2}$	252
$BaBi_3$	740	$Mg_{\sim 0.47}Tl_{\sim 0.53}$	220
Bi_2Pt	10	$Mo_{0.16}Ti_{0.84}$	<985
Bi_3Sr	530	$NbSn_2$	620
Bi_5Tl_3	>400	$PbTl_{0.27}$	756
CdSn	>266	$PbTl_{0.17}$	796
$CoSi_2$	105	$PbTl_{0.12}$	849
$Cr_{0.1}Ti_{0.3}V_{0.6}$	1,360	$PbTl_{0.075}$	880
$In_{1-0.86}Mg_{0-0.14}$	272.4–259.2	$PbTl_{0.04}$	864

From Roberts, B. W., Properties of Selected Superconductive Materials, National Bureau of Standards Technical Notes 482 and 724, U.S. Government Printing Office, Washington, D.C., 1969, 1972.

Key to Crystal Structure Types

"*Strukturbericht*" *type*[c]	*Example*	*Class*	"*Strukturbericht*" *type*[c]	*Example*	*Class*
A1	Cu	Cubic, f.c.	$C15_b$	$AuBe_5$	Cubic
A2	W	Cubic, b.c.	C16	$CuAl_2$	Tetragonal, b.c.
A3	Mg	Hexagonal, close packed	C18	FeS_2	Orthorhombic
A4	Diamond	Cubic, f.c.	C22	Fe_2P	Trigonal
A5	White Sn	Tetragonal, b.c.	C23	$PbCl_2$	Orthorhombic
A6	In	Tetragonal, b.c. (f.c. cell usually used)	C32	AlB_2	Hexagonal
			C36	$MgNi_2$	Hexagonal
A7	As	Rhombohedral	C37	Co_2Si	Orthorhombic
A8	Se	Trigonal	C49	$ZrSi_2$	Orthorhombic
A10	Hg	Rhombohedral	C54	$TiSi_2$	Orthorhombic
A12	α-Mn	Cubic, b.c.	C_c	Si_2Th	Tetragonal, b.c.
A13	β-Mn	Cubic	DO_3	BiF_3	Cubic, f.c.
A15	"β-W" (WO_3)	Cubic	DO_{11}	Fe_3C	Orthorhombic
B1	NaCl	Cubic, f.c.	DO_{18}	Na_3As	Hexagonal
B2	CsCl	Cubic	DO_{19}	Ni_3Sn	Hexagonal
B3	ZnS	Cubic	DO_{20}	$NiAl_3$	Orthorhombic
B4	ZnS	Hexagonal	DO_{22}	$TiAl_3$	Tetragonal
$B8_1$	NiAs	Hexagonal	DO_e	Ni_3P	Tetragonal, b.c.
$B8_2$	Ni_2In	Hexagonal	$D1_3$	Al_4Ba	Tetragonal, b.c.
B10	PbO	Tetragonal	$D1_c$	$PtSn_4$	Orthorhombic
B11	γ-CuTi	Tetragonal	$D2_1$	CaB_6	Cubic
B17	PtS	Tetragonal	$D2_c$	MnU_6	Tetragonal, b.c.
B18	CuS	Hexagonal	$D2_d$	$CaZn_5$	Hexagonal
B20	FeSi	Cubic	$D5_2$	La_2O_3	Trigonal
B27	FeB	Orthorhombic	$D5_8$	Sb_2S_3	Orthorhombic
B31	MnP	Orthorhombic	$D7_3$	Th_3P_4	Cubic, b.c.
B32	NaTl	Cubic, f.c.	$D7_b$	Ta_3B_4	Orthorhombic
B34	PdS	Tetragonal	$D8_1$	Fe_3Zn_{10}	Cubic, b.c.
B_f	δ-CrB	Orthorhombic	$D8_2$	Cu_5Zn_8	Cubic, b.c.
B_g	MoB	Tetragonal, b.c.	$D8_3$	Cu_9Al_4	Cubic
B_h	WC	Hexagonal	$D8_8$	Mn_5Si_3	Hexagonal
B_i	γ-MoC	Hexagonal	$D8_b$	CrFe	Tetragonal
C1	CaF_2	Cubic, f.c.	$D8_i$	Mo_2B_5	Rhombohedral
$C1_b$	MgAgAs	Cubic, f.c.	$D10_2$	Fe_3Th_7	Hexagonal
C2	FeS_2	Cubic	$E2_1$	$CaTiO_3$	Cubic
C6	CdI_2	Trigonal	$E9_3$	Fe_3W_3C	Cubic, f.c.
C11b	$MoSi_2$	Tetragonal, b.c.	$L1_0$	CuAu	Tetragonal
C12	$CaSi_2$	Rhombohedral	$L1_2$	Cu_3Au	Cubic
C14	$MgZn_2$	Hexagonal	L'_{2b}	ThH_2	Tetragonal, b.c.
C15	Cu_2Mg	Cubic, f.c.	L'_3	Fe_2N	Hexagonal

[a] See key at end of table.
[b] *w/o* denotes weight percent.
[c] See original source.

Table 2–7 (continued)
SELECTED SUPERCONDUCTIVE COMPOUNDS AND ALLOYS

From Pearson, W. B., *Handbook of Lattice Spacing and Structure of Metals,* Vol. 1, 1958, 79; Vol. 2, 1967, 3, Pergamon Press, Oxford, Engl. Reprinted with permission of Pergamon Press Ltd.

Table 2–8
HIGH CRITICAL MAGNETIC-FIELD SUPERCONDUCTIVE COMPOUNDS AND ALLOYS

With Critical Temperatures, H_{c1}, H_{c2}, H_{c3}, and the Temperature of Field Observations, T_{obs}

Substance	T_c, *K*	H_{c1}, *kg*	H_{c2}, *kg*	H_{c3}, *kg*	T_{obs}, *K*[a]
Al_2CMo_3	9.8–10.2	0.091	156		1.2
$AlNb_3$		0.375			
$Ba_xO_3Sr_{1-x}Ti$	<0.1–0.55	0.0039 max.			
$Bi_{0.5}Cd_{0.1}Pb_{0.27}Sn_{0.13}$			>24		3.06
Bi_xPb_{1-x}	7.35–8.4	0.122 max.	~30 max.		4.2
$Bi_{0.56}Pb_{0.44}$	8.8		15		4.2
$Bi_{7.5\ w/o}Pb_{92.5\ w/o}$[b]			2.32		
$Bi_{0.099}Pb_{0.901}$		0.29	2.8		
$Bi_{0.02}Pb_{0.98}$		0.46	0.73		
$Bi_{0.53}Pb_{0.32}Sn_{0.16}$			>25		3.06
$Bi_{1-0.93}Sn_{0-0.07}$			0–0.032		3.7
Bi_5Tl_3	6.4		>5.56		3.35
C_8K (excess K)	0.55		0.160 (H⊥c)		0.32
			0.730 (H‖c)		0.32
C_8K	0.39		0.025 (H⊥c)		0.32
			0.250 (H‖c)		0.32
$Co_{0.44}Mo_{0.56}$	12.5–13.5	0.087	98.5		1.2
CNb	8–10	0.12	16.9		4.2
$CNb_{0.4}Ta_{0.6}$	10–13.6	0.19	14.1		1.2
CTa	9–11.4	0.22	4.6		1.2
$Ca_xO_3Sr_{1-x}Ti$	<0.1–0.55	0.002–0.004			
$Cd_{0.1}Hg_{0.9}$ (by weight)		0.23	0.34		2.04
$Cd_{0.05}Hg_{0.95}$		0.28	0.31		2.16
$Cr_{0.10}Ti_{0.30}V_{0.60}$	5.6	0.071	84.4		0
GaN	5.85	0.725			4.2
Ga_xNb_{1-x}			>28		4.2
GaSb (annealed)	4.24		2.64		3.5
$GaV_{1.95}$	5.3		73[c]		
$GaV_{2.1-3.5}$	6.3–14.45		230–300[d]		0
GaV_3		0.4	350[c]		0
			500[d]		
$GaV_{4.5}$	9.15		121[e]		0
Hf_xNb_y			>52–>102		1.2
Hf_xTa_y			>28–>86		1.2
$Hg_{0.05}Pb_{0.95}$		0.235	2.3		
$Hg_{0.101}Pb_{0.899}$		0.23	4.3		4.2
$Hg_{0.15}Pb_{0.85}$	~6.75		>13		2.93
$In_{0.98}Pb_{0.02}$	3.45	0.1		0.12	2.76
$In_{0.96}Pb_{0.04}$	3.68	0.1	0.12	0.25	2.94
$In_{0.94}Pb_{0.06}$	3.90	0.095	0.18	0.35	3.12
$In_{0.913}Pb_{0.087}$	4.2	~0.17	0.55	2.65	
$In_{0.316}Pb_{0.684}$		0.155	3.7		4.2
$In_{0.17}Pb_{0.83}$			2.8	5.5	4.2
$In_{1.000}Te_{1.002}$	3.5–3.7		1.2[e]		0
$In_{0.95}Tl_{0.05}$		0.263	0.263		3.3
$In_{0.90}Tl_{0.10}$		0.257	0.257		3.25
$In_{0.83}Tl_{0.17}$		0.242	0.39		3.21
$In_{0.75}Tl_{0.25}$		0.216	0.50		3.16
LaN	1.35	0.45			0.76
La_3S_4	6.5	≈0.15	>25		1.3
La_3Se_4	8.6	≈0.2	>25		1.25
$Mo_{0.52}Re_{0.48}$	11.1		14–21	22–33	4.2
			18–28	37–43	1.3

Table 2–8 (continued)
HIGH CRITICAL MAGNETIC-FIELD SUPERCONDUCTIVE COMPOUNDS AND ALLOYS

With Critical Temperatures, H_{c1}, H_{c2}, H_{c3}, and the Temperature of Field Observations, T_{obs} (continued)

Substance	T_c, *K*	H_{c1}, *kg*	H_{c2}, *kg*	H_{c3}, *kg*	T_{obs}, *K*[a]
$Mo_{0.6\pm0.05}Re_{0.395}$	10.6		14–20	20–37	4.2
			19–26	26–37	1.3
$Mo_{\sim0.5}Tc_{\sim0.5}$			~75[e]		0
$Mo_{0.16}Ti_{0.84}$	4.18	0.028	98.7[e]		0
			36–38		3.0
$Mo_{0.913}Ti_{0.087}$	2.95	0.060	~15		4.2
$Mo_{0.1-0.3}U_{0.9-0.7}$	1.85–2.06		>25		
$Mo_{0.17}Zr_{0.83}$			~30		
$N_{(12.8\ w/o)}Nb$	15.2		>9.5		13.2
NNb (wires)	16.1		153[e]		0
			132		4.2
			95		8
			53		12
NNb_xO_{1-x}	13.5–17.0		~38		
NNb_xZr_{1-x}	9.8–13.8		4–>130		4.2
$N_{0.93}Nb_{0.85}Zr_{0.15}$	13.8		>130		4.2
$Na_{0.086}Pb_{0.914}$		0.19	6.0		
$Na_{0.016}Pb_{0.984}$		0.28	2.05		
Nb	9.15		2.020		1.4
			1.710		4.2
Nb		0.4–1.1	3–5.5		4.2
Nb (unstrained)		1.1–1.8	3.40	6–9.1	4.2
Nb (strained)		1.25–1.92	3.44	6.0–8.7	4.2
Nb (cold-drawn wire)		2.48	4.10	≈10	4.2
Nb (film)			>25		4.2
NbSc			>30		
Nb_3Sn		0.170	221		4.2
			70		14.15
			54		15
			34		16
			17		17
$Nb_{0.1}Ta_{0.9}$		0.084	0.154		4.195
$Nb_{0.2}Ta_{0.8}$			10		4.2
$Nb_{0.65-0.73}Ta_{0.02-0.10}Zr_{0.25}$			>70–>90		4.2
Nb_xTi_{1-x}			148 max.		1.2
			120 max.		4.2
$Nb_{0.222}U_{0.778}$		1.98	23		1.2
Nb_xZr_{1-x}			127 max.		1.2
			94 max.		4.2
O_3SrTi	0.43	.0049[e]	.504[e]		0
O_3SrTi	0.33	.00195[e]	.420[e]		0
$PbSb_{1\ w/o}$ (quenched)			>1.5		4.2
$PbSb_{1\ w/o}$ (annealed)			>0.7		4.2
$PbSb_{2.8\ w/o}$ (quenched)			>2.3		4.2
$PbSb_{2.8\ w/o}$ (annealed)			>0.7		4.2
$Pb_{0.871}Sn_{0.129}$		0.45	1.1		
$Pb_{0.965}Sn_{0.035}$		0.53	0.56		
$Pb_{1-0.26}Tl_{0-0.74}$	7.20–3.68		2–6.9[e]		0
$PbTl_{0.17}$	6.73		4.5[e]		0
$Re_{0.26}W_{0.74}$			>30		
$Sb_{0.93}Sn_{0.07}$			0.12		3.7
SiV_3	17.0	0.55	156[c]		
Sn_xTe_{1-x}		0.00043–0.00236	0.005–0.0775		0.012–0.079
Ta (99.95%)		0.425	1.850		1.3
		0.325	1.425		2.27
		0.275	1.175		2.66
		0.090	0.375		3.72
$Ta_{0.5}Nb_{0.5}$			3.55		4.2
$Ta_{0.65-0}Ti_{0.35-1}$	4.4–7.8		>14–138		1.2
$Ta_{0.5}Ti_{0.5}$			138		1.2
Te	~3.3	0.25[e]			0
Tc_xW_{1-x}	5.75–7.88		8–44		4.2
Ti				2.7	4.2
$Ti_{0.75}V_{0.25}$	5.3	0.029[c]	199[e]		0
$Ti_{0.775}V_{0.225}$	4.7	0.024[c]	172[e]		0
$Ti_{0.615}V_{0.385}$	7.07	0.050	~34		4.2
$Ti_{0.516}V_{0.484}$	7.20	0.062	~28		4.2
$Ti_{0.415}V_{0.585}$	7.49	0.078	~25		4.2

Table 2–8 (continued)
HIGH CRITICAL MAGNETIC-FIELD SUPERCONDUCTIVE COMPOUNDS AND ALLOYS

With Critical Temperatures, H_{c1}, H_{c2}, H_{c3}, and the Temperature of Field Observations, T_{obs} (continued)

Substance	T_c, *K*	H_{c1}, *kg*	H_{c2}, *kg*	H_{c3}, *kg*	T_{obs}, *K*[a]
$Ti_{0.12}V_{0.88}$			17.3	28.1	4.2
$Ti_{0.09}V_{0.91}$			14.3	16.4	4.2
$Ti_{0.06}V_{0.94}$			8.2	12.7	4.2
$Ti_{0.03}V_{0.97}$			3.8	6.8	4.2
Ti_xV_{1-x}			108 max.		1.2
V	5.31	~0.8	~3.4		1.79
		~0.75	~3.15		2
		~0.45	~2.2		3
		~0.30	~1.2		4
$V_{0.26}Zr_{0.74}$	≈5.9	0.238			1.05
		0.227			1.78
		0.185			3.04
		0.165			3.5
W (film)	1.7–4.1		>34		1

[a]Temperature of critical field measurement.
[b]*w/o* denotes weight percent.
[c]Parabolic extrapolation.
[d]Linear extrapolation.
[e]Extrapolated.

From Roberts, B. W., in *Handbook of Chemistry and Physics,* 55th ed., Weast R. C., Ed., CRC Press, Cleveland, 1974, E-97.

Table 2–9
PROPERTIES OF DIELECTRICS

In most cases properties were determined by ASTM (American Society for Testing and Materials) test methods at room temperature under standard conditions. Values will, in general, change considerably with temperature.

Plastics

Material	*Dielectric constant, 10^6 cycles*	*Dielectric strength, volts/mil*	*Volume resistivity, ohms-cm, 23°C*	*Loss factor*[a]
Allyl resin, cast	3.6–4.5	380	$>4 \times 10^{14}$	0.028–0.06
Aniline formadehyde resin, no filler	3.5–3.6	600–650	10^{16}–10^{17}	0.006–0.008
Casein	6.1–6.8	400–700		0.052
Cellulose acetate, molding	3.2–7.0	250–365	10^{10}–10^{13}	0.01–1.0
Cellulose acetate, sheet	4.0–5.5	250–300	10^{11}–10^{13}	0.04–0.06
Cellulose acetate butyrate	3.2–6.2	250–400	10^{10}–10^{12}	0.01–0.04
Cellulose nitrate (proxylin)	6.4	300–600	$(10–15) \times 10^{10}$	0.06–0.09
Cold-molded compound, inorganic, refractory		45		
Cold-molded compound, organic, non-refractory	6.0	85–115	1.3×10^{12}	0.07
Ethyl cellulose	2.8–3.9	350–500	10^{12}–10^{14}	0.01–0.06
Epoxy cast resin	3.62	400	10^{16}–10^{17}	0.019
Glycerol phthalate, cast, alkyd	3.7–4.0	300–350	$>10^{14}$	0.025–0.035
Melamine formaldehyde resins, molding				
Alpha cellulose filler	7.2–8.2	300–400	$10^{12} \times 10^{14}$	0.027–0.045
Asbestos filler	6.1–6.7	350–400	2.4×10^{11}	0.041–0.050
Cellulose filler	4.7–7.0	350–400		0.032–0.060
Flock filler		300–330		
Macerated fabric filler	6.5–6.9	250–350	10^{9}–10^{10}	0.036
Methyl methacrylate, cast	2.7–3.2	450–500	$>10^{15}$	0.02–0.03
Methyl methacrylate, molding	2.7–3.2	450–500	$>10^{14}$	0.02–0.03
Mica, glass-bonded, compression	7.4–7.85		10^{14}–10^{15}	0.0015–0.002
Mica, glass-bonded, injection	6.9–9.2		10^{14}–10^{17}	0.0015–0.012
Nylon (F.M. 3001)	3.5	470	4×10^{14}	0.03
Phenol formaldehyde resins, molding				
Asbestos filler	5.0–7.0	100–350	10^{10}–10^{12}	0.10–0.50
Glass fiber	6.6	140–370	7×10^{12}	0.02
Mica filler	4.2–5.2	300–460	10^{12}–10^{14}	0.005–0.04
No filler	4.5–5.0	300–400	10^{11}–10^{12}	0.015–0.03
Sisal felt	3–5	250–400	10^{11}–10^{12}	0.3–0.5
Wood flour filler	4.–7.	200–425	10^{9}–10^{13}	0.03–0.07
Phenol formaldehyde resins, cast				
Mineral filler	9–15	100–250	10^{9}–10^{12}	0.07–0.2
No filler	4.0–5.5	350–400	10^{12}–10^{13}	0.04–0.05
Polyacrylic ester (filled and vulcanized)		400–700	2×10^{11} at 70°C	
Polyester, cast resin, rigid	2.8–4.1	380–500	10^{14}	0.006–0.026
Polyester, cast resin, flexible	4.1–5.2	250–500		0.023–0.052
Polyester molding materials (glass fiber filler)	4.0–4.5	150–400	10^{12}–10^{14}	0.015–0.020
Polyethylene	2.3	460	1.6×10^{13}	<0.0005
Polymonochlorodifluoroethylene	2.5	400	1.2×10^{18}	0.010
Polystyrene molding	2.4–2.65	500–700	10^{17}–10^{19}	0.0001–0.0004
Polytetrafluoroethylene	2.0	480	$>10^{15}$	<0.0002–0.0005
Rubber, hard	2.8	470	2×10^{15}	0.06
Rubber, chlorinated	3 approx.		1.5×10^{13}	0.006
Rubber, modified, isomerized	2.4–2.7	620	5×10^{16}	0.0008–0.002
Shellac		200–600	1.8×10^{9}	
Silicone molding compound (glass-fiber-filled)	3.7	185	10^{11}–10^{13}	0.0017
Styrene, modified, molding (shock-resistant type)	2.4–3.8	300–600	10^{12}–10^{17}	0.0004–0.02
Urea formaldehyde resin, alpha cellulose filler	6.4–6.9	300–400	10^{12}–10^{13}	0.028–0.032
Vinyl butyral, flexible, unfilled	3.92	350	5×10^{10}	0.061
Vinyl butyral, rigid	3.33	400	$>10^{14}$	0.0065
Vinyl chloride, rigid	2.8–3.0	700–1300	$>10^{16}$	0.006–0.014
Vinyl chloride, flexible, filled	3.5–4.5	600–800	5×10^{14}	0.09–0.10
Vinyl chloride, flexible, unfilled	3.5–4.5	800–1000	5×10^{12}	0.09–0.10

Table 2–9 (continued)
PROPERTIES OF DIELECTRICS

Plastics (continued)

Material	*Dielectric constant, 10^6 cycles*	*Dielectric strength, volts/mil*	*Volume resistivity, ohms-cm, 23°C*	*Loss factor*[a]
Vinyl chloride acetate, rigid	3.0–3.1	425	10^{16}	0.018–0.019
Vinyl chloride acetate, flexible, unfilled	3.3–4.3	300–400	10^{11}–10^{13}	0.04–0.14
Vinyl chloride acetate, flexible, filled	3.3–4.3	250–350	10^{11}–10^{13}	0.04–0.14
Vinyl formal molding compound	3.0	490		0.023
Vinylidene chloride	3.0–4.0	350	10^{14}–10^{16}	0.05–0.08

[a]Power factor × dielectric constant equals loss factor.

From Bolz, R. E. and Tuve, G. L., Eds., *Handbook of Tables for Applied Engineering Science,* 2nd ed., CRC Press, Cleveland, 1973, 243.

Ceramics

Material	*Dielectric constant, 10^6 cycles*	*Dielectric strength, volts/mil*	*Volume resistivity, ohms-cm, 23°C*	*Loss factor*[a]
Alumina	4.5–8.4	40–160	10^{11}–10^{14}	0.0002–0.01
Corderite	4.5–5.4	40–250	10^{12}–10^{14}	0.004–0.012
Forsterite	6.2	240	10^{14}	0.0004
Porcelain (dry process)	6.0–8.0	40–240	10^{12}–10^{14}	0.003–0.02
Porcelain (wet process)	6.0–7.0	90–400	10^{12}–10^{14}	0.006–0.01
Porcelain, zircon	7.1–10.5	250–400	10^{13}–10^{15}	0.0002–0.008
Steatite	5.5–7.5	200–400	10^{13}–10^{15}	0.0002–0.004
Titanates (Ba, Sr, Ca, Mg, and Pb)	15–12,000	50–300	10^{8}–10^{15}	0.0001–0.02
Titanium dioxide	14–110	100–210	10^{12}–10^{18}	0.0002–0.005

[a]Power factor × dielectric constant equals loss factor.

Glasses

Type	*Dielectric constant, 100 mc, 20°C*	*Volume resistivity, megohm-cm, 350°C*	*Loss factor*[a]
Corning 0010	6.32	10	0.015
Corning 0080	6.75	0.13	0.058
Corning 0120	6.65	100	0.012
Pyrex 1710	6.00	2,500	0.025
Pyrex 3320	4.71		0.019
Pyrex 7040	4.65	80	0.013
Pyrex 7050	4.77	16	0.017
Pyrex 7052	5.07	25	0.019
Pyrex 7060	4.70	13	0.018
Pyrex 7070	4.00	1,300	0.0048
Vycor 7230	3.83		0.0061
Pyrex 7720	4.50	16	0.014
Pyrex 7740	5.00	4	0.040
Pyrex 7750	4.28	50	0.011
Pyrex 7760	4.50	50	0.0081
Vycor 7900	3.9	130	0.0023
Vycor 7910	3.8	1,600	0.00091
Vycor 7911	3.8	4,000	0.00072
Corning 8870	9.5	5,000	0.0085
G. E. Clear (silica glass)	3.81	4,000–30,000	0.00038
Quartz (fused)	3.75–4.1 (1 mc)		0.0002 (1 mc)

[a]Power factor × dielectric constant equals loss factor.

From Weast, R. C., Ed., *Handbook of Chemistry and Physics*, 55th ed., CRC Press, Cleveland, 1974, E-60.

Table 2–10
DIELECTRIC CONSTANTS FOR VARIOUS SOLIDS

Effect of High Frequencies at Room Temperature

Material	*Frequency, cps*			*Material*	*Frequency, cps*		
	10^3	10^6	10^9		10^3	10^6	10^9
Alkyd isocyanate foam	1.223	1.218	1.20	Nylon 66	3.75	3.33	3.16
Aluminum oxide	8.83	8.80	8.80	Nylon 610	3.50	3.14	3.0
Asbestos fiber	4.80	3.1	—	Plexiglas	3.12	2.76	—
Asphalt	2.66	2.58	2.55 (1×10^{10})	Polycarbonate	3.17	3.02	2.96
Balata	2.50	2.50	2.42	Polyether, chlorinated	3.1	3.0	2.9
Beeswax, yellow	2.66	2.53	2.45	Polyethylene	2.37	2.35	2.33
Buna S	2.66	2.56	2.52	Polyisobutylene	2.23	a	a
Butyl rubber compound	2.42	2.40	2.39	Polypropylene	2.25	a	a
Calcium titanate	167.7	a	a	Polyvinyliden fluoride	8.4	8.0	6.6
Cellulose nitrate	8.4	6.6	5.2	Porcelain (dry process)	5.36	5.08	5.04
Dichloronaphthalene	3.04	2.98	2.93	Rutile	100.0	a	a
Gutta percha	2.60	2.53	2.47	Selenium, amorphous	6.00	a	a
Hevea compound	36.	9.0	6.8	Shellac, natural	3.81	3.47	3.10
Lucite	2.84	2.63	2.58	Strontium titanate	233.0	232.0	232.0
Magnesium oxide	9.65	a	a	Teflon FEP	2.1	a	a
Magnesium silicate	5.98	5.97	5.96	Teflon PTFE	2.0	a	a
Magnesium titanate	13.9	a	a	Urea-formaldehyde	6.7	6.0	5.2
Methyl cellulose	6.8	5.7	4.3	Vinylite QYNA	3.10	2.88	2.85
Mica, ruby	5.4	a	a	Vinylite VYHH	3.12	2.91	2.83
Neoprene	6.60	6.26	4.5	Vinylite 5544	7.20	4.13	3.05

[a]No appreciable variation with frequency in the range $10^3 - 10^9$ cps.

From Bolz, R. E. and Tuve, G. L., Eds., *Handbook of Tables for Applied Engineering Science,* 2nd ed., CRC Press, Cleveland, 1973, 244.

Table 2–11
RESISTOR COLOR CODE AND STANDARD VALUES

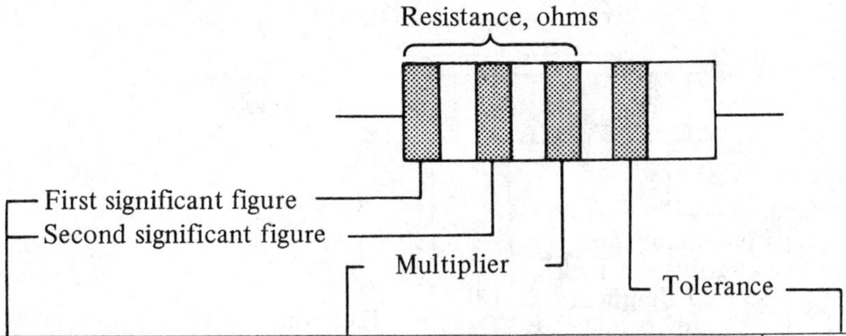

Color	*Value*	*Color*	*Multiplier*	*Color*	*Tolerance, percent*
Black	0	Black	1	No color	±20
Brown	1	Brown	10	Black	±20
Red	2	Red	100	Silver	±10
Orange	3	Orange	1,000	Gold	± 5
Yellow	4	Yellow	10,000	White	±10[a]
Green	5	Green	100,000	Green	± 5[a]
Blue	6	Blue	1,000,000		
Violet	7	Violet	10,000,000		
Gray	8	Silver	0.01		
White	9	Gold	0.1		
		Gray	0.01[a]		
		White	0.1[a]		

[a]Optional.

Standard Resistor Values, Significant Figures

±20%: 10, 15, 22, 33, 47, 68 (Series 6).[b]
±10%: 10, 12, 15, 18, 22, 27, 33, 39, 47, 56, 68, 82 (Series 12).[b]
±5%: 10, 11, 12, 13, 15, 16, 18, 20, 22, 24, 27, 30, 33, 36, 39, 43, 47, 51, 56, 62, 68, 75, 82, 91 (Series 24).[b]

[b]For series specifications see Bolz, R. E. and Tuve, G. L., Eds., *Handbook of Tables for Applied Engineering Science,* 2nd ed., CRC Press, Cleveland, 1973, 246.

From Bolz, R. E. and Tuve, G. L., Eds., *Handbook of Tables for Applied Engineering Science,* 2nd ed., CRC Press, Cleveland, 1973, 247.

Table 2–12
CAPACITOR COLOR CODE

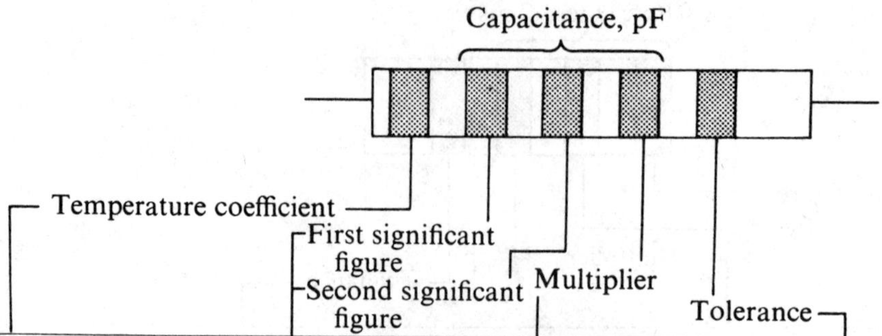

Color	*Temp. coeff., ppm/°C*	*Color*	*Value*	*Color*	*Multiplier*	*Color*	*More than 10 pF (in %)*	*Less than 10 pF*
Black	0	Black	0	Black	1	Black	±20	2.0
Brown	−30	Brown	1	Brown	10	Brown	±1	0.1
Red	−80	Red	2	Red	100	Red	±2	
Orange	−150	Orange	3	Orange	1,000	Green	±5	0.5
Yellow	−220	Yellow	4	Gray	0.01	Gray		0.25
Green	−330	Green	5	White	0.1	White	±10	1.0
Blue	−470	Blue	6					
Violet	−750	Violet	7					
Gray	+30	Gray	8					
White	+500	White	9					

From Bolz, R. E. and Tuve, G. L., Eds., *Handbook of Tables for Applied Engineering Science,* 2nd ed., CRC Press, Cleveland, 1973, 247.

Table 2–13
THERMAL EXPANSION COEFFICIENTS FOR MATERIALS USED IN INTEGRATED CIRCUITS

Coefficient of Linear Thermal Expansion of Selected Materials per K

Material	*Temperature, K*					
	300	*400*	*500*	*600*	*700*	*800*
Aluminum	23.2	24.9	26.4	28.3	30.7	33.8
Beryllium oxide	4.7	—	6.0	—	7.0	—
Copper	16.8	17.7	18.3	18.9	19.4	20.0
Germanium	5.7	6.2	6.5	6.7	6.9	7.2
Gold	14.1	14.5	15.	15.4	15.9	16.5
Indium	31.9	38.5	—	—	—	—
Lead	28.9	29.8	32.1	—	—	—
Molybdenum	5.0	5.2	5.3	5.4	5.5	5.7
Nickel	12.7	13.8	15.2	17.2	16.4	16.8
Platinum	8.9	9.2	9.5	9.7	10.0	10.2
Silicon	2.5	3.1	3.5	3.8	4.1	4.3
Silver	19.2	20.0	20.6	21.4	22.3	23.4
Tantalum	6.5	6.6	6.8	6.9	7.0	7.1
Tin	21.2	24.2	27.5	—	—	—
Tungsten	4.5	4.6	4.6	4.7	4.7	4.8
Vitreous silica	.42	.56	.56	.55	.54	.54

Material	*Coefficient range*	*Temperature range, °C*
Aluminum oxide ceramic	6.0–7.0	25–300
Brass	17.7–21.2	25–300
Kanthal A	13.9–15.1	20–900
Kovar	5.0	25–300
Pyrex glass	3.2	25–300
Pyroceram (#9608)	4–20	25–300
Pyroceram cement		
Vitreous (#45)	4	0–300
Devitrified	2.4	25–300
Pyroceram cement (#89, #95)	8–10	—
Silicon carbide	4.8	0–1,000
Silicon nitride		
α	2.9	25–1,000
β	2.25	25–1,000
Solder glass (Kimble CV-101)	809	0–300

Note: Multiply all values by 10^{-6}.

From Beadles, R. L., Interconnections and Encapsulation, *Integrated Silicon Device Technology,* Vol. 14, Research Triangle Institute, Research Triangle Park, N. C., 1967. With permission.

Table 2–14
DIFFUSION COEFFICIENT D

For Self-diffusion and Diffusion of Foreign Atoms

$$D = D_\sigma e^{-\Delta E/kT}$$

Semiconductor and diffusing element	*D_o, cm^2/s*	*ΔE, eV*	*Temperature, deg C*
Aluminum antimonide (AlSb)			
Al		≈1.8	
Cu	3.5×10^{-3}	0.36	150–500
Sb		≈1.5	
Zn	0.33 ± 0.15	1.93 ± 0.04	660–860
Cadmium selenide (CdSe)			
Se	2.6×10^{-3}	1.55	700–1800
Cadmium sulfide (CdS)			
Ag	$2.5 \times 10^{+1}$	1.2	250–500
Cd	3.4	2.0	750–1000
Cu	1.5×10^{-3}	0.76	450–750
Cadmium telluride (CdTe)			
Au	$6.7 \times 10^{+1}$	2.0	600–1000
In	4.1×10^{-1}	1.6	450–1000
Calcium ferrate (III) ($CaFe_2O_4$)			
Ca	30	3.7	
Fe	0.4	3.1	

Table 2–14 (continued)
DIFFUSION COEFFICIENT D

For Self-diffusion and Diffusion of Foreign Atoms (continued)

Semiconductor and diffusing element	D_o, cm^2/s	ΔE, eV	*Temperature, deg C*
α-Calcium metasilicate ($CaSiO_3$)			
Ca	7.4×10^4	4.8	
Gallium antimonide (GaSb)			
Ga	3.2×10^3	3.15	650–700
In	1.2×10^{-7}	0.53	400–650
Sb	3.4×10^4	3.44	650–700
	$8.7 \times 10^{+2}$	1.13	470–570
Sn	2.4×10^{-5}	0.80	320–570
Te	3.8×10^{-4}	1.2	400–650
Gallium arsenide (GaAs)			
Ag	2.5×10^{-3}	1.5	
	3.9×10^{-11}	0.33	
	4×10^{-4}	0.8 ± 0.05	500–1160
As	4×10^{21}	10.2 ± 1.2	1200–1250
Au	10^{-3}	1.0 ± 0.2	740–1024
Cd	0.05 ± 0.04	2.43 ± 0.06	868–1149
	[a]5.0×10^{-2}	2.8[a]	
Cu	0.03	0.52	100–600
Ga	1×10^7	5.60 ± 0.32	1125–1250
Li	0.53	1.0	250–400
Mg	1.4×10^{-4}	1.89	
	2.3×10^{-2}	2.6	740–1024
	[a]2.6×10^{-2}	2.7[a]	
Mn	[a]6.5×10^{-1}	2.49[a]	
	8.5×10^{-3}	1.7	740–1024
S	1.2×10^{-4}	1.8	
	[a]1.6×10^{-5}	1.63[a]	
	2.6×10^{-5}	1.86	
	4×10^3	4.04 ± 0.15	1000–1200
Se	3×10^3	4.16 ± 0.16	1000–1200
Sn	[a]3.8×10^{-2}	2.7	
	6×10^{-4}	2.5	1069–1215
Zn	[a]2.5×10^{-1}	3.0[a]	
	3.0×10^{-7}	1.0	
	6.0×10^{-7}	0.6	
	15 ± 7	2.49 ± 0.05	800
Gallium phosphide (GaP)			
Zn	1.0	2.1	700–1300
Germanium (Ge)			
Ag	4.4×10^{-2}	1.0	700–900
As	6.3	2.4	600–850
Au	2.2×10^{-2}	2.5	
B	1.6×10^{-9}	4.6	600–850
Cu	1.9×10^{-4}	0.18	600–850
Fe	1.3×10^{-1}	1.1	750–850
Ga	$4.0 \times 10^{+1}$	3.1	600–850
Ge	$8.7 \times 10^{+1}$	3.2	750–920
He	6.1×10^{-3}	0.69	750–850
In	3×10^{-2}	2.4	600–850
Li	1.3×10^{-4}	0.47	200–600
Ni	8×10^{-1}	0.9	700–875
P	2.5	2.5	600–850
Pb	—	3.6	600–850
Sb	4.0	2.4	600–850
Sn	1.7×10^{-2}	1.9	600–850
Zn	$1.0 \times 10^{+1}$	2.5	600–850
Indium antimonide (InSb)			
Ag	1.0×10^{-7}	0.25	
Au	[a]7×10^{-4}	0.32[a]	140–510

Table 2–14 (continued)
DIFFUSION COEFFICIENT D

For Self-diffusion and Diffusion of Foreign Atoms (continued)

Semiconductor and diffusing element	D_o, cm^2/s	ΔE, eV	*Temperature, deg C*
Cd	[a]1.0×10^{-5}	1.1[a]	250–500
	1.23×10^{-9}	0.52	442–519
	1.26	1.75	
	1.3×10^{-4}	1.2	360–500
Co	2.7×10^{-11}	0.39	
	10^{-7}	0.25	440–510
Cu	3.0×10^{-5}	0.37	
	[a]9.0×10^{-4}	1.08[a]	
Fe	10^{-7}	0.25	440–510
Hg	[a]4.0×10^{-6}	1.17[a]	
In	0.05	1.81	450–500
	1.8×10^{-9}	0.28	
Ni	10^{-7}	0.25	440–510
Sb	0.05	1.94	450–500
	1.4×10^{-6}	0.75	
Sn	5.5×10^{-8}	0.75	390–512
Te	1.7×10^{-7}	0.57	300–500
Zn	0.5	1.35	360–500
	1.6×10^{-6}	2.3 ± 0.3	360–500
	5.5	1.6	360–500
(Polycrystal)	1.7×10^{-7}	0.85	390–512
	[a]5.3×10^{7}	2.61	
(High concentration)	6.3×10^{8}	2.61	
	[a]8.7×10^{-10}	0.7[a]	
(Conc. = 2.2×10^{20} cm^{-3})	9.0×10^{-10}	0	
(Single crystal)	1.4×10^{-7}	0.86	390–512
Indium arsenide (InAs)			
Cd	4.35×10^{-4}	1.17	600–900
Cu		0.52[a]	
Ge	3.74×10^{-6}	1.17	600–900
Mg	1.98×10^{-6}	1.17	600–900
S	6.78	2.20	600–900
Se	12.55	2.20	600–900
Sn	1.49×10^{-6}	1.17	600–900
Te	3.43×10^{-5}	1.28	600–900
Zn	3.11×10^{-3}	1.17	600–900
Indium phosphide (InP)			
In	1×10^{5}	3.85 ± 0.0	850–1000
P	7×10^{10}	5.65 ± 0.0	850–1000
Iron oxide (Fe_3O_4)			
Fe	5.2	2.4	
Lead metasilicate ($PbSiO_3$)			
Pb	85	2.6	
Lead orthosilicate ($PbSiO_4$)			
Pb	8.2	2.0	
Mercury selenide (HgSe)			
Sb	6.3×10^{-5}	0.85	540–630
Nickel aluminate ($NiAl_2O_4$)			
Cr	1.17×10^{-3}	2.2	
Fe	1.33	3.5	
Nickel chromate (III) ($NiCr_2O_4$)			
Cr	0.74	3.1	
Cr	2.03×10^{-5}	1.9	
Fe	1.35×10^{-3}	2.6	
Ni	0.85	3.2	
Selenium (Se) (amorphous)			
Fe	1.1×10^{-5}	0.38	300–400
Ge	9.4×10^{-6}	0.39	300–400
In	5.2×10^{-6}	0.32	300–400
Sb	2.8×10^{-8}	0.29	300–400

Table 2–14 (continued)
DIFFUSION COEFFICIENT D

For Self-diffusion and Diffusion of Foreign Atoms (continued)

Semiconductor and diffusing element	D_o, cm^2/s	ΔE, eV	*Temperature, deg C*
Se	7.6×10^{-10}	0.14	300–400
Sn	4.8×10^{-8}	0.39	300–400
Te	5.4×10^{-6}	0.53	300–400
Tl	1.4×10^{-6}	0.35	300–400
Zn	3.8×10^{-7}	0.29	300–400
Silicon (Si)			
Al	8.0	3.5	1100–1400
Ag	2×10^{-3}	1.6	1100–1350
As	3.2×10^{-1}	3.5	1100–1350
Au	1.1×10^{-3}	1.1	800–1200
B	$1.0 \times 10^{+1}$	3.7	950–1200
Bi	$1.04 \times 10^{+3}$	4.6	1100–1350
Cu	4×10^{-2}	1.0	800–1100
Fe	6.2×10^{-3}	0.86	1000–1200
Ga	3.6	3.5	1150–1350
H_2	9.4×10^{-3}	0.47	1000–1200
He	1.1×10^{-1}	0.86	1000–1200
In	$1.65 \times 10^{+1}$	3.9	1100–1350
Li	9.4×10^{-3}	0.78	100–800
P	$1.0 \times 10^{+1}$	3.7	1100–1350
Sb	5.6	3.9	1100–1350
Tl	$1.65 \times 10^{+1}$	3.9	1100–1350
Silicon carbide (SiC)			
Al	2.0×10^{-1}	4.9	1800–2250
B	$1.6 \times 10^{+2}$	5.6	1850–2250
Cr	2.3×10^{-1}	4.8	1700–1900
Sulfur (S)			
S	2.8×10^{13}	2.0	> 100
Tin zinc oxide ($SnZn_2O_4$)			
Sn	2×10^5	4.7	
Zn	37	3.3	
Zinc aluminate ($ZnAl_2O_4$)			
Zn	2.5×10^2	3.4	
Zinc chromate (III) ($ZnCr_2O_4$)			
Cr	8.5	3.5	
Zn	60	3.7	
Zinc ferrate (III) ($ZnFe_2O_4$)			
Fe	8.5×10^2	3.5	
Zn	8.8×10^2	3.7	
Zinc selenide (ZnSe)			
Cu	1.7×10^{-5}	0.56	200–570
Zinc sulfide (ZnS)			
Zn	$1.0 \times 10^{+16}$	6.50	> 1030
	$1.5 \times 10^{+4}$	3.25	940–1030
	3.0×10^{-4}	1.52	< 940

[a]Values obtained at the low concentration limit.

REFERENCES

Boltaks, B. I. (Carasso, J. I., translator), *Diffusion in Semiconductors,* Infosearch, Ltd., London (distributed by Pion, Ltd.; in the U.S.A. by Academic Press, New York), 1963.
Madelung, O. (Meyerhofer, D., translator), *Physics of III-V Compounds,* John Wiley & Sons, New York, 1964.
Willardson, R. K., and Beer, A. C., *Semiconductors and Semimetals,* Vol. 4, Academic Press, New York, 1968.

From Bolz, R. E. and Tuve, G. L., Eds., *Handbook of Tables for Applied Engineering Science,* 2nd ed., CRC Press, Cleveland, 1973, 251.

Table 2–15
TYPICAL FILM MATERIALS AND PROCESSES

Application	*Material*	*Process*
Thin-film resistors	Nichrome (Ni–Cr)	Vacuum evaporation
	Tin oxide (SnO_2)	Vapor plating
	Tantalum–tantalum nitride (Ta–TaN)	Sputtering
	Cermet chromium-silicon monoxide (Cr–SiO)	Vacuum evaporation
Thick-film resistors	Palladium–silver Palladium oxide–silver and glass	Silk screen
Thin-film dielectrics	Silicon monoxide (SiO)	Vacuum evaporation
	Silicon dioxide (SiO_2)	Vapor plating, high temperature, steam oxidation, reactive sputtering
	Aluminum oxide (Al_2O_3)	Vapor plating, anodization of aluminum films
	Tantalum oxide (Ta_2O_5)	Anodization of tantalum, reactive sputtering
Thick-film dielectrics	$BaTiO_3$ or TiO_2 and glass mixtures	Silk screen
Thin-film conductors	Chromium–gold	Vacuum evaporation
	Chromium–copper	Vacuum evaporation
	Aluminum	Vacuum evaporation
	Nickel on ceramic	Electroless plating
Thick-film conductors	Gold–platinum–glass Gold–glass Silver–glass	Silk screen

REFERENCE

Schwartz, B. and Schwartz, N., Eds., *Measurement Techniques for Thin Films,* The Electrochemical Society, New York, 1967.

From Westman, H. P., Ed., *Reference Data for Radio Engineers,* 5th ed., Howard W. Sams & Co., Indianapolis, 1968. With permission.

Table 2–16
THIN-FILM DEPOSITION DATA

NAME	SYMBOL	MELTING POINT °C	DENSITY g/cm3	TEMPERATURE (°C) @ VAP. PRESS. 10-8 TORR	10-6 TORR	10-4 TORR	EVAPORATION TECHNIQUES ELECTRON BEAM	CRUCIBLE	COIL	BOAT	BASKET	SPUTTER*	REMARKS n = Index of refraction
Aluminum	Al	660	2.70	677	821	1010	Xlnt.	TiB_2-BN ZrB_2, BN	W	TiB_2-BN Al_2O_3	W	RF DC Sputtergun	Alloys and wets; tungsten-stranded superior
Aluminum Antimonide	AlSb	1080	4.3	—	—	—	—	—	—	—	—	RF	—
Aluminum Arsenide	AlAs	1600	3.7	—	—	~1300	—	—	—	—	—	RF	—
Aluminum Bromide	$AlBr_3$	97	3.01	—	—	~50	—	Graphite	—	Mo	—	RF	—
Aluminum Carbide	Al_4C_3	1400	2.36	—	—	~800	Fair	—	—	—	—	RF	n = 2.7
Aluminum 2% Copper	Al2%Cu	640	2.82	—	—	—	—	—	—	—	—	RF DC Sputtergun	Wire feed and flash. Difficult from dual sources.
Aluminum Fluoride	AlF_3	1257 Subl.	3.07	410 Sublimes.	490	700	Poor	Graphite	—	Mo W Ta	—	RF	—
Aluminum Nitride	AlN	Subl.	3.26	—	—	~1750	Fair	—	—	—	—	RF RF-Reactive	Decomposes. Reactive evaporate in 10^{-3} N_2 with glow discharge.
Aluminum Oxide (a) (alumina)	Al_2O_3	2045	3.97	—	—	1550	Xlnt.	—	—	W	W	RF-Reactive	Sapphire xlnt in EB, forms smooth hard films. n = 1.66
Aluminum Phosphide	AlP	2000	2.42	—	—	—	—	—	—	—	—	RF	—
Aluminum 2% Silicon	Al2%Si	640	2.69	—	—	1010	—	TiB_2-BN	—	—	—	RF DC Sputtergun	Wire feed and flash. Difficult from dual sources.
Antimony	Sb	630	6.68	279	345	425	Poor	BN, C Al_2O_3	Mo Ta	Mo, Ta Al_2O_3 Coated	Mo Ta	RF DC Sputtergun	Toxic. Evaporates well.
Antimony Telluride	Sb_2Te_3	619	6.50	—	—	600	—	Carbon	—	—	—	RF	Decomposes over 750°C
Antimony Trioxide	Sb_2O_3	656	5.2 or 5.76	—	—	~300	Good	BN, Al_2O_3	—	Pt	Pt	RF-Reactive	Toxic. Decomposes on W. n = 1.85
Antimony Triselenide	Sb_2Se_3	611	—	—	—	—	—	Carbon	—	Ta	—	RF	Stoichiometry variable.
Antimony Trisulphide	Sb_2S_3	550	4.64	—	—	~200	—	—	—	Mo Ta	Mo Ta	—	n = 3.01. No decomposition.
Arsenic	As	814	5.73	107 Sublimes.	150	210	Poor	Al_2O_3 BeO Vit. Carbon	—	C	—	—	Toxic. Sublimes rapidly at low temperature.
Arsenic Selenide	As_2Se_3	360	4.75	—	—	—	—	Quartz	—	—	—	RF	—
Arsenic Trisulphide	As_2S_3	300	3.43	—	—	~400	Fair	Quartz	—	Mo	—	RF	n = 2.7
Barium	Ba	710	3.5	545	627	735	Fair	Metals	W	W Ta Mo	W	RF	Wets w/o alloying — reacts with ceramics.

Table 2–16 (continued)
THIN-FILM DEPOSITION DATA

NAME	SYMBOL	MELTING POINT °C	DENSITY g/cm³	TEMPERATURE (°C) @ VAP. PRESS.			EVAPORATION TECHNIQUES					SPUTTER*	REMARKS n = Index of refraction
				10^{-8} TORR	10^{-6} TORR	10^{-4} TORR	ELECTRON BEAM	CRUCIBLE	COIL	BOAT	BASKET		
Barium Chloride	$BaCl_2$	962	3.86	—	—	~650	—	—	—	Ta Mo	—	RF	Use gentle preheat to outgas.
Barium Fluoride	BaF_2	1280	4.83	—	—	~700	—	—	—	—	—	RF	n = 1.47
Barium Oxide	BaO	1923	5.72 or 5.32	—	—	~1300	Poor	Al_2O_3	—	Pt	Pt	RF RF-Reactive	Decomposes slightly. n = 1.98
Barium Sulphide	BaS	—	4.25	—	—	1100	—	—	—	Mo	—	RF	n = 2.16
Barium Titanate	$BaTiO_3$	Dec.	6.0	Decomposes			—	—	—	—	—	RF	Decomposes, yields free Ba from single source; sputtering preferred; or co-evaporate from 2 sources
Beryllium	Be	1278	1.85	710	878	1000	Xlnt.	BeO C Vit. Carbon	W	W Ta	W	RF DC Sputtergun	Wets W/Mo/Ta. Metal powder and oxides are toxic. Evaporates easily.
Beryllium Chloride	$BeCl_2$	440	1.90	—	—	~150	—	—	—	—	—	RF	—
Beryllium Fluoride	BeF_2	Subl.	1.99	—	—	~200	Good	—	—	—	—	—	Toxic.
Beryllium Oxide	BeO	2530	3.01	—	—	1900	Good	—	—	—	W	RF RF-Reactive	Powders toxic. No decomposition from EB guns. n = 1.72
Bismuth	Bi	271	9.80	330	410	520	Good	Al_2O_3 Vit. Carbon	W	W, Mo Al_2O_3 Ta	W	DC RF	Vapors are toxic. High resistivity. No shorting of baskets.
Bismuth Fluoride	BiF_3	727	8.75	— Sublimes	—	~300	—	—	—	—	—	RF	—
Bismuth Oxide	Bi_2O_3	820	8.9	—	—	~1400	Poor	—	—	Pt	Pt	RF RF-Reactive	Vapors are toxic. n = 1.97
Bismuth Selenide	Bi_2Se_3	710	7.66	—	—	~650	Good	Graphite Quartz	—	—	—	RF	Sputtering preferred; or co-evaporate from 2 sources.
Bismuth Telluride	Bi_2Te_3	585	7.85	—	—	~600	—	Graphite Quartz	—	W Mo	—	RF	Sputtering preferred; or co-evaporate from 2 sources.
Bismuth Titanate	$Bi_2Ti_2O_7$	—	—	Decomposes			—	—	—	—	—	RF	Decomposes. Sputtering preferred; or co-evaporate from 2 sources in 10^{-2} O_2
Bismuth Trisulphide	Bi_2S_3	685	7.39	—	—	—	—	—	—	—	—	RF	n ≃ 1.5
Boron	B	2100	2.36	1278 Sublimes	1548	1797	Xlnt.	C Vit. Carbon	—	C	—	RF	Material explodes with rapid cooling. Forms carbide with container.
Boron Carbide	B_4C	2350	2.50	2500	2580	2650	Xlnt.	—	—	—	—	RF	Similar to chromium.

*Materials marked RF or RF-reactive may be sputtered with Sloan RF Sputtergun.

Table 2–16 (continued)
THIN-FILM DEPOSITION DATA

NAME	SYMBOL	MELTING POINT °C	DENSITY g/cm³	TEMPERATURE (°C) @ VAP. PRESS.			EVAPORATION TECHNIQUES					SPUTTER*	REMARKS n = Index of refraction
				10-8 TORR	10-6 TORR	10-4 TORR	ELECTRON BEAM	CRUCIBLE	COIL	BOAT	BASKET		
Boron Nitride	BN	2300	2.20	— Sublimes	—	~1600	Poor	—	—	—	—	RF RF-Reactive	Sputtering preferred; Decomposes.
Boron Oxide	B_2O_3	460	1.82	—	—	~1400	Good	—	—	Pt Mo	—	—	n = 1.46
Boron Trisulphide	B_2S_3	310	1.55	—	—	800	—	Graphite	—	—	—	RF	—
Cadmium	Cd	321	8.64	64	120	180	Poor	Al_2O_3 Quartz	—	W, Cb Mo Ta	W Mo Ta	DC RF	Poisons vacuum systems, low sticking coefficient.
Cadmium Arsenide	Cd_3As_2	721	6.21	—	—	—	—	Quartz	—	—	—	RF	—
Cadmium Fluoride	CdF_2	1070	6.64	—	—	~500	—	—	—	—	—	RF	n = 1.56
Cadmium Oxide	CdO	900	6.95	—	—	~530	—	—	—	—	—	RF-Reactive	Disproportionates. n = 2.49
Cadmium Selenide	CdSe	1264	5.81	—	—	540	Good	Al_2O_3 Quartz	—	Mo Ta	—	RF	Evaporates easily. n = 2.4
Cadmium Sulphide	CdS	1750	4.82	— Sublimes	—	550	Fair	Al_2O_3 Quartz	—	W Mo Ta	W	RF	Sticking co-efficient strongly affected by substrate temperature. Stoichiometry variable. n = 2.4
Cadmium Telluride	CdTe	1098	6.20	—	—	450	—	—	W	W Mo Ta	Ta Mo	RF	Stoichiometry depends on substrate temperature n ≅ 2.6
Calcium	Ca	842	1.55	272 Sublimes	357	459	Poor	Al_2O_3 Quartz	W	W	W	—	Corrodes in air.
Calcium Fluoride	CaF_2	1360	3.18	—	—	~1100	—	Quartz	W Mo Ta	W Mo Ta	W Mo Ta	RF	Rate control important. Use gentle preheat to outgas. n = 1.23
Calcium Oxide	CaO	2580	3.35	—	—	~1700	—	ZrO_2	—	W Mo	—	RF RF-Reactive	Forms volatile oxides with W and Mo. n = 1.84
Calcium Silicate	$CaO\text{-}SiO_2$	1540	2.90	—	—	—	Good	Quartz	—	—	—	RF	n = 1.61
Calcium Sulphide	CaS	—	2.18	—	—	1100	—	—	—	Mo	—	RF	Decomposes. n = 2.14
Calcium Titanate	$CaTiO_3$	1975	4.10	1490	1600	1690	Poor	—	—	—	—	RF	Disproportionates except in sputtering.
Calcium Tungstate	$CaWO_4$	—	6.06	—	—	—	Good	—	—	W	—	RF	n = 1.92
Carbon	C	Subl.	1.8 – 2.3	1657 Sublimes	1867	2137	Xlnt.	—	—	—	—	RF	EB preferred. Arc evaporation. Poor film adhesion.
Cerium	Ce	795	8.23	970	1150	1380	Good	Al_2O_3 BeO Vit. Carbon	W	W Ta	W	DC RF	—

*Materials marked RF or RF-reactive may be sputtered with Sloan RF Sputtergun.

Table 2–16 (continued)
THIN-FILM DEPOSITION DATA

NAME	SYMBOL	MELTING POINT °C	DENSITY g/cm³	TEMPERATURE (°C) @ VAP. PRESS.			EVAPORATION TECHNIQUES					SPUTTER*	REMARKS n = Index of refraction
				10^{-8} TORR	10^{-6} TORR	10^{-4} TORR	ELECTRON BEAM	CRUCIBLE	COIL	BOAT	BASKET		
Ceric Oxide	CeO_2	2600	7.3	1890 Sublimes	2000	2310	Good	—	—	W	—	RF RF-Reactive	Very little decomposition.
Cerium Fluoride	CeF_3	1418	6.16	—	—	~900	Good	—	—	W Mo Ta	Mo Ta	RF	Use gentle preheat to outgas. n ≅ 1.7
Cerium Oxide	Ce_2O_3	1692	6.87	—	—	—	Fair	—	—	W	—	—	Alloys with source; use .015-.020 W boat. n = 1.95
Cesium	Cs	28	1.87	−16	+22	+30	—	Quartz	—	S.S.	—	—	—
Cesium Bromide	CsBr	636	4.44	—	—	~400	—	—	—	W	—	RF	n = 1.70
Cesium Chloride	CsCl	646	3.97	—	—	600	—	—	—	W	—	RF	n = 1.64
Cesium Fluoride	CsF	684	3.59	—	—	~500	—	—	—	W	—	RF	—
Cesium Hydroxide	CsOH	272	3.67	—	—	550	—	—	—	Pt	—	—	—
Cesium Iodide	CsI	621	4.51	—	—	~500	—	—	—	W	—	RF	n = 1.79
Chiolote	$Na_5Al_3F_{14}$	—	2.9	—	—	~800	—	—	—	Mo W	—	RF	n = 1.33
Chromium	Cr	1890	7.20	837 Sublimes	977	1157	Good	Vit. Carbon	W	Cr-plated rod or strip	W	RF DC	Films very adherent. High rates possible.
Chromium Boride	CrB	2760	6.17	—	—	—	—	—	—	—	—	RF DC Sputtergun	—
Chromium Bromide	$CrBr_2$	842	4.36	—	—	550	—	—	—	Inconel	—	RF	—
Chromium Carbide	Cr_3C_2	1890	6.68	—	—	~2000	Fair	—	—	W	—	RF DC Sputtergun	—
Chromium Chloride	$CrCl_2$	824	2.75	—	—	550	—	—	—	Fe Inconel	—	RF	Sublimes easily.
Chromium Oxide	Cr_2O_3	2435	5.21	—	—	~2000	Good	—	—	W Mo	W	RF RF-Reactive	Disproportionates to lower oxides, reoxidizes @ 600°C in air. n = 2.4
Chromium Silicide	Cr_3Si_2	—	5.5	—	—	—	—	—	—	—	—	RF DC Sputtergun	—
Chromium-Silicon Monoxide	Cr-SiO	Influenced by composition					Good	—	—	W	W	DC RF	Flash.
Cobalt	Co	1495	8.90	850	990	1200	Xlnt.	Al_2O_3 BeO	—	W Cb	W	DC RF	Alloys with refractory metals.

*Materials marked RF or RF-reactive may be sputtered with Sloan RF Sputtergun.

Table 2–16 (continued)
THIN-FILM DEPOSITION DATA

NAME	SYMBOL	MELTING POINT °C	DENSITY g/cm³	TEMPERATURE (°C) @ VAP. PRESS.			EVAPORATION TECHNIQUES					SPUTTER*	REMARKS n = Index of refraction
				10^{-8} TORR	10^{-6} TORR	10^{-4} TORR	ELECTRON BEAM	CRUCIBLE	COIL	BOAT	BASKET		
Cobalt Bromide	$CoBr_2$	678	4.91	— Sublimes	—	400	—	—	—	Inconel	—	RF	—
Cobalt Chloride	$CoCl_2$	740	3.36	— Sublimes	—	472	—	—	—	Inconel	—	RF	—
Cobalt Oxide	CoO	1935	6.4	—	—	—	—	—	—	—	—	DC-Reactive RF-Reactive	Sputtering preferred.
Copper	Cu	1083	8.92	727	857	1017	Xlnt.	Al_2O_3 TiB_2BN	W	Mo	W	DC RF	Films do not adhere well. Use intermediate layer, e.g., chromium. Evaporates from any source material.
Copper Chloride	CuCl	422	3.53	—	—	~600	—	—	—	—	—	RF	n = 1.93
Copper Oxide	Cu_2O	1235	6.0	— Sublimes	—	~600	Good	Al_2O_3	—	Ta	—	DC-Reactive RF-Reactive	n = 2.70
Cryolite	Na_3AlF_6	1000	2.9	1020	1260	1480	Xlnt.	Vit. Carbon	—	W Mo Ta	W Mo Ta	RF	Large chunks reduce spitting. Little decomposition.
Dysprosium	Dy	1409	8.54	625 Sublimes	750	900	Good	—	—	Ta	—	RF DC Sputtergun	—
Dysprosium Fluoride	DyF_3	1360	—	— Sublimes	—	~800	Good	—	—	Ta	—	RF	—
Dysprosium Oxide	Dy_2O_3	2340	8.16	—	—	~1400	—	—	—	Ir	—	RF RF-Reactive	Loses oxygen.
Erbium	Er	1497	9.06	650 Sublimes	775	930	Good	—	—	W Ta	—	DC RF	—
Erbium Fluoride	ErF_3	1350	—	—	—	~750	—	—	—	—	—	RF	—
Erbium Oxide	Er_2O_3	2400	8.64	—	—	~1600	—	—	—	Ir	—	RF RF-Reactive	Loses oxygen.
Europium	Eu	822	5.26	280	360	480	Fair	Al_2O_3	—	W Ta	—	RF DC	Low tantalum solubility.
Europium Fluoride	EuF_2	1380	6.5	—	—	~950	—	—	—	Mo	—	RF	—
Europium Oxide	Eu_2O_3	2056	8.18	—	—	~1600	Good	ThO_2	—	Ir Ta W	—	RF RF-Reactive	Loses oxygen; films clear and hard.
Europium Sulphide	EuS	—	—	—	—	—	Good	—	—	—	—	RF	—
Gadolinium	Gd	1312	7.89	760	900	1175	Xlnt.	Al_2O_3	—	Ta	—	RF DC Sputtergun	High Ta solubility.
Gadolinium Carbide	GdC_2	—	—	—	—	1500	—	Carbon	—	—	—	RF	Decomposes.

*Materials marked RF or RF-reactive may be sputtered with Sloan RF Sputtergun.

Table 2–16 (continued)
THIN-FILM DEPOSITION DATA

NAME	SYMBOL	MELTING POINT °C	DENSITY g/cm³	TEMPERATURE (°C) @ VAP. PRESS.			EVAPORATION TECHNIQUES					SPUTTER*	REMARKS n = Index of refraction
				10^{-8} TORR	10^{-6} TORR	10^{-4} TORR	ELECTRON BEAM	CRUCIBLE	COIL	BOAT	BASKET		
Gadolinium •Oxide	Gd_2O_3	2310	7.41	—	—	—	Fair	—	—	Ir	—	RF RF-Reactive	Loses oxygen.
Gallium	Ga	30	5.90	619	742	907	Good	Al_2O_3 BeO Quartz	—	—	—	—	Alloys with refractory metals. Use EB gun.
Gallium Antimonide	GaSb	710	5.6	—	—	—	Fair	—	—	W Ta	—	RF	Flash evaporate.
Gallium Arsenide	GaAs	1238	5.3	—	—	—	Good	Carbon	—	W Ta	—	RF	Flash evaporate.
Gallium Nitride	GaN	Subl.	6.1	—	—	~200	—	Al_2O_3	—	—	—	RF RF-Reactive	Evaporate Ga in 10^{-3} N_2.
Gallium Oxide (β)	Ga_2O_3	1900	5.88	—	—	—	—	—	—	Pr W	—	RF	Loses oxygen.
Gallium Phosphide	GaP	1540	4.1	—	770	920	Xlnt.	Quartz	—	W Ta	W	RF	Does not decompose. Rate control important.
Germanium	Ge	937	5.35	812	957	1167	Xlnt.	Quartz Al_2O_3	—	W C Ta	—	DC RF	Excellent films from EB guns.
Germanium Monoxide	GeO	—	—	—	—	550	—	Quartz	—	—	—	RF	—
Germanium Nitride	Ge_3N_2	—	5.2	— Sublimes	—	~650	—	—	—	—	—	RF-Reactive	Sputtering preferred.
Germanium Oxide	GeO_2	1086	6.24	—	—	~625	Good	Quartz Al_2O_3	—	Ta Mo	W Mo	RF-Reactive	Similar to SiO, film predominantly GeO.
Germanium Telluride	GeTe	725	6.20	—	—	381	—	Quartz Al_2O_3	—	W Mo	W	RF	—
Glass, Schott 8329	—	—	2.20	—	—	—	Xlnt.	—	—	—	—	RF	Evaporable alkali glass. Melt in air before evaporating.
Gold	Au	1062	19.32	807	947	1132	Xlnt.	Al_2O_3 BN Vit. Carbon	W	W, Mo Coated Al_2O_3	W	DC RF	Films soft, not very adherent.
Hafnium	Hf	2230	13.09	2160	2250	3090	Good	—	—	—	—	DC RF	—
Hafnium Boride	HfB_2	3250	10.5	—	—	—	—	—	—	—	—	DC RF	—
Hafnium Carbide	HfC	4160	12.2	— Sublimes	—	~2600	—	—	—	—	—	DC RF	—
Hafnium Nitride	HfN	2852	13.8	—	—	—	—	—	—	—	—	RF RF-Reactive	—
Hafnium Oxide	HfO_2	2812	9.68	—	—	~2500	Fair	—	—	W	—	DC RF RF-Reactive	Film HfO

*Materials marked RF or RF-reactive may be sputtered with Sloan RF Sputtergun.

Table 2–16 (continued)
THIN-FILM DEPOSITION DATA

NAME	SYMBOL	MELTING POINT °C	DENSITY g/cm³	TEMPERATURE (°C) @ VAP. PRESS. 10-8 TORR	10-6 TORR	10-4 TORR	ELECTRON BEAM	EVAPORATION TECHNIQUES CRUCIBLE	COIL	BOAT	BASKET	SPUTTER*	REMARKS n = Index of refraction
Hafnium Silicide	$HfSi_2$	1750	7.2	—	—	—	—	—	—	—	—	RF	—
Holmium	Ho	1470	8.80	650 Sublimes	770	950	Good	—	W	W Ta	W	—	—
Holmium Fluoride	HoF_3	1143	—	—	—	~800	—	Quartz	—	—	—	DC RF	—
Holmium Oxide	Ho_2O_3	2370	8.41	—	—	—	—	—	—	Ir	—	RF RF-Reactive	Loses oxygen.
Inconel	Ni/Cr/Fe	1425	8.5	—	—	—	Good	—	W	W	W	DC RF	Use fine wire pre-wrapped on W. Low rate req'd for smooth films.
Indium	In	157	7.30	487	597	742	Xlnt.; Mo Liner req'd	Graphite Al_2O_3	—	W Mo	W	DC RF	Wets W and Cu; use Mo liner in guns.
Indium Antimonide	InSb	535	5.8	500	—	~400	—	—	—	W	—	RF	Decomposes; sputtering preferred; or co-evaporate from 2 sources; flash.
Indium Arsenide	InAs	943	5.7	780	870	970	—	—	—	W	—	RF	Sputtering preferred; or co-evaporate from 2 sources; flash.
Indium Nitride	InN	1200	7.0	—	—	—	—	—	—	—	—	RF	—
Indium Oxide	In_2O	—	6.99	— Sublimes	—	650	—	—	—	—	—	RF	Decomposes.
Indium Oxide	In_2O_3	1565	7.18	—	—	~600	Poor	Al_2O_3	—	W Pt	W	RF RF-Reactive	Film In_2O
Indium Phosphide	InP	1058	4.8	—	630	730	—	Graphite	—	W Ta	W Ta	RF	Deposits P rich. Flash evaporate.
Indium Selenide	In_2Se_3	890	5.7	—	—	—	—	—	—	—	—	RF	Sputtering preferred; or co-evaporate from 2 sources; flash.
Indium Sesquisulphide	In_2S_3	1050	4.90	— Sublimes	—	850	—	Graphite	—	—	—	RF	Film In_2S
Indium Sulphide	In_2S	653	5.87	—	—	650	—	Graphite	—	—	—	RF	—
Indium Telluride	In_2Te_3	667	5.8	—	—	—	—	—	—	—	—	RF	Sputtering preferred; or co-evaporate from 2 sources; flash.
Iridium	Ir	2459	22.4	1850	2080	2380	Fair	ThO_2	—	—	—	DC RF	—
Iron	Fe	1535	7.86	858	998	1180	Xlnt.	Al_2O_3 BeO	W	W	W	DC RF	Attacks W. Films hard, smooth. Use gentle preheat to outgas.
Iron Bromide	$FeBr_2$	689	4.64	—	—	561	—	Fe	—	—	—	RF	—
Iron Chloride	$FeCl_2$	670	2.98	— Sublimes	—	300	—	Fe	—	—	—	RF	—

*Materials marked RF or RF-reactive may be sputtered with Sloan RF Sputtergun.

Table 2–16 (continued)
THIN-FILM DEPOSITION DATA

NAME	SYMBOL	MELTING POINT °C	DENSITY g/cm3	TEMPERATURE (°C) @ VAP. PRESS. 10-8 TORR	10-6 TORR	10-4 TORR	EVAPORATION TECHNIQUES ELECTRON BEAM	CRUCIBLE	COIL	BOAT	BASKET	SPUTTER*	REMARKS n = Index of refraction
Iron Iodide	FeI_2	592	5.31	—	—	400	—	Fe	—	—	—	RF	—
Iron Oxide	FeO	1425	5.7	—	—	—	Poor	—	—	—	—	RF RF-Reactive	Decomposes; sputtering preferred.
Iron Oxide	Fe_2O_3	1565	5.24	—	—	—	Good	—	—	W	W	—	Disproportionates to Fe_3O_4 at 1530°C, $n \cong 3.0$
Iron Sulphide	FeS	1195	4.84	—	—	—	—	Al_2O_3	—	—	—	RF	Decomposes.
Kanthal	FeCrAl	—	7.1	—	—	—	—	—	W	W	—	DC RF	—
Lanthanum	La	920	6.17	990	1212	1388	Xlnt.	Al_2O_3	—	W Ta	—	RF	Films will burn in air if scraped.
Lanthanum Boride	LaB_6	2210	2.61	—	—	—	Good	—	—	—	—	RF	—
Lanthanum Fluoride	LaF_3	1490	~6	—	—	900	Good	—	—	Ta Mo	Ta	RF	No decomposition. $n \cong 1.6$
Lanthanum Oxide	La_2O_3	2250	5.84	—	—	1400	Good	—	—	W Ta	—	RF	Loses oxygen. $n \cong 1.73$
Lead	Pb	328	11.34	342	427	497	Xlnt.	Al_2O_3 Quartz	W	W Mo	W Ta	DC RF	Toxic. Carefully controlled rates req'd for superconductors.
Lead Chloride	$PbCl_2$	678	5.85	—	—	~325	—	Al_2O_3	—	Pt	—	RF	Little decomposition. $n = 2.2$
Lead Fluoride	PbF_2	822	8.24	—	—	~400	—	Al_2O_3	—	W Pt, Mo	—	RF	$n = 1.75$
Lead Oxide	PbO	890	9.53	—	—	~550	—	Quartz Al_2O_3	—	Pt	—	RF-Reactive	No decomposition. $n \cong 2.6$
Lead Stannate	$PbSnO_3$	1115	8.1	670	780	905	Poor	Al_2O_3	—	Pt	Pt	RF	Disproportionates.
Lead Selenide	PbSe	1065	8.10	— Sublimes	—	~500	—	Graphite Al_2O_3	—	W Mo	W	RF	—
Lead Sulphide	PbS	1114	7.5	— Sublimes	—	550	—	Quartz Al_2O_3	—	W	W Mo	RF	Little decomposition. $n = 3.91$
Lead Telluride	PbTe	917	8.16	780	910	1050	—	Al_2O_3 Graphite	—	Pt	—	RF	Vapors toxic. Deposits Te rich. Sputtering preferred, or co-evaporate from 2 sources.
Lead Titanate	$PbTiO_3$	—	7.52	—	—	—	—	—	—	Ta	—	RF	—
Lithium	Li	179	0.53	227	307	407	Good	Al_2O_3 BeO	—	Ta S.S.	—	—	Metal reacts violently in air.

*Materials marked RF or RF-reactive may be sputtered with Sloan RF Sputtergun.

Table 2–16 (continued)
THIN-FILM DEPOSITION DATA

NAME	SYMBOL	MELTING POINT °C	DENSITY g/cm^3	TEMPERATURE (°C) @ VAP. PRESS.			EVAPORATION TECHNIQUES					SPUTTER*	REMARKS n = Index of refraction
				10^{-8} TORR	10^{-6} TORR	10^{-4} TORR	ELECTRON BEAM	CRUCIBLE	COIL	BOAT	BASKET		
Lithium Bromide	LiBr	547	3.46	—	—	~500	—	—	—	Ni	—	RF	n = 1.78
Lithium Chloride	LiCl	613	2.07	—	—	400	—	—	—	Ni	—	RF	Use gentle preheat for outgas. n = 1.66
Lithium Fluoride	LiF	870	2.60	875	1020	1180	Good	—	—	Ni, Ta Mo W	—	RF	Rate control important for optical films. Use gentle preheat for outgas. n = 1.33
Lithium Iodide	LiI	446	4.06	—	—	400	—	—	—	Mo W	—	RF	—
Lithium Oxide	Li_2O	>1700	2.01	—	—	850	—	—	—	Pt Ir	—	RF	n = 1.64
Lutetium	Lu	1652	9.84	— Sublimes	—	1300	Xlnt.	Al_2O_3	—	Ta	—	RF DC Sputtergun	—
Lutetium Oxide	Lu_2O_3	—	9.41	—	—	1400	—	—	—	Ir	—	RF	Decomposes.
Magnesium	Mg	651	1.74	185 Sublimes	247	327	Good	Al_2O_3 Vit. Carbon	W	W Mo Ta, Cb	W	DC RF	Extremely high rates possible.
Magnesium Aluminate	$MgAl_2O_4$	2135	3.6	—	—	—	Good	—	—	—	—	RF	Natural spinel.
Magnesium Bromide	$MgBr_2$	195	3.72	—	—	250	—	—	—	Ni	—	RF	Decomposes.
Magnesium Chloride	$MgCl_2$	116	2.32	—	—	400	—	—	—	Ni	—	RF	Decomposes. n = 1.6
Magnesium Fluoride	MgF_2	1266	2.9-3.2	—	—	925	Xlnt.	Al_2O_3	—	—	—	RF	Rate control and substrate heat important for optical films. Reacts with W. Excellent with Mo. n = 1.38
Magnesium Iodide	MgI_2	>700	4.24	—	—	200	—	—	—	Pr	—	RF	—
Magnesium Oxide	MgO	2800	3.58	—	—	1300	Good	Carbon Al_2O_3	—	—		RF RF-Reactive	Evaporate in 10^{-3} of O_2 to retain stoichiometry. W produces volatile oxides. n ≅ 1.7
Manganese	Mn	1244	7.20	507 Sublimes	572	647	Good	Al_2O_3 BeO	W	W Ta Mo	W	DC RF	—
Manganese Bromide	$MnBr_2$	695	4.38	—	—	500	—	—	—	Inconel	—	RF	—
Manganese Chloride	$MnCl_2$	650	2.98	—	—	450	—	—	—	Inconel	—	RF	—
Manganese Dioxide	MnO_2	—	5.02	—	—	—	Poor	—	—	W	W	RF-Reactive	Loses O_2 @ 535°C.
Manganese Sulphide	MnS	—	3.99	—	—	1300	—	—	—	Mo	—	RF	Decomposes. n = 2.7

*Materials marked RF or RF-reactive may be sputtered with Sloan RF Sputtergun.

Table 2–16 (continued)
THIN-FILM DEPOSITION DATA

NAME	SYMBOL	MELTING POINT °C	DENSITY g/cm³	TEMPERATURE (°C) @ VAP. PRESS.			EVAPORATION TECHNIQUES					SPUTTER*	REMARKS n = Index of refraction
				10-8 TORR	10-6 TORR	10-4 TORR	ELECTRON BEAM	CRUCIBLE	COIL	BOAT	BASKET		
Mercury	Hg	–39	13.55	–68	–42	–6	—	—	—	—	—	—	—
Mercury Sulphide	HgS	Subl.	8.10	— Sublimes	—	250	—	Al_2O_3	—	—	—	RF	Decomposes.
Molybdenum	Mo	2610	10.22	1592	1822	2117	Xlnt.	—	—	—	—	DC RF	Films smooth, hard. Careful degas req'd.
Molybdenum Boride	MoB_2	2100	7.12	—	—	—	Poor	—	—	—	—	RF DC Sputtergun	—
Molybdenum Carbide	Mo_2C	2687	9.18	—	—	—	Fair	—	—	—	—	RF DC Sputtergun	Evaporation of $Mo(CO)_6$ yields Mo_2C.
Molybdenum Disulphide	MoS_2	1185	4.80	—	—	~50	—	—	—	—	—	RF	—
Molybdenum Silicide	$MoSi_2$	—	6.3	—	—	—	—	—	—	W	—	RF	Decomposes.
Molybdenum Trioxide	MoO_3	795	4.70	—	—	~900	—	Al_2O_3 BN	—	Mo Pt	Mo	RF	Slight O_2 loss. n ≅ 1.9
Neodymium	Nd	1024	7.00	731	871	1062	Xlnt.	Al_2O_3	—	Ta	—	DC RF	Low Ta solubility.
Neodymium Fluoride	NdF_3	1410	6.5	—	—	~900	Good	Al_2O_3	—	Mo W	Mo Ta	RF	Very little decomposition. n ≅ 1.6
Neodymium Oxide	Nd_2O_3	2272	7.24	—	—	~1400	Good	ThO_2	—	Ta W	—	RF RF-Reactive	Loses oxygen, films clear, EB preferred. n = 1.79
Nichrome IV	Ni/Cr	1395	8.50	847	987	1217	Xlnt.	Al_2O_3 Vit. Carbon BeO	W	Al_2O_3 Coated	W Ta	DC RF	Alloys with refractory metals.
Nickel	Ni	1453	8.90	927	1072	1262	Xlnt.	Al_2O_3 BeO Vit. Carbon	W	W	W	DC RF	Alloys with refractory metals. Forms smooth adherent films.
Nickel Bromide	$NiBr_2$	963	4.64	— Sublimes	—	362	—	—	—	Inconel	—	RF	—
Nickel Chloride	$NiCl_2$	1001	3.55	— Sublimes	—	444	—	—	—	Inconel	—	RF	—
Nickel Oxide	NiO	1990	7.45	—	—	~1470	—	Al_2O_3	—	—	—	RF-Reactive	Dissociates upon heating. n = 2.18
Niobium (Columbium)	Nb	2468	8.55	1728	1977	2287	Xlnt.	—	—	W	—	DC RF	Attacks W source.
Niobium Boride	NbB_2	3050	6.97	—	—	—	—	—	—	—	—	RF DC Sputtergun	—
Niobium Carbide	NbC	3800	7.82	—	—	—	Fair	—	—	—	—	RF DC Sputtergun	—

*Materials marked RF or RF-reactive may be sputtered with Sloan RF Sputtergun.

Table 2–16 (continued)
THIN-FILM DEPOSITION DATA

NAME	SYMBOL	MELTING POINT °C	DENSITY g/cm³	TEMPERATURE (°C) @ VAP. PRESS. 10-8 TORR	10-6 TORR	10-4 TORR	EVAPORATION TECHNIQUES ELECTRON BEAM	CRUCIBLE	COIL	BOAT	BASKET	SPUTTER*	REMARKS n = Index of refraction
Niobium Nitride	NbN	2573	8.4	—	—	—	—	—	—	—	—	RF RF-Reactive	Reactive, evaporate Nb in 10^{-3} N_2.
Niobium Oxide	NbO	—	6.27	—	—	1100	—	—	—	Pt	—	RF	—
Niobium Pentoxide	Nb_2O_5	1530	4.47	—	—	—	—	—	—	W	W	RF RF-Reactive	n ≅ 1.95
Niobium Telluride	$NbTe_x$	—	7.6	—	—	—	—	—	—	—	—	RF	Composition variable.
Niobium-Tin	Nb_3Sn	—	—	—	—	—	Xlnt.	—	—	—	—	RF DC Sputtergun	Co-evaporate from 2 sources.
Niobium Trioxide	Nb_2O_3	1780	7.5	—	—	—	—	—	—	W	W	RF	—
Osmium	Os	1700	22.5	2170	2430	2760	Fair	—	—	—	—	DC RF	—
Osmium Oxide	Os_2O_3	—	—	—	—	—	—	—	—	—	—	RF	Decomposes. Deposit Os in 10^{-3} O_2.
Palladium	Pd	1550	12.40	842 Sublimes	992	1192	Xlnt.	Al_2O_3 BeO	W	W	W	DC RF	Alloys with refractory metals; rapid evaporation suggested.
Palladium Oxide	PdO	870	8.31	—	—	575	—	Al_2O_3	—	—	—	RF-Reactive	Decomposes.
Parylene (Union Carbide)	C_8H_8	300—400	1.1	—	—	—	—	—	—	—	—	—	Vapor depositable plastic.
Permalloy	Ni/Fe	1395	8.7	947	1047	1307	Good	Al_2O_3 Vit. Carbon	—	W	—	DC RF	Film low in Ni content. Use 84% Ni source. JVST Vol. 7, No. 6, p. 573
Phosphorus	P	41.4	1.82	327	361	402	—	Al_2O_3	—	—	—	—	Metal reacts violently in air.
Phosphorus Nitride	P_3N_5	—	2.51	—	—	—	—	—	—	—	—	RF RF-Reactive	—
Platinum	Pt	1769	21.45	1292	1492	1747	Xlnt.	C ThO_2	W Pt	W	—	DC RF	Alloys with metals. Films soft, poor adhesion.
Platinum Oxide	PtO_2	450	10.2	—	—	—	—	—	—	—	—	RF-Reactive	—
Plutonium	Pu	635	—	—	—	—	—	—	—	W	—	DC RF	Toxic, radioactive.
Polonium	Po	254	9.4	117	170	244	—	Quartz	—	—	—	—	Radioactive
Potassium	K	64	0.86	23	60	125	—	Quartz	—	Mo	—	—	Metal reacts violently in air. Use gentle preheat to outgas.

*Materials marked RF or RF-reactive may be sputtered with Sloan RF Sputtergun.

Table 2–16 (continued)
THIN-FILM DEPOSITION DATA

NAME	SYMBOL	MELTING POINT °C	DENSITY g/cm³	TEMPERATURE (°C) @ VAP. PRESS.			EVAPORATION TECHNIQUES					SPUTTER*	REMARKS n = Index of refraction
				10^{-8} TORR	10^{-6} TORR	10^{-4} TORR	ELECTRON BEAM	CRUCIBLE	COIL	BOAT	BASKET		
Potassium Bromide	KBr	730	2.75	—	—	~450	—	Quartz	—	Ta Mo	—	RF	Use gentle preheat to outgas. n = 1.56
Potassium Chloride	KCl	776	1.98	—	—	510	—	—	—	Ta Ni	—	RF	Use gentle preheat to outgas. n = 1.49
Potassium Fluoride	KF	880	2.48	—	—	~450	—	Quartz	—	—	—	RF	Use gentle preheat to outgas. n = 1.35
Potassium Hydroxide	KOH	360	2.04	—	—	~400	—	—	—	Pt	—	—	Use gentle preheat to outgas.
Potassium Iodide	KI	723	3.13	—	—	~500	—	—	—	Ta	—	RF	Use gentle preheat to outgas. n = 1.68
Praseodymium	Pr	931	6.78	800	950	1150	—	—	—	Ta	—	RF DC Sputtergun	—
Praseodymium Oxide	Pr_2O_3	2125	6.88	—	—	1400	Good	ThO_2	—	Ir	—	RF RF-Reactive	Loses oxygen.
Radium	Ra	700	5.0	246	320	416	—	—	—	—	—	—	—
Rhenium	Re	3180	20.53	1928	2207	2571	Xlnt.	—	—	—	—	DC RF	Fine wire will self-evaporate.
Rhenium Oxide	ReO_3	—	~7	—	—	—	—	—	—	—	—	RF	Evaporate Re in 10^{-3} O_2.
Rhodium	Rh	1966	12.4	1277	1472	1707	Good	ThO_2 Vit. Carbon	W	W	W	DC RF	EB gun preferred.
Rubidium	Rb	38.5	1.47	−3	37	111	—	Quartz	—	—	—	DC RF	—
Rubidium Chloride	RbCl	715	2.76	—	—	~550	—	Quartz	—	—	—	RF	n = 1.49
Rubidium Iodide	RbI	642	3.55	—	—	~400	—	Quartz	—	—	—	RF	—
Ruthenium	Ru	2700	12.6	1780	1990	2260	Fair	—	—	W	—	DC RF	—
Samarium	Sm	1072	7.54	373	460	573	Good	Al_2O_3	—	Ta	—	RF DC Sputtergun	—
Samarium Oxide	Sm_2O_3	2350	7.43	—	—	—	Good	ThO_2	—	Ir	—	RF RF-Reactive	Loses O_2. Films smooth, clear.
Scandium	Sc	1539	2.99	714 Sublimes	837	1002	Xlnt.	Al_2O_3 BeO	—	W	—	RF	Alloys with Ta
Scandium Oxide	Sc_2O_3	2300	3.86	—	—	—	—	—	—	—	—	RF RF-Reactive	—
Selenium	Se	217	4.2-4.8	89	125	170	Good	Al_2O_3 Vit. Carbon	W Mo	W Mo	W Mo	RF DC Sputtergun	Toxic. Poisons vacuum systems.

*Materials marked RF or RF-reactive may be sputtered with Sloan RF Sputtergun.

Table 2–16 (continued)
THIN-FILM DEPOSITION DATA

NAME	SYMBOL	MELTING POINT °C	DENSITY g/cm³	TEMPERATURE (°C) @ VAP. PRESS. 10-8 TORR	10-6 TORR	10-4 TORR	EVAPORATION TECHNIQUES ELECTRON BEAM	CRUCIBLE	COIL	BOAT	BASKET	SPUTTER*	REMARKS n = Index of refraction
Silicon	Si	1410	2.42	992	1147	1337	Xlnt.	BeO Ta Vit. Carbon	—	W Ta	—	DC RF	Alloys with W; use heavy W boat. SiO produced above 4 x 10^{-6} Torr. EB best.
Silicon Boride	SiB_6	—	—	—	—	—	Poor	—	—	—	—	RF	—
Silicon Carbide	SiC	2700	3.22	—	—	1000	—	—	—	—	—	RF	Sputtering preferred.
Silicon Dioxide	SiO_2	1610—1710	2.2-2.7	— influenced by composition	—	~1025	Xlnt.	Al_2O_3	—	—	—	RF	Quartz xlnt in EB. n = 1.47
Silicon Monoxide	SiO	1702	2.1	— Sublimes	—	850	Xlnt.	Ta	W	Ta	W	RF RF-Reactive	Baffle box source best for resistance evaporation. Low rate suggested. n = 1.6
Silicon Nitride	Si_3N_4	Subl.	3.44	—	—	~800	—	—	—	—	—	RF RF-Reactive	—
Silicon Selenide	SiSe	—	—	—	—	550	—	Quartz	—	—	—	RF	—
Silicon Sulphide	SiS	Subl.	1.85	—	—	450	—	Quartz	—	—	—	RF	—
Silicon Telluride	$SiTe_2$	—	4.39	—	—	550	—	Quartz	—	—	—	RF	—
Silver	Ag	961	10.49	847	958	1105	Xlnt.	Al_2O_3 Mo	W	Ta Mo	W	DC RF	Evaporates well from any source.
Silver Bromide	AgBr	432	6.47	—	—	~380	—	Quartz	—	Ta	—	RF	n = 2.25
Silver Chloride	AgCl	455	5.56	—	—	~520	—	Quartz	—	Mo Pt	Mo	RF	n = 2.07
Silver Iodide	AgI	558	5.67	—	—	~900	—	—	—	Ta	—	RF	n = 2.21
Sodium	Na	97	0.97	74	124	192	—	Quartz	—	Ta S.S.	—	—	Use gentle preheat to outgas. Metal reacts violently in air.
Sodium Bromide	NaBr	755	3.20	—	—	~400	—	Quartz	—	—	—	RF	Use gentle preheat to outgas. n = 1.64
Sodium Chloride	NaCl	801	2.16	—	—	530	Poor	Quartz	—	Ta W Mo	—	RF	Cu ovens, little decomposition. Use gentle preheat to outgas. n = 1.54
Sodium Cyanide	NaCN	563	—	—	—	~550	—	—	—	Ag	—	RF	Use gentle preheat to outgas. n = 1.45
Sodium Fluoride	NaF	988	2.79	945	1080	1200	Poor	BeO	—	Mo Ta W	—	RF	Use gentle preheat to outgas. No decomposition. n = 1.30
Sodium Hydroxide	NaOH	318	2.13	—	—	~470	—	—	—	Pt	—	—	Use gentle preheat to outgas. n = 1.36

*Materials marked RF or RF-reactive may be sputtered with Sloan RF Sputtergun.

Table 2–16 (continued)
THIN-FILM DEPOSITION DATA

NAME	SYMBOL	MELTING POINT °C	DENSITY g/cm³	TEMPERATURE (°C) @ VAP. PRESS.			EVAPORATION TECHNIQUES					SPUTTER*	REMARKS n = Index of refraction
				10^{-8} TORR	10^{-6} TORR	10^{-4} TORR	ELECTRON BEAM	CRUCIBLE	COIL	BOAT	BASKET		
Spinel	MgO_3 $5Al_2O_3$	—	8.0	—	—	—	Good	—	—	—	—	RF	n = 1.72
Strontium	Sr	769	2.6	239	309	403	Good	Vit. Carbon	W	W Ta Mo	W	RF DC Sputtergun	Wets but does not alloy with refractory metals. May react violently in air.
Strontium Fluoride	SrF_2	1450	4.24	—	—	~1000	—	Al_2O_3	—	—	—	RF	n = 1.44
Strontium Oxide	SrO	2460	4.7	— Sublimes	—	1500	—	Al_2O_3	—	Mo	—	RF	Reacts with Mo and W; n = 1.87
Strontium Sulphide	SrS	—	3.70	—	—	—	—	—	—	Mo	—	RF	Decomposes. n = 2.11
Sulphur	S_8	115	2.0	13	19	57	—	Quartz	—	W	W	—	Poisons vacuum system.
Supermalloy	Ni/Fe/Mo	1410	8.9	—	—	—	Good	—	—	—	—	RF DC	Sputtering preferred; or co-evaporate from 2 sources, Permalloy and Mo.
Tantalum	Ta	2996	16.6	1960	2240	2590	Xlnt.	—	—	—	—	DC RF	Forms good films.
Tantalum Boride	TaB_2	3000	12.38	—	—	—	—	—	—	—	—	RF DC Sputtergun	—
Tantalum Carbide	TaC	3880	14.65	—	—	~2500	—	—	—	—	—	RF DC Sputtergun	—
Tantalum Nitride	TaN	3360	16.30	—	—	—	—	—	—	—	—	RF RF-Reactive DC Sputtergun	Reactive; evaporate Ta in 10^{-3} N_2.
Tantalum Pentoxide	Ta_2O_5	1800	8.74	1550	1780	1920	Good	Vit. Carbon	W	Ta	W	RF RF-Reactive	Slight decomposition; evaporate in 10^{-3} Torr of O_2. n ≅ 2.6
Tantalum Sulphide	TaS_2	1300	—	—	—	—	—	—	—	—	—	RF	—
Technetium	Tc	2200	11.5	1570	1800	2090	—	—	—	—	—	—	—
Teflon	PTFE	330	2.9	—	—	—	—	—	—	W	—	RF	Baffled source. Film structure doubtful.
Tellurium	Te	452	6.25	157	207	277	Poor	Al_2O_3 Quartz	W	W Ta	W	RF	Wets w/o alloying. Toxic.
Terbium	Tb	1357	8.27	800	950	1150	Xlnt.	Al_2O_3	—	Ta	—	RF	—
Terbium Fluoride	TbF_3	1176	—	—	—	~800	—	—	—	—	—	RF	—
Terbium Oxide	Tb_2O_3	2387	7.87	—	—	1300	—	—	—	Ir	—	RF	Partially decomposes.

*Materials marked RF or RF-reactive may be sputtered with Sloan RF Sputtergun.

Table 2–16 (continued)
THIN-FILM DEPOSITION DATA

NAME	SYMBOL	MELTING POINT °C	DENSITY g/cm³	TEMPERATURE (°C) @ VAP. PRESS.			EVAPORATION TECHNIQUES					SPUTTER*	REMARKS n = Index of refraction
				10-8 TORR	10-6 TORR	10-4 TORR	ELECTRON BEAM	CRUCIBLE	COIL	BOAT	BASKET		
Terbium Oxide	Tb_4O_7	—	—	—	—	—	—	—	—	Ta	—	RF	Films TbO.
Thallium	Tl	302	11.85	280	360	470	Poor	Al_2O_3 Quartz	—	W Ta	W	DC Sputtergun	Wets freely, very toxic.
Thallium Bromide	TlBr	480	7.56	— Sublimes	—	~250	—	Quartz	—	—	—	RF	Toxic
Thallium Chloride	TlCl	430	7.00	— Sublimes	—	~150	—	Quartz	—	—	—	RF	n = 2.25
Thallium Iodide (B)	TlI	440	7.09	— Sublimes	—	~250	—	Quartz	—	—	—	RF	n = 2.78
Thallium Oxide	Tl_2O_3	717	9.65	—	—	350	—	—	—	—	—	RF	Goes to Tl_2O at 850°C.
Thorium	Th	1700	11.7	1430 Sublimes	1660	1925	Xlnt.	—	W	W Ta Mo	W	—	Toxic, radioactive.
Thorium Carbide	ThC_2	2773	8.96	—	—	~2300	—	Carbon	—	—	—	RF DC Sputtergun	Radioactive.
Thorium Dioxide	ThO_2	3050	10.03	—	—	~2100	Good	—	—	W	—	RF RF-Reactive	Radioactive. n = 1.86, 2.20
Thorium Fluoride	ThF_4	900	6.3	—	—	~750	Fair	Vit. Carbon	—	Ni	W	RF	Radioactive
Thorium Oxyfluoride	$ThOF_2$	900	9.1	—	—	—	—	—	—	Mo Ta	--	—	Radioactive. n = 1.52
Thorium Sulphide	ThS_2	—	6.80	—	—	—	—	—	—	—	—	RF	Sputtering preferred; or co-evaporate from 2 sources.
Thulium	Tm	1545	9.32	461 Sublimes	554	680	—	Al_2O_3	—	Ta	—	DC Sputtergun	—
Thulium Oxide	Tm_2O_3	—	8.90	—	—	1500	—	—	—	Ir	—	RF	Decomposes.
Tin	Sn	232	7.75	682	807	997	Xlnt.	Al_2O_3	W	Mo	W	DC RF	Wets Mo; use Ta liner in EB guns.
Tin Oxide	SnO_2	1127	6.95	— Sublimes	—	~350	Xlnt.	Quartz Al_2O_3	W	W	W	RF RF-Reactive	Films from W oxygen deficient, oxidize in air. n = 2.0
Tin Selenide	SnSe	861	6.18	—	—	~400	—	Quartz	—	—	—	RF	—
Tin Sulphide	SnS	882	5.08	—	—	~450	—	Quartz	—	—	—	RF	—
Tin Telluride	SnTe	780	6.44	—	—	~450	—	Quartz	—	—	—	RF	—

*Materials marked RF or RF-reactive may be sputtered with Sloan RF Sputtergun.

Table 2–16 (continued)
THIN-FILM DEPOSITION DATA

NAME	SYMBOL	MELTING POINT °C	DENSITY g/cm³	TEMPERATURE (°C) @ VAP. PRESS. 10^{-8} TORR	10^{-6} TORR	10^{-4} TORR	EVAPORATION TECHNIQUES ELECTRON BEAM	CRUCIBLE	COIL	BOAT	BASKET	SPUTTER*	REMARKS n = Index of refraction
Titanium	Ti	1675	4.50	1067	1235	1453	Xlnt.	Vit. Carbon TiC	—	Ta	—	DC RF	Alloys with refractory metals; evolves gas on first heating.
Titanium Boride	TiB_2	2980	4.50	—	—	—	Poor	—	—	—	—	RF DC Sputtergun	—
Titanium Carbide	TiC	3140	4.93	—	—	~2300	—	—	—	—	—	RF DC Sputtergun	—
Titanium Dioxide (rutile)	TiO_2	1640	4.29	—	—	~1300	Fair	—	—	—	W	RF RF-Reactive	Sub-oxide, must be reoxidized to rutile. Ta reduces TiO_2 to TiO and Ti. $n = 2.4$
Titanium Monoxide	TiO	1750	4.93	—	—	~1500	Good	Vit. Carbon	—	W Ta	—	RF	Use gentle preheat to outgas. $n \cong 2.2$
Titanium Nitride	TiN	2930	5.43	—	—	—	Good	—	—	Mo	—	RF RF-Reactive DC Sputtergun	Sputtering preferred. Decomposes with thermal evaporation.
Titanium Sesquioxide	Ti_2O_3	2130	4.6	—	—	—	Good	—	—	W	—	RF	Decomposes.
Titanium Silicide	$TiSi_2$	—	—	—	—	—	—	—	—	—	—	RF DC Sputtergun	—
Tungsten	W	3410	19.3	2117	2407	2757	Good	—	—	—	—	RF DC Sputtergun	Forms volatile oxides. Films hard & adherent.
Tungsten Boride	WB_2	2900	12.75	—	—	—	Poor	—	—	—	—	RF	—
Tungsten Carbide	W_2C	2860	17.15	1480	1720	2120	Xlnt.	—	—	C	—	RF DC Sputtergun	—
Tungsten Disulphide	WS_2	1250	7.5	—	—	—	—	—	—	—	—	RF	—
Tungsten Selenide	WSe_2	—	9.0	—	—	—	—	—	—	—	—	RF	—
Tungsten Silicide	WSi_2	2900	9.4	—	—	—	—	—	—	—	—	RF DC Sputtergun	—
Tungsten Telluride	WTe_3	—	9.49	—	—	—	—	Quartz	—	—	—	RF	—
Tungsten Trioxide	WO_3	1473	7.16	—	—	980	Good	—	—	W Pt	—	RF-Reactive	Use gentle preheat to outgas. W reduces oxide slightly. $n = 1.68$
Uranium	U	1132	18.7	1132	1327	1582	Good	—	W	Mo W	W	—	Films oxidize.
Uranium Carbide	UC_2	2260	11.28	—	—	2100	—	Carbon	—	—	—	RF	Decomposes.
Uranium Dioxide	UO_2	2176	10.9	—	—	—	—	—	—	W	W	RF	Ta causes decomposition.

*Materials marked RF or RF-reactive may be sputtered with Sloan RF Sputtergun.

Table 2–16 (continued)
THIN-FILM DEPOSITION DATA

NAME	SYMBOL	MELTING POINT °C	DENSITY g/cm³	TEMPERATURE (°C) @ VAP. PRESS.			EVAPORATION TECHNIQUES					SPUTTER*	REMARKS n = Index of refraction
				10^{-8} TORR	10^{-6} TORR	10^{-4} TORR	ELECTRON BEAM	CRUCIBLE	COIL	BOAT	BASKET		
Uranium Fluoride	UF_4	~1000	—	—	—	300	—	—	—	Ni	—	RF	—
Uranium Oxide	U_3O_8	dec	8.30	—	—	—	—	—	—	W	W	RF-Reactive	Decomposes at 1300°C to UO_2.
Uranium Phosphide	UP_2	—	8.57	—	—	1200	—	—	—	Ta	—	RF	Decomposes.
Uranium Sulphide	U_2S_3	—	—	—	—	1400	—	—	—	W	—	RF	Slight decomposition.
Vanadium	V	1890	5.96	1162	1332	1547	Xlnt.	—	—	Mo	—	DC RF	Wets Mo. EB evaporated films preferred.
Vanadium Boride	VB_2	2400	5.10	—	—	—	—	—	—	—	—	RF DC Sputtergun	—
Vanadium Carbide	VC	2810	5.77	—	—	~1800	—	—	—	—	—	RF DC Sputtergun	—
Vanadium Dioxide	VO_2	1967	4.34	— Sublimes	—	~575	—	—	—	—	—	RF RF-Reactive	Sputtering preferred.
Vanadium Nitride	VN	2320	6.13	—	—	—	—	—	—	—	—	RF RF-Reactive DC Sputtergun	—
Vanadium Pentoxide	V_2O_5	690	3.36	—	—	~500	—	Quartz	—	—	—	RF	—
Vanadium Silicide	VSi_2	—	4.42	—	—	—	—	—	—	—	—	RF	—
Ytterbium	Yb	824	6.98	520	590	690	Good	—	—	Ta	—	DC RF	—
Ytterbium Fluoride	YbF_3	1157	—	—	—	~800	—	—	—	—	—	RF	—
Ytterbium Oxide	Yb_2O_3	2346	9.17	—	—	~1500	—	—	—	Ir	—	RF RF-Reactive	Loses oxygen.
Yttrium	Y	1509	4.48	830	973	1157	Xlnt.	Al_2O_3	—	W Ta	W	RF DC Sputtergun	High Ta solubility.
Yttrium Aluminum Oxide	$Y_3Al_5O_{12}$	1990	—	—	—	—	Good	—	W	—	—	RF	Films not ferroelectric.
Yttrium Fluoride	YF_3	1387	4.0	—	—	—	—	—	—	—	—	RF	—
Yttrium Oxide	Y_2O_3	2680	4.84	—	—	~2000	Good	C	—	W	—	RF RF-Reactive	Loses oxygen, films smooth and clear. n = 1.79
Zinc	Zn	419	7.14	127	177	250	Xlnt.	Al_2O_3 Quartz	W	Mo W Ta	W	DC RF	Evaporates well under wide range of conditions.

*Materials marked RF or RF-reactive may be sputtered with Sloan RF Sputtergun.

Table 2–16 (continued)
THIN-FILM DEPOSITION DATA

NAME	SYMBOL	MELTING POINT °C	DENSITY g/cm³	TEMPERATURE (°C) @ VAP. PRESS. 10-8 TORR	10-6 TORR	10-4 TORR	EVAPORATION TECHNIQUES: ELECTRON BEAM	CRUCIBLE	COIL	BOAT	BASKET	SPUTTER*	REMARKS n = Index of refraction
Zinc Antimonide	Zn_3Sb_2	546	6.3	—	—	—	—	—	—	—	—	RF	—
Zinc Bromide	$ZnBr_2$	394	4.22	—	—	~300	—	Carbon	—	W	—	RF	Decomposes.
Zinc Fluoride	ZnF_2	872	4.84	790	905	1035	—	Quartz	—	Pt Ta	—	RF	—
Zinc Nitride	Zn_3N_2	—	—	—	—	—	—	—	—	Mo	—	RF	Decomposes.
Zinc Oxide	ZnO	1975	5.61	—	—	~1800	—	—	—	—	—	RF RF-Reactive	—
Zinc Selenide	ZnSe	1526	5.42	—	—	660	—	Quartz	W Mo	Ta W Mo	W	RF	Use gentle preheat to outgas. Evaporates well. n = 2.6
Zinc Sulphide	ZnS	1830	4.09	— Sublimes	—	~800	Good	Quartz	—	Ta Mo	—	RF	Use gentle preheat to outgas. Films partially decompose. Sticking coefficient varies with substrate temperature. n ≅ 2.3
Zinc Telluride	ZnTe	1238	6.0	—	—	~600	—	—	—	Mo Ta	—	RF	Use gentle preheat to outgas. n = 2.8
Zircon	$ZrSiO_4$	2550	4.56	—	—	—	—	—	—	—	—	RF	—
Zirconium	Zr	1852	6.40	1477	1702	1987	Xlnt.	—	—	W	—	RF DC	Alloys with W. Films oxidize readily.
Zirconium Boride	ZrB_2	3040	6.08	—	—	—	Good	—	—	—	—	RF DC Sputtergun	—
Zirconium Carbide	ZrC	3540	6.73	—	—	~2500	—	—	—	—	—	RF DC-Sputtergun	—
Zirconium Nitride	ZrN	2980	7.09	—	—	—	—	—	—	—	—	RF RF-Reactive DC Sputtergun	Reactively evaporate in 10^{-3} N_2 atmosphere.
Zirconium Oxide	ZrO_2	2700	5.6	—	—	~2200	Good	—	—	W	—	RF RF-Reactive	Films oxygen deficient, clear and hard. n = 2.05
Zirconium Silicide	$ZrSi_2$	—	4.88	—	—	—	—	—	—	—	—	RF DC Sputtergun	—

*Materials marked RF or RF-reactive may be sputtered with Sloan RF Sputtergun.

Table furnished through the courtesy of Sloan Technology Corporation.

Table 2–17
PROPERTIES OF COMMONLY USED GLASS SUBSTRATE MATERIALS

For conversion factors see Table 2–18.

	Soda lime	*Alkali zinc borosilicate (microsheet)*	*Lime alumino-silicate (alkali-free)*	*Lime alumino-silicate (alkali-free)*	*Barium alumino-silicate (alkali-free)*	*Alkali borosilicate*	*96% silica*	*Fused silica*	*Fotoceram*[a]	*Synthetic sapphire*[b]
Code number[c]	0080	0211	1715	1723	7059	7740	7900	7940	—	—
Annealing point, deg C	512	542	866	710	650	565	910	1 050	—	—
Softening point, deg C	696	720	1 060	910	872	820	1 500	1 580	700	2 040
Thermal exp coef, 10^{-6}/deg C										
0–300 deg C	9.2	7.2	3.5	4.6	4.5	3.25	0.8	0.56	10.4	5.8
Thermal conductivity										
cal/cm·s·deg C at 25 deg C	0.002 3	—	—	0.003 2	—	0.002 7	0.003 8	0.003 4	0.005	0.098
cal/cm·s·deg C at 300 deg C	0.003 2	—	—	—	0.003 6	—	—	0.004 2	—	—
Density, g/cm^3	2.47	2.57	2.48	2.63	2.76	2.23	2.18	2.20	2.46	3.98
Dielectric constant, 100 hz										
25 deg C	8.3	6.8	6.0	6.4	5.9	4.9	3.9	3.9		
200 deg C	—	—	6.1	6.6	6.1	—	4.1	3.9		
400 deg C	—	—	7.2	7.3	7.4	—	—	—		
Dielectric constant, 1 Mhz										
25 deg C	6.9	6.6	5.9	6.4	5.8	4.6	3.9	3.9	5.6	9.4–11.5
200 deg C	9.3	7.4	6.1	6.5	5.9	5.1	3.9	3.9		
400 deg C	—	—	6.3	6.7	6.1	—	3.9	3.9		
Loss tangent, 100 hz										
25 deg C	0.078	0.01	0.001 8	0.000 8	0.001 1	0.027	0.000 6	0.000 06		
200 deg C	—	—	0.004 7	0.002 4	0.006 2	—	0.07	0.000 3		
400 deg C	—	—	0.2	0.09	0.19	—	—	—		
Loss tangent, 1 Mhz										
25 deg C	0.01	0.004 7	0.002 4	0.001 3	0.001 1	0.006 2	0.000 6	0.000 02	0.006	—
200 deg C	0.17	0.032	0.002 8	0.001 4	0.001 8	0.03	0.001	0.000 02		
400 deg C	—	—	0.006 1	0.003 4	0.007 1		0.026	0.000 3		
Log volume resistivity										
ohm-cm at 25 deg C	6.4	8.3	13.6	14.1	13.5	8.1	9.7	11.8	14.0	12.0
ohm-cm at 350 deg C	5.1	6.7	11.3	11.8	11.3	6.6	8.1	10.2	—	—
Dielectric strength,										
kV rms at 25 deg C	0.35	2.0	>10	>10	>10	2.0	7.0	>10	—	—
Weatherability, g/cm^2	>5	0.05–0.25	<0.01	<0.01	<0.01	0.05–0.25	<0.01	<0.01	—	Unaffected
Chemical durability, mg/cm^2										
5% HCl, 24 hr	0.02	0.03	0.1	0.4	5.5	0.005	0.001	0.001	—	Unaffected
5% NaOH, 6 hr	0.5	2.0	1.2	0.3	3.7	1.1	1.1	0.7	—	Unaffected
0.02*N* Na_2CO_3, 6 hr	0.1	0.1	0.15	0.1	0.3	0.1	0.03	0.03	—	Unaffected

[a]Courtesy of Adolf Meller Company, Providence, R. I.
[b]Code numbers of Corning Glass Works, Corning, N.Y.
[c]Trade name, Corning Glass Works, Corning, N.Y.

From Westman, H. P., Ed., *Reference Data for Radio Engineers,* 5th ed., Howard W. Sams & Co., Indianapolis, 1968, and Bulletin CEP-2/5M/9-62, Corning Glass Works, Corning, N.Y., 1962. With permission.

Table 2–18
PROPERTIES OF COMMONLY USED CERAMIC SUBSTRATE MATERIALS

Conversion Factors:

For temperatures in K, add 273 to temperatures in °C.
For thermal conductivity in W/m·K, multiply cal/cm sec°C by 418.4.
For density in kg/m^3, multiply g/cm^3 by 1000.

Ceramic type	*Alumina ($96\% Al_2O_3$)*	*Alumina ($99.7\% Al_2O_3$)*[b]	*Dense alumina ($85\% Al_2O_3$)*	*Dense alumina ($94\% Al_2O_3 + CaO + SiO_2$)*	*Dense alumina ($96\% Al_2O_3 + MgO + SiO_2$)*	*Beryllia (BeO)*	*Dense beryllia (98% BeO)*	*Dense beryllia (99.5% BeO)*	*Titania (TiO_2)*[a]	*Barium titanate*	*Magnesium titanate*[b]
Code number[c]	—	—	576	719	614	—	735	754	192	—	—
Softening temp. (°C)	1650	>1650	1100	1500	1550	—	1600	1600	1600	1550	—
Melting point (°C)	2050	>2050	—	—	—	2550	—	—	1920	1700	—
Thermal exp. coef. (10^{-6}/°C)	6.4	5.0	6.5	6.2	6.4	7.5	6.1	6.0	8.3	8.1	7.5
Thermal conductivity cal/cm/sec/°C at 25°C	0.50	0.045	0.060	0.073	0.084	0.20	0.50	0.55	0.012	0.003	0.1
Density (gm/cm^3)	3.7	3.9	3.4	3.58	3.7	3.08	2.9	2.88	4.0	5.5	3.6
Dielectric constant (1 MHz) at 25°C	10.0	9.5 (10 GHz)	8.3	8.9	9.3	7.1	6.3	6.4	85	10 to 10^4	16 (10 GHz)
Loss tangent (1 MHz) at 25°C	0.0002	0.0001	0.0058	0.0018	0.0028	0.0006	0.0006	0.0006	0.0002	>0.02	0.0002
Log volume resistivity (ohm-cm) at 30°C	14.0	>14	10.7	12.8	10.0	>14	13.8	>14	14.0	12.0	>14
Dielectric strength at 25°C and 60 Hz (volts/mil)	—	—	230	230	230	—	255	260	100	—	—
Surface smoothness (microinches)[d]	10 to 30	10 to 30	10 to 30	10 to 30	10 to 30	10 to 30	10 to 30	10 to 30	0.1 to 30	10 to 30	10 to 30
Chemical resistance	excellent	excellent	excellent	excellent	excellent	excellent	excellent	excellent	good	good	good

[a] Also known as Rutile.
[b] Substrates used as dielectrics for microminiature microwave strip lines.
[c] Code nubmers of ALSIMAG brand ceramics, American Lava Corp., Chattanooga, Tenn.
[d] 1 microinch = 250 Å, or 0.025 μm.

From Westman, H. P., Ed., *Reference Data for Radio Engineers*, 5th ed., Howard W. Sams & Co., Indianapolis, 1968. With permission.

Table 2–19
DIELECTRIC STRENGTH OF VARIOUS INSULATING MATERIALS

In rms V/mil at Frequencies from 60 hz to 100 Mhz

Material	*Thickness, mils*	*Frequency*						
		60 hz	*1 khz*	*38 khz*	*180 khz*	*2 Mhz*	*18 Mhz*	*100 Mhz*
Polystyrene, unpigmented	30	3 174	2 400	1 250	977	725	335	220
Polyethylene, unpigmented	30	1 091	965	500	460	343	180	132
Polytetrafluoroethylene (Teflon[a])	30	850	808	540	500	375	210	143
Monochlorotrifluoroethylene (Kel-F[b])	20	2 007	1 478	1 054	600	354	129	29[d]
Glass-bonded mica	32	712	643	—	360	207	121	76
Soda-lime glass	32	1 532	1 158	—	230	90	55	20[d]
Dry-process porcelain	32	232	226	—	90	83	71	60[d]
Steatite	32	523	427	—	300	80	58	56[d]
Forsterite (AlSiMag-243)	65	499	461	455	365	210	112	74
Alumina, 85% (AlSiMag-576[c])	55	298	298	253	253	178	112	69

[a]Trademark of E. I. du Pont de Nemours & Co., Inc., Wilmington, Del.
[b]Trademark of Minnesota Mining and Manufacturing Co. (3M Co.), St. Paul, Minn.
[c]Trademark of American Lava Corp., Ridgefield, N. J.
[d]Puncture with attendant volume heating effect.

From Frisco, L. J., Dependence of electric field strength, *Electro-Tech.*, 68, 110, 1961. With permission.

Table 2–20
THIN FILMS FOR INTEGRATED CIRCUITS

See Table 2–16 for other properties of thin films.

A. Physical and Electrical Properties of Various Films

Conversion Factors:

For temperatures in K, add 273 to temperatures in °C.
For thermal conductivity in W/m · K, multiply cal/cm sec °C by 418.4.
For density in kg/m^3, multiply g/cm^3 by 1,000.

Characteristics	*Ta*	*Ta_2O_5*	*Ti*	*TiO_2*	*SiO*	*Si*	*SiO_2*	*Al_2O_3*	*$Al_2Si_2O_7$*
Density, g/cm^3	16.6	8.75	4.54	4.24	2.15	2.42	2.32	3.96	Unknown
Melting point, °C	2996	1470	1668	1920	—	1420	1728	2040	Unknown
Thermal conductivity, cal/°C sec cm	0.13	Unknown	0.041	0.01	Unknown	0.20	0.003	0.01	Unknown
Thermal coefficient of expansion, per °C	6.46×10^{-6}	0.55×10^{-6}	8.4×10^{-6}	9.1×10^{-6}	Unknown	7.6×10^{-6}	7.0×10^{-6}	8.0×10^{-6}	Unknown
Resistivity, ohm-cm	12.5×10^{-6}	Unknown	42×10^{-6}	10^{14} to 10^{25}	10^9	10^{-2} to 10^5	10^{14}	10^{11}	$>10^{14}$
Relative dielectric constant	—	25	—	100	1.8–5.8	12.0	3.78	10	8–9

From *Integrated Silicon Device Technology*, Vol. 2, Capacitance, Research Triangle Institute, Research Triangle Park, N.C., 1963. With permission.

B. Resistivity and Temperature Coefficient of Resistance of Various Metal Films

Symbols:

ρ_f = electrical resistivity in microhm-cm
TCR = temperature coefficient of resistance per °C

Metal	*ρ_f before annealing, microhm-cm*	*ρ_f after annealing at 600°C, microhm-cm*	*Temperature coefficient of resistance*	*Bulk TCR, 0–100°C*	*Ratio of film TCR to bulk TCR*
Au	22.2	4.95	$2,800 \times 10^{-6}$	3,400	.82
Pt	8.7	15.65	$2,500 \times 10^{-6}$	3,900	.64
Ir	12.8	42.5	$1,800 \times 10^{-6}$	4,000	.45
Rh	17.3	15.8	$2,000 \times 10^{-6}$	4,600	.43
Pd	20.3	20.8	$2,300 \times 10^{-6}$	3,700	.62
Ni	28.5	41.0	$5,000 \times 10^{-6}$	6,400	.78
Cr	172.5	62.0	600×10^{-6}	—	—
Ti	67.1	59.9	700×10^{-6}	5,400	.13
Zr	134.0	—	$<100 \times 10^{-6}$	4,400	.02
Mo	99.5	49.0	200×10^{-6}	3,300	.06
Ta	768.0	—	$<100 \times 10^{-6}$	3,100	.03
W	4,390.0	422.5	$<100 \times 10^{-6}$	4,800	.02
Al	.41	.36	$2,800 \times 10^{-6}$	4,300	.65

From *Integrated Silicon Technology*, Vol. 1, Resistance, Research Triangle Institute, Research Triangle Park, N.C., 1963. With permission.

Table 2–21
SPUTTERING RATES FOR METALS USED IN THIN-FILM DEPOSITION

Sputtering Yields in Atoms/Ion

DIRECT-CURRENT SPUTTERING YIELDS FOR MATERIALS BOMBARDED BY Ar^+ IONS

Target	Bombarding energy, volts					
	200	600	1000	2000	5000	10 kv
Ag	1.6	3.4				8.8
Al	0.35	1.2				
Au	1.1	2.8				
Co	0.6	1.4				
Cr	0.7	1.3				
Cu	1.1	2.3	3.2	4.3	5.5	6.6
Fe	0.5	1.3	1.4	2.0[a]	2.5[a]	
Ge	0.5	1.2	1.5	2.0	3.0	
Mo	0.4	0.9	1.1			2.2
Nb	0.25	0.65				
Ni	0.7	1.5	2.1			
Os	0.4	0.95				
Pd	1.0	2.4				
Pt	0.6	1.6				
Re	0.4	0.9				
Rh	0.55	1.5				
Si	0.2	0.5	0.6	0.9	1.4	
Ta	0.3	0.6				
Th	0.3	0.7				
Ti	0.2	0.6				
U	0.35	1.0				
W	0.3	0.6				
Zr	0.3	0.75				
GaSb (111)	0.4	0.9	1.2			
SiC		1.8				

[a]Type 304 stainless steel.

DIRECT-CURRENT SPUTTERING YIELDS FOR MATERIALS BOMBARDED BY Ne^+ IONS

Target	Bombarding energy, volts		
	200	600	1000
Ag	1.0	2.0	
Al	0.2	0.8	
Au	0.6	1.2	
Co	0.4	1.0	
Cr	0.5	1.05	
Cu[a]	0.8	2.0 (1.5)	1.8
Fe	0.4	1.0 (0.7)	0.8
Ge	0.3	0.8	
Mo	0.2	0.5 (0.3)	0.5
Nb	0.2	0.4	
Ni	0.5	1.3 (1.1)	1.2
Pd	0.6	1.3	
Pt	0.3	0.7	
Re	0.15	0.4	
Rh	0.4	0.8	
Si	0.0	0.5	
Ta	0.1	0.3	
Ti	0.2	0.45	
U	0.2	0.5	
V	0.2	0.55	
W	0.1	0.3	
Zr	0.2	0.4	
SiC		0.55	

[a]2.6 at 5 kev and 3.0 at 10 kev.

Table 2–21 (continued)
SPUTTERING RATES FOR METALS USED IN THIN-FILM DEPOSITION

Sputtering Yields in Atoms/Ion (continued)

DIRECT-CURRENT SPUTTERING YIELDS FOR MATERIALS BOMBARDED BY Hg^+ IONS

Target	*Bombarding energy, volts*			
	200	*400*	*5 kv*	*10 kv*
Ag	0.85		14	22.5
Al	0.15	0.45		
Au	1.0		13.5	21
C	0.05	0.15		
Co	0.2	0.6	4	6
Cr	0.2	0.8		
Cu	0.55	1.3	8.5	11.5
Fe	0.15	0.5	2.5	4
Ge	0.2	0.6	1.5 (1k)	
Mo	0.1	0.45	2.5	5
Nb	0.1	0.3		
Ni	0.25	0.7	5.5	7
Pd	0.5	1.3	7	9
Pt	0.5	1.4	7.5	10
Rh	0.35	1.0	5	7
Si	0.05	0.15		
Ta	0.2	0.5	2.5	3
Ti	0.1	0.4	2	2.5
V	0.1	0.3	2.5	3.5
W	0.15	0.55	2.5	2.5
Zr	0.15	0.35		

DIRECT-CURRENT SPUTTERING YIELDS FOR MATERIALS BOMBARDED BY Kr^+ IONS

Target	*Bombarding energy, volts*		
	200	*600*	*1000*
Ag	1.4	3.9	6.4 (1075 volts)
Al	0.3	1.1	
Au	1.1	3.4	
C	0.05	0.2	
Co	0.3	1.3	
Cr	0.6	1.6	
Cu	0.9	2.8	3.4
Fe	0.4	1.2	1.4
Ge	0.4	1.4	
Mo	0.3	1.1	1.2
Nb	0.2	0.7	
Ni	0.5	1.5	1.7
Pd	1.0	2.6	
Pt	0.7	2.1	
Rh	0.5	1.7	
Si	0.1	0.6	
Ta	0.3	1.0	
Ti	0.2	0.5	
W	0.4	1.1	
Zr	0.2	0.7	
SiC		1.6	

Table 2–21 (continued)
SPUTTERING RATES FOR METALS USED IN THIN-FILM DEPOSITION

Sputtering Yields in Atoms/Ion (continued)

DIRECT-CURRENT SPUTTERING YIELDS FOR MATERIALS BOMBARDED BY Xe^+ IONS

Target	Bombarding energy, volts		
	200	600	1000
Ag	1.1	4.2	
Al	0.2	1.0	
Au	1.0	3.1	
C	0.04	0.2	
Co	0.4	1.3	
Cr	0.4	1.9	
Cu	0.8	2.4	3.6
Fe	0.3	1.2	1.8
Ge	0.3	1.2	1.6
Mo	0.3	1.1	1.6
Nb	0.2	0.6	
Ni	0.4	1.5	2.2
Pd	0.9	2.5	
Pt	0.7	2.2	
Rh	0.5	1.6	
Si	0.08	0.5	
Ta	0.3	1.0	
Ti	0.1	0.5	
W	0.4	1.2	
Zr	0.2	0.7	
SiC		1.6	

DIRECT-CURRENT SPUTTERING YIELDS FOR MATERIALS BOMBARDED BY He^+ IONS

Target	Bombarding energy, volts	
	200	600
Ag	0.08	0.2
Al	0.005	0.02
Au	0.02	0.08
C	0.02	0.09
Co	0.04	0.2
Cr	0.07	0.2
Cu	0.1	0.3
Fe	0.07	0.2
Ge	0.03	0.08
Mo	0.005	0.04
Nb	0.005	0.03
Ni	0.06	0.2
Pd	0.06	0.2
Pt	0.004	0.04
Rh	0.02	0.07
Si	0.05	0.2
Ta	0.002	0.01
Ti	0.04	0.08
W	0.001	0.008
SiC		0.16

Table 2–21 (continued)
SPUTTERING RATES FOR METALS USED IN THIN-FILM DEPOSITION

Sputtering Yields in Atoms/Ion (continued)

SPUTTERING YIELDS IN MOLECULES PER ION FOR INSULATORS BOMBARDED BY ARGON USING RF

Target	*Mean ion energy, kv*		
	1.1	*2.0*	*2.9*
SiO_2	0.16	0.39	0.50
Pyrex 7740	0.15	0.33	0.43
Al_2O_3	0.05	0.12	0.17

SPUTTERING YIELDS FOR MATERIALS BOMBARDED BY N^+ IONS

Target	*Bombarding energy, kv*		
	1	*3*	*8*
Cu	1.5	2.5	2.5
Fe	0.6	1.0	1.2
Mo	0.15	0.35	0.4
Ni	0.7	1.2	1.1
W	0.15	0.3	0.45

DIRECT-CURRENT SPUTTERING YIELDS FOR MATERIALS BOMBARDED BY N_2^+ IONS

Target	*Bombarding energy, kv*		
	1	*3*	*8*
Cu	2.0	3.8	5.1
Fe	0.75	1.5	1.75
Mo	0.3	0.6	0.8
Ni	1.1	2.0	2.2
W	0.2	0.5	0.7

From Maissel, L. I., Deposition of thin films by cathodic sputtering, *Physics of Thin Films,* Vol. 3, Hass, G. and Thun, R. E., Eds., Academic Press, New York, 1966. With permission.

Table 2–22
PROPERTIES OF SEMICONDUCTORS

The term *semiconductor* is applied to a material in which electric current is carried by electrons or holes; its electrical conductivity when extremely pure rises exponentially with temperature and may be increased from this low "intrinsic" value by many orders of magnitude by "doping" with electrically active impurities.

Semiconductors are characterized by an energy gap in the allowed energies of electrons in the material that separates the normally filled energy levels of the *valence band* (where "missing" electrons behave as positively charged current carriers or "holes") and the *conduction band* (where electrons behave somewhat as a gas of free negatively charged carriers with an effective mass dependent on the material and the direction of the electrons' motion). This energy gap depends on the nature of the material and varies with direction in anisotropic crystals. It is slightly dependent on temperature and pressure, and this dependence is usually almost linear at normal temperatures and pressures.

I. General Properties of Semiconductors[a]

Substance	*Lattice parameters at room temperature, Å*	*Density, g/cc*	*Melting point, K*	*Minimum room temperature energy gap, ev*	*Comparative thermal conductivity*	*Heat of formation, kcal/mole*	*Mobility (room temperature)* Electrons cm²/volt-sec	Holes cm²/volt-sec	*Remarks*
Part A. Tetrahedral Semiconductors									
§ A1 Diamond Structure Elements (Strukturbericht symbol A4, Space Group Fd3m-O_h^7)									
C	3.5597	3.51	4300	5.4	2000	161	1800	1400	
Si	5.43072	2.3283	1685	1.107	1240	77.5	1900	500	
Ge	5.65754	5.3234	1231	0.67	640	69.5	3800	1820	
α-Sn	6.4912	5.765	503	0.08		64	2500	2400	
§ A2 Sphalerite (Zinc Blende) Structure Compounds (Strukturbericht symbol B3 Space Group F$\bar{4}$3m-T_d^2)									
I-VII Compounds									
CuF	4.255								
CuCl	5.4057	3.53	695						
CuBr	5.6905	4.72	770	2.94		115			
CuI	6.0427	5.63	878			105			
AgBr						102	4000		
AgI	6.473	5.67				93	30		
II-VI Compounds									
BeS	4.865	2.36							
BeSe	5.139	4.315							
BeTe	5.626	5.090							
BePo	8.38	7.3							
ZnO	4.63								See § A3
ZnS	5.4093	4.079	1920	3.54	140	114	180	5(400°C)	See also § A3
ZnSe	5.6676	5.42	1790	2.58	140	101	540	28	
ZnTe	6.101	5.72	1510	2.26	140	90	340	100	
ZnPo									
CdS	5.5818								See also § A3
CdSe	6.05								See § A3
CdTe	6.477	5.86	1370	1.44	55	81	1200	50	
CdPo									
HgS	5.8517	7.73	~2020						
HgSe	6.084	8.25	1070	0.30	10	59	20000		
HgTe	6.429	8.17	943	0.15	20	58	25000	350	
III-V Compounds									
BN	3.615	3.49	3000	~4	200	195			
BP(L.T.)	4.538	2.9		~6			500	70	
BAs	4.777								
AlP	5.451	2.85	1770	2.5					
AlAs	5.6622	3.81	1870	2.16		150	1200	420	
AlSb	6.1355	4.218	1330	1.60	600	140	200–400	550	
GaP	5.4505	4.13	1750	2.24	1100	152	300	100	
GaAs	5.65315	5.316	1510	1.35	370	128	8800	400	
GaSb	6.0954	5.619	980	0.67	270	118	4000	1400	
InP	5.86875	4.787	1330	1.27	800	134	4600	150	
InAs	6.05838	5.66	1215	0.36	290	114	33000	460	
InSb	6.47877	5.775	798	0.165	160	107	78000	750	

Table 2–22 (continued)
PROPERTIES OF SEMICONDUCTORS

I. General Properties of Semiconductors

Substance	Lattice parameters at room temperature, Å		Density, g/cc	Melting point, K	Minimum room temperature energy gap, ev	Comparative thermal conductivity	Heat of formation, kcal/mole	Mobility (room temperature) cm²/volt-sec Electrons	Holes	Remarks
Other Sphalerite Structure Compounds										
β-SiC	4.348		3.21	3070	2.3			4000		
Ga_2Te_3	5.899		5.75	1063	~1.0	~14	65			
In_2Te_3(H.T.)	6.150		5.8	940	~1.0	~8	47.4	~10		
$MgGeP_2$	5.652									
$ZnSnP_2$	5.65				2.1					
$ZnSnAs_2$(H.T.)	5.851		5.53	1050	~0.7	70				
§ A3 Wurtzite (Zincite) Structure Compounds (Strukturbericht symbol B4, Space Group P 6_3 mc-C_{6v}^4)										
I-VII Compounds										
CuCl	3.91	6.42		T_c 680°K						
CuBr	4.06	6.66		T_c 658°K						
CuI	4.31	7.09								
AgI	4.580	7.494			2.63					
II-VI Compounds										
BeO	2.698	4.380		2800						
MgTe	4.54	7.39	3.85	~2800						
ZnO	3.24950	5.2069	5.66	2250	3.2	6	154	180		
ZnS	3.8140	6.2576	4.1	2100	3.67		110			
ZnSe	3.996	6.626								
ZnTe	4.27	6.99								
CdS	4.1348	6.7490	4.82	2020	2.42		96	400		
CdSe	4.299	7.010	5.66	1530	1.74		90	650		
CdTe	4.57	7.47			1.50					
III-V Compounds										
BP(H.T.)	3.562	5.900								
AlN	3.111	4.978	3.26	~2500	6.02		197			
GaN	3.180	5.166	6.10	1500	3.34		157			
InN	3.533	5.693	6.88	1200	2.0		133			
Other Wurtzite Structure Compounds										
SiC	3.076	5.048								
MnTe	4.078	6.701			~1.0					
Al_2S_3	3.579	5.829	2.55		4.1		426			
Al_2Se_3	3.890	6.30	3.91		3.1		367			
§ A4 Chalcopyrite Structure Compounds (Strukturbericht symbol $E1_1$, Space Group I $\bar{4}$ 2d $-D_{2d}^{12}$)										
I-III-VI_2 Compounds										
$CuAlS_2$	5.323	10.44	3.47		2.5					
$CuAlSe_2$	5.617	10.92	4.70	1270	1.1					
$CuAlTe_2$	5.976	11.80	5.50	1160	0.88					
$CuGaS_2$	5.360	10.49	4.35							
$CuGaSe_2$	5.618	11.01	5.56	1310	0.96, 1.63					
$CuGaTe_2$	6.013	11.93	5.99	1150	0.82, 1.0					
$CuInS_2$	5.528	11.08	4.75		1.2					
$CuInSe_2$	5.785	11.57	5.77	1250	0.86, 0.92	37				
$CuInTe_2$	6.179	12.365	6.10	970	0.95	49				
$CuTlS_2$	5.580	11.17	6.32							
$CuTlSe_2$(L.T.)	5.844	11.65	7.11	680	1.07					
$CuFeS_2$	5.25	10.32		1150	0.53					
$CuFeSe_2$				850	0.16					
$CuLaS_2$	5.65	10.86								
$AgAlS_2$	5.707	10.28	3.94							
$AgAlSe_2$	5.968	10.77	5.07	1220	0.7					
$AgAlTe_2$	6.309	11.85	6.18	1000	0.56					
$AgGaS_2$	5.755	10.28	4.72							
$AgGaSe_2$	5.985	10.90	5.84	1120	1.66					
$AgGaTe_2$	6.301	11.96	6.05	990	1.1	10				
$AgInS_2$(L.T.)	5.828	11.19	5.00		1.9					
$AgInSe_2$	6.102	11.69	5.81	1053	1.18	30				
$AgInTe_2$	6.42	12.59	6.12	965	0.96, 0.52					
$AgFeS_2$	5.66	10.30	4.53							

Table 2–22 (continued)
PROPERTIES OF SEMICONDUCTORS

1. General Properties of Semiconductors (continued)

Substance	Lattice parameters at room temperature, Å		Density, g/cc	Melting point, K	Minimum room temperature energy gap, ev	Comparative thermal conductivity	Heat of formation, kcal/mole	Mobility (room temperature) cm²/volt-sec Electrons	Holes	Remarks
II-IV-V_2 Compounds										
$ZnSiP_2$	5.400	10.441	3.39	1640	2.3			1000		
$ZnGeP_2$	5.465	10.771	4.17	1295	2.2					
$CdSiP_2$	5.678	10.431	4.00	~1470	2.2			1000		
$CdGeP_2$	5.741	10.775	4.48	~1060	1.8					
$CdSnP_2$	5.900	11.518			1.5					
$ZnSiAs_2$	5.61	10.88	4.70	~1350	1.7				50	
$ZnGeAs_2$	5.672	11.153	5.32	~1150	0.85	110				
$ZnSnAs_2$	5.8515	11.704	5.53	~ 910	0.65	150			300	Disorders at 910°K
$CdSiAs_2$	5.884	10.882								
$CdGeAs_2$	5.9427	11.2172	5.60	~ 903	0.53	40		70	25	Disorders at 903°K
$CdSnAs_2$	6.0944	11.9182	5.72	880	0.26	70		22000	250	
§ A5 "Defect Chalcopyrite" Structure Compounds (Strukturbericht symbol E3, Space Group I $\bar{4}$ $- S_4^2$)										
$ZnAl_2Se_4$	5.503	10.90	4.37							
$ZnAl_2Te_4$(?)	5.104	12.05	4.95							
$ZnGa_2S_4$(?)	5.274	10.44	3.80							
$ZnGa_2Se_4$(?)	5.496	10.99	5.21							
$ZnGa_2Te_4$(?)	5.937	11.87	5.67		1.35					
$ZnIn_2Se_4$	5.711	11.42	5.44	1250	2.6					
$ZnIn_2Te_4$	6.122	12.24	5.83	1075	1.2					
$CdAl_2S_4$	5.564	10.32	3.06							
$CdAl_2Se_4$	5.747	10.68	4.54							
$CdAl_2Te_4$(?)	6.011	12.21	5.10							
$CdGa_2S_4$	5.577	10.08	4.03							
$CdGa_2Se_4$	5.743	10.73	5.32							
$CdGa_2Te_4$	6.093	11.81	5.77							
$CdIn_2Te_4$	6.205	12.41	5.9	1060	(1.26 or 0.9)	4000				
$HgAl_2S_4$	5.488	10.26	4.11							
$HgAl_2Se_4$	5.708	10.74	5.05							
$HgAl_2Te_4$(?)	6.004	12.11	5.81							
$HgGa_2S_4$	5.507	10.23	5.00							
$HgGa_2Se_4$	5.715	10.78	6.18							
$HgIn_2Se_4$	5.764	11.80	6.3	1100	0.6					
$HgIn_2Te_4$(?)	6.186	12.37	6.3	980	0.86		200			
§ A6 Other Tetrahedral Compounds										
α-SiC	3.0817 15.1183		3.21	3070	2.86			400		6H structure
$Hg_5Ga_2Te_8$	6.235									B3 with super lattice
$Hg_5In_2Te_8$	6.328				0.7			2000		B3 with super lattice
Part B. Octahedral Semiconductors										
Halite Structure Semiconductors (Strukturbericht symbol B1, Space Group Fm3m $- O_h^5$)										
SnSe	6.020			1133						
SnTe	6.313			1080 (max)	0.5	91				
PbS	5.9362		7.61	1390	0.37	23	104	600	600	
PbSe	6.1243		8.15	1340	0.26	17	94	1000	900	
PbTe	6.454		8.16	1180	0.25	23	94	1600	600	
Selected Other Binary Chalcides										
BiSe	5.99		7.98	880	0.4					
BiTe	6.47									
EuSe	6.191			2300	1.8	2.4				
GdSe	5.771			2400						
NiD	4.1684		6.6	2260	2.0 or 3.7			4		
CdO	4.6953			1700	2.5	7	127	100		
SrS	6.0199		3.643	3000	4.1					

Table 2–22 (continued)
PROPERTIES OF SEMICONDUCTORS

I. General Properties of Semiconductors (continued)[a]

Substance	Lattice parameters at room temperature, Å	Density, g/cc	Melting point, K	Minimum room temperature energy gap, ev	Comparative thermal conductivity	Heat of formation, kcal/mole	Mobility (room temperature) cm³/volt-sec: Electrons	Holes	Remarks
Selected Ternary Compounds									
$AgSbSe_2$	5.786	6.60	910	0.58	10.5				
$AgSbTe_2$ (or $Ag_{19}Sb_{29}Te_{52}$)	6.078	7.12	830	0.7, 0.27	8.6, 0.3				
$AgBiS_2$(H.T.)	5.648								
$AgBiSe_2$(H.T.)	5.82								
$AgBiTe_2$(H.T.)	6.155								

[a]Listed by crystal structure.

II. Semiconducting Properties of Selected Materials

Substance	Minimum energy gap, ev: Room temperature	0 K	$\frac{dE_g}{dT}$ × 10⁴ ev/°C	$\frac{dE_g}{dP}$ × 10⁶ ev cm²/kg	Density of states electron effective mass, m_{d_n} (m_o)	Electron mobility and temperature dependence: μ_n cm²/volt-sec	$-x$	Density of states hole effective mass, m_{d_p} (m_o)	Hole mobility and temperature dependence: μ_p cm²/volt-sec	$-x$	Dominant emission wavelength, 77 K, microns
Si	1.107	1.153	−2.3	−2.0	0.58	1,900	2.6	1.06	500	2.3	1.274
Ge	0.67	0.744	−3.7	+7.3	0.35	3,800	1.66	0.56	1,820	2.33	1.770
α-Sn	0.08	0.094	−0.5		0.02	2,500	1.65	0.3	2,400	2.0	—
Te	0.33				0.68	1,100		0.19	560		—
III-V Compounds											
AlAs	2.2	2.3				1,200			420		—
AlSb	1.6	1.7	−3.5	−1.6	0.09	200	1.5	0.4	500	1.8	—
GaP	2.24	2.40	−5.4	−1.7	0.35	300	1.5		150	1.5	0.59
GaAs	1.35	1.53	−5.0	+9.4	0.068	9,000	1.0	0.5	500	2.1	0.84
GaSb	0.67	0.78	−3.5	+12	0.050	5,000	2.0	0.23	1,400	0.9	1.6
InP	1.27	1.41	−4.6	+4.6	0.067	5,000	2.0		200	2.4	0.91
InAs	0.36	0.43	−2.8	+8	0.022	33,000	1.2	0.41	460	2.3	3.1
InSb	0.165	0.23	−2.8	+15	0.014	78,000	1.6	0.4	750	2.1	5.2
II-VI Compounds											
ZnO	3.2		−9.5	+0.6	0.38	180	1.5				0.37
ZnS	3.54		−5.3	+5.7		180			5 (400°C)		0.33
ZnSe	2.58	2.80	−7.2	+6		540			28		—
ZnTe	2.26			+6		340			100		—
CdO	2.5±.1		−6		0.1	120					—
CdS	2.42		−5	+3.3	0.165	400		0.8			0.49
CdSe	1.74	1.85	−4.6		0.13	650	1.0	0.6			0.68
CdTe	1.44	1.56	−4.1	+8	0.14	1,200		0.35	50		0.78
HgSe	0.30				0.030	20,000	2.0				—
HgTe	0.15		−1		0.017	25,000		0.5	350		—
Halite Structure Compounds											
PbS	0.37	0.28	+4		0.16	800		0.1	1,000	2.2	4.3
PbSe	0.26	0.16	+4		0.3	1,500		0.34	1,500	2.2	8.5
PbTe	0.25	0.19	+4	−7	0.21	1,600		0.14	750	2.2	6.5

Table 2–22 (continued)
PROPERTIES OF SEMICONDUCTORS

II. Semiconducting Properties of Selected Materials (continued)

Substance	Minimum energy gap, ev: Room temperature	Minimum energy gap, ev: 0 K	$\frac{dE_g}{dT}$ × 10^4 ev/°C	$\frac{dE_g}{dP}$ × 10^6 ev cm²/kg	Density of states electron effective mass, m_{d_n} (m_o)	Electron mobility and temperature dependence: μ_n cm²/volt-sec	Electron mobility and temperature dependence: $-x$	Density of states hole effective mass, m_{d_p} (m_o)	Hole mobility and temperature dependence: μ_p cm²/volt-sec	Hole mobility and temperature dependence: $-x$	Dominant emission wavelength, 77 K, microns
Others											
ZnSb	0.50	0.56			0.15	10				1.5	
CdSb	0.45	0.57	−5.4		0.15	300			2,000	1.5	
Bi_2S_3	1.3					200			1,100		
Bi_2Se_3	0.27					600			675		
Bi_2Te_3	0.13		−0.95		0.58	1,200	1.68	1.07	510	1.95	
Mg_2Si		0.77	−6.4		0.46	400	2.5		70		
Mg_2Ge		0.74	−9			280	2		110		
Mg_2Sn	0.21	0.33	−3.5		0.37	320			260		
Mg_3Sb_2		0.32				20			82		
Zn_3As_2	0.93					10	1.1		10		
Cd_3As_2	0.13				0.046	15,000	0.88				
GaSe	2.05		3.8						20		
GaTe	1.66	1.80	−3.6			14	−5				
InSe	1.8					900					
TlSe	0.57		−3.9		0.3	30		0.6	20	1.5	
$CdSnAs_2$	0.23				0.05	25,000	1.7				
Ga_2Te_3	1.1	1.55	−4.8								
α-In_2Te_3	1.1	1.2			0.7				50	1.1	
β-In_2Te_3	1.0								5		
$Hg_5In_2Te_8$	0.5								11,000		
SnO_2									78		

III. Valence Bands of Semiconductors[a]

Semiconductors with Valence Band Maximum at Center of Brillouin Zone ("Γ")

Substance	Band curvature effective mass (Expressed as fraction of free electron mass): Heavy holes	Light holes	"Split-off" band holes	Energy separation of "split-off" band, ev	Measured (light) hole mobility, cm²/volt-sec
Si	0.52	0.16	0.25	0.044	500
Ge	0.34	0.043	0.08	0.3	1,820
Sn	0.3				2,400
AlAs					
AlSb	0.4			0.7	550
GaP				0.13	100
GaAs	0.8	0.12	0.20	0.34	400
GaSb	0.23	0.06		0.7	1,400
InP				0.21	150
InAs	0.41	0.025	0.083	0.43	460
InSb	0.4	0.015		0.85	750
CdTe	0.35				50
HgTe	0.5				350

Semiconductors with Multiple Valence Band Maxima

Substance	Number of equivalent valleys and direction	Curvature effective masses: Longitudinal, m_L	Curvature effective masses: Transverse, m_T	Anisotropy, $K = \frac{m_L}{m_T}$	Measured (light) hole mobility, cm²/volt-sec
PbSe	4 "L" [111]	0.095	0.047	2.0	1,500
PbTe	4 "L" [111]	0.27	0.02	10	750
Bi_2Te_3	6	0.207	~0.045	4.5	515

Table 2–22 (continued)
PROPERTIES OF SEMICONDUCTORS

IV. Conduction Bands of Semiconductors[a]

Single Valley Semiconductors

Substance	*Energy gap, ev*	*Effective mass, m_o*	*Mobility, cm^2/volt-sec*	*Comments*
GaAs	1.35	0.067	8,500	3(or 6?) equivalent [100] valleys 0.36 ev above this maximum with a mobility of ~50.
InP	1.27	0.067	5,000	3(or 6?) equivalent [100] valleys 0.4 ev above this minimum.
InAs	0.36	0.022	33,000	Equivalent valleys ~1.0 ev above this minimum.
InSb	0.165	0.014	78,000	
CdTe	1.44	0.11	1,000	4(or 8?) equivalent [111] valleys 0.51 ev above this minimum.

Multivalley Semiconductors

Substance	*Energy gap*	*Number of equivalent valleys and direction*	*Band curvature effective mass*		*Anisotropy,*
			Longitudinal, m_L	*Transverse,* m_T	$K = \frac{m_L}{m_T}$
Si	1.107	6 in [100] "Δ"	0.90	0.192	4.7
Ge	0.67	4 in [111] at "L"	1.588	0.0815	19.5
GaSb	0.67	as Ge (?)	~1.0	~0.2	~5
PbSe	0.26	4 in [111] at "L"	0.085	0.05	1.7
PbTe	0.25	4 in [111] at "L"	0.21	0.029	5.5
Bi_2Te_3	0.13	6			~0.05

[a]Room temperature data.

From Pamplin, B. R., in *Handbook of Chemistry and Physics,* 55th ed., Weast, R. C., Ed., CRC Press, Cleveland, 1974, E-99.

Table 2–23
ELASTICITY PARAMETERS FOR SOME CUBIC CRYSTALS

Symbols:

γt = transverse frequencies γe = longitudinal frequencies c = velocity of light

Material	*Stiffness constants in 10^{12} dynes/cm² at 300 K*			*Debye temperature, K, (referred to 0 K)*	*Transverse and longitudinal lattice vibration frequencies*		*KNOOP hardness, kg/mm²*
	C_{11}	C_{12}	C_{44}		γ_t/c	γ_e/c	
Diamond	10.76	1.25	5.76	—	—	—	7000
Na	0.073	0.062	0.042	156	—	—	—
Li	0.135	0.114	0.088	335	—	—	—
Ge	1.285	0.483	0.680	374	—	—	700–880
Si	1.66	0.639	0.796	648	—	—	1100–1400
GaSb	0.885	0.404	0.433	266	—	—	450
InSb	0.672	0.367	0.302	203	185	197	220
MgO	2.86	0.87	1.48	—	—	—	370
NaCl	0.487	0.124	0.126	—	—	—	—
RbBr	0.317	0.042	0.039	—	—	—	—
RbI	0.256	0.031	0.029	—	—	—	—
CsBr	0.300	0.078	0.076	—	—	—	—
CsI	0.246	0.067	0.062	—	—	—	—
InP	—	—	—	322	307	351	540
InAs	0.833	0.453	0.396	249	219	243	380
AlSb	0.894	0.443	0.416	292	318	345	360
AlP	—	—	—	588	—	—	—
AlAs	—	—	—	417	—	—	—
GaAs	1.19	0.538	0.598	344	373	297	750
GaP	—	—	—	435	—	—	950
GaSb	8.849	4.037	4.325	266	231	240	—
LiF	1.112	0.420	0.628	—	—	—	—
KCl	0.403	0.066	0.063	—	—	—	—
BaF_2	0.891	0.400	0.254	—	—	—	—
K	0.037	0.031	0.019	91.1	—	—	—
Al	1.068	0.607	0.282	428	—	—	32
Au	1.923	1.631	0.420	162.4	—	—	60

REFERENCES

Madelung, O., *Physics of III-V Compounds* (Transl. by Meyerhofer, D.), John Wiley & Sons, New York, 1964.
Willardson, R. K. and Beer, A. C., Eds., *Semiconductors and Semimetals,* Vol. 2, Physics of III-V Compounds, Academic Press, New York, 1966.
Hilsum, C. and Rose-Innes, A. C., *Semiconducting III-V Compounds,* Pergamon Press, Elmsford, N.Y., 1961.

From Bolz, R. E. and Tuve, G. L., Eds., *Handbook of Tables for Applied Engineering Science,* 2nd ed., CRC Press, Cleveland, 1973, 278.

Table 2–24
ELECTRON PARAMAGNETIC RESONANCE

Summary of Electron Paramagnetic Resonance Results in Compound Semiconductors

Host lattice	*Paramagnetic atom*	*Spin S*	*g-value*	*Fine structure constant |a|, gauss*	*Hyperfine interaction A, gauss*	*Remarks*
GaAs	Manganese	$\frac{5}{2}$	2.003	15.5	−57	Fine structure not resolved
GaAs	Iron	$\frac{5}{2}$	2.046	374	—	Hyperfine structure not resolved
GaAs	Iron or defects	$\frac{5}{2}$	2.046	374	—	Spin density 10^2 times greater than iron density
GaAs	Nickel	$\frac{3}{2}$	2.106	—	—	One very broad line, about 130 G in half-width
GaAs	Zinc	$\frac{1}{2}$	8.1	—	—	Bound hole detected by uniaxial stress
GaAs	Cadmium	$\frac{1}{2}$	6.7	—	—	Bound hole detected by uniaxial stress
GaAs	Conduction elec.	$\frac{1}{2}$	0.52	—	—	About 100 G in half-width
InSb	Conduction elec.	$\frac{1}{2}$	50.7–48.8	—	—	g-value and line width change with concentration
GaP	Manganese	$\frac{5}{2}$(?)	2.002	—	60.5	Only one line observed
GaP	Iron	$\frac{5}{2}$	2.023	429	—	Hyperfine structure not resolved

From Goldstein, B., Electron paramagnetic resonance, in *Semiconductors and Semimetals,* Vol. 2, Physics of III-V compounds, Willardson, R. K. and Beer, A. C., Eds., Academic Press, New York, 1966. With permission.

Table 2–25
LASER LINES STRONGLY ABSORBED BY THE ATMOSPHERE

Laser	*λ, microns*	*Absorber*
Atomic krypton	1.7843	H_2O
Atomic krypton	1.9211	H_2O
Tm^{+3}–$CaWO_4$	1.911	H_2O
Tm^{+3}–$CaWO_4$	1.916	H_2O
U^{+3}–SrF_2	2.472	H_2O
U^{+3}–CaF_2	2.511	H_2O
Atomic krypton	2.5234	H_2O
U^{+3}–BaF_2	2.556	H_2O
U^{+3}–CaF_2	2.613	H_2O
Atomic neon	3.391317	CH_4
Carbon monoxide	5.2 to 7	H_2O
Cesium	7.1821	H_2O
Atomic neon	18.3040	H_2O
Atomic neon	20.351	H_2O

From Eppers, W. C., Atmospheric transmission, in *Handbook of Lasers,* Pressley, R. J., Ed., The Chemical Rubber Co., Cleveland, 1971, 40.

Table 2–26
LASER LINES WITH WEAK TO MODERATE ABSORPTION BY THE ATMOSPHERE

Laser	*λ*	*Comment*
Ionized argon	4880 Å, 5145 Å	
Atomic neon (He–Ne)	6328 Å	
GaAs	8300 Å, 9200 Å	Close attention must be paid to temperature of operation; increased absorption occurs from approx. 8600 Å to 9250 Å
Ruby	6934 → 6945 Å	Strong H_2O absorptions can occur.
Nd^{+3}	≈1.06 μ	Very low absorption
Atomic neon (He–Ne)	1.1523 μ 5 lines	Moderate H_2O absorption
CH_4 Raman shift of 1.06 μ	1.53 μ	
Er^{+3} (CaF_2) (glass)	1.55 → 1.65 μ	
Ho^{+3}–$CaWO_4$	2.04 μ	Mostly clear
Ho^{+3}–YAG	to	
Ho^{+3}–CaF_2	2.128 μ	
Atomic Xe	3.50704 μ	
DF	3.8 μ	
CO_2	10.6 μ	CO_2 Water absorption

From Eppers, W. C., Atmospheric transmission, in *Handbook of Lasers,* Pressley, R. J., Ed., The Chemical Rubber Co., Cleveland, 1971, 40.

Table 2–27
SPECTRAL RANGE COVERED BY SEMICONDUCTOR LASERS

			Mode of excitation and reference			
	λ (μ)	$h\nu$ (eV)	*Injection*	*Electron beam*	*Optical*	*Avalanche*
ZnS	0.33	3.8		1	2	
ZnO[a]	0.37	3.4		3,4		
$Zn_{1-x}Cd_xS$	0.49–0.32	2.5–3.82			5	
ZnSe	0.46	2.7		6		
CdS[a]	0.49	2.5		4,7–11	12	
ZnTe	0.53	2.3		13		
GaSe	0.59	2.1		14		
$CdSe_{1-x}S_x$	0.49–0.68	2.5–1.8		15	16	
$CdSe_{0.95}S_{0.05}$	0.675	1.8			16	
CdSe	0.675	1.8		4,8,17	18,19	
$Al_{1-x}Ga_xAs$[a]	0.63–.90	2.0–1.4	20–29			
$GaAs_{1-x}P_x$[a]	0.61–0.90	2.0–1.4	30–34	35,36		
CdTe	0.785	1.6		37–39		
GaAs[a]	0.83–.91[b]	1.50–1.38	40–42	43–45	46,47	48
InP	0.91	1.36	49–50			51
$GaAs_{1-x}Sb_x$	0.95–1.5	1.4–0.83	[c]			
$CdSnP_2$	1.01	1.25		52		
$InAs_{1-x}P_x$	0.9–3.2	1.4–0.39	[c]			
$InAs_{0.94}P_{0.06}$	0.942	1.32	53			
$InAs_{0.51}P_{0.49}$	1.6	0.78	53			
GaSb	1.55	0.80	54–56	57		
$In_{1-x}Ga_xAs$	0.58–3.1	2.14–0.4	58			
$In_{0.65}Ga_{0.35}As$	1.77	0.70	58			
$In_{0.75}Ga_{0.25}As$	2.07	0.60	58			
Cd_3P_2	2.1	0.58			59	
InAs	3.1	0.39	60–63	64	65	
$InAs_{1-x}Sb_x$	3.1–5.4	0.39–0.23	66			
$InAs_{0.98}Sb_{0.02}$	3.19	0.39	66			
$Cd_{1-x}Hg_xTe$	3–15	0.41–0.08		67	67	
$Cd_{0.32}Hg_{0.68}Te$	3.8	0.33			67	
Te	3.72	0.334		68		
PbS	4.3	0.29	69	70		
InSb	5.2	0.236	71–73	74	75,76	77,78
PbTe	6.5	0.19	79	70		
$PbS_{1-x}Se_x$	3.9–8.5	0.32–0.146	80	81		
PbSe	8.5	0.146	82–85	70		
PbSnTe	28	0.045	86			
PbSnSe	8–31.2	0.155–0.040	87–89			

[a]Have lased at room temperature.
[b]Depending on temperature and doping.
[c]Expected, but not observed.

Table 2–27 (continued)
SPECTRAL RANGE COVERED BY SEMICONDUCTOR LASERS

REFERENCES

1. **Hurwitz, C. E.,** *Appl. Phys. Lett.,* 9, 116, 1966.
2. **Wang, S. and Chang, C. C.,** *Appl. Phys. Lett.,* 12, 193, 1968.
3. **Nicoll, F. H.,** *Appl. Phys. Lett.,* 9, 13, 1966; *J. Appl. Phys.,* 39, 4469, 1968.
4. **Packard, J. R., Campbell, D. A., and Tait, W. C.,** *J. Appl. Phys.,* 38, 5255, 1967.
5. **Brodin, N. S., Budnick, P. I., Vitrikhovskii, B. L., and Zakrevskii, S. V.,** Proceedings, International Conference on Physics of Semiconductors, Moscow, 1968, p. 610.
6. **Bogdankevich, O. V., Zverev, M. M., Krasilnikov, A. I., and Pechenov, A. N.,** *Phys. Status Solidi,* 19, K5, 1967.
7. **Basov, N. G., Bogdankevich, O. V., and Deviatkov, A. G.,** *Zh. Eksp. Teor. Fiziol.,* 47, 1588, 1964; *Sov. Phys. JETP,* 20, 1067, 1965.
8. **Hurwitz, C. E.,** *Appl. Phys. Lett.,* 8, 121, 1966.
9. **Nicoll, F. H.,** *Appl. Phys. Lett.,* 10, 69, 1967.
10. **Nicoll, F. H.,** *RCA Rev.,* 29, 379, 1968.
11. **Brewster, J. L.,** *Appl. Phys. Lett.,* 13, 385, 1968.
12. **Basov, N. G., Grasyuk, A. G., Zubarev, I. G., and Katulin, V. A.,** *Sov. Phys. Solid State,* 7, 2932, 1966.
13. **Hurwitz, C. E.,** *IEEE J. Quant. Electron.,* 3, 333, 1967.
14. **Basov, N. G., Bogdankevich, O. V., and Pechenov, A. N.,** *Sov. Phys. Dokl.,* 10, 329, 1965.
15. **Hurwitz, C. E.,** *Appl. Phys. Lett.,* 8, 121, 1966.
16. **Keune, D. L., Rossi, J. A., Gaddy, O. L., Merkelo, H., and Holonyak, N., Jr.,** *Appl. Phys. Lett.,* 14, 99, 1969.
17. **Vavilov, V. S. and Nolle, E. L.,** Proceedings, International Conference on Physics of Semiconductors, Moscow, 1968, p. 600.
18. **Stillman, G. E., Sirkis, M. D., Rossi, J. A., Johnson, M. R., and Holonyak, N., Jr.,** *Appl. Phys. Lett.,* 9, 268, 1966.
19. **Holonyak, N., Jr., Johnson, M. R., and Keune, D. L.,** *IEEE J. Quant. Electron.,* 4, 199, 1968.
20. **Rupprecht, H., Woodall, J. M., and Pettit, G. D.,** *IEEE J. Quant. Electron.,* 4, 35, 1968.
21. **Susaki, W., Sogo, T., and Oku, T.,** *IEEE J. Quant. Electron.,* 4, 423, 1968.
22. **Nelson, H. and Kressel, H.,** *Appl. Phys. Lett.,* 15, 7, 1969.
23. **Kressel, H. and Nelson, H.,** *RCA Rev.,* 30, 106, 1969.
24. **Hayashi, I., Panish, M. B., and Foy, P.,** *IEEE J. Quant. Electron.,* 5, 211, 1969.
25. **Panish, M. B., Hayaski, I., and Sumski, S.,** *IEEE J. Quant. Electron.,* 5, 210, 1969.
26. **Panish, M. B., Hayaski, I., and Sumski, S.,** *Appl. Phys. Lett.,* 16, 326, 1970.
27. **Hayashi, I. and Panish, M. B.,** *J. Appl. Phys.,* 41, 150, 1970.
28. **Alfreov, Zh. I., Andreev, V. M., Korolkov, V. I., Portnoi, E. L., and Tretyakov, D. N.,** *Sov. Phys. Semiconductors,* 2, 1289, 1969.
29. **Alferov, Zh. I., Andreev, V. M., Portnoi, E. L., and Trukan, M. K.,** *Sov. Phys. Semiconductors,* 3, 1107, 1969.
30. **Holonyak, N., Jr., and Bevacqua, S. F.,** *Appl. Phys. Lett.,* 1, 82, 1962.
31. **Holonyak, N., Jr.,** *Trans. Met. Soc. AIME,* 230, 276, 1964.
32. **Pilkuhn, M., and Rupprecht, H.,** *J. Appl. Phys.,* 36, 684, 1965.
33. **Pankove, J. I., Nelson, H., Tietjen, J. J., Hegyi, I. J., and Maruska, H. P.,** *RCA Rev.,* 28, 560, 1967.
34. **Pankove, J. L. and Hegyi, I. J.,** *Proc Inst. Elec. Electron. Eng.,* 56, 324, 1968.
35. **Basov, N. G., Bogdankevich, O. V., Eliseev, P. G., and Lavrushin, B. M.,** *Sov. Phys. Solid State,* 8, 1073, 1966.
36. **Kurbatov, L. N., Mashchenko, V. E., Mochalkin, N. N., Britov, A. D., and Dirochka, A. I.,** Proceedings, International Conference on Physics of Semiconductors, Moscow, 1968, p. 587.
37. **Vavilov, V. S. and Nolle, E. L.,** *Dokl. Akad. Nauk SSSR,* 164, 73, 1965.
38. **Vavilov, V. S., Nolle, E. L., and Egorov, V. D.,** *Sov. Phys. Solid State,* 7, 749, 1965; 9, 657, 1967.
39. **Basov, N. G., Bogdankevich, O. V., Eliseev, P. G., and Lavrushin, B. M.,** *Sov. Phys. Solid State,* 8, 1073, 1966.
40. **Nathan, M. I., Dumke, W. P., Burns, G., Dill, F. H., Jr., and Lasher, G. J.,** *Appl. Phys. Lett.,* 1, 62, 1962.
41. **Hall, R. N., Fenner, G. E., Kingsley, J. D., Soltys, T. J., and Carlson, R. O.,** *Phys. Ref. Lett.,* 9, 366, 1962.
42. **Quist, T. M., Rediker, R. H., Keyes, R. J., Krag, W. E., Lax, B., McWhorter, A. L., and Zeiger, H. J.,** *Appl. Phys. Lett.,* 1, 91, 1962.
43. **Hurwitz, C. E. and Keyes, R. J.,** *Appl. Phys. Lett.,* 5, 139, 1964.
44. **Cusano, D. A.,** *Appl. Phys. Lett.,* 7, 151, 1965.
45. **Bogdankevich, O. V., Borisov, N. A., Kryukova, I. V., and Lavrushin, B. M.,** *Sov. Phys. Semiconductors,* 2, 845, 1969.
46. **Basov, N. G., Gasyuk, A. Z., and Katulin, V. A.,** *Sov. Phys. Dokl.,* 10, 343, 1965.
47. **Dapkus, P. D., Holonyak, N., Jr., Rossi, J. A., Williams, F. V., and High, D. A.,** *J. Appl. Phys.,* 40, 3300, 1969.
48. **Southgate, P. D.,** *IEEE J. Quant. Electron.,* 4, 179, 1968.
49. **Weiser, K. and Levitt, R. S.,** *Appl. Phys. Lett.,* 2, 176, 1963.
50. **Burns, G., Levitt, R. S., Nathan, M. I., and Weiser, K.,** *Proc. Inst. Elec. Electron. Eng.,* 51, 1148, 1963.

Table 2–27 (continued)
SPECTRAL RANGE COVERED BY SEMICONDUCTOR LASERS

51. **Southgate, P. D. and Mazzochi, R. T.,** *Phys. Rev. Lett.,* A28, 216, 1968.
52. **Berkovskii, F. M., Goryunova, N. A., Orlov, V. M., Ryvkin, S. M., Sokolova, V. I., Tsevkova, E. V., and Shpenkov, G. P.,** *Sov. Phys. Semiconductors,* 2, 1027, 1969.
53. **Alexander, F. B., Bird, V. R., Carpenter, D. R., Manley, G. W., McDermott, P. S., Peloke, J. R., Quinn, H. F., Riley, R. J., and Yetter, L. R.,** *Appl. Phys. Lett.,* 4, 13, 1964.
54. **Chipaux, C. and Eymard, E.,** *Phys. Status Solidi,* 10, 165, 1965.
55. **Kryukova, I. V. et al.,** *Sov. Phys. Solid State,* 8, 822, 1538, 1966.
56. **Pistoulet, B. and Mathieu, H.,** Proceedings, International Conference on Physics of Semiconductors, Moscow, 1968, P. 352.
57. **Benoit-à-la Guillaume, C. and Debever, J. M.,** *Compt. Rend.,* 258, 2200, 1964.
58. **Melngailis, I., Strauss, A. J., and Rediker, R. H.,** *Proc. Elec. Electron. Eng.,* 51, 1154, 1963.
59. **Bishop, S. G., Moore, W. J., and Swiggard, E. M.,** *Appl. Phys. Lett.,* 15, 12, 1969.
60. **Melngailis, I.,** *Appl. Phys. Lett.,* 2, 176, 1963.
61. **Rodot, M., Leroux Hugon, P., Besson, J., and Lebloch, H.,** *Onde Elec.,* 45, 1197, 1965.
62. **Melngailis, I. and Rediker, R. H.,** *J. Appl. Phys.,* 37, 899, 1966.
63. **Rodot, M., Vérié, C., Marfaing, Y., Besson, J., and Lebloch, H.,** *IEEE J. Quant. Electron.,* 2, 586, 1966.
64. **Benoit-à-la Guillaume, C. and Debever, J. M.,** *Solid State Commun.,* 1, 10, 1965.
65. **Melngailis, I.,** *IEEE J. Quant. Electron.,* 1, 104, 1965.
66. **Basov, N. G., Dudenkova, A. V., Krasil'nikov, A. I., Nikitin, V. V., and Fedoseev, K. P.,** *Sov. Phys. Solid State,* 8, 847, 1966.
67. **Melngailis, I. and Strauss, A. J.,** *Appl. Phys. Lett.,* 8, 179, 1966.
68. **Benoit-à-la Guillaume, C. and Debever, J. M.,** *Solid State Commun.,* 3, 19, 1965.
69. **Butler, J. F. and Calawa, A. R.,** *J. Electrochem. Soc.,* 54, 1056, 1965.
70. **Hurwitz, C. E., Calawa, A. R., and Rediker, R. H.,** *IEEE J. Quant. Electron.,* 1, 102, 1965.
71. **Benoit-à-la Guillaume, C. and Lavallard, P.,** *Solid State Commun.,* 1, 148, 1963.
72. **Melngailis, I., Phelan, R. J., and Rediker, R. H.,** *Appl. Phys. Lett.,* 5, 99, 1964.
73. **Melngailis, I.,** *Appl. Phys. Lett.,* 6, 59, 1965.
74. **Benoit-à-la Guillaume, C. and Debever, J. M.,** *Solid State Commun.,* 2, 145, 1964.
75. **Phelan, R. J. and Rediker, R. H.,** *Appl. Phys. Lett.,* 6, 70, 1965.
76. **Phelan, R. J., Calawa, A. R., Rediker, R. H., Keyes, R. J., and Lax, B.,** *Appl. Phys. Lett.,* 3, 143, 1963.
77. **Shotov, A. P., Grishechkina, S. P., Kopilovskii, B. D., and Muminov, R. A.,** *Sov. Phys. Solid State,* 8, 865, 1998, 1966.
78. **Shotov, A. P., Grishechkina, S. P., and Muminov, R. A.,** Proceedings International Conference on Physics of Semiconductors, Moscow, 1968, p. 539.
79. **Butler, J. F., Calawa, A. R., Phelan, R. J., Jr., Harman, T. C., Strauss, A. J., and Rediker, R. H.,** *Appl. Phys. Lett.,* 5, 75, 1964.
80. **Krubatov, L. N., Britov, A. D., Aver'yanov, I. S., Mashchenko, V. E., Mochalkin, N. N., and Dirochka, A. I.,** *Sov. Phys. Semiconductors,* 2, 1008, 1969.
81. **Krubatov, L. N., Mashchenkov, V. E., Mochalkin, N. N., Britov, A. D., and Dirochka, A. I.,** Proceedings, International Conference on Physics of Semiconductors, Moscow, 1968, p. 587.
82. **Butler, J. F., Calawa, A. R., Phelan, R. J., Jr., Strauss, A. J., and Rediker, R. H.,** *Solid State Commun.,* 2, 303, 1964.
83. **Butler, J. F., Calawa, A. R., and Rediker, R. H.,** *IEEE J. Quant. Electron.,* 1, 4, 1965.
84. **Besson, J. M., Butler, J. F., Calawa, A. R., Paul, W., and Rediker, R. H.,** *Appl. Phys. Lett.,* 7, 206, 1965.
85. **Chambouleyron, I., Besson, J. M., Balkanski, M., Rodot, H., and Abrales, H.,** Proceedings, International Conference on Physics of Semiconductors, Moscow, 1968, p. 546.
86. **Butler, J. F. and Harman, T. C.,** *Appl. Phys. Lett.,* 12, 347, 1968; *IEEE J. Quant. Electron.,* 5, 50, 1969.
87. **Butler, J. F., Calawa, A. R., and Harman, T. C.,** *Appl. Phys. Lett.,* 9, 427, 1966.
88. **Harman, T. C., Calawa, A. R., Melngailis, I., and Dimmock, J. O.,** *Appl. Phys. Lett.,* 14, 333, 1969.
89. **Calawa, A. R., Dimmock, J. O., Harman, T. C., and Melngailis, I.,** *Phys. Rev. Lett.,* 23, 7, 1969.

From Pankove, J. I., Injection lasers, in *Handbook of Lasers,* Pressley, R. J., Ed., The Chemical Rubber Co., Cleveland, 1971, 368.

Table 2–28
CONTINUOUS-WAVE INSULATING CRYSTAL LASERS

Laser ion	*Host*	*Sensitizer ion(s)*	*Wave-length, μm*	*Tempera-ture, K*	*Optical pump*	*Power, W*	*Efficiency,[a] percent*	*Refer-ence*
IRON GROUP IONS								
Cr^{3+}	Al_2O_3		0.694	300	Hg	2.4	~0.1	1,2
			0.694[b]	4.2, 77	Ar laser			3
Ni^{2+}	MgF_2		1.67	85	W	1	0.2	4
Ni^{2+}	MnF_2		1.93	85	W			4
DIVALENT RARE EARTH IONS								
Dy^{2+}	CaF_2		2.36	77	W	1.2	0.06	5
			2.36	27	Sunlight			6
Tm^{2+}	CaF_2		1.12	27	Hg			7
TRIVALENT RARE EARTH IONS								
Nd^{3+}	$Ca(NbO_3)_2$		1.06	300	Xe	0.12	0.05	8
Nd^{3+}	$Ca_5(PO_4)_3F$		1.06	300	W	1.3	0.2	9
Nd^{3+}	$CaWO_4$		1.06	300	Xe	<0.1	~0.01	10
			1.06	300	Hg	~0.01	~0.01	11
			1.06	85	Hg	0.5	0.03	12
Nd^{3+}	LaF_3		1.04	300				13
Nd^{3+}	$YAlO_3$ (b axis)		1.06	300	Kr	35	0.8	14
			1.08	300	Kr	100	1.8	15
Nd^{3+}	$YAlO_3$ (c axis)		1.06	300	Kr	6.5	0.3	16
Nd^{3+}	$Y_3Al_5O_{12}$		1.06	300	W	~25	1.0	17
			1.06	300	Kr	250	2.1	18
			1.06	300	Kr	1 100[c]	2.0	19
			1.06	300	Plasma arc	200	0.2	20
			1.06	300	Na-doped Hg	0.5	0.2	21
			1.32	300	W	0.03	~0.01	22
		Cr^{3+}	1.06	300	Hg	10	0.4	23
Ho^{3+}	CaF_2	Er^{3+}, Tm^{3+}, Yb^{3+}	2.1	77	Xe			24
Ho^{3+}	Er_2O_3	Er^{3+}	2.12	77	W			25
Ho^{3+}	$Er_{1.5}Y_{1.5}Al_5O_{12}$	Er^{3+}	2.10	85	Hg, W			26
Ho^{3+}	$Y_3Al_5O_{12}$	Cr^{3+}	2.10	85	W			26
		Cr^{3+}	2.12	85	Hg, W			26
		Er^{3+}, Tm^{3+}	2.12	85	W	15	5	28
			2.12	77	W-I	20	3.5	27
Tm^{3+}	CaF_2	Er^{3+}	1.9	77	Xe			24
Tm^{3+}	Er_2O_3	Er^{3+}	1.93	77	W			29

[a]Overall efficiency: laser output/electrical energy input to pump lamp. Because the efficiency depends on a number of factors, such as rod quality, pump cavity efficiency, and output coupling efficiency, the original references should be consulted when comparing values.
[b]Nonspiking, single-mode operation.
[c]Multiple laser rods in series inside one resonant cavity.

Table 2–28 (continued)
CONTINUOUS-WAVE INSULATING CRYSTAL LASERS

REFERENCES

1. **Evtuhov, V. and Neeland, J. K.,** *J. Appl. Phys.*, 38, 4051, 1967.
2. **Roess, D.,** *IEEE J. Quant. Electron.*, 2, 208, 1966.
3. **Birnbaum, M., Wendzikowski, P. H., and Fincher, C. L.,** *Appl. Phys. Lett.*, 1b, 436, 1970.
4. **Johnson, L. F., Guggenheim, H. J., and Thomas, R. A.,** *Phys. Rev.*, 149, 179, 1966.
5. **Pressley, R. J. and Wittke, J. P.,** *IEEE J. Quant. Electron.*, 3, 1966, 1967.
6. **Kiss, Z. J., Lewis, H. R., and Duncan, R. C.,** *Appl. Phys. Lett.*, 2, 93, 1963.
7. **Duncan, R. C. and Kiss, Z. J.,** *Appl. Phys. Lett.*, 3, 23, 1963.
8. **Bagdasarov, Kh. S., Gritsenko, M. M., Zubkova, F. M., Kaminskii, A. A., Kevorkov, A. M., and Li, L.,** *Sov. Phys. Crystallogr.*, 15, 323, 1970.
9. **Hopkins, R. H.** et al., Technical Report AFAL-TR-69-239, Air Force Avionics Laboratory, Wright-Patterson Air Force Base, Ohio, 1969.
10. **Kaminskii, A. A., Kornienko, L. S., Maksimova, G. V., Osiko, V. V., Prokhorov, A. M., and Shipulo, G. P.,** *Sov. Phys. JETP*, 22, 22, 1966.
11. **Johnson, L. F., Boyd, G. D., Nassau, K., and Soden, R. R.,** *Phys. Rev.*, 126, 1406, 1962.
12. **Johnson, L. F.,** in *Quantum Electrics Proceedings of the Third International Congress*, Grivet, P. and Bloembergen, N., Eds., Columbia University Press, New York, 1964, 1021.
13. **Voronko, Yu. K., Dmitruk, M. V., Kaminskii, A. A., Osiko, V. V., and Shpakov, V. N.,** *Sov. Phys. JETP*, 27, 400, 1968.
14. **Massey, G. A. and Yarborough, J. M.,** *Appl. Phys. Lett.*, 18, 576, 1971.
15. **Bass, M. and Weber, M. J.,** *Laser Focus*, 7, 34, 1971.
16. **Weber, M. J., Bass, M., Monchamp, R. R., and Comperchio, E.,** Technical Report AFML-TR-70-258, Air Force Materials Laboratory, Wright-Patterson Air Force Base, Ohio, 1970.
17. **Geusic, J. E.,** *NEREM Rec.*, 8, 192, 1966.
18. **Koechner, W.,** *Laser Focus*, 5, 29, 1969.
19. **Erickson, K.,** *Laser Focus*, 6, 16, 1970; private communication.
20. **Jackson, J. E. and Yenni, D. M.,** Second Interim Technical Report Contract SRCR-66-4, 1966.
21. **Schlecht, R. G., Church, C. H., and Larson, D. A.,** (abstract), *IEEE J. Quant. Electron.*, 2, 48, 1966.
22. **Smith, R. G.,** *IEEE J. Quant. Electron.*, 4, 505, 1968.
23. **Pressley, R. J., Collard J. R., Goedertier, P. V., Sterzer, F., and Zernik, W.,** Technical Report AFAL-TR-66-129, Air Force Avionics Laboratory, Wright-Patterson Air Force Base, Ohio, 1966.
24. **Voronko, Yu. K., Dmitruk, M. V., Murina, T. M., and Osiko, V. V.,** *Izv. Akad. Nauk SSSR Neorgan. Materialy*, 5, 506, 1969.
25. **Soffer, B. H. and Hoskins, R. H.,** *IEEE J. Quant. Electron.*, 2, 253, 1966.
26. **Johnson, L. F., Geusic, J. E., and Van Uitert, L. G.,** *Appl. Phys. Lett.*, 7, 127, 1965.
27. **Devor, D. P., Soffer, B. H., and Moss, G. E.,** Technical Report AFAL-TR-71-181, Air Force Avionics Laboratory, Wright-Patterson Air Force Base, Ohio, 1971.
28. **Johnson, L. F., Geusic, J. E., and Van Uitert, L. G.,** *Appl. Phys. Lett.*, 8, 200, 1966.
29. **Soffer, B. H. and Hoskins, R. H.,** *Appl. Phys. Lett.*, 6, 200, 1965.
30. **Boyd, G. D., Collins, R. J., Porto, S. P. S., Yariv, A., and Hargreaves, W. A.,** *Phys. Rev. Lett.*, 8, 269, 1962.

From Weber, M. J., Insulating crystal lasers, in *Handbook of Lasers*, Pressley, R. J., Ed., The Chemical Rubber Co., Cleveland, 1971, 393.

Table 2–29
PROPERTIES OF SOLID-LASER MATERIALS

The host crystal in which the best laser results were obtained is discussed in detail. Other host crystals are listed in the "Remarks" column.

Note: Numbers in parentheses designate temperatures in Kelvin.

Laser material	*Output wavelength, microns*	*Operating mode (and temperature, K)*	t_{spont}, *milli-seconds*	*Pulse threshold in joules (and temperature, K)*	*Useful absorption regions, microns*	*Position of terminal level, E_1 (cm^{-1})*	*Laser transition*	*Remarks*
$Cr^{3+}:Al_2O_3$ (ruby) 10^{19} atoms/cc	0.6934 (R_1, 77°)	cw (77) pulsed (350)	3	~800 (77)	0.5–0.6 0.32–0.44	0	$^2E(\bar{E})\rightarrow{}^4A_2$	"Spiking" observed in both pulsed and cw operation
	0.6929 (R_2, 290°)	pulsed			0.5–0.6 0.32–0.44	0	$^2E(2\bar{A})\rightarrow{}^4A_2$	
$Cr^{3+}:Al_2O_3$ $n \sim 10^{20}$	0.701, 0.704	pulsed (77)				~100		Due to paired chromium ions
$U^{3+}:CaF_2$	2.613	pulsed (300) cw (~100)	0.13 (77)	1 (77)	~0.9	609	$^4I_{11/2}\rightarrow{}^4I_{9/2}$	"Spiking" present in pulsed but not in cw operation. Pulsed emission also observed in BaF_2 at 2.556 μ and in SrF_2 at 2.472 μ, 2.408 μ
	2.438	pulsed (77)						
	2.511	pulsed (77)		6 (77)				
	2.223	pulsed (77)		2000 (77)		0		
$Nd^{3+}:CaWO_4$	1.065	cw (300)		~1 (77)	0.57–0.6	~2000	$^4F_{3/2}\rightarrow{}^4I_{11/2}$	"Spiking" present in pulsed but not in cw operation. Laser emission from Nd^{3+} also observed in $SrMO_4$, $SrWO_4$, $CaMO_4$, $PbMO_4$, CaF_2, SrF_2, BaF_2, and LaF_3
	1.063	pulsed		14 (77)				
	1.066	pulsed	~0.1 (77)	6 (77)				
	1.058	pulsed		80 (77)				
	1.064	pulsed		7 (77)				
Nd^{3+}:Glass	1.06	pulsed (300)		~50		~2000	$^4F_{3/2}\rightarrow{}^4I_{11/2}$	
$Pr^{3+}:CaWO_4$	1.047	pulsed (77)	0.05 (77)	15 (77)	0.45–0.5	377	$^1G_4\rightarrow{}^3H_4$	Pulsed laser emission of Pr^{3+} was also detected in $SrMO_4$

Table 2–29 (continued)
PROPERTIES OF SOLID-LASER MATERIALS

Laser material	*Output wavelength, microns*	*Operating mode (and temperature, K)*	*t_{spont}, milli-seconds*	*Pulse threshold in joules (and temp-erature, K)*	*Useful absorption regions, microns*	*Position of termi-nal level, E_1 (cm^{-1})*	*Laser transition*	*Remarks*
$Dy^{2+}:CaF_2$	2.36	cw (90)	~10 (77)	20 (77)	0.8–1 0	35	$^5I_7 \rightarrow {}^5I_8$	"Spiking" in pulsed operation but not in continuous
$Tm^{3+}:CaWO_4$	1.911 1.916	pulsed (77) pulsed (77)		60 (77) 73 (77)	{0.46–0.48 {1.7–1 8	~325	$^3H_4 \rightarrow {}^3H_6$	1.918 μ emission also observed. Laser emission also observed in SrF_2
$Er^{3+}:CaWO_4$	1.612	pulsed (77)		800 (77)	0.38 0.52	375	$^4I_{13/2} \rightarrow {}^4I_{15/2}$	
$Ho^{3+}:CaWO_4$	2.046 2.059	pulsed (77) pulsed (77)		80 (77) 250 (77)	0.44–0.46	~230	$^5I_7 \rightarrow {}^5I_8$	
$Tm^{2+}:CaF_2$	1.116	pulsed (~4)	4	50 (4)	0.28–0.34 0.39–0.46 0.53–0.63	0	$^2F_{5/2} \rightarrow {}^2F_{7/2}$	
$Sm^{2+}:CaF_2$	0.708	pulsed (20)	0.002	0.01 (20)	0.425–0.5 0.59–0.65	263	$^5D_0 \rightarrow {}^7F_1$	No "spiking" in pulsed operation. Laser action at 0.6969 μ also observed in $SrF_2:Sm^{2+}$
Yb^{3+}:Glass	1.015	pulsed (77)	1.5	1300	~0.91 ~0.95 ~0.98		$^2F_{5/2} \rightarrow {}^2F_{7/2}$	
Gd^{3+}:Glass	0.3125	pulsed (77)	4 (300)		0.274 0.277		$^6P_{7/2} \rightarrow {}^8S_{7/2}$	
Ho^{3+}:Glass	$\lambda > 1.95\ \mu$	pulsed (77)	~0.7 (77)	3600 (77)	0.44–0.46		$^5I_7 \rightarrow {}^5I_8$	

From Yariv, A. and Gordon, J. P., *Proc. Inst. Elec. Electron. Eng.*, 51(1), 13, 1963. With permission.

Table 2–30
PROPERTIES OF LINEAR ELECTROOPTICAL MATERIALS

The refractive index of a crystal is described by an ellipsoid (indicatrix),

$$B_{ij}X_iX_j = 1 \equiv B_{11}X_1^2 + B_{22}X_2^2 + B_{33}X_3^2 + 2B_{23}X_2X_3 + 2B_{13}X_1X_3 + 2B_{12}X_1X_2,$$

in which summation over repeated indices is understood and $B_{ij} = B_{ji}$. By definition,

$$B_{ij} = \xi_0\, \partial E_i/\partial D_j = \frac{1}{\epsilon}$$

where ξ_0 is the vacuum permittivity and ξ is the relative dielectric constant.

The electrooptic coefficient $r_{ij,k} \equiv r_{lk}$ is is defined by

$$\Delta B_{ij} = r_{ij,k}E_k$$

$$\Delta B_l = r_{lk}E_k,$$

in which the indices i, j, and k each cover the rectangular coordinate axes 1, 2, and 3, and $l = (ij)$ refers to the six reduced combinations 1 = (11), 2 = (22), 3 = (33), 4 = (23), 5 = (13), and 6 = (12).

If $r_{ij,k}$ is determined at constant strain – for example, by making a measurement at high frequencies well above acoustic resonances of the sample—the crystal is clamped, at indicated by the letter (S) or by $r^S_{ij,k}$. If $r_{ij,k}$ is determined at constant stress—for example, at low frequencies well below the acoustic resonances of the sample—the crystal is free, as indicated by the letter T or by $r^T_{ij,k}$.

In a principal axis system where $B_{ij} = 0$ for $i \neq j$, and $B_{ii} \equiv B_i \equiv 1/\epsilon$,

$$\Delta B_{ij} = -\Delta\epsilon_{ij}/\epsilon_i\epsilon_j,$$

for $\Delta\epsilon_{ij} \ll \epsilon_i, \epsilon_j$)

Typical accuracies for r_{lk} are ±15%. References containing more extensive wavelength and temperature dependence are indicated by λ and t, respectively. Electrooptic coefficients are stated in SI units (m/volt); to convert CGS units (cm/statvolt), multiply by 3 × 10^4. Unless stated explicitly, the signs of r_{lk} have not been determined.

A. ABO_3-type Compounds

Material and symmetry (critical temp, K)	*Electrooptic coefficients*			*Refractive index*		*Dielectric constant*
	r_{l3}, 10^{-12} m/V	r_{lk}, 10^{-12} m/V	λ, μm	n_i	λ, μm	ε_i
$LiNbO_3$, *3m*	$(T)r_c = 19$	$(T)r_{22} = 7$	.633[23]	$n_1 = n_2 = 2.378\,0$	.45[32,λ]	$(T)\varepsilon_1 = \varepsilon_2 = 78$[31,t]
(1 470)	$(T)r_c = 17.4$	$(T)r_{22} = 3.2$	.633[24]	2.271 6	.70[33,t,λ]	$(T)\varepsilon_3 = 32$
		$(T)r_{51} = 32$		2.237 0	1.00[35,t,λ]	$(S)\varepsilon_1 = \varepsilon_2 = 43$[42]
	$(T)r_{33} = +32.2$	$(T)r_{22} = 6.8$	.633[25,t]	2.197 4	2.00	$(S)\varepsilon_3 = 28$
	$(T)r_{13} = +10$			2.115 5	4.00	
	$(T)r_c = 18$	$(T)r_{22} = 6.7$	.633[28]	$n_3 = 2.277\,2$	.45	
	$(T)r_c = 17$	$(T)r_{22} = 5.7$	1.15[28]	2.187 4	.70	
	$(T)r_c = 16$	$(T)r_{22} = 3.1$	3.39[28]	2.156 7	1.00	
	All$(T)r_{ij} > 0$		.633[26]	2.125 0	2.00	
	$(S)r_{33} = +30.8$	$(S)r_{22} = 3.4$	.633[29]	2.055 3	4.00	
	$(S)r_{13} = +8.6$	$(S)r_{51} = +28$				
	$(S)r_{33} = 28$	$(S)r_{22} = 3.1$	3.39[30]			
	$(S)r_{13} = 6.5$	$(S)r_{51} = 23$				
$LiTaO_3$, *3m*	$(T)r_c = 22$		.633[34]	$n_1 = n_2 = 2.183\,4$	.60[36,λ]	$(T)\varepsilon_2 = \varepsilon_1 = 51$[37]
(890)	$(S)r_{33} = 30.3$	$(S)r_{51} = 20$	.633[34]	2.130 5	1.20	$(T)\varepsilon_3 = 45$
	$(S)r_{13} = 7$	$(S)r_{22} \approx 1$		2.033 5	4.00	$(S)\varepsilon_2 = \varepsilon_1 = 41$
	$(S)r_{33} = 27$	$(S)r_{51} = 15$	3.39[30]	$n_3 = 2.187\,8$	.60	$(S)\varepsilon_3 = 43$
	$(S)r_{13} = 4.5$	$(S)r_{22} \approx .3$		2.134 1	1.20	
	All$(T)r_{ij} > 0$		.633[27]	2.037 7	4.00	
$Ba_2NaNb_5O_{15}$, *mm2*	$(T)r_c = 34$		.633[40]	$n_1 = 2.322$	.633[39,λ]	$(T)\varepsilon_1 = 235$[41]
(833)	$(T)r_{33} = 48$	$(T)r_{42} = 92$	.633[39]	$n_2 = 2.321$		$(T)\varepsilon_2 = 247$
	$(T)r_{13} = 15$	$(T)r_{51} = 90$		$n_3 = 2.218$		$(T)\varepsilon_3 = 51$
	$(T)r_{23} = 13$					$(S)\varepsilon_1 = 222$
	$(S)r_{33} = +29$	$(S)r_{42} = 75$	.633[38]			$(S)\varepsilon_2 = 227$
	$(S)r_{23} = 8$	$(S)r_{51} = 88$				$(S)\varepsilon_3 = 32$
	$(S)r_{13} = 7$					

Table 2–30 (continued)
PROPERTIES OF LINEAR ELECTROOPTICAL MATERIALS

B. KDP- and ADP-type Crystals

Material	T_c, *K* [*1*]	*Electrooptic coefficients*		*Refractive index*		*Dielectric constant*	
		r_{63}, 10^{-12} *m/V*	r_{41}, 10^{-12} *m/V*	n_3	n_1	ε_3	ε_1
KH_2PO_4 (KDP)	123	(*T*)−10.5[2]; [3,λ; 4,*t*] (*T*)9.37[5] (*S*)8.8[11]; [3,λ] (*S*)8.15[12]; $r_{63}<0$[6]	+8.6[2] $r_{41}<0$[7]	1.47	1.51[8,9—λ,*t*]	(*T*)21 (*S*)21	42[10] 44[13]
KD_2PO_4 (DKDP)	222	(*T*)26.4[14]; [3,λ; 4,*t*] (*S*)24.0[15]; .93r_{63}^T [3,λ]	8.8[16]	1.47	1.51[9—λ,*t*]	(*T*)50[14] (*S*)48	58[17]
$NH_4H_2PO_4$ (ADP)	148[a]	(*T*)−8.5[2]; [19,3,λ] (*S*)5.5[2]; 4.1[21]; [3,λ]	24.5[16], 23.1[20] $r_{41}<0$[7]	1.48	1.53[8,9—λ,*t*]	(*T*)15 (*S*)14	56[10] 58[17]
$ND_4D_2PO_4$ (DADP)	242[a]	(*T*)11.9[18,λ]; [22,*t*]			1.52[18]		

[a] Antiferroelectric transition.

C. AB-type Compounds

Material and symmetry	*Electrooptic coefficients*		*Refractive index*		*Dielectric constant,* ε_i
	r_{lk}, 10^{-12} *m/V*	λ, *μm*	n_i	λ, *μm*	
GaAs, *$\bar{4}3m$*	(*S*)r_{41} = 1.2 (*S*)r_{41} = −1.5 (*S*+*T*)r_{41} = 1.2 to 1.6 (*T*)r_{41} = 1.0–1.2 (*T*)r_{41} = 1.6 (*S*)r_{41} = 1.5	.9–1.08[43] 3.39[44] 1.0–3.0[45] 4.0–12.0[45] 10.6[47,48] Raman scat. [48,49]	n_o = 3.60 n_o = 3.50 n_o = 3.42 n_o = 3.30	.9[50] 1.02[50] 1.25[50] >5.0[51]	(*S*)ε = 13.2[52] (*S*)ε = 12.3[53] (*T*)ε = 12.5[51]
CdTe, *$\bar{4}3m$*	(*T*)r_{41} = 6.8 (*T*)r_{41} = 6.8 (*T*)r_{41} = 5.5 (*T*)r_{41} = 5.0	3.39[59] 10.6[59] 23.35[60] 27.95[60]	n_o = 2.82 n_o = 2.60 n_o = 2.58 n_o = 2.53	1.3[54] 10.6[61] 23.34[61] 27.95[61]	(*S*)ε = 9.4[53]
CdS, *6mm*	(*T*)r_c = 4 (*T*)r_{51} = 3.7 (*T*)r_c = 5.5 (*S*)r_{33} = 2.4 (*S*)r_{13} = 1.1	.589[62] .589[62] 10.6[47] .633[58] .633[58]	n_3 = 2.48 n_1 = n_2 = 2.46 n_3 = 2.3	.63[55] .63[55] 10.0[55]	(*T*)ε_3 = 10.33[56] (*T*)ε_1 = 9.35[56] (*S*)ε_1 = 9.02[56]; 8.7[57] (*S*)ε_3 = 9.53[56]; 9.25[57]

Condensed from Kaminow, I. P. and Turner, E. H., Linear electrooptical materials, in *Handbook of Lasers,* Pressley, R. J., Ed., The Chemical Rubber Co., Cleveland, 1971, 447.

Table 2–30 (continued)
PROPERTIES OF LINEAR ELECTROOPTICAL MATERIALS

REFERENCES

1. **Jona, F. and Shirane**, in *Ferroelectric Crystals,* Macmillan Company, Riverside, N.J., 1962.
2. **Carpenter, R. O'B.,** The electrooptic effect in crystals of the dihydrogen phosphate type, Part III, Measurement of coefficients, *J. Opt. Soc. Am.,* 40, 225, 1950; Electrooptic sound-on-film modulator, *J. Opt. Soc. Am.,* 25, 1145, 1953.
3. **Vasilevskaya, A. S.,** The electrooptic properties of crystals of KDP-type, *Sov. Phys. Crystallogr.,* 11, 644, 1967; Ward, J. F. and Frankcen, P. A., Structure of nonlinear optical phenomena in potassium dihydrogen phosphate, *Phys. Rev.,* 133, A-183, 1964.
4. **Sonin, A. S., Vasilevskaya, A. S., and Strukov, B. A.,** Electrooptic properties of potassium dihydrogen phosphate and deuterated potassium dihydrogen phosphate in the region of their phase transitions, *Sov. Phys. Solid State,* 8, 2758, 1967.
5. **Blokh, O. G.,** Dispersion of r^{63} for crystals of ADP and KDP, *Sov. Phys. Crystallogr.,* 7, 509, 1962.
6. **Turner, E. H.,** unpublished data.
7. **Ward, J. F. and New, G. H. C.,** Optical rectification in ammonium dihydrogen phosphate, potassium dihydrogen phosphate, and quartz, *Proc. R. Soc.,* 4299, 238, 1967.
8. **Zernike, F., Jr.,** Refractive indices of ADP and KDP between 0.2 and 1.5 μ, *J. Opt. Soc. Am.,* 54, 1215, 1964; Correction, *J. Opt. Soc. Am.,* 55, 210E, 1965; Vishnevskii, V. N. and Stefanski, I. V., Temperature dependence on the dispersion of the refractivity of ADP and KDP single crystals, *Opt. Spektrosk.,* 20, 195, 1966.
9. **Yamazaki, M. and Ogawa, T.,** Temperature dependences of the refractive indices of $NH_4H_2PO_4$, and partially deuterated KH_2PO_4, *J. Opt. Soc. Am.,* 56, 1407, 1966; Phillips, R. A., Temperature variation of the index of refraction of ADP, KDP, and deuterated KDP, *J. Opt. Soc. Am.,* 56, 629, 1966.
10. **Berlincourt, D. A., Curran, D. R., and Jaffe,** in *Physical Acoustics,* Vol. 1. Part A, Mason, W. P., Ed., Academic Press, New York, 1964, 169.
11. **Rosner, R. D., Turner, E. H., and Kaminow, I. P.,** Clamped electrooptic coefficients of KDP and quartz, *Appl. Opt.,* 6, 778, 1967.
12. **Ohm, E. A.,** A linear optical modulator with high FM sensitivity, *Appl. Opt.,* 6, 1233, 1967.
13. **Kaminow, I. P. and Harding, G. O.,** Complex dielectric constant of KH_2PO_4 at 9.2 Gc/sec., *Phys. Rev.,* 129, 1562, 1963.
14. **Sliker, T. R. and Burlage, S. R.,** Some dielectric and optical properties of KD_2PO_4 *J. Appl. Phys.,* 34, 1837, 1963.
15. **Christmas, T. M. and Wildey, C. G.,** Pulse measurement of r_{63}, in KD*P, *Electron. Lett.,* 6, 152, 1970.
16. **Ott, J. H. and Sliker, T. R.,** Linear electrooptic effects in KH_2PO_4, and its isomorph, *J. Opt. Soc. Am.,* 54, 1442, 1964.
17. **Kaminow, I. P.,** Microwave dielectric properties of $NH_4H_2PO_4$, KH_2AsO_4, and partially deuterated KH_2PO_4, *Phys. Rev.,* 138, A1539, 1965.
18. **Adhav, R. S.,** Linear electro-optic effects in tetragonal phosphates and arsenates, *J. Opt. Soc. Am.,* 59, 414, 1969.
19. **Koetser, H.,** Measurement of r_{63} for ADP up to electric breakdown, *Electron. Lett.,* 3, 52, 1967.
20. **Ley, J. M.,** Low-voltage light-amplitude modulation, *Electron. Lett.,* 2, 12, 1966.
21. **Silverstein, L., and Sucher, M.,** Determination of the Pockels electro-optic coefficient in ADP at 5.5 Ghz, *Electron. Lett.,* 2, 437, 1966.
22. **Vasilevskaya, A. S.,** Electrooptic and elastooptical properties of deuterated ammonium dihydrogen phosphate crystals, *Sov. Phys. Solid State,* 8, 2756, 1967.
23. **Lenzo, P. V., Spencer, E. G., and Nassau, K.,** Electrooptic coefficient in lithium niobate, *J. Opt. Soc. Am.,* 56, 633, 1966.
24. **Bernal, E., Chen, G. D., and Lee, T. C.,** Low frequency electrooptic and dielectric constants of lithium niobate, *Phys. Lett.,* 21(3), 259, 1966.
25. **Zook, J. D., Chen, D., and Otto, G. N.,** Temperature dependence and model of the electrooptic effect in $LiNbO_3$, *Appl. Phys. Lett.,* 11(5), 159, 1967.
26. **Hulme, K. F., Davies, P. H., and Cound, V. M.,** The signs of the electrooptic coefficients for lithium niobate, *J. Phys. C (Solid State Phys.,)* 2, 855, 1969.
27. **Luther-Davies, B.** et al., The signs of the electrooptic coefficients for lithium tantalate, *J. Phys. C (Solid State Phys.),* 3, L106, 1970.
28. **Smakula, P. H. and Clapsy, P. C.,** The electrooptic effect in $LiNbO_3$ and KTN, *Trans. AIME,* 239, 421, 1967.
29. **Turner, E. H.,** High-frequency electrooptic coefficient of lithium niobate, *Appl. Phys. Lett.,*8(11), 303, 1966.Signs of coefficients were determined later; determination to be published.
30. **Turner, E. H.,** Electrooptic coefficients of some crystals at 3.39 microns, Paper Th A13, Optical Society of America, San Francisco, 1966.
31. **Nassau, K., Levinstein, H. J., and Loiacono, G. M.,** Ferroelectric lithium niobate, 2. Preparation of single domain crystals, *J. Phys. Chem. Solids,* 27, 989, 1966.

Table 2–30 (continued)
PROPERTIES OF LINEAR ELECTROOPTICAL MATERIALS

32. **Boyd, G. D.** et al., $LiNbO_3$: an efficient phase matchable nonlinear optical material, *Appl. Phys. Lett.*, 15(11), 234, 1964.
33. **Iwasaki, H.**, et al., Piezoelectric and optical properties of $LiNbO_3$ single crystals, *Rev. Elec. Commun. Lab.*, 16(5), 385, 1968.
34. **Lenzo, P. V.** et al., Electrooptic coefficients and elastic wave propagation in single domain ferroelectric lithium tantalate, *Appl. Phys. Lett.*, 8(4), 81, 1966.
35. **Boyd, G. D., Bond, W. L., and Carter, H. L.**, Refractive index as a function of temperature in $LiNbO_3$, *J. Appl. Phys.*, 38(4), 1941, 1967.
36. **Bond, W. L.**, Measurement of the refractive indices of several crystals, *J. Appl. Phys.*, 36, 1674, 1965.
37. **Warner, A. W., Onoe, M., and Coquin, G. A.**, Determination of elastic and piezoelectric constants for crystals in class (3m), *J. Acoust. Soc. Am.*, 42, 1223, 1967.
38. **Turner, E. H.**, unpublished data.
39. **Singh, S., Draegert, D. A., and Geusic, J. E.**, Optical and ferroelectric properties of barium sodium niobate, *Phys. Rev.*, 2B, 2709, 1970.
40. **Byer, R. L.**, et al., Nonlinear optical properties of $Ba_2NaNb_5O_{15}$ in the tetragonal phase, *J. Appl. Phys.*, 40(1), 444, 1969.
41. **Warner, A. W., Coquin, G. A., and Fink, J. L.**, Elastic and piezoelectric constants of $Ba_2NaNb_5O_{15}$, *J. Appl. Phys.*, 40, 4353, 1969.
42. **Kaminow, I. P., and Turner, E. H.**, Electrooptic light modulators, *Proc. Inst. Elect. Electron. Eng.*, 54, 1374, 1966.
43. **Singh, S.**, private communication.
44. **Turner, E. H. and Kaminow, I. P.**, Electrooptical effect in gallium arsenide, *J. Opt. Soc. Am.*, 53, 523, 1963.
45. **Turner, E. H.**, to be published.
46. **Walsh, T. E.**, Gallium arsenide electrooptic modulators, *RCA Rev.*, 27, 323, 1966.
47. **Yariv, A., Mead, C. A., and Parker, J. V.**, GaAs as an electrooptic modulator at 10.6 microns, *IEEE J. Quant. Electron.*, 2, 243, 1966.
48. **Kaminow, I. P.**, Measurements of the electrooptic effect in CdS, ZnTe, and GaAs at 10.6 microns, *IEEE J. Quant. Electron.*, 4, 23, 1968.
49. **Mooradian, A. and McWhorter, A. L.**, Light scattering from plasmons and phonons in GaAs, in *Light Scattering Spectra of Solids*, Wright, G. B., Ed., Springer-Verlag, New York, 1969, 297.
50. **Johnston, W. D., Jr. and Kaminow, I. P.**, Contributions to optical nonlinearity in GaAs as determined from Raman scattering efficiencies, *Phys. Rev.*, 188, 1209, 1969.
51. **Marpel, D. T. F.**, Refractive index of GaAs, *J. Appl. Phys.*, 35, 1241, 1964.
52. **Hambleton, K. G., Hilsum, C., and Holeman, B. R.**, Determination of the effective ionic charge of gallium arsenide from direct measurement of the dielectric constant, *Proc. Phys. Soc.*, 77, 1147, 1961.
53. **Jones, S. and Mao, S.**, Further investigation of the dielectric constant of gallium arsenide, *J. Appl. Phys.*, 39, 4038, 1968.
54. **Johnson, C. J., Sherman, G. H., and Weil, R.**, Far infrared measurement of the dielectric properties of GaAs and CdTe at 300k and 8K, *J. Appl. Opt.*, 8, 1667, 1969.
55. **Marple, D. T. F.**, Refractive index of ZnSe, ZnTe, and CdTe, *J. Appl. Phys.*, 35(3), 539, 1964.
56. **Shiozawa, L. R. and Jost, J. M.**, Research on II–VI Compound Semiconductors, Report AD 620297, Clevite Corporation, Palo Alto, Calif., 1965.
57. **Berlincourt, D., Jaffe, H., and Shiozawa, L. R.**, Electroelastic properties of the sulfides, selenides, and tellurides of zinc and cadmium, *Phys. Rev.*, 129, 1009, 1963.
58. **Barker, A. S., Jr. and Summers, C. J.**, Infrared dielectric function of CdS, *J. Appl. Phys.*, 41, 3552, 1970.
59. **Turner, E. H.**, to be published.
60. **Kiefer, J. E. and Yariv, A.**, Electrooptic characteristics of CdTe at 3.39 and 10.6 μ, *Appl. Phys. Lett.*, 15(1), 26, 1969.
61. **Johnson, C. J.**, Electrooptic effect in CdTe at 23.35 and 27.95 microns, *Proc. Inst. Elect. Electron. Eng.*, 56, 1719, 1968.
62. **Lorimer, O. G. and Spitzer, W. G.**, Infarared refractive index and absorption of InAs and absorption of InAs and CdTe, *J. Appl. Phys.*, 36, 1841, 1965.
63. **Gainon, D. J. A.**, Linear electrooptic effect in CdS, *J. Opt. Soc. Am.*, 54, 270, 1964.

Table 2–31
MAGNETIC MATERIALS AND UNITS

A. Units and Conversions

Name	*Symbol*	*cgs unit*	*Conversion, SI to cgs*
Force	F	dyne	1 newton = 100,000 dynes
Magnetomotive force	$\mathcal{F}$	gilbert	1 ampere-turn = 0.4π gilbert
Power	P	ergs/sec	1 watt = 10^7 ergs/sec
Flux	Φ	maxwell	1 weber = 10^8 maxwell
Flux density[a]	B	gauss	1 tesla = 1 weber/meter^2 = 10,000 gauss
Magnetizing force[b]	H	oersted	1 ampere-turn/meter = 4π oersted/1,000

[a]Sometimes called intensity of field or magnetic induction. (Retentivity of magnetization is also expressed in gauss.)
[b]Sometimes called magnetic field strength.

B. Properties of Magnetic Materials

Name	*Symbols*	*Definition*	*Examples*
Saturation flux density	B_s	Field intensity when material is saturated	Ferrite: 3,500 Cobalt-iron: 22,000
Remanence or retentivity	B_r	Magnetization remaining after mmf is removed	Ferrite: 1,000 Pure iron: 10,000 Co-Fe-V: 14,000
Coercivity	H_c	Coercive or magnetizing force to remove remanence	Mild steel: 50 Alnico 4: 700
Susceptibility	M/H	Ratio of magnetization to magnetizing force	
Permeability	μ, B/H	Ratio of flux density to field intensity, usually relative	Air: 1 Iron: 1,000 Nickel-iron: 100,000
Energy product	BH_{max}	Criterion of internal losses	Alnico 1: 1.4 Alnico 5: 5.0
Curie point	T_c	Temperature of high-permeability material above which thermal agitation overcomes molecular magnetic field	Ferrite: 150°C Silicon-iron: 700°C
Neel point	T_n	Temperature below which antiferro-magnetics become spontaneously magnetized	Cesium: 13K Chromium: 312K

Table 2–31 (continued)
MAGNETIC MATERIALS AND UNITS

C. Classification and Uses of Magnetic Materials

Class or use	*Properties*	*Typical examples*
Ferromagnetic material	High permeability, 10 to 10^6; spontaneous magnetization below Curie point; magnetization in proportion to field strength	Iron (pure): μ = 8,000 Silicon iron: μ = 7,000 Mild steel: μ = 450 Cast iron: μ = 90 Cobalt: μ = 60
Antiferromagnetic material; (or ferrimagnetic)	Antiparallel spin; transition at Curie point	Ferrous oxide; manganese oxide
Paramagnetic material	Magnetization in proportion to field strength. Permeability slightly above unity	Chromium; aluminum; beryllium; some salts
Diamagnetic material	Permeability slightly less than unity. Induced magnetism direction opposite to field	Copper; bismuth; zinc; silver; paraffin; wood
Ferrites (and garnets)	High permeability, 1 to 10; high resistivity; low power loss; high coercivity; low saturation flux density	High-frequency components; computer memories; magnetic ceramics
Electromagnet and transformer cores (soft)	High permeability; high resistivity; low energy product; low coercivity	Silicon iron
Permanent magnets (hard magnets)	High retentivity; high coercivity; low energy product if a-c	Iron, pure: B_r = 20,000 Co-Fe-V: B_r = 14,000 Cast iron: B_r = 5,000 MnZn ferrite: B_r = 1,000

From Bolz, R. E. and Tuve, G. L., Eds., *Handbook of Tables for Applied Engineering Science,* 2nd ed., CRC Press, Cleveland, 1973, 304.

Table 2–32
PROPERTIES OF HIGH-PERMEABILITY MAGNETIC MATERIALS

Properties are expressed in the cgs electromagnetic system (typical values).[a]

Name	*Composition, %*	*Permeability*		*Coercivity, H_c, Oe*	*Retentivity, B_r, G*	*B, max, G*	*Resistivity, microhm-cm*
		Initial	*Maximum*				
Iron, pure (laboratory conditions)	Annealed	25,000	350,000	0.05	12,000	14,000	9.7
Iron, Swedish		250	5,500	1.0	13,000	21,000	10
Iron, cast		100	600	4.5	5,300	20,000	30
Iron, silicon	4 Si, bal. Fe (hot-rolled)	500	7,000	0.3	7,000	20,000	50
Rhometal	36 Ni, bal. Fe	1,000	5,000	0.5	3,600	10,000	90
Permalloy 45	45 Ni, bal. Fe	2,500	25,000	0.3		16,000	45
Mumetal	71–78 Ni, 4.3–6 Cu, 0–2 Cr, bal. Fe	20,000	100,000	0.05	6,000	7,200	25–50
Supermalloy	79 Ni, 5 Mo, bal. Fe	100,000	1,000,000	0.002		8,000	60
HyMu80	80 Ni, bal. Fe	20,000	100,000	0.05		8,700	57
Alfenol	16 Al, bal. Fe	3,450	116,000	0.025	3,800	7,825	150
Permendure	50 Co, 1–2 V, bal. Fe	800	4,500	2.0	14,000	24,000	26
Sendust	10 Si, 5 Al, bal. Fe (cast)	30,000	120,000	0.05	5,000	10,000	60–80
Ferroxcube 3	Mn-Zn-Ferrite	1,000	1,500	0.1	1,000	3,000	$>10^6$
Ferroxcube 101	Ni-Zn-Ferrite	1,100		0.18	1,100	2,300	$>10^5$

REFERENCES

Bozorth, R. N., *Ferromagnetism,* D. Van Nostrand Co., New York, 1951.

Bardell, P. R., *Magnetic Materials in the Electrical Industry,* Philosphical Library, New York, 1955.

Technical catalog data by Allegheny Ludlum Steel Corporation, Pittsburgh, and Ferroxcube Corporation of America, Saugerties, N.Y.

[a]Values are approximate only and vary with heat treatment and mechanical working of the material.

From Westman, H. P., *Reference Data for Radio Engineers,* 5th ed., Howard W. Sams & Co., Indianapolis, 1968. With permission.

Table 2–33
DATA ON METALLIC MAGNETIC-CORE MATERIALS

Class or trade name	*Composition, % (remainder is iron)*	*Characteristic property or application*	*Permeability*		*Direct-current saturation, kilogauss*	*Residual induction, kilogauss*	*Coercive force, oersteds*	*Resistivity, microhm-cm*	*Curie temperature, °C*
			Initial	*Maximum*					
SILICON-IRON[a]									
Silicon-Iron	4 Si	Transformer	400	7,000	20	12	0.5	60	690
Hypersil							0.1		
Trancor 3X							to		
Silectron	3.5 Si	Grain oriented	1,500	35,000	20	13.7	0.3	50	750
Sendust	9.5 Si, 5.5 Al	High frequency, powder	30,000	120,000	10	5	0.05	80	—
COBALT-IRON[a]									
Hyperco	35 Co, 0.5 Cr	High saturation	650	10,000	24	> 13	> 1	28	970
Permendur 2V	49 Co, 2 V		800	4,500	24	14	2	25	980
NICKEL-IRON[a]									
Perminvar 45–25	45 Ni, 25 Co		400	2,000	15.5	3.3	1.2	20	715
Perminvar 7–70	70 Ni, 7 Co	"Constant"	850	4,000	12.5	2.4	0.6	15	650
Conpernik	50 Ni	permeability	1,500	2,000	16	—	—	45	—
Isoperm 36	36 Ni, 9 Cu		60	65	—	—	—	70	300
Isoperm 50	50 Ni	High frequency	90	100	16	—	—	40	500
Permalloy 45	45 Ni		2,700	23,000	16.5	8	0.3	45	440
Allegheny 4750	47 to 50 Ni		9,000	50,000	16	6.2[a]	0.08[a]	52	430
Armco 48	48 Ni	Combine good	—	—	16	—	—	—	—
Nicaloi	49 Ni	permeability	—	—	16	—	—	—	—
High Perm 49	49 Ni	and flux	5,000	50,000	16	6.5	0.03	43	475
Hipernik	50 Ni, Si, Mn	density	4,000	100,000	16	8[a]	0.03[a]	45	500
Monimax	47 Ni, 3 Mo		2,000	38,000	15	—	0.06	80	390
Sinimax	42 Ni, 3 Si	High resistivity	3,500	30,000	11	—	0.1	90	290
Permenorm 5000Z									
Permenite									
Deltamax									
Hypernik V			400	40,000	15.5	13	0.2	40	450
Orthonik			to	to	to	to	to	to	to
Orthonol	45 to 50 Ni		1,700	100,000	16	15	0.4	50	500
Permalloy 65	65 to 68 Ni	Rectangular hysteresis loop	1,500	250,000 to 600,000	13	13	0.03	20	600
Alloy 1040	72 Ni, 14 Cu, 3 Mo		40,000	100,000	6	2.5	0.02	55	290
Mumetal	77 Ni, 5 Cu, 2 Cr		20,000	100,000	8	6	0.05	60	400
Permalloy 78	78 Ni, 0.6 Mn		9,000	100,000	10.7	6	0.05	16	580
Mo-Permalloy 4–79	79 Ni, 4 Mo		20,000	75,000	8	5.5	0.05	55	—
Supermalloy	79 Ni, 5 Mo		55,000	500,000	6.8	—	0.002	65	400
		Highest perme-	to	to	to		to		
		ability, low	150,000	1,000,000	7.8		0.05		
Hymu 80	80 Ni	saturation	10,000	100,000	8	—	0.06	58	460
FERRITES[b]									
3C3	MnZi	High-frequency transformers	2,200 ± 20%		4.6	3.5	0.1	60×10^6	150
3B7 and 3B9	MnZi	High-Q coils	2,300 ± 20%		4.6	3.0	0.2	100×10^6	170
3D3	NiZi	High-frequency	750 ± 20%		4.7	3.0	0.5	150×10^6	150
4C4	NiZi	High-Q coils	125 ± 20%		4.1	2.0	4.5	10×10^9	300

[a] B_m = 10,000 gauss.
[b] Data furnished by Ferroxcube Corporation of America, Saugerties, N.Y.

Note: The table shows characteristics as listed by the manufacturers. The parameters of different lots of material may vary considerably from the above values. In cases of residual induction and coercive force, the difference may amount to 50%.

From Hoh, S. R., Evaluation of high-performance core materials (Part I), *Tele-Tech Electron. Ind.*, 12, 86–89, 154–156, 1953. With permission.

Table 2–34
MAGNETIC PROPERTIES AND COMPOSITION OF PERMANENT MAGNETIC ALLOYS[a]

Name	*Composition,*[b] *weight percent*					*Remanence, B_r, gauss*	*Coercive force, H_c, oersteds*	*Maximum energy product, $(BH)_{max}$, gauss-oersteds × 10^{-6}*
	Al	*Ni*	*Co*	*Cu*	*Other*			
U.S.A.								
Alnico I	12	20–22	5			7,100	440	1.4
Alnico II	10	17	12.5	6		7,200	540	1.6
Alnico III	12	24–26		3		6,900	470	1.35
Alnico IV	12	27–28	5			5,500	700	1.3
Alnico V[c]	8	14	24	3		12,500	600	5.0
Alnico V DG[c]	8	14	24	3		13,100	640	6.0
Alnico VI[c]	8	15	24	3	1.25 Ti	10,500	750	3.75
Alnico VII[c]	8.5	18	24	3	5 Ti	7,200	1,050	2.75
Alnico XII	6	18	35		8 Ti	5,800	950	1.6
Carbon steel					1 Mn 0.9 C	10,000	50	0.2
Chromium steel					3.5 Cr 0.9 C 0.3 Mn	9,700	65	0.3
Cobalt steel			17		2.5 Cr 8 W 0.75 C	9,500	150	0.65
Cunico		21	29	50		3,400	660	0.80
Cunife		20		60		5,400	550	1.5
Ferroxdur 1			$BaFe_{12}O_{19}$			2,200	1,800	1.0
Ferroxdur 2			$BaF_{12}O_{19}$ (oriented)			3,840	2,000	3.5
Platinum-Cobalt			23		77 Pt	6,000	4,300	7.5
Remalloy			12		17 Mo	10,500	250	1.1
Silmanol	4.4				86.6 Ag 8.8 Mn	550	6,000	0.075
Tungsten steel					5 W 0.3 Mn 0.7 C	10,300	70	0.32
Vicalloy I			52		10 V	8,800	300	1.0
Vicalloy II (wire)			52		14 V	10,000	510	3.5
Germany								
Alni 90	12	21				8,000	350	1.2
Alni 120	13	27				6,000	570	1.2
Alnico 130	12	23	5			6,300	620	1.4
Alnico 160	11	24	12	4		6,200	700	1.6
Alnico 190	12	21	15	4		7,000	700	1.8
Alnico 250	8	19	23	4	6 Ti	6,500	1,000	2.2
Alnico 400[c]	9	15	23	4		12,000	650	4.8
Alnico 580[c] (semicolumnar)	9	15	23	4		13,000	700	6.0
Oerstit 800	9	18	19	4	4 Ti	6,600	750	1.95
Great Britain								
Alcomax I	7.5	11	25	3	1.5 Ti	12,000	475	3.5
Alcomax II	8	11.5	24	4.5		12,400	575	4.7
Alcomax IISC (semicolumnar)	8	11	22	4.5		12,800	600	5.15
Alcomax III	8	13.5	24	3	0.8 Nb	12,500	670	5.10
Alcomax IIISC (semicolumnar)	8	13.5	24	3	0.8 Nb	13,000	700	5.80
Alcomax IV	8	13.5	24	3	2.5 Nb	11,200	750	4.30
Alcomax IVSC (semicolumnar)	8	13.5	24	3	2.5 Nb	11,700	780	5.10
Alni, high B_r	13	24		3.5		6,200	480	1.25
Alni, normal						5,600	580	1.25
Alni, high H_c	12	32			0–0.5 Ti	5,000	680	1.25
Alnico, high B_r	10	17	12	6		8,000	500	1.70

Table 2–34 (continued)
MAGNETIC PROPERTIES AND COMPOSITION OF PERMANENT MAGNETIC ALLOYS[a]

Name	Composition,[b] weight percent					Remanence, B_r, gauss	Coercive force, H_c, oersteds	Maximum energy product, $(BH)_{max}$, gauss-oersteds × 10^{-6}
	Al	Ni	Co	Cu	Other			
Alnico, normal						7,250	560	1.70
Alnico, high H_c	10	20	13.5	6	0.25 Ti	6,600	620	1.70
Columax (columnar)	similar to Alcomax III or IV					13,000–14,000	700–800	7.0–8.5
Hycomax	9	21	20	1.6		9,500	830	3.3

[a]See Table 2–31 for conversion factors.
[b]Unlisted remainder of composition is either iron or iron plus trace impurities.
[c]Alloys so designated are cast anisotropic; all others are cast isotropic.

From Weast, R. C., Ed., *Handbook of Chemistry and Physics,* 55th ed., CRC Press, Clevleand, 1974, E-115.

Table 2–35
CAST PERMANENT MAGNETIC ALLOYS[a]

Name	Composition,[b] weight percent	Specific gravity, g/cc	Thermal expansion: Cm × 10^{-6} / cm × °C	Thermal expansion: Between °C	Tensile strength: Kg[c] / mm^2	Tensile strength: Form	Remarks[d]	Use
U.S.A.								
Alnico I	Al 12; Ni 20–22; Co 5	6.9	12.6	20–300	2.9	Cast	i	Permanent magnets
Alnico II	Al 10; Ni 17; Cu 6; Co 12.5	7.1	12.4	20–300	2.1 45.7	Cast Sintered	i	Temperature controls, magnetic toys, and novelties
Alnico III	Al 12; Ni 24–26; Cu 3	6.9	12	20–300	8.5	Cast	i	Tractor magnetos
Alnico IV	Al 12; Ni 27–28; Co 5	7.0	13.1	20–300	6.3 42.1	Cast Sintered	i	Application requiring high coercive force
Alnico V	Al 8; Ni 14; Co 24; Cu 3	7.3	11.3		3.8 35	Cast Sintered	a	Application requiring high energy
Alnico V DG	Al 8; Ni 14; Co 24; Cu 3	7.3	11.3				a, c	
Alnico VI	Al 8; Ni 15; Co 24; Cu 3; Ti 1.25	7.3	11.4		16.1	Cast	a	Application requiring high energy
Alnico VII	Al 8.5; Ni 18; Cu 3; Co 24; Ti 5	7.17	11.4				a	
Alnico XII	Al 6; Ni 18; Co 35; Ti 8	7.2	11	20–300				Permanent magnets
Comol	Co 12; Mo 17	8.16	9.3	20–300	88.6			Permanent magnets
Cunife	Cu 60; Ni 20	8.52			70.3			Permanent magnets
Cunico	Cu 50; Ni 21	8.31			70.3			Permanent magnets
Barium ferrite Feroxdur	$BaFe_{12}O_{19}$	4.7	10		70.3			Ceramics
Great Britain								
Alcomax I	Al 7.5; Ni 11; Co 25; Cu 3; Ti 1.5						a	Permanent magnets
Alcomax II	Al 8; Ni 11.5; Co 24; Cu 4.5						a	Permanent magnets

Table 2–35 (continued)
CAST PERMANENT MAGNETIC ALLOYS[a]

Name	Composition,[b] weight percent	Specific gravity, g/cc	Thermal expansion: $\frac{Cm \times 10^{-6}}{cm \times °C}$	Thermal expansion: Between °C	Tensile strength: $\frac{Kg}{mm^2}$ [c]	Tensile strength: Form	Remarks[d]	Use
Alcomax II SC	Al 8; Ni 11; Co 22; Cu 4.5	7.3					a, sc	
Alcomax III	Al 8; Ni 13.5; Co 24; Nb 0.8	7.3					a	Magnets for motors, loudspeakers
Alcomax IV	Al 8; Ni 13.5; Cu 3; Co 24; Nb 2.5							Magnets for cycle-dynamos
Columax	Similar to Alcomax III or IV						a, sc	Permanent magnets, heat-treatable
Hycomax	Al 9; Ni 21; Co 20; Cu 1.6						a	Permanent magnets
Alnico (high H_c)	Al 10; Ni 20; Co 13.5; Cu 6; Ti 0.25	7.3					i	
Alnico (high B_r)	Al 10; Ni 17; Co 12; Cu 6	7.3					i	
Alni (high H_c)	Al 12; Ni 32; Ti 0–0.5	6.9					i	
Alni (high B_r)	Al 13; Ni 24; Cu 3.5						i	
Germany								
Alnico 580	Al 9; Ni 15; Co 23; Cu 4						i	
Alnico 400	Al 9; Ni 15; Co 23; Cu 4						a	
Oerstit 800	Al 9; Ni 18; Co 19; Cu 4; Ti 4						i	Permanent magnets
Alnico 250	Al 8; Ni 19; Co 23; Cu 4; Ti 6						i	
Alnico 190	Al 12; Ni 21; Cu 4; Co 15						i	
Alnico 130	Al 12; Ni 23; Co 5						i	
Alni 120	Al 13; Ni 27						i	
Alni 90	Al 12; Ni 21						i	
Austria								
Alnico 160	Al 11; Ni 24; Co 12; Cu 4						i	Permanent magnets, sintered

[a]For properties of permanent magnetic alloys see Table 2–34.
[b]The additional alloying metal for each of the magnets listed is iron.
[c]$kg/mm^2 \times 1422.3 = lb/in.^2$; $kg/mm^2 \times 9.807 = MN/m^2$.
[d]i = isotropic; a = anisotropic; c = columnar; sc = semicolumnar.

From Weast, R. C., Ed., *Handbook of Chemistry and Physics,* 55th ed., CRC Press, Cleveland, 1974, E-117.

Table 2–36
FERROMAGNETIC MATERIALS – CURIE TEMPERATURES[a]
Upper Transition Temperature

Material	*Curie temp, K*	*Material*	*Curie temp, K*	*Material*	*Curie temp, K*
AuFe	300	EuO	69.5	$GdNi_2$	85
Au_4Mn	263	EuS	16	$GdOs_2$	66
Au_4V	43	EuSe	4.58	$GdPd_2$	335
$BaFeO_3$	180	FeAl	923	GdZn	~280
$BiMnO_3$	103	Fe_3Al	773	$HgCr_2S_4$	36
$CdFe_2$	782	FeB	598	$HgCr_2Se_4$	120
$CeAl_2$	8	Fe_3C	483	Ho	20
$CeCo_3$	78	Fe_3Cr	1273	$HoAl_2$	27
$CeCo_5$	687	Fe-Ni		$HoCo_5$	1025
	737	4.5% Fe	683.0	$HoFe_2$	608
	464	19% Fe	834.0	$HoIr_2$	12
$CeFe_2$	235	23% Fe	876.0	HoNi	31
	878	50% Fe	786.1	$HoNi_2$	23
$CeFe_5$	228	Fe_2O_3	848	$HoNi_3$	66
Co	1390	ε-Fe_2O_3	483	$HoOs_2$	9
Co_2B	429	γ-Fe_2O_3	743	$LaCrO_3$	300
	433	Fe_3O_4	848–858	Mn (45% Al)	~653
Co_3B	747	FeP	215	Mn-Al-Co	Co rich: 370
$CoFeCoO_4$	450	Fe_2P (hexagonal)	278		Mn rich: 466
$CoFe_2O_4$	673–769	FePt	743	$MnAu_4$	360
CoPt	813	Fe_3Pt	453	MnB	578
CoS_2	110	FeRh	668	MnBi	633
Co_2VO_4	160	Fe-Si	1043.9–1012.6	Mn_3Ge	28
Cr	311	0.9–7.4 at. % Si		Mn_3Ge_2	300
CrO_2	391	Fe_3Si	808	Mn_3In	583
CuMnAl	433	Fe_3Sn	743	$MnNi_3$	750 (ordered)
Dy	85	Fe_2TiO_4	142		132 (dis-
$DyAl_2$	53	Fe-53.3% V	280		ordered)
	62	Fe_2Zr	628	Mn_3O_4	45
$DyCo_5$	1125	$Ga_{2-x}Fe_xO_3$		MnO-35.5%,	467.5
	966	$x = 1.08$	350	ZnO-15%,	
$DyCo_3$	450	$x = 1.20$	305	Fe_2O_3-49.5%	
$DyFe_2$	638	$x = 0.80$	205	MnO-26%,	368.5
	663	Gd	294	ZnO-24%,	
$DyIr_2$	23	$GdAl_2$	176	Fe_2O_3-50%	
DyN	22	$GdCl_3$	2.20	MnTe	260
DyNi	48	$GdCo_2$	404	Mn_2TiO_4	~77
$DyOs_2$	15	$GdCo_3$	612	Mn_2VO_4	62
Er	20	Gd-Dy		$NdAl_2$	65
$ErAl_2$	21	% Dy: 10	285	$NdCo_2$	116
$ErCo_2$	39, 36	% Dy: 50	226	NdGe	28
$ErCo_3$	401	% Dy: 61	193	$NdGe_2$	3.6
$ErFe_2$	473	% Dy: 87.5	120	NdNi	35
$ErIr_2$	3	$GdFe_2$	782	Ni	627
ErN	6		813		628.3
	5	$Gd_3Fe_5O_{12}$	564		633
ErNi	10		574.6	Ni-Cr	324
$ErNi_2$	14	$GdIr_2$	88	5.6 at. % Cr	
	21		90	$NiFe_2O_4$	858
$ErNi_3$	62	GdN	69–72	$NiMnO_3$	437
EuH_2	24	GdNi	73	Pd_3Fe	540

Table 2–36 (continued)
FERROMAGNETIC MATERIALS – CURIE TEMPERATURES[a]

Upper Transition Temperature (continued)

Material	*Curie temp, K*	*Material*	*Curie temp, K*	*Material*	*Curie temp, K*
$PrAl_2$	34	SmIG	562	UFe_2	172
$PrCo_3$	349	SmNi	45	USe	160.5
PrGe	39	Tb	210	USe_2	13.1
$PrIr_2$	16	$TbAl_2$	121	UTe	103
PrNi	20	$TbCo_3$	506	YCo_3	301
$PrNi_2$	8	TbGa	155	YIG: Nd	548–568
$SmAl_2$	122	TbN	40	0–40% Nd_2O_3	
$SmCo_5$	1020	TbNi	50	$3Y_2O_3 \cdot Al_2O_3 \cdot 4Fe_2O_3$	415
	747	$TbNi_2$	46	$YbCo_5$	973
	1015			$ZrFe_2$	633

[a]Curie temperature is defined as the point (temperature increasing) at which a material ceases to be ferromagnetic and becomes paramagnetic, i.e., the saturation decreases to zero. Pure iron is magnetic up about 1,455°F (790°C); above this temperature it is non-magnetic. Materials gain ferromagnetism on cooling, but not always at the same temperature.

Note: The values listed in this table represent a small and somewhat random sample of those reported in the quoted source (see below). Users should consult the original source, which includes a large bibliography and keyed references.

REFERENCES

Weast, R. C., Ed., Magnetic susceptibility of the elements and inorganic compounds, in *Handbook of Chemistry and Physics,* 55th ed., CRC Press, Cleveland, 1974, E-121.

Bozorth, R. M., McGuire, T. R., and Hudson, R. P., Magnetic properties of materials, in *American Institute of Physics Handbook,* 2nd ed., Sect. 5g, Gray, D. E., Ed., McGraw-Hill, New York, 1963, 5-164.

From ORNL-RMIC-7 (Rev.), Research Materials Information Center, Solid State Division, Oak Ridge National Laboratory, Oak Ridge, Tenn., 1969.

Table 2–37
ANTIFERROMAGNETIC MATERIALS – NEEL TEMPERATURE[a]

Upper Transition Temperature

Material	*Neel temp,*[b] *K*	*Material*	*Neel temp.*[b] *K*	*Material*	*Neel temp.*[b] *K*	*Material*	*Neel temp,*[b] *K*
AuMn	493	CrGe	62	$FeWO_4$	66	β-MnS	110
Au_3Mn	140	$CrVO_4$	50	GdAs	25	(hexagonal)	
Ba_2CoWO_6	17	Cr_2WO_6	69	GdBi	28	$MnSO_4$	11.5
$BaFe_{12}O_{19}$	709.5	$CuCl_2$	70	GdP	15	MnTe	306.7
	(for H⊥c)	CuF_2	68.7	$GdPO_4$	225	$MnTe_2$	80
$CdCr_2O_4$	9	CuHo	27	GdSb	28	$MnUO_4$	12
Ce	13		(elec. res.)	Ho	133	$MnWO_4$	16
CeS	7	CuO	230	HoGe	18	Nd	12
CeSb	18	$CuSO_4$	34.5	$HoGe_2$	11	$NdSn_3$	4.7
CeSe	12	$CuWO_4$	90	HoSb	9	NdTe	13
CeTe	10	DyCu	64	$KCoF_3$	135	$NiCl_2$	50
$CoBr_2$	19	$DyCu_2$	24	$KCrF_3$	40	NiF_2	~73.2
$CoCO_3$	18	DyH_2	8	$KCuF_3$	243	$NiWO_4$	67
$CoCl_2$	25	Er	84	$KFeF_3$	113	$RbMnF_3$	54.5
CoF_2	37–45	Eu	103	$KMnF_3$	88		66
CoF_3	460	Fe_3Al	750	$LaVO_3$	137		82
$CoMoO_3$	391	$FeCl_2$	23.5	α-Mn	~100		83
$CoMoO_4$	5	$FeCl_3$	15	$MnCl_2$	1.96		~100
CoO	292	FeF_2	78.11	MnF_2	66.2	Sm	15
$CoSO_4$	12	FeO	186	MnF_3	47	Tb	230
$CoUO_4$	12	Fe_2O_3	259	MnO	118	Tm	51
$CoWO_4$	55	α-Fe_2O_3	963	MnO_2	92	UO_2	~30
Cr	312	FeS	593	Mn_2O_3	80	$ZnFe_2O_4$	15
CrF_2	53	$FeSO_4$	21	MnS	165		9
CrF_3	80	FeSn	373	β-MnS	160		
CrFe	308	$FeTiO_3$	68	(cubic)			

[a]Neel temperature is defined as the point above which an antiferromagnetic material becomes paramagnetic; above the Neel point the susceptibility decreases with increasing temperature.
[b]No attempt has been made to choose a best value where more than one temperature is reported.

Note: The values listed in this table represent a small and somewhat random sample of those reported in the quoted source (see below). Users should consult the original source, which includes a large bibliography and keyed references.

From ORNL-RMIC-7 (Rev.), Research Materials Information Center, Solid State Division, Oak Ridge National Laboratory, Oak Ridge, Tenn., 1969.

Table 2–38
NUMBER μ_B OF BOHR MAGNETONS PER MAGNETIC ATOM AND DATA ON SATURATION MAGNETIZATION AND CURIE POINTS

Substance	*Saturation magnetization M_s, gauss*		*$\mu_B(0\ K)$, per formula unit*	*Ferromagnetic Curie temperature, K*
	Room temperature	*0 K*		
Fe	1707	1740	2.22	1043
Co	1400	1446	1.72	1400
Ni	485	510	0.606	631
Gd	—	2010	7.10	292
Dy	—	2920	10.0	85
Cu_2MnAl	500	(550)	(4.0)	710
MnAs	670	870	3.4	318
MnBi	620	680	3.52	630
Mn_4N	183	—	1.0	743
MnSb	710	—	3.5	587
MnB	152	163	1.92	578
CrTe	247	—	2.5	339
$CrBr_3$	—	—	—	37
CrO_2	515	—	2.03	392
$MnOFe_2O_3$	410	—	5.0	573
$FeOFe_2O_3$	480	—	4.1	858
$CoOFe_2O_3$	400	—	3.7	793
$NiOFe_2O_3$	270	—	2.4	858
$CuOFe_2O_3$	135	—	1.3	728
$MgOFe_2O_3$	110	—	1.1	713
UH_3	—	230	0.90	180
EuO	—	1920	6.8	69
$GdMn_2$	—	215	2.8	303
$Gd_3Fe_5O_{12}$	0	605	16.0	564
$Y_3Fe_5O_{12}$ (YIG)	130	200	5.0	560

REFERENCES

Bozorth, R. M., McGuire, T. R., and Hudson, R. P., Magnetic properties of materials, in *American Institute of Physics Handbook,* 2nd ed., Sect. 5g, Gray, D. E., Ed., McGraw-Hill, New York, 1963, 5-164.

Hellwege, K. H., Ed., *Numerical Data and Functional Relationships in Physics, Chemistry, Astronomy, Geophysics, and Technology,* 6th ed., Springer-Verlag, New York, 1962.

Table 2–39
FERROMAGNETIC CURIE POINTS AND SATURATION MOMENTS OF AB_5 COMPOUNDS OF RARE-EARTH ELEMENTS (A), HAVING HEXAGONAL $CaCu_5$ STRUCTURE

Per Formula Unit, at Temperatures Indicated

Rare-earth elements, A		*Compounds, B*		
Name	*Symbol*	*Mn*[1]	*Co*[2,3]	*Ni*[3]
Yttrium	Y	490K 2.21 μ_B (78K) 1.38 μ_B (298K)	995K 8.2 μ_B (1.4K)	~0 μ_B (1.4K)
Cerium	Ce		687K 7.4 μ_B (1.4K)	
Praseodymium	Pr		9.9 μ_B (1.4K)	
Neodymium	Nd		10.5 μ_B (1.4K)	2.2 μ_B (1.4K)
Samarium	Sm	440K 1.72 μ_B (78K) 1.40 μ_B (298K)	1015K 8.6 μ_B(1.4K)	0.7 μ_B (1.4K)
Gadolinium	Gd	465K 6.23 μ_B (78K) 2.89 μ_B (298K)	1030K 1.3 μ_B (1.4K)	6.1 μ_B (1.4K)
Terbium	Tb	445K 6.18 μ_B (78K) 2.66 μ_B (298K)	0.7 μ_B (1.4K)	7.0 μ_B (1.4K)
Dysprosium	Dy	430K 5.34 μ_B (78K) 2.49 μ_B (298K)	1125K 1.5 μ_B (1.4K)	7.7 μ_B (1.4K)
Holmium	Ho	425K 5.12 μ_B (78K) 1.99 μ_B (298K)	1025K 1.9 μ_B (1.4K)	7.2μ_B (1.4K)
Erbium	Er	415K 3.74 μ_B (78K) 1.63 μ_B (298K)	1.3 μ_B (1.4K)	7.7 μ_B (1.4K)
Thulium	Tm		1.5 μ_B (1.4K)	

REFERENCES

1. **Cherry, L. T. and Wallace, W. E.,** Magnetic characteristics of some intermetallic compounds between manganese and the lanthanide metals, *J. Appl. Phys. Suppl.,* 32, 340, 1961.
2. **Nassau, K., Cherry, L. V., and Wallace, W. E.,** Intermetallic compounds between lanthanons and transmission metals of the first long period II – ferrimagnetism of AB_5 cobalt compounds, *J. Phys. Chem. Solids,* 16, 131, 1960.
3. **Nesbitt, E. A., Williams, H. J., Wernick, J. H., and Sherwood, R. C.,** Magnetic moments of intermetallic compounds of transition and rare-earth elements, *J. Appl. Phys.,* 33, 1674, 1932.

From Westbrook, J. T., Magnetic properties of some rare-earth compounds, in *Intermetallic Compounds,* John Wiley & Sons, New York, 1967. With permission.

Table 2–40
PHYSICAL AND MAGNETIC CONSTANTS OF RARE-EARTH METALS

Element	Atomic No., Z	Mol. weight	Density, g/cc	Cryst. form at room temp	Transition point, °C	M.P., °C	Curie point, θ_f, K	Neel point, θ_N, K	Asymptotic Curie point, θ_a, K	Electronic state of 3+ ion				Magnetic moment in Bohr magnetons		
										Number of 4f electrons	S	L	J	Effective moment		Saturation moment, gJ
														3+ ion	Metal	
Sc	21	44.96	2.992	h.c.p.	1335	1539										
Y	39	88.92	4.478	h.c.p.	1459	1509										
La	57	138.92	6.174	h.c.p.	310	920				0	0	0	0	0	0	
			6.186	f.c.c.	868											
Ce	58	140.13	6.771	f.c.c.	725	795		12.5	−46	1	$\frac{1}{2}$	3	$2\frac{1}{2}$	2.52	2.51	
Pr	59	140.92	6.782	hex.	798	935			−21	2	1	5	4	3.60	2.56	
Nd	60	144.27	7.004	hex.	862	1024		7.5	−16	3	$\frac{3}{2}$	6	$4\frac{1}{2}$	3.50	3.3–3.71	
Pm	61	(147)				1035				4	2	6	4			
Sm	62	150.35	7.536	rhomb.	917	1072		14.8		5	$\frac{5}{2}$	5	$2\frac{1}{2}$		1.74	
Eu	63	152.0	5.259	b.c.c.		826		(90)	15	6	3	3	0		8.3	
Gd	64	157.26	7.895	h.c.p.	1264	1312	289		310	7	$\frac{7}{2}$	0	$3\frac{1}{2}$	7.80	7.93	7.12
Tb	65	158.93	8.272	h.c.p.	1317	1356	218	230	236	8	3	3	6	9.74	9.62	9.25
Dy	66	162.51	8.536	h.c.p.		1407	90	179	151	9	$\frac{5}{2}$	5	$7\frac{1}{2}$	10.5	10.67	10.2
Ho	67	164.94	8.803	h.c.p.		1461	20	133	87	10	2	6	8	10.6	10.9	9.7[a]
Er	68	167.27	9.051	h.c.p.		1497	20	80 (53)	41.6	11	$\frac{3}{2}$	6	$7\frac{1}{2}$	9.6	10.0	8.3[a]
Tm	69	168.94	9.332	h.c.p.		1545	22	53	20	12	1	5	6	7.1	7.56	
Yb	70	173.04	6.977	f.c.c.	798	824				13	$\frac{1}{2}$	3	$3\frac{1}{2}$	4.4	0.0	
Lu	71	174.99	9.842	h.c.p.		1652				14	0	0	0	0	0	

[a]Determined by neutron diffraction.

From Chickazumi, S., *Physics of Magnetism*, John Wiley & Sons, New York, 1964. With permission.

Table 2–41
INSTRUMENT TRANSDUCERS

Table A classifies instrument transducers, giving the common or descriptive name for each, a brief statement of the principle and nature of the device, and a statement of the basic quantity measured. Table B lists almost one hundred properties or characteristics to be measured, with typical transducers used for each measurement.

Almost all transducers (except counters) respond to one of four basic inputs, viz., displacement, force, temperature, or radiation (and derivatives). Each of these four inputs may, in turn, be utilized to change the resistance, inductance, or capacitance of a passive electric circuit or the output of a voltage generator. Hence, a number of choices are possible for measuring any one property or quantity, especially with the inclusion of special circuits for differentiating, integrating, and damping. Any particular requirements of range, sensitivity, accuracy, and dynamic response can almost always be met by the instrument designer.

Only common applications and examples for guidance in studying the extensive technical literature and for interpreting suppliers' literature are listed here. No distinction is made between equilibrium measurements and the great variety of dynamic measurements. It should also be emphasized that most transducers permit a choice of readout, i.e., indicating, digital, recording, integrating, or combinations of these.

This versatility and the ready availability of transducer instruments in a variety of grades have resulted in almost complete displacement of other instrumentation by electrical transducers.

Of commercial instrument transducers, about one half of the market is represented either by measurements of displacement (or dimension) or measurements of force (or pressure). The two other large uses are those of temperature measurement and of fluid-flow measurement.

Transducer Development

New and improved transducers are in the process of rapid development, with emphasis on compactness, ease of calibration, and improvement in the ranges available in a given type or design. Environmental specifications are becoming more demanding, especially in regard to vibration. There are many new developments in radiation measurement.

New circuitry is available for dealing with zero shift and predictable error. Printed circuits, rigid mounting and encapsulation, and contributions from solid-state science are greatly improving the transducer; at the same time there are parallel improvements in readout instrumentation.

Table 2–41 (continued)
INSTRUMENT TRANSDUCERS

Table A. Classification of Transducers

Name or class	*Nature and principle*	*Basic measurement*
	EXTERNALLY POWERED TRANSDUCERS (PASSIVE)	
VARIABLE RESISTANCE		
1. Variable resistor	Slider or contact varies resistance in potentiometer or bridge circuit	Displacement, linear or angular
2. Resistance thermometer	Wire or thermistor, with large temperature-coefficient of resistivity	Temperature
3. Resistance strain gage	Resistance of a wire grid; foil or semiconductor changed by stress	Displacement, strain
4. Hot-wire meter	Heated wire or film (constant temperature or constant current) in fluid stream	Temperature (fluid velocity inferred)
5. Radiation bolometer	Radiation focused on resistance-thermometer sensor	Temperature (total radiation inferred)
6. Thermistor radiometer	Radiation focused on thermistor bolometer	Temperature (total radiation inferred)
7. Thickness gage	Resistance between contacts depends on thickness and resistivity of separating material	Dimension
8. Photoconductive cell	Radiation on photoresistive element	Radiation
9. Photoemissive or photomultiplier tube	Radiation causes electron emission and current (amplification available)	Radiation (illumination)
10. Ionization gage	Glow-discharge tube in high-frequency field; asymmetry generates voltage	Displacement
11. Resistance hygrometer	Resistivity of conductive strip changed by moisture	Partial pressure (humidity)
VARIABLE CAPACITANCE		
12. Adjustable capacitor	Capacitance varied by changing distance between plates or area of plates	Displacement
13. Capacitance bridge pickup	Modification of No. 12 using a-c bridge; high sensitivity	Displacement
14. Dielectric gage	Capacitance varied by changing position or thickness of dielectric	Displacement, dimension
15. Dielectric thermometer	Variation of capacitance with temperature of dielectric	Temperature
16. Condenser microphone	Capacitance between diaphragm and fixed electrode varied by sound pressure	Displacement
VARIABLE INDUCTANCE		
17. Air-gap gage	Self-inductance or mutal inductance changed by varying the magnetic path	Displacement, thickness
18. Differential transformer	Transformer with differential secondaries and movable magnetic core	Displacement
19. Reluctance pickup	Reluctance of magnetic circuit varied by positioning or core material	Displacement
20. Eddy-current gage	Inductance of a-c coil varied by position of eddy-current plate	Displacement
21. Magnetostriction gage	Magnetic properties varied by pressure and stress	Force
22. Hall-effect transducer	Magnetic field interacts with current through semiconductor to produce voltage at right angle	Field strength
23. Inductance bridge pick up	Modification of No. 17 using inductance bridge	Displacement
	SELF-GENERATING TRANSDUCERS	
24. Moving magnet-and-coil generator	Relative movement of coil and magnet varies output voltage	Displacement velocity, linear or angular
25. Thermocouple and thermopile	Pairs of dissimilar metals or semiconductors generate voltage if terminals not at same temperature	Temperature
26. Piezoelectric pickup	Quartz or other crystal mounted in compression, bending, or twisting	Force
27. Photovoltaic cell	Layer-built semiconductor cell or transistor generates voltage from radiation	Radiation, light
28. Radiation counter (special class)	Gas counters collect charge released by ionizing radiation	Radiation, radioactive or nuclear

Table 2–41 (continued)
INSTRUMENT TRANSDUCERS

Table B. Examples of Transducer Applications

Table B lists some typical applications of instrument transducers and common types of transducers used for each measurement. Since the properties of materials play a large part in all measurement techniques, tables that give such properties will suggest other applications of transducers. Another major field of transducer use not listed here is in energy measurements.

Quantity to be measured	*Transducers* — *Common examples*	*Transducers* — *See Table A numbers*
Acceleration, angular	Unbonded strain gage; force balance	3, 12, 19, 26
Acceleration, linear	Seismic potentiometer; piezoelectric accelerometer	3, 12, 19, 26
Altitude	Capsule or bellows altimeter	1, 18
Angle	Variable reluctance pickup	18, 19
Blast pressure	Piezoelectric pickup	3, 12, 26
Count, events	Stroboscope; electronic-pulse counter	8, 27
Count, particles	Photoconductive cell; photovoltaic element	
Current, stream	Rotating-current meter; impact-tube meter	24
Density, gas	Hot-wire meter	4
Dewpoint	Photovoltaic cell	11, 27
Dielectric constant	Dielectric gage	14
Dimension, linear	Differential transformer; slide-wire resistor	1, 3, 7
Dimension, micrometer	Capacitance gage; unbonded strain gage	3, 12, 13, 14
Displacement, angular	Slide-wire potentiometer; inductance gage; eddy-current gage; electrolytic	1, 3, 17, 19, 20
Displacement, linear	Differential transformer; slide-wire resistor; reluctance pickup	1, 3, 12, 17, 18, 19, 20
Distance	Slide-wire resistor	1, 3, 12, 17, 20
Duration, time	Tuning fork	
Emissivity	Thermopile radiometer	5, 6, 25
Field strength, magnetic	Hall-effect pickup	22
Film thickness	Dielectric gage	7, 12, 14, 17
Flow, gas or vapor	Differential head meter (orifice, nozzle, pitot); hot-wire or thermocouple anemometer	3, 4, 18, 19, 25
Flow, liquid	Differential head meter (orifice, venturi); turbine flowmeter	1, 3, 18
Flow, open-channel	Rotating-current meter	24
Force	Reluctance pickup; carbon pile; strain gage; magnetostrictive gage	3, 12, 17, 18, 19, 21, 23, 26
Frequency	Moving-coil generator; stroboscope	17, 19, 24
Gamma rays	Geiger counter	28
Hardness	Indenter (displacement)	3, 12, 17, 18
Head		(See pressure)
Heat flow	Thermopile sandwich	25
Humidity, air	Resistance hygrometer; thermocouple physchrometer	2, 11, 27
Infrared	Thermistor bolometer; thermopile photoconductive cell	5, 6, 8, 25
Ionization		10
Jerk		3, 12, 26
Level, liquid	Dielectric gage; capacitance gage	12, 14
Light intensity	Photovoltaic cell	5, 8, 25, 27
Load, force or weight	Bonded strain gage (strut); inductor (elastic element); proving ring (displacement)	1, 3, 12, 13, 17, 18, 19, 21, 26

Table 2–41 (continued)
INSTRUMENT TRANSDUCERS

Table B. Examples of Transducer Applications (continued)

Quantity to be measured	Transducers: Common examples	Transducers: See Table A numbers
Moisture, in solids	Resistance gage; dielectric gage; nuclear magnetic resonance	1, 14
Noise	Condenser microphone; piezoelectric crystal	16, 26
Nuclear radiation	Geiger counter	28
Particle counting	Ionization gage	10, 28
Position, angular	Resistance gage; differential transformer	1, 12, 18, 19
Pressure, absolute		
Position, linear	Contact potentiometer	1, 12, 18, 19
	Bourdon-tube potentiometer	1, 3, 18, 19
Pressure, differential	Bellows gage	1, 12, 18, 19
Pressure, dynamic	Piezoelectric pickup; strain gage	3, 12, 16, 19, 26
Pressure, gage	Reluctance gage; strain-gage pickup; capacitor (elastic element)	1, 3, 12, 18
Pressure, impact	Pitot and bellows with differential transformer	12, 18, 19
Radiation, light (optical)	Optical-target thermopile; photoresistive gage	8, 9, 27
Radiation, nuclear	Geiger counter	10, 28
Radiation, total	Photomultiplier tube; bolometer	5, 6, 8, 25
Reflectivity	Radiometer	5, 6, 25
Rotational speed	Reluctance pickup	19, 24
Rugosity	Moving-coil tracer	19, 24
Shock	Ceramic crystal	

Quantity to be measured	Transducers: Common examples	Transducers: See Table A numbers
Sound pressure	Piezoelectric; condenser microphone	16, 26
Speed, rotational	Moving-coil tachometer; pulse counter; stroboscopic counter	19, 24
Strain	Wire or foil strain gage	3
Temperature	Thermocouple; wire resistance; thermistor	2, 5, 6, 8, 15, 25
Thickness, metal	Eddy-current gage; capacitance pick-up; contact gage (resistance); ultrasonic probe; isotope gage	1, 3, 12, 13, 14, 17
Time	Synchronous motor; tuning fork	
Torque	Strain gage	3, 12, 18, 26
Turbulence, fluid	Hot-wire pickup	4
Vacuum, high	Ionization gage	10
Vacuum, low	Corrugated diaphragm	
Velocity, linear	Moving-coil generator	
Vibration acceleration	Piezoelectric crystal; strain gage (force)	3, 12, 18, 26
Vibration displacement	Reluctance gage; seismic vibrometer	18, 19
Vibration frequency	Calibrated oscilloscope; stroboscopic counter	
Vibration velocity	Magnet and coil	24
Viscosity	Drag-cup torque meter; falling-ball displacement gage	3, 18
Voltage	Moving-coil meter or galvanometer	19, 20, 24
Weight	Strain gage; force balance	3, 12, 17, 18, 26

REFERENCE

Minnar, E. J., Ed., *ISA Transducer Compendium,* Plenum Press, New York, 1963.

From Bolz, R. E. and Tuve, G. L., Eds., *Handbook of Tables for Applied Engineering Science,* 2nd ed., CRC Press, Cleveland, 1973, 975.

Table 2–42
THERMOCOUPLE CALIBRATION

The thermocouple temperature-emf tables on the following pages are condensed from USA Standard for Temperature Measurement Thermocouples, C96.1–1969, published by the Instrument Society of America, 530 William Penn Place, Pittsburgh, Pa. This code was approved by the American National Standards Institute and prepared with the assistance and approval of representatives of the major technical societies and Federal Departments in the United States. The code includes specifications for thermocouples and extension wires and for their fabrication, protection, and installation. Suppliers will furnish calibrated thermocouples and color-coded wire conforming with these specifications.

The table below gives standard and special limits of error for protected thermocouples; these apply only within the limits of temperatures recommended for the various wire sizes, as shown in the table.

Temperature and Error Limits for Thermocouples

Thermocouple type	*Name (positive wire first)*	*Thermocouple limits of error, °F*			*Upper temp limits, °F, for AWG wire size*				
		Range, °F	*Standard*	*Special*	*8*	*14*	*20*	*24*	*28*
E	Chromel-constantan	32 to 600 600 to 1600	±3 ±.50%	— —	1600	1200	1000	800	800
J	Iron-constantan	32 to 530 530 to 1400	±4 ±.75%	±2 ±.38%	1400	1100	900	700	700
K	Chromel-Alumel®[a]	32 to 530 530 to 2300	±4 ±.75%	±2 ±.38%	2300	2000	1800	1600	1600
R, S	Platinum-rhodium	32 to 1000 1000 to 2700	±5 ±.50%	±2.5 ±.25%	—	—	—	2700	—
T	Copper-constantan	−150 to −75 −75 to +200 200 to 700	±2.0% ±1.5 ±.75%	±1% ±.75 ±.38%	—	700	500	400	400

[a]Chromel–Alumel Thermocouple Alloys, Hoskins Manufacturing Co., Detroit, Mich.

From USA Standard for Temperature Measurement Thermocouples, C96.1–1969, Instrument Society of America, Pittsburgh. With permission.

Table 2–43
COPPER-CONSTANTAN THERMOCOUPLE CALIBRATION

Type T Thermocouples; Electromotive Force in Absolute Millivolts; Reference Junction at 32°F

Temp. °F	*0°*	*10°*	*20°*	*30°*	*40°*	*50°*	*60°*	*70°*	*80°*	*90°*
					Millivolts					
−200	−4.111	−4.246	−4.377	−4.504	−4.627	−4.747	−4.863	−4.974	−5.081	−5.185
−100	−2.559	−2.730	−2.897	−3.062	−3.223	−3.380	−3.533	−3.684	−3.829	−3.972
(−)0	−0.670	−0.872	−1.072	−1.270	−1.463	−1.654	−1.842	−2.026	−2.207	−2.385
(+)0	−0.670	−0.463	−0.254	−0.042	+0.171	0.389	0.609	0.832	1.057	1.286
100	1.517	1.751	1.987	2.226	2.467	2.711	2.958	3.207	3.458	3.712
200	3.967	4.225	4.486	4.749	5.014	5.280	5.550	5.821	6.094	6.370
300	6.647	6.926	7.208	7.491	7.776	8.064	8.352	8.642	8.935	9.229
400	9.525	9.823	10.123	10.423	10.726	11.030	11.336	11.643	11.953	12.263
500	12.575	12.888	13.203	13.520	13.838	14.157	14.477	14.799	15.122	15.447
600	15.773	16.101	16.429	16.758	17.089	17.421	17.754	18.089	18.425	18.761
700	19.100	19.439	19.779	20.120	20.463	20.805	—	—	—	—

Correction Table for Reference Junction Other Than 32°F

Note: Correction should be added to observe emf before entering the above table.

Reference junction, °F	35	40	45	50	55	60	65	70	75	80	85	90	95
Correction, millivolts	.064	.171	.280	.389	.499	.609	.720	.832	.944	1.057	1.171	1.286	1.401

REFERENCES

Weast, R. C., Ed., *Handbook of Chemistry and Physics,* 55th ed., CRC Press, Cleveland, 1974, E-111.
NBS Circular 561, National Bureau of Standards, Washington, D.C., 1955.

From USA Standard for Temperature Measurement Thermocouples, C96.1–1969, Instrument Society of America, Pittsburgh. With permission.

Table 2–44
IRON-CONSTANTAN THERMOCOUPLE CALIBRATION

Type J Thermocouples; Electromotive Force in Absolute Millivolts; Reference Junction at 32°F

Temp. °F	*0°*	*10°*	*20°*	*30°*	*40°*	*50°*	*60°*	*70°*	*80°*	*90°*
	Millivolts									
−300	−7.52	−7.66	−7.79	—	—	—	—	—	—	—
−200	−5.76	−5.96	−6.16	−6.35	−6.53	−6.71	−6.89	−7.06	−7.22	−7.38
−100	−3.49	−3.73	−3.97	−4.21	−4.44	−4.68	−4.90	−5.12	−5.34	−5.55
(−)0	−0.89	−1.16	−1.43	−1.70	−1.96	−2.22	−2.48	−2.74	−2.99	−3.24
(+)0	−0.89	−0.61	−0.34	−0.06	+0.22	0.50	0.79	1.07	1.36	1.65
100	1.94	2.23	2.52	2.82	3.11	3.41	3.71	4.01	4.31	4.61
200	4.91	5.21	5.51	5.81	6.11	6.42	6.72	7.03	7.33	7.64
300	7.94	8.25	8.56	8.87	9.17	9.48	9.79	10.10	10.41	10.72
400	11.03	11.34	11.65	11.96	12.26	12.57	12.88	13.19	13.50	13.81
500	14.12	14.42	14.73	15.04	15.34	15.65	15.96	16.26	16.57	16.88
600	17.18	17.49	17.80	18.11	18.41	18.72	19.03	19.34	19.64	19.95
700	20.26	20.56	20.87	21.18	21.48	21.79	22.10	22.40	22.71	23.01
800	23.32	23.63	23.93	24.24	24.55	24.85	25.16	25.47	25.78	26.09
900	26.40	26.70	27.02	27.33	27.64	27.95	28.26	28.58	28.89	29.21
1000	29.52	29.84	30.16	30.48	30.80	31.12	31.44	31.76	32.08	32.40
1100	32.72	33.05	33.37	33.70	34.03	34.36	34.68	35.01	35.35	35.68
1200	36.01	36.35	36.69	37.02	37.36	37.71	38.05	38.39	38.74	39.08
1300	39.43	39.78	40.13	40.48	40.83	41.19	41.54	41.90	42.25	42.61
1400	42.96	43.32	43.68	44.03	44.39	44.75	45.10	45.46	45.82	46.18

Correction Table for Reference Junction Other Than 32°F

Note: Correction should be added to observed emf before entering the above table.

Reference junction, °F	35	40	45	50	55	60	65	70	75	80	85	90	95
Correction, millivolts	0.08	0.22	0.36	0.50	0.65	0.79	0.93	1.07	1.22	1.36	1.51	1.65	1.80

From USA Standard for Temperature Measurement Thermocouples, C96.1 – 1969, Instrument Society of America, Pittsburgh. With permission.

REFERENCES

Weast, R. C., Ed., *Handbook of Chemistry and Physics,* 55th ed., CRC Press, Cleveland, 1974, E-109.

NBS Circular 561, National Bureau of Standards, Washington, D.C., 1955.

Table 2–45
CHROMEL-CONSTANTAN THERMOCOUPLE CALIBRATION

Type E Thermocouples; Electromotive Force in Absolute Millivolts; Reference Junction at 32°F

Temp. °F	0°	10°	20°	30°	40°	50°	60°	70°	80°	90°
	Millivolts									
−200	−6.40	−6.62	−6.83	−7.04	−7.24	−7.44	−7.62	−7.80	−7.97	−8.14
−100	−3.94	−4.21	−4.47	−4.73	−4.98	−5.23	−5.48	−5.72	−5.95	−6.18
(−)0	−1.02	−1.33	−1.64	−1.94	−2.24	−2.54	−2.83	−3.11	−3.39	−3.67
(+)0	−1.02	−0.71	−0.39	−0.07	0.26	0.59	0.92	1.26	1.59	1.93
100	2.27	2.62	2.97	3.32	3.68	4.04	4.40	4.77	5.13	5.50
200	5.87	6.25	6.62	7.00	7.38	7.76	8.15	8.54	8.93	9.32
300	9.71	10.11	10.51	10.91	11.31	11.71	12.11	12.52	12.93	13.34
400	13.75	14.17	14.59	15.00	15.42	15.84	16.26	16.68	17.10	17.52
500	17.95	18.38	18.81	19.23	19.66	20.09	20.52	20.95	21.39	21.82
600	22.25	22.69	23.13	23.57	24.00	24.44	24.88	25.32	25.76	26.20
700	26.65	27.09	27.53	27.97	28.42	28.86	29.31	29.75	30.19	30.64
800	31.09	31.54	31.98	32.43	32.87	33.32	33.77	34.22	34.67	35.12
900	35.57	36.02	36.47	36.92	37.37	37.82	38.26	38.71	39.16	39.61
1000	40.06	40.51	40.96	41.41	41.86	42.31	42.76	43.21	43.66	44.11
1100	44.56	45.01	45.46	45.91	46.36	46.81	47.26	47.71	48.15	48.60
1200	49.04	49.49	49.93	50.37	59.82	51.27	51.72	52.16	52.61	53.05
1300	53.50	53.94	54.38	54.83	55.27	55.71	56.15	56.59	57.03	57.48
1400	57.92	58.36	58.80	59.24	59.68	60.11	60.55	60.99	61.43	61.86
1500	62.30	62.74	63.17	63.60	64.04	64.47	64.90	65.34	65.77	66.20
1600	66.63	67.03	67.48	67.91	68.34	68.76	69.19	69.62	70.05	70.47

Note: Correction should be added to observed emf before entering the above table.

Correction Table for Reference Junction Other Than 32°F

Note: Correction should be added to observed emf before entering the above table.

Reference junction, °F	35	40	45	50	55	60	65	70	75	80	85	90	95
Correction, millivolts	0.10	0.26	0.42	0.59	0.76	0.92	1.09	1.26	1.42	1.59	1.76	1.93	2.10

From USA Standard for Temperature Measurement Thermocouples, C96.1–1969, Instrument Society of America, Pittsburgh. With permission.

Table 2–46
CHROMEL-ALUMEL THERMOCOUPLE CALIBRATION

Type K Thermocouples; Electromotive Force in Absolute Millivolts; Reference Junction at 32°F

Temp. °F	*0°*	*10°*	*20°*	*30°*	*40°*	*50°*	*60°*	*70°*	*80°*	*90°*
	Millivolts									
−300	−5.51	−5.60	—	—	—	—	—	—	—	—
−200	−4.29	−4.44	−4.58	−4.71	−4.84	−4.96	−5.08	−5.20	−5.30	−5.41
−100	−2.65	−2.84	−3.01	−3.19	−3.36	−3.52	−3.69	−3.84	−4.00	−4.15
(−)0	−0.68	−0.89	−1.10	−1.30	−1.50	−1.70	−1.90	−2.09	−2.28	−2.47
(+)0	−0.68	−0.47	−0.26	−0.04	+0.18	0.40	0.62	0.84	1.06	1.29
100	1.52	1.74	1.97	2.20	2.43	2.66	2.89	3.12	3.36	3.59
200	3.82	4.05	4.28	4.51	4.74	4.97	5.20	5.42	5.65	5.87
300	6.09	6.31	6.53	6.76	6.98	7.20	7.42	7.64	7.87	8.09
400	8.31	8.54	8.76	8.98	9.21	9.43	9.66	9.88	10.11	10.34
500	10.57	10.79	11.02	11.25	11.48	11.71	11.94	12.17	12.40	12.63
600	12.86	13.09	13.32	13.55	13.78	14.02	14.25	14.48	14.71	14.95
700	15.18	15.41	15.65	15.88	16.12	16.35	16.59	16.82	17.06	17.29
800	17.53	17.76	18.00	18.23	18.47	18.70	18.94	19.18	19.41	19.65
900	19.89	20.13	20.36	20.60	20.84	21.07	21.31	21.54	21.78	22.02
1000	22.26	22.49	22.73	22.97	23.20	23.44	23.68	23.91	24.15	24.39
1100	24.63	24.86	25.10	25.34	25.57	25.81	26.05	26.28	26.52	26.75
1200	26.98	27.22	27.45	27.69	27.92	28.15	28.39	28.62	28.86	29.09
1300	29.32	29.56	29.79	30.02	30.25	30.49	30.72	30.95	31.18	31.42
1400	31.65	31.88	32.11	32.34	32.57	32.80	33.02	33.25	33.48	33.71
1500	33.93	34.16	34.39	34.62	34.84	35.07	35.29	35.52	35.75	35.97
1600	36.19	36.42	36.64	36.87	37.09	37.31	37.54	37.76	37.98	38.20
1700	38.43	38.65	38.87	39.09	39.31	39.53	39.75	39.96	40.18	40.40
1800	40.62	40.84	41.05	41.27	41.49	41.70	41.92	42.14	42.35	42.57
1900	42.78	42.99	43.21	43.42	43.63	43.85	44.06	44.27	44.49	44.70
2000	44.91	45.12	45.33	45.54	45.75	45.96	46.17	46.38	46.58	46.79
2100	47.00	47.21	47.41	47.62	47.82	48.03	48.23	48.44	48.64	48.85
2200	49.05	49.25	49.45	49.65	49.86	50.06	50.26	50.46	50.65	50.85
2300	51.05	51.25	51.45	51.64	51.84	52.03	52.23	52.42	52.62	52.81

Correction Table for Reference Junction Other Than 32°F

Note: Correction should be added to observed emf before entering the above table.

Reference junction, °F	35	40	45	50	55	60	65	70	75	80	85	90	95
Correction, millivolts	0.07	0.18	0.29	0.40	0.51	0.62	0.73	0.84	0.95	1.06	1.18	1.29	1.40

REFERENCES

Weast, R. C., Ed., *Handbook of Chemistry and Physics,* 55th ed., CRC Press, Cleveland, 1974, E-106.
NBS Circular 561, National Bureau of Standards, Washington, D.C., 1955.

Table 2–47
PLATINUM-RHODIUM THERMOCOUPLE CALIBRATION

Electromotive Force in Absolute Millivolts; Reference Junction at 32°F

A. Type R Thermocouples, Platinum-13% Rhodium

Temp. °F	0°	10°	20°	30°	40°	50°	60°	70°	80°	90°
	Millivolts									
0	—	—	—	—	0.024	0.056	0.087	0.120	0.153	0.187
100	0.221	0.256	0.291	0.327	0.364	0.401	0.439	0.477	0.516	0.555
200	0.595	0.635	0.676	0.717	0.758	0.800	0.843	0.886	0.929	0.973
300	1.017	1.061	1.106	1.151	1.196	1.242	1.287	1.334	1.380	1.427
400	1.474	1.521	1.569	1.616	1.664	1.712	1.761	1.809	1.858	1.907
500	1.956	2.005	2.055	2.105	2.155	2.205	2.255	2.306	2.357	2.407
600	2.458	2.510	2.561	2.613	2.664	2.716	2.768	2.820	2.872	2.924
700	2.977	3.029	3.082	3.135	3.188	3.240	3.293	3.347	3.400	3.453
800	3.506	3.560	3.614	3.667	3.721	3.775	3.829	3.883	3.937	3.991
900	4.046	4.100	4.155	4.210	4.264	4.319	4.374	4.430	4.485	4.540
1000	4.596	4.651	4.707	4.763	4.818	4.874	4.930	4.987	5.043	5.099
1100	5.156	5.212	5.269	5.326	5.383	5.440	5.497	5.555	5.612	5.669
1200	5.726	5.784	5.842	5.899	5.957	6.015	6.073	6.131	6.190	6.248
1300	6.307	6.365	6.424	6.483	6.542	6.601	6.660	6.719	6.778	6.838
1400	6.897	6.957	7.017	7.076	7.136	7.196	7.257	7.317	7.377	7.438
1500	7.498	7.559	7.620	7.681	7.742	7.803	7.864	7.925	7.987	8.048
1600	8.110	8.172	8.234	8.296	8.358	8.420	8.482	8.545	8.607	8.670
1700	8.732	8.795	8.858	8.921	8.984	9.048	9.111	9.174	9.238	9.302
1800	9.365	9.429	9.493	9.557	9.621	9.686	9.750	9.815	9.879	9.944
1900	10.009	10.074	10.139	10.204	10.269	10.334	10.400	10.465	10.531	10.597
2000	10.662	10.728	10.794	10.860	10.926	10.992	11.058	11.124	11.190	11.257
2100	11.323	11.389	11.456	11.522	11.589	11.655	11.722	11.789	11.855	11.922
2200	11.989	12.055	12.122	12.189	12.256	12.322	12.389	12.456	12.523	12.590
2300	12.657	12.724	12.790	12.857	12.924	12.991	13.058	13.124	13.191	13.258
2400	13.325	13.391	13.458	13.525	13.591	13.658	13.725	13.791	13.858	13.924
2500	13.991	14.058	14.124	14.191	14.257	14.324	14.390	14.457	14.523	14.589
2600	14.656	14.722	14.789	14.855	14.921	14.988	15.054	15.120	15.186	15.253
2700	15.319	15.385	15.451	15.517	15.583	15.649	15.715	15.781	15.847	15.913

B. Type S Thermocouples, Platinum-10% Rhodium

Temp. °F	0°	10°	20°	30°	40°	50°	60°	70	80°	90°
	Millivolts									
0	—	—	—	—	0.024	0.055	0.086	0.119	0.152	0.186
100	0.220	0.255	0.291	0.327	0.363	0.400	0.438	0.476	0.516	0.556
200	0.596	0.637	0.678	0.721	0.763	0.807	0.850	0.894	0.939	0.984
300	1.030	1.075	1.121	1.167	1.214	1.261	1.309	1.357	1.406	1.455
400	1.504	1.553	1.603	1.653	1.703	1.754	1.805	1.856	1.908	1.960
500	2.012	2.065	2.117	2.170	2.223	2.277	2.330	2.384	2.438	2.493
600	2.547	2.602	2.657	2.712	2.768	2.823	2.879	2.935	2.991	3.047
700	3.103	3.160	3.217	3.273	3.330	3.387	3.445	3.502	3.560	3.618
800	3.677	3.735	3.794	3.852	3.911	3.970	4.029	4.087	4.146	4.205
900	4.264	4.324	4.384	4.443	4.503	4.563	4.624	4.685	4.746	4.807
1000	4.868	4.930	4.991	5.053	5.115	5.176	5.238	5.301	5.363	5.426
1100	5.488	5.551	5.614	5.677	5.741	5.805	5.869	5.933	5.996	6.060
1200	6.125	6.188	6.252	6.317	6.381	6.446	6.511	6.577	6.642	6.706
1300	6.773	6.838	6.904	6.970	7.037	7.103	7.169	7.235	7.302	7.369
1400	7.436	7.503	7.571	7.639	7.706	7.774	7.842	7.911	7.979	8.047
1500	8.116	8.184	8.253	8.322	8.391	8.460	8.530	8.599	8.669	8.739
1600	8.809	8.879	8.949	9.019	9.090	9.161	9.232	9.303	9.374	9.445
1700	9.516	9.587	9.659	9.730	9.802	9.874	9.946	10.019	10.092	10.164
1800	10.237	10.310	10.383	10.456	10.529	10.603	10.676	10.749	10.823	10.898
1900	10.973	11.048	11.122	11.197	11.273	11.348	11.424	11.499	11.575	11.651
2000	11.726	11.802	11.878	11.954	12.029	12.105	12.182	12.258	12.335	12.411
2100	12.488	12.564	12.641	12.718	12.795	12.871	12.948	13.025	13.102	13.178
2200	13.255	13.332	13.409	13.486	13.564	13.641	13.718	13.795	13.872	13.949
2300	14.027	14.104	14.181	14.258	14.335	14.412	14.490	14.567	14.644	14.721
2400	14.798	14.875	14.952	15.029	15.107	15.184	15.261	15.338	15.415	15.492
2500	15.568	15.645	15.722	15.800	15.877	15.954	16.031	16.108	16.185	16.263
2600	16.340	16.417	16.494	16.571	16.648	16.725	16.802	16.880	16.957	17.033
2700	17.110	17.186	17.263	17.340	17.416	17.493	17.569	17.646	17.723	17.799

Table 2–47 (continued)
PLATINUM-RHODIUM THERMOCOUPLE CALIBRATION

Correction Table for Reference Junction Other Than 32°F for Pt-Rh Couples

Note: Correction should be added to observed emf before entering the above tables.

Reference junction, °F	35	40	45	50	55	60	65	70	75	80	85	90	95
Correction, millivolts	.009	.024	.038	.055	.070	.086	.102	.119	.134	.152	.168	.186	.202

REFERENCES

Zysk, E. D. and Robertson, A. R., Thermocouples for temperatures above 1500°C, *Instrum. Technol.,* 8(11), 30, 1961.
Weast, R. C., Ed., *Handbook of Chemistry and Physics,* 55th ed., CRC Press, Cleveland, 1974, E-105.
NBS Circular 561, National Bureau of Standards, Washington, D.C., 1955.

From USA Standard for Temperature Measurement Thermocouples, C96.1–1969, Instrument Society of America, Pittsburgh. With permission.

Table 2–48
PROPERTIES OF PIEZOELECTRIC CERAMICS

The main properties of representative piezoelectric ceramic transducer compositions are listed. Generally the notation follows that of the IRE Standards.[1] Various electromechanical coupling factors (k), free and clamped relative permittivities (ϵ/ϵ_0), and piezoelectric d and g constants are given. The four coupling factors evaluate the ability of the material to convert energy from electrical to mechanical form (or vice versa) in the planar, transverse, parallel, and shear modes, respectively. The free (superscript T) and clamped (superscript S) permittivities govern impedance parallel and transverse to the field direction at frequencies away from the electromechanical resonances. The piezoelectric d and g constants measure charge density and field generated by an applied stress, respectively, in the parallel (33), transverse (31), and shear (15) directions of a transducer. A high d constant is valuable for generating motion and a high g constant for generating electrical signals. The table also includes tan δ, mechanical Q, and the frequency constants for a bar poled in a thin dimension (N_1) and a thin plate (N_3). The properties are intimately tied to crystal structure and chemical composition of the ceramic.

Barium titanate was the original piezoelectric ceramic. Both unmodified and modified compositions are represented. It has been largely supplanted by the lead titanate-zirconates which have higher coupling factors and can operate over a wide temperature range. The first two compositions predominate in present usage.

The last two compositions in the table have specialized uses. Lead metaniobate has strongly anisotropic piezoelectric response and very low mechanical Q, both assets in ultrasonic flaw detection. Sodium-potassium niobate, unlike the other compositions, is hot pressed rather than kiln fired. Its low relative permittivity and high value of thickness frequency constant are good for delay line transducers.

A few other modified lead titanate-zirconate compositions find substantial use. Their compositions are proprietary and they are not included here. Data on their properties may be obtained from manufacturers and elsewhere.[2] Recent sources of information[2-4] are available for a more thorough review of materials and applications.

Table 2–48 (continued)
PROPERTIES OF PIEZOELECTRIC CERAMICS

Quantity	Material			
	$Pb_{.94}Sr_{.06}(Ti_{.48}Sr_{.52})O_3$	$Pb_{.988}(Ti_{.48}Zr_{.52})_{.976}Nb_{.024}O_3$	$Pb(Ti_{.48}Zr_{.52})O_3$	$Pb(Ti_{.46}Zr_{.54})O_3$
k_p	0.58	0.60	0.53	0.47
k_{31}	0.334	0.344	0.313	0.280
k_{33}	0.70	0.705	0.67	0.626
k_{15}	0.71	0.685	0.694	0.701
$\epsilon_{33}^T/\epsilon_o$	1300	1700	730	450
$\epsilon_{33}^S/\epsilon_o$	635	830	399	260
$\epsilon_{11}^T/\epsilon_o$	1475	1730	1180	990
$\epsilon_{11}^S/\epsilon_o$	730	916	612	504
d_{33}, 10^{-12} C/N	289	374	223	152
d_{31}	-123	-171	-94	-60
d_{15}	496	584	494	440
g_{33}, 10^{-3} Vm/N	26.1	24.8	34.5	38.1
g_{31}	-11.1	-11.4	-14.5	-15.1
g_{15}	39.4	38.2	47.2	50.3
Tan δ	0.004	0.02	0.004	0.003
Q_m	500	75	500	680
Density, 10^3 kg/m^3	7.6	7.75	7.6	7.6
N_1, $H_z \cdot m$	1650	1400	–	1680
N_3 (thin plate)	2000	1770	–	2090
Curie point	328°C	365°C	386°C	370°C
Structure	Perovskite	Perovskite	Perovskite	Perovskite
Symmetry	Tetr.	Tetr.	Tetr.	Rhomb.
Commercial use	Ultrasonics, sonar	Microphones, hydrophones		

Table 2–48 (continued)
PROPERTIES OF PIEZOELECTRIC CERAMICS

	Material				
Quantity	$BaTiO_3$	95 wt % $BaTiO_3$ 5 wt % $CaTiO_3$	80 wt % $BaTiO_3$ 12 wt % $PbTiO_3$ 8 wt % $CaTiO_3$	$PbNb_2O_6$	$Na_{.5}K_{.5}NbO_3$
k_p	0.36	0.33	0.19	0.07	0.45
k_{31}	0.212	0.194	0.113	0.045	0.27
k_{33}	0.50	0.48	0.34	0.38	0.53
k_{15}	0.48	0.491	0.30	–	–
$\epsilon^T_{33}/\epsilon_o$	1700	1200	450	225	420
$\epsilon^S_{33}/\epsilon_o$	1260	910	395	–	–
$\epsilon^T_{11}/\epsilon_o$	1450	1300	–	–	–
$\epsilon^S_{11}/\epsilon_o$	1115	1000	–	–	–
d_{33}, 10^{-12} C/N	190	149	60	85	160
d_{31}	-78	-58	-20	$\sim$-9	-49
d_{15}	260	242	–	–	–
g_{33}, 10^{-3} Vm/N	12.6	14.1	15.1	42.5	43
g_{31}	-5.2	-5.5	-5.0	-4.5	-13.1
g_{15}	20.2	21.0	–	–	–
Tan δ	0.01	0.006	0.006	0.01	0.014
Q_m	300	400	1200	11	240
Density, 10^3 kg/m^3	5.7	5.55	5.4	6.0	4.46
N_1, H_z·m	2200	2290	2430		2540
N_3 (thin plate)	2520	2740	–		3470
Curie point	115°C	115°C	140°C	570°C	420°C
Structure	Perovskite	Perovskite	Perovskite	K-tungsten bronze	Perovskite
Symmetry	Tetr.	Tetr.	Tetr.	Orth.	Orth.
Commercial use				Flaw detectors	Delay lines

REFERENCES

1. IRE standards on piezoelectric crystals: measurements of piezoelectric ceramics, 1961, *Proc. IRE*, 49, 1161, 1961.
2. **Berlincourt, D.,** Piezoelectric crystals and ceramics, in *Ultrasonic Transducer Materials,* Mattiat, O. E., Ed., Plenum Press, New York, 1971, chap. 2, p. 63.
3. **Berlincourt, D. A., Curran, D. R., and Jaffe, H.,** Piezoelectric and piezomagnetic materials, in *Physical Acoustics,* Vol. 1, Part A, Mason, W. P., Ed., Academic Press, New York, 1964, chap. 3, p. 169.
4. **Jaffe, B., Cook, W. R., Jr., and Jaffe, H.,** *Piezoelectric Ceramics,* Academic Press, New York, 1971.

Table compiled by B. Jaffe.

Table 2–49
COMPARISON OF BATTERY TYPES

For conversion of temperature to K, see Table 4–13 of *CRC Handbook of Materials Science,* Volume I. For output in kJ/kg, multiply the values in watt-hr/lb by 7.9367.

Name	*Type*	*Anode*	*Cathode*	*Electrolyte*	*Nominal cell voltage*	*Temp range, °F*	*Typical output, watt-hr/lb*	*Cycle life if recharged (50% discharge)*
Leclanché or carbon-zinc	Primary	Zinc	C + MnO_2	$NH_4Cl-ZnCl_2$	1.5	40–130	2–30	–
Mercury	Primary	Zinc	HgO	$KOH-K_2Zn_2O_3$	1.35 and 1.4	40–130	50	–
Silver oxide	Primary	Zinc	Ag_2O or AgO	KOH or NaOH	1.5	0–130	30–80	100–300
Alkaline or manganese zinc	Primary and rechargeable	Zinc or zinc–Hg	MnO_2	KOH or NaOH	1.5	0–130		50–100
Lalande	Primary	Zinc	CuO	NaOH	0.65		20	–
Nickel-cadmium	Rechargeable (secondary)	Cadmium	$Ni(OH)_2$	KOH	1.25	0–115	6–15	100–2,000
Silver-cadmium	Rechargeable	Cadmium	Ag_2O_2 or AgO	KOH	1.1	-40–100	10–40	300–1,000
Lead-acid or Planté	Rechargeable	Lead	PbO_2	H_2SO_4	2.0	-40–120	7–26	100–400
Nickel-iron or Edison	Rechargeable	Iron	NiO_2	KOH	1.2		10–15	100–3,000
Cuprous chloride	Activated	Magnesium	Cu_2Cl_2	Sea water	1.2	-80–150	20–40	–
Silver chloride	Activated	Magnesium	AgCl	Sea water	1.4	-80–150	20–80	–

Notes: **Mercury oxide and zinc batteries** are important commercially (called RM batteries for Signal Corps walkie-talkies). They have the following characteristics: very flat voltage curve; good heavy-drain characteristics; high capacity per unit volume and weight; long dry-storage life (90% at 4 years); suited for continuous service; available in miniature (button-type); usable, at light loads, down to -40°F and up to 160°F; withstand pressure, vibration, acceleration, impact; and often used as voltage-reference sources, 1.35 V/cell.

Silver oxide and zinc batteries may be dry-stored and activated immediately prior to use. They have a high capacity per unit weight and are non-magnetic. Special batteries may be charged, but their high-current performance is inferior to the primary type.

Manganese oxide and zinc batteries sustain voltage at high current drain. They are usable to -5°F, are inexpensive, and have a long shelf-life.

Lalande copper oxide and zinc batteries are commercially used for crossing and semaphore signals, approach lighting, etc., in sizes of 75–1,000 Ahr (ampere hour). They have a flat voltage-time curve and allow easy field replacement. KOH electrolyte is substituted when service is much below atmospheric freezing (but above 0°F).

Nickel-cadmium batteries may be recharged many times and tolerate overcharge.

Silver-cadmium batteries have low cell voltage but high output per unit volume and weight. Their chief characteristics are long charge-cycle life, rapid charge, non-magnetic, and no residual field.

From Bolz, R. E. and Tuve, G. L., Eds., *Handbook of Tables for Applied Engineering Science,* 2nd ed., CRC Press, Cleveland, 1973, 554.

Table 2–50
PROBABLE EFFICIENCIES OF FUEL CELLS AT VARIOUS LOADS

Percentage of maximum power	*Efficiencies in percent*			
	Present dissolved methanol type cells	*Future dissolved methanol type cells*	*Present hydrazine type cells*	*Diesel engine*
10	43	72	68	22
20	36	67	65	27
40	28	60	58	29
60	23	56	50	31
80	19	49	43	28
100	12	35	30	25

From Berger, C., Ed., *Handbook of Fuel Cell Technology,* Prentice-Hall, Englewood Cliffs, N. J., © 1968. With permission.

Table 2–51
PROTOTYPE BATTERIES

For energy density in kJ/kg, multiply the values in Whr/lb by 7.9367. For energy density in MJ/m^3, multiply the values in $Whr/in.^3$ by 219.69. For power density in W/kg, multiply the values in W/lb by 2.2046.

Type	*Composition (charged)*			*Average energy density*			*Maximum power density, W/lb*	*Cell potential, v*		*Cell life*	*Remarks*	*Manufacturer or developer*
	+Cathode	*−Anode*	*Electrolyte*	*Whr/lb Theoretical*	*Whr/lb Actual*	*$Whr/in.^3$*		*Open*	*Discharging*			
Zinc-nickel-S[a]	Ni	Zn	KOH	220	40–50							Yardney Electric
Zinc-air	Air	Zn	KOH	464	50–60 to 100	2–3	30 (40 est)	1.65	0.9–1.2 to 1.4	Several hundred cycles	Several approaches to avoid dendritic zinc	GE, General Atomics (Gulf), Yardney, ESB
Zinc-air-P[b]	Air	Zn	KOH	464	150		20 (40 est.)	1.65	0.9–1.2			Leesona-Moos
Magnesium-air-P	Air	Mg	Aqueous solutions of NaCl, CaCl, LiCl, MgCl	345	500 with local H_2O refills				1.3			GE
Sodium-sulfur-S	S	Na	Ceramic	345	84–100 est. to 150	4 est. 8	100 est. to 200	2	1.75	Indefinite	Operates at high temperature (250–300°C), self-maintaining once brought to operating temperature	Ford
Sodium-air-S	Air	Na	NaOH	930	160–215 (4-hr discharge)	5	40–55 (4-hr discharge)	2.60	2.3–2.4	Indefinite	2-step process involving complex sodium amalgam—operates at 130°C	Atomic International (Northern American Aviation)

Table 2–51 (continued)
PROTOTYPE BATTERIES

Type	Composition (charged)			Average energy density			Maximum power density, W/lb	Cell potential, v		Cell life	Remarks	Manufacturer or developer
				Whr/lb		Whr/in.^{3}						
	+ Cathode	− Anode	Electrolyte	Theoretical	Actual			Open	Discharging			
Sodium-air-P	Air	Na	NaOH	930			~40	1.9–2.0	1.4		Sodium-mercury amalgam, ambient temperature	Western Reserve
Calcium-air-P	Air	Ca	KOH									Yardney
Lithium-moist air-S	Air + H_2O	Li	Non-aqueous	2566								Globe-Union
Aluminum-air-P	Air	Al	KOH		240–400		30 av. 48–75 peak	2.7	1.1		Power controlled by adjustment of liquid level	Zaromb Research Corp.
Iron-air-S	Air	Fe		650							Proprietary development	Westinghouse
Dry tape-P	Halogen systems	Li	Organic		200 (ex hardware)				3.0–3.1			Monsanto
		Mg	Aqueous		100 (ex hardware)				2.0			
Dry tape-P	Ag_2O_2	Zn	KOH		25–30							Monsanto
Lithium-chlorine-S	Cl_2	Li	Lithium-chloride	1200	100 est. to 300		75 est. to 150	3.5	3.2	Indefinite	Operates at 650°C	GM
Lithium-tellurium-S	Li_2Te	Li	LiCl-LiF eutectic	270	90		140	1.94	1.67–1.79	Indefinite	Operates at 471°C	Argonne National Laboratory
Lithium-nickel chloride-S	$NiCl_2$	Li	Propylene carbonate or K-phosphofluoride	435	est. 100		Low <20	2.50			Room temperature	Gulton

Table 2–51 (continued)
PROTOTYPE BATTERIES

Type	Composition (charged)			Average energy density			Maximum power density, W/lb	Cell potential, v		Cell life	Remarks	Manufacturer or developer
				Whr/lb								
	+ Cathode	− Anode	Electrolyte	Theoretical	Actual	$Whr/in.^3$		Open	Discharging			
Lithium-nickel fluoride-S	NiF_2	Li	Same as above	620	90–100		Low <70	2.9			Room temperature	Gulton
Lithium-silver difluoride-S	AgF_2	Li	Butyrolactone, KPF_6	678			Low			Has undergone 50 cycles at 90% depth of discharge	Room temperature	Whittaker
Lithium-silver chloride-S	AgCl	Li	Li salts in PC, other organics	231	30 est. 90		Low		2.3–2.8	50–200 cycles	Room temperature	Lockheed, Electrochimica
Lithium-copper fluoride-P	CuF_2	Li	Li salts, in PC, other organics	746	57–80 110 est. 200		Low	3.6	2.3–2.8		Room temperature	Lockheed, ESB, Electrochimica
Lithium-copper chloride-S	$CuCl_2$	Li	Li salts in PC, other organics	503	25 est. 175		Low	3.1	2.3–2.8		Room temperature	Mallory, Electrochimica

[a] S – secondary.
[b] P – primary.

From *SAE J.*, 76(12), 64, 1968. With permission.

Table 2–52
TYPICAL COMPLETE FUEL-CELL SYSTEMS

For mass in kg, multiply values in pounds by 0.4536. For volume in m^3, multiply values in ft^3 by 0.028 317.

Output, kW	*Mass (excluding fuel), lb*	*Volume, cu ft*	*Fuel and oxidant*	*Electrode and electrolyte*	*Conditions*	*Operating application*
UNION CARBIDE						
0.3[a]	33 (including fuel)	0.87	Hydrazine, air	Carbon plated on nickel circulating KOH	Ambient	Commercial
1.0–2.5[b]	42 + auxiliaries	3.36	Hydrogen, oxygen		120°F–150°F	Commercial
3.74–9.4[c]		28	Hydrogen, oxygen		120°F–150°F	
94	3650		Hydrogen, oxygen		120°F–150°F	GM Electrovan
ALLIS-CHALMERS						
0.5			Methanol, oxygen	Porous nickel electrodes, platinized	Ambient	Demonstration
3				anode, silver cathode, asbestos	Ambient	Demonstration
4.5	1190	34	Hydrazine, air JP–150 (reformer)	matrix held KOH	Reformer 1450°F Cell 160–180°F	U.S. Army
2.5 (overload)	169 (complete)	< 5.3	Hydrogen, oxygen		190°F	Space
15		20 + auxiliaries	Propane, oxygen		Ambient	Mounted on tractor
MONSANTO RESEARCH						
0.06	14.5	0.35	Hydrazine, air	Circulating KOH	200°F	
5	200		Hydrazine, air		200°F	
60	1080		Hydrazine, air		200°F	U.S. Army truck
PRATT & WHITNEY (UNITED AIRCRAFT)						
0.5	82.75	2.63	JP–4 (reformer)	Asbestos matrix, KOH		Army (battery charger)
1			Hydrogen, oxygen		~500°F, 55 psi	NASA (LM)
2			Hydrogen, oxygen		~500°F, 55 psi	NASA (Apollo)
GENERAL ELECTRIC						
0.03	7	0.3	Lithium hydride, air	Ion-exchange membrane, sulfonic acid	35–110°F	Military field use
0.06	10	0.7	Active metal/ hydrogen, air		35–110°F	Military field use
0.2	60	—	Hydrogen, air		35–110°F	
1.0	70 + auxiliaries		Hydrogen, oxygen		35–110°F	NASA (Gemini type)
1.5	140	8	JP–4 reformer, air		35–110°F	Battery charger
ENERGY CONVERSION LTD. (U.K.)						
6	1300	2813	Methanol, air	—	—	Electric truck
CHLORIDE GROUP (U.K.)						
—	—	—	Hydrazine, air	—	—	Demonstration
SHELL (U.K.)						
—	—	—	Methanol, air	Acid electrolyte	—	Demonstration
ASEA (SWEDEN)						
50	—	—	Ammonia (reformer)		—	Submarine
VARTA A.G. (GERMANY)						
2	—	—	Hydrogen, oxygen		—	Fork-lift truck

[a]H2R–1–278.
[b]E2F–1–468.
[c]E2F–4–864.

From *SAE J.*, 76(12), 73, 1968. With permission.

Table 2–53
FUEL-CELL CHARACTERISTICS

For current density in A/m^2, multiply values in amp/sq ft by 10.764. For output in m^3/kW, multiply values in ft^3/kW by 0.0283. For output in g/W, multiply values by lb/kW by 0.4536.

Type	*Fuel*	*Oxidant*	*Electrode*	*Electrolyte*	*Operating temperature*	*Operating pressure*	*Open-circuit voltage*		*Current density, amp/sq ft*	*Output, Whr/lb*	*Output*		*% efficiency thermal to electric*
							Theoretical	*Measured*			ft^3/kW	lb/kW	
Ion-exchange membrane	Hydrogen	Oxygen or air	Activated metal	(Solid) ion-exchange membrane	50°F above ambient −65 to +165°F	Atmospheric	1.1	1 at 0.8 volt and 10 ma	22 at 0.8 volt	100 (measured)	3.5–5	250–500	60 at 15–20 amp/sq ft
Redox	Liquefied fuel	Oxygen or air	Porous metal	Liquids	70–85°C	Approx atmospheric	~1		200	1,200 with air 1,600 with pure oxygen	5	50–75	
Carbox	HCO-petroleum hydrocarbons (kerosene)	Oxygen or air	Porous metal	Fused carbonate	500–800°C	Atmospheric	~1	0.7	60				64
Hydrox	Hydrogen	Oxygen	Porous metal		200–250°C, 400–500°F	10–55 atm, 400–600 psi	1.1	1.0	Up to 1,000		0.25–1.2	40–90	
Thermal regenerative	Hydrogen	Group I metals	Fuel metal and nickel	Fused group I chlorides	608 or 1004°F	200–500 mm, Hg abs	0.75	0.72	245 at 0.72 volt with lithium	~5			Carnot: 40
Solar regenerative	Nitric oxide	Chlorine	Carbon-disk	Liquid nitrosyl chloride	70°F	15 psig	0.21	0.21	2 at 0.1 volt				
Low temp-pressure	Hydrogen	Oxygen	Specially processed carbon	12 molar solution of KOH	70–150°F	1–5 atm	1.2	1.12	100 at 0.95 volt and 104°F and oxygen at 5 atm	1,620	3.5–5	250–500	75

From Bolz, R. E. and Tuve, G. L., Eds., *Handbook of Tables for Applied Engineering Science,* 2nd ed., CRC Press, 1973, 559.

Table 2–54
ELECTROMOTIVE FORCE AND COMPOSITION OF VOLTAIC CELLS

Standard Cells

Name of cell	*Negative pole*	*Solution*	*Positive pole*	*Depolarizer*	*Electromotive force, volts*
Weston normal	Cadmium amalgam	Saturated solution of $CdSO_4$	Mercury	Paste of Hg_2SO_4 and $CdSO_4$	1.0183 at 20°C
Clark standard	Zinc amalgam	Saturated solution of $ZnSO_4$	Mercury	Paste of Hg_2SO_4 and $ZnSO_4$	1.4328 at 15°C

Temperature equations (temperature in °C):

Clark cell: $E_t = 1.4328[1 - 0.00119(t - 15) - 0.000007(t - 15)^2]$ volt

Weston cell: $E_t = 1.0183[1 - 0.0000406(t - 20) - 0.00000095(t - 20)^2 + 0.00000001(t - 20)^3]$ volt

Double Fluid Cells

Name of cell	*Negative pole*	*Solution*	*Positive pole*	*Solution*	*Electromotive force, volts*
Bunsen	Amal. zinc	1 part H_2SO_4 to 12 parts H_2O	Carbon	Fuming nitric acid	1.94
Bunsen	Amal. zinc	1 part H_2SO_4 to 12 parts H_2O	Carbon	HNO_3, density 1.38	1.86
Bichromate	Amal. zinc	12 parts $K_2Cr_2O_7$ to 25 parts H_2SO_4 and 100 parts H_2O	Carbon	1 part H_2SO_4 to 12 parts H_2O	2.00
Bichromate	Amal. zinc	1 part H_2SO_4 to 12 parts H_2O	Carbon	12 parts $K_2Cr_2O_7$ to 100 parts H_2O	2.03
Daniell	Amal. zinc	1 part H_2SO_4 to 4 parts H_2O	Copper	Saturated solution of $CuSO_4 + 5H_2O$	1.06
Daniell	Amal. zinc	5% solution of $ZnSO_4 + 6H_2O$	Copper	Saturated solution of $CuSO_4 + 5H_2O$	1.08
Daniell	Amal. zinc	1 part NaCl to 4 parts H_2O	Copper	Saturated solution of $CuSO_4 + 5H_2O$	1.05
Grove	Amal. zinc	1 part H_2SO_4 to 12 parts H_2O	Platinum	Fuming nitric acid	1.93
Grove	Amal. zinc	Solution of $ZnSO_4$	Platinum	HNO_3, density 1.33	1.66

From Forsythe, W. E., Ed., *Smithsonian Physical Tables,* 9th ed., The Smithsonian Institution, Washington, D.C., 1956, 377.

Table 2–55
CHARACTERISTICS OF NUCLEAR BATTERIES

	Constant-current charging			Contact-potential difference	Junction	Photo-junction	Thermo-junction
Radioactive material	Sr^{90}	H^3	Kr^{85}	H^3	Sr^{90}	Pm^{147}	Po^{210}
Half-life	25 yr	12 yr	10 yr	12 yr	25 yr	2.6 yr	138 days
Quantity	10 Mc	1 curie	1 curie	1.5 Mc/cell	50 Mc	4.5 curies	3,000 curies
Size	1 cu in.	1 cu in.	5 cu in.	1 cu in.		0.2×0.7 in. diam	5.5×4.75 in. diam
Weight	6 oz	1 oz	14 oz	1.5 oz		0.6 oz less shielding	5 lb
Current amp	10^{-12}	6×10^{-10}	10^{-9}	10^{-10}	5×10^{-6}	0.25–1 volt, 20μw	5 watts
Voltage	14 kv	1 kv	1 kv	100 volts (66 cells)	0.2 volt		
Development status	Sr batteries in production; prototypes of H, Kr batteries under test			Development complete, but not in production	Development complete, but not in production	Development complete	Prototypes completed; larger units being investi-gated
Manufacturers	Radiation Research Corp.; Patterson Moos Div.; Universal Winding Co.			Tracerlab, Inc.	RCA	Elgin National Watch Co.	Mound Laboratory; Martin Company

From *Electronics*, March 20, 1959; copyright McGraw-Hill, Inc., 1959. With permission.

Section 3

Nuclear Materials

3.1 RADIATION DOSE AND RISK DETERMINATION

Allen Brodsky
Mercy Hospital
Duquesne University
Pittsburg, Pennsylvania

RELATIONSHIP BETWEEN PHYSICAL QUANTITIES AND BIOLOGICAL EFFECTS

A number of acute and long-term effects of radiation on animal and human species have been related to the physical energy absorbed from various types of ionizing radiation. However, the relative effectiveness of each type of radiation per unit energy absorbed in biological tissue has been found to vary not only with the type of radiation and its quantum energy, but also with the rate at which the energy is delivered, the kind of tissue, age and species of animal, the biological effect under consideration, and other experimental and epidemiologic variables. For mammals, beta radiation and the recoil electrons ejected from atoms by X- or gamma radiation generally produce approximately the same order of magnitude of biological effects. On the other hand, heavier particles such as the alpha particles emitted by certain radionuclides lose their energy at higher rates of linear energy transfer (LET) along their paths and seem to produce somewhat higher damage per unit energy absorbed. In the case of alpha-emitting radionuclides, of course, since the characteristic alpha particles emitted do not have sufficient range to penetrate the dead layer of skin, biological damage is produced only when the radioactive material itself is distributed within an organ by inhalation or ingestion.

Thus, for purposes of radiation hazard evaluation, several units of radiation exposure and dose must be introduced to account for the several methods of measuring and assessing the effects of different types of radiation. Since most radiations of a given type, as emitted by most radionuclides, have average LET's within a narrow range, and consequently seem to have relative biological effectivenesses (RBE's) within a narrow range, characteristic simplifications can be introduced to limit the number of new units and the definitions required for most radiation protection applications. In this section, only the most useful definitions and data are presented for use in radiation dose measurements and calculations for external sources of radiation. A short summary of available quantitative data for risk determination is included, but no summary of the complex data of radiation therapy or experimental radiobiology is within the scope of this section. Elementary introductions to radiobiology are contained in References 13, 34, and 39 for purposes of radiation therapy, low-level radiation risk evaluation, and management of radiation accident cases, respectively. Reference 39 also gives an extensive compilation of data and methods for evaluating exposure from inhaled and environmental sources of radioactive material.

BASIC UNITS OF RADIOACTIVITY, RADIATION MEASUREMENT, AND RADIATION DOSE

The following definitions of quantities and units will suffice in dealing with most problems in radiation protection (health physics) and dosimetry:

Roentgen (R) – a unit for expressing exposure from X- or gamma radiation in terms of the ionization produced in air, which can be measured by appropriate air ionization chambers and electrical instruments. It has been defined as an exposure "such that the associated corpuscular emission per 0.001293 g of air produces, in air, ions carrying one electrostatic unit of quantity of electricity of either sign."[1] More recent definitions have been given in units of coulombs per kilogram, as 2.58×10^{-4} C/kg of air.[1] This simply means that an exposure (formerly called "exposure dose"[2]) of 1 R generates recoil electrons per 0.001293 g of dry air within a small volume surrounding the point of measurement to the extent that these electrons produce 1 esu (electrostatic unit) of positive charge and 1 esu of negative charge (as ion pairs) in air. Since many electrons will leave the element of volume before losing all of their energy, a "standard air chamber" is necessary in order to most accurately measure the roentgen (see Figure 3.1–1). This chamber is specially designed to achieve an "electronic equilibrium" condition in which the ionization lost by electrons leaving that volume is compen-

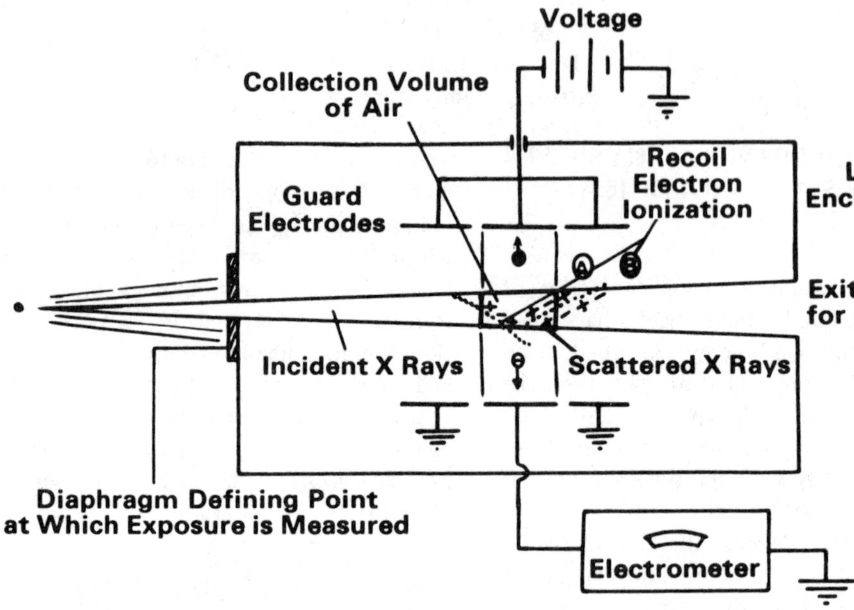

FIGURE 3.1–1. Schematic diagram of a "standard air chamber." (From Brodsky, A., in *CRC Handbook of Radioactive Nuclides,* Wang, Y., Ed., The Chemical Rubber Co., Cleveland, 1969, 574.

sated for by an equilibrium number of electrons entering the volume from a preceding volume of air.[1] Thus, the roentgen is a unit that expresses a point quantity, essentially the ionization density produced near the point of measurement in air. When applied to human exposure in a large beam or field of radiation, either the field must be uniform or the exposure in roentgens must be measured at each point in the field. In summary, the roentgen may be remembered for practical purposes as the X- or gamma exposure 1 esu (+ or −) per cubic centimeter of air (at STP 0°C, 760 mm Hg).

rad – a unit of "absorbed dose" (D) in any medium for any kind of ionizing radiation.[1,2] It is simply defined as:

$$1 \text{ rad} = 100 \text{ ergs/gram}.$$

Integral absorbed dose – the integral $\int D dm$ over an organ or the whole body, where D is the variable absorbed dose within mass element dm. It is useful in obtaining an idea of the total energy absorbed by a region of tissue, which is more closely related to certain macroscopic biological changes than the dose only near one point. Common units are gram-rads, and 1 gram-rad = 100 ergs. The integral absorbed dose may be divided by the total mass of tissue in the region of interest to obtain the average tissue dose $\bar{D}$ in rads.

rep – a unit formerly used for similar purposes as the rad, but defined for any type of radiation in terms of the energy absorbed in tissue equivalent to that which would be absorbed from 1 R of X- or gamma radiation. One rep has been defined variously as the absorption of energy ranging from 84 erg/cm^3 to 93 erg/gram of tissue.[3,4]

rem – a unit of "RBE dose" or "dose equivalent" (DE), used to express the estimated equivalent of any type of radiation that would produce the same biological end point as 1 rad delivered by X- or gamma radiation. Thus,

$$\text{DE (in rem)} = \text{Dose (in rad)} \times \text{QF} \times \text{DF},$$

where QF (formerly called RBE) accounts for the relative biological effectiveness of the radiation compared to X-radiation for radiation protection purposes, and DF (formerly designated as n) is the "relative damage factor"[5] or "distribution factor"[1,2] used to account for differences in the distribution of the rad dose to the organ of concern as a result of uneven uptake of the radionuclide, etc., as opposed to the QF or RBE factors reserved for more intrinsic biological characteristics of the emitted radiations.

gram-rem – the unit for integral absorbed dose when several types of radiation are involved and the equivalent doses of each are to be added after multiplication by appropriate relative biological effectiveness factors.

man-rem – a term used in estimating expected frequencies of disease in a population by determining the equivalent integral absorbed dose over the population. For somatic effects, the number of persons in the exposed population might be multiplied by the average dose equivalent to each person; for genetic effects, the average gonadal dose would be multiplied by the number of people exposed.

kerma (*K*) – the quotient E_K/m, where E_K is the sum of all kinetic energies of charged particles liberated by indirectly ionizing particles (e.g., neutrons) in a volume element, and *m* is the mass of matter in that volume element.[6]

Energy fluence (*F*) – as defined by the ICRU,[6] the quotient $\Delta E_F/\Delta a$, where ΔE_F is the sum of all the energies, exclusive of rest energies, entering a sphere of cross-sectional area Δa (e.g., in units of MeV/cm^2).

Energy flux density (intensity *I*) – the quotient $\Delta F/\Delta t$, where ΔF is the energy fluence in a small time interval Δt.

Particle fluence (Θ) – the quotient $\Delta N/\Delta a$, the number of particles flowing into a small sphere per unit cross-sectional area. This is the same quantity as nvt, used in the case of neutron-diffusion theory, where n = neutron density in $neutrons/cm^3$, v = average neutron velocity in cm/sec, and t = time in seconds over which the fluence is integrated. Another way of envisioning the meaning of nvt would be to imagine the total distance in centimeters traveled by all the neutrons present at a given moment per cubic centimeter of volume.

Particle flux density (ϕ) – the quotient $\Delta\Phi/\Delta t$, where $\Delta\Phi$ is the particle fluence in time Δt; units may be $particles/cm^2$-sec.

Mass attenuation coefficient (μ/ρ) – the fractional number of incident particles (or photons) interacting with a given material per unit mass thickness that they pass through; i.e., $\mu/\rho = dN/N\rho dl$, where dN/N is the probability of interaction per unit thickness dl of density ρ. Common units are cm^2/g (i.e., fractional number/g-cm^{-2}).

Mass energy-absorption coefficient (μ_{en}/ρ) – the fractional energy removed from incident indirectly ionizing particles (or photons) per unit mass thickness; i.e., $\mu_{en}/\rho = dE/E\rho dl$, where dE is the energy removed from the incident particles (not including rest energy or energy reirradiation as bremsstrahlung), E is the sum of the energies (excluding rest energies) of the incident particles, and ρdl is the mass thickness in g/cm^2, as above. Common units are again cm^2/g (i.e., fractional energy/g-cm^{-2}). (See Reference 33.)

Curie (Ci)– the most common unit used to express the radioactivity (A) of a material; it is the amount of any radionuclide (or combination of radionuclides) in which there are 3.7×10^{10} nuclear transformations or disintegrations per second, or 2.22×10^{12} disintegrations per minute (dpm). Combined with appropriate prefixes, the symbol is commonly used for the following other scientific units:

1 megacurie (MCi)	$= 10^6$ Ci
1 kilocurie (kCi)	$= 10^3$ Ci
1 millicurie (mCi)	$= 10^{-3}$ Ci
1 microcurie (μCi)	$= 10^{-6}$ Ci
1 nanocurie (nCi)	$= 10^{-9}$ Ci
1 picocurie (pCi)	$= 10^{-12}$ Ci = $\mu\mu$Ci.

The above designations are widely used, since we deal with a wide range of quantities of radioactivity in evaluating radiation exposure potentials.

In addition to the basic quantities and units defined above, the derived units will be introduced later in this section. Table 3.1–1 presents the other fundamental quantities and units listed by the ICRU.[2,6]

PHYSICAL MEASUREMENTS OF DOSE AND THE BRAGG-GRAY PRINCIPLE

Since the standard air chamber is a relatively large, expensive, and sensitive instrument, and since measurements of the roentgen apply directly only to X- or gamma rays, many other field and laboratory instruments have been devised for measuring the absorbed dose or dose equivalent from various types of radiation more directly, although in some cases less precisely. Instruments have also been designed to measure over wide ranges of intensity and energy. Some of the methodology has been reviewed in References 7 to 12.

Small-cavity chambers with air- or tissue-equivalent walls have often been used as secondary-standard instruments for measuring radiation exposure or absorbed dose. By use of the modified Bragg-Gray principle,[11] the ionization

Table 3.1–1
QUANTITIES AND UNITS

No.	Name	Symbol	Dimensions	Units: mksa	Units: cgs	Units: Special
4	Energy imparted (integral absorbed dose)	–	E	J	erg	g-rad
5	Absorbed dose	D	EM^{-1}	J kg^{-1}	erg g^{-1}	rad
6	Absorbed-dose rate	–	$EM^{-1}T^{-1}$	J kg^{-1} s^{-1}	erg g^{-1} s^{-1}	rad s^{-1}, etc.
7	Particle fluence or fluence	Φ	L^{-2}	m^{-2}	cm^{-2}	
8	Particle flux density	φ	$L^{-2}T^{-1}$	m^{-2} s^{-1}	cm^{-2} s^{-1}	
9	Energy fluence	F	EL^{-2}	J m^{-2}	erg cm^{-2}	
10	Energy flux density or intensity	I	$EL^{-2}T^{-1}$	J m^{-2} s^{-1}	erg cm^{-2} s^{-1}	
11	Kerma	K	EM^{-1}	J kg^{-1}	erg g^{-1}	
12	Kerma rate	–	$EM^{-1}T^{-1}$	J kg^{-1} s^{-1}	erg g^{-1} s^{-1}	
13	Exposure	X	–	–	–	R (roentgen)
14	Exposure rate	–	$QM^{-1}T^{-1}$	C kg^{-1} s^{-1}	esu g^{-1} s^{-1}	R s^{-1}, etc.
15	Mass attenuation coefficient	$\frac{\mu}{\rho}$	L^2M^{-1}	m^2 kg^{-1}	cm^2 g^{-1}	
16	Mass energy-transfer coefficient	$\frac{\mu K}{\rho}$	L^2M^{-1}	m^2 kg^{-1}	cm^2 g^{-1}	
17	Mass energy-absorption coefficient	$\frac{\mu_{en}}{\rho}$	L^2M^{-1}	m^2 kg^{-1}	cm^2 g^{-1}	
18	Mass stopping power	$\frac{S}{\rho}$	EL^2M^{-1}	J m^2 kg^{-1}	erg cm^2 g^{-1}	
19	Linear energy transfer	LET	EL^{-1}	J m^{-1}	erg m^{-1}	keV (um)$^{-1}$
20	Average energy per ion pair	W	E	J	erg	eV
22	Activity	A	T^{-1}	s^{-1}	s^{-1}	Ci (curie)
23	Specific gamma-ray constant	Γ	QL^2M^{-1}	C m^2 kg^{-1}	esu cm^2 g^{-1}	R m^2 h^{-1}, etc.
	Dose equivalent	DE	–	–	–	rem

From *NBS Handbook 87,* U.S. Government Printing Office, Washington, D.C., 1963, 43.

collected by an electrode within a small cavity in a suitably designed chamber can be related to the energy deposited per gram of wall material; also, the relationship may hold constant for a wide range of energies as long as the Bragg-Gray conditions are fulfilled. In its simplest form, the Bragg-Gray principle states that the ratio of the energy absorbed per gram of a medium to the energy absorbed per gram of gas in a small cavity in the medium is constant (almost independent of the initial energy of the recoil electrons produced in the medium). Since the energy to produce an ion pair in a gas (W_g) is relatively independent of energy, we have the Bragg-Gray principle[13]

$$E_m = S_g^m \times W_g \times J_g,$$

where E_m is the energy absorbed per gram of wall material, S_g^m represents the relative mass stopping power ratio $(dE/\rho dl)_m/(dE/\rho dl)_g$ for electrons in the material and in the gas (i.e., the ratio of the rates of energy loss per unit path measured in g/cm^2), W_g expresses the average energy required to produce an ion pair in the gas (now usually taken as 33.7 eV/ion pair[13]), and J_g is the number of ion pairs produced per gram of gas in the cavity. Average mass stopping power ratios are given in Table 3.1–2 for electrons of various initial kinetic energies.[13,14]

The conditions to be met for the Bragg-Gray relation to hold[13] are briefly:

1. Cavity dimensions must be small compared to the ranges of most secondary ionizing particles;
2. Most ionizing particles should originate in

Table 3.1–2
MEAN MASS STOPPING POWER RATIOS RELATIVE TO AIR (S_m)

$$S_m = \frac{1}{T}\int_0^{T_0} S_m \, dT$$

Table A

Initial electron kinetic energy (T_0), MeV	Including density effect		
	C	Water	Tissue
0.002	1.070	1.238	1.216
0.003	1.064	1.226	1.216
0.004	1.060	1.220	1.199
0.005	1.058	1.215	1.195
0.006	1.055	1.212	1.191
0.007	1.054	1.208	1.188
0.008	1.052	1.206	1.186
0.009	1.051	1.203	1.183
0.01	1.050	1.202	1.182
0.02	1.044	1.191	1.172
0.03	1.041	1.185	1.166
0.04	1.039	1.181	1.163
0.05	1.038	1.179	1.160
0.06	1.037	1.177	1.159
0.07	1.036	1.175	1.157
0.08	1.035	1.174	1.156
0.09	1.034	1.173	1.155
0.1	1.034	1.172	1.154
0.2	1.030	1.166	1.148
0.3	1.027	1.163	1.145
0.4	1.024	1.161	1.143
0.5	1.022	1.159	1.141
0.6	1.020	1.158	1.140
0.7	1.017	1.156	1.138
0.8	1.016	1.154	1.136
0.9	1.014	1.152	1.134
1	1.012	1.150	1.132
2	1.001	1.139	1.121
3	0.985	1.121	1.103
4	0.976	1.110	1.093
5	0.968	1.108	1.084
6	0.961	1.093	1.076
8	0.950	1.080	1.063
10	0.940	1.069	1.052

Note: More recent data are given below in Tables 3.1–2B and 3.1–2C.

Table 3.1–2 (continued)
MEAN MASS STOPPING POWER RATIOS RELATIVE TO AIR (S_m)

Table B

Mean mass stopping power ratios, $\bar{S}^{m}_{air}$, relative to air, corrected for polarization, for electronic equilibrium spectra generated by monoenergetic initial electrons and by Compton electrons produced by monoenergetic gammas.

	Element and state of molecular binding									
Initial electron energy	**H sat**	**H unsat**	**C sat**	**C unsat**	**N amines, nitrates**	**N ring**	**–O–**	**O=**	**Graphite**	**Al**
.1 MeV	2.52	2.59	1.016	1.021	.976	1.018	.978	.994	1.014	.859
.2 MeV	2.52	2.59	1.015	1.019	.978	1.016	.979	.995	1.013	.870
.3 MeV	2.48	2.55	1.014	1.018	.979	1.016	.981	.995	1.011	.876
.4 MeV	2.46	2.53	1.014	1.018	.980	1.015	.981	.996	1.009	.879
.5 MeV	2.44	2.51	1.013	1.017	.980	1.015	.982	.996	1.007	.881
.6 MeV	2.44	2.50	1.012	1.016	.980	1.013	.981	.995	1.005	.882
.7 MeV	2.42	2.48	1.010	1.013	.978	1.011	.980	.993	1.003	.883
.8 MeV	2.40	2.46	1.009	1.012	.978	1.010	.979	.992	1.001	.884
1.0 MeV	2.39	2.44	1.004	1.008	.975	1.005	.977	.988	.998	.885
1.2 MeV	2.37	2.42	1.001	1.004	.973	1.002	.974	.985	.995	.885
1.5 MeV	2.35	2.39	.995	.998	.967	.996	.969	.980		
Gammas	**Stopping powers for electrons set in motion by gamma rays**									
.15 MeV	2.73	2.85	1.020	1.027	.970	1.022	.972	.992	1.017	.835
.25 MeV	2.62	2.72	1.017	1.022	.974	1.019	.976	.994	1.015	.853
.4 MeV	2.55	2.63	1.016	1.020	.977	1.017	.978	.995	1.013	.866
.6 MeV	2.50	2.57	1.014	1.018	.979	1.016	.980	.995	1.011	.874
1.0 MeV	2.44	2.50	1.008	1.012	.977	1.000	.978	.991	1.005	.881
1.5 MeV	2.39	2.45	1.001	1.005	.972	1.003	.973	.985	.999	.883
2.0 MeV	2.36	2.42	.994	.997	.996	.995	.967	.978		
2.5 MeV	2.32	2.37	.987	.990	.960	.988	.962	.973		

Table C

Mean mass stopping power ratios, $\bar{S}^{m}_{air}$, relative to air, corrected for polarization for electronic equilibrium spectra generated by Compton electrons produced by monoenergetic gamma rays for composite materials.

Gamma ray emitter	Energy of radiation	Polyethylene	Water	Tissue (muscle)	Polystyrene	Lucite®	Graphite
^{198}Au	.41 MeV	1.233	1.149	1.149	1.139	1.124	1.013
^{137}Cs	.67 MeV	1.225	1.145	1.145	1.133	1.120	1.010
^{60}Co	1.25 MeV	1.209	1.135	1.133	1.120	1.109	1.002

Table compiled from Johns, H. E., *The Physics of Radiology,* 2nd ed., Charles C Thomas, Springfield, Ill., 1964, 293, and *ICRU Handbook 85,* International Commission on Radiation Units and Measurements, Washington, D.C. With permission.

the chamber walls, and very few primary interactions should occur in the gas;

3. The fluence of primary and secondary particles should be nearly uniform across the cavity;

4. The wall of the cavity should be thick enough so that all recoil charged particles traversing the cavity originate within the wall material, but the wall must not be so thick that it attenuates the primary radiation appreciably.

Cavity chambers may be made with walls having an average atomic number $\bar{Z}$ simulating that of soft tissue, so that they can measure the rad dose more directly. More often, however, R chambers having air-equivalent walls are used, with appropriate wall thicknesses for the X- or gamma-ray energies to be measured (see Table 3.1–3 and Figure 3.1–2). Then the exposure or "air dose" measured in roentgens can be converted to the appropriate absorbed dose in tissue, expressed in rads, by the equation[13]

$$D_{tissue}\ (\text{rads}) = 0.869 \times \frac{(\mu_{en}/\rho)_{tissue}}{(\mu_{en}/\rho)_{air}} \times R = f \times R.$$

Some values of μ_{en}/ρ for various materials and gamma-ray energies are given in Table 3.1–4. Values of f relative to air for water, compact bone, and muscle are given in Table 3.1–5.

When exposure to X- or gamma radiation is measured at or near the surface of the body with an air wall chamber of "equilibrium" thickness, the dose in rads at various depths in tissue must be corrected for attenuation by the use of depth dose curves or tables. However, the fractional dose at each depth depends not only on the quantum energy of the radiation, but on the area of the beam at the body surface, on the distance from source to skin, and on other factors.[2,13] A few representative depth-dose and tissue-air ratio data are presented in Tables 3.1–6 through 3.1–13. The equivalent kilovoltage (effective keV) corresponding to some of the HVL specifications may be obtained from Figure 3.1–3. However, these data should be checked by suitable measurements[36] when accuracy such as that required for therapeutic purposes is needed. Table 3.1–14 provides data for converting photon fluences to exposures expressed in roentgen units.

Figures 3.1–4 through 3.1–6 shows graphs of mass-energy absorption coefficients (μ_{en}/ρ) given as total absorption, μ_a/ρ, vs. photon energy for air, copper, and sodium iodide. Total attenuation coefficients and component contributions of individual interactions are also plotted for comparison. Figure 3.1–7 gives the differential scattering cross-sections[41] for the total number of photons (solid curves) and for the incident energy fluence (dashed curves). These curves give the fractional amount of photons or energy, respectively, incident upon 1 cm^2 of a "thin" absorber that is scattered into unit solid angle (per steradian) at scattering angle ϕ per electron contained in the 1 cm^2 of absorber. These data are useful for radiation fluence and dose measurements and calculations. Their detailed derivation and qualifications, and much additional information on radiation dosimetry, may be found in the references at the end of this section. These references are only representative and will hopefully lead the interested reader to the wealth of additional literature on these subjects.

Table 3.1–3
EQUILIBRIUM WALL THICKNESSES FOR SEVERAL PHOTON SPECTRA UP TO 3 MeV

Radiation	Peak energy, MeV	Mean energy, MeV	Equilibrium wall, cm
^{137}Cs	0.66	0.66	0.2–0.3
^{60}Co	1.25	1.25	0.4–0.6
X-rays	2.00	0.67	0.3–0.4
X-rays	3.00	1.00	0.4–0.6

From Johns, H. E., *The Physics of Radiology,* 2nd ed., 1964, 293. Courtesy of Charles C Thomas, Publisher, Springfield, Ill.

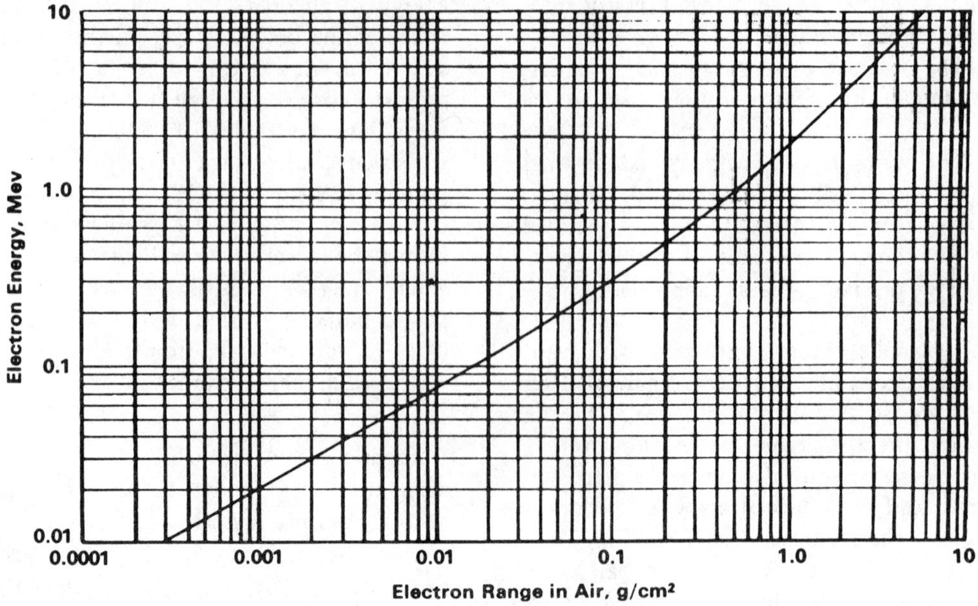

FIGURE 3.1–2. Electron range in air (in g/cm^2) as a function of the electron energy. (From Johns, H. E., *The Physics of Radiology;* 2nd ed., 1964, 291. Courtesy of Charles C Thomas, Publisher, Springfield, Ill.)

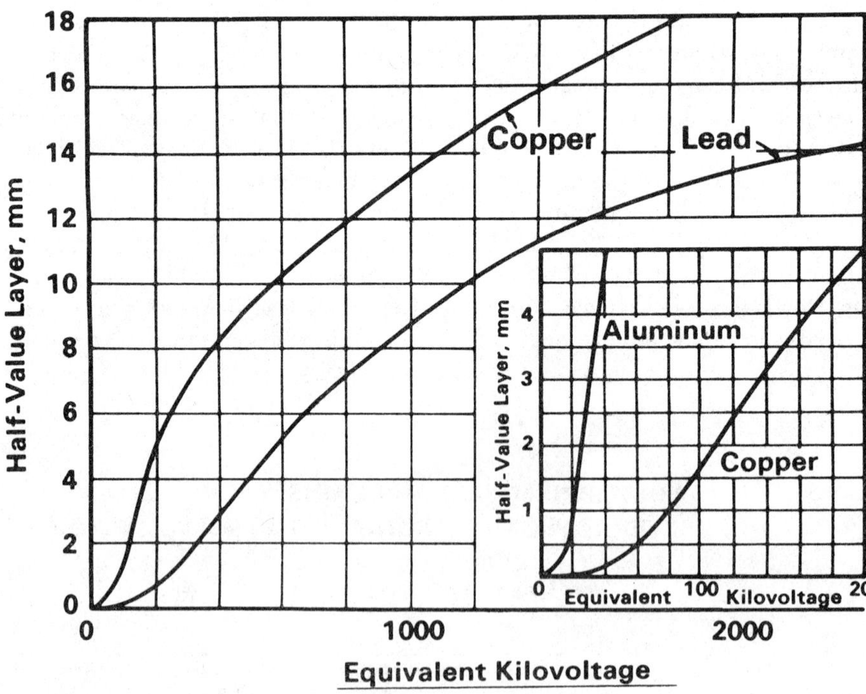

FIGURE 3.1–3. Relationships between HVL and effective kilovoltage. (From Johns, H. E., *The Physics of Radiology,* 2nd ed., 1964, 256. Courtesy of Charles C Thomas, Publisher, Springfield, Ill.)

Table 3.1–4
VALUES OF THE (a) MASS ENERGY-TRANSFER COEFFICIENT, μ_K/ρ (cm^2/g), AND (b) MASS ENERGY-ABSORPTION COEFFICIENT, μ_{en}/ρ (cm^2/g)

Hydrogen through Sodium

	$_1$H		$_4$Be		$_5$B		$_6$C		$_7$N		$_8$O		$_{11}$Na	
$h\nu$(MeV)	(a)	(b)	(a)	(b)	(a)	(b)	(a)	(b)	(a)	(b)	(a)	(b)	(a)	(b)
0.01	0.00986	0.00986	0.368	0.368	0.911	0.911	1.97	1.97	3.38	3.38	5.39	5.39	14.9	14.9
0.015	0.0110	0.0110	0.104	0.104	0.248	0.248	0.536	0.536	0.908	0.908	1.44	1.44	4.20	4.20
0.02	0.0135	0.0135	0.0469	0.0469	0.0998	0.0998	0.208	0.208	0.362	0.362	0.575	0.575	1.70	1.70
0.03	0.0185	0.0185	0.0195	0.0195	0.0338	0.0338	0.0594	0.0594	0.105	0.105	0.165	0.165	0.475	0.475
0.04	0.0231	0.0231	0.0146	0.0146	0.0210	0.0210	0.0306	0.0306	0.0493	0.0493	0.0733	0.0733	0.199	0.199
0.05	0.0271	0.0271	0.0142	0.0142	0.0175	0.0175	0.0233	0.0233	0.0319	0.0319	0.0437	0.0437	0.106	0.106
0.06	0.0306	0.0306	0.0147	0.0147	0.0170	0.0170	0.0211	0.0211	0.0256	0.0256	0.0322	0.0322	0.0668	0.0668
0.08	0.0362	0.0362	0.0166	0.0166	0.0179	0.0179	0.0205	0.0205	0.0223	0.0223	0.0249	0.0249	0.0382	0.0382
0.10	0.0406	0.0406	0.0184	0.0184	0.0194	0.0194	0.0215	0.0215	0.0224	0.0224	0.0237	0.0237	0.0297	0.0297
0.15	0.0481	0.0481	0.0216	0.0216	0.0226	0.0226	0.0245	0.0245	0.0247	0.0247	0.0251	0.0251	0.0260	0.0260
0.2	0.0525	0.0525	0.0235	0.0235	0.0245	0.0245	0.0265	0.0265	0.0267	0.0267	0.0268	0.0268	0.0264	0.0264
0.3	0.0569	0.0569	0.0255	0.0255	0.0266	0.0266	0.0287	0.0287	0.0287	0.0287	0.0288	0.0288	0.0277	0.0277
0.4	0.0586	0.0586	0.0262	0.0262	0.0273	0.0273	0.0295	0.0295	0.0295	0.0295	0.0295	0.0295	0.0284	0.0284
0.5	0.0593	0.0593	0.0265	0.0265	0.0276	0.0276	0.0297	0.0297	0.0297	0.0296	0.0297	0.0297	0.0285	0.0285
0.6	0.0587	0.0587	0.0263	0.0263	0.0274	0.0273	0.0296	0.0295	0.0296	0.0295	0.0296	0.0296	0.0284	0.0284
0.8	0.0574	0.0574	0.0257	0.0256	0.0268	0.0267	0.0289	0.0288	0.0289	0.0289	0.0289	0.0289	0.0277	0.0275
1.0	0.0555	0.0555	0.0248	0.0248	0.0259	0.0258	0.0279	0.0279	0.0280	0.0279	0.0280	0.0278	0.0268	0.0266
1.5	0.0507	0.0507	0.0227	0.0227	0.0237	0.0236	0.0256	0.0255	0.0256	0.0255	0.0256	0.0254	0.0245	0.0243
2	0.0465	0.0464	0.0208	0.0208	0.0218	0.0217	0.0235	0.0234	0.0236	0.0234	0.0236	0.0234	0.0227	0.0225
3	0.0399	0.0398	0.0181	0.0180	0.0190	0.0188	0.0206	0.0204	0.0207	0.0205	0.0208	0.0206	0.0202	0.0199
4	0.0353	0.0352	0.0163	0.0161	0.0172	0.0170	0.0187	0.0185	0.0189	0.0186	0.0191	0.0188	0.0188	0.0184
5	0.0319	0.0317	0.0149	0.0148	0.0158	0.0156	0.0174	0.0171	0.0177	0.0173	0.0179	0.0175	0.0179	0.0174
6	0.0292	0.0290	0.0140	0.0138	0.0149	0.0146	0.0164	0.0161	0.0167	0.0163	0.0171	0.0166	0.0173	0.0161
8	0.0253	0.0252	0.0126	0.0123	0.0135	0.0132	0.0151	0.0147	0.0156	0.0151	0.0160	0.0155	0.0167	0.0159
10	0.0227	0.0225	0.0117	0.0114	0.0127	0.0123	0.0143	0.0138	0.0149	0.0143	0.0154	0.0148	0.0161	0.0155

Table 3.1–4 (continued)
VALUES OF THE (a) MASS ENERGY-TRANSFER COEFFICIENT, μ_K/ρ (cm^2/g), AND (b) MASS ENERGY-ABSORPTION COEFFICIENT, μ_{en}/ρ (cm^2/g)

Magnesium through Potassium

	$_{12}$Mg		$_{13}$Al		$_{14}$Si		$_{15}$P		$_{16}$S		$_{18}$Ar		$_{19}$K	
$h\nu$(MeV)	(a)	(b)	(a)	(b)	(a)	(b)	(a)	(b)	(a)	(b)	(a)	(b)	(a)	(b)
0.01	20.1	20.1	25.5	25.5	33.3	33.3	39.8	39.8	49.7	49.7	62.3	62.3	77.6	77.6
0.015	5.80	5.80	7.47	7.47	9.75	9.75	11.8	11.8	14.9	14.9	19.1	19.1	23.9	23.9
0.02	2.38	2.38	3.06	3.06	4.01	4.01	4.91	4.91	6.21	6.21	8.02	8.02	10.2	10.2
0.03	0.671	0.671	0.868	0.868	1.14	1.14	1.39	1.39	1.77	1.77	2.31	2.31	2.94	2.94
0.04	0.276	0.276	0.357	0.357	0.472	0.472	0.572	0.572	0.727	0.727	0.962	0.962	1.23	1.23
0.05	0.144	0.144	0.184	0.184	0.241	0.241	0.293	0.293	0.372	0.372	0.488	0.488	0.623	0.623
0.06	0.0888	0.0888	0.111	0.111	0.144	0.144	0.173	0.173	0.218	0.218	0.284	0.284	0.366	0.366
0.08	0.0475	0.0475	0.0562	0.0562	0.0700	0.0700	0.0820	0.0820	0.101	0.101	0.128	0.128	0.162	0.162
0.10	0.0346	0.0346	0.0386	0.0386	0.0459	0.0459	0.0511	0.0511	0.0609	0.0609	0.0735	0.0735	0.0913	0.0913
0.15	0.0279	0.0279	0.0285	0.0285	0.0312	0.0312	0.0322	0.0322	0.0357	0.0357	0.0377	0.0377	0.0442	0.0442
0.2	0.0277	0.0277	0.0276	0.0276	0.0292	0.0292	0.0293	0.0293	0.0311	0.0311	0.0304	0.0304	0.0343	0.0343
0.3	0.0288	0.0288	0.0282	0.0282	0.0294	0.0294	0.0288	0.0288	0.0299	0.0299	0.0278	0.0278	0.0304	0.0304
0.4	0.0294	0.0294	0.0287	0.0287	0.0298	0.0298	0.0291	0.0291	0.0301	0.0301	0.0275	0.0275	0.0298	0.0298
0.5	0.0294	0.0294	0.0287	0.0286	0.0298	0.0298	0.0291	0.0291	0.0300	0.0300	0.0272	0.0272	0.0294	0.0293
0.6	0.0293	0.0293	0.0286	0.0286	0.0296	0.0295	0.0288	0.0288	0.0297	0.0297	0.0296	0.0269	0.0291	0.0290
0.8	0.0286	0.0285	0.0279	0.0277	0.0289	0.0288	0.0280	0.0278	0.0290	0.0288	0.0262	0.0261	0.0283	0.0282
1.0	0.0276	0.0275	0.0270	0.0269	0.0279	0.0277	0.0272	0.0270	0.0280	0.0278	0.0253	0.0251	0.0273	0.0270
1.5	0.0253	0.0251	0.0247	0.0245	0.0255	0.0253	0.0248	0.0246	0.0256	0.0253	0.0232	0.0229	0.0250	0.0247
2	0.0234	0.0232	0.0229	0.0226	0.0237	0.0234	0.0231	0.0228	0.0238	0.0235	0.0215	0.0212	0.0233	0.0229
3	0.0210	0.0206	0.0206	0.0202	0.0214	0.0210	0.0209	0.0204	0.0216	0.0211	0.0198	0.0192	0.0214	0.0208
4	0.0196	0.0191	0.0193	0.0188	0.0202	0.0196	0.0198	0.0192	0.0205	0.0199	0.0189	0.0182	0.0206	0.0198
5	0.0187	0.0181	0.0185	0.0179	0.0194	0.0187	0.0191	0.0184	0.0200	0.0192	0.0185	0.0177	0.0202	0.0193
6	0.0182	0.0175	0.0181	0.0172	0.0191	0.0182	0.0188	0.0179	0.0197	0.0188	0.0184	0.0174	0.0202	0.0190
8	0.0177	0.0168	0.0177	0.0168	0.0187	0.0177	0.0187	0.0175	0.0197	0.0184	0.0186	0.0173	0.0205	0.0190
10	0.0175	0.0164	0.0176	0.0165	0.0188	0.0175	0.0188	0.0174	0.0200	0.0184	0.0190	0.0174	0.0210	0.0191

Table 3.1–4 (continued)

VALUES OF THE (a) MASS ENERGY-TRANSFER COEFFICIENT, μ_K/ρ (cm^2/g), AND (b) MASS ENERGY-ABSORPTION COEFFICIENT, μ_{en}/ρ (cm^2/g)

Calcium through Copper

	$_{20}$Ca		$_{26}$Fe		$_{27}$Co		$_{28}$Ni		$_{29}$Cu	
$h\nu$(MeV)	(a)	(b)	(a)	(b)	(a)	(b)	(a)	(b)	(a)	(b)
0.01	91.6	91.6	142.	142.	148.6	148.6	161.	161.	160.	160.
0.015	28.6	28.6	49.3	49.3	52.5	52.5	58.2	58.2	59.4	59.4
0.02	12.2	12.2	22.8	22.8	24.4	24.4	27.4	27.4	28.2	28.2
0.03	3.60	3.60	7.28	7.28	7.97	7.97	9.06	9.06	9.50	9.50
0.04	1.50	1.50	3.17	3.17	3.49	3.49	4.03	4.03	4.24	4.24
0.05	0.764	0.764	1.64	1.64	1.64	1.64	2.15	2.15	2.22	2.22
0.06	0.444	0.444	0.961	0.961	1.08	1.08	1.25	1.25	1.32	1.32
0.08	0.196	0.196	0.414	0.414	0.454	0.454	0.528	0.528	0.573	0.573
0.10	0.109	0.109	0.219	0.219	0.243	0.243	0.284	0.284	0.302	0.302
0.15	0.0497	0.0497	0.0814	0.0814	0.0887	0.0887	0.101	0.101	0.106	0.106
0.2	0.0371	0.0371	0.0495	0.0495	0.0523	0.0523	0.0582	0.0582	0.0597	0.0597
0.3	0.0318	0.0318	0.0335	0.0335	0.0347	0.0347	0.0374	0.0374	0.0370	0.0370
0.4	0.0309	0.0309	0.0308	0.0308	0.0308	0.0308	0.0326	0.0326	0.0318	0.0318
0.5	0.0304	0.0304	0.0295	0.0295	0.0294	0.0294	0.0309	0.0309	0.0298	0.0298
0.6	0.0300	0.0299	0.0287	0.0286	0.0284	0.0283	0.0298	0.0297	0.0287	0.0286
0.8	0.0291	0.0289	0.0275	0.0273	0.0272	0.0270	0.0284	0.0282	0.0272	0.0271
1.0	0.0280	0.0278	0.0264	0.0262	0.0260	0.0258	0.0272	0.0269	0.0261	0.0258
1.5	0.0257	0.0254	0.0241	0.0237	0.0237	0.0234	0.0247	0.0243	0.0237	0.0233
2	0.0240	0.0236	0.0225	0.0220	0.0222	0.0217	0.0232	0.0227	0.0222	0.0217
3	0.0220	0.0214	0.0212	0.0204	0.0202	0.0202	0.0219	0.0211	0.0211	0.0202
4	0.0213	0.0205	0.0209	0.0199	0.0208	0.0197	0.0218	0.0207	0.0211	0.0200
5	0.0211	0.0200	0.0211	0.0198	0.0210	0.0196	0.0222	0.0207	0.0214	0.0200
6	0.0211	0.0198	0.0215	0.0199	0.0215	0.0198	0.0227	0.0209	0.0220	0.0202
8	0.0215	0.0198	0.0226	0.0204	0.0226	0.0203	0.0239	0.0215	0.0234	0.0209
10	0.0222	0.0201	0.0238	0.0209	0.0239	0.0210	0.0254	0.0222	0.0248	0.0215

Table 3.1–4 (continued)
VALUES OF THE (a) MASS ENERGY-TRANSFER COEFFICIENT, μ_K/ρ (cm^2/g), AND (b) MASS ENERGY-ABSORPTION COEFFICIENT, μ_{en}/ρ (cm^2/g)

Tin

	$_{50}$Sn	
$h\nu$(MeV)	(a)	(b)
0.0010	11110	11110
0.0015	3950	3950
0.0020	1954	1954
0.0030	705	705
0.0039288	360	360
L_3 edge		
0.0039288	1067	1067
0.0040	1019	1019
0.0041573	930	930
L_2 edge		
0.0041573	1187	1187
0.0044648	971	971
L_1 edge		
0.0044648	1207	1207
0.005	880	880
0.006	540	539
0.008	250	249
0.010	136.5	136.4
0.015	43.7	43.6
0.020	19.83	19.81
0.0291947	6.83	6.82
K edge		
0.0291947	16.70	16.69
0.030	16.18	16.17
0.04	9.97	9.96
0.05	6.25	6.24
0.06	4.20	4.19
0.08	2.19	2.18
0.10	1.257	1.250
0.15	0.446	0.442
0.20	0.211	0.209
0.30	0.0853	0.0843
0.4	0.0536	0.0530
0.5	0.0423	0.0416
0.6	0.0358	0.0353
0.8	0.0301	0.0294
1.0	0.0270	0.0264
1.5	0.0233	0.0226
2.0	0.0220	0.0210
3.0	0.0219	0.0205
4	0.0232	0.0212
5	0.0247	0.0221
6	0.0262	0.0230
8	0.0292	0.0245
10	0.0319	0.0258

Table 3.1–4 (continued)
VALUES OF THE (a) MASS ENERGY-TRANSFER COEFFICIENT, μ_K/ρ (cm^2/g), AND (b) MASS ENERGY-ABSORPTION COEFFICIENT, μ_{en}/ρ (cm^2/g)

Lead

$h\nu$(MeV)	$_{82}$Pb (a)	(b)
M_1 edge		
0.003854	1454	1453
0.004	1298	1297
0.005	747	747
0.006	479	479
0.008	230	230
0.010	131.0	130.7
0.0130406	66.2	66.0
L_3 edge		
0.0130406	128.8	128.8
0.015	91.7	91.7
0.0152053	89.6	89.6
L_2 edge		
0.0152053	113.0	113.0
0.015855	101.7	101.6
L_1 edge		
0.015855	123.0	123.0
0.02	69.2	69.1
0.03	24.6	24.6
0.04	11.83	11.78
0.05	6.57	6.54
0.06	4.11	4.08
0.08	1.924	1.908
0.088005	1.494	1.481
K edge		
0.088005	2.47	2.47
0.10	2.28	2.28
0.15	1.164	1.154
0.2	0.637	0.629
0.3	0.265	0.259
0.4	0.1474	0.1432
0.5	0.0984	0.0951
0.6	0.0737	0.0710
0.8	0.0503	0.0481
1.0	0.0396	0.0377
1.5	0.0288	0.0271
2	0.0259	0.0240
3	0.0260	0.0234
4	0.0281	0.0245
5	0.0306	0.0259
6	0.0331	0.0272
8	0.0378	0.0294
10	0.0419	0.0310

Table 3.1–4 (continued)
VALUES OF THE (a) MASS ENERGY-TRANSFER COEFFICIENT, μ_K/ρ (cm^2/g), AND (b) MASS ENERGY-ABSORPTION COEFFICIENT, μ_{en}/ρ (cm^2/g)

Uranium

	$_{92}$U	
hν(MeV)	(a)	(b)
N_1 edge		
0.001439	4460	4460
0.0015	4050	4050
0.002	2110	2110
0.003	818	818
0.003545	554	554
M_5 edge		
0.003545	1191	1191
0.003720	1055	1055
M_4 edge		
0.03720	1520	1520
0.004	1258	1258
0.004299	1046	1046
M_3 edge		
0.004299	1242	1242
0.005	851	850
0.005179	781	780
M_2 edge		
0.005179	839	839
0.005546	707	706
M_1 edge		
0.005546	739	739
0.006	612	611
0.008	300	299
0.010	170.5	170.2
0.015	59.6	59.4
0.017165	42.2	42.1
L_3 edge		
0.017165	80.1	80.1
0.020	55.5	55.5
0.020945	49.4	49.4

	$_{99}$U	
hν(MeV)	(a)	(b)
L_2 edge		
0.020945	66.9	66.9
0.021771	61.2	61.1
L_1 edge		
0.021771	69.6	69.6
0.03	33.0	33.0
0.04	16.33	16.27
0.05	9.30	9.25
0.06	5.82	5.78
0.08	2.76	2.73
0.10	1.535	1.517
0.011562	1.049	1.036
K edge		
0.11562	1.660	1.658
0.15	1.120	1.193
0.20	0.712	0.704
0.30	0.322	0.315
0.4	0.1835	0.1779
0.5	0.1222	0.1175
0.6	0.0902	0.0864
0.8	0.0599	0.0569
1.0	0.0458	0.0432
1.5	0.0317	0.0295
2	0.0278	0.0255
3	0.0276	0.0246
4	0.0298	0.0256
5	0.0324	0.0270
6	0.0348	0.0281
8	0.0392	0.0298
10	0.0432	0.0312

Table 3.1–4 (continued)
VALUES OF THE (a) MASS ENERGY-TRANSFER COEFFICIENT, μ_K/ρ (cm^2/g), AND (b) MASS ENERGY-ABSORPTION COEFFICIENT, μ_{en}/ρ (cm^2/g)

Some Mixtures and Compounds

	Air		Water		0.8N(0.4M) H_2SO_4 solution		Compact bone (femur)		Muscle (striated)		Polystyrene, $(C_8H_8)_n$	
$h\nu$(MeV)	(a)	(b)	(a)	(b)	(a)	(b)	(a)	(b)	(a)	(b)	(a)	(b)
0.01	4.61	4.61	4.79	4.79	5.36	5.36	19.2	19.2	4.87	4.87	1.82	1.82
0.015	1.27	1.27	1.28	1.28	1.45	1.45	5.84	5.84	1.32	1.32	0.495	0.495
0.02	0.511	0.511	0.512	0.512	0.585	0.585	2.46	2.46	0.533	0.533	0.193	0.193
0.03	0.148	0.148	0.149	0.149	0.169	0.169	0.720	0.720	0.154	0.154	0.0562	0.0562
0.04	0.0668	0.0668	0.0677	0.0677	0.0761	0.0761	0.304	0.304	0.0701	0.0701	0.0300	0.0300
0.05	0.0406	0.0406	0.0418	0.0418	0.0460	0.0460	0.161	0.161	0.0431	0.0431	0.0236	0.0236
0.06	0.0305	0.0305	0.0320	0.0320	0.0344	0.0344	0.0998	0.0998	0.0328	0.0328	0.0218	0.0218
0.08	0.0243	0.0243	0.0262	0.0262	0.0271	0.0271	0.0537	0.0537	0.0264	0.0264	0.0217	0.0217
0.10	0.0234	0.0234	0.0256	0.0256	0.0260	0.0260	0.0387	0.0387	0.0256	0.0256	0.0231	0.0231
0.15	0.0250	0.0250	0.0277	0.0277	0.0277	0.0277	0.0305	0.0305	0.0275	0.0275	0.0263	0.0263
0.2	0.0268	0.0268	0.0297	0.0297	0.0296	0.0296	0.0301	0.0301	0.0294	0.0294	0.0286	0.0286
0.3	0.0287	0.0287	0.0319	0.0319	0.0319	0.0319	0.0310	0.0310	0.0317	0.0317	0.0309	0.0309
0.4	0.0295	0.0295	0.0328	0.0328	0.0327	0.0327	0.0315	0.0315	0.0325	0.0325	0.0318	0.0318
0.5	0.0297	0.0296	0.0330	0.0330	0.0330	0.0330	0.0317	0.0317	0.0328	0.0328	0.0321	0.0321
0.6	0.0296	0.0295	0.0329	0.0329	0.0328	0.0328	0.0315	0.0314	0.0326	0.0325	0.0319	0.0318
0.8	0.0289	0.0289	0.0321	0.0321	0.0320	0.0320	0.0307	0.0306	0.0318	0.0318	0.0311	0.0310
1.0	0.0280	0.0278	0.0311	0.0309	0.0310	0.0308	0.0297	0.0295	0.0308	0.0306	0.0300	0.0300
1.5	0.0256	0.0254	0.0284	0.0282	0.0283	0.0281	0.0272	0.0270	0.0282	0.0280	0.0275	0.0275
2	0.0236	0.0234	0.0262	0.0260	0.0261	0.0259	0.0251	0.0249	0.0259	0.0257	0.0253	0.0252
3	0.0207	0.0205	0.0229	0.0227	0.0229	0.0227	0.0221	0.0219	0.0227	0.0225	0.0221	0.0219
4	0.0189	0.0186	0.0209	0.0206	0.0209	0.0206	0.0204	0.0200	0.0207	0.0204	0.0200	0.0198
5	0.0178	0.0174	0.0195	0.0191	0.0194	0.0191	0.0192	0.0187	0.0193	0.0189	0.0185	0.0182
6	0.0168	0.0164	0.0185	0.0180	0.0184	0.0180	0.0184	0.0178	0.0183	0.0178	0.0174	0.0171
8	0.0157	0.0152	0.0170	0.0166	0.0171	0.0166	0.0173	0.0167	0.0169	0.0164	0.0159	0.0155
10	0.0151	0.0145	0.0162	0.0157	0.0162	0.0157	0.0168	0.0159	0.0160	0.0155	0.0150	0.0145

Table 3.1–4 (continued)
VALUES OF THE (a) MASS ENERGY-TRANSFER COEFFICIENT, μ_K/ρ (cm^2/g), AND (b) MASS ENERGY-ABSORPTION COEFFICIENT, μ_{en}/ρ (cm^2/g)

Some Mixtures and Compounds (continued)

	Methyl methacrylate (Perspex®, plexiglass, Lucite®), $C_5H_8O_2$		Polyethylene, $(CH_2)_n$		Bakelite® (typical composition), $C_{45}H_{38}O_7$		Pyrex® glass (Corning #7740)		Concrete (typical composition)	
hν(MeV)	(a)	(b)	(a)	(b)	(a)	(b)	(a)	(b)	(a)	(b)
0.01	2.91	2.91	1.69	1.69	2.43	2.43	16.5	16.5	25.5	25.5
0.015	0.783	0.783	0.461	0.461	0.658	0.0658	4.75	4.75	7.66	7.66
0.02	0.310	0.310	0.180	0.180	0.258	0.258	1.94	1.94	3.22	3.22
0.03	0.0899	0.0899	0.0535	0.0535	0.0748	0.0748	0.554	0.554	0.936	0.936
0.04	0.0437	0.0437	0.0295	0.0295	0.0374	0.0374	0.232	0.232	0.393	0.393
0.05	0.0301	0.0301	0.0238	0.0238	0.0269	0.0269	0.122	0.122	0.204	0.204
0.06	0.0254	0.0254	0.0225	0.0225	0.0235	0.0235	0.0768	0.0768	0.124	0.124
0.08	0.0232	0.0232	0.0228	0.0228	0.0221	0.0221	0.0428	0.0428	0.0625	0.0625
0.10	0.0238	0.0238	0.0243	0.0243	0.0230	0.0230	0.0325	0.0325	0.0424	0.0424
0.15	0.0266	0.0266	0.0279	0.0279	0.0260	0.0260	0.0274	0.0274	0.0290	0.0290
0.2	0.0287	0.0287	0.0303	0.0303	0.0281	0.0281	0.0276	0.0276	0.0290	0.0290
0.3	0.0310	0.0310	0.0328	0.0328	0.0303	0.0303	0.0289	0.0289	0.0295	0.0295
0.4	0.0318	0.0318	0.0337	0.0337	0.0312	0.0312	0.0295	0.0295	0.0298	0.0298
0.5	0.0322	0.0322	0.0340	0.0340	0.0315	0.0315	0.0297	0.0297	0.0300	0.0300
0.6	0.0319	0.0319	0.0338	0.0337	0.0313	0.0312	0.0295	0.0294	0.0297	0.0297
0.8	0.0312	0.0311	0.0330	0.0329	0.0305	0.0305	0.0288	0.0287	0.0290	0.0289
1.0	0.0302	0.0301	0.0319	0.0319	0.0295	0.0295	0.0279	0.0277	0.0281	0.0279
1.5	0.0276	0.0275	0.0292	0.0291	0.0270	0.0269	0.0254	0.0252	0.0256	0.0254
2	0.0254	0.0253	0.0268	0.0267	0.0248	0.0247	0.0235	0.0233	0.0237	0.0235
3	0.0222	0.0220	0.0234	0.0232	0.0217	0.0215	0.0209	0.0207	0.0212	0.0209
4	0.0202	0.0199	0.0211	0.0209	0.0197	0.0195	0.0194	0.0190	0.0198	0.0193
5	0.0187	0.0184	0.0195	0.0192	0.0183	0.0180	0.0184	0.0179	0.0188	0.0182
6	0.0177	0.0173	0.0182	0.0180	0.0173	0.0169	0.0178	0.0171	0.0183	0.0176
8	0.0162	0.0158	0.0166	0.0162	0.0158	0.0154	0.0170	0.0163	0.0176	0.0168
10	0.0153	0.0148	0.0155	0.0151	0.0150	0.0145	0.0166	0.0157	0.0174	0.0163

Note: See original source for data references.

From Hubbell, J. H., *Photon Cross Sections, Attenuation Coefficients, and Energy Absorption Coefficients from 10 KeV to 100 GeV*, NSRDS-NBS-29, U.S. Government Printing Office, Washington, D.C., August 1969.

Table 3.1–5
f FACTOR TO CONVERT FROM ROENTGENS TO RADS

Values of f Relative to Air for Water, Compact Bone, and Muscle; $f = 0.869 \frac{(\mu_{en}/\rho)_{med}}{(\mu_{en}/\rho)_{air}}$

	Photon energy, MeV	Water to air	Compact bone to air	Muscle to air
	.01	.903	3.62	.918
	.015	.876	4.00	.907
	.02	.871	4.18	.903
	.03	.874	4.23	.904
	.04	.883	3.95	.912
	.05	.895	3.45	.922
	.06	.912	2.84	.933
	.08	.937	1.92	.944
	.10	.950	1.44	.951
	.15	.961	1.06	.955
	.2	.964	.976	.957
	.3	.966	.939	.958
	.4	.967	.931	.959
	.5	.968	.927	.959
	.6	.968	.924	.959
^{137}Cs	.662	.968	.923	.958
	.8	.968	.922	.958
	1.0	.967	.922	.957
^{60}Co	1.25	.966	.922	.957
	1.5	.966	.923	.956
	2	.965	.925	.954
	3	.962	.928	.951
	4	.958	.933	.948
	5	.956	.938	.945
	6	.954	.943	.943
	8	.949	.955	.937
	10	.945	.956	.929

From Johns, H. E. and Cunningham, J. R., *The Physics of Radiology,* 3rd ed., 1971, 736. Courtesy of Charles C Thomas, Publisher, Springfield, Ill.

STANDARDS FOR MEASURING ABSORBED DOSE OF X- OR GAMMA RAYS IN RADIATION THERAPY

Recent recommendations[36,38] have been made for purposes of radiation therapy dosimetry, and these reports provide helpful data and information for obtaining accurate X- or gamma-ray dose measurements. The original reports should be read for details, qualifications, and further literature reference. The following section is quoted from ICRU Report 23 to illustrate some of the basic considerations involved in radiation dose measurements with secondary thimble ionization chambers, and to provide ready reference to some of the more frequently used methods and data.

Determination of Absorbed-dose Rates

First, attention must be directed towards the measurement of the absorbed-dose rate at the calibration point and towards the derivation of the peak (or the surface) absorbed-dose rates. For this purpose a number of simple recommendations are made. These concern:

1. the measuring system to be used,
2. the desirable features of the instrument based on that system and
3. the way in which the instrument is used.

The System of Measurement

When making measurements of absorbed dose or of any other physical quantity, a distinction must be drawn between the instrument which is used for measurements under working conditions and the standard instrument against which it has been calibrated. These latter devices are generally designed to operate under strictly defined laboratory conditions and would normally be inconvenient – or even completely unsuitable – for the type of measurement required in practice. For example, a free-air ionization chamber would not give an accurate measurement of the exposure rate at the end of the applicator of a typical 250 kV x-ray apparatus. For other reasons, neither calorimetric nor chemical methods are appropriate for regular clinical measurements. It is preferable to use one of the dosimeters which have been designed for this type of measurement. Although a number of physical effects (e.g., thermoluminescence, fluorescence, photoconductivity) have been used in the design of dosimeters – and indeed have advantages for special purposes – there is no doubt that the calibrated ionization chamber should remain the basis of clinical dosimetry. An important practical advantage of this recommendation is that most radiotherapy departments are already equipped with a suitable dosimeter of this type. The specifications for this instrument are discussed in the following section.

For x rays generated at potentials of 50–250 kV, exposure standards, using the free-air ionization chamber, are well established and a vast amount of radiotherapeutic experience is based upon them. In this quality range clinical dosimeters should always be calibrated directly or indirectly against one of these national standard instruments, as described below. Essentially the same considerations apply to cobalt-60 gamma rays or to the x rays of similar quality generated at 2 MV, except that for these qualities the standard instruments are graphite cavity chambers.

For x rays generated at potentials above 2 MV, the problem is more difficult. Although several alternative methods are available for measurement of absorbed dose, none has so far been adopted by any standards laboratory. Each individual user of a dosimeter must, therefore, calibrate it either by carrying out his own absolute

(continued on page 263)

Table 3.1–6
TISSUE-AIR RATIOS FOR RECTANGULAR FIELDS[a]

Dose to Small Mass of Tissue at Depth d for a Dose of 1 rad to the Same Mass of Tissue at the Same Point in Free Space

HVL 1.0 mm Cu

	Field size (cm × cm) at depth d, or axis of rotation														
d, cm	0 × 0	4 × 4	4 × 6	4 × 8	4 × 10	4 × 15	6 × 6	6 × 8	6 × 10	6 × 15	8 × 8	8 × 10	8 × 15	10 × 10	10 × 15
0[b]	1.00	1.17	1.20	1.22	1.24	1.25	1.25	1.28	1.30	1.32	1.31	1.34	1.36	1.36	1.39
2	.682	1.04	1.10	1.14	1.16	1.19	1.19	1.24	1.26	1.30	1.29	1.32	1.37	1.36	1.42
4	.472	.798	.850	.905	.926	.956	.945	1.00	1.03	1.09	1.06	1.11	1.17	1.15	1.23
5	.393	.692	.750	.796	.818	.855	.836	.890	.925	.980	.950	1.00	1.06	1.05	1.13
6	.330	.600	.656	.694	.723	.761	.734	.788	.822	.871	.851	.901	.959	.952	1.03
7	.277	.514	.567	.601	.630	.666	.638	.692	.723	.770	.750	.794	.850	.845	.920
8	.233	.440	.490	.520	.548	.584	.555	.604	.630	.676	.658	.700	.757	.750	.820
9	.195	.375	.420	.449	.472	.509	.476	.520	.548	.591	.570	.611	.665	.656	.725
10	.163	.321	.357	.384	.407	.439	.410	.449	.474	.516	.494	.533	.584	.573	.637
11	.135	.270	.304	.327	.348	.379	.350	.387	.408	.449	.426	.462	.510	.500	.560
12	.114	.230	.259	.280	.299	.327	.300	.334	.352	.388	.368	.402	.445	.436	.493
13	.094	.195	.220	.239	.255	.282	.257	.285	.304	.340	.317	.346	.388	.378	.430
14	.079	.165	.187	.204	.219	.243	.219	.245	.260	.292	.273	.301	.338	.327	.375
15	.066	.140	.159	.174	.186	.209	.186	.210	.225	.256	.235	.260	.294	.284	.328
16	.056	.119	.135	.148	.161	.180	.160	.180	.194	.219	.203	.225	.256	.248	.287
17	.047	.101	.115	.127	.136	.155	.136	.154	.165	.189	.175	.105	.221	.214	.248
18	.039	.085	.098	.108	.117	.134	.116	.133	.142	.163	.150	.167	.191	.185	.215
19	.032	.073	.083	.092	.100	.116	.099	.114	.122	.140	.129	.144	.166	.160	.187
20	.027	.062	.071	.079	.086	.100	.085	.097	.105	.121	.111	.124	.144	.138	.163

Table 3.1–6 (continued)
TISSUE-AIR RATIOS FOR RECTANGULAR FIELDS

Dose to Small Mass of Tissue at Depth d for a Dose of 1 rad to the Same Mass of Tissue at the Same Point in Free Space (continued)

HVL 2.0 mm Cu

	Field size (cm × cm) at depth d, or axis of rotation														
d, cm	0 × 0	4 × 4	4 × 6	4 × 8	4 × 10	4 × 15	6 × 6	6 × 8	6 × 10	6 × 15	8 × 8	8 × 10	8 × 15	10 × 10	10 × 15
0[b]	1.00	1.14	1.16	1.18	1.19	1.21	1.19	1.22	1.24	1.25	1.25	1.27	1.30	1.29	1.32
2	.720	1.02	1.07	1.10	1.12	1.15	1.14	1.17	1.20	1.23	1.22	1.25	1.29	1.29	1.33
4	.515	.800	.846	.884	.908	.937	.918	.967	1.00	1.04	1.02	1.07	1.12	1.10	1.17
5	.438	.700	.754	.786	.810	.841	.822	.870	.902	.950	.927	.972	1.02	1.01	1.08
6	.371	.612	.665	.697	.723	.757	.735	.783	.812	.857	.839	.882	.936	.927	.995
7	.316	.534	.582	.616	.640	.669	.643	.691	.725	.765	.745	.789	.840	.832	.900
8	.268	.461	.505	.538	.558	.594	.567	.615	.640	.684	.663	.706	.760	.748	.815
9	.228	.399	.440	.469	.490	.520	.493	.539	.565	.605	.584	.625	.673	.670	.730
10	.193	.341	.378	.404	.426	.457	.430	.470	.493	.533	.513	.552	.600	.590	.651
11	.165	.296	.329	.351	.372	.400	.372	.412	.435	.470	.450	.485	.530	.526	.580
12	.140	.256	.284	.303	.323	.351	.322	.357	.376	.412	.394	.425	.467	.460	.515
13	.119	.222	.246	.262	.282	.308	.280	.312	.330	.362	.343	.372	.412	.406	.455
14	.102	.192	.212	.227	.245	.269	.243	.269	.285	.318	.300	.325	.363	.354	.402
15	.086	.166	.185	.198	.213	.236	.214	.236	.250	.280	.261	.285	.320	.312	.356
16	.073	.143	.160	.172	.185	.205	.184	.204	.218	.245	.228	.250	.283	.273	.315
17	.063	.124	.139	.150	.163	.180	.160	.178	.190	.216	.199	.218	.250	.240	.279
18	.054	.107	.121	.131	.141	.158	.140	.155	.167	.190	.174	.190	.219	.210	.246
19	.046	.093	.106	.114	.124	.139	.122	.135	.146	.178	.151	.166	.193	.185	.219
20	.040	.081	.092	.100	.108	.123	.106	.118	.128	.147	.132	.147	.172	.163	.195

Table 3.1–6 (continued)
TISSUE-AIR RATIOS FOR RECTANGULAR FIELDS

Dose to Small Mass of Tissue at Depth d for a Dose of 1 rad to the Same Mass of Tissue at the Same Point in Free Space (continued)

HVL 3.0 mm Cu

d, cm	Field size (cm × cm) at depth d, or axis of rotation														
	0 × 0	4 × 4	4 × 6	4 × 8	4 × 10	4 × 15	6 × 6	6 × 8	6 × 10	6 × 15	8 × 8	8 × 10	8 × 15	10 × 10	10 × 15
0[b]	1.00	1.12	1.14	1.15	1.16	1.18	1.16	1.18	1.20	1.22	1.21	1 23	1.24	1.24	1.26
1	.856	1.10	1.12	1.14	1.16	1.17	1.15	1.18	1.20	1.22	1.21	1.23	1.25	1.25	1.28
2	.736	1.00	1.05	1.07	1.09	1.11	1.10	1.13	1.15	1.18	1.17	1.20	1.23	1.22	1.26
3	.632	.895	.946	.964	.988	1.01	1.00	1.04	1.07	1.10	1.07	1.13	1.16	1.15	1.20
4	.541	.798	.844	.868	.888	.917	.898	.936	.970	1.00	.988	1.03	1.07	1.06	1.12
5	.467	.705	.752	.776	.798	.830	.809	.850	.877	.915	.902	.943	.990	.978	1.04
6	.401	.624	.668	.697	.721	.750	.728	.771	.798	.836	.820	.860	.905	.902	.959
7	.345	.549	.588	.615	.641	.670	.645	.686	.712	.750	.731	.776	.822	.815	.876
8	.296	.476	.516	.543	.565	.595	.566	.607	.634	.674	.655	.700	.745	.736	.796
9	.256	.419	.450	.479	.500	.529	.500	.539	.565	.600	.580	.622	.669	.660	.722
10	.222	.363	.394	.417	.438	.468	.439	.473	.497	.535	.513	.549	.596	.587	.645
11	.190	.317	.345	.366	.384	.412	.382	.417	.436	.473	.450	.489	.531	.520	.574
12	.164	.273	.299	.319	.337	.364	.335	.365	.384	.417	.398	.434	.474	.461	.512
13	.143	.239	.261	.279	.296	.320	.294	.320	.337	.370	.350	.381	.419	.407	.454
14	.123	.206	.228	.244	.259	.281	.257	.281	.298	.326	.308	.335	.371	.360	.405
15	.106	.182	.199	.214	.228	.248	.226	.247	.260	.288	.270	.295	.329	.318	.361
16	.092	.158	.174	.187	.198	.218	.198	.217	.230	.255	.238	.259	.290	.281	.320
17	.080	.138	.152	.165	.175	.192	.174	.190	.202	.225	.210	.228	.258	.249	.287
18	.068	.120	.133	.143	.153	.169	.151	.166	.177	.200	.185	.200	.228	.220	.256
19	.059	.105	.116	.126	.135	.149	.133	.146	.156	.176	.163	.176	.202	.194	.226
20	.051	.092	.102	.110	.113	.132	.117	.129	.137	.156	.143	.157	.180	.172	.201

Table 3.1–6 (continued)
TISSUE-AIR RATIOS FOR RECTANGULAR FIELDS

Dose to Small Mass of Tissue at Depth d for a Dose of 1 rad to the Same Mass of Tissue at the Same Point in Free Space (continued)

Cobalt 60

	Field size (cm × cm) at depth d, or axis of rotation													
d, cm	0 × 0	4 × 4	4 × 6	4 × 8	4 × 10	4 × 15	5 × 5	6 × 6	6 × 8	6 × 10	6 × 15	7 × 7	8 × 8	8 × 10
0.5[b]	1.000	1.015	1.018	1.020	1.022	1.025	1.018	1.022	1.025	1.027	1.031	1.025	1.029	1.032
1	.965	.996	1.001	1.005	1.008	1.012	1.003	1.009	1.014	1.018	1.023	1.015	1.021	1.025
2	.905	.956	.965	.970	.973	.978	.967	.976	.983	.988	.994	.985	.992	.997
3	.845	.915	.926	.932	.936	.942	.928	.940	.948	.954	.961	.950	.959	.966
4	.792	.872	.885	.893	.897	.903	.888	.902	.912	.918	.926	.914	.924	.931
5	.742	.829	.843	.852	.856	.863	.847	.862	.873	.880	.889	.875	.887	.895
6	.694	.786	.801	.810	.815	.823	.805	.821	.833	.840	.851	.835	.847	.856
7	.650	.743	.758	.767	.773	.781	.762	.778	.791	.799	.810	.793	.807	.819
8	.608	.700	.715	.725	.731	.740	.719	.736	.749	.757	.769	.751	.765	.775
9	.570	.659	.674	.684	.689	.700	.677	.695	.708	.716	.730	.710	.724	.734
10	.534	.620	.635	.644	.650	.661	.638	.655	.668	.677	.691	.671	.685	.695
11	.501	.581	.596	.606	.612	.623	.600	.616	.630	.639	.652	.632	.647	.658
12	.469	.546	.560	.570	.576	.587	.563	.580	.594	.603	.617	.596	.611	.622
13	.440	.513	.527	.537	.544	.555	.530	.547	.561	.570	.584	.563	.578	.589
14	.412	.482	.496	.505	.512	.523	.499	.515	.528	.538	.552	.531	.545	.557
15	.386	.454	.467	.476	.483	.494	.470	.485	.498	.507	.522	.501	.515	.526
16	.361	.427	.440	.449	.455	.466	.443	.458	.470	.479	.494	.472	.485	.496
17	.338	.402	.414	.423	.429	.440	.417	.431	.443	.452	.467	.445	.458	.469
18	.317	.378	.390	.398	.404	.415	.393	.406	.418	.426	.441	.420	.433	.443
19	.297	.355	.366	.375	.381	.391	.369	.383	.394	.403	.417	.396	.409	.420
20	.278	.333	.344	.353	.358	.369	.347	.361	.372	.380	.394	.374	.386	.396
22	.246	.293	.304	.312	.317	.327	.306	.318	.328	.336	.350	.330	.342	.352
24	.215	.258	.268	.275	.280	.290	.270	.281	.290	.298	.311	.292	.303	.312
26	.187	.228	.236	.243	.248	.257	.238	.249	.258	.264	.277	.259	.270	.278
28	.164	.200	.210	.215	.219	.228	.210	.221	.228	.234	.246	.230	.239	.246
30	.144	.178	.185	.190	.194	.202	.186	.195	.202	.208	.218	.205	.212	.219

Table 3.1–6 (continued)
TISSUE-AIR RATIOS FOR RECTANGULAR FIELDS

Dose to Small Mass of Tissue at Depth d for a Dose of 1 rad to the Same Mass of Tissue at the Same Point in Free Space (continued)

Cobalt 60 (continued)

	Field size (cm × cm) at depth d, or axis of rotation												
d, cm	**8 × 15**	**8 × 20**	**10 × 10**	**10 × 15**	**10 × 20**	**12 × 12**	**15 × 15**	**15 × 20**	**20 × 20**	**20 × 30**	**25 × 25**	**30 × 30**	**35 × 35**
0.5[b]	1.037	1.041	1.035	1.042	1.046	1.041	1.051	1.056	1.063	1.071	1.073	1.080	1.084
1	1.032	1.035	1.031	1.038	1.043	1.038	1.048	1.054	1.062	1.069	1.072	1.079	1.084
2	1.005	1.009	1.004	1.013	1.018	1.014	1.025	1.032	1.040	1.049	1.052	1.059	1.065
3	.975	.980	.974	.985	.990	.985	.999	1.006	1.016	1.026	1.029	1.038	1.044
4	.942	.947	.940	.952	.959	.953	.968	.977	.987	.999	1.002	1.014	1.021
5	.907	.913	.905	.918	.925	.919	.936	.946	.957	.971	.974	.988	.998
6	.869	.876	.867	.882	.890	.883	.902	.912	.925	.940	.944	.959	.970
7	.830	.837	.827	.844	.853	.845	.866	.878	.893	.909	.913	.929	.941
8	.790	.798	.787	.805	.815	.806	.830	.843	.859	.877	.881	.899	.912
9	.751	.760	.747	.767	.778	.768	.793	.808	.825	.845	.849	.869	.882
10	.713	.722	.709	.729	.741	.730	.756	.771	.790	.811	.816	.337	.852
11	.675	.685	.672	.692	.704	.692	.719	.736	.755	.777	.782	.803	.820
12	.640	.650	.636	.657	.670	.658	.685	.702	.722	.744	.750	.772	.790
13	.607	.618	.603	.625	.638	.626	.653	.670	.690	.713	.720	.743	.762
14	.575	.586	.571	.593	.606	.594	.622	.639	.660	.684	.691	.715	.734
15	.545	.556	.540	.563	.576	.563	.593	.610	.633	.656	.662	.687	.706
16	.516	.527	.510	.533	.547	.533	.564	.582	.605	.628	.634	.660	.679
17	.488	.499	.483	.506	.519	.506	.536	.554	.577	.601	.608	.633	.653
18	.462	.474	.457	.479	.493	.479	.509	.528	.551	.575	.582	.607	.627
19	.438	.449	.433	.455	.469	.455	.485	.503	.526	.550	.557	.583	.603
20	.415	.426	.410	.431	.445	.431	.461	.479	.502	.527	.534	.560	.580
22	.369	.380	.364	.385	.398	.384	.413	.431	.456	.481	.488	.515	.535
24	.329	.340	.324	.345	.358	.345	.373	.390	.412	.438	.446	.471	.492
26	.294	.304	.290	.309	.322	.308	.336	.352	.373	.398	.405	.431	.451
28	.263	.270	.257	.276	.288	.276	.302	.320	.339	.362	.368	.393	.413
30	.233	.242	.288	.245	.257	.244	.268	.286	.305	.328	.335	.358	.377

Table 3.1–6 (continued)
TISSUE-AIR RATIOS FOR RECTANGULAR FIELDS

Dose to Small Mass of Tissue at Depth d for a Dose of 1 rad to the Same Mass of Tissue at the Same Point in Free Space (continued)

[a] A "tissue-air ratio" (TAR) is the dose at a point in a medium divided by a dose to a "small volume" of tissue in air; it had advantages in rotational radiation therapy calculations since it is relatively independent of the distances between point source of radiation and skin surface.

[b] This entry is also the backscatter factor (BSF). The BSF is defined as the ratio of dose in the medium at depth of maximum electron build-up (close to zero here) divided by the dose within a small volume of tissue in air at the same point; a "small volume" would be a volume just large enough to provide electron build-up – about dimensions equal to twice the maximum secondary electron range in tissue.

From Johns, H. E. and Cunningham, J. R., *The Physics of Radiology,* 3rd ed., 1971, 748. Courtesy of Charles C Thomas, Publisher, Springfield, Ill.

Table 3.1–7

BACKSCATTER FACTOR FOR CIRCULAR FIELDS

Dose to Small Mass of Tissue at d_m, the Depth of Maximum Electronic Build-up for a Dose of 1 rad to the Same Mass of Tissue at the Same Point in Space

Half-value layer	Area, cm² 10	16	20	25	35	50	64	80	100	150	200	300	400
	Radius, cm												
	1.78	**2.26**	**2.52**	**2.82**	**3.34**	**3.99**	**4.51**	**5.05**	**5.64**	**6.77**	**7.98**	**9.75**	**11.3**
Al, mm													
1.0	1.108	1.128	1.138	1.148	1.164	1.179	1.189	1.197	1.205	1.218	1.229		
2.0	1.118	1.143	1.154	1.168	1.190	1.211	1.225	1.238	1.250	1.266	1.279		
3.0	1.134	1.164	1.179	1.194	1.217	1.240	1.256	1.270	1.283	1.302	1.318		
4.0	1.141	1.174	1.190	1.208	1.236	1.265	1.283	1.299	1.314	1.334	1.350		
Cu, mm													
0.25	1.174	1.205	1.220	1.237	1.263	1.292	1.312	1.330	1.348	1.374	1.395	1.424	1.450
0.5	1.186	1.220	1.235	1.254	1.282	1.314	1.336	1.357	1.376	1.406	1.430	1.463	1.492
1.0	1.150	1.184	1.200	1.221	1.252	1.288	1.314	1.338	1.360	1.393	1.420	1.458	1.490
1.5	1.138	1.169	1.184	1.201	1.230	1.262	1.284	1.306	1.327	1.361	1.391	1.428	1.460
2.0	1.119	1.145	1.160	1.176	1.201	1.230	1.250	1.269	1.288	1.320	1.348	1.385	1.418
3.0	1.098	1.120	1.130	1.144	1.164	1.188	1.205	1.222	1.238	1.266	1.289	1.316	1.340
4.0	1.076	1.094	1.104	1.114	1.132	1.152	1.168	1.182	1.197	1.220	1.240	1.264	1.280

From Johns, H. E. and Cunningham, J. R., *The Physics of Radiology*, 3rd ed., 1971, 751. Courtesy of Charles C Thomas, Publisher, Springfield, Ill.

Table 3.1–8
BACKSCATTER FACTOR FOR RECTANGULAR FIELDS

Dose to Small Mass of Tissue at d_m, the Depth of Maximum Electronic Build-up for a Dose of 1 rad to the Same Mass of Tissue at the Same Point in Space

Field size (cm × cm)

Half-value layer	4 × 4	4 × 6	4 × 8	4 × 10	4 × 15	4 × 20	6 × 6	6 × 8	6 × 10	6 × 15	6 × 20
Cu, mm											
0.5	1.214	1.244	1.261	1.272	1.285	1.292	1.283	1.306	1.321	1.340	1.350
1.0	1.180	1.211	1.230	1.243	1.258	1.266	1.252	1.279	1.297	1.318	1.330
1.5	1.166	1.193	1.210	1.222	1.237	1.245	1.230	1.253	1.269	1.291	1.303
2.0	1.144	1.169	1.184	1.194	1.208	1.216	1.201	1.222	1.237	1.257	1.269
3.0	1.116	1.137	1.149	1.158	1.170	1.176	1.164	1.182	1.194	1.211	1.221

Field size (cm × cm)

	8 × 8	8 × 10	8 × 15	8 × 20	10 × 10	10 × 15	10 × 20	15 × 15	15 × 20	20 × 20
0.5	1.334	1.352	1.376	1.390	1.373	1.401	1.418	1.439	1.462	1.489
1.0	1.311	1.333	1.360	1.375	1.357	1.389	1.407	1.430	1.456	1.487
1.5	1.282	1.302	1.330	1.345	1.324	1.357	1.376	1.400	1.426	1.457
2.0	1.248	1.265	1.292	1.307	1.286	1.317	1.335	1.358	1.384	1.415
3.0	1.204	1.219	1.241	1.253	1.237	1.262	1.277	1.296	1.315	1.337

From Johns, H. E. and Cunningham, J. R., *The Physics of Radiology,* 3rd ed., 1971, 751. Courtesy of Charles C Thomas, Publisher, Springfield, Ill.

Table 3.1–9
PERCENTAGE DEPTH DOSE[a] FOR RECTANGULAR FIELDS

HVL 1.0 mm Cu, FSD 50 cm

Depth, cm	Open-ended applicator (cm× cm)										
	0 × 0	4 × 4	4 × 6	4 × 8	4 × 10	4 × 15	4 × 20	6 × 6	6 × 8	6 × 10	6 × 15
b	100.0	118.0	121.1	123.0	124.3	125.8	126.6	125.2	127.9	129.7	131.8
0	100.0	100.0	100.0	100.0	100.0	100.0	100.0	100.0	100.0	100.0	100.0
1	79.0	92.9	94.7	95.5	96.0	96.6	96.8	96.9	97.9	98.5	99.3
2	63.0	81.3	84.3	85.7	86.5	87.4	87.7	87.9	89.6	90.6	91.7
3	50.5	70.3	73.5	75.3	76.5	77.8	78.3	77.4	79.8	81.3	82.9
4	40.5	60.0	63.2	65.1	66.4	68.0	68.7	67.2	60.7	71.3	73.4
5	32.5	50.7	53.8	55.8	57.1	58.7	59.5	57.7	60.2	61.9	64.2
6	26.3	42.7	45.5	47.4	48.8	50.4	51.3	49.2	51.6	53.4	55.7
7	21.3	35.8	38.3	40.1	41.5	43.1	44.0	41.7	44.1	45.8	48.1
8	17.3	29.9	32.2	33.9	35.2	36.8	37.7	35.3	37.6	39.2	41.4
9	14.0	25.0	27.1	28.7	29.8	31.4	32.2	29.9	32.0	33.5	35.6
10	11.3	20.9	22.8	24.2	25.2	26.7	27.5	25.3	27.2	28.6	30.6
11	9.1	17.4	19.2	20.4	21.3	22.7	23.5	21.4	23.1	24.3	26.2
12	7.4	14.6	16.2	17.3	18.1	19.4	20.1	18.1	19.6	20.7	22.5
13	5.9	12.2	13.6	14.6	15.4	16.5	17.1	15.3	16.6	17.6	19.3
14	4.8	10.2	11.4	12.3	13.0	14.0	14.6	12.9	14.1	15.0	16.5
15	3.9	8.6	9.6	10.4	11.0	11.9	12.5	10.9	12.0	12.8	14.1
16	3.2	7.2	8.1	8.7	9.3	10.1	10.7	9.2	10.2	10.9	12.0
17	2.6	6.0	6.8	7.3	7.8	8.6	9.1	7.8	8.6	9.2	10.3
18	2.1	5.0	5.7	6.2	6.6	7.3	7.7	6.6	7.3	7.8	8.8
19	1.7	4.2	4.8	5.2	5.6	6.2	6.6	5.6	6.2	6.7	7.5
20	1.4	3.5	4.0	4.4	4.8	5.3	5.6	4.7	5.3	5.7	6.4

Table 3.1–9 (continued)
PERCENTAGE DEPTH DOSE[a] FOR RECTANGULAR FIELDS

HVL 1.0 mm Cu, FSD 50 cm (continued)

	Open-ended applicator (cm × cm)										
Depth, cm	6 × 20	8 × 8	8 × 10	8 × 15	8 × 20	10 × 10	10 × 15	10 × 20	15 × 15	15 × 20	20 × 20
b	133.0	131.1	133.3	136.0	137.5	135.7	138.9	140.7	143.0	145.6	148.7
0	100.0	100.0	100.0	100.0	100.0	100.0	100.0	100.0	100.0	100.0	100.0
1	99.5	99.1	99.8	100.7	100.9	100.6	101.5	101.8	102.6	102.8	103.0
2	92.1	91.6	92.8	94.0	94.5	94.0	95.4	95.9	97.0	97.6	98.4
3	83.6	82.5	84.2	86.2	87.0	86.1	88.3	89.2	90.9	92.0	93.4
4	74.4	72.6	74.5	77.0	78.1	76.6	79.4	80.7	82.8	84.4	86.1
5	65.2	63.2	65.2	67.8	69.1	67.3	70.3	71.8	74.0	75.8	77.8
6	56.8	54.6	56.6	59.4	60.6	58.8	61.9	63.4	65.6	67.5	69.7
7	49.2	46.9	48.9	51.7	52.9	51.1	54.2	55.7	57.9	59.9	62.2
8	42.5	40.2	42.1	44.8	46.1	44.2	47.2	48.8	50.9	53.0	55.3
9	36.7	34.4	36.2	38.7	40.1	38.2	41.1	42.6	44.6	46.7	49.1
10	31.6	29.4	31.1	33.4	34.8	32.9	35.6	37.1	39.0	41.0	43.4
11	27.2	25.1	26.6	28.8	30.2	28.8	30.8	32.3	34.0	35.9	38.2
12	23.4	21.4	22.7	24.9	26.1	24.3	26.6	28.1	29.6	31.4	33.5
13	20.1	18.3	19.4	21.4	22.5	20.8	23.0	24.3	25.8	27.4	29.3
14	17.2	15.6	16.6	18.4	19.4	17.8	19.8	21.0	22.4	23.9	25.7
15	14.8	13.2	14.2	15.8	16.7	15.3	17.1	18.2	19.4	20.8	22.5
16	12.7	11.2	12.1	13.5	14.4	13.1	14.8	15.8	16.9	18.2	19.7
17	10.9	9.5	10.3	11.6	12.4	11.2	12.7	13.7	14.7	15.9	17.2
18	9.4	8.1	8.8	10.0	10.7	9.6	10.9	11.8	12.7	13.8	15.1
19	8.1	7.0	7.6	8.6	9.3	8.3	9.4	10.2	11.0	12.0	13.2
20	6.9	6.0	6.5	7.4	8.0	7.1	8.1	8.8	9.5	10.4	11.5

Table 3.1–9 (continued)
PERCENTAGE DEPTH DOSE[a] FOR RECTANGULAR FIELDS

HVL 1.5 mm Cu, FSD 50 cm

	Open ended applicator (cm × cm)										
Depth, cm	0 × 0	4 × 4	4 × 6	4 × 8	4 × 10	4 × 15	4 × 20	6 × 6	6 × 8	6 × 10	6 × 15
b	100.0	116.6	119.3	121.0	122.2	123.7	124.5	123.0	125.3	126.9	129.1
0	100.0	100.0	100.0	100.0	100.0	100.0	100.0	100.0	100.0	100.0	100.0
1	80.8	94.0	95.3	96.0	96.4	97.0	97.2	97.0	97.8	98.4	99.1
2	65.2	83.2	85.5	86.8	87.5	88.4	88.8	88.5	90.0	91.0	92.1
3	52.7	72.0	74.9	76.6	77.6	78.8	79.3	78.5	80.8	82.1	83.8
4	43.0	61.3	64.4	66.3	67.6	69.0	69.7	68.2	70.8	72.4	74.4
5	35.0	52.2	55.1	57.1	58.4	60.0	60.8	58.9	61.5	63.2	65.4
6	23.4	44.2	47.0	48.9	50.2	52.0	52.8	50.6	53.2	54.9	57.2
7	23.2	37.3	40.0	41.8	43.1	44.8	45.7	43.4	45.8	47.5	49.8
8	18.8	31.4	33.9	35.6	36.8	38.5	39.4	37.0	39.3	40.9	43.2
9	15.3	26.4	28.6	30.2	31.4	33.0	33.9	31.5	33.6	35.2	37.4
10	12.4	22.3	24.2	25.6	26.7	28.2	29.1	26.9	28.8	30.2	32.3
11	10.2	18.8	20.5	21.8	22.8	24.2	25.0	22.9	24.6	25.9	27.8
12	8.3	15.8	17.4	18.5	19.4	20.7	21.5	19.5	21.0	22.2	24.0
13	6.7	13.3	14.7	15.7	16.5	17.7	18.5	16.6	17.9	19.0	20.7
14	5.5	11.2	12.4	13.4	14.1	15.2	15.9	14.1	15.3	16.3	17.8
15	4.5	9.4	10.5	11.4	12.0	13.1	13.6	12.0	13.1	14.0	15.4
16	3.7	7.9	8.9	9.6	10.2	11.2	11.7	10.2	11.2	12.0	13.3
17	3.1	6.7	7.5	8.2	8.7	9.6	10.1	8.7	9.6	10.3	11.5
18	2.5	5.7	6.4	7.0	7.4	8.2	8.7	7.4	8.2	8.9	10.0
19	2.1	4.8	5.4	5.9	6.3	7.1	7.5	6.3	7.0	7.6	8.6
20	1.7	4.0	4.6	5.0	5.4	6.1	6.5	5.3	6.0	6.5	7.4

Table 3.1–9 (continued)
PERCENTAGE DEPTH DOSE[a] FOR RECTANGULAR FIELDS

HVL 1.5 mm Cu, FSD 50 cm (continued)

	Open-ended applicator (cm × cm)										
Depth, cm	6 × 20	8 × 8	8 × 10	8 × 15	8 × 20	10 × 10	10 × 15	10 × 20	15 × 15	15 × 20	20 × 20
b	130.3	128.2	130.2	133.0	134.5	132.4	135.7	137.6	140.0	142.6	145.7
0	100.0	100.0	100.0	100.0	100.0	100.0	100.0	100.0	100.0	100.0	100.0
1	99.4	98.8	99.5	100.3	100.6	100.2	101.2	101.5	102.3	102.5	102.7
2	92.6	91.8	92.9	94.3	94.9	94.1	95.6	96.3	97.4	98.3	99.3
3	84.5	83.2	84.8	86.8	87.7	86.6	88.8	89.8	91.4	92.7	94.2
4	75.3	73.7	75.6	78.0	79.0	77.7	80.4	81.6	83.5	85.1	86.9
5	66.4	64.5	66.5	69.1	70.3	68.7	71.7	73.1	75.3	77.1	79.2
6	58.3	56.1	58.1	60.9	62.2	60.4	63.5	65.0	67.3	69.3	71.7
7	51.0	48.6	50.6	53.4	54.8	52.9	56.0	57.7	59.9	62.0	64.6
8	44.4	42.0	43.9	46.6	48.1	46.1	49.2	50.9	53.0	55.3	57.9
9	38.6	36.2	38.0	40.6	42.1	40.1	43.1	44.8	46.8	49.1	51.7
10	33.5	31.1	32.8	35.3	36.7	34.7	37.6	39.2	41.2	43.4	45.9
11	29.0	26.7	28.3	30.6	32.0	30.0	32.7	34.3	36.1	38.2	40.6
12	25.1	22.9	24.3	26.5	27.8	25.9	28.4	30.0	31.6	33.6	35.8
13	21.7	19.7	20.9	22.9	24.1	22.4	24.7	26.1	27.6	29.4	31.5
14	18.7	16.9	18.0	19.8	21.0	19.4	21.5	22.8	24.2	25.8	27.7
15	16.2	14.5	15.5	17.2	18.3	16.8	18.7	19.9	21.2	22.7	24.3
16	14.0	12.4	13.4	14.9	15.9	14.5	16.3	17.4	18.6	19.9	21.4
17	12.1	10.7	11.6	13.0	13.9	12.5	14.2	15.2	16.4	17.5	18.8
18	10.5	9.2	10.0	11.3	12.1	10.8	12.4	13.3	14.4	15.4	16.6
19	9.1	7.9	8.6	9.8	10.5	9.3	10.8	11.6	12.6	13.6	14.6
20	7.9	6.7	7.4	8.5	9.1	8.1	9.4	10.1	11.0	11.9	12.8

Table 3.1–9 (continued)
PERCENTAGE DEPTH DOSE[a] FOR RECTANGULAR FIELDS

HVL 2.0 mm Cu, FSD 50 cm

Depth, cm	Open-ended applicator (cm × cm)										
	0 × 0	4 × 4	4 × 6	4 × 8	4 × 10	4 × 15	4 × 20	6 × 6	6 × 8	6 × 10	6 × 15
b	100.0	114.4	116.9	118.4	119.4	120.8	121.6	120.1	122.2	123.7	125.7
0	100.0	100.0	100.0	100.0	100.0	100.0	100.0	100.0	100.0	100.0	100.0
1	81.4	93.8	95.2	95.8	96.2	96.7	96.9	96.8	97.7	98.2	98.8
2	66.5	83.9	85.9	87.0	87.7	88.5	88.9	88.4	89.8	90.7	91.8
3	54.0	72.5	75.2	76.7	77.7	78.9	79.4	78.6	80.6	81.9	83.5
4	44.2	62.1	65.0	66.7	67.9	69.4	70.0	68.7	71.0	72.5	74.5
5	36.2	52.9	55.7	57.6	58.8	60.5	61.2	59.5	61.9	63.5	65.6
6	29.6	44.9	47.6	49.5	50.7	52.4	53.2	51.3	53.7	55.3	57.5
7	24.3	38.0	40.6	42.4	43.6	45.3	46.1	44.1	46.4	48.0	50.3
8	19.9	32.1	34.6	36.3	37.4	39.1	39.0	37.8	40.1	41.6	43.8
9	16.4	27.1	29.4	31.0	32.1	33.7	34.5	32.4	34.5	36.0	38.1
10	13.4	22.9	25.0	26.5	27.5	29.0	29.8	27.7	29.7	31.1	33.1
11	11.1	19.4	21.3	22.6	23.6	25.0	25.8	23.6	25.5	26.8	28.7
12	9.1	16.5	18.1	19.3	20.2	21.5	22.3	20.2	21.9	23.1	24.9
13	7.5	14.0	15.4	16.5	17.3	18.5	19.3	17.3	18.8	19.9	21.6
14	6.2	11.9	13.1	14.1	14.8	15.9	16.7	14.8	16.1	17.1	18.7
15	5.1	10.1	11.2	12.1	12.7	13.7	14.4	12.7	13.8	14.7	16.2
16	4.2	8.5	9.5	10.3	10.9	11.8	12.4	10.9	11.8	12.6	14.0
17	3.5	7.2	8.1	8.8	9.3	10.2	10.7	9.3	10.1	10.9	12.1
18	2.9	6.1	6.9	7.5	8.0	8.8	9.3	7.9	8.7	9.4	10.5
19	2.4	5.2	5.9	6.4	6.8	7.6	8.0	6.7	7.5	8.1	9.1
20	2.0	4.4	4.9	5.4	5.8	6.5	6.9	5.7	6.4	6.9	7.9

Table 3.1–9 (continued)
PERCENTAGE DEPTH DOSE[a] FOR RECTANGULAR FIELDS

HVL 2.0 mm Cu, FSD 50 cm (continued)

Depth, cm	Open-ended applicator (cm × cm)										
	6 × 20	8 × 8	8 × 10	8 × 15	8 × 20	10 × 10	10 × 15	10 × 20	15 × 15	15 × 20	20 × 20
b	126.9	124.8	126.5	129.2	130.7	128.6	131.7	133.5	135.8	138.4	141.5
0	100.0	100.0	100.0	100.0	100.0	100.0	100.0	100.0	100.0	100.0	100.0
1	99.0	98.6	99.3	100.0	100.3	99.9	100.8	101.0	101.7	102.0	102.4
2	92.3	91.4	92.5	93.8	94.4	93.6	95.1	95.8	96.9	97.8	98.9
3	84.2	83.0	84.5	86.4	87.3	86.1	88.3	89.3	90.9	92.1	93.6
4	75.4	73.6	75.4	77.8	78.9	77.4	80.1	81.4	83.4	85.0	86.8
5	66.6	64.7	66.6	69.1	70.4	68.7	71.6	73.0	75.1	77.0	79.1
6	58.6	56.5	58.5	61.1	62.5	60.6	63.6	65.2	67.3	69.3	71.6
7	51.4	49.2	51.1	53.8	55.3	53.2	56.3	57.9	60.1	62.2	64.6
8	45.0	42.7	44.5	47.2	48.7	46.6	49.7	51.3	53.5	55.6	58.1
9	39.3	37.0	38.7	41.4	42.8	40.7	43.8	45.3	47.5	49.6	52.1
10	34.2	32.0	33.6	36.2	37.5	35.5	38.4	40.0	42.0	44.1	46.5
11	29.8	27.6	29.1	31.6	32.8	30.9	33.6	35.2	37.1	39.1	41.4
12	25.9	23.8	25.2	27.5	28.7	26.9	29.4	30.9	32.7	34.6	36.7
13	22.5	20.5	21.8	23.9	25.1	23.4	25.7	27.1	28.7	30.5	32.5
14	19.6	17.6	18.9	20.8	21.9	20.3	22.4	23.8	25.2	26.9	28.7
15	17.0	15.2	16.3	18.1	19.1	17.6	19.6	20.8	22.2	23.7	25.4
16	14.8	13.1	14.1	15.7	16.7	15.2	17.1	18.2	19.5	20.9	22.5
17	12.9	11.3	12.2	13.7	14.6	13.2	15.0	16.0	17.2	18.4	19.9
18	11.2	9.8	10.5	12.0	12.3	11.5	13.2	14.0	15.2	16.3	17.6
19	9.7	8.4	9.1	10.5	11.2	10.0	11.5	12.3	13.4	14.4	15.6
20	8.4	7.2	7.9	9.1	9.7	8.7	10.0	10.8	11.7	12.7	13.7

Table 3.1–9 (continued)
PERCENTAGE DEPTH DOSE[a] FOR RECTANGULAR FIELDS

HVL 3.0 mm Cu, FSD 50 cm

	Open-ended applicator (cm × cm)										
Depth, cm	**0 × 0**	**4 × 4**	**4 × 6**	**4 × 8**	**4 × 10**	**4 × 15**	**4 × 20**	**6 × 6**	**6 × 8**	**6 × 10**	**6 × 15**
b	100.0	111.6	113.7	114.9	115.8	117.0	117.6	116.4	118.2	119.4	121.1
0	100.0	100.0	100.0	100.0	100.0	100.0	100.0	100.0	100.0	100.0	100.0
1	82.3	93.9	95.1	95.6	95.9	96.3	96.5	96.5	97.1	97.5	98.1
2	68.0	84.6	86.2	87.1	87.6	88.4	88.7	88.3	89.4	90.1	91.2
3	56.2	73.7	76.0	77.3	78.1	79.2	79.7	78.8	80.5	81.6	83.1
4	46.4	63.1	65.8	67.4	68.4	69.6	70.3	69.0	71.0	72.4	74.1
5	38.6	54.2	56.7	58.4	59.5	60.9	61.6	60.1	62.2	63.7	65.6
6	32.0	46.3	48.7	50.4	51.5	53.1	53.8	52.0	54.2	55.7	57.8
7	26.5	39.3	41.7	43.4	44.5	46.1	46.8	44.9	47.1	48.6	50.8
8	22.0	33.4	35.7	37.3	38.5	40.0	40.7	38.7	40.9	42.4	44.5
9	18.4	28.5	30.6	32.1	33.2	34.7	35.4	33.4	35.4	36.9	38.9
10	15.4	24.3	26.2	27.6	28.6	30.0	30.8	28.7	30.6	32.0	34.0
11	12.8	20.7	22.4	23.7	24.6	26.0	26.7	24.7	26.4	27.7	29.6
12	10.7	17.6	19.1	20.4	21.2	22.5	23.1	21.2	22.8	24.0	25.7
13	9.0	15.1	16.3	17.5	18.3	19.5	20.0	18.2	19.7	20.8	22.4
14	7.5	12.9	14.0	15.0	15.7	16.9	17.4	15.7	17.0	18.0	19.5
15	6.3	11.0	12.0	12.8	13.5	14.6	15.1	13.5	14.6	15.5	17.0
16	5.3	9.4	10.3	11.0	11.6	12.6	13.1	11.6	12.6	13.4	14.8
17	4.5	8.0	8.8	9.4	10.0	10.9	11.4	10.0	10.9	11.6	12.9
18	3.7	6.8	7.5	8.1	8.6	9.4	9.9	8.6	9.4	10.0	11.2
19	3.1	5.8	6.4	7.0	7.4	8.1	8.6	7.4	8.1	8.7	9.8
20	2.6	4.9	5.5	6.0	6.4	7.1	7.5	6.3	7.0	7.6	8.5

Table 3.1–9 (continued)
PERCENTAGE DEPTH DOSE[a] FOR RECTANGULAR FIELDS

HVL 3.0 mm Cu, FSD 50 cm (continued)

	Open-ended applicator (cm × cm)										
Depth, cm	6 × 20	8 × 8	8 × 10	8 × 15	8 × 20	10 × 10	15 × 15	10 × 20	10 × 15	15 × 20	20 × 20
b	122.1	120.4	121.9	124.1	125.3	123.7	126.2	127.7	129.6	131.5	133.7
0	100.0	100.0	100.0	100.0	100.0	100.0	100.0	100.0	100.0	100.0	100.0
1	98.3	97.9	98.3	99.1	99.3	98.9	99.7	100.0	100.6	101.0	101.4
2	91.5	90.7	91.6	92.8	93.3	92.6	93.9	94.5	95.6	96.2	96.8
3	82.5	82.4	83.7	85.4	86.3	85.1	87.0	88.1	89.5	90.8	92.3
4	75.0	73.4	74.9	77.1	78.1	76.7	79.1	80.3	82.1	83.8	85.7
5	66.6	64.8	66.5	68.8	70.0	68.4	71.1	72.5	74.5	76.2	78.3
6	58.8	56.8	58.6	61.1	62.3	60.6	63.5	65.0	67.0	68.9	71.3
7	51.8	49.7	51.5	54.1	55.3	53.6	56.5	58.1	60.1	62.0	64.2
8	45.6	43.4	45.2	47.7	49.0	47.2	50.1	51.7	53.7	55.6	57.9
9	40.0	37.8	39.6	42.0	43.3	41.5	44.3	45.9	47.9	49.8	52.0
10	35.0	32.8	34.5	36.9	38.1	36.3	39.1	40.6	42.6	44.5	46.7
11	30.6	28.5	30.0	32.3	33.5	31.7	34.4	35.8	37.7	39.6	41.7
12	26.7	24.8	26.1	28.3	29.4	27.7	30.3	31.5	33.3	35.1	37.2
13	23.3	21.5	22.7	24.7	25.8	24.2	26.6	27.8	29.4	31.1	33.1
14	20.3	18.6	19.7	21.6	22.6	21.1	23.3	24.5	25.9	27.5	29.4
15	17.7	16.1	17.1	18.9	19.8	18.4	20.4	21.6	22.9	24.4	26.2
16	15.5	13.9	14.9	16.5	17.4	16.0	17.9	19.0	20.2	21.6	23.2
17	13.5	12.0	13.0	14.4	15.3	13.9	15.7	16.7	17.8	19.1	20.7
18	11.8	10.4	11.3	12.6	13.4	12.2	13.8	14.7	15.7	16.9	18.5
19	10.3	9.0	9.8	11.1	11.8	10.7	12.1	13.0	13.9	15.0	16.4
20	9.0	7.8	8.5	9.7	10.3	9.3	10.6	11.4	12.3	13.3	14.5

Table 3.1–9 (continued)
PERCENTAGE DEPTH DOSE[a] FOR RECTANGULAR FIELDS

Cobalt 60 – SSD 50 cm, 1.25 MeV, HVL 11 mm Pb

	Open-ended applicator (cm × cm)										
Depth, cm	**0 × 0**	**4 × 4**	**4 × 6**	**4 × 8**	**4 × 10**	**4 × 15**	**4 × 20**	**6 × 6**	**6 × 8**	**6 × 10**	**6 × 15**
c	100.0	101.5	101.8	102.0	102.2	102.5	102.7	102.2	102.5	102.7	103.1
0	Surface dose 30–50% depending upon collimator										
0.5	100.0	100.0	100.0	100.0	100.0	100.0	100.0	100.0	100.0	100.0	100.0
1	94.6	96.0	96.3	96.5	96.6	96.6	96.6	96.7	96.9	97.0	97.1
2	85.2	88.7	89.3	89.6	89.8	89.9	89.9	90.1	90.5	90.6	90.8
3	76.8	81.6	82.5	82.9	83.1	83.3	83.4	83.6	84.1	84.4	84.7
4	69.3	75.0	76.0	76.5	76.7	77.0	77.1	77.3	77.9	78.3	78.7
5	92.6	68.8	70.0	70.4	70.7	71.1	71.2	71.3	72.0	72.5	73.0
6	56.4	63.0	64.1	64.7	65.1	65.5	65.6	65.6	66.4	66.9	67.5
7	51.0	57.6	58.7	59.4	59.8	60.2	60.4	60.2	61.1	61.6	62.2
8	46.1	52.6	53.7	54.4	54.8	55.3	55.5	55.2	56.1	56.6	57.3
9	41.7	48.0	49.1	49.8	50.1	50.7	51.0	50.5	51.4	52.0	52.7
10	37.8	43.8	44.9	45.5	45.9	46.5	46.8	46.2	47.1	47.7	48.5
11	34.3	40.0	41.0	41.6	42.0	42.6	43.0	42.3	43.2	43.8	44.7
12	31.1	36.5	37.5	38.1	38.5	39.1	39.5	38.8	39.7	40.2	41.1
13	28.2	33.3	34.3	34.9	35.3	35.9	36.3	35.6	36.4	37.0	37.9
14	25.6	30.5	31.4	32.0	32.4	33.0	33.4	32.6	33.4	34.0	34.9
15	23.3	27.9	28.7	29.3	29.7	30.3	30.7	29.9	30.6	31.2	32.1
16	21.1	25.5	26.2	26.8	27.2	27.9	28.2	27.4	28.1	28.7	29.6
17	19.3	23.3	24.0	24.6	24.9	25.6	26.0	25.1	25.8	26.4	27.3
18	17.5	21.3	22.0	22.6	22.9	23.5	24.0	23.0	23.7	24.3	25.2
19	15.9	19.5	20.2	20.7	21.0	21.6	22.1	21.1	21.8	22.4	23.3
20	14.5	17.8	18.5	19.0	19.3	19.9	20.3	19.4	20.0	20.6	21.5

Table 3.1–9 (continued)
PERCENTAGE DEPTH DOSE[a] FOR RECTANGULAR FIELDS

Cobalt 60 – SSD 50 cm, 1.25 MeV, HVL 11 mm Pb (continued)

	Open-ended applicator (cm × cm)										
Depth, cm	6 × 20	8 × 8	8 × 10	8 × 15	8 × 20	10 × 10	10 × 15	10 × 20	15 × 15	15 × 20	20 × 20
c	103.3	102.9	103.2	103.7	104.1	103.5	104.2	104.6	105.1	105.6	106.3
0	Surface dose 30–50% depending upon collimator										
0.5	100.0	100.0	100.0	100.0	100.0	100.0	100.0	100.0	100.0	100.0	100.0
1	97.1	97.1	97.3	97.4	97.4	97.5	97.6	97.6	97.7	97.7	97.7
2	90.9	90.9	91.1	91.3	91.4	91.4	91.6	91.7	91.9	92.0	92.1
3	84.8	84.7	85.0	85.4	85.5	85.4	85.8	86.0	86.2	86.5	86.7
4	78.9	78.7	79.1	79.6	79.8	79.6	80.1	80.4	80.7	81.1	81.5
5	73.2	72.9	73.4	74.0	74.3	74.0	74.6	75.0	75.4	75.9	76.4
6	67.8	67.4	67.9	63.6	69.0	68.6	69.4	69.8	70.3	70.9	71.6
7	62.6	62.1	62.7	63.5	64.0	63.4	64.4	64.9	65.5	66.2	67.0
8	57.7	57.1	57.7	58.6	59.2	58.5	59.6	60.2	60.9	61.7	62.6
9	53.2	52.4	53.1	54.1	54.7	53.9	55.1	55.8	56.6	57.5	58.5
10	49.0	48.1	48.8	49.9	50.5	49.7	50.9	51.7	52.5	53.6	54.7
11	45.1	44.2	44.9	46.0	46.7	45.8	47.1	47.9	48.7	49.9	51.1
12	41.6	40.7	41.4	42.5	43.2	42.2	43.6	44.4	45.2	46.4	47.7
13	38.4	37.4	38.1	39.2	39.9	39.0	40.3	41.1	42.0	43.2	44.5
14	35.4	34.4	35.1	36.2	36.9	36.0	37.3	38.1	39.0	40.2	41.6
15	32.7	31.6	32.3	33.4	34.1	33.2	34.5	35.3	36.3	37.5	38.8
16	30.2	29.1	29.7	30.9	31.6	30.6	32.0	32.8	33.8	35.0	36.3
17	27.9	26.8	27.4	28.6	29.3	28.2	29.7	30.5	31.5	32.7	34.0
18	25.8	24.7	25.3	26.5	27.2	26.1	27.5	28.3	29.3	30.5	31.8
19	23.8	22.7	23.4	24.5	25.2	24.1	25.5	26.3	27.3	28.5	29.8
20	22.0	20.9	21.6	22.7	23.3	22.2	23.7	24.4	25.4	26.6	27.9

Table 3.1–9 (continued)
PERCENTAGE DEPTH DOSE[a] FOR RECTANGULAR FIELDS

Cobalt 60 – SSD 60 cm

Depth, cm	Open-ended applicator (cm × cm)										
	0 × 0	4 × 4	4 × 6	4 × 8	4 × 10	4 × 15	4 × 20	6 × 6	6 × 8	6 × 10	6 × 15
c	100.0	101.5	101.8	102.0	102.2	102.5	102.7	102.2	102.5	102.7	103.1
0	Surface dose 30–50% depending upon collimator										
0.5	100.0	100.0	100.0	100.0	100.0	100.0	100.0	100.0	100.0	100.0	100.0
1	95.0	96.5	96.7	96.8	96.9	97.0	97.0	97.0	97.2	97.3	97.4
2	86.0	89.7	90.2	90.5	90.7	90.8	90.8	90.8	91.2	91.4	91.6
3	77.9	83.2	83.9	84.3	84.6	84.8	84.8	84.8	85.3	85.6	85.9
4	70.7	77.0	77.8	78.3	78.6	78.9	79.0	78.9	79.6	80.0	80.3
5	64.2	71.0	71.9	72.5	72.8	73.1	73.4	73.2	74.0	74.5	74.8
6	58.3	65.4	66.4	67.0	67.3	67.7	68.0	67.7	68.6	69.1	69.6
7	53.0	60.1	61.2	61.8	62.1	62.6	62.8	62.5	63.4	63.9	64.5
8	48.2	55.1	56.2	56.8	57.2	57.7	57.9	57.6	58.4	59.0	59.7
9	43.9	50.4	51.5	52.1	52.6	53.1	53.4	53.0	53.8	54.4	55.1
10	39.9	46.1	47.2	47.8	48.3	48.8	49.2	48.7	49.5	50.1	50.9
11	36.3	42.2	43.3	43.9	44.4	44.9	45.3	44.8	45.6	46.2	47.0
12	33.1	38.7	39.8	40.4	40.9	41.4	41.8	41.2	42.0	42.6	43.4
13	30.2	35.5	36.5	37.2	37.6	38.2	38.5	37.9	38.7	39.3	40.1
14	27.5	32.5	33.5	34.2	34.6	35.2	35.5	34.8	35.7	36.3	37.1
15	25.1	29.8	30.8	31.4	31.8	32.4	32.8	32.0	32.9	33.5	34.3
16	22.9	27.4	28.3	28.9	29.3	29.9	30.3	29.4	30.3	30.0	31.7
17	20.9	25.2	26.1	26.6	27.0	27.6	28.0	27.1	28.0	28.5	29.3
18	19.1	23.2	24.0	24.5	24.9	25.5	25.9	25.0	25.8	26.3	27.2
19	17.4	21.3	22.1	22.6	22.9	23.6	23.9	23.0	23.8	24.3	25.2
20	15.9	19.5	20.3	20.8	21.0	21.8	22.0	21.1	21.9	22.4	23.3

Table 3.1–9 (continued)
PERCENTAGE DEPTH DOSE[a] FOR RECTANGULAR FIELDS

Cobalt 60 – SSD 60 cm (continued)

Open-ended applicator (cm × cm)

Depth, cm	6 × 20	8 × 8	8 × 10	8 × 15	8 × 20	10 × 10	10 × 15	10 × 20	15 × 15	15 × 20	20 × 20
c	103.3	102.9	103.2	103.7	104.1	103.5	104.2	104.6	105.1	105.6	106.3
0	Surface dose 30–50% depending upon collimator										
0.5	100.0	100.0	100.0	100.0	100.0	100.0	100.0	100.0	100.0	100.0	100.0
1	97.4	97.4	97.5	97.7	97.7	97.7	97.8	97.9	98.0	98.0	98.1
2	91.7	91.7	91.9	92.1	92.2	92.1	92.4	92.5	92.7	92.8	93.0
3	86.1	86.0	86.3	86.6	86.8	86.6	87.1	87.2	87.5	87.7	88.0
4	80.5	80.3	80.7	81.2	81.4	81.2	81.8	82.0	82.4	82.7	83.1
5	75.1	74.8	75.3	75.9	76.2	75.9	76.6	76.9	77.4	77.8	78.3
6	69.9	69.5	70.0	70.7	71.1	70.7	71.5	71.9	72.5	73.0	73.7
7	64.9	64.4	64.9	65.7	66.2	65.6	66.6	67.0	67.8	68.4	69.2
8	60.1	59.5	60.1	61.0	61.5	60.8	61.9	62.4	63.3	64.0	64.9
9	55.6	54.9	55.6	56.6	57.1	56.3	57.5	58.1	59.0	59.8	60.9
10	51.4	50.6	51.3	52.4	52.9	52.1	53.4	54.0	54.9	55.8	57.0
11	47.5	46.7	47.4	48.5	49.1	48.2	49.5	50.2	51.1	52.1	53.4
12	44.0	43.1	43.8	44.9	45.6	44.6	45.9	46.7	47.6	48.7	50.0
13	40.7	39.8	40.5	41.6	42.3	41.3	42.6	43.4	44.4	45.5	46.9
14	37.7	36.7	37.5	38.5	39.2	38.2	39.6	40.4	41.4	42.5	43.9
15	34.9	33.9	34.6	35.7	36.4	35.4	36.8	37.6	38.6	39.7	41.1
16	32.3	31.3	32.0	33.1	33.8	32.8	34.2	35.0	36.0	37.1	38.5
17	29.9	28.9	29.6	30.7	31.4	30.4	31.8	32.6	33.6	34.8	36.1
18	27.8	26.7	27.4	28.5	29.2	28.2	29.6	30.4	31.4	32.6	33.9
19	25.8	24.7	25.4	26.5	27.2	26.2	27.5	28.4	29.3	30.5	31.8
20	23.9	22.8	23.4	24.6	25.3	24.2	25.5	26.4	27.3	28.5	29.8

Table 3.1–9 (continued)
PERCENTAGE DEPTH DOSE[a] FOR RECTANGULAR FIELDS

Cobalt 60 – SSD 80 cm

Depth, cm	Open-ended applicator (cm × cm)										
	0 × 0	4 × 4	4 × 6	4 × 8	4 × 10	4 × 15	4 × 20	6 × 6	6 × 8	6 × 10	6 × 15
c	100.0	101.5	101.8	102.0	102.2	102.5	102.7	102.2	102.5	102.7	103.1
0	Surface dose 30–50% depending upon collimator										
0.5	100.0	100.0	100.0	100.0	100.0	100.0	100.0	100.0	100.0	100.0	100.0
1	95.4	96.8	97.0	97.2	97.3	97.4	97.4	97.4	97.6	97.7	97.8
2	87.1	90.6	91.2	91.5	91.6	91.8	91.8	91.9	92.2	92.5	92.7
3	79.5	84.7	85.5	85.9	86.1	86.4	86.4	86.5	86.9	87.3	87.6
4	72.7	79.0	79.9	80.4	80.6	81.0	81.1	81.1	81.7	82.1	82.5
5	66.5	73.5	74.5	75.1	75.3	75.7	75.9	75.9	76.6	77.0	77.5
6	60.8	68.1	69.2	69.9	70.1	70.5	70.7	70.7	71.5	71.9	72.5
7	55.6	62.9	64.1	64.8	65.1	65.5	65.7	65.7	66.5	67.0	67.6
8	50.9	58.0	59.2	59.9	60.3	60.8	61.0	60.8	61.7	62.2	62.9
9	46.6	53.5	54.7	55.3	55.8	56.3	56.6	56.2	57.1	57.7	58.5
10	42.7	49.3	50.5	51.1	51.6	52.2	52.5	52.0	52.9	53.5	54.4
11	39.2	45.5	46.6	47.3	47.8	48.4	48.6	48.1	49.0	49.6	50.5
12	35.9	41.9	43.0	43.7	44.2	44.8	45.1	44.5	45.4	46.0	46.9
13	32.9	38.6	39.7	40.4	40.9	41.4	41.8	41.1	42.0	42.7	43.6
14	30.2	35.6	36.6	37.3	37.8	38.4	38.7	38.0	38.9	39.6	40.5
15	27.7	32.9	33.8	34.5	35.0	35.6	35.9	35.2	36.1	36.7	37.6
16	25.4	30.4	31.3	32.0	32.4	33.1	33.4	32.6	33.5	34.1	35.0
17	23.3	28.1	29.0	29.6	30.0	30.7	31.0	30.2	31.1	31.6	32.6
18	21.4	26.0	26.9	27.4	27.9	28.5	28.8	28.0	28.8	29.4	30.3
19	19.6	24.0	24.9	25.4	25.9	26.5	26.8	26.0	26.7	27.4	28.2
20	18.0	22.1	22.9	23.5	23.9	24.5	24.8	24.0	24.8	25.4	26.2

Table 3.1–9 (continued)
PERCENTAGE DEPTH DOSE[a] FOR RECTANGULAR FIELDS

Cobalt 60 – SSD 80 cm (continued)

Depth, cm	Open-ended applicator (cm × cm) 6 × 20	8 × 8	8 × 10	8 × 15	8 × 20	10 × 10	10 × 15	10 × 20	15 × 15	15 × 20	20 × 20
c	103.3	102.9	103.2	103.7	104.1	103.5	104.2	104.6	105.1	105.6	106.3
0	Surface dose 30–50% depending upon collimator										
0.5	100.0	100.0	100.0	100.0	100.0	100.0	100.0	100.0	100.0	100.0	100.0
1	97.8	97.8	98.0	98.1	98.1	98.2	98.3	98.3	98.4	98.4	98.4
2	92.8	92.7	93.0	93.2	93.3	93.3	93.6	93.6	93.9	93.9	94.0
3	87.7	87.6	87.9	88.3	88.5	88.3	88.8	88.9	89.3	89.4	89.6
4	82.7	82.5	82.9	83.4	83.6	83.4	84.0	84.2	84.7	84.9	85.2
5	77.7	77.4	77.9	78.5	78.8	78.5	79.2	79.5	80.1	80.4	80.8
6	72.7	72.4	73.0	73.7	74.0	73.6	74.4	74.7	75.4	75.8	76.4
7	67.9	67.5	68.1	68.9	69.2	68.8	69.8	70.1	70.8	71.4	72.1
8	63.3	62.7	63.4	64.3	64.7	64.1	65.2	65.7	66.5	67.2	68.0
9	58.9	58.2	58.9	59.9	60.4	59.7	60.9	61.4	62.3	63.1	64.0
10	54.8	54.0	54.8	55.8	56.3	55.6	56.9	57.4	58.4	59.2	60.2
11	51.0	50.1	50.9	52.0	52.5	51.7	53.1	53.7	54.7	55.6	56.6
12	47.4	46.5	47.3	48.4	49.0	48.1	49.5	50.2	51.2	52.1	53.2
13	44.1	43.2	44.0	45.1	45.7	44.8	46.2	46.9	47.9	48.8	50.0
14	41.0	40.1	40.9	42.0	42.6	41.8	43.1	43.9	44.9	45.8	47.0
15	38.1	37.2	38.0	39.2	39.8	38.9	40.3	41.0	42.0	43.0	44.2
16	35.5	34.5	35.3	36.5	37.1	36.2	37.6	38.3	39.3	40.3	41.5
17	33.1	32.1	32.8	34.0	34.6	33.7	35.1	35.8	36.8	37.8	39.0
18	30.8	29.8	30.5	31.7	32.3	31.4	32.8	33.5	34.5	35.5	36.7
19	28.7	27.7	28.4	29.6	30.2	29.2	30.7	31.4	32.3	33.4	34.6
20	26.8	25.7	26.4	27.6	28.2	27.2	28.6	29.4	30.3	31.4	32.6

Table 3.1–9 (continued)
PERCENTAGE DEPTH DOSE[a] FOR RECTANGULAR FIELDS

Cobalt 60– SSD 100 cm

	Open-ended applicator (cm × cm)										
Depth, cm	0 × 0	4 × 4	4 × 6	4 × 8	4 × 10	4 × 15	4 × 20	6 × 6	6 × 8	6 × 10	6 × 15
c	100.0	101.5	101.8	102.0	102.2	102.5	102.7	102.2	102.5	102.7	103.1
0	Surface dose 30–50% depending upon collimator										
0.5	100.0	100.0	100.0	100.0	100.0	100.0	100.0	100.0	100.0	100.0	100.0
1	95.9	97.1	97.3	97.5	97.6	97.7	97.7	97.7	97.9	98.0	98.2
2	87.9	91.4	91.9	92.2	92.4	92.5	92.6	92.6	92.6	93.1	93.4
3	80.7	85.8	86.5	86.9	87.2	87.3	87.5	87.5	87.9	88.2	88.6
4	73.8	80.2	81.2	81.7	82.0	82.2	82.4	82.4	83.0	83.4	83.8
5	67.8	74.8	76.0	76.6	76.9	77.2	77.4	77.3	78.1	78.6	79.0
6	62.3	69.7	70.9	71.6	71.9	72.3	72.5	72.4	73.2	73.8	74.3
7	57.3	64.8	66.0	66.7	67.1	67.5	67.7	67.6	68.4	69.0	69.6
8	52.7	60.1	61.3	62.0	62.4	62.9	63.1	62.9	63.8	64.4	65.1
9	48.5	55.7	56.9	57.6	58.0	58.5	58.8	58.4	59.4	60.0	60.7
10	44.7	51.5	52.7	53.4	53.8	54.4	54.7	54.2	55.2	55.8	56.6
11	41.2	47.7	48.8	49.5	49.9	50.5	50.8	50.3	51.3	51.9	52.7
12	38.0	44.1	45.2	45.9	46.3	46.9	47.2	46.7	47.7	48.2	49.1
13	35.0	40.8	41.9	42.6	43.0	43.6	43.9	43.3	44.3	44.9	45.8
14	32.2	37.8	38.9	39.5	40.0	40.6	40.9	40.2	41.2	41.8	42.7
15	29.6	35.0	36.1	36.7	37.2	37.8	38.1	37.4	38.3	38.9	39.9
16	27.2	32.5	33.5	34.1	34.5	35.2	35.5	34.8	35.6	36.3	37.2
17	25.0	30.1	31.1	31.7	32.1	32.8	33.1	32.3	33.1	33.8	34.7
18	23.0	27.9	28.8	29.4	29.8	30.5	30.8	30.0	30.8	31.5	32.4
19	21.2	25.8	26.7	27.3	27.7	28.4	28.7	27.9	28.7	29.3	30.2
20	19.5	23.8	24.7	25.3	25.7	26.4	26.7	25.9	26.7	27.3	28.2

Table 3.1–9 (continued)
PERCENTAGE DEPTH DOSE[a] FOR RECTANGULAR FIELDS

Cobalt 60 – SSD 100 cm (continued)

	Open-ended applicator (cm × cm)										
Depth, cm	6 × 20	8 × 8	8 × 10	8 × 15	8 × 20	10 × 10	10 × 15	10 × 20	15 × 15	15 × 20	20 × 20
c	103.3	102.9	103.2	103.7	104.1	103.5	104.2	104.6	105.1	105.6	106.3
0	Surface dose 30– 50% depending upon collimator										
0.5	100.0	100.0	100.0	100.0	100.0	100.0	100.0	100.0	100.0	100.0	100.0
1	98.2	98.1	98.3	98.5	98.5	98.6	98.8	98.8	99.0	98.9	98.9
2	93.4	93.3	93.6	93.9	93.9	93.9	94.3	94.3	94.6	94.6	94.7
3	88.6	88.5	88.9	89.3	89.3	89.3	89.8	89.8	90.2	90.3	90.5
4	83.9	83.7	84.2	84.7	84.8	84.7	85.3	85.4	85.9	86.1	86.3
5	79.2	78.9	79.6	80.1	80.3	80.1	80.8	81.0	81.6	81.9	82.2
6	74.5	74.2	74.9	75.6	75.8	75.5	76.3	76.6	77.3	77.7	78.1
7	69.9	69.5	70.2	71.0	71.3	70.9	71.8	72.2	73.0	73.5	74.0
8	65.4	64.9	65.6	66.5	66.9	66.4	67.4	67.9	68.7	69.3	70.0
9	61.1	60.5	61.2	62.1	62.6	62.0	63.1	63.7	64.5	65.2	66.1
10	57.0	56.3	57.0	58.0	58.6	57.3	59.0	59.7	60.6	61.3	62.3
11	53.2	52.4	53.1	54.2	54.8	53.9	55.2	55.9	56.9	57.7	58.7
12	49.6	48.7	49.5	50.7	51.2	50.3	51.7	52.4	53.4	54.3	55.3
13	46.3	45.4	46.1	47.3	47.9	47.0	48.4	49.1	50.2	51.1	52.1
14	43.2	42.3	43.0	44.2	44.8	43.9	45.3	46.0	47.1	48.1	49.1
15	40.3	39.4	40.1	41.3	41.9	41.0	42.4	43.1	44.2	45.2	46.2
16	37.7	36.7	37.4	38.6	39.2	38.3	39.7	40.4	41.5	42.5	43.5
17	35.2	34.2	34.9	36.1	36.7	35.8	37.2	37.9	39.0	40.0	41.0
18	32.9	31.9	32.6	33.8	34.4	33.5	34.9	35.6	36.7	37.6	38.6
19	30.7	29.7	30.5	31.6	32.3	31.3	32.7	33.4	34.5	35.4	36.4
20	28.7	27.7	28.5	29.6	30.2	29.3	30.6	31.3	32.4	33.3	34.4

Table 3.1–9 (continued)
PERCENTAGE DEPTH DOSE[a] FOR RECTANGULAR FIELDS

4 MV Linear Accelerator– SSD 100 cm

Depth, cm	Open-ended applicator (cm × cm) 4 × 4	5 × 5	6 × 6	7 × 7	8 ×8	10 × 10	12 × 12	15 × 15	20 × 20
1	100.0	100.0	100.0	100.0	100.0	100.0	100.0	100.0	100.0
2	96.5	96.8	97.2	97.4	97.5	97.6	97.7	97.9	98.2
3	91.3	92.0	92.5	92.9	93.0	93.3	93.5	93.7	94.3
4	85.9	86.8	87.4	87.8	88.2	88.7	89.2	89.7	90.5
5	80.4	81.3	82.2	82.9	83.4	84.0	84.4	84.9	85.7
6	74.9	76.1	77.0	77.9	78.6	79.5	80.1	80.7	81.5
7	69.9	71.3	72.2	73.2	74.0	75.0	75.8	76.4	77.3
8	65.6	66.9	68.0	69.1	69.8	71.0	71.7	72.3	73.4
9	61.2	62.8	63.7	64.7	65.5	66.6	67.7	68.5	69.8
10	57.2	58.5	59.8	60.8	61.7	62.7	63.7	64.7	66.2
11	53.4	54.9	56.0	57.0	57.8	59.0	60.0	61.4	62.7
12	49.8	51.1	52.3	53.4	54.3	55.7	56.8	58.1	59.5
13	46.6	47.8	49.0	50.0	51.0	52.4	53.5	54.9	56.3
14	43.5	44.7	45.8	46.9	47.8	49.3	50.3	51.7	53.4
15	40.7	41.9	43.0	44.0	45.0	46.4	47.5	48.8	50.3
16	37.9	39.2	40.3	41.3	42.1	43.6	44.7	46.0	47.8
17	35.4	36.6	37.8	38.7	39.5	40.9	42.0	43.4	45.0
18	33.0	34.2	35.2	36.2	37.0	38.5	39.6	40.9	42.6
19	30.9	32.0	32.9	34.0	34.8	36.0	37.2	38.5	40.1
20	28.9	29.9	30.8	31.8	32.7	34.0	35.1	36.3	38.1
22	25.3	26.2	27.1	27.9	28.7	30.0	31.0	32.4	34.2
24	22.1	23.0	23.9	24.6	25.4	26.5	27.6	28.9	30.6
26	(19.4)	(20.2)	(20.9)	(21.6)	(22.3)	(23.5)	(24.5)	(25.8)	(27.4)
28	(16.9)	(17.7)	(18.3)	(19.0)	(19.7)	(20.7)	(21.7)	(22.9)	(24.6)
30	(14.9)	(15.5)	(16.1)	(16.8)	(17.4)	(18.4)	(19.4)	(20.4)	(21.9)

Table 3.1–9 (continued)
PERCENTAGE DEPTH DOSE[a] FOR RECTANGULAR FIELDS

6 MV Linear Accelerator – SSD 100 cm

	Open-ended applicator (cm × cm)								
Depth, cm	4 × 4	5 × 5	6 × 6	7 × 7	8 × 8	10 × 10	12 × 12	15 × 15	20 × 20
1.5	100	100	100	100	100	100	100	100	100
2.0	99.3	99.3	99.3	99.2	99.2	99.2	99.1	99.0	99.0
3	94.6	94.8	95.0	95.0	95.1	95.2	95.2	95.2	95.2
4	89.6	90.1	90.4	90.6	90.9	91.1	91.2	91.3	91.5
5	84.8	85.3	85.8	86.2	86.5	86.9	87.1	87.4	87.8
6	79.8	80.6	81.2	81.8	82.3	83.0	83.3	83.6	84.1
7	75.2	76.0	76.7	77.3	77.9	78.8	79.4	79.8	80.5
8	71.0	71.8	72.6	73.3	73.9	74.9	75.6	76.3	77.0
9	66.8	67.7	68.6	69.3	70.0	71.2	72.1	73.0	73.8
10	62.8	63.8	64.6	65.5	66.3	67.5	68.4	69.2	70.2
11	59.2	60.2	61.2	62.1	62.8	64.0	65.0	65.9	67.0
12	55.8	56.7	57.7	58.6	59.4	60.6	61.8	62.7	63.9
13	52.5	53.5	54.4	55.3	56.1	57.4	58.5	59.7	61.0
14	49.5	50.5	51.4	52.3	53.1	54.4	55.5	56.7	58.1
15	46.6	47.6	48.6	49.4	50.2	51.6	52.8	54.0	55.4
16	43.9	44.9	45.8	46.8	47.6	49.0	50.0	51.3	52.8
17	41.4	42.4	43.4	44.2	45.0	46.3	47.4	48.8	50.2
18	39.0	40.0	40.9	41.8	42.6	44.0	45.1	46.2	47.9
19	36.7	37.6	38.6	39.4	40.2	41.6	42.8	44.1	45.7
20	34.7	35.6	36.4	37.3	38.1	39.5	40.6	41.8	43.6
22	30.8	31.7	32.5	33.3	34.0	35.4	36.6	37.8	39.5
24	27.4	28.2	29.0	29.8	30.4	31.7	32.8	34.2	35.6
26	24.3	25.2	25.8	26.6	27.2	28.4	29.4	30.6	32.2
28	21.6	22.4	23.1	23.8	24.4	25.6	26.6	27.8	29.2
30	19.3	20.0	20.7	21.4	22.0	23.0	23.9	25.1	26.6

Table 3.1–9 (continued)
PERCENTAGE DEPTH DOSE[a] FOR RECTANGULAR FIELDS

[a]The "percentage depth dose" (%D.D.) is a ratio for fixed source skin distance (SSD) of the dose at a depth in a medium to the dose at depth of maximum electron build-up, for a fixed field size at the surface. It depends on SSD and takes into account measured distance as well as material attenuation.
[b]The first line gives the dose to a small mass of tissue at the surface for 100 rad of primary radiation to the same mass of tissue. It is also 100 times the backscatter factor.
[c]The first line gives the dose at the maximum (0.5 cm) for 100 rad at the same point in free space. This entry divided by 100 is the backscatter factor.

From Johns, H. E. and Cunningham, J. R., *The Physics of Radiology*, 3rd ed., 1971. Courtesy of Charles C Thomas, Publisher, Springfield, Ill.

Table 3.1–10
PERCENTAGE DEPTH DOSE DATA FOR BETATRON RADIATION

15 MV, SSD = 100 cm

Depth, cm	Area of field, cm^2: 25	100	200–400
0.5	61.5	64.5	68.0
1	81.0	82.0	86.5
2	98.5	98.0	98.5
3	100.0	100.0	100.0
4	98.0	98.0	97.5
5	94.5	94.5	94.0
6	90.5	91.0	90.5
7	86.5	87.0	87.5
8	83.0	84.0	84.0
10	76.0	77.0	77.5
12	69.0	71.0	71.5
14	62.5	64.5	65.5
16	57.0	59.0	60.0
18	52.0	54.0	55.5
20	47.5	50.0	51.0

22 MV, SSD = 100 cm

Depth, cm	Field sizes larger than 50 cm^2
0.5	50.0
1	70.0
2	90.1
3	98.0
4	100.0
5	99.5
6	96.6
7	93.0
8	89.1
10	81.9
12	75.5
14	69.9
16	64.2
18	59.1
20	54.5

31 MV, SSD = 100 cm

Depth, cm	Field sizes larger than 50 cm^2
2	83.5
3	94.0
4	99.5
5	100.0
6	99.0
7	96.0
8	93.0
10	88.0
12	82.0
14	76.5
16	71.5
18	66.0
20	62.0

From Johns, H. E. and Cunningham, J. R., *The Physics of Radiology*, 3rd ed., 1971, 767. Courtesy of Charles C Thomas, Publisher, Springfield, Ill.

Table 3.1–11
DEPTH DOSE DATA FOR HIGH-ENERGY ELECTRONS

15 MeV Electrons

Percentage depth dose	Field, cm × cm			
	2 × 2	4 × 4	6 × 6	10 × 10
100	1.0	1.6	1.7	1.8
95	1.9	2.7	2.9	3.1
90	2.3	3.5	3.7	4.0
80	2.9	4.0	4.3	4.6
70	3.4	4.5	4.9	5.2
60	4.0	4.9	5.3	5.5
50	4.5	5.4	5.7	6.0
40	4.9	5.8	6.1	6.4
30	5.4	6.2	6.4	6.7
20	5.9	6.5	6.9	7.1
10	6.3	7.3	7.4	7.5

20 MeV Electrons

Percentage depth dose	Field, cm × cm			
	4 × 4	6 × 6	10 × 10	15 × 15
100	1.8	1.9	2.1	2.1
95	3.1	3.5	4.0	4.0
90	4.0	4.4	4.9	5.0
80	4.8	5.5	5.8	5.9
70	5.4	6.2	6.6	6.7
60	6.1	6.9	7.2	7.3
50	6.7	7.4	7.8	7.9
40	7.3	8.0	8.3	8.4
30	7.8	8.4	8.7	8.8
20	8.3	8.7	9.0	9.0
10	9.7	9.8	9.8	9.8

30 MeV Electrons

Percentage depth dose	Field, cm × cm			
	4 × 4	6 × 6	10 × 10	15 × 15
100	2.3	2.4	2.6	2.7
95	3.8	4.5	5.5	5.7
90	5.1	5.9	6.3	7.0
80	6.3	7.5	8.3	8.7
70	7.5	8.7	9.5	9.9
60	8.5	9.8	10.5	11.0
50	9.4	10.8	11.3	11.8
40	10.4	11.6	12.2	12.6
30	11.3	12.5	13.0	13.4
20	12.5	13.3	13.7	14.1
10	14.2	14.5	14.9	15.0

From Johns, H. E. and Cunningham, J. R., *The Physics of Radiology*, 3rd ed., 1971, 767. Courtesy of Charles C Thomas, Publisher, Springfield, Ill.

Table 3.1–12
4 MV (CLINAC®) X-RAY CENTRAL AXIS PERCENTAGE DEPTH DOSES – 80-cm TSD (LEAD FLATNESS FILTER)

	Field size, cm									
Depth, cm	3 × 3	4 × 4	5 × 5	6 × 6	8 × 8	10 × 10	12 × 12	15 × 15	20 × 20	25 × 25	30 × 30
1.2	100.0	100.0	100.0	100.0	100.0	100.0	100.0	100.0	100.0	100.0	100.0
2	95.8	96.3	96.6	96.7	96.7	96.8	96.8	96.8	96.8	96.9	97.0
3	89.4	90.3	90.9	91.2	91.6	92.2	92.4	92.6	93.0	93.2	93.5
4	83.1	84.3	85.3	86.0	86.5	87.4	87.8	88.1	88.6	88.9	89.3
5	77.2	78.8	79.9	80.6	81.6	82.4	83.0	83.7	84.3	84.8	85.3
6	71.5	73.3	74.5	75.5	76.8	77.9	78.8	79.5	80.3	80.9	81.6
7	66.4	68.2	69.7	70.7	72.1	73.8	74.5	75.2	76.2	77.0	77.8
8	61.7	63.2	64.7	65.8	67.6	68.8	70.0	71.2	72.2	73.0	73.9
9	57.1	58.7	60.1	61.3	63.0	64.6	66.0	67.2	68.4	69.2	70.1
10	53.0	54.6	56.0	57.2	59.1	60.8	62.2	63.6	64.8	65.7	66.8
11	49.3	50.8	52.2	53.3	55.1	57.0	58.6	59.9	61.2	62.4	63.3
12	45.7	47.0	48.3	49.8	51.7	53.4	55.0	56.2	58.0	59.0	60.1
13	42.4	43.7	45.0	46.3	48.2	50.0	51.8	52.9	54.8	56.0	57.2
14	39.3	40.5	41.7	43.1	45.1	46.7	48.5	49.9	52.0	53.0	54.2
15	36.5	37.8	38.9	40.1	42.1	43.9	45.5	47.0	48.9	50.2	51.2
16	33.9	35.1	36.2	37.4	39.3	41.0	42.7	44.2	46.0	47.3	48.5
17	31.4	32.6	33.7	34.9	36.7	38.3	40.0	41.6	43.3	44.5	45.7
18	29.3	30.4	31.4	32.5	34.3	35.7	37.5	39.0	40.9	42.1	43.3
19	27.2	28.2	29.2	30.3	32.0	33.5	35.0	36.8	38.4	39.7	41.0
20	25.2	26.2	27.2	28.3	30.0	31.4	32.7	34.4	36.3	37.8	38.8
21	23.4	24.3	25.3	26.3	28.0	29.3	30.8	32.4	34.1	35.4	36.4
22	21.7	22.6	23.5	24.5	26.3	27.4	28.7	30.3	32.1	33.3	34.4
23	20.2	21.1	22.0	22.9	24.5	25.7	26.9	28.5	30.2	31.2	32.4
24	18.8	19.6	20.4	21.4	22.9	24.1	25.4	26.7	28.5	29.5	30.5
25	17.4	18.2	19.1	19.8	21.4	22.6	23.8	25.0	26.8	27.8	28.8
26	16.2	16.9	17.8	18.6	20.0	21.1	22.2	23.5	25.3	26.2	27.1
27	14.9	15.7	16.5	17.3	18.6	19.7	20.9	22.0	23.8	24.7	25.6
28	13.8	14.6	15.4	16.2	17.4	18.6	19.6	20.7	22.4	23.2	24.1
29	12.9	13.5	14.4	15.0	16.3	17.3	18.3	19.4	21.2	21.9	22.8
30	11.9	12.6	13.4	14.0	15.2	16.2	17.2	18.2	19.9	20.6	21.4

From Peterson, M. and Golden, R., *Radiology*, 103(3), 675, 1972. With permission.

Table 3.1–13
TISSUE-AIR RATIOS[a]

4 MV, Lead Flatness Filter

Depth, cm	Field size, cm: 0	3 × 3	4 × 4	5 × 5	6 × 6	8 × 8	10 × 10	12 × 12	15 × 15	20 × 20	25 × 25	30 × 30
1.2	1.000	1.011	1.015	1.018	1.022	1.030	1.037	1.044	1.055	1.066	1.071	1.074
2	.952	.982	.993	.998	1.003	1.012	1.021	1.030	1.042	1.054	1.060	1.064
3	.899	.940	.954	.962	.971	.985	.998	1.007	1.019	1.033	1.040	1.044
4	.847	.898	.910	.923	.933	.951	.966	.979	.995	1.008	1.017	1.026
5	.799	.850	.867	.883	.897	.917	.933	.949	.967	.984	.995	1.004
6	.753	.810	.827	.842	.856	.881	.900	918	.937	.955	.968	.979
7	.710	.767	.785	.801	.819	.845	.866	.884	.908	.928	.941	.954
8	.670	.727	.746	.764	.782	.809	.832	.852	.875	.900	.916	.929
9	.632	.689	.708	.727	.744	.775	.800	.818	.844	.873	.889	.903
10	.598	.656	.671	.690	.710	.740	.765	.786	.812	.843	.862	.876
11	.564	.620	.637	.655	.673	.706	.731	.752	.780	.813	.834	.850
12	.529	.588	.601	.620	.641	.673	.698	.720	.749	.786	.808	.824
13	.500	.555	.570	.590	.608	.639	.666	.690	.718	.757	.780	.796
14	.471	.525	.542	.560	.577	.609	.636	.661	.689	.728	.752	.768
15	.444	.495	.512	.529	.547	.578	.605	.630	.660	.701	.725	.743
16	.419	.471	.487	.502	.519	.549	.576	.601	.631	.673	.698	.717
17	.395	.445	.461	.476	.493	.521	.549	.574	.605	.645	.672	.693
18	.373	.421	.436	.451	.465	.494	.521	.545	.576	.618	.646	.669
19	.351	.399	.412	.428	.444	.471	.496	.520	.550	.593	.620	.644
20	.331	.376	.392	.406	.421	.448	.473	.496	.527	.568	.597	.620
21	.312	.356	.371	.385	.399	.424	.450	.473	.503	.543	.573	.595
22	.294	.337	.351	.364	.378	.404	.428	.450	.481	.521	.549	.572
23	.277	.319	.332	.346	.359	.383	.408	.429	.459	.499	.525	.549
24	.262	.302	.315	.328	.340	.365	.383	.408	.438	.477	.504	.528
25	.246	.285	.298	.310	.323	.346	.368	.389	.419	.455	.482	.505

[a]These are typical values measured in water for the Varian CLINAC-4® at Rosewood General Hospital, Houston, Texas.

From Peterson, M. and Golden, R., unpublished data, 1971.

Table 3.1–14
MASS ABSORPTION COEFFICIENTS FOR AIR

Energy Fluence per Roentgen, Photon Fluence per Roentgen, and Exposure per Millicurie as a Function of Photon Energy

	Photon energy, MeV	Mass coefficients for air, cm^2/g — Attenuation $\frac{\mu'}{\rho}$	Attenuation $\frac{\mu}{\rho}$	Energy transfer, $\frac{\mu k}{\rho}$	Energy absorption, $\frac{\mu en}{\rho}$	Energy fluence per roentgen, erg/cm^2-R	Photon fluence per roentgen, N/cm^2-R	Specific gamma ray constant (roentgen per hour per millicurie at 1 cm[a])
	.010	4.82	5.04	4.61	4.61	18.8	11.8×10^8	9.00
	.015	1.45	1.56	1.27	1.27	68.4	28.5	3.72
	.02	.691	.758	.511	.511	170	53.1	2.00
	.03	.318	.350	.148	.148	587	122	.866
	.04	.229	.248	.0668	.0668	1301	203	.521
	.05	.196	.206	.0406	.0406	2140	267	.396
	.06	.179	.187	.0305	.0305	2849	297	.357
	.08	.162	.167	.0243	.0243	3576	279	.379
	.10	.151	.155	.0234	.0234	3714	232	.457
	.15	.134	.136	.0250	.0250	3476	145	.732
	.2	.123	.124	.0268	.0268	3243	101	1.05
	.3	.106	.107	.0287	.0287	3028	63.1	1.68
	.4	.0954	.0954	.0295	.0295	2946	46.0	2.30
	.5	.0868		.0298	.0296	2936	36.7	2.89
	.6	.0804		.0296	.0295	2946	30.7	3.45
^{137}Cs	.661[b]	.0772		.0294	.0294	2956	27.9	
	.80	.0706		.0289	.0289	3007	23.5	4.51
	1.0	.0635		.0280	.0278	3126	19.5	5.43
^{60}Co	1.25[b]	.0572		.0268	.0266	3267	16.3	
	1.5	.0517		.0256	.0254	3421	14.3	7.44
	2	.0444		.0236	.0234	3714	11.6	9.13
	3	.0358		.0207	.0205	4239	8.83	12.0
	4	.0308		.0189	.0186	4672	7.30	14.5
	5	.0276		.0178	.0174	4994	6.24	17.0
	6	.0252		.0168	.0164	5299	5.52	19.2
	8	.0223		.0157	.0152	5717	4.47	23.7
	10	.0204		.0151	.0145	5993	3.75	28.3

Table 3.1–14 (continued)
MASS ABSORPTION COEFFICIENTS FOR AIR

Energy Fluence per Roentgen, Photon Fluence per Roentgen, and Exposure per Millicurie as a Function of Photon Energy (continued)

Equations:

$$\text{Energy fluence per roentgen} = \frac{86.9}{(\mu_{en}/\rho)_{air}} \cdot \frac{\text{ergs}}{\text{cm}^2\ \text{R}}$$

$$\text{Photon fluence per roentgen} = \frac{86.9}{(\mu_{en}/\rho)_{air}} \cdot \frac{1}{1.6 \times 10^{-6}} \cdot \frac{1}{h\nu\ (\text{in MeV})} \frac{\text{photons}}{\text{cm}^2\ \text{R}}$$

$$r = \frac{3.700 \times 10^7 \times 3600 \times (\mu_{en}/\rho)_{air}}{4\pi} \cdot \frac{h\nu\ (\text{in MeV})}{86.9} \times 1.6 \times 10^{-6} \frac{\text{R cm}^2}{\text{hr} \cdot \text{mCi}}$$

$$W_{air} = 33.7\ \text{eV/ion pair}$$

[a]Calculated on the basis of 3.700×10^7 gamma rays emitted per sec/mCi.
[b]To calculate the value for ^{137}Cs or ^{60}Co, one also requires information concerning the disintegration scheme.

From Johns, H. E. and Cunningham, J. R., *The Physics of Radiology,* 3rd ed., 1971, 735. Courtesy of Charles C Thomas, Publisher, Springfield, Ill.

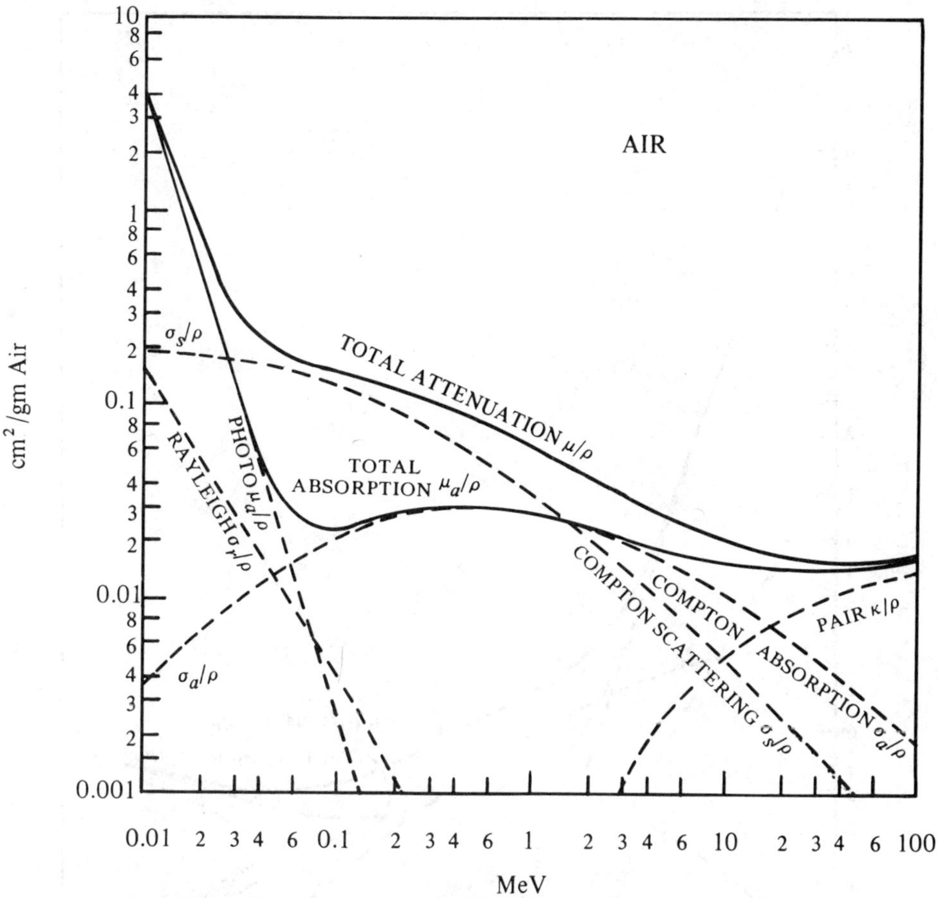

FIGURE 3.1–4. Mass attenuation coefficients for photons in "air" taken as 78.04 vol % nitrogen, 21.02 vol % oxygen, and 0.94 vol % argon. At 0°C and 760 mm Hg pressure, the density of air is ρ = 0.001293 g/cm³. (From Evans, R. D., in *Radiation Dosimetry*, 3rd ed., Attix, F. H. and Roesch, W. C., Eds., Academic Press, New York, 1968. With permission.)

(continued from page 229).

measurements or by making use of the cobalt-60 gamma ray, or 2 MV x ray, exposure calibration. The latter alternative involves an additional factor appropriate to the energy of the radiation being used. This procedure, which is described in more detail in subsequent sections, represents the best method available at the present time and, in the interests of consistency, should be generally adopted until cavity chambers can also be calibrated for higher energy photons.

Despite these differences in *principle* between measurements of high energy and medium energy radiations, in practice a single *technique* is applicable to nearly the whole range of radiation qualities. The only exception occurs with the relatively low voltage x rays which are used for superficial therapy.

The Working Instrument

Three main features of an ionization chamber have to be considered, namely its size, the materials used in its construction, and the thickness of its walls. Since the aim is to measure the exposure at a point, the ionization chamber must be small. It is satisfactory if the internal diameter is about 5 mm and the length about 15 mm. Dimensions twice as great as these should never be exceeded.

It is necessary that the chamber response should be as independent as possible of radiation energy. The chamber should be constructed of suitable materials so that it is effectively "air-equivalent", i.e., its response (scale divisions per roentgen) to a given exposure should not vary with energy by more than 5%, at least for x rays generated between 100 and 300 kV. Strict "air-equivalence" is difficult to achieve with lower energies, but no chamber should be used for such purposes that cannot satisfy the stated criterion in the medium energy range. If it is satisfactory in that range, it can generally be relied on also to have satisfactory characteristics for higher energy radiations, though this should not be taken to imply that the medium energy calibration factor can be extrapolated directly to higher energies.

For any particular photon radiation, an ionization chamber should be used for the measurement of exposure only if its wall thickness is such that there is negligible contribution to the ionization within the chamber from secondary electrons produced outside. The requisite thickness increases with photon energy. It is about 50

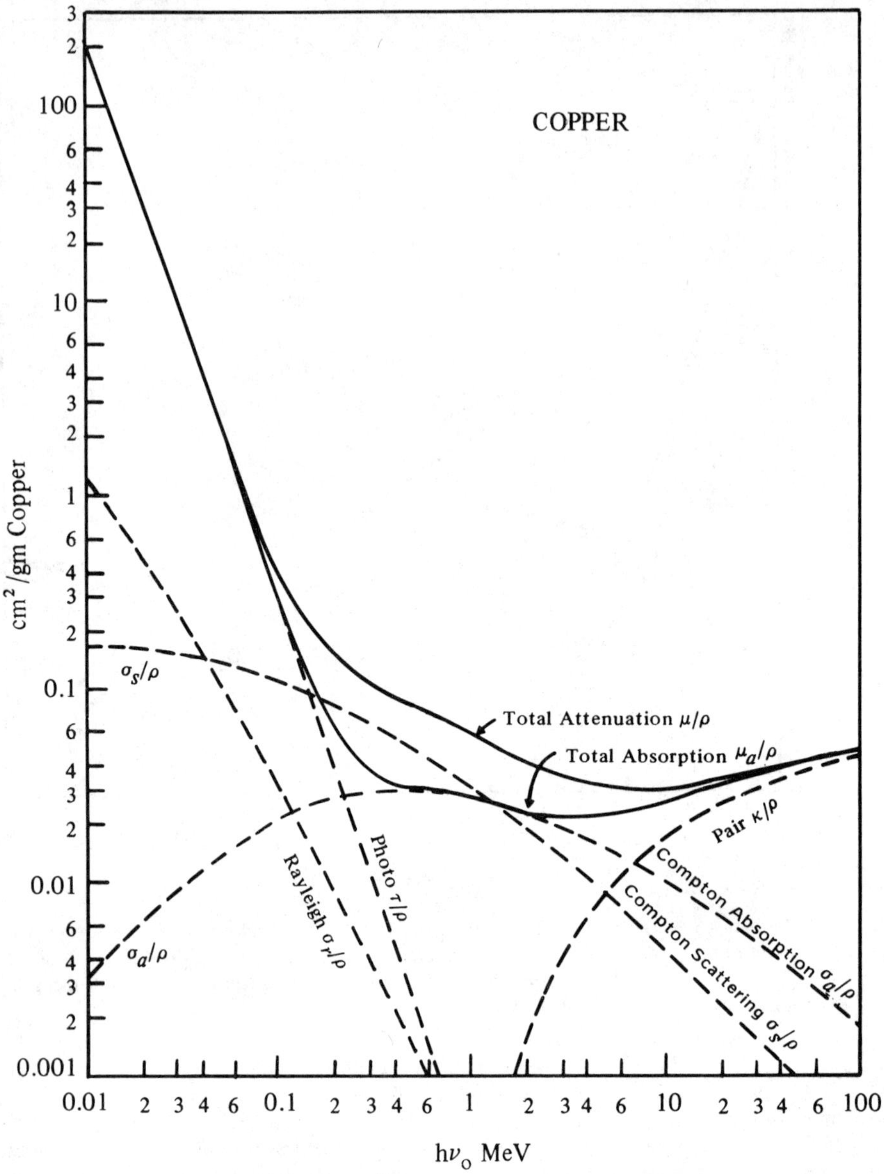

FIGURE 3.1–5. Mass attenuation coefficients for photons in copper (Z = 29). The dashed branch on the μ_a/ρ curve shows the effect of excluding the annihilation photons. The corresponding linear coefficients for copper may be obtained by multiplying all curves by ρ = 8.92 g/cm³ Cu. (From Evans, R. D., in *Radiation Dosimetry,* 3rd ed., Attix, F. H. and Roesch, W. C., Eds., Academic Press, New York, 1968. With permission.)

mg/cm² (0.5 mm of material of density 1 g cm^{-3}) for x rays generated at 300 kV. Since, for reasons of strength, most clinical dosimeters have chamber walls of about this thickness, they are directly applicable up to this generating potential. For higher energy radiations, thicker walls are needed and it is usual to supplement the chamber wall with close-fitting caps of Perspex (Lucite or Plexiglas). An extra thickness of 4–5 mm of this material is needed for measuring cobalt-60 gamma rays or 2 MV x rays.

A number of modern commercial instruments fulfill the desired conditions very satisfactorily, and in what follows, it is assumed that one of these is being used and that it has a thimble chamber whose cavity diameter is less that 8 mm and whose length is about 1.5 cm. It is also assumed that the instrument satisfies some further criteria. The first of these is that the chamber does not have a metal stem which produces marked attenuation of the scattered radiation when used in a phantom. The second is that it does not exhibit the stem leakage effects of some earlier instruments. Thirdly, there must be negligible ion recombination in the chamber.

A test to determine whether there is any stem leakage is quite easy to perform, whilst the ion collection efficiency can be checked by making comparative

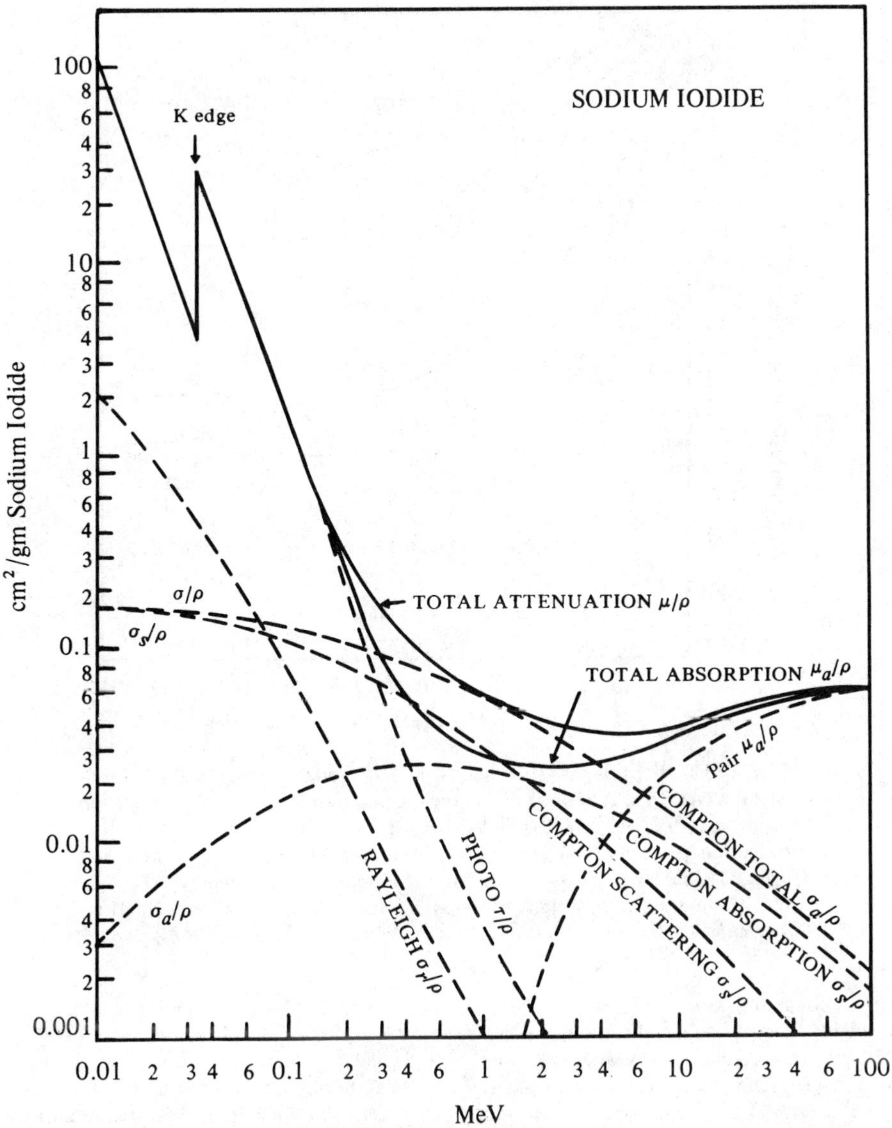

FIGURE 3.1–6. Mass attenuation coefficients for photons in pure NaI. The "Compton total" attenuation coefficient $(\sigma/\rho) = (\sigma_a/\rho) + (\sigma_s/\rho)$ is shown explicitly, because of its usefulness in predicting the behavior of NaI(Tl) scintillators. The 0.1 to 0.2% thallium activator in NaI(Tl) scintillators is ignored here. The dashed branch on the μ_a/ρ curve shows the effect of excluding annihilation photons. Linear attenuation coefficients for NaI may be obtained using $\rho = 3.67$ g/cm³ NaI. (From Evans, R. D., in *Radiation Dosimetry*, 3rd ed., Attix, F. H. and Roesch, W. C., Eds., Academic Press, New York, 1968. With permission.)

measurements with different collecting potentials. However, it is strongly recommended that the manufacturers of dosimeters should specify the exposure rate at which the loss of charge due to recombination becomes 1%. No correction for the effect is likely to be needed unless the mean exposure rate exceeds 200 R/min, but special care is needed with pulsed radiation emission and especially when the pulses are very short (less than a few microseconds) so that the instantaneous exposure rates may be very high. About 1 R per pulse of 1 microsecond duration can be recorded in chambers of the dimensions given above with an error due to recombination of about 5%, which can be reduced if a higher than usual collecting potential is applied.

Before an instrument is used it should be calibrated against a national standard at a number of appropriate radiation qualities. If direct access to a national standard is not possible, the instrument may, alternatively, be compared with another instrument which has been calibrated against a national standard.

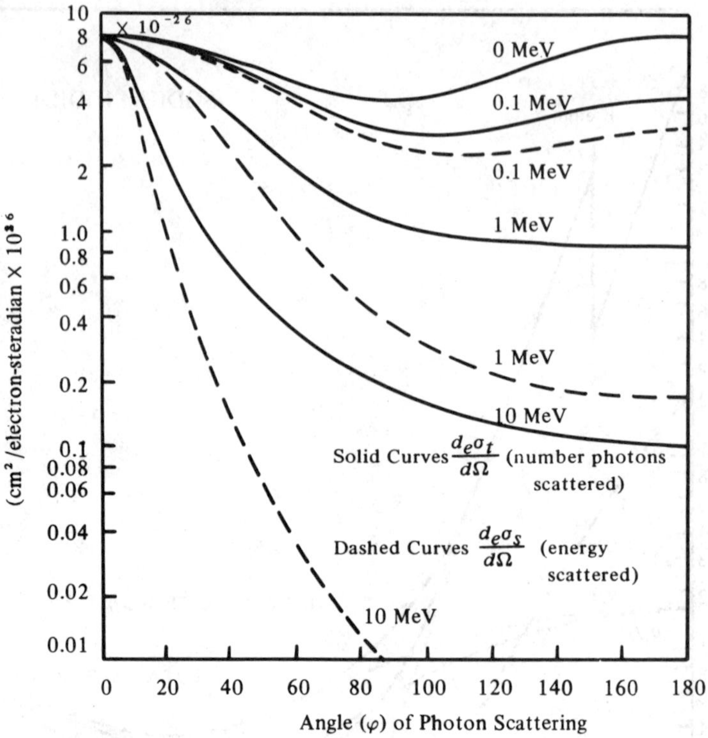

FIGURE 3.1–7. Graphs showing $d_e\sigma_t/d\Omega$ (solid curves) and $d_e\sigma_s/d\Omega$ (dashed curves) as a function of the angle of photon scattering, φ, for photon energies of 0, 0.1, 1.0, and 10 MeV. $d_e\sigma_t/d\Omega$ is the differential cross section per electron per unit solid angle for the number of photons scattered, and $d_e\sigma_s/d\Omega$ is the same cross section for the fraction of the energy scattered. (From Klein-Nishina equations as given in Hine, G. J. and Brownell, G. L., Eds., *Radiation Dosimetry,* 1st ed., Academic Press, New York, 1956.)

The Technique of Measurement

Since it is highly desirable that different workers should use the same measurement technique, it is very convenient that a single technique is applicable to nearly the whole range of x-ray and gamma-ray energies. Despite the differences in fundamental methods mentioned earlier, the technique to be described is applicable to a very large proportion of the radiations in regular use for radiotherapy. X rays for superficial therapy form a minor exception which is dealt with in a subsequent section.

X Rays Generated at Potentials Above 150 kV and High Energy Gamma Rays

This category includes radiations which are used when the region of chief clinical interest lies several centimeters below the skin. For this reason, and for others that have been extensively discussed in other publications, it is recommended that the calibration measurement be carried out with an ionization chamber positioned on the central axis of the beam, at a depth, *d*, below the surface of a water phantom. The values of *d* recommended for various radiation qualities are given in Table A. The chamber should be protected from the water by enclosure in a water-tight Perspex (Lucite or Plexiglas) tube. Figure A shows the general design of the phantom with the protective tube. The constructional material is Perspex, Lucite, Plexiglas or similar material and the phantom should have a cross-sectional area of 30 cm × 30 cm and be 20 cm deep. A smaller phantom may suffice if only small beams are under study.

The absorbed dose, expressed in rads, can be calculated from the equation

$$D = R \cdot k_1 \cdot k_2 \cdot N \cdot F$$

where D is the absorbed dose at depth d in the undisturbed water phantom with the chamber removed, and the meanings of the other symbols are as follows:

R is the instrument reading;

k_1 is a factor to correct for any difference in temperature and pressure at the time of measurement from those prevailing when the instrument was calibrated;

k_2 is a factor to correct for differences, such as quality, between the radiation field used for calibration and that being used;

N is the calibration factor, determined by the standardizing laboratory at a stated quality of radiation, and under stated conditions of temperature and pressure, for the conversion of the instrument reading into a statement of exposure, expressed in roentgens;

F is a composite coefficient relating the exposure in

Table A
RECOMMENDED VALUES OF THE CALIBRATION DEPTH (*d*) IN WATER

Radiation	*d*/cm
150 keV–10 MeV X-rays	5
Cesium-137, cobalt-60 gamma rays	5
11–25 MeV X-rays	7
26–50 MeV X-rays	10

From ICRU Report 23, Measurement of Absorbed Dose in a Phantom Irradiated by a Single Beam of X or Gamma Rays, International Commission on Radiation Units and Measurements, Washington, D.C., January 15, 1973, 4. With permission.

Table B
THE CONVERSION COEFFICIENT, *F*

Primary beam quality (HVL, MV, or nuclide)	F/rad · R^{-1}
0.5 mm Al	0.89
1 mm Al	0.88
2 mm Al	0.87
4 mm Al	0.87
6 mm Al	0.88
8 mm Al	0.89
0.5 mm Cu	0.89
1.0 mm Cu	0.91
1.5 mm Cu	0.93
2.0 mm Cu	0.94
3.0 mm Cu	0.95
4.0 mm Cu	0.96
^{137}Cs, ^{60}Co	0.95
2 MV	0.95
4 MV	0.94
6 MV	0.94
8 MV	0.93
10 MV	0.93
12 MV	0.92
14 MV	0.92
16 MV	0.91
18 MV	0.91
20 MV	0.90
25 MV	0.90
30 MV	0.89
35 MV	0.88

From ICRU Report 23, Measurement of Absorbed Dose in a Phantom Irradiated by a Single Beam of X or Gamma Rays, International Commission on Radiation Units and Measurements, Washington, D.C., January 15, 1973, 4. With permission.

roentgens to the absorbed dose in water expressed in rads. It incorporates a "displacement correction" and its precise significance is different when applied to medium energy radiations and when applied to high energy radiations. The product ($N \cdot F$) may be regarded as the absorbed dose calibration coefficient of the instrument for the specified measurement conditions. The value of F depends on the radiation quality (Table B).

X Rays Generated at Potentials Between 40 and 150 kV

The limits of this radiation category are not intended to be rigid. The essential feature is that this section applies to radiations of low penetrating power, and treatments in which the absorbed doses of interest are in or near the surface. Under these circumstances it is recommended that the calibration measurement should be made with the chamber positioned free in air on the central ray of the beam, as close as possible to the eventual position of the treated surface. The surface absorbed dose, D, in a water phantom, is then related to the ionization chamber reading R by

$$D = R \cdot k_1 \cdot k_2 \cdot N \cdot F \cdot \left(\frac{s + x}{s}\right)^2 \cdot B$$

where:

k_1, k_2, N, and F have the same meanings as above;

s is the source-surface distance used in treatments;

x is the distance between the locations of the surface and of the chamber center, e.g., the distance between the end of the applicator (treatment cone) and the chamber center. The sign of x is positive when the chamber is farther from the source. If no applicator is used, it is desirable to center the chamber at the treatment distance, in which case $x = 0$;

B is the back scatter factor appropriate to the field size and radiation quality.

Commentary on Numerical Values of the Factors Involved in the Determination of the Absorbed-dose Rate

The Temperature and Pressure Correction Factor, k_1

If the ionization chamber is unsealed, the value of the correction factor, N, given by the standardizing laboratory applies only to measurements made at a specific temperature (t_0), usually 20, 22 or 25°C, though sometimes 0°C, and pressure (p_0), usually 760 mm Hg pressure. Frequently measurements are carried out with the air ambient temperature (t) and pressure (p) different from those specified, and allowance must be made for this by use of the factor which is given by:

$$k_1 = \frac{273 + t}{273 + t_0} \cdot \frac{p_0}{p}$$

The temperature is that of the air in the ionization

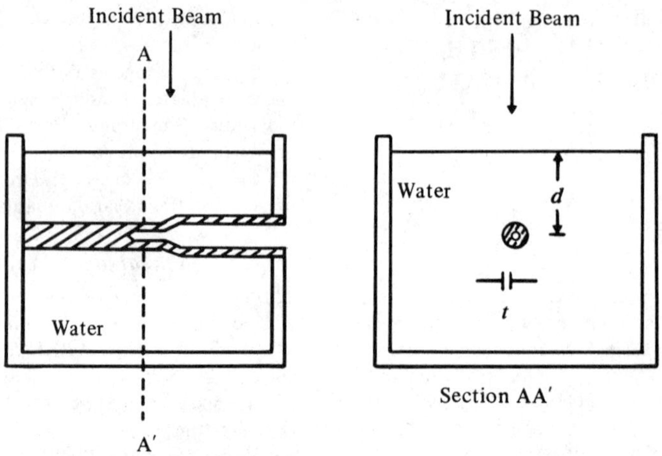

FIGURE A. Phantom with protective tube for measurements at a depth in water. Phantom dimensions: perpendicular to beam 30 × 30 cm, depth 20 cm. Alternatively, the phantom should extend laterally to leave a margin of at least 5 cm around the beam. The wall thickness, *t*, of the Perspex® sleeve should be approximately 2 mm, but it is not critical. (From Measurement of Absorbed Dose in a Phantom Irradiated by a Single Beam of X or Gamma Rays, ICRU Report 23, International Commission on Radiation Units and Measurements, Washington, D.C., January 15, 1973, 4. With permission.)

chamber, and this is only the same as the room temperature – which is the temperature usually measured – if adequate time is allowed for the chamber, and any phantom in which it is to be used, to come to room temperature.

The Exposure Calibration Factor, N

For radiations generated at potentials up to 400 kV, the value of *N* is found by calibrating the ionization chamber against a standard free-air chamber. *N* is the number by which the instrument reading (corrected as above for the ambient temperature and pressure) must be multiplied to yield the exposure in roentgens. Since, especially for low energy x rays, the value of *N* changes with changes of quality, the half value layer of the beam must be determined with appropriate accuracy.

For higher energy radiations, the only exposure calibration facilities available at standards laboratories are for cobalt-60 gamma rays or 2 MV x rays. Consequently, the value of *N* for gamma ray beams from cesium-137 and cobalt-60, and for x-ray beams generated at or above 2 MV, will be that obtained for cobalt-60 gamma rays or for 2 MV x rays. Calibration of the chamber at these qualities requires that it should have a cap of sufficient size to bring the total thickness to 500 – 800 mg per cm^2, in order to obtain electronic equilibrium.

This cap, which is usually of Perspex, Lucite, Plexiglas or similar material, attenuates and scatters the beam to some extent, and this is allowed for in the factor *N*, the use of which gives the exposure at the location of the center of the chamber, *in the absence of the chamber and its extra cap.* When, as here recommended, the chamber is used for the measurement of exposure in a phantom, allowance must be made for the effect of the material which the chamber and its calibration cap displace. An appropriate factor, which is about 0.98 for a chamber with an air cavity of diameter 6 mm, is incorporated in the value of *F* quoted for the higher radiation energies. It should be noted that the reading of the chamber is virtually unaffected if the chamber is used in the phantom without its calibration cap, whose thickness is then replaced by an equal thickness of water.

For the lower energy radiations, for which no added cap is necessary, there will still be some perturbation due to the displacement of phantom material by the different materials of the chamber system. Therefore, in principle, some correction is necessary. In practice, for the type of ionization chamber envisaged in these recommendations, the magnitude of this correction is much less than 1%.

The Conversion Coefficient, F

In the 40 – 400 kV range, *F* is essentially the "*f*-factor" which is appropriate for calculating the absorbed dose from the exposure under conditions of electron equilibrium. This coefficient takes account of *W*, the average energy necessary to produce an ion pair in air, and of the relative energy absorptions in air and water. For monoenergetic radiation, *F* is proportional to $(\mu en/\rho)$ water/$(\mu en/\rho)$air, where $(\mu en/\rho)$ is the mass energy absorption coefficient.

For the heterogeneous beams encountered in practice, the value of *F* has been obtained using an equivalent photon energy, defined as the energy of a monoenergetic beam which has the same half value layer in aluminium or copper as the radiation being considered. The relevant half value layer is not that of the primary beam, but one which takes into account the fact that the quality of the radiation inside the phantom is different from that of the primary beam because of filtration and scattering in the phantom. Data are available which enable this quality to be deduced with sufficient accuracy from knowledge of the primary beam quality. Due allowance has been made for this quality change in the calculation of the *F* values given in Table B, in which the stated beam qualities are those of the primary beam. The values were obtained for a field size 10 cm × 10 cm. If a particular value of *F* is applied to a different field size, the error introduced will be less than 2%.

In the case of measurements of cesium-137 and cobalt-60 gamma rays, and of x rays of higher energy (for which the factor *N* is obtained either with cobalt-60 gamma rays, or with x rays generated at a potential of 2 MV), the tabulated values of *F* embody all the necessary factors (stopping power ratios, including allowance for the polarization or density effect; and allowance for the perturbation produced by the presence of the measuring device in the phantom) for converting the corrected instrument reading into the absorbed dose in water expressed in rads. In this quality range *F* is identical with the factor usually known as $C\lambda$.

3.4 The Back Scatter Factor, B

Unless special facilities for the measurement of back scatter are available, standard backscatter factors should be used.

3.5 The Depth of Measurement, d

To avoid ambiguity it is desirable to specify a definite depth of measurement. The criteria which guide the choice are firstly that the result should not depend on very exact positioning of the chamber, secondly that there should be electron equilibrium and thirdly, that the measurement should be made within the region of interest to the radiotherapist. For generating potentials in the range 150 kV to 10 MV, the recommended depth of 5 cm essentially fulfills all of these criteria. At higher generating potentials, the recommended depth is 7 cm (11 – 25 MV) or 10 cm (26 – 50 MV). The precise value of the depth is not critical, provided that it is known.

3.6 The Inverse-Square Factor, $(s+x)^2/s^2$

Use of the factor $(s + x)^2/s^2$ implies that the inverse-square law is obeyed, but, in practice, this may not be so. To minimize the resulting error, x should be, if possible, zero and, in any case, sufficiently small that the correction to be applied is less than 5%. If this condition cannot be met, it is necessary to make measurements for various values of x, using the smallest available chamber, and to extrapolate the readings to $x = 0$.[a]

Some physical constants and conversion data useful in radiation dosimetry are given in Tables 3.1–15 through 3.1–18.

QUANTITATION OF DOSE-EFFECT RELATIONSHIPS

The analysis of possible mechanisms relating radiation dose delivered to tissue and the observed biological effects has been reviewed by Burch,[15] who emphasizes the need for quantitative methods of predicting long-term as well as early biological effects resulting from radiation exposure. Upton and Kimball[16] reviewed data indicating the relationships between exposure and effects in animals, and Wald[17] reviewed the various biomedical effects observed in humans exposed to radiation. Many other comprehensive reviews of radiobiological effects are available; they are cited in References 15 to 18, 34, and 40. The information now available shows certain definite biomedical effects of radiation on humans at high doses and dose rates, but continuing investigation is underway to provide improved understanding and predictability of the shape of the probabilistic dose-response relationships at low doses and dose rates. Therefore, recommended limits of long-term occupational and population exposure must be based on estimated upper limits of the probabilities of disease determined by conservative extrapolation of dose-response curves to low dose or dose rate regions. Since a comprehensive treatment of dose-response relations is beyond the scope of this handbook, only some examples of possible types of response relationships are presented below.

Cell survival after irradiation is often plotted as the logarithm of the surviving fraction vs. dose (see Figure 3.1–8). These curves are often fitted to more or less smoothly varying data, showing that for large numbers of experimental units such as bacterial or mammalian cells an underlying probabilistic dose-response relationship can be explicitly and accurately defined. These types of curves often are fitted by the n-target model

$$S_D = 1 - (1 - e^{-\lambda D})^n,$$

when n is termed the extrapolation (or target) number, and λ the probability that unit radiation dose D will "hit" any of the n targets.

Although useful for describing much of the accumulated data, this model is of limited use in

[a]From Measurement of Absorbed Dose in a Phantom Irradiated by a Single Beam of X or Gamma Rays, ICRU Report 23, International Commission on Radiation Units and Measurements, Washington, D.C., January 15, 1973. With permission.

Table 3.1–15
PHYSICAL CONSTANTS

c	Velocity of light	2.998×10^{8} m/sec
h	Planck's Constant	6.625×10^{-34} J sec
N	Avogadro's Number	6.023×10^{23} molecules per mole
e	Electronic charge	4.803×10^{-10} esu = 1.602×10^{-19} C
eV	Electron volt	1.602×10^{-19} J = 1.602×10^{-12} erg
MeV	Million electron volt	1.602×10^{-13} J = 1.602×10^{-6} erg
	Mass of electron	9.109×10^{-31} kg = 0.5110 MeV
	Mass of proton	1.6724×10^{-27} kg = 938.2 MeV
	Mass of neutron	1.6747×10^{-27} kg = 939.5 MeV
amu	Atomic mass unit	931.14 MeV

1 curie (Ci) = 3.700×10^{10} dis/sec   1 year = 5.260×10^{5} min = 3.156×10^{7} sec

1 millicurie (mCi) = 3.700×10^{7} dis/sec   1 day = 1.44×10^{3} min = 8.64×10^{4} sec

1 microcurie (μCi) = 3.700×10^{4} dis/sec

W_{air} = 33.7 eV/ion pair = 54.0×10^{-12} erg/ion pair = 0.1124 erg/esu

1 esu = 2.082×10^{9} ion pairs = 3.336×10^{-10} C   1 C = 6.242×10^{18} ion pairs

$$1\,\text{Roentgen (R)} = \frac{1\ \text{esu}}{\text{cm}^3\ \text{air(NTP)}} = \frac{1\ \text{esu}}{.001293\ \text{g air}} = 773.4\,\frac{\text{esu}}{\text{g air}} = 2.58 \times \frac{10^{-4}\ \text{C}}{\text{kg air}}$$

$$= 2.082 \times \frac{10^{9}\ \text{ion pairs}}{\text{cm}^3\ \text{air(NTP)}} = 1.610 \times 10^{12}\,\frac{\text{ion pairs}}{\text{g air}}$$

$$= 86.9\,\frac{\text{ergs}}{\text{g air}} = 8.69 \times 10^{-3}\,\frac{\text{J}}{\text{kg air}} = 0.869\ \text{rad in air}$$

1 rad = 100 ergs/g = 10^{-2} Joules/kg = 6.24×10^{7} MeV/g

From Johns, H. E. and Cunningham, J. R., *The Physics of Radiology,* 3rd ed., 1971. Courtesy of Charles C Thomas, Publisher, Springfield, Ill.

describing the influence of many factors on radiosensitivity, as pointed out by Burch,[15] who has developed more elaborate multiple-hit models. Dose-response curves determined in terms of chromosome breaks in the plants *Tradescantia* and *Vicia* show that two-break aberrations increase nearly as the square of the dose for radiation of low LET at high intensity, but increase at a lower rate and almost linearly with the dose at low intensity.[16] For high LET radiation (see the neutron curve of Figure 3.1–9), the two-break aberrations increase almost linearly with dose and independently of intensity.

Figure 3.1–10 illustrates the type of dose-response relationships that may be observed in studying genetic mutations at the animal level. One may note that, although large confidence intervals must be attached to each data point as a result of statistical limitations in observing the relatively rare mutations with an animal population of limited size, a definite increase in mutation rate with dose is, nevertheless, observed. Also, high dose rates again appear more effective than low dose rates.

When a response such as mortality (1 – survival) of animals is plotted as a function of radiation dose in rad, curves such as those in Figure 3.1–11 are obtained. The S-shaped curve typical of many chemically toxic agents[19] is observed here for radiation.

When plotting this type of mortality data for mice exposed to high-energy protons (440 and 730 MeV) and X-rays, Bradley et al.[20] found that a probit mortality (or probability) vs. log (dose) scale linearized the dose-response curves and made them approximately parallel for given strains exposed to different LET radiations (see Figure 3.1–12). The point where each line crosses the 50% mortality level is called the LD_{50} for that type of radiation, that age, strain, and species of animal, and that time interval of observation. Thus, whatever the basic mechanisms at the cellular and subcellular levels, they may often combine to produce an effect represented at the animal or human level by the empirical log-normal dose-response function.[19,21] This is the functional shape that the probit-type plot is designed to linearize for convenience in bioassay

Table 3.1–16
VALUES OF THE ATOMIC WEIGHT M, THE FACTOR $(N_A/M) \times 10^{-24}$ FOR CONVERTING ATTENUATION DATA FROM b/ATOM TO $cm^2\ g^{-1}$, AND TYPICAL DENSITIES ρ FOR CONVERTING FROM $cm^2\ g^{-1}$ TO cm^{-1}

Element Z	Element Symbol	M,[a] atomic weight, g/g-atom	$\frac{N_A}{M} \times 10^{-24}, b\ \frac{cm^2}{g} / \frac{b}{atom}$	ρ,[c] g/cm³
1	H	1.00797	0.5975	0.00008988 g, (H_2)
2	He	4.0026	.1505	.0001785 g
3	Li	6.939	.08679	.534
4	Be	9.0122	.06683	1.85
5	B	10.811	.05571	2.535
6	C	12.01115	.05014	2.25 (graphite)[d]
7	N	14.0067	.04300	.001250 g, (N_2)
8	O	15.9994	.03764	.001429 g, (O_2)
9	F	18.9984	.03170	.001696 g, (F_2)
10	Ne	20.183	.02984	.0008999 g
11	Na	22.9898	.02620	.971
12	Mg	24.312	.02477	1.74
13	Al	26.9815	.02232	2.70
14	Si	28.086	.02144	2.42
15	P	30.9738	.01944	1.8–2.7
16	S	32.064	.01878	1.96–2.07
17	Cl	35.453	.01699	.003214 g, (Cl_2)
18	Ar	39.948	.01508	.001784 g
19	K	39.102	.01540	.87
20	Ca	40.08	.01503	1.55
21	Sc	44.956	.01340	3.02
22	Ti	47.90	.01257	4.5
23	V	50.942	.01182	5.87
24	Cr	51.996	.01158	7.14
25	Mn	54.9380	.01096	7.3
26	Fe	55.847	.01078	7.86
27	Co	58.9332	.01022	8.71
28	Ni	58.71	.01026	8.8
29	Cu	63.54	.009478	8.93
30	Zn	65.37	.009213	6.92
31	Ga	69.72	.008638	5.93
32	Ge	72.59	.008297	5.46
33	As	74.9216	.008038	5.73
34	Se	78.96	.007627	4.82
35	Br	79.909	.007537	3.121
36	Kr	83.80	.007187	.003743 g
37	Rb	85.47	.007046	1.53
38	Sr	87.62	.006873	2.6
39	Y	88.905	.006774	3.8
40	Zr	91.22	.006602	6.44

Table 3.1–16 (continued)
VALUES OF THE ATOMIC WEIGHT M, THE FACTOR $(N_A/M) \times 10^{-24}$ FOR CONVERTING ATTENUATION DATA FROM b/ATOM TO $cm^2\ g^{-1}$, AND TYPICAL DENSITIES ρ FOR CONVERTING FROM $cm^2\ g^{-1}$ TO cm^{-1}

Element		M,[a] atomic		
Z	Symbol	weight, g/g-atom	$\frac{N_A}{M} \times 10^{-24}$,[b] $\frac{cm^2}{g} / \frac{b}{atom}$	ρ,[c] g/cm³
41	Nb	92.906	.006482	8.4
42	Mo	95.94	.006277	9.01
43	Tc	(99)	(.006083)	[11.50]
44	Ru	101.07	.005959	12.1
45	Rh	102.905	.005853	12.44
46	Pd	106.4	.005660	12.25
47	Ag	107.870	.005583	10.49
48	Cd	112.40	.005358	8.65
49	In	114.82	.005245	7.43
50	Sn	118.69	.005074	5.75–7.29
51	Sb	121.75	.004947	6.62
52	Te	127.60	.004720	6.25
53	I	126.9044	.004746	4.94
54	Xe	131.30	.004587	.005896 g
55	Cs	132.905	.004531	1.873
56	Ba	137.34	.004385	3.5
57	La	138.91	.004336	6.15
58	Ce	140.12	.004298	6.90
59	Pr	140.907	.004274	6.48
60	Nd	144.24	.004175	7.00
61	Pm	(145)	(.004153)	[7.22][e]
62	Sm	150.35	.004006	7.7–7.8
63	Eu	151.96	.003963	[5.259]
64	Gd	157.25	.003830	[7.948]
65	Tb	158.924	.003790	[8.272]
66	Dy	162.50	.003706	[8.536]
67	Ho	164.930	.003652	[8.803]
68	Er	167.26	.003601	[9.051][f]
69	Tm	168.934	.003565	[9.332]
70	Yb	173.04	.003480	[6.977]
71	Lu	174.97	.003442	[9.872]
72	Hf	178.49	.003374	13.3
73	Ta	180.948	.003328	17.1
74	W	183.85	.003276	19.3
75	Re	186.2	.003234	20.53
76	Os	190.2	.003166	22.8
77	Ir	192.2	.003133	22.42–22.8
78	Pt	195.09	.003087	21.4
79	Au	196.967	.003058	19.3
80	Hg	200.59	.003002	13.55 1

Table 3.1–16 (continued)
VALUES OF THE ATOMIC WEIGHT M, THE FACTOR $(N_A/M) \times 10^{-24}$ FOR CONVERTING ATTENUATION DATA FROM b/ATOM TO $cm^2\ g^{-1}$, AND TYPICAL DENSITIES ρ FOR CONVERTING FROM $cm^2\ g^{-1}$ TO cm^{-1}

Element Z	Element Symbol	M,[a] atomic weight, g/g-atom	$\frac{N_A}{M} \times 10^{-24}$,[b] $\frac{cm^2}{g} / \frac{b}{atom}$	ρ,[c] g/cm³
81	Tl	204.37	.002947	11.86
82	Pb	207.19	.002907	11.34
83	Bi	208.980	.002882	9.78
84	Po	(210)	(.002868)	[9.32]
85	At	(210)	(.002868)	–
86	Rn	(222)	(.002713)	0.00996 g
87	Fr	(223)	(.002701)	–
88	Ra	(226)	(.002665)	5(?)
89	Ac	(227)	(.002653)	[10.07]
90	Th	232.038	.002595	11.0
91	Pa	(231)	(.002607)	[15.37]
92	U	238.03	.002530	18.7
93	Np	(237)	(.002541)	[19.36]
94	Pu	(242)	(.002489)	[19.84]
95	Am	(243)	(.002478)	[11.7]
96	Cm	(247)	(.002438)	[~7]
97	Bk	(249)	(.002419)	–
98	Cf	(251)	(.002399)	–
99	Es	(254)	(.002371)	–
100	Fm	(253)	(.002380)	–

[a]Atomic weights are those recommended in 1961 by the International Union of Pure and Applied Chemistry, based on M = 12.0000 for C^{12}. These values are based on average isotropic abundances and ignore natural and artificial variations. In practice, for example, M can vary from 6.01513 for pure $_3Li^7$, with a corresponding variation in the conversion factor from 0.1001 to 0.08584.

[b]N_A = Avogadro's Number = 6.02252×10^{23} atoms/g-atom, C^{12} scale.

[c]Densities are for common solids, except as denoted liquid (l) or gas (g) at 20°C, and in square brackets ([ ]). Gas densities are at S.T.P.: 0°C, 76 cm Hg.

[d]Graphite theoretical density, based on X-ray diffraction data. Commercial-grade pile graphite has a density range of 1.5 to 1.9 g/cm³.

[e]The density of Pm^{147} is 7.22 g/cm³.

[f]Density values of erbium in the literature vary from 4.77 to the values 9.05 and 9.16.

From Hubbell, J. H., *Photon Cross Sections, Attenuation Coefficients, and Energy Absorption Coefficients, from 10 KeV to 100 GeV,* NSRDS-NBS-29, U.S. Government Printing Office, Washington, D.C., August 1969, 6.

Table 3.1–17
CONVERSION FACTORS AND DENSITIES FOR A FEW COMPOUNDS AND MIXTURES

Substance	Conversion factor, $\frac{cm^2}{g} / \frac{b}{molecule}$	ρ, g/cm³
H_2O	0.03344	1.00 1, 0.917 (ice)
SiO_2	.01002	2.32
NaI	.004019	3.667
Air (20°C, 76 cm Hg)	–	0.001205 g
Concrete	–	2.2–2.4
0.8 *N* H_2SO_4 solution	–	1.0494
Bone	–	1.7–2.0
Muscle	–	~ 1
Polystyrene, $(C_8H_8)_n$	–	1.05–1.07
Polyethylene, $(CH_2)_n$	–	0.92
Polymethyl methacrylate (Lucite®), $(C_5H_8O_2)_n$	–	1.19
Bakelite® (typical), $C_{43}H_{38}O_7$	–	1.20–1.70
Pyrex® glass (Corning No. 7740)	–	2.23

From Hubbell, J. H., *Photon Cross Sections, Attenuation Coefficients, and Energy Absorption Coefficients, from 10 KeV to 100 GeV*, NSRDS-NBS-29, U.S. Government Printing Office, Washington, D.C., August 1969, 7.

of toxic agents or pharmaceuticals, and statistical methods of curve fitting and analysis are available for these functions.[19] Other types of functions have also been tried for fitting radiation dose-response data of varying kinds. With present information, no unique theoretical basis is available for choosing specific functions, so mathematical functions are chosen either for convenience, for investigating the consistency of interim theoretical predictions, or for assistance in estimating upper limits of risk from environmental radiation sources.[34]

The log-normal function is briefly described here since a summary of the log-normal function, its properties, and simplified procedures for its application are presented in Reference 21. Briefly, suppose that x is the dose of a population of animals that $P(\ln X \leqq \ln x)$ is the probability of animals having "sensitivities" or lethal effects at doses $X \leqq x$; then P may be considered as the (expected) fraction of animals dying in a group administered a dose x, and a log-normally distributed P would be given by the function

$$P(\ln X \leqslant \ln x) = \frac{1}{\sqrt{2\pi}\sigma_g} \int_{\ln X = \ln 0 = -\infty}^{\ln X = \ln x} \exp\left[-(\ln X - \ln \mu_g)^2/2\sigma_g^2\right] d(\ln X).$$

Here, σ_g is the standard deviation ($\sqrt{\mathrm{Var}(\ln X)}$) in the frequency distribution of $\ln X$, $\ln \mu g$ is the mean value of $\ln X$, and μ_g is the "geometric mean of X." The above expression is the form for the cumulative integral over a normal distribution of $\ln X$, and has been shown to be a special case of a family of logarithmic distributions.[22] The LD_{50} value where a probit-type mortality plot (Figure 3.1–12) crosses the 50% mortality line would be an estimate of μ_g of the log-normal dose-response function (if the data indicate that a straight line on such a plot is applicable). The parameter σ_g is the standard deviation in $\ln X$ (the natural logarithm of the randomly distributed dose required to kill an animal), and this parameter may also be easily estimated from the probit plot. The estimated value of σ_g can be obtained from the equation

$$\sigma_g = \ln x_{0.8413} - \ln x_{0.50} = \ln(x_{0.8413}/x_{0.50}),$$

or

$$\sigma_g = \ln x_{0.50} - \ln x_{0.1587} = \ln(x_{0.50}/x_{0.1587}),$$

Table 3.1–18

FRACTIONS BY WEIGHT, w_j OF ELEMENTS IN SOME MIXTURES AND COMPOUNDS

Element														
Z	Symbol	H_2O	SiO_2	NaI	Air	Concrete	0.8 *N* H_2SO_4 solution[a]	Bone, compact	Muscle, striated	Polystyrene, $(C_8H_8)_n$	Lucite®, $(C_5H_8O_2)_n$	Polyethylene, $(CH_2)_n$	Bakelite®, $C_{45}H_{38}O_7$	Pyrex® glass[b]
1	H	0.1119	–	–	–	0.0056	0.1084	0.064	0.102	0.0774	0.0805	0.1437	0.0574	–
5	B	–	–	–	–	–	–	–	–	–	–	–	–	0.0401
6	C	–	–	–	–	–	–	0.278	0.123	0.9226	0.5999	0.8563	0.7746	–
7	N	–	–	–	0.755	–	–	0.027	0.035	–	–	–	–	–
8	O	0.8881	0.5326	–	0.232	0.4983	0.8791	0.410	0.72893	–	0.3196	–	0.1680	0.5396
11	Na	–	–	0.1534	–	0.0171	–	–	0.0008	–	–	–	–	0.0282
12	Mg	–	–	–	–	0.0024	–	0.002	0.0002	–	–	–	–	–
13	Al	–	–	–	–	0.0456	–	–	–	–	–	–	–	0.0116
14	Si	–	0.4674	–	–	0.3158	–	–	–	–	–	–	–	0.3772
15	P	–	–	–	–	–	–	0.070	0.002	–	–	–	–	–
16	S	–	–	–	–	0.0012	0.0125	0.002	0.005	–	–	–	–	–
18	Ar	–	–	–	0.013	–	–	–	–	–	–	–	–	–
19	K	–	–	–	–	0.0192	–	–	0.003	–	–	–	–	0.0033
20	Ca	–	–	–	–	0.0826	–	0.147	0.00007	–	–	–	–	–
26	Fe	–	–	–	–	0.0122	–	–	–	–	–	–	–	–
53	I	–	–	0.8466	–	–	–	–	–	–	–	–	–	–

[a] 3.832% H_2SO_4, 96.168% H_2O by weight.

[b] SiO_2, 80.7%; B_2O_3, 12.9%, Na_2O_3, 3.8%; Al_2O_3, 2.2%; K_2O, 0.4% by weight. 0.2% has been added to SiO_2 to make total = 100%. Otherwise, these percentages, with 80.5% SiO_2, are those given for Pyrex glass (Corning 7740).

From Hubbell, J. H., *Photon Cross Sections, Attenuation Coefficients, and Energy Absorption Coefficients, from 10 KeV to 100 GeV*, NSRDS-NBS-29, U.S. Government Printing Office, Washington, D.C., August 1969, 8.

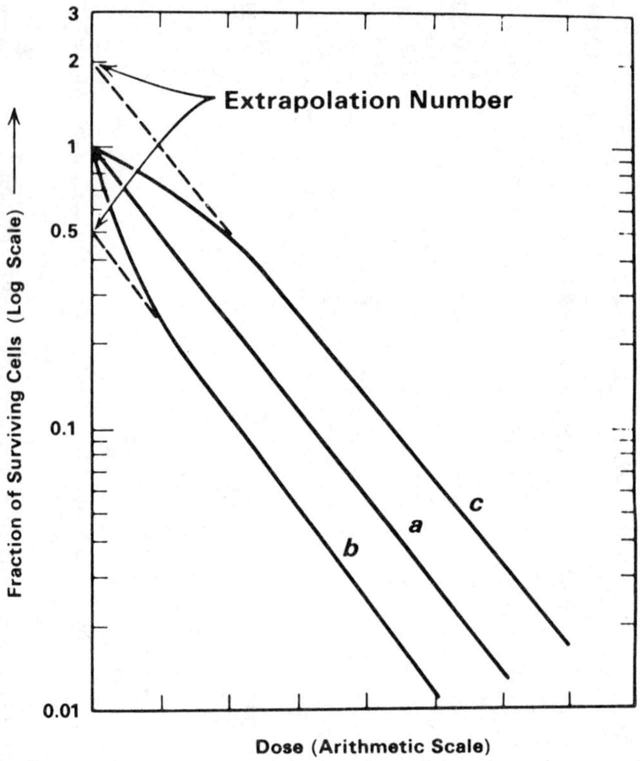

FIGURE 3.1–8. Types of mammalian cell survival curve (logarithm of fraction of cells, S, surviving the [acute] irradiation against dose D). Curve *a*: $S = 2\ D$; observed when homogeneous cells are irradiated by particles (such as low-energy natural α particles) of high LET. Curve *b* gives an extrapolation number of less than unity (0.5 in example), observed when heterogeneous, mixtures of cells (of low and high radiosensitivity) are irradiated. Curve *c* gives an extrapolation number of more than unity (2 in example), observed when most types of mammalian cells are irradiated by, for example, ^{60}Co gamma rays – that is, by radiation of low average LET. (From Morgan, K. Z. and Turner, J. E., Eds., *Principles of Radiation Protection – A Textbook of Health Physics,* John Wiley & Sons, New York, 1967, 377. With permission.)

where $x_{0.8413}$ is the value of x where the "% mortality vs. log X line" crosses the 84.13% mortality value, and $x_{0.50} = \mu_g$ is the value of x where the line crosses the 50% mortality value. Now the "standard geometric deviation"[21] or geometric mean standard deviation[22] is defined as

$$s_g = x_{0.8413}/x_{0.50} = x_{0.50}/x_{0.1587}),$$

and the average dose $\bar{x}$ required to kill an animal is then given by the relationship[21]

$$\log_{10}\bar{x} = \log \mu_g + 1.1513(\log_{10} s_g)^2 .$$

The particular significance of μ_g and s_g for a log-normal dose-response relationship is that $\mu \stackrel{\times}{\div} s_g$, and not $\bar{x} \pm \sigma_x$, gives the interval in which 68% of the population "sensitivities" will lie, if X is interpreted as the dose to which a fraction of the population is sensitive enough to die (or manifest some other biological end point). The concept of s_g may also be useful, for example, in predicting that less than 2.5% of the population are likely to require doses greater than $s_g^2\ \mu_g$ to die; in other words, if the log-normal function holds at higher doses, less than about 2.5% of the population would survive more than s_g^2 times the LD_{50}. On the other hand, 2.5% of the population would be affected by doses less than the LD_{50} divided by s_g^2. Thus, the S-shaped or log-normal dose-response relationship would predict that at low doses some finite, although small, effect is still possible.

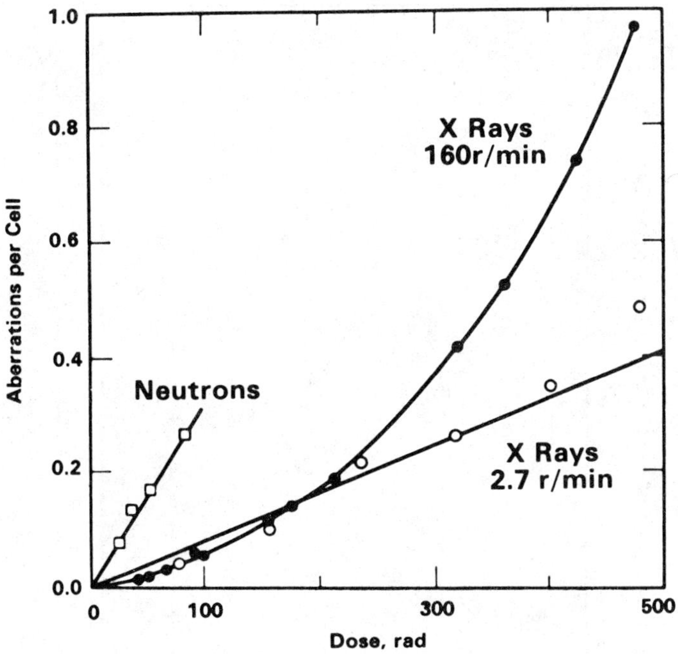

FIGURE 3.1–9. Dose curves for two-break chromosomal aberrations. The X-ray data are replotted from Sax (1941), the neutron data from Giles (1943). The neutron doses were given in *n* units, and have been converted to rad by multiplication with the factor 2.5 (this conversion is very approximate and should not be relied upon). (From Morgan, K. Z. and Turner, J. E., Eds., *Principles of Radiation Protection – A Textbook of Health Physics,* John Wiley & Sons, New York, 1967, 404. With permission.)

The S-shaped curve has also been suggested by some of the animal data on carcinogenesis. However, when dose levels are so high that cell lethality is produced, the rising portion of the S-shaped curve may not be evident (see Figure 3.1–13) However, a stochastic two-stage model of carcinogenesis,[23] involving a second event conditioned on a specific prior event followed by the cell growth stage of Arley,[24] predicts a dose-response curve shape so similar to the log-normal that it could not be distinguished from it by any reasonably sized experiment. Thus, the variation in the dose X at which the biological end point (e.g., an observed tumor by time T) may occur could be interpreted as resulting largely from ordinary chance variations in the action of a radiation dose on an individual animal rather than from real individual variations in sensitivity between animals. Some purely stochastic variation of this kind, as well as real variations in individual susceptibility, may also be expected to contribute to the range of doses over which lethality may occur. In any case, however, the type of dose-response relationship represented by any of these S-shaped curves would yield a predicted incidence of cancer at low doses or dose rates that would be lower[23] than the incidence obtained by linear extrapolations through the origin from cancer incidences observed at higher doses or dose rates. Leukemogenesis, on the other hand, may indeed be linearly related to dose.[18,34]

Thus, from such considerations, conservative estimates of biomedical effects at the population level may be made for purposes of setting standards for radiation protection. Some order-of-magnitude estimates of the biomedical effects of radiation per unit dose are given below for use in assessing the potential hazards of "maximum credible accidents" or inadvertent exposures deemed possible in individual applications.

ESTIMATION OF HUMAN RISKS OF LOW-LEVEL RADIATION EXPOSURE

Lethal dose (whole body) – LD_{50}^{30} = 250 to 500 rem when received within a short period of time (say, less than 1 day) and without therapeutic

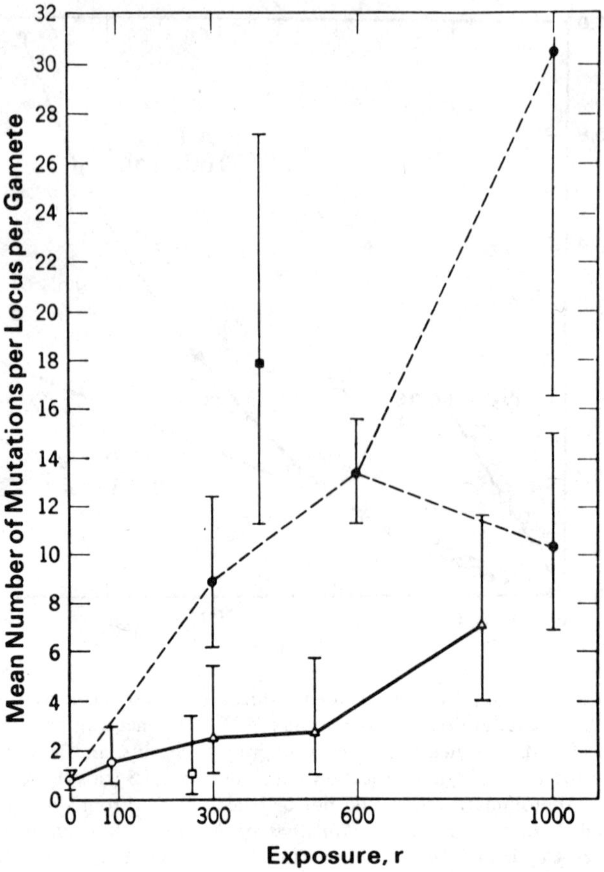

FIGURE 3.1–10. Exposure curves for specific-locus mutation in the mouse. 90% confidence intervals shown. Solid points represent results with acute X-rays (80 to 90 r/min). Open points are chronic ν-ray results (triangles and square, 90/r week; circles, 10 r/week). Squares are mutation rates in females; all other points are mutation rates in males. The point for zero dose is the sum of all male controls. The top 1,000-r point represents results of successive exposures to 600 and 400 r, respectively, separated by an interval of 15 weeks (Russel et al., 1960). (From Morgan, K. Z. and Turner, J. E., Eds., *Principles of Radiation Protection – A Textbook of Helath Physics,* John Wiley & Sons, New York, 1967, 416. With permission.)

treatment.[17] (Consider also biological variability or chance nature of some radiation effects as discussed above.)

Shortening of life span – Estimated at 1 to 4 days per roentgen of exposure early in life, but the exact magnitude is not established.[17,18]

Bone cancer – Above 4 to 8 cases per million population per 70-year period per rem dose received over a 70-year life span, assuming 10% of the natural incidence is a result of background radiation;[15,18] the frequency at low doses may actually be much less than this upper-limit estimate. These estimates are not inconsistent with the risk estimate of Table 3.1–19.

Leukemia – An average of about 2 cases per million adults per rem average exposure to the entire population considered per year-at-risk following the exposure, although the incidence-vs.-time curve peaks after a latent period that probably varies according to dose level;[23,25] however, for children irradiated in utero or in preconception gametal stages, recent evidence of Graham et al.[26] tends to confirm the earlier data of McMahon,[27,28] and Stewart[37] indicating a

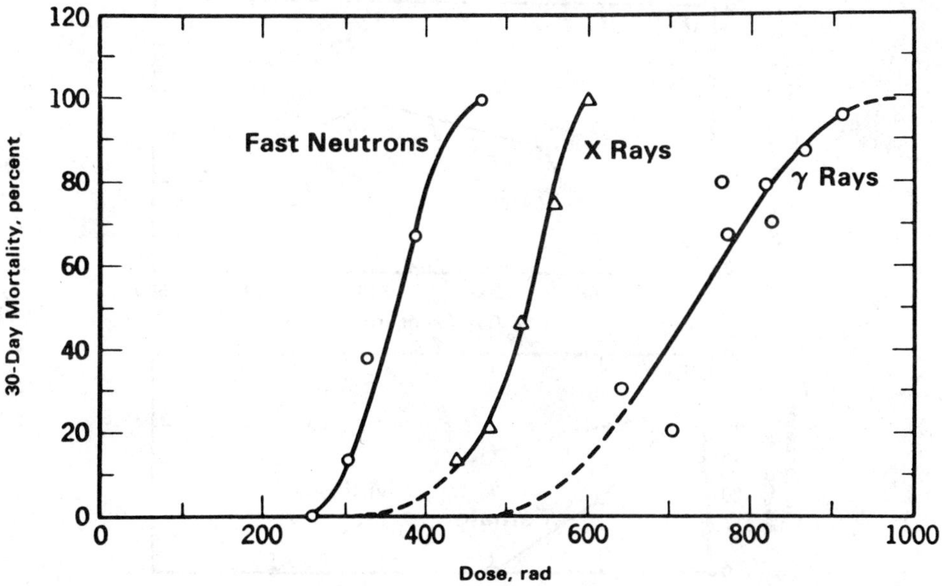

FIGURE 3.1–11. Thirty-day mortality of mice exposed to X-rays, fast neutrons, and gamma rays. The mice were 9 to 12 weeks old at the time of whole-body irradiation. (From Morgan, K. Z. and Turner, J. E., Eds., *Principles of Radiation Protection – A Textbook of Health Physics,* John Wiley & Sons, New York, 1967, 429. With permission.)

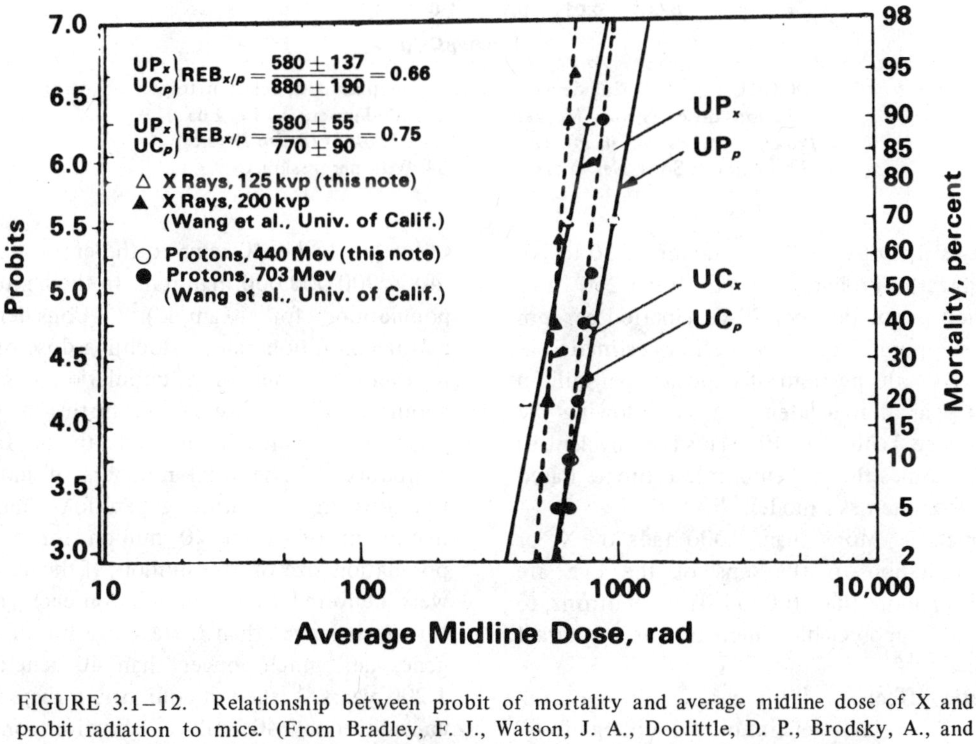

FIGURE 3.1–12. Relationship between probit of mortality and average midline dose of X and probit radiation to mice. (From Bradley, F. J., Watson, J. A., Doolittle, D. P., Brodsky, A., and Sutton, R. B., *Health Phys.*, 10, 72, 1964. Reprinted with permission from Pergamon Press.)

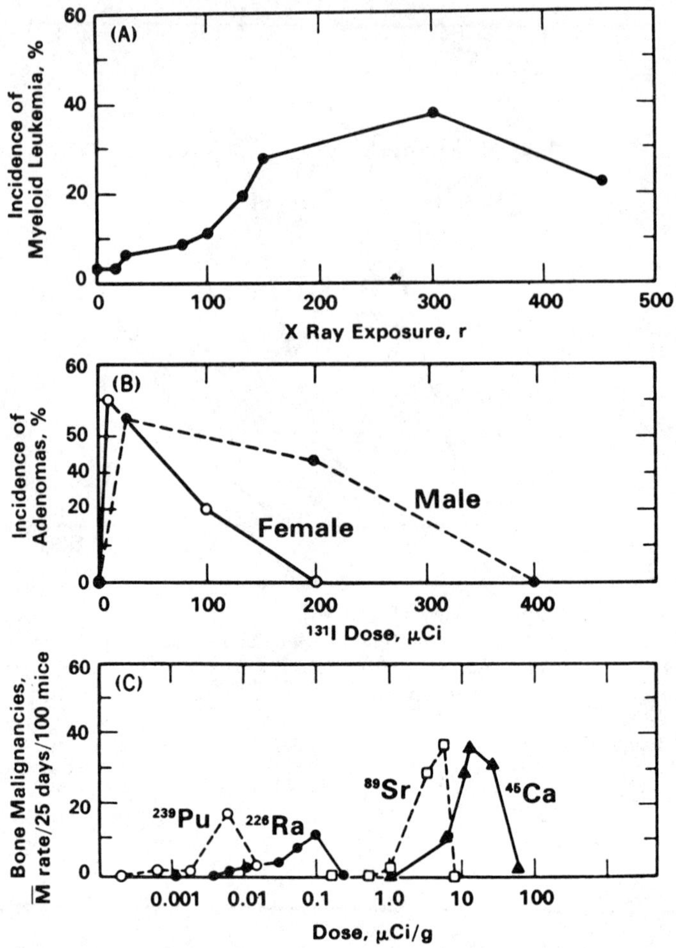

FIGURE 3.1–13. Incidence of various neoplasms in relation to radiation dose. (From Morgan, K. Z. and Turner, J. E., Eds., *Principles of Radiation Protection – A Textbook of Health Physics*, John Wiley & Sons, New York, 1967, 434. With permission.)

higher sensitivity by perhaps a factor of 50 to 70.

All cancer – Probably no more than 250 cases per million population per 70-year period per rem dose to the population.[24] Sometimes estimated as about five deaths per rem of exposure per million for each year after a latent period following the exposure (see Table 3.1–19). This is equivalent to about five times the leukemia rate estimate based on the "absolute risk" model.[34]

Cataracts – More than 1,000 rads of X- or gamma radiation to the lens of the eye are required, or more than 100 rads from neutrons, to cause an appreciable increase in cataract incidence.[17,18]

Genetic effects – With a long range of uncertainty,[34] a single-generation "doubling dose" to double the natural mutation rate has been estimated to be 40 rems to the entire population (40 × 200,000,000 man-rems to the United States population, for example).[29] Considering the natural mutation rate, a doubling dose of 40 rem to every member of a population might cause about 1 out of every 200 births in the next generation to result in a death or failure to reproduce.[24] The total numbers of induced extinctions in all future generations might be a maximum of about 40 million for a constant population size of 200 million. If the 40 rem dose were delivered to the population each generation for much longer than the average life of a mutant gene, i.e., much longer than 40 generations or 1,200 years,[30] then at equilibrium there would be an additional 40 million abortive conceptions, stillbirths, neonatal deaths, or failures to

Table 3.1–19
ASSUMED VALUES USED IN CALCULATING ESTIMATES OF RISK

Age at irradiation	Type of cancer	Duration of latent period, years	Duration of plateau region, years[a]	Risk estimate: Absolute risk,[b] deaths/10^6 per year per rem	Risk estimate: Relative risk, % increment in deaths per rem
In utero	Leukemia	0	10	25	50
	All other cancer	0	10	25	50
0–9 years	Leukemia	2	25	2.0	5.0
	All other cancer	15	(a)30 (b)Life	1.0	2.0
10+ years	Leukemia	2	25	1.0	2.0
	All other cancer	15	(a)30 (b)Life	5.0	0.2

[a]Plateau region = interval following latent period during which risk remains elevated. Risk estimate is an average rate over this period of years.
[b]The absolute risk for those aged 10 years or more at the time of irradiation for all cancer excluding leukemia can be broken down into the respective sites as follows:

Type of cancer	Deaths/10^6 per year per rem
Breast	1.5[c]
Lung	1.3
GI, including stomach	1.0
Bone	0.2
All other cancer	1.0
Total	5.0

[c]This is derived from the value of 6.0 quoted in the original source (Appendix II, Section A1e) corrected for a 50% cure rate and the inclusion of males as well as females in the population.

From BEIR, The Effects on Populations of Exposure to Low Levels of Ionizing Radiation, Report of the Advisory Committee on the Biological Effects of Ionizing Radiation (BEIR), Division of Medical Sciences, National Academy of Sciences, Washington, D.C., November 1972.

reproduce per generation per 200,000,000 population. Of course, the proportion of each of these is uncertain. In addition, however, the surviving population would presumably carry a large pool of radiation-caused mutants who might have varying degrees of health impairment, as well as a possible small – but increasing – percentage of individuals evolving with superior characteristics as a result of selection in favor of the beneficial long-surviving mutant fraction.

Fertility – The sterilizing dose to the gonads for men is a minimum of 300 to 400 rem in a single exposure; for women, it is 1,000 to 2,000 rem fractionated over 10 to 24 days.[34] Conception should be deferred after irradiation for about 2 months for the male; the corresponding period is not known for females, and recovery may not exist regarding genetic effects.[34]

Other radiation effects – Some additional nonspecific health impairment or loss of vitality might result from cell loss following somatic mutations, although, if the destruction of cells is at a low rate, regeneration may prevent organ failure or ill health.[31] These additional effects, particularly by degenerative cardiovascular and renal diseases,[15] may have a smaller relative increase per rem above natural incidence levels, but, since they are more prevalent, they may somewhat exceed in absolute numbers the excess deaths from radiation-induced cancer. Nevertheless, the total excess mortality induced by radiation exposure would not be expected to be more than the order of magnitude produced by genetic and carcinogenic effects, and can be maintained lower than other normally accepted risks for the benefits of nuclear energy and its medical,

Table 3.1–20
DOSE-LIMITING RECOMMENDATIONS[a]

Maximum permissible dose equivalent for occupational exposure	
Combined whole body occupational exposure	
Prospective annual limit – paragraphs 229, 233	5 rems in any one year
Retrospecive annual limit – paragraphs 230, 233	10–15 rems in any one year
Long-term accumulation to age *N* years – paragraphs 231	$(N - 18) \times 5$ rems
Skin – paragraphs 234, 235	15 rems in any one year
Hands – paragraphs 236, 237	75 rems in any one year (25/quarter)
Forearms – paragraphs 236, 237	30 rems in any one year (10/quarter)
Other organs, tissues and organ systems – paragraphs 238, 239	15 rems in any one year (5/quarter)
Fertile women (with respect to fetus) – paragraphs 240, 241	0.5 rem in gestation period
Dose limits for the public, or occasionally exposed individuals	
Individual or occasional – paragraphs 245, 246, 253, 254	0.5 rem in any one year
Students – paragraphs 255, 256	0.1 rem in any one year
Population dose limits	
Genetic – paragraphs 247, 248	0.17 rem average per year
Somatic – paragraphs 250, 251	0.17 rem average per year
Emergency dose limits – life saving	
Individual (older than 45 years if possible) – paragraph 258	100 rems
Hands and forearms – paragraphs 258	200 rems, additional (300 rems, total)
Emergency dose limits – less urgent	
Individual – paragraph 259	25 rems
Hands and forearms – paragraph 259	100 rems, total
Family of radioactive patients	
Individual (under age 45) – paragraphs 267, 268	0.5 rem in any one year
Individual (over age 45) – paragraphs 267, 268	5 rems in any one year

[a]Paragraph citations refer to original source.

From NCRP Report No. 39, Basic Radiation Protection Criteria, National Council on Radiation Protection and Measurements, Washington, D.C., January 15, 1971. With permission.

industrial, and consumer by-products.[24] Further details on biological effects and literature references may be obtained from References 34 and 40, and supplementary reports as they are published by the respective agencies.

Exposure limits – Recommended exposure limits[40] of the National Council on Radiation Protection and Measurements are shown in Table 3.1–20. The original report[40] should be consulted for a thorough discussion of radiobiological knowledge and radiation protection philosophy.

REFERENCES

1. International Commission on Radiological Units and Measurements, Report of the ICRU, *NBS Handbook 78,* U.S. Government Printing Office, Washington, D.C., 1959.
2. International Commission on Radiological Units and Measurements, Clinical Dosimetry – Recommendations of the ICRU, *NBS Handbook 87,* U.S. Government Printing Office, Washington, D.C., 1963, 38.
3. **Parker, H. M.,** Health physics, instrumentation, and radiation protection, *Adv. Biol. Med. Phys.,* 1, 243, 1948.
4. **Roesch, W. C. and Attix, H. F., Eds.,** *Radiation Dosimetry,* 2nd ed., Vol. 1, Academic Press, New York, 1968, chap. 1.
5. International Commission on Radiological Protection, Report of Committee 2, *Health Phys.,* 3, 1, 1960.
6. International Commission on Radiological Units and Measurements, Report 10a, Radiation Quantities and Units, *NBS Handbook 84,* U.S. Government Printing Office, Washington, D.C., 1962.
7. **Attix, F. H. and Roesch, W. C., Eds.,** Instrumentation, *Radiation Dosimetry,* Vol. 2, Academic Press, New York, 1966.
8. **Hine, G. J. and Brownell, G. L., Eds.,** *Radiation Dosimetry,* Academic Press, New York, 1956.
9. **Price, W. J.,** *Nuclear Radiation Detection,* 2nd ed., McGraw-Hill, New York, 1966.
10 **Morgan, K. Z. and Turner, J. E., Eds.,** *Principles of Radiation Protection – A Textbook of Health Physics,* John Wiley & Sons, New York, 1967.
11. National Committee on Radiation Protection and Measurements, NCRP Report No. 27, Stopping Powers for Use with Cavity Chambers, *NBS Handbook 79,* U.S. Government Printing Office, Washington, D.C., 1961.
12. National Committee on Radiation Protection and Measurements, Measurement of Neutron Flux and Spectra for Physical and Biological Applications, Recommendations of the NCRP, *NBS Handbook 72,* U.S. Government Printing Office, Washington, D.C., 1960.
13. **Johns, H. E. and Cunningham, J. R.,** *The Physics of Radiology,* 3rd ed., Charles C Thomas, Springfield, Ill., 1971.
14. International Commission on Radiological Units and Measurements, Report of the ICRU, *NBS Handbook 62,* U.S. Government Printing Office, Washington, D.C., 1956.
15. **Burch, P. R. J.,** Radiation physics, in *Principles of Radiation Protection – A Textbook of Health Physics,* Morgan, K. Z. and Turner, J. E., Eds., John Wiley & Sons, New York, 1967, 366.
16. **Upton, A. C. and Kimball, R. F.,** Radiation biology, in *Principles of Radiation Protection – A Textbook of Health Physics,* Morgan, K. Z. and Turner, J. E., Eds., John Wiley & Sons, New York, 1967, 398.
17. **Wald, N.,** Evaluation of human exposure data, in *Principles of Radiation Protection – A Textbook of Health Physics,* Morgan, K. Z. and Turner, J. E., Eds., John Wiley & Sons, New York, 1967, 448.
18. United National Scientific Committee on the Effects of Atomic Radiation, Various Reports, United Nations, New York, 1958, 1962, 1964, 1966. (Continuing reviews are underway.)
19. **Finney, D. J.,** *Probit Analysis,* Cambridge University Press, New York, 1962.
20. **Bradley, F. J., Watson, J. A., Doolittle, D. P., Brodsky, A., and Sutton, R. B.,** *Health Phys.,* 10, 71, 1964.
21. **Schubert, J., Brodsky, A., and Tyler, S.,** The log-normal function as a stochastic model of the distribution of strontium-90 and other fission products in humans, *Health Phys.,* 13, 1187, 1967.
22. **Espenscheid, W. F., Kerker, M., and Marijevic, E.,** *J. Phys. Chem.,* 68, 3093, 1964.
23. **Brodsky, A.,** A Stochastic Model of Carcinogenesis and Its Implications in the Dose-Response Plane, presented at the Health Physics Society Meeting in Los Angeles, 1966. Abstract in *Health Phys.,* 12, 1176, 1966. Detailed treatment of the model in Brodsky, A., A Stochastic Model of Carcinogenesis as Applied to Skin Tumors in Mice, dissertation, University of Pittsburgh, 1966.
24. **Brodsky, A.,** *Am. J. Public Health,* 55(12), 1971, 1965.
25. **Cobb, S., Miller, M., and Wald, N.,** On the estimation of the incubation period in malignant disease, I, *J. Chronic Dis.,* 9(4), 385, 1959.
26. **Graham, L. S., Levine, M. L., Lilienfeld, A. M., Schuman, L., Gibson, R., David, J. E., and Hempleman, L. H.,** Preconception Intrauterine and Postnatal Irradiation as Related to Leukemia, presented at the Meeting of the American Public Health Association, New York, October 8, 1964.
27. **McMahon, B.,** *J. Natl. Cancer Inst.,* 28, 1173, 1962.
28. United Nations Scientific Committee on the Effects of Atomic Radiation, Report, United Nations, New York, 1962.
29. National Academy of Sciences, The Biological Effects of Atomic Radiation, *Summary Reports,* National Research Council, Washington, D.C., 1956. (See also more recent report, Reference 34.)
30. **Muller, H. J.,** Radiation and human mutations, *Sci. Am.,* 193, 58, 1955.
31. **Henshaw, P. S.,** *Health Phys.,* 1, 141, 1958.
32. International Commission on Radiological Protection, *Radiosensitivity and Spatial Distribution of Dose,* ICRP Publication 14, Pergamon Press, Oxford, Engl., 1969.
33. **Evans, R. D.,** X-ray and gamma-ray interactions, in *Radiation Dosimetry,* 2nd ed., Attix, F. H. and Roesch, W. C., Eds., Academic Press, New York, 1968, 93.
34. **BEIR,** The Effects on Populations of Exposure to Low Levels of Ionizing Radiation, Report of the Advisory Committee on the Biological Effects of Ionizing Radiation (BEIR), Division of Medical Sciences, National Academy of Sciences, Washington, D.C., November 1972.

35. **Hubbell, J. H.,** *Photon Cross Sections, Attenuation Coefficients, and Energy Absorption Coefficients, from 10 KeV to 100 GeV,* NSRDS-NBS-29, U.S. Government Printing Office, Washington, D.C., August 1969 (available from Superintendent of Documents).
36. ICRU Report 23, Measurement of Absorbed Dose in a Phantom Irradiated by a Single Beam of X or Gamma Rays, International Commission on Radiation Units and Measurements, 7910 Woodmont Ave., Washington, D.C., 20014, January 15, 1973.
37. **Stewart, A. M., Webb, J., Giles, D., and Hewitt, D.,** Preliminary communication: malignant disease in childhood and diagnostic irradiation in utero, *Lancet,* ii, 447, 1956; see also Stewart, A. M., Low dose radiation cancers in man, in *Advances in Cancer Research,* Klein, G. and Weinhouse, S., Eds., Academic Press, New York, 1971, 359.
38. Scientific Committee on Radiation Dosimetry, AAPM, Protocol for the dosimetry of X- and gamma- ray beams with maximum energies between 0.6 and 50 MeV, *Phys. Med. Biol.,* 16, 397, 1971.
39. **Wang, Y., Ed.,** *CRC Handbook of Radioactive Nuclides,* The Chemical Rubber Co., Cleveland, 1969.
40. NCRP Report No. 39, Basic Radiation Protection Criteria, National Council on Radiation Protection and Measurements, 4201 Connecticut Ave., N. W., Washington, D.C., 20008, January 15, 1971.
41. **Hine, G. J. and Browell, G. L., Eds.,** *Radiation Dosimetry,* 1st ed., Academic Press, New York, 1956, 64.
42. **Cember, H.,** *Introduction to Health Physics,* Pergamon Press, London, 1969 (general introductory text).

3.2 DOSE TO VARIOUS BODY ORGANS FROM INHALATION OR INGESTION OF SOLUBLE RADIONUCLIDES[a]

D. F. Bunch

At the National Reactor Testing Station (NRTS) a great number of calculations are performed each year by the various contractors to estimate the radiological consequences from operational releases, minor and major incidents, and potential accidental releases. Much of this is repetitious in that each person making the calculation must perform the necessary mathematics to solve the various equations used in calculating dose. In almost all cases, the mathematical and biological parameters are those recommended for the "standard man" by the International Commission on Radiological Protection (ICRP).[1,2] To eliminate the need for this repetition and to establish more uniform practices in calculations, a computer program was written and estimates of dose were prepared for the isotopes and major organs listed in the ICRP reports. These estimates take the form of dose conversion factors, such that a rapid and reasonable estimate of dose may be made. This section gives information for dose from the ingestion or inhalation of soluble radionuclides.

Calculation of Dose

Since, for the most part, close accuracy is not desired or even warranted, the parameters used are those recommended by ICRP for continuous exposure. The general expression used is

$$\text{Dose} = \frac{AfET_E}{m}\left(1 - \exp\frac{-1.26 \times 10^4}{T_E}\right) \times \frac{1.6 \times 10^{-6} \times 3.2 \times 10^{15}}{0.693 \times 10^2} \tag{3.2–1}$$

where f = fractional uptake by ingestion or inhalation to the organ of interest, E = effective energy = $\Sigma EF(\text{RBE})n$, T_E = effective half-time of material in organ of interest, m = mass of organ, 1.6×10^{-6} = erg/MeV, 3.2×10^{15} = dis/day/curie, and 0.693×10^2 = erg/g/rad.

The exponential term assumes a 50-year post-exposure period to correct for certain isotopes that do not reach equilibrium in this time. The term A is defined as:

$A = 1$ to calculate rem per curie (Ci) inhaled or ingested (3.2–2)

$A = 1\,B$ to calculate rem/Ci-; sec/m^3 where B breathing rate in m^3/sec (Ci-sec/m^3 is the time integrated concentration of airborne radioactivity) (3.2–3)

$A = 1/f$ to calculate rem/Ci in the organ. (3.2–4)

The derivation, assumptions, and limitation of Equation 3.2–1 have been discussed in detail in References 1, 3, and 4, and these should be referred to for more detailed information (see especially Reference 3). It should be emphasized that these calculations should not be applied to the general population and, further, that derived doses are only approximations.

Application

Sample Calculation 1. If it is known that 1 μCi of I-131 has been inhaled, the estimation of dose may be made as follows:

1. If the thyroid is the organ of interest, the conversion factor from curies inhaled to dose in rem (from Table 3.2–1) is 1.48 E + 06 or 1.48×10^6 rem/Ci inhaled.

[a]This section is reprinted from *Dose to Various Body Organs from Inhalation or Ingestion of Soluble Radionuclides* which was prepared in the Health and Safety Division of the U.S. Atomic Energy Commission, Idaho Operations Office, AEC Research and Development Report, Health and Safety, TID-4500, August 1966.

Table 3.2–1
DOSE CONVERSION FACTORS

Notes:

Column 1: The identification format is Z.A. as in tritium 01.003. (Z = atomic number; A = atomic weight; M = metastable)
Column 2: This is $\Sigma EF(\text{RBE})n$ in MeV.
Column 3: This is T_E in days.
Column 4: Weight of organ of interest, assuming standard-man parameters; e.g., bone = 700 g.
Column 5: rem/Ci inhaled.
Column 6: rem/Ci-sec/m^3 for breathing rate typical of active portion of day, 10 m^3/8 hr.
Column 7: rem/Ci-sec/m^3 for average breathing rate, 20 m^3/24 hr.
Column 8: rem/Ci in organ.
Column 9: rem/Ci ingested.

Identification (Z.A.)	Energy (MeV)	T(1/2) Days	Weight in Grams	Inh. Dose (rem/Ci)	Dose/Conc rem/Ci-s/m³ Hi-Rate	Dose/Conc rem/Ci-s/m³ Lo-Rate	Organ Dose (rem/Ci)	Ing. Dose (rem/Ci)
THYROID 20 GRAMS								
24.051	8.40E−03	2.66E+01	20.0	1.90E+02	6.57E−02	4.29E−02	8.26E+05	3.72E−00
49.114M	9.20E−01	7.20E−00	20.0	2.45E+03	8.47E−01	5.53E−01	2.45E+07	1.96E+01
49.115M	1.60E−01	1.90E−01	20.0	1.12E+01	3.89E−03	2.53E−03	1.12E+05	8.99E−02
49.115	1.70E−01	8.40E−00	20.0	5.28E+02	1.82E−01	1.19E−01	5.28E+06	4.22E−00
50.113	1.60E−01	4.30E+01	20.0	7.12E+02	2.46E−01	1.60E−01	2.54E+07	1.27E+02
50.125	9.30E−01	8.40E−00	20.0	8.09E+02	2.79E−01	1.82E−01	2.89E+07	1.44E+02
51.122	5.90E−01	1.60E−00	20.0	2.79E+01	9.66E−03	6.30E−03	3.49E+06	3.14E−00
51.124	5.70E−01	3.80E−00	20.0	6.41E+01	2.21E−02	1.44E−02	8.01E+06	7.21E−00
51.125		4.00E−00	20.0	1.66E+01	5.76E−03	3.75E−03	2.08E+06	1.86E−00
52.125	1.10E−01	7.80E−00	20.0	1.20E+03	4.17E−01	2.72E−01	3.17E+06	7.93E+02
52.127M	3.00E−01	8.30E−00	20.0	3.50E+03	1.21E−00	7.90E−01	9.21E+06	2.30E+03
52.127	2.40E−01	3.70E−01	20.0	1.24E+02	4.31E−02	2.81E−02	3.28E+05	8.21E+01
52.129M	6.80E−01	7.10E−00	20.0	6.78E+03	2.34E−00	1.53E−00	1.78E+07	4.46E+03
52.129	6.00E−01	5.10E−02	20.0	4.30E+01	1.48E−02	9.70E−03	1.13E+05	2.83E+01
52.131M	6.90E−01	1.10E−00	20.0	1.06E+03	3.69E−01	2.40E−01	2.80E+06	7.02E+02
52.132	7.40E−01	2.40E−00	20.0	2.49E+03	8.63E−01	5.63E−01	6.57E+06	1.64E+03
53.126	1.60E−01	1.21E+01	20.0	1.64E+06	5.69E+02	3.71E+02	7.16E+06	2.14E+06
53.129	6.80E−02	1.38E+02	20.0	7.98E+06	2.76E+03	1.80E+03	3.47E+07	1.04E+07
53.131	2.30E−01	7.60E−00	20.0	1.48E+06	5.14E+02	3.35E+02	6.46E+06	1.94E+06
53.132	6.50E−01	9.70E−02	20.0	5.36E+04	1.85E+01	1.21E+01	2.33E+05	6.99E+04
53.133	5.40E−01	8.70E−01	20.0	3.99E+05	1.38E+02	9.02E+01	1.73E+06	5.21E+05
53.134	8.20E−01	3.60E−02	20.0	2.51E+04	8.69E−00	5.66E−00	1.09E+05	3.27E+04
53.135	5.20E−01	2.80E−01	20.0	1.23E+05	4.28E+01	2.79E+01	5.38E+05	1.61E+05
75.183	3.40E−02	2.90E−00	20.0	1.27E+03	4.41E−01	2.88E−01	3.64E+05	1.27E+03
75.186	3.60E−01	1.70E−00	20.0	7.92E+03	2.74E−00	1.78E−00	2.26E+06	7.92E+03
75.187	1.20E−02	3.00E−00	20.0	4.66E+02	1.61E−01	1.05E−01	1.33E+05	4.66E+02
75.188	8.00E−01	5.70E−01	20.0	5.90E+03	2.04E−00	1.33E−00	1.68E+06	5.90E+03
85.211	6.10E+01	3.00E−01	20.0	1.80E+06	6.22E+02	4.05E+02	7.82E+07	2.34E+06
KIDNEY 300 GRAMS								
4.007	1.20E−02	3.70E+01	300.0	8.21E+02	2.84E−01	1.85E−01	1.09E+05	6.57E−00
21.046	5.00E−01	4.00E+01	300.0	2.46E+04	8.53E−00	5.56E−00	4.93E+06	9.86E−00
21.047	2.00E−01	3.30E−00	300.0	8.14E+02	2.81E−01	1.83E−01	1.62E+05	3.25E−01
21.048	1.10E−00	1.80E−00	300.0	2.44E+03	8.44E−01	5.51E−01	4.88E+05	9.76E−01
23.048	7.00E−01	1.32E+01	300.0	2.27E+04	7.88E−00	5.14E−00	2.27E+06	1.82E+03

Table 3.2–1 (continued)
DOSE CONVERSION FACTORS

Identification (Z. A.)	Energy (MeV)	T(1/2) Days	Weight in Grams	Inh. Dose (rem/Ci)	Dose/Conc rem/Ci-s/m³ Hi-Rate	Dose/Conc rem/Ci-s/m³ Lo-Rate	Organ Dose (rem/Ci)	Ing. Dose (rem/Ci)
KIDNEY (Continued)								
24.051	1.20E−02	2.66E+01	300.0	5.35E+01	1.85E−02	1.20E−02	7.87E+04	1.02E−00
27.057	4.50E−02	9.20E−00	300.0	8.16E+01	2.82E−02	1.84E−02	1.02E+05	6.12E+01
27.058M	4.80E−02	3.70E−01	300.0	3.50E−00	1.21E−03	7.90E−04	4.38E+03	2.62E−00
27.058	2.20E−01	8.40E−00	300.0	3.64E+02	1.26E−01	8.22E−02	4.55E+05	2.73E+02
27.060	5.60E−01	9.50E−00	300.0	1.04E+03	3.63E−01	2.36E−01	1.31E+06	7.87E+02
29.064	1.70E−01	5.10E−01	300.0	4.27E+02	1.47E−01	9.65E−02	2.13E+04	2.13E+02
30.065	1.10E−01	9.30E+01	300.0	3.02E+04	1.04E+01	6.83E−00	2.52E+06	1.00E+04
30.069M	4.70E−01	5.80E−01	300.0	8.06E+02	2.79E−01	1.82E−01	6.72E+04	2.68E+02
30.069	3.70E−01	3.60E−02	300.0	3.94E+01	1.36E−02	8.89E−03	3.28E+03	1.31E+01
31.072	8.90E−01	5.50E−01	300.0	6.03E+02	2.08E−01	1.36E−01	1.20E+05	2.41E−00
32.071	1.00E−02	6.00E−00	300.0	1.18E+02	4.09E−02	2.67E−02	1.48E+04	4.44E−00
33.073	3.60E−02	6.70E+01	300.0	1.60E+03	5.55E−01	3.62E−01	5.94E+05	1.78E+02
33.074	3.40E−01	1.70E+01	300.0	3.84E+03	1.33E−00	8.68E−01	1.42E+06	4.27E+02
33.076	1.10E−00	1.10E−00	300.0	8.05E+02	2.78E−01	1.81E−01	2.98E+05	8.95E+01
33.077	2.40E−01	1.60E−00	300.0	2.55E+02	8.84E−02	5.77E−02	9.47E+04	2.84E+01
34.075	7.20E−02	1.01E+01	300.0	5.38E+03	1.86E−00	1.21E−00	1.79E+05	7.17E+03
40.093	1.90E−02	9.00E+02	300.0	2.37E+04	8.24E−00	5.37E−00	4.76E+06	9.53E−00
40.095	4.60E−01	5.90E+01	300.0	3.34E+04	1.15E+01	7.55E−00	6.69E+06	1.33E+01
40.097	1.50E−00	7.10E−01	300.0	1.31E+03	4.54E−01	2.96E−01	2.62E+05	5.25E−01
41.093M	3.80E−02	6.40E+02	300.0	2.99E+04	1.03E+01	6.76E−00	5.99E+06	1.19E+01
41.095	2.00E−02	3.35E+01	300.0	8.26E+02	2.85E−01	1.86E−01	1.65E+05	3.30E−01
41.097	6.00E−01	5.10E−02	300.0	3.77E+01	1.30E−02	8.51E−03	7.54E+03	1.50E−02
42.099	4.80E−01	1.50E−00	300.0	9.31E+03	3.21E−00	2.09E−00	1.70E+05	1.02E+04
43.096M	4.20E−01	3.60E−02	300.0	1.86E+01	6.45E−03	4.20E−03	3.72E+03	1.86E+01
43.096	4.70E−01	3.50E−00	300.0	2.02E+03	7.01E−01	4.57E−01	4.05E+05	2.02E+03
43.097M	9.00E−02	1.60E+01	300.0	1.77E+03	6.14E−01	4.00E−01	3.55E+05	1.77E+03
43.097	2.00E−02	2.00E+01	300.0	4.93E+02	1.70E−01	1.11E−01	9.86E+04	4.93E+02
43.099M	2.60E−02	2.50E−01	300.0	8.01E−00	2.77E−03	1.80E−03	1.60E+03	8.01E−00
43.099	9.40E−02	2.00E+01	300.0	2.31E+03	8.02E−01	5.23E−01	4.63E+05	2.31E+03
44.097	7.70E−02	1.30E−00	300.0	1.23E+03	4.27E−01	2.78E−01	2.46E+04	1.48E+02
44.103	2.20E−01	2.40E−00	300.0	6.51E+03	2.25E−00	1.46E−00	1.30E+05	7.81E+02
44.105	8.40E−01	1.80E−01	300.0	1.86E+03	6.45E−01	4.20E−01	3.72E+04	2.23E+02
44.106	1.30E−00	2.48E−00	300.0	3.97E+04	1.37E+01	8.97E−00	7.95E+05	4.77E+03
45.103M	5.40E−02	3.80E−02	300.0	5.06E−00	1.75E−03	1.14E−03	5.06E+02	3.03E−00
45.105	1.90E−01	1.44E−00	300.0	6.74E+02	2.33E−01	1.52E−01	6.74E+04	4.04E+02
46.103	6.10E−02	1.10E+01	300.0	4.96E+03	1.71E−00	1.12E−00	1.65E+05	3.31E+03
46.109	4.20E−01	5.60E−01	300.0	1.74E+03	6.02E−01	3.92E−01	5.80E+04	1.16E+03
47.105	2.20E−01	8.00E−00	300.0	2.17E+03	7.50E−01	4.89E−01	4.34E+05	8.68E+01
47.110M	6.50E−01	1.00E+01	300.0	8.01E+03	2.77E−00	1.80E−00	1.60E+06	3.20E+02
47.111	3.70E−01	4.00E−00	300.0	1.82E+03	6.31E−01	4.11E−01	3.65E+05	7.30E+01
48.109	9.80E−02	1.84E+02	300.0	1.11E+05	3.84E+01	2.50E+01	4.44E+06	1.11E+03
48.115M	6.10E−01	3.80E+01	300.0	1.42E+05	4.94E+01	3.22E+01	5.71E+06	1.42E+03
48.115	5.60E−01	2.20E−00	300.0	7.59E+03	2.62E−00	1.71E−00	3.03E+05	7.59E+01
49.113M	1.90E−01	7.30E−00	300.0	3.42E+03	1.18E−00	7.72E−01	3.42E+05	2.73E+01
49.114M	9.30E−01	2.70E+01	300.0	6.19E+04	2.14E+01	1.39E+01	6.19E+06	4.95E+02
49.115M	1.90E−01	1.90E−01	300.0	8.90E+01	3.08E−02	2.00E−02	8.90E+03	7.12E−01
49.115	1.70E−01	6.00E+01	300.0	2.51E+04	8.70E−00	5.67E−00	2.51E+06	2.01E+02
52.125M	1.40E−01	2.00E+01	300.0	2.07E+04	7.16E−00	4.67E−00	6.90E+05	1.38E+04

Table 3.2–1 (continued)
DOSE CONVERSION FACTORS

Identification (Z.A.)	Energy (MeV)	T(1/2) Days	Weight in Grams	Inh. Dose (rem/Ci)	Dose/Conc rem/Ci-s/m³ Hi-Rate	Dose/Conc rem/Ci-s/m³ Lo-Rate	Organ Dose (rem/Ci)	Ing. Dose (rem/Ci)
KIDNEY (Continued)								
52.127M	3.20E−01	2.30E+01	300.0	5.44E+04	1.88E+01	1.22E+01	1.81E+06	3.63E+04
52.127	2.40E−01	3.90E−01	300.0	6.92E+02	2.39E−01	1.56E−01	2.30E+04	4.61E+02
52.129M	7.80E−01	1.60E+01	300.0	9.23E+04	3.19E+01	2.08E+01	3.07E+06	6.15E+04
52.129	6.80E−01	5.10E−02	300.0	2.56E+02	8.87E−02	5.79E−02	8.55E+03	1.71E+02
52.131M	8.10E−01	1.20E−00	300.0	7.19E+03	2.48E−00	1.62E−00	2.39E+05	4.79E+03
52.132	9.60E−01	2.90E−00	300.0	2.06E+04	7.12E−00	4.64E−00	6.86E+05	1.37E+04
55.131	2.10E−01	8.00E−00	300.0	3.10E+03	1.07E−00	7.01E−01	4.14E+05	4.14E+03
55.134M	1.10E−01	1.30E−01	300.0	2.64E+01	9.15E−03	5.97E−03	3.52E+03	3.52E+01
55.134	4.60E−01	4.00E+01	300.0	3.40E+04	1.17E+01	7.68E−00	4.53E+06	4.53E+04
55.135	6.60E−02	4.20E+01	300.0	5.12E+03	1.77E−00	1.15E−00	6.83E+05	6.83E+03
55.136	2.90E−01	1.00E+01	300.0	5.36E+03	1.85E−00	1.21E−00	7.15E+05	7.15E+03
55.137	3.70E−01	4.20E+01	300.0	2.87E+04	9.94E−00	6.48E−00	3.83E+06	3.83E+04
56.131	1.40E−01	4.90E−00	300.0	4.73E−00	1.63E−03	1.06E−03	1.69E+05	8.46E−01
56.140	1.20E−00	5.10E−00	300.0	4.22E+01	1.46E−02	9.53E−03	1.50E+06	7.54E−00
58.141	1.80E−01	3.00E+01	300.0	6.66E+03	2.30E−00	1.50E−00	1.33E+06	2.66E−00
58.143	8.20E−01	1.33E−00	300.0	1.34E+03	4.65E−01	3.03E−01	2.69E+05	5.38E−01
58.144	1.30E−00	1.91E+02	300.0	3.06E+05	1.05E+02	6.91E+01	6.12E+07	1.22E+02
59.142	8.10E−01	8.00E−01	300.0	7.99E+02	2.76E−01	1.80E−01	1.59E+05	3.19E−01
59.143	3.20E−01	1.35E+01	300.0	5.32E+03	1.84E−00	1.20E−00	1.06E+06	2.13E−00
60.144	2.00E+01	6.56E+02	300.0	3.23E+07	1.11E+04	7.30E+03	3.23E+09	1.61E+04
60.147	3.10E−01	1.11E+01	300.0	8.48E+03	2.93E−00	1.91E−00	8.48E+05	4.24E−00
60.149	9.70E−01	8.30E−02	300.0	1.98E+02	6.87E−02	4.48E−02	1.98E+04	9.92E−02
61.147	6.90E−02	3.83E+02	300.0	3.25E+04	1.12E+01	7.35E−00	6.51E+06	1.30E+01
61.149	4.20E−01	2.20E−00	300.0	1.13E+03	3.94E−01	2.57E−01	2.27E+05	4.55E−01
62.147	2.30E+01	6.56E+02	300.0	1.86E+07	6.43E+03	4.19E+03	3.72E+09	7.44E+03
62.151	4.20E−02	6.45E+02	300.0	3.34E+04	1.15E+01	7.54E−00	6.68E+06	1.33E+01
62.153	2.50E−01	1.95E−00	300.0	6.01E+02	2.08E−01	1.35E−01	1.20E+05	2.40E−01
63.152	7.10E−01	3.80E−01	300.0	4.99E+02	1.72E−01	1.12E−01	6.65E+04	1.99E−01
63.152	2.50E−01	1.13E+03	300.0	5.22E+05	1.80E+02	1.17E+02	6.96E+07	2.09E+02
63.154	7.60E−01	1.18E+03	300.0	1.65E+06	5.73E+02	3.74E+02	2.21E+08	6.63E+02
63.155	8.30E−02	4.38E+02	300.0	6.72E+04	2.32E+01	1.51E+01	8.96E+06	2.69E+01
65.160	4.00E−01	6.60E+01	300.0	4.88E+04	1.68E+01	1.10E+01	6.51E+06	1.95E+01
67.166	6.90E−01	1.10E−00	300.0	9.36E+02	3.23E−01	2.11E−01	1.87E+05	3.74E−01
68.169	1.90E−01	9.30E−00	300.0	2.17E+03	7.53E−01	4.91E−01	4.35E+05	8.71E−01
68.171	4.60E−01	3.10E−01	300.0	1.75E+02	6.08E−02	3.96E−02	3.51E+04	7.03E−02
69.170	3.40E−01	9.20E+01	300.0	3.85E+04	1.33E+01	8.70E−00	7.71E+06	1.54E+01
69.171	3.00E−02	2.26E+02	300.0	8.36E+03	2.89E−00	1.88E−00	1.67E+06	3.34E−00
70.175	1.50E−01	4.10E−00	300.0	1.97E+03	6.82E−01	4.45E−01	1.51E+05	7.58E−01
71.177	1.60E−01	6.70E−00	300.0	6.61E+02	2.28E−01	1.49E−01	2.64E+05	2.64E−01
72.181	2.50E−01	4.30E+01	300.0	1.32E+04	4.58E−00	2.99E−00	2.65E+06	5.30E−00
73.182	4.50E−01	8.80E+01	300.0	7.32E+04	2.53E+01	1.65E+01	9.76E+06	2.93E+01
76.185	2.50E−01	4.80E−00	300.0	5.92E+03	2.04E−00	1.33E−00	2.96E+05	1.48E+03
76.191M	3.90E−02	5.20E−01	300.0	1.00E+02	3.46E−02	2.25E−02	5.00E+03	2.50E+01
76.191	1.10E−01	3.80E−00	300.0	2.06E+03	7.13E−01	4.65E−01	1.03E+05	5.15E+02
76.193	3.80E−01	1.00E−00	300.0	1.87E+03	6.48E−01	4.23E−01	9.37E+04	4.68E+02
77.190	1.20E−01	9.70E−00	300.0	4.01E+03	1.39E−00	9.07E−01	2.87E+05	1.29E+03
77.192	5.00E−01	3.00E+01	300.0	5.18E+04	1.79E+01	1.16E+01	3.70E+06	1.66E+04
77.194	8.10E−01	7.80E−01	300.0	2.18E+03	7.54E−01	4.92E−01	1.55E+05	7.01E+02

Table 3.2–1 (continued)
DOSE CONVERSION FACTORS

Identification (Z. A.)	Energy (MeV)	T(1/2) Days	Weight in Grams	Inh. Dose (rem/Ci)	Dose/Conc rem/Ci-s/m³ Hi-Rate	Dose/Conc rem/Ci-s/m³ Lo-Rate	Organ Dose (rem/Ci)	Ing. Dose (rem/Ci)
KIDNEY (Continued)								
78.191	2.20E−01	2.90E−00	300.0	4.72E+03	1.63E−00	1.06E−00	1.57E+05	1.57E+03
78.193M	2.30E−02	3.20E−00	300.0	5.44E+02	1.88E−01	1.22E−01	1.81E+04	1.81E+02
78.193	1.40E−02	6.00E+01	300.0	6.21E+03	2.15E−00	1.40E−00	2.07E+05	2.07E+03
78.197	5.00E−01	5.60E−02	300.0	2.07E+02	7.16E−02	4.67E−02	6.90E+03	6.90E+01
78.197	2.30E−01	7.40E−01	300.0	1.25E+03	4.35E−01	2.84E−01	4.19E+04	4.19E+02
79.196	1.50E−01	5.50E−00	300.0	1.83E+03	6.33E−01	4.13E−01	2.03E+05	6.10E+02
79.198	4.10E−01	2.70E−00	300.0	2.45E+03	8.50E−01	5.54E−01	2.73E+05	8.19E+02
79.199	1.20E−01	3.10E−00	300.0	8.25E+02	2.85E−01	1.86E−01	9.17E+04	2.75E+02
80.197M	1.80E−01	9.40E−01	300.0	9.18E+03	3.17E−00	2.07E−00	4.17E+04	1.08E+04
80.197	4.30E−02	2.30E−00	300.0	5.36E+03	1.85E−00	1.21E−00	2.43E+04	6.34E+03
80.203	1.50E−01	1.10E+01	300.0	8.95E+04	3.09E+01	2.02E+01	4.07E+05	1.05E+05
81.200	1.30E−01	9.70E−01	300.0	7.46E+02	2.58E−01	1.68E−01	3.11E+04	7.15E+02
81.201	1.10E−01	2.10E−00	300.0	1.36E+03	4.73E−01	3.08E−01	5.69E+04	1.31E+03
81.202	2.40E−01	4.40E−00	300.0	6.25E+03	2.16E−00	1.41E−00	2.60E+05	5.99E+03
81.204	2.50E−01	7.00E−00	300.0	1.03E+04	3.58E−00	2.33E−00	4.31E+05	9.92E+03
82.203	6.90E−02	2.16E−00	300.0	1.47E+03	5.08E−01	3.31E−01	3.67E+04	3.67E+02
82.210	1.00E+01	4.94E+02	300.0	4.87E+07	1.68E+04	1.09E+04	1.21E+09	1.21E+07
82.212	8.10E+01	4.40E−01	300.0	3.51E+05	1.21E+02	7.93E+01	8.79E+06	8.79E+04
83.206	5.80E−01	3.10E−00	300.0	3.54E+04	1.22E+01	8.00E−00	4.43E+05	1.33E+03
83.207	3.30E−01	6.00E−00	300.0	3.90E+04	1.35E+01	8.81E−00	4.88E+05	1.46E+03
83.210	1.90E+01	3.00E−00	300.0	1.12E+06	3.89E+02	2.53E+02	1.40E+07	4.21E+04
83.212	8.20E+01	4.20E−02	300.0	6.79E+04	2.35E+01	1.53E+01	8.49E+05	2.54E+03
84.210	5.50E+01	4.60E+01	300.0	1.24E+07	4.31E+03	2.81E+03	6.24E+08	2.49E+06
89.227	6.20E+01	6.00E+03	300.0	2.41E+08	8.35E+04	5.45E+04	8.05E+10	8.05E+04
89.228	5.50E+01	2.60E−01	300.0	1.05E+04	3.66E−00	2.38E−00	3.52E+06	3.52E−00
90.227	6.10E+01	1.84E+01	300.0	2.76E+06	9.57E+02	6.24E+02	2.76E+08	1.38E+03
90.228	5.60E+01	6.78E+02	300.0	9.36E+07	3.23E+04	2.11E+04	9.36E+09	4.68E+04
90.230	4.80E+01	2.20E+04	300.0	1.13E+09	3.92E+05	2.56E+05	1.13E+11	5.67E+05
90.231	1.40E−01	1.07E−00	300.0	3.69E+02	1.27E−01	8.33E−02	3.69E+04	1.84E−01
90.232	4.10E+01	2.20E+04	300.0	9.70E+08	3.35E+05	2.18E+05	9.70E+10	4.85E+05
90.234	9.00E−01	2.41E+01	300.0	5.35E+04	1.85E+01	1.20E+01	5.35E+06	2.67E+01
91.230	1.50E+02	1.77E+01	300.0	1.31E+06	4.52E+02	2.94E+02	1.31E+08	5.22E+02
91.231	7.90E+01	5.10E+04	300.0	2.17E+09	7.52E+05	4.90E+05	2.17E+11	8.70E+05
91.233	1.50E−01	2.74E+01	300.0	1.01E+04	3.50E−00	2.28E−00	1.01E+06	4.05E−00
92.230	3.50E+02	8.70E−00	300.0	2.10E+07	7.27E+03	4.74E+03	7.51E+08	8.26E+03
92.232	1.10E+02	1.50E+01	300.0	1.13E+07	3.94E+03	2.57E+03	4.07E+08	4.47E+03
92.233	5.00E+01	1.50E+01	300.0	5.18E+06	1.79E+03	1.16E+03	1.85E+08	2.03E+03
92.234	4.90E+01	1.50E+01	300.0	5.07E+06	1.75E+03	1.14E+03	1.81E+08	1.99E+03
92.235	4.60E+01	1.50E+01	300.0	4.76E+06	1.64E+03	1.07E+03	1.70E+08	1.87E+03
92.236	4.70E+01	1.50E+01	300.0	4.86E+06	1.68E+03	1.09E+03	1.73E+08	1.91E+03
92.238	4.30E+01	1.50E+01	300.0	4.45E+06	1.54E+03	1.00E+03	1.59E+08	1.75E+03
92.240	1.10E−00	5.66E−01	300.0	4.30E+03	1.48E−00	9.70E−01	1.53E+05	1.68E+02
93.237	4.90E+01	6.40E+04	300.0	1.03E+09	3.58E+05	2.33E+05	1.38E+11	4.14E+05
93.239	2.10E−01	2.33E−00	300.0	9.05E+02	3.13E−01	2.04E−01	1.20E+05	3.62E−01
94.238	5.70E+01	1.60E+04	300.0	1.22E+10	4.24E+06	2.76E+06	1.22E+11	4.90E+06
94.239	5.30E+01	3.20E+04	300.0	1.36E+10	4.71E+06	3.07E+06	1.36E+11	5.44E+06
94.240	5.30E+01	3.20E+04	300.0	1.36E+10	4.71E+06	3.07E+06	1.36E+11	5.44E+06
94.241	2.50E−00	4.20E+03	300.0	2.46E+08	8.51E+04	5.55E+04	2.46E+09	9.84E+04

Table 3.2–1 (continued)
DOSE CONVERSION FACTORS

Identification (Z. A.)	Energy (MeV)	T(1/2) Days	Weight in Grams	Inh. Dose (rem/Ci)	Dose/Conc rem/Ci-s/m³ Hi-Rate	Dose/Conc rem/Ci-s/m³ Lo-Rate	Organ Dose (rem/Ci)	Ing. Dose (rem/Ci)
KIDNEY (Continued)								
94.242	5.10E+01	3.20E+04	300.0	1.31E+10	4.53E+06	2.95E+06	1.31E+11	5.24E+06
94.243	3.70E−01	2.08E−01	300.0	9.49E+01	3.28E−02	2.14E−02	1.89E+04	1.13E−02
94.244	4.80E+01	3.20E+04	300.0	7.52E+08	2.60E+05	1.70E+05	1.50E+11	9.02E+04
95.241	5.70E+01	2.30E+04	300.0	1.02E+09	3.53E+05	2.30E+05	1.36E+11	4.09E+05
95.242	7.20E+01	1.80E+04	300.0	1.01E+09	3.50E+05	2.28E+05	1.34E+11	4.05E+05
95.242	7.80E+01	6.67E−01	300.0	8.08E+04	2.79E+01	1.82E+01	1.08E+07	3.23E+01
95.243	5.40E+01	2.70E+04	300.0	1.00E+09	3.47E+05	2.26E+05	1.34E+11	4.02E+05
95.244	4.40E+01	1.80E−02	300.0	1.46E+03	5.06E−01	3.30E−01	1.95E+05	5.86E−01
96.242M	7.80E+01	1.61E+02	300.0	1.54E+07	5.35E+03	3.49E+03	3.09E+09	6.19E+03
96.243	6.00E+01	8.40E+03	300.0	4.82E+08	1.67E+05	1.08E+05	9.65E+10	1.93E+05
96.244	6.00E+01	5.20E+03	300.0	3.50E+08	1.21E+05	7.91E+04	7.01E+10	1.40E+05
96.245	5.60E+01	2.40E+04	300.0	6.77E+08	2.34E+05	1.52E+05	1.35E+11	2.70E+05
96.246	5.60E+01	2.40E+04	300.0	6.77E+08	2.34E+05	1.52E+05	1.35E+11	2.70E+05
96.247	5.50E+01	2.40E+04	300.0	6.64E+08	2.30E+05	1.50E+05	1.32E+11	2.65E+05
96.248	5.20E+01	2.40E+04	300.0	5.46E+09	1.89E+06	1.23E+06	1.09E+12	2.18E+06
SKELETON 7000 GRAMS								
4.007	8.50E−03	4.80E+01	7000.0	3.45E+02	1.19E−01	7.78E−02	4.31E+03	2.76E−00
6.014	2.70E−01	4.00E+01	7000.0	2.28E+03	7.89E−01	5.15E−01	1.14E+05	2.85E+03
9.018	1.40E−00	7.80E−02	7000.0	4.61E+02	1.59E−01	1.04E−01	1.15E+03	6.11E+02
15.032	3.50E−00	1.41E+01	7000.0	1.66E+05	5.77E+01	3.76E+01	5.21E+05	1.98E+05
16.035	2.80E−01	7.61E+01	7000.0	4.50E+03	1.55E−00	1.01E−00	2.25E+05	6.75E+03
20.045	4.30E−01	1.62E+02	7000.0	3.68E+05	1.27E+02	8.30E+01	7.36E+05	3.97E+05
20.047	2.60E−00	4.90E−00	7000.0	6.73E+04	2.32E+01	1.51E+01	1.34E+05	7.27E+04
21.046	9.00E−01	2.40E+01	7000.0	1.14E+04	3.94E−00	2.57E−00	2.28E+05	4.56E−00
21.047	8.90E−01	3.10E−00	7000.0	1.45E+03	5.04E−01	3.29E−01	2.91E+04	5.83E−01
21.048	1.60E−00	1.70E−00	7000.0	1.43E+03	4.97E−01	3.24E−01	2.87E+04	5.75E−01
23.048	1.20E−00	1.44E+01	7000.0	7.30E+03	2.52E−00	1.64E−00	1.82E+05	5.11E+02
26.055	6.50E−03	6.65E+02	7000.0	1.37E+03	4.74E−01	3.09E−01	4.56E+04	4.56E+02
28.059	7.70E−03	8.00E+02	7000.0	1.30E+04	4.50E−00	2.93E−00	6.51E+04	9.76E+03
28.063	1.10E−01	7.79E+02	7000.0	1.81E+05	6.26E+01	4.08E+01	9.05E+05	1.35E+05
28.065	5.30E−00	1.10E−01	7000.0	1.23E+03	4.26E−01	2.78E−01	6.16E+03	9.24E+02
30.065	9.40E−02	2.06E+02	7000.0	9.21E+03	3.18E−00	2.07E−00	2.04E+05	3.07E+03
30.069M	2.10E−00	5.80E−01	7000.0	5.79E+02	2.00E−01	1.30E−01	1.28E+04	1.93E+02
30.069	1.90E−00	3.60E−02	7000.0	3.25E+01	1.12E−02	7.34E−03	7.23E+02	1.08E+01
31.072	2.60E−00	5.60E−01	7000.0	1.15E+03	3.99E−01	2.60E−01	1.53E+04	4.61E−00
38.085M	3.40E−02	4.90E−02	7000.0	1.69E−01	5.85E−03	3.82E−03	6.04E+01	1.26E−01
38.085	9.10E−02	6.48E+01	7000.0	1.74E+04	6.03E−00	3.93E−00	6.23E+04	1.30E+04
38.089	2.80E−00	5.04E+01	7000.0	4.17E+05	1.44E+02	9.42E+01	1.49E+06	3.13E+05
38.090	5.50E−00	6.40E+03	7000.0	3.84E+07	1.32E+04	8.67E+03	3.20E+08	2.88E+07
38.091	7.40E−00	4.00E−01	7000.0	7.53E+03	2.61E−00	1.69E−00	2.68E+04	5.65E+03
38.092	8.00E−00	1.10E−01	7000.0	2.60E+03	9.01E−01	5.87E−01	9.30E+03	1.95E+03
39.090	4.40E−00	2.68E−00	7000.0	2.36E+04	8.19E−00	5.34E−00	1.24E+05	9.34E−00
39.091M	3.00E−00	3.50E−02	7000.0	2.10E+02	7.29E−02	4.75E−02	1.11E+03	8.32E−02
39.091	2.90E−00	5.80E+01	7000.0	3.37E+05	1.16E+02	7.62E+01	1.77E+06	1.33E+02
39.092	6.90E−00	1.50E−01	7000.0	2.07E+03	7.19E−01	4.69E−01	1.09E+04	8.20E−01

Table 3.2–1 (continued)
DOSE CONVERSION FACTORS

Identification (Z.A.)	Energy (MeV)	T(1/2) Days	Weight in Grams	Inh. Dose (rem/Ci)	Dose/Conc rem/Ci-s/m³ Hi-Rate	Dose/Conc rem/Ci-s/m³ Lo-Rate	Organ Dose (rem/Ci)	Ing. Dose (rem/Ci)
SKELETON (Continued)								
39.093	6.50E−00	4.20E−01	7000.0	5.48E+03	1.89E−00	1.23E−00	2.88E+04	2.16E−00
40.093	9.50E−02	1.00E+03	7000.0	9.75E+04	3.37E+01	2.19E+01	1.08E+06	3.90E+01
40.095	1.10E−00	5.95E+01	7000.0	6.22E+04	2.15E+01	1.40E+01	6.91E+05	2.49E+01
40.097	6.20E−00	7.10E−01	7000.0	4.18E+03	1.44E−00	9.45E−01	4.65E+04	1.67E−00
41.093M	1.20E−01	7.87E+02	7000.0	9.98E+04	3.45E+01	2.25E+01	9.98E+05	3.79E+01
41.095	3.70E−01	3.38E+01	7000.0	1.32E+04	4.57E−00	2.98E−00	1.32E+05	5.02E−00
41.097	2.40E−00	5.10E−02	7000.0	1.29E+02	4.47E−02	2.92E−02	1.29E+03	4.91E−02
43.096M	3.90E−01	3.60E−02	7000.0	1.48E−01	5.13E−05	3.34E−05	1.48E+02	1.48E−01
43.096	3.50E−01	3.70E−00	7000.0	1.36E+01	4.73E−03	3.08E−03	1.36E+04	1.36E+01
43.097M	3.70E−01	2.00E+01	7000.0	7.82E+01	2.70E−02	1.76E−02	7.82E+04	7.82E+01
43.097	1.90E−02	2.50E+01	7000.0	5.02E−00	1.73E−03	1.13E−03	5.02E+03	5.02E−00
43.099M	2.00E−02	2.50E−01	7000.0	5.28E−02	1.82E−05	1.19E−05	5.28E+01	5.28E−02
43.099	4.70E−01	2.50E+01	7000.0	1.24E+02	4.29E−02	2.80E−02	1.24E+05	1.24E+02
44.097	1.30E−01	2.40E−00	7000.0	6.95E+01	2.28E−02	1.48E−02	3.29E+03	7.91E−00
44.103	6.20E−01	1.20E+01	7000.0	1.57E+03	5.44E−01	3.54E−01	7.86E+04	1.88E+02
44.105	3.50E−00	1.90E−01	7000.0	1.40E+02	4.86E−02	3.17E−02	7.03E+03	1.68E+01
44.106	6.50E−00	1.50E+01	7000.0	2.06E+04	7.13E−00	4.65E−00	1.03E+06	2.47E+03
45.103M	1.90E−01	3.80E−02	7000.0	1.52E−00	5.28E−04	3.44E−04	7.63E+01	7.63E−01
45.105	9.50E−01	1.39E−00	7000.0	2.79E+02	9.65E−02	6.30E−02	1.39E+04	1.39E+02
47.105	1.60E−01	1.70E+01	7000.0	3.73E+02	1.29E−01	8.43E−02	2.87E+04	1.43E+01
47.110M	1.10E−00	2.60E+01	7000.0	3.93E+03	1.35E−00	8.87E−01	3.02E+05	1.51E+02
47.111	1.80E−00	6.00E−00	7000.0	1.48E+03	5.13E−01	3.34E−01	1.14E+05	5.70E+01
49.113M	6.80E−01	7.30E−02	7000.0	2.09E+01	7.26E−03	4.73E−03	5.24E+02	1.78E−01
49.114M	4.50E−00	2.60E+01	7000.0	4.94E+04	1.71E+01	1.11E+01	1.23E+06	4.20E+02
49.115M	7.40E−01	1.90E−01	7000.0	5.94E+01	2.05E−02	1.34E−02	1.48E+03	5.05E−01
49.115	8.50E−01	5.70E+01	7000.0	2.04E+04	7.08E−00	4.62E−00	5.12E+05	1.74E+02
50.113	7.00E−01	5.30E+01	7000.0	3.13E+04	1.08E+01	7.08E−00	3.92E+05	7.84E+03
50.125	4.80E−00	8.70E−00	7000.0	3.53E+04	1.22E+01	7.97E−00	4.41E+05	8.82E+03
51.122	2.30E−00	2.70E−00	7000.0	1.96E+03	6.81E−01	4.44E−01	6.56E+04	1.96E+02
51.124	6.90E−01	3.80E+01	7000.0	8.31E+03	2.87E−00	1.87E−00	2.77E+05	8.31E+02
51.125	2.80E−01	9.00E+01	7000.0	1.77E+03	6.12E−00	4.00E−00	5.89E+05	1.77E+03
52.125M	5.10E−01	2.00E+01	7000.0	3.66E+03	1.26E−00	8.27E−01	1.07E+05	2.48E+03
52.127M	1.50E−00	2.30E+01	7000.0	1.24E+04	4.28E−00	2.79E−00	3.64E+05	8.38E+03
52.127	1.20E−00	3.90E−01	7000.0	1.68E+02	5.81E−02	3.79E−02	4.94E+03	1.13E+02
52.129M	3.20E−00	1.60E+01	7000.0	1.84E+04	6.36E−00	4.15E−00	5.41E+05	1.24E+04
52.129	2.80E−00	5.10E−02	7000.0	5.13E+01	1.77E−02	1.15E−02	1.50E+03	3.47E+01
52.131M	2.60E−00	1.20E−00	7000.0	1.12E+03	3.87E−01	2.53E−01	3.29E+04	7.58E+02
52.132	3.10E−00	2.90E−00	7000.0	3.23E+03	1.11E−00	7.29E−01	9.50E+04	2.18E+03
55.134M	1.70E−02	9.30E−00	7000.0	5.01E+01	1.73E−02	1.13E−02	1.67E+03	6.68E+01
55.131	4.90E−01	1.30E−01	7000.0	2.02E+01	6.98E−03	4.55E−03	6.73E+02	2.69E+01
55.134	9.90E−01	1.20E+02	7000.0	3.76E+04	1.30E+01	8.50E−00	1.25E+06	5.02E+04
55.135	3.30E−01	1.40E+02	7000.0	1.46E+04	5.06E−00	3.30E−00	4.88E+05	1.95E+04
55.136	7.20E−01	1.19E+01	7000.0	2.71E+03	9.40E−01	6.13E−01	9.05E+04	3.62E+03
55.137	1.40E−00	1.38E+02	7000.0	6.12E+04	2.11E+01	1.38E+01	2.04E+06	8.16E+04
56.131	1.10E−01	9.80E−00	7000.0	2.16E+03	7.49E−01	4.88E−01	1.13E+04	3.98E+02
56.140	4.20E−00	1.07E+01	7000.0	9.02E+04	3.12E+01	2.03E+01	4.75E+05	1.66E+04
57.140	2.70E−00	1.68E−00	7000.0	4.79E+03	1.65E−00	1.08E−00	4.79E+04	1.91E+04
58.141	8.10E−01	3.10E+01	7000.0	1.99E+04	6.88E−00	4.49E−00	2.65E+05	7.96E−00

Table 3.2–1 (continued)
DOSE CONVERSION FACTORS

Identification (Z. A.)	Energy (MeV)	T(1/2) Days	Weight in Grams	Inh. Dose (rem/Ci)	Dose/Conc rem/Ci-s/m³ Hi-Rate	Dose/Conc rem/Ci-s/m³ Lo-Rate	Organ Dose (rem/Ci)	Ing. Dose (rem/Ci)
SKELETON (Continued)								
58.143	3.80E−00	1.33E−00	7000.0	4.00E+03	1.38E−00	9.04E−01	5.34E+04	1.60E−00
58.144	6.30E−00	2.43E+02	7000.0	1.21E+06	4.19E+02	2.73E+02	1.61E+07	4.85E+02
59.142	3.90E−00	8.00E−01	7000.0	3.29E+03	1.14E−00	7.44E−01	3.29E+04	1.31E−00
59.143	1.60E−00	1.36E+01	7000.0	2.30E+04	7.95E−00	5.19E−00	2.30E+05	9.20E−00
,60.144	1.00E+02	1.50E+03	7000.0	1.42E+08	4.93E+04	3.21E+04	1.58E+09	5.54E+04
60.147	1.40E−00	1.12E+01	7000.0	1.49E+04	5.16E−00	3.36E−00	1.65E+05	5.80E−00
60.149	4.70E−00	8.30E−02	7000.0	3.71E+02	1.28E−01	8.37E−02	4.12E+03	1.44E−01
61.147	3.50E−01	5.70E+02	7000.0	1.89E+05	6.56E+01	4.28E+01	2.10E+06	7.38E+01
61.149	1.90E−00	2.20E−00	7000.0	3.97E+03	1.37E−00	8.97E−01	4.41E+04	1.54E−00
62.147	1.15E+02	1.50E+03	7000.0	1.64E+08	5.67E+04	3.70E+04	1.82E+09	6.38E+04
62.151	1.30E−01	1.44E+03	7000.0	1.78E+05	6.16E+01	4.01E+01	1.97E+06	6.92E+01
62.153	1.10E−00	1.96E−00	7000.0	2.05E+03	7.09E−01	4.62E−01	2.27E+04	7.97E−01
63.152	2.90E−00	3.80E−01	7000.0	1.04E+03	3.62E−01	2.36E−01	1.16E+04	4.19E−01
63.152	4.50E−01	1.14E+03	7000.0	4.88E+05	1.68E+02	1.10E+02	5.42E+06	1.95E+02
63.154	2.70E−00	1.19E+03	7000.0	3.05E+06	1.05E+03	6.89E+02	3.39E+07	1.22E+03
63.155	2.80E−01	4.39E+02	7000.0	1.16E+05	4.04E+01	2.63E+01	1.29E+06	4.67E+01
64.153	2.30E−01	1.91E+02	7000.0	5.10E+04	1.76E+01	1.15E+01	4.64E+05	2.08E+01
64.159	7.50E−01	7.50E−01	7000.0	6.54E+02	2.26E−01	1.47E−01	5.94E+03	2.67E−01
65.160	1.10E−00	6.80E+01	7000.0	1.18E+05	4.10E+01	2.67E+01	7.90E+05	4.74E+01
66.165	1.50E−00	9.70E−02	7000.0	2.30E+02	7.98E−02	5.20E−02	1.53E+03	9.22E−02
66.166	3.90E−00	3.40E−00	7000.0	2.10E+04	7.27E−00	4.74E−00	1.40E+05	8.41E−00
67.166	3.40E−00	1.10E−00	7000.0	6.32E+03	2.18E−00	1.42E−00	3.95E+04	2.53E−00
68.169	5.80E−01	9.30E−00	7000.0	8.55E+03	2.95E−00	1.93E−00	5.70E+04	3.42E−00
68.171	2.00E−00	3.10E−01	7000.0	9.83E+02	3.40E−01	2.21E−01	6.55E+03	3.93E−01
69.170	1.70E−00	1.13E+02	7000.0	3.24E+05	1.12E+02	7.33E+01	2.03E+06	1.32E+02
69.171	1.50E−00	4.10E+02	7000.0	1.04E+06	3.59E+02	2.34E+02	6.50E+06	4.22E+02
70.175	7.10E−01	4.10E−00	7000.0	4.61E+03	1.59E−00	1.04E−00	3.07E+04	1.78E−00
71.177	7.60E−01	6.75E−00	7000.0	9.21E+03	3.18E−00	2.08E−00	5.42E+04	3.68E−00
72.181	7.40E+01	4.30E+01	7000.0	1.34E+06	4.65E+02	3.03E+02	3.36E+07	5.04E+02
73.182	1.00E−00	8.20E+01	7000.0	4.33E+04	1.49E+01	9.78E−00	8.66E+05	1.73E+01
74.181	4.70E−02	8.50E−00	7000.0	8.44E+01	2.92E−02	1.90E−02	4.22E+03	2.95E+01
74.185	6.80E−01	8.00E−00	7000.0	1.15E+03	3.97E−01	2.59E−01	5.75E+04	4.02E+02
74.187	1.40E−00	9.00E−01	7000.0	2.66E+02	9.21E−02	6.01E−02	1.33E+04	9.32E+01
75.183	5.50E−02	3.30E−00	7000.0	9.59E−00	3.31E−03	2.16E−03	1.91E+03	9.59E−00
75.186	1.80E−00	1.82E−00	7000.0	1.73E+02	5.99E−02	3.90E−02	3.46E+04	1.73E+02
75.187	6.20E−02	3.50E−00	7000.0	1.14E+01	3.96E−03	2.58E−03	2.29E+03	1.14E+01
75.188	3.90E−00	5.90E−01	7000.0	1.21E+02	4.20E−02	2.74E−02	2.43E+04	1.21E+02
81.200	9.50E−02	9.70E−01	7000.0	2.53E+01	8.76E−03	5.71E−03	9.74E+02	2.43E+01
81.201	4.40E−01	2.10E−00	7000.0	2.53E+02	8.78E−02	5.73E−02	9.76E+03	2.44E+02
81.202	9.40E−01	4.40E−00	7000.0	1.13E+03	3.93E−01	2.56E−01	4.37E+04	1.09E+03
81.204	1.30E−00	7.00E−00	7000.0	2.50E+03	8.65E−01	5.64E−01	9.62E+04	2.40E+03
82.203	5.10E−02	2.17E−00	7000.0	9.35E+01	3.23E−02	2.11E−02	1.16E+03	2.33E+01
82.210	2.90E+01	2.40E+03	7000.0	5.85E+07	2.02E+04	1.32E+04	7.31E+08	1.46E+07
82.212	4.10E+02	4.40E−01	7000.0	1.52E+05	5.27E+01	3.44E+01	1.90E+06	3.81E+04
83.206	4.30E−01	4.30E−00	7000.0	1.50E+02	5.20E−02	3.39E−02	1.95E+04	5.86E−00
83.207	2.40E−01	1.32E+01	7000.0	2.57E+02	8.92E−02	5.81E−02	3.34E+04	1.00E+01
83.210	4.00E+01	3.60E−00	7000.0	1.17E+04	4.05E−00	2.64E−00	1.52E+06	4.56E+02
83.212	4.11E+02	4.20E−02	7000.0	1.40E+03	4.86E−01	3.17E−01	1.82E+05	5.47E+01

Table 3.2–1 (continued)
DOSE CONVERSION FACTORS

Identification (Z. A.)	Energy (MeV)	T(1/2) Days	Weight in Grams	Inh. Dose (rem/Ci)	Dose/Conc rem/Ci-s/m³ Hi-Rate	Dose/Conc rem/Ci-s/m³ Lo-Rate	Organ Dose (rem/Ci)	Ing. Dose (rem/Ci)
SKELETON (Continued)								
84.210	2.80E+02	2.00E+01	7000.0	1.77E+06	6.14E+02	4.00E+02	5.92E+07	3.55E+05
88.223	2.80E−00	1.17E+01	7000.0	6.92E+04	2.39E+01	1.56E+01	3.46E+05	5.19E+04
88.224	2.80E−00	3.64E−00	7000.0	2.15E+04	7.45E−00	4.86E−00	1.07E+05	1.61E+04
88.226	1.10E−00	1.60E+03	7000.0	7.43E+05	2.57E+02	1.67E+02	1.85E+07	5.57E+05
88.228	1.90E−00	2.10E+03	7000.0	1.68E+06	5.82E+02	3.79E+02	4.20E+07	1.26E+06
89.227	1.00E+03	7.20E+03	7000.0	5.03E+09	1.74E+06	1.13E+06	6.28E+10	1.88E+06
89.228	9.70E+02	2.60E−01	7000.0	2.13E+05	7.37E+01	4.81E+01	2.66E+06	7.99E+01
90.227	9.90E+02	1.84E+01	7000.0	3.46E+07	1.19E+04	7.82E+03	1.92E+08	1.34E+04
90.228	9.70E+02	6.93E+02	7000.0	1.27E+09	4.42E+05	2.88E+05	7.10E+09	4.97E+05
90.230	2.40E+02	7.30E+04	7000.0	5.28E+09	1.82E+06	1.19E+06	2.93E+10	2.05E+06
90.231	5.60E−01	1.07E−00	7000.0	1.14E+03	3.94E−01	2.57E−01	6.33E+03	4.43E−01
90.232	2.70E+02	7.30E+04	7000.0	5.94E+09	2.05E+06	1.34E+06	3.30E+10	2.31E+06
90.234	4.50E−00	2.41E+01	7000.0	2.06E+05	7.13E+01	4.65E+01	1.14E+06	8.02E+01
91.230	1.60E+03	1.77E+01	7000.0	6.58E+06	2.26E+03	1.49E+03	5.98E+07	2.68E+05
91.231	7.50E+02	7.30E+04	7000.0	1.00E+10	3.49E+06	2.27E+06	9.17E+10	4.12E+07
91.233	4.10E−01	2.74E+01	7000.0	1.30E+04	4.51E−00	2.94E−00	1.18E+05	5.34E+01
92.230	1.80E+03	1.95E+01	7000.0	1.03E+07	3.59E+03	2.34E+03	3.71E+08	4.08E+03
92.232	1.20E+03	3.00E+02	7000.0	1.06E+08	3.68E+04	2.40E+04	3.80E+09	4.18E+04
92.233	2.50E+02	3.00E+02	7000.0	2.22E+07	7.68E+03	5.01E+03	7.92E+08	8.72E+03
92.234	2.40E+02	3.00E+02	7000.0	2.13E+07	7.37E+03	4.80E+03	7.61E+08	8.37E+03
92.235	2.30E+02	3.00E+02	7000.0	2.04E+07	7.06E+03	4.60E+03	7.29E+08	8.02E+03
92.236	2.30E+02	3.00E+02	7000.0	2.04E+07	7.06E+03	4.60E+03	7.29E+08	8.02E+03
92.238	2.20E+02	3.00E+02	7000.0	1.95E+07	6.75E+03	4.40E+03	6.97E+08	7.67E+03
92.240	5.30E−00	5.87E−01	7000.0	9.20E+02	3.18E−01	2.07E−01	3.28E+04	3.61E+01
93.237	2.50E+02	7.30E+04	7000.0	3.36E+09	1.16E+06	7.59E+05	3.05E+10	1.37E+06
93.239	9.80E−01	2.33E−00	7000.0	2.65E+03	9.18E−01	5.99E−01	2.41E+04	1.08E−00
94.238	2.80E+02	2.30E+04	7000.0	2.15E+09	7.45E+05	4.86E+05	2.87E+10	8.61E+05
94.239	2.70E+02	7.20E+04	7000.0	2.47E+09	8.56E+05	5.58E+05	3.29E+10	9.89E+05
94.240	2.70E+02	7.10E+04	7000.0	2.47E+09	8.55E+05	5.57E+05	3.29E+10	9.88E+05
94.241	1.40E+01	4.50E+03	7000.0	4.69E+07	1.62E+04	1.05E+04	6.25E+08	1.87E+04
94.242	2.50E+02	7.30E+04	7000.0	2.29E+09	7.93E+05	5.17E+05	3.05E+10	9.17E+05
94.243	2.00E−00	2.08E−01	7000.0	8.79E+02	3.04E−01	1.98E−01	4.39E+03	1.05E−01
94.244	2.40E+02	7.30E+04	7000.0	7.16E+09	2.48E+06	1.61E+06	3.57E+10	8.59E+05
95.241	2.80E+02	5.10E+04	7000.0	2.08E+09	7.20E+05	4.69E+05	3.30E+10	8.26E+05
95.242M	3.60E+02	3.20E+04	7000.0	2.09E+09	7.25E+05	4.73E+05	3.33E+10	8.32E+05
95.242	4.00E+02	6.67E−01	7000.0	1.49E+05	5.16E+01	3.37E+01	2.37E+06	5.92E+01
95.243	2.70E+02	7.10E+04	7000.0	2.07E+09	7.18E+05	4.68E+05	3.29E+10	8.23E+05
95.244	2.50E+02	1.80E−02	7000.0	2.99E+03	1.03E−00	6.76E−01	4.75E+04	1.18E−00
96.242	4.00E+02	1.62E+02	7000.0	5.13E+07	1.77E+04	1.15E+04	6.85E+08	2.05E+04
96.243	3.00E+02	1.10E+04	7000.0	1.78E+09	6.17E+05	4.02E+05	2.37E+10	7.13E+05
96.244	3.00E+02	6.10E+03	7000.0	1.26E+09	4.38E+05	2.85E+05	1.68E+10	5.06E+05
96.245	2.80E+02	7.20E+04	7000.0	2.56E+09	8.87E+05	5.79E+05	3.42E+10	1.02E+06
96.246	2.80E+02	7.10E+04	7000.0	2.56E+09	8.86E+05	5.78E+05	3.41E+10	1.02E+06
96.247	2.70E+02	7.30E+04	7000.0	2.47E+09	8.57E+05	5.59E+05	3.30E+10	9.90E+05
96.248	2.60E+02	7.30E+04	7000.0	2.07E+10	7.18E+06	4.68E+06	2.77E+11	8.30E+06
96.249	2.70E+01	4.40E−02	7000.0	9.41E+02	3.25E−01	2.12E−01	1.25E+04	3.76E−01
97.249	2.00E+01	2.89E+02	7000.0	1.22E+07	4.22E+03	2.75E+03	6.11E+07	1.46E+03
97.250	2.90E+01	1.34E−01	7000.0	8.21E+03	2.84E−00	1.85E−00	4.10E+04	9.85E−01

Table 3.2–1 (continued)
DOSE CONVERSION FACTORS

Identification (Z.A.)	Energy (MeV)	T(1/2) Days	Weight in Grams	Inh. Dose (rem/Ci)	Dose/Conc rem/Ci-s/m³ Hi-Rate	Dose/Conc rem/Ci-s/m³ Lo-Rate	Organ Dose (rem/Ci)	Ing. Dose (rem/Ci)
SKELETON (Continued)								
98.249	3.00E+02	5.10E+04	7000.0	7.08E+09	2.44E+06	1.59E+06	3.54E+10	8.49E+05
98.250	3.10E+02	3.50E+03	7000.0	2.23E+09	7.71E+05	5.03E+05	1.11E+10	2.67E+05
98.251	2.90E+02	5.80E+04	7000.0	6.94E+09	2.40E+06	1.56E+06	3.47E+10	8.33E+05
98.252	1.10E+03	7.94E+02	7000.0	1.84E+09	6.38E+05	4.16E+05	9.23E+09	2.21E+05
98.253	3.70E+02	1.80E+01	7000.0	1.40E+07	4.87E+03	3.17E+03	7.04E+07	1.68E+03
98.254	1.89E+04	5.60E+01	7000.0	2.23E+09	7.74E+05	5.05E+05	1.11E+10	2.68E+05
99.253	3.70E+02	2.00E+01	7000.0	1.56E+07	5.41E+03	3.53E+03	7.82E+07	1.87E+03
99.254M	6.60E+02	1.60E−00	7000.0	2.23E+06	7.72E+02	5.03E+02	1.11E+07	2.67E+02
99.254	6.20E+02	4.77E+02	7000.0	6.25E+08	2.16E+05	1.41E+05	3.12E+09	7.50E+04
99.255	3.80E+02	3.00E+01	7000.0	2.41E+07	8.33E+03	5.43E+03	1.20E+08	2.89E+03
100.254	6.60E+02	1.35E−01	7000.0	1.88E+05	6.51E+01	4.25E+01	9.41E+05	2.26E+01
100.255	3.80E+02	8.96E−01	7000.0	7.19E+05	2.49E+02	1.62E+02	3.59E+06	8.63E+01
100.256	1.80E+04	1.11E−01	7000.0	4.22E+06	1.46E+03	9.53E+02	2.11E+07	5.06E+02
LIVER 1700 GRAMS								
4.007	1.60E−02	4.50E+01	1700.0	7.83E+02	2.71E−01	1.76E−01	3.13E+04	6.26E−00
15.032	6.90E−01	8.00E−00	1700.0	9.61E+03	3.32E−00	2.16E−00	2.40E+05	1.20E+04
21.046	6.40E−01	2.50E+01	1700.0	2.78E+04	9.63E−00	6.28E−00	6.96E+05	1.04E+01
21.047	2.10E−01	3.10E−00	1700.0	1.13E+03	3.92E−01	2.55E−01	2.83E+04	4.25E−01
21.048	1.10E−00	1.70E−00	1700.0	3.25E+03	1.12E−00	7.34E−01	8.14E+04	1.22E−00
23.048	9.00E−01	1.31E+01	1700.0	1.02E+04	3.55E−00	2.31E−00	5.13E+05	6.15E+02
25.052	9.60E−01	4.50E−00	1700.0	1.31E+04	4.55E−00	2.97E−00	1.88E+05	3.76E+03
25.054	2.30E−01	2.30E+01	1700.0	1.61E+04	5.57E−00	3.63E−00	2.30E+05	4.60E+03
25.056	1.30E−00	1.10E−01	1700.0	4.35E+02	1.50E−01	9.83E−02	6.22E+03	1.24E+02
26.055	6.50E−03	3.68E+02	1700.0	4.16E+03	1.44E−00	9.39E−01	1.04E+05	1.35E+03
26.059	4.20E−00	4.17E+01	1700.0	3.04E+05	1.05E+02	6.88E+01	7.62E+06	9.91E+04
27.057	5.30E−02	9.20E−00	1700.0	4.24E+02	1.46E−01	9.57E−02	2.12E+04	1.48E+02
27.058M	5.90E−02	3.70E−01	1700.0	1.90E+01	6.57E−03	4.28E−03	9.50E+02	6.65E−00
27.058	2.90E−01	8.40E−00	1700.0	2.12E+03	7.33E−01	4.78E−01	1.06E+05	7.42E+02
27.060	7.20E−01	9.50E−00	1700.0	5.95E+03	2.06E−00	1.34E−00	2.97E+05	2.08E+03
28.059	7.70E−03	5.00E+02	1700.0	5.02E+03	1.73E−00	1.13E−00	1.67E+05	3.35E+03
28.063	2.10E−02	4.92E+02	1700.0	1.34E+04	4.66E−00	3.04E−00	4.49E+05	8.99E+03
28.065	1.20E−00	1.10E−01	1700.0	1.72E+02	5.96E−02	3.89E−02	5.74E+03	1.14E+02
29.064	1.90E−01	5.30E−01	1700.0	1.31E+02	4.54E−02	2.96E−02	4.38E+03	8.76E+01
30.065	1.50E−01	6.60E+01	1700.0	4.74E+04	1.63E+01	1.06E+01	4.30E+05	1.50E+04
30.069M	5.00E−01	5.80E−01	1700.0	1.38E+03	4.80E−01	3.13E−01	1.26E+04	4.41E+02
30.069	3.70E−01	3.60E−02	1700.0	6.37E+01	2.20E−02	1.43E−02	5.79E+02	2.02E+01
31.072	1.10E−00	5.30E−01	1700.0	1.59E+03	5.53E−01	3.60E−01	2.53E+04	6.34E−00
32.071	1.00E−02	4.60E−00	1700.0	1.00E+01	3.46E−03	2.25E−03	2.00E+03	4.00E−01
33.073	4.10E−02	6.70E+01	1700.0	9.56E+02	3.30E−01	2.15E−01	1.19E+05	1.07E+02
33.074	3.80E−01	1.70E+01	1700.0	2.24E+03	7.78E−01	5.07E−01	2.81E+05	2.53E+02
33.076	1.10E−00	1.10E−00	1700.0	4.21E+02	1.45E−01	9.50E−02	5.26E+04	4.74E+01
33.077	2.40E−01	1.60E−00	1700.0	1.33E+02	4.62E−02	3.01E−02	1.67E+04	1.50E+01
34.075	9.40E−02	2.00E+01	1700.0	4.09E+03	1.41E−00	9.23E−01	8.18E+04	4.91E+03
37.086	6.60E−01	1.44E+01	1700.0	1.65E+04	5.72E−00	3.73E−00	4.13E+05	2.06E+04
37.087	9.00E−02	6.30E+01	1700.0	9.87E+03	3.41E−00	2.22E−00	2.46E+05	1.23E+04
40.093	1.90E−02	3.20E+02	1700.0	6.14E+03	2.12E−00	1.38E−00	3.06E+05	2.15E−00
40.095	5.70E−01	5.30E+01	1700.0	2.63E+04	9.09E−00	5.93E−00	1.31E+06	9.20E−00

Table 3.2–1 (continued)
DOSE CONVERSION FACTORS

Identification (Z.A.)	Energy (MeV)	T(1/2) Days	Weight in Grams	Inh. Dose (rem/Ci)	Dose/Conc rem/Ci-s/m³ Hi-Rate	Dose/Conc rem/Ci-s/m³ Lo-Rate	Organ Dose (rem/Ci)	Ing. Dose (rem/Ci)
LIVER (continued)								
40.097	1.60E−00	7.10E−01	1700.0	9.88E+02	3.42E−01	2.23E−01	4.94E+04	3.46E−01
41.093M	3.80E−02	6.88E+02	1700.0	2.27E+04	7.87E−00	5.13E−00	1.13E+06	1.02E+01
41.095	2.60E−01	3.36E+01	1700.0	7.60E+03	2.63E−00	1.71E−00	3.80E+05	3.42E−00
41.097	6.40E−01	5.10E−02	1700.0	2.84E+01	9.83E−03	6.41E−03	1.42E+03	1.27E−02
42.099	4.80E−01	2.66E−00	1700.0	3.53E+03	1.22E−00	7.97E−01	5.43E+04	4.34E+03
43.096M	6.00E−01	3.60E−02	1700.0	1.41E−00	4.87E−04	3.18E−04	9.40E+02	1.41E−00
43.096	6.40E−02	3.80E−00	1700.0	1.58E+01	5.49E−03	3.58E−03	1.05E+04	1.58E+01
43.097M	9.00E−02	2.30E+01	1700.0	1.35E+02	4.67E−02	3.05E−02	9.01E+04	1.35E+02
43.097	2.00E−02	3.00E+01	1700.0	3.91E+01	1.35E−02	8.84E−03	2.61E+04	3.91E+01
43.099M	9.40E−02	3.00E+01	1700.0	1.84E+02	6.36E−02	4.15E−02	1.22E+05	1.84E+02
43.099	3.50E−02	2.50E−01	1700.0	5.71E−01	1.97E−04	1.28E−04	3.80E+02	5.71E−01
45.103M	5.50E−02	3.80E−02	1700.0	1.27E−00	4.40E−04	2.87E−04	9.09E+01	7.27E−01
45.105	1.90E−01	1.40E−00	1700.0	1.62E+02	5.60E−02	3.65E−02	1.15E+04	9.26E+01
46.103	6.30E−02	9.00E−00	1700.0	7.40E+02	2.56E−01	1.67E−01	2.46E+04	4.93E+02
46.109	4.20E−01	5.50E−01	1700.0	3.01E+02	1.04E−01	6.80E−02	1.00E+04	2.01E+02
47.105	2.90E−01	1.10E+01	1700.0	1.06E+03	3.69E−01	2.41E−01	1.38E+05	4.16E+01
47.110M	8.40E−01	1.42E+01	1700.0	3.99E+03	1.38E−00	9.02E−01	5.19E+05	1.55E+02
47.111	3.80E−01	5.00E−00	1700.0	6.36E+02	2.20E−01	1.43E−01	8.27E+04	2.48E+01
48.109	1.00E−01	1.40E+02	1700.0	1.15E+05	4.00E+01	2.61E+01	6.09E+05	1.15E+03
48.115	6.10E−01	3.50E+01	1700.0	1.76E+05	6.10E+01	3.98E+01	9.29E+05	1.76E+03
48.115	5.80E−01	2.20E−00	1700.0	1.05E+04	3.65E−00	2.38E−00	5.55E+04	1.05E+02
49.113M	2.10E−01	7.30E−02	1700.0	2.66E+01	9.23E−03	6.02E−03	6.67E+02	1.86E−01
49.114M	9.40E−01	2.60E+01	1700.0	4.25E+04	1.47E+01	9.60E−00	1.06E+06	2.97E+02
49.115M	2.00E−01	1.90E−01	1700.0	6.61E+01	2.28E−02	1.49E−02	1.65E+03	4.63E−01
49.115	1.70E−01	5.80E+01	1700.0	1.71E+04	5.93E−00	3.87E−00	4.29E+05	1.20E+02
50.113	2.30E−01	4.30E+01	1700.0	1.20E+03	4.17E−01	2.72E−01	4.30E+05	2.15E+02
50.125	9.40E−01	8.40E−00	1700.0	9.62E+02	3.32E−01	2.17E−01	3.43E+05	1.71E+02
51.122	6.70E−01	2.60E−00	1700.0	3.79E+01	1.31E−02	8.55E−03	7.58E+04	4.54E−00
51.124	9.20E−01	2.30E+01	1700.0	4.60E+02	1.59E−01	1.03E−01	9.21E+05	5.52E+01
51.125	2.10E−01	3.60E+01	1700.0	1.77E+02	6.14E−02	4.00E−02	3.55E+05	2.12E+01
52.125M	1.40E−01	2.00E+01	1700.0	2.43E+03	8.43E−01	5.50E−01	1.21E+05	1.21E+03
52.127M	3.20E−01	2.30E+01	1700.0	6.40E+03	2.21E−00	1.44E−00	3.20E+05	3.20E+03
52.127	2.40E−01	3.90E−01	1700.0	8.14E+01	2.81E−02	1.83E−02	4.07E+03	4.07E+01
52.129M	8.30E−01	1.60E+01	1700.0	1.15E+04	3.99E−00	2.60E−00	5.78E+05	5.78E+03
52.129	7.30E−01	5.10E−02	1700.0	3.24E+01	1.12E−02	7.31E−03	1.62E+03	1.62E+01
52.131M	9.70E−01	1.20E−00	1700.0	1.01E+03	3.50E−01	2.28E−01	5.06E+04	5.06E+02
52.132	1.10E−00	2.90E−00	1700.0	2.77E+03	9.60E−01	6.26E−01	1.38E+05	1.38E+03
55.131	2.40E−02	9.00E−00	1700.0	4.70E+02	1.62E−01	1.06E−01	9.40E+03	6.58E+02
55.134M	1.50E−01	1.30E−01	1700.0	4.24E+01	1.46E−02	9.57E−03	8.48E+02	5.94E+01
55.134	5.70E−01	8.10E+01	1700.0	1.00E+05	3.47E+01	2.26E+01	2.00E+06	1.40E+05
55.135	6.60E−02	9.00E+01	1700.0	1.29E+04	4.47E−00	2.91E−00	2.58E+05	1.80E+04
55.136	3.50E−01	1.14E+01	1700.0	8.68E+03	3.00E−00	1.95E−00	1.73E+05	1.21E+04
55.137	4.10E−01	8.90E+01	1700.0	7.94E+04	2.74E+01	1.79E+01	1.58E+06	1.11E+05
56.131	1.90E−01	1.15E+01	1700.0	1.61E+01	5.59E−03	3.64E−03	9.51E+04	2.85E−00
56.140	1.40E−00	1.26E+01	1700.0	1.30E+02	4.51E−02	2.94E−02	7.67E+05	2.30E+01
57.140	1.10E−00	1.68E−00	1700.0	3.21E+03	1.11E−00	7.26E−01	8.04E+04	1.20E+04
58.141	1.80E−01	2.90E+01	1700.0	1.36E+04	4.71E−00	3.07E−00	2.27E+05	5.68E−00
58.143	8.50E−01	1.32E−00	1700.0	2.93E+03	1.01E−00	6.61E−01	4.88E+04	1.22E−00

Table 3.2–1 (continued)
DOSE CONVERSION FACTORS

Identification (Z.A.)	Energy (MeV)	T(1/2) Days	Weight in Grams	Inh. Dose (rem/Ci)	Dose/Conc rem/Ci-s/m³ Hi-Rate	Dose/Conc rem/Ci-s/m³ Lo-Rate	Organ Dose (rem/Ci)	Ing. Dose (rem/Ci)
LIVER (continued)								
58.144	1.30E−00	1.46E+02	1700.0	4.95E+05	1.71E+02	1.11E+02	8.26E+06	2.06E+02
59.142	8.10E−01	8.00E−01	1700.0	1.41E+03	4.87E−01	3.18E−01	2.82E+04	5.64E−01
59.143	3.20E−01	1.32E+01	1700.0	9.19E+03	3.18E−00	2.07E−00	1.83E+05	3.67E−00
60.144	2.00E+01	1.31E+02	1700.0	1.48E+07	5.12E+03	3.34E+03	1.14E+08	5.70E+03
60.147	3.20E−01	1.00E+01	1700.0	1.81E+04	6.26E−00	4.08E−00	1.39E+05	6.96E−00
60.149	9.90E−01	8.30E−02	1700.0	4.64E+02	1.60E−01	1.04E−01	3.57E+03	1.78E−01
61.147	6.90E−02	3.83E+02	1700.0	2.30E+04	7.95E−00	5.19E−00	1.15E+06	6.90E−00
61.149	4.40E−01	2.20E−00	1700.0	8.42E+02	2.91E−01	1.90E−01	4.21E+04	2.52E−01
62.147	2.30E+01	1.87E+02	1700.0	1.68E+07	5.82E+03	3.80E+03	1.87E+08	6.55E+03
62.151	4.20E−02	1.86E+02	1700.0	3.06E+04	1.05E+01	6.90E−00	3.40E+05	1.19E+01
62.153	2.60E−01	1.94E−00	1700.0	1.97E+03	6.83E−01	4.45E−01	2.19E+04	7.68E−01
63.152	7.10E−01	3.80E−01	1700.0	7.04E+02	2.43E−01	1.59E−01	1.17E+04	2.93E−01
63.152	3.30E−01	1.24E+02	1700.0	1.06E+05	3.69E+01	2.41E+01	1.78E+06	4.45E+01
63.154	8.60E−01	1.24E+02	1700.0	2.78E+05	9.63E+01	6.28E+01	4.64E+06	1.16E+02
63.155	9.50E−02	1.05E+02	1700.0	2.60E+04	9.01E−00	5.87E−00	4.34E+05	1.08E+01
64.153	9.90E−02	1.56E+02	1700.0	2.01E+04	6.97E−00	4.55E−00	6.72E+05	8.06E−00
64.159	3.30E−01	7.50E−01	1700.0	3.23E+02	1.11E−01	7.29E−02	1.07E+04	1.29E−01
66.165	3.90E−01	9.70E−02	1700.0	3.29E+01	1.13E−02	7.43E−03	1.64E+03	9.88E−03
66.166	7.80E−01	3.40E−00	1700.0	2.30E+03	7.98E−01	5.21E−01	1.15E+05	6.92E−01
67.166	6.90E−01	1.10E−00	1700.0	6.60E+02	2.28E−01	1.49E−01	3.30E+04	1.98E−01
68.169	2.20E−01	9.20E−00	1700.0	6.60E+02	2.28E−01	1.49E−01	8.81E+04	2.64E−01
72.181	2.90E−01	4.30E+01	1700.0	5.97E+04	2.06E+01	1.34E+01	5.42E+05	2.44E+01
73.182	5.60E−01	8.80E+01	1700.0	1.71E+05	5.93E+01	3.87E+01	2.14E+06	6.43E+01
74.181	8.70E−02	3.90E−00	1700.0	2.95E+02	1.02E−01	6.66E−02	1.47E+04	8.86E+01
74.185	1.40E−01	3.80E−00	1700.0	4.63E+02	1.60E−01	1.04E−01	2.31E+04	1.38E+02
74.187	4.40E−01	8.00E−01	1700.0	3.06E+02	1.06E−01	6.91E−02	1.53E+04	9.19E+01
75.183	1.00E−01	1.17E+01	1700.0	2.54E+02	8.80E−02	5.74E−02	5.09E+04	2.54E+02
75.186	3.70E−01	2.98E−00	1700.0	2.39E+02	8.30E−02	5.41E−02	4.79E+04	2.39E+02
75.187	1.20E−02	1.40E+01	1700.0	3.65E+01	1.26E−02	8.25E−03	7.31E+03	3.65E+01
75.188	8.50E−01	6.80E−01	1700.0	1.25E+02	4.35E−02	2.83E−02	2.51E+04	1.25E+02
76.185	2.90E−01	5.20E−00	1700.0	6.56E+02	2.27E−01	1.48E−01	6.56E+04	2.62E+02
76.191	4.90E−02	5.20E−01	1700.0	1.10E+01	3.83E−03	2.50E−03	1.10E+03	4.43E−00
76.191	1.20E−01	4.10E−00	1700.0	2.14E+02	7.40E−02	4.83E−02	2.14E+04	8.56E+01
76.193	3.80E−01	1.00E−00	1700.0	1.65E+02	5.72E−02	3.73E−02	1.65E+04	6.61E+01
77.190	1.60E−01	8.30E−00	1700.0	4.04E+03	1.39E−00	9.13E−01	5.78E+04	1.32E+03
77.192	6.00E−01	2.00E+01	1700.0	3.65E+04	1.26E+01	8.25E−00	5.22E+05	1.20E+04
77.194	8.10E−01	7.70E−01	1700.0	1.90E+03	6.57E−01	4.28E−01	2.71E+04	6.24E+02
78.191	3.10E−01	2.60E−00	1700.0	8.42E+01	2.91E−02	1.90E−02	3.50E+04	2.80E+01
78.193M	3.20E−02	3.20E−00	1700.0	1.06E+01	3.70E−03	2.41E−03	4.45E+03	3.56E−00
78.193	1.90E−02	2.00E+01	1700.0	3.96E+01	1.37E−02	8.95E−03	1.65E+04	1.32E+01
78.197M	5.10E−01	5.60E−02	1700.0	2.98E−00	1.03E−03	6.73E−04	1.24E+03	9.94E−01
78.197	2.40E−01	7.20E−01	1700.0	1.80E+01	6.24E−03	4.07E−03	7.52E+03	6.01E−00
79.196	2.10E−01	5.50E−00	1700.0	5.02E+02	1.73E−01	1.13E−01	5.02E+04	2.01E+02
79.198	4.40E−01	2.70E−00	1700.0	5.17E+02	1.78E−01	1.16E−01	5.17E+04	2.06E+02
79.199	1.30E−01	3.10E−00	1700.0	1.75E+02	6.06E−02	3.95E−02	1.75E+04	7.01E+01
80.197M	1.90E−01	9.30E−01	1700.0	6.92E+02	2.39E−01	1.56E−01	7.69E+03	8.46E+02
80.197	5.20E−02	2.30E−00	1700.0	4.68E+02	1.62E−01	1.05E−01	5.20E+03	5.72E+02
80.203	1.70E−01	1.04E+01	1700.0	6.92E+03	2.39E−00	1.56E−00	7.69E+04	8.46E+03

Table 3.2–1 (continued)
DOSE CONVERSION FACTORS

Identification (Z.A.)	Energy (MeV)	T(1/2) Days	Weight in Grams	Inh. Dose (rem/Ci)	Dose/Conc rem/Ci-s/m³ Hi-Rate	Dose/Conc rem/Ci-s/m³ Lo-Rate	Organ Dose (rem/Ci)	Ing. Dose (rem/Ci)
LIVER (continued)								
81.200	1.80E−01	9.20E−01	1700.0	1.44E+02	4.98E−02	3.25E−02	7.20E+03	1.44E+02
81.201	1.20E−01	1.90E−00	1700.0	1.98E+02	6.86E−02	4.47E−02	9.92E+03	1.98E+02
81.202	2.70E−01	3.50E−00	1700.0	8.22E+02	2.84E−01	1.85E−01	4.11E+04	8.22E+02
81.204	2.50E−01	5.00E−00	1700.0	1.08E+03	3.76E−01	2.45E−01	5.44E+04	1.08E+03
82.203	9.40E−02	2.17E−00	1700.0	2.04E+02	7.06E−02	4.60E−02	8.87E+03	5.68E+01
82.210	1.00E+01	1.50E+03	1700.0	1.50E+07	5.19E+03	3.38E+03	6.52E+08	4.17E+06
82.212	8.30E+01	4.40E−01	1700.0	3.65E+04	1.26E+01	8.25E−00	1.58E+06	1.01E+04
83.206	8.00E−01	4.50E−00	1700.0	6.26E+03	2.16E−00	1.41E−00	1.56E+05	2.35E+02
83.207	4.50E−01	1.49E+01	1700.0	1.16E+04	4.03E−00	2.63E−00	2.91E+05	4.37E+02
83.210	1.30E+01	3.80E−00	1700.0	8.60E+04	2.97E+01	1.94E+01	2.15E+06	3.22E+03
83.212	8.30E+01	4.20E−01	1700.0	6.06E+04	2.09E+01	1.36E+01	1.51E+06	2.27E+03
84.210	5.50E+01	3.20E+01	1700.0	3.83E+06	1.32E+03	8.64E+02	7.66E+07	7.66E+05
89.227	6.20E+01	1.90E+03	1700.0	6.65E+08	2.30E+05	1.50E+05	5.12E+09	2.56E+05
89.228	5.60E+01	2.60E−01	1700.0	8.23E+04	2.85E+01	1.85E+01	6.33E+05	3.16E+01
90.227	6.10E+01	1.84E+01	1700.0	4.88E+05	1.69E+02	1.10E+02	4.88E+07	2.44E+02
90.228	5.60E+01	6.91E+02	1700.0	1.68E+07	5.82E+03	3.80E+03	1.68E+09	8.42E+03
90.230	4.80E+01	5.70E+04	1700.0	2.36E+08	8.17E+04	5.33E+04	2.36E+10	1.18E+05
90.231	1.60E−01	1.07E−00	1700.0	7.45E+01	2.57E−02	1.68E−02	7.45E+03	3.72E−02
90.232	4.10E+01	5.70E+04	1700.0	2.01E+08	6.97E+04	4.55E+04	2.01E+10	1.00E+05
90.234	9.00E−01	2.41E+01	1700.0	9.44E+03	3.26E−00	2.13E−00	9.44E+05	4.72E−00
91.231	6.30E+01	5.80E+04	1700.0	4.03E+08	1.39E+05	9.11E+04	3.10E+10	1.55E+05
91.233	1.80E−01	2.74E+01	1700.0	2.79E+03	9.65E−01	6.29E−01	2.14E+05	1.07E−00
93.237	4.90E+01	5.40E+04	1700.0	3.11E+08	1.07E+05	7.03E+04	2.39E+10	1.19E+05
93.239	2.20E−01	2.33E−00	1700.0	2.90E+02	1.00E−01	6.54E−02	2.23E+04	1.11E−01
94.238	5.70E+01	1.60E+04	1700.0	1.08E+08	3.74E+04	2.44E+04	2.16E+10	4.32E+04
94.239	5.30E+01	3.00E+04	1700.0	1.18E+08	4.10E+04	2.67E+04	2.37E+10	4.74E+04
94.240	5.30E+01	3.00E+04	1700.0	1.18E+08	4.10E+04	2.67E+04	2.37E+10	4.74E+04
94.241	1.00E−00	4.10E+03	1700.0	8.51E+05	2.94E+02	1.92E+02	1.70E+08	3.40E+02
94.242	5.10E+01	3.00E+04	1700.0	1.14E+08	3.95E+04	2.57E+04	2.28E+10	4.56E+04
94.243	2.50E−01	2.08E−01	1700.0	8.60E+01	2.97E−02	1.94E−02	2.26E+03	1.01E−02
95.241	5.70E+01	3.40E+03	1700.0	7.24E+08	2.50E+05	1.63E+05	8.22E+09	2.88E+05
95.242M	6.80E+01	3.30E+03	1700.0	7.06E+08	2.44E+05	1.59E+05	8.02E+09	2.81E+05
95.242	7.50E+01	6.67E−01	1700.0	1.60E+05	5.56E+01	3.63E+01	1.82E+06	6.40E+01
95.243	5.40E+01	3.50E+03	1700.0	7.04E+08	2.43E+05	1.58E+05	8.00E+09	2.80E+05
95.244	2.00E+01	1.80E−02	1700.0	1.37E+03	4.77E−01	3.11E−01	1.56E+04	5.48E−01
96.242	7.80E+01	1.54E+02	1700.0	5.22E+07	1.80E+04	1.18E+04	5.22E+08	2.90E+04
96.243	6.00E+01	2.50E+03	1700.0	6.48E+08	2.24E+05	1.46E+05	6.48E+09	2.59E+05
96.244	6.00E+01	2.10E+03	1700.0	5.47E+08	1.89E+05	1.23E+05	5.47E+09	2.18E+05
96.245	5.60E+01	3.00E+03	1700.0	7.20E+08	2.49E+05	1.62E+05	7.20E+09	2.88E+05
96.246	5.60E+01	3.00E+03	1700.0	7.20E+08	2.49E+05	1.62E+05	7.20E+09	2.88E+05
96.247	5.50E+01	3.00E+03	1700.0	7.07E+08	2.44E+05	1.59E+05	7.07E+09	2.82E+05
96.248	5.20E+01	3.00E+03	1700.0	5.81E+08	2.01E+05	1.31E+05	5.81E+09	2.32E+05
BRAIN 1500 GRAMS								
15.032	6.90E−01	1.35E+01	1500.0	2.02E+03	6.99E−01	4.56E−01	4.59E+05	2.43E+03
29.064	2.10E−01	5.30E−01	1500.0	2.19E+01	7.59E−03	4.95E−03	5.49E+03	1.64E+01

Table 3.2–1 (continued)
DOSE CONVERSION FACTORS

Identification (Z.A.)	Energy (MeV)	T(1/2) Days	Weight in Grams	Inh. Dose (rem/Ci)	Dose/Conc rem/Ci-s/m³ Hi-Rate	Dose/Conc rem/Ci-s/m³ Lo-Rate	Organ Dose (rem/Ci)	Ing. Dose (rem/Ci)
LUNGS 1000 GRAMS (SOL.)								
14.031	5.90E−01	1.10E−01	1000.0	3.36E+02	1.16E−01	7.58E−02	4.80E+03	4.32E+02
24.051	1.40E−02	2.66E+01	1000.0	2.75E+02	9.53E−02	6.21E−02	2.75E+04	5.51E−00
26.055	6.50E−03	8.19E+02	1000.0	2.36E+03	8.17E−01	5.33E−01	3.93E+05	7.87E+02
26.059	4.20E−01	4.45E+01	1000.0	8.29E+03	2.87E−00	1.87E−00	1.38E+06	2.76E+03
43.096M	3.80E−01	3.60E−02	1000.0	4.55E−01	1.57E−04	1.02E−04	1.01E+03	4.55E+01
43.096	6.40E−01	2.30E−00	1000.0	4.90E+01	1.69E−02	1.10E−02	1.08E+05	4.90E+01
43.097M	9.00E−02	4.70E−00	1000.0	1.40E+01	4.87E−03	3.17E−03	3.13E+04	1.40E+01
43.097	2.00E−02	5.00E−00	1000.0	3.33E−00	1.15E−03	7.51E−04	7.40E+03	3.33E+00
43.099M	3.50E−02	2.40E−01	1000.0	2.79E−01	9.67E−05	6.31E−05	6.21E+02	2.79E−01
43.099	9.40E−02	5.00E−00	1000.0	1.56E+01	5.41E−03	3.53E−03	3.47E+04	1.56E+01
51.122	6.70E−01	2.70E−00	1000.0	1.07E+03	3.70E−01	2.41E−01	1.33E+05	1.20E+02
51.124	9.20E−01	3.80E+01	1000.0	2.06E+04	7.15E−00	4.67E−00	2.58E+06	2.32E+03
51.125	2.10E−01	9.00E+01	1000.0	1.21E+04	4.17E−00	2.72E−00	1.51E+06	1.35E+03
55.131	2.40E−02	9.30E−00	1000.0	3.77E+01	1.31E−02	8.50E−03	1.64E+04	4.92E+01
55.134M	1.70E−01	1.30E−01	1000.0	3.76E−00	1.30E−03	8.48E−04	1.63E+03	4.90E−00
55.134	5.70E−01	1.20E+02	1000.0	1.16E+04	4.02E−00	2.62E−00	5.06E+06	1.51E+04
55.135	6.60E−02	1.40E+02	1000.0	1.57E+03	5.44E−01	3.54E−01	6.83E+05	2.05E+03
55.136	3.50E−01	1.91E+01	1000.0	7.08E+02	2.45E−01	1.59E−01	3.08E+05	9.24E+02
55.137	4.10E−01	1.38E+02	1000.0	9.62E+03	3.33E−00	2.17E−00	4.18E+06	1.25E+04
56.131	1.90E−01	1.16E+01	1000.0	8.97E−00	3.10E−03	2.02E−03	1.63E+05	1.63E−00
56.140	1.40E−00	1.28E+01	1000.0	7.29E+01	2.52E−02	1.64E−02	1.32E+06	1.32E+01
81.200	1.80E−01	9.50E−01	1000.0	3.66E+01	1.26E−02	8.28E−03	1.26E+04	3.41E+01
81.201	1.20E−01	2.00E−00	1000.0	5.15E+01	1.78E−02	1.16E−02	1.77E+04	4.79E+01
81.202	2.70E−01	4.00E−00	1000.0	2.31E+02	8.01E−02	5.23E−02	7.99E+04	2.15E+02
81.204	2.50E−01	6.00E−00	1000.0	3.21E+02	1.11E−01	7.26E−02	1.11E+05	2.99E+02
HEART 300 GRAMS								
29.064	1.70E−01	5.30E−01	300.0	8.88E+01	3.07E−02	2.00E−02	2.22E+04	6.66E+01
SPLEEN 150 GRAMS								
4.007	1.20E−02	4.90E+01	150.0	1.45E+02	5.01E−02	3.27E−02	2.90E+05	1.16E−00
23.048	7.00E−01	1.37E+01	150.0	1.23E+04	4.25E−00	2.77E−00	4.73E+06	9.46E+02
26.055	6.50E−03	3.88E+02	150.0	7.46E+03	2.58E−00	1.68E−00	1.24E+06	2.48E+03
26.059	3.40E−01	4.19E+01	150.0	4.21E+04	1.45E+01	9.51E−00	7.02E+06	1.40E+04
27.057	4.50E−02	9.20E−00	150.0	1.14E+02	3.95E−02	2.58E−02	2.04E+05	8.57E+01
27.058	4.80E−02	3.70E−01	150.0	4.90E−00	1.69E−03	1.10E−03	8.76E+03	3.67E−00
27.058	2.20E−01	8.40E−00	150.0	5.10E+02	1.76E−01	1.15E−01	9.11E+05	3.82E+02
27.060	5.60E−01	9.50E−00	150.0	1.46E+03	5.08E−01	3.31E−01	2.62E+06	1.10E+03
29.064	1.70E−01	4.20E−01	150.0	1.05E+03	3.65E−01	2.38E−01	3.52E+04	7.04E+02
31.072	8.90E−01	5.40E−01	150.0	5.92E+02	2.05E−01	1.33E−01	2.37E+05	2.37E−00
34.075	7.20E−02	1.60E+01	150.0	1.98E+03	6.88E−01	4.48E−01	5.68E+05	2.55E+03
37.086	6.60E−01	1.32E+01	150.0	1.28E+04	4.46E−00	2.90E−00	4.29E+06	1.71E+04
37.087	9.00E−02	4.50E+01	150.0	5.99E+03	2.07E−00	1.35E−00	1.99E+06	7.99E+03
40.093	1.90E−02	9.00E+02	150.0	1.46E+04	5.07E−00	3.31E−00	9.78E+06	5.87E−00
40.095	4.60E−01	5.90E+01	150.0	2.00E+04	6.94E−00	4.53E−00	1.33E+07	8.03E−00
40.097	1.50E−00	7.10E−01	150.0	7.88E+02	2.72E−01	1.77E−01	5.25E+05	3.15E−01
41.093M	3.80E−02	7.56E+02	150.0	2.83E+04	9.80E−00	6.39E−00	1.41E+07	1.13E+01

Table 3.2–1 (continued)
DOSE CONVERSION FACTORS

Identification (Z.A.)	Energy (MeV)	T(1/2) Days	Weight in Grams	Inh. Dose (rem/Ci)	Dose/Conc rem/Ci-s/m³ Hi-Rate	Dose/Conc rem/Ci-s/m³ Lo-Rate	Organ Dose (rem/Ci)	Ing. Dose (rem/Ci)
SPLEEN (continued)								
41.095	2.00E−01	3.38E+01	150.0	6.66E+03	2.30E−00	1.50E−00	3.33E+06	2.66E−00
41.097	6.00E−01	5.10E−02	150.0	3.01E+01	1.04E−02	6.81E−03	1.50E+04	1.20E−02
45.103M	5.40E−02	3.80E−02	150.0	3.54E−00	1.22E−03	7.99E−04	1.01E+03	2.02E−00
45.105	1.90E−01	1.42E−00	150.0	4.65E+02	1.61E−01	1.05E−01	1.33E+05	2.66E+02
46.103	6.10E−02	8.00E−00	150.0	8.42E+02	2.91E−01	1.90E−01	2.40E+05	4.81E+02
46.109	4.20E−01	5.50E−01	150.0	3.98E+02	1.37E−01	9.00E−02	1.13E+05	2.27E+02
49.113M	1.90E−01	7.30E−02	150.0	3.42E+01	1.18E−02	7.72E−03	6.84E+03	2.73E−01
49.114	9.30E−01	2.40E+01	150.0	5.50E+04	1.90E+01	1.24E+01	1.10E+07	4.40E+02
49.115M	1.90E−01	1.90E−01	150.0	8.90E+01	3.08E−02	2.00E−02	1.78E+04	7.12E−01
49.115	1.70E−01	4.80E+01	150.0	2.01E+04	6.96E−00	4.54E−00	4.02E+06	1.61E+02
52.125M	1.40E−01	2.00E+01	150.0	5.24E+03	1.81E−00	1.18E−00	1.38E+06	3.45E+03
52.127M	3.20E−01	2.30E+01	150.0	1.37E+04	4.77E−00	3.11E−00	3.63E+06	9.07E+03
52.127	2.40E−01	3.90E−01	150.0	1.75E+02	6.07E−02	3.95E−02	4.61E+04	1.15E+02
52.129M	7.80E−01	1.60E+01	150.0	2.33E+04	8.09E−00	5.27E−00	6.15E+06	1.53E+04
52.129	6.80E−01	5.10E−02	150.0	6.50E+01	2.24E−02	1.46E−02	1.71E+04	4.27E+01
52.131M	8.00E−01	1.20E−00	150.0	1.79E+03	6.22E−01	4.06E−01	4.73E+05	1.18E+03
52.132	9.60E−01	2.90E−00	150.0	5.21E+03	1.80E−00	1.17E−00	1.37E+06	3.43E+03
55.131	2.10E−02	9.10E−00	150.0	3.58E+02	1.23E−01	8.08E−02	9.42E+04	4.71E+02
55.134	1.30E−01	1.30E−01	150.0	3.16E+01	1.09E−02	7.14E−03	8.33E+03	4.16E+01
55.134	4.60E−01	8.80E+01	150.0	7.58E+04	2.62E+01	1.71E+01	1.99E+07	9.98E+04
55.135	6.60E−02	9.80E+01	150.0	1.21E+04	4.19E−00	2.73E−00	3.19E+06	1.59E+04
55.136	2.90E−01	1.15E+01	150.0	6.25E+03	2.16E−00	1.41E−00	1.64E+06	8.22E+03
55.137	3.70E−01	9.70E+01	150.0	6.72E+04	2.32E+01	1.51E+01	1.77E+07	8.85E+04
56.131	1.40E−01	6.10E−00	150.0	5.89E−00	2.04E−03	1.33E−03	4.21E+05	1.05E−00
56.140	1.20E−00	6.40E−00	150.0	5.30E+01	1.83E−02	1.19E−02	3.78E+06	9.47E−00
72.181	2.50E−01	4.10E+01	150.0	1.51E+05	5.24E+01	3.42E+01	5.05E+06	6.57E+01
73.182	4.50E−01	7.60E+01	150.0	4.21E+04	1.45E+01	9.51E−00	1.68E+07	1.68E+01
77.190	1.20E−01	9.70E−00	150.0	3.44E+03	1.19E−00	7.77E−01	5.74E+05	1.14E+03
77.192	5.00E−01	3.00E+01	150.0	4.44E+04	1.53E+01	1.00E+01	7.40E+06	1.48E+04
77.194	8.10E−01	7.80E−01	150.0	1.87E+03	6.46E−01	4.22E−01	3.11E+05	6.23E+02
78.191	2.20E−01	2.90E−00	150.0	7.55E+02	2.61E−01	1.70E−01	3.14E+05	2.51E+02
78.193M	2.30E−02	3.20E−00	150.0	8.71E+01	3.01E−02	1.96E−02	3.63E+04	2.90E+01
78.193	1.40E−02	6.00E+01	150.0	9.94E+02	3.44E−01	2.24E−01	4.14E+05	3.31E+02
78.197	5.00E−02	5.60E−02	150.0	3.31E−00	1.14E−03	7.48E−04	1.38E+03	1.10E−00
78.197	2.30E−01	7.40E−01	150.0	2.01E+02	6.97E−02	4.54E−02	8.39E+04	6.71E+01
79.196	1.50E−01	5.50E−00	150.0	6.10E+02	2.11E−01	1.37E−01	4.07E+05	2.03E+02
79.198	4.10E−01	2.70E−00	150.0	8.19E+02	2.83E−01	1.84E−01	5.46E+05	2.73E+02
79.199	1.20E−01	3.10E−00	150.0	2.75E+02	9.52E−02	6.21E−02	1.83E+05	9.17E+01
80.197M	1.70E−01	9.00E−01	150.0	7.54E+02	2.61E−01	1.70E−01	7.54E+04	1.50E+03
80.197	4.30E−02	2.10E−00	150.0	4.45E+02	1.54E−01	1.00E−01	4.45E+04	8.90E+02
80.203	1.50E−01	8.20E−00	150.0	6.06E+03	2.09E−00	1.36E−00	6.06E+05	1.21E+04
83.206	5.80E−01	3.90E−00	150.0	2.90E+03	1.00E−00	6.54E−01	1.11E+06	1.11E+02
83.207	3.30E−01	1.00E+01	150.0	4.23E+03	1.46E−00	9.55E−01	1.62E+06	1.62E+02
83.210	1.70E+01	3.00E−00	150.0	6.54E+04	2.26E+01	1.47E+01	2.51E+07	2.51E+03
83.212	8.20E+01	4.20E−02	150.0	4.41E+03	1.52E−00	9.96E−01	1.69E+06	1.69E+02
84.210	5.50E+01	4.20E+01	150.0	1.13E+07	3.94E+03	2.57E+03	1.13E+09	2.27E+06
85.211	6.10E+01	3.00E−01	150.0	2.08E+05	7.15E+01	4.70E+01	1.04E+07	3.12E+05

Table 3.2–1 (continued)
DOSE CONVERSION FACTORS

Identification (Z.A.)	Energy (MeV)	T(1/2) Days	Weight in Grams	Inh. Dose (rem/Ci)	Dose/Conc rem/Ci-s/m³ Hi-Rate	Dose/Conc rem/Ci-s/m³ Lo-Rate	Organ Dose (rem/Ci)	Ing. Dose (rem/Ci)
PANCREAS 70 GRAMS								
25.052	5.60E−01	2.80E−00	70.0	1.49E+04	5.16E−00	3.36E−00	1.65E+06	4.97E+03
25.054	1.30E−01	5.60E−00	70.0	6.92E+03	2.39E−00	1.56E−00	7.69E+05	2.30E+03
25.056	1.10E−00	1.10E−01	70.0	1.15E+03	3.98E−01	2.59E−01	1.27E+05	3.83E+02
27.057	4.00E−02	9.20E−00	70.0	3.11E+02	1.07E−01	7.02E−02	3.89E+05	2.33E+02
27.058M	3.90E−02	3.70E−01	70.0	1.22E+01	4.22E−03	2.75E−03	1.52E+04	9.15E−00
27.058	1.70E−01	8.40E−00	70.0	1.20E+03	4.17E−01	2.72E−01	1.50E+06	9.05E+02
27.060	4.40E−01	9.50E−00	70.0	3.53E+03	1.22E−00	7.97E−01	4.41E+06	2.65E+03
30.065	8.40E−02	2.30E+01	70.0	1.83E+04	6.35E−00	4.14E−00	2.04E+06	6.12E+03
30.069M	4.50E−01	5.70E−01	70.0	2.44E+03	8.44E−01	5.50E−01	2.71E+05	8.13E+02
30.069	3.70E−01	3.60E−02	70.0	1.26E+02	4.38E−02	2.85E−02	1.40E+04	4.22E+01
37.086	6.50E−01	1.43E+01	70.0	2.26E+04	7.81E−00	5.10E−00	9.82E+06	2.94E+04
37.087	9.00E−02	6.00E+01	70.0	1.31E+04	4.54E−00	2.96E−00	5.70E+06	1.71E+04
TESTES 40 GRAMS								
14.031	5.90E−01	1.10E−01	40.0	4.08E+01	1.41E−02	9.21E−03	1.20E+05	5.16E+01
16.035	5.60E−02	7.64E+01	40.0	7.75E+03	2.68E−00	1.75E−00	7.91E+06	1.02E+04
30.065	5.60E−02	1.28E+02	40.0	3.58E+03	1.23E−00	8.08E−01	1.32E+07	1.19E+03
30.069M	4.30E−01	5.80E−01	40.0	1.24E+02	4.30E−02	2.81E−02	4.61E+05	4.15E+01
52.125M	1.10E−01	2.00E+01	40.0	4.47E+03	1.54E−00	1.01E−00	4.07E+06	3.05E+03
52.127M	3.10E−01	2.30E+01	40.0	1.45E+04	5.01E−00	3.27E−00	1.31E+07	9.89E+03
52.127	2.40E−01	3.90E−01	40.0	1.90E+02	6.58E−02	4.29E−02	1.73E+05	1.29E+02
52.129M	6.90E−01	1.60E+01	40.0	2.24E+04	7.77E−00	5.07E−00	2.04E+07	1.53E+04
52.129	6.00E−01	5.10E−02	40.0	6.22E+01	2.15E−02	1.40E−02	5.66E+04	4.24E+01
52.132	7.30E−01	2.90E−00	40.0	4.30E+03	1.49E−00	9.72E−01	3.91E+06	2.93E+03
30.069	3.70E−01	3.60E−02	40.0	5.06E+04	1.76E+01	1.15E+01	3.38E+07	6.76E+04
PROSTATE 20 GRAMS								
24.051	8.40E−03	2.66E+01	20.0	1.90E+02	6.57E−02	4.29E−02	8.26E+05	3.72E−00
30.065	5.60E−02	1.30E+01	20.0	5.38E+04	1.86E+01	1.21E+01	2.69E+06	1.61E+04
30.069M	4.30E−01	5.60E−01	20.0	1.78E+04	6.16E−00	4.02E−00	8.90E+05	5.34E+03
30.069	3.70E−01	3.60E−02	20.0	9.85E+02	3.40E−01	2.22E−01	4.92E+04	2.95E+02
ADRENALS 20 GRAMS								
14.031	5.90E−01	1.10E−01	20.0	1.68E+02	5.81E−02	3.79E−02	2.40E+05	2.04E+02
OVARIES 8 GRAMS								
14.031	5.90E−01	1.10E−01	8.0	2.40E+01	8.30E−03	5.41E−03	6.00E+05	3.00E+01
30.065	5.60E−02	7.40E+01	8.0	4.59E+03	1.59E−00	1.03E−00	3.83E+07	1.53E+03
30.069M	4.30E−01	5.80E−01	8.0	2.76E+02	9.57E−02	6.24E−02	2.30E+06	9.22E+01
30.069	3.70E−01	3.60E−02	8.0	1.47E+01	5.11E−03	3.33E−03	1.23E+05	4.92E−00
85.211	6.10E+01	3.00E−01	8.0	2.93E+05	1.01E−02	6.62E+01	1.95E+08	3.91E+05

Table 3.2–1 (continued)
DOSE CONVERSION FACTORS

Identification (Z.A.)	Energy (MeV)	T(1/2) Days	Weight in Grams	Inh. Dose (rem/Ci)	Dose/Conc rem/Ci-s/m³ Hi-Rate	Dose/Conc rem/Ci-s/m³ Lo-Rate	Organ Dose (rem/Ci)	Ing. Dose (rem/Ci)
SKIN 2000 GRAMS								
14.031	5.90E−01	1.10E−01	2000.0	7.20E+01	2.49E−02	1.62E−02	2.40E+03	7.20E+01
16.035	5.60E−02	8.24E+01	2000.0	1.28E+03	4.42E−01	2.88E−01	1.70E+05	1.70E+03
43.096M	2.10E−02	3.60E−02	2000.0	1.39E−01	4.83E−05	3.15E−05	2.79E+01	1.39E−01
43.096	8.30E−03	3.00E−00	2000.0	4.60E−00	1.59E−03	1.03E−03	9.21E+02	4.60E−00
43.097M	7.10E−02	9.00E−00	2000.0	1.18E+02	4.08E−02	2.66E−02	2.36E+04	1.18E+02
43.097	1.10E−03	1.00E+01	2000.0	2.03E−00	7.04E−04	4.59E−04	4.07E+02	2.03E−00
43.099M	2.20E−03	2.40E−01	2000.0	9.76E−02	3.37E−05	2.20E−05	1.95E+01	9.76E−02
43.099	9.40E−02	1.00E+01	2000.0	1.73E+02	6.01E−02	3.92E−02	3.47E+04	1.73E+02
49.113M	1.30E−01	7.30E−02	2000.0	1.75E+01	6.07E−03	3.96E−03	3.51E+02	1.26E−01
49.114M	9.00E−01	2.60E+01	2000.0	4.32E+04	1.49E+01	9.76E−00	8.65E+05	3.11E+02
49.115M	1.40E−01	1.90E−01	2000.0	4.92E+01	1.70E−02	1.11E−02	9.84E+02	3.54E−01
49.115	1.70E−01	6.70E+01	2000.0	2.10E+04	7.28E−00	4.75E−00	4.21E+05	1.51E+02
75.183	1.20E−03	1.90E+01	2000.0	1.09E+02	3.79E−02	2.47E−02	8.43E+02	1.09E+02
75.186	3.60E−01	3.30E−00	2000.0	5.71E+03	1.97E−00	1.28E−00	4.39E+04	5.71E+03
75.187	1.20E−02	2.50E+01	2000.0	1.44E+03	4.99E−01	3.25E−01	1.11E+04	1.44E+03
75.188	7.80E−01	6.90E−01	2000.0	2.58E+03	8.95E−01	5.84E−01	1.99E+04	2.58E+03
FAT 10000 GRAMS								
6.014	5.40E−02	1.20E+01	10000.0	1.82E+03	6.30E−01	4.11E−01	4.79E+03	2.39E+03
55.131	2.40E−02	9.30E−00	10000.0	3.79E−00	1.31E−03	8.57E−04	1.65E+03	4.95E−00
MUSCLE 30000 GRAMS								
30.065	3.20E−01	2.18E+02	30000.0	1.54E+04	5.35E−00	3.49E−00	1.72E+05	5.16E+03
30.069	6.40E−01	5.80E−01	30000.0	8.24E+01	2.85E−02	1.85E−02	9.15E+02	2.74E+01
30.069	3.70E−01	3.60E−02	30000.0	2.95E−00	1.02E−03	6.67E−04	3.28E+01	9.85E−01
37.086	7.00E−01	1.51E+01	30000.0	8.86E+03	3.06E−00	2.00E−00	2.60E+04	1.17E+04
37.087	9.00E−02	8.00E+01	30000.0	6.03E+03	2.08E−00	1.36E−00	1.77E+04	7.99E+03
55.134M	2.60E−01	1.30E−01	30000.0	2.50E+01	8.65E−03	5.64E−03	8.33E+01	.00E−99
55.134	1.10E−00	1.20E+02	30000.0	9.76E+04	3.37E+01	2.20E+01	3.25E+05	.00E−99
55.135	6.60E−02	1.40E+02	30000.0	6.83E+03	2.36E−00	1.54E−00	2.27E+04	.00E−99
55.136	6.50E−01	1.19E+01	30000.0	5.72E+03	1.98E−00	1.29E−00	1.90E+04	.00E−99
55.137	5.90E−01	1.38E+02	30000.0	6.02E+04	2.08E+01	1.35E+01	2.00E+05	.00E−99
56.131	3.80E−01	1.15E+01	30000.0	8.94E−00	3.09E−03	2.01E−03	1.07E+04	1.61E−00
56.140	2.30E−00	1.27E+01	30000.0	5.98E+01	2.06E−02	1.34E−02	7.20E+04	1.08E+01
81.200	4.00E−01	9.30E−01	30000.0	2.38E+02	8.25E−02	5.38E−02	9.17E+02	2.20E+02
81.201	1.70E−01	1.90E−00	30000.0	2.07E+02	7.16E−02	4.67E−02	7.96E+02	1.91E+02
81.202	3.80E−01	3.80E−00	30000.0	9.26E+02	3.20E−01	2.08E−01	3.56E+03	8.54E+02
81.204	2.50E−01	5.50E−00	30000.0	8.81E+02	3.05E−01	1.99E−01	3.39E+03	8.13E+02
TISSUE 43000 GRAMS								
1.003	1.00E−02	1.20E+01	43000.0	2.06E+02	7.14E−02	4.66E−02	2.06E+02	2.06E+02

Table 3.2–1 (continued)
DOSE CONVERSION FACTORS

Identification (Z.A.)	Energy (MeV)	T(1/2) Days	Weight in Grams	Inh. Dose (rem/Ci)	Dose/Conc rem/Ci-s/m³ Hi-Rate	Dose/Conc rem/Ci-s/m³ Lo-Rate	Organ Dose (rem/Ci)	Ing. Dose (rem/Ci)
TOTAL BODY 70000 GRAMS								
1.003	1.00E-02	1.20E+01	70000.0	1.26E+02	4.38E-02	2.86E-02	1.26E+02	1.26E+02
4.007	3.50E-02	4.10E+01	70000.0	3.79E+02	1.31E-01	8.55E-02	1.51E+03	3.03E-00
6.014	5.40E-02	1.00E+01	70000.0	4.28E+02	1.48E-01	9.66E-02	5.70E+02	5.70E+02
9.018	8.90E-01	7.80E-02	70000.0	5.50E+01	1.90E-02	1.24E-02	7.33E+01	7.33E+01
11.022	1.60E-00	1.10E+01	70000.0	1.39E+04	4.82E-00	3.14E-00	1.86E+04	1.86E+04
11.024	2.70E-00	6.00E-01	70000.0	1.28E+03	4.44E-01	2.89E-01	1.71E+03	1.71E+03
14.031	5.90E-01	1.10E-01	70000.0	4.66E+01	1.61E-02	1.05E-02	6.86E+01	5.83E+01
15.032	6.90E-01	1.35E+01	70000.0	6.20E+03	2.14E-00	1.40E-00	9.84E+03	7.38E+03
16.035	5.60E-02	4.43E+01	70000.0	1.96E+03	6.80E-01	4.43E-01	2.62E+03	2.62E+03
17.036	2.60E-01	2.90E+01	70000.0	5.97E+03	2.06E-00	1.34E-00	7.97E+03	7.97E+03
17.038	2.30E-00	2.60E-02	70000.0	4.74E+01	1.64E-02	1.06E-02	6.32E+01	6.32E+01
20.045	8.60E-02	1.62E+02	70000.0	8.10E+03	2.80E-00	1.82E-00	1.47E+04	7.36E+03
20.047	1.40E-00	4.90E-00	70000.0	3.98E+03	1.37E-00	9.00E-01	7.25E+03	3.62E+03
21.046	1.30E-00	2.20E+01	70000.0	7.55E+03	2.61E-00	1.70E-00	3.02E+04	3.02E-00
21.047	2.60E-01	3.10E-00	70000.0	2.13E+02	7.36E-02	4.80E-02	8.52E+02	8.52E-02
21.048	2.20E-00	1.70E-00	70000.0	9.88E+02	3.41E-01	2.23E-01	3.95E+03	3.95E-01
23.048	1.90E-00	1.16E+01	70000.0	6.05E+03	2.09E-00	1.36E-00	2.32E+04	4.65E+02
24.051	2.50E-02	2.66E+01	70000.0	1.75E+02	6.08E-02	3.96E-02	7.03E+02	3.51E-00
25.052	2.10E-00	4.20E-00	70000.0	2.79E+03	9.67E-01	6.31E-01	9.32E+03	9.32E+02
25.054	5.10E-01	1.62E+01	70000.0	2.62E+03	9.06E-01	5.91E-01	8.73E+03	8.73E+02
25.056	1.90E-00	1.10E-01	70000.0	6.62E+01	2.29E-02	1.49E-02	2.20E+02	2.20E+01
26.055	6.50E-03	4.63E+02	70000.0	9.54E+02	3.30E-01	2.15E-01	3.18E+03	3.18E+02
26.059	8.10E-01	4.27E+01	70000.0	1.09E+04	3.79E-00	2.47E-00	3.65E+04	3.65E+03
27.057	9.00E-02	9.20E-00	70000.0	3.50E+02	1.21E-01	7.90E-02	8.75E+02	2.62E+02
27.058M	9.90E-02	3.70E-01	70000.0	1.54E+01	5.35E-03	3.49E-03	3.87E+01	1.16E+01
27.058	6.10E-01	8.40E-00	70000.0	2.16E+03	7.49E-01	4.88E-01	5.41E+03	1.62E+03
27.060	1.50E-00	9.50E-00	70000.0	6.02E+03	2.08E-00	1.35E-00	1.50E+04	4.51E+03
28.059	7.70E-03	6.67E+02	70000.0	2.17E+03	7.51E-01	4.90E-01	5.42E+03	1.62E+03
28.063	2.10E-02	6.52E+02	70000.0	5.78E+03	2.00E-00	1.30E-00	1.44E+04	4.34E+03
28.065	1.40E-00	1.10E-01	70000.0	6.51E+01	2.25E-02	1.46E-02	1.62E+02	4.88E+01
29.064	2.50E-01	5.30E-01	70000.0	5.46E+01	1.88E-02	1.23E-02	1.40E+02	3.92E+01
30.065	3.20E-01	1.94E+02	70000.0	1.96E+04	6.81E-00	4.44E-00	6.56E+04	6.56E+03
30.069M	6.40E-01	5.80E-01	70000.0	1.17E+02	4.07E-02	2.65E-02	3.92E+02	3.92E+01
30.069	3.70E-01	3.60E-02	70000.0	4.22E-00	1.46E-03	9.53E-04	1.40E+01	1.40E-00
31.072	1.80E-00	5.40E-01	70000.0	2.56E+02	8.88E-02	5.79E-02	1.02E+03	1.02E-00
32.071	1.00E-02	9.20E-01	70000.0	2.52E-00	8.74E-04	5.70E-04	9.72E-00	9.72E-02
33.073	6.10E-02	6.00E+01	70000.0	1.04E+03	3.61E-01	2.35E-01	3.86E+03	1.16E+02
33.074	5.60E-01	1.65E+01	70000.0	2.63E+03	9.12E-01	5.95E-01	9.76E+03	2.93E+02
33.076	1.30E-00	1.10E-00	70000.0	4.08E+02	1.41E-01	9.21E-02	1.51E+03	4.53E+01
33.077	2.40E-01	1.60E-00	70000.0	1.09E+02	3.79E-02	2.47E-02	4.05E+02	1.21E+01
34.075	2.00E-01	1.01E+01	70000.0	1.49E+03	5.17E-01	3.37E-01	2.13E+03	1.92E+03
35.082	1.80E-00	1.30E-00	70000.0	1.85E+03	6.41E-01	4.18E-01	2.47E+03	2.47E+03
37.086	7.00E-01	1.32E+01	70000.0	7.32E+03	2.53E-00	1.65E-00	9.76E+03	9.76E+03
37.087	9.00E-02	4.50E+01	70000.0	3.21E+03	1.11E-00	7.24E-01	4.28E+03	4.28E+03
38.085M	9.80E-02	4.90E-02	70000.0	8.65E-00	3.00E-03	1.94E-03	2.16E+01	6.49E-00
38.085	3.30E-01	6.47E+01	70000.0	9.02E+03	3.12E-00	2.03E-00	2.25E+04	6.77E+03
38.089	5.50E-01	5.03E+01	70000.0	1.16E+04	4.04E-00	2.64E-00	2.92E+04	8.77E+03

Table 3.2–1 (continued)
DOSE CONVERSION FACTORS

Identification (Z.A.)	Energy (MeV)	T(1/2) Days	Weight in Grams	Inh. Dose (rem/Ci)	Dose/Conc rem/Ci-s/m³ Hi-Rate	Dose/Conc rem/Ci-s/m³ Lo-Rate	Organ Dose (rem/Ci)	Ing. Dose (rem/Ci)
TOTAL BODY (continued)								
38.090	1.10E−00	5.70E+03	70000.0	2.36E+06	8.16E+02	5.32E+02	5.90E+06	1.77E+06
38.091	1.90E−00	4.00E−01	70000.0	3.55E+02	1.23E−01	8.04E−02	8.88E+02	2.67E+02
38.092	2.60E−00	1.10E−01	70000.0	1.20E+02	4.18E−02	2.72E−02	3.02E+02	9.07E+01
39.090	8.90E−01	2.68E−00	70000.0	6.30E+02	2.18E−01	1.42E−01	2.52E+03	2.52E−01
39.091	9.30E−01	3.50E−02	70000.0	8.60E−00	2.97E−03	1.94E−03	3.44E+01	3.44E−03
39.091	5.90E−01	5.80E+01	70000.0	9.04E+03	3.12E−00	2.04E−00	3.61E+04	3.61E−00
39.092	1.60E−00	1.50E−01	70000.0	6.34E+01	2.19E−02	1.43E−02	2.53E+02	2.53E−02
39.093	1.70E−00	4.20E−01	70000.0	1.88E+02	6.52E−02	4.25E−02	7.54E+02	7.54E−02
40.093	1.90E−02	4.50E+02	70000.0	2.54E+03	8.82E−01	5.75E−01	1.02E+04	1.02E−00
40.095	1.10E−00	5.55E+01	70000.0	1.61E+04	5.58E−00	3.64E−00	6.45E+04	6.45E−00
40.097	2.10E−00	7.10E−01	70000.0	3.94E+02	1.36E−01	8.89E−02	1.57E+03	1.57E−01
41.093M	3.80E−02	6.30E+02	70000.0	6.32E+03	2.18E−00	1.42E−00	2.53E+04	2.53E−00
41.095	5.10E−01	3.35E+01	70000.0	4.51E+03	1.56E−00	1.01E−00	1.80E+04	1.80E−00
41.097	8.70E−01	5.10E−02	70000.0	1.17E+01	4.05E−03	2.64E−03	4.69E+01	4.69E−03
42.099	5.10E−01	1.80E−00	70000.0	6.62E+02	2.29E−01	1.50E−01	1.02E+03	8.15E+02
43.096M	3.00E−01	3.60E−02	70000.0	5.70E−00	1.97E−03	1.28E−03	1.14E+01	5.70E−00
43.096	1.40E−00	8.00E−01	70000.0	5.92E+02	2.04E−01	1.33E−01	1.18E+03	5.92E+02
43.097M	9.00E−02	9.90E−01	70000.0	4.70E+01	1.62E−02	1.06E−02	9.41E+01	4.70E+01
43.097	2.00E−02	1.00E−00	70000.0	1.05E+01	3.65E−03	2.38E−03	2.11E+01	1.05E+01
43.099M	8.00E−02	2.00E−01	70000.0	8.45E−00	2.92E−03	1.90E−03	1.69E+01	8.45E−00
43.099	9.40E−02	1.00E−00	70000.0	4.96E+01	1.71E−02	1.12E−02	9.93E+01	4.96E+01
44.097	1.50E−01	2.00E−00	70000.0	8.56E+01	2.96E−02	1.93E−02	3.17E+02	9.51E−00
44.103	4.40E−01	6.20E−00	70000.0	7.78E+02	2.69E−01	1.75E−01	2.88E+03	8.65E+01
44.105	1.20E−00	1.90E−01	70000.0	6.50E+01	2.25E−02	1.46E−02	2.41E+02	7.23E−00
44.106	1.40E−00	7.20E−00	70000.0	2.87E+03	9.95E−01	6.49E−01	1.06E+04	3.19E+02
45.103M	5.50E−02	3.80E−02	70000.0	7.73E−01	2.67E−04	1.74E−04	2.20E−00	4.41E−01
45.105	2.00E−01	1.33E−00	70000.0	9.84E+01	3.40E−02	2.22E−02	2.81E+02	5.62E+01
46.103	6.40E−02	3.90E−00	70000.0	9.23E+01	3.19E−02	2.08E−02	2.63E+02	5.27E+01
46.109	4.20E−01	5.10E−01	70000.0	7.92E+01	2.74E−02	1.78E−02	2.26E+02	4.52E+01
47.105	6.30E−01	4.40E−00	70000.0	7.61E+02	2.63E−01	1.71E−01	2.93E+03	2.93E+01
47.110	1.70E−00	4.90E−00	70000.0	2.28E+03	7.92E−01	5.16E−01	8.80E+03	8.80E+01
47.111	4.00E−01	3.00E−00	70000.0	3.29E+02	1.14E−01	7.44E−02	1.26E+03	1.26E+01
48.109	1.10E−01	1.40E+02	70000.0	4.07E+03	1.40E−00	9.18E−01	1.62E+04	4.07E+01
48.115M	6.10E−01	3.50E+01	70000.0	5.64E+03	1.95E−00	1.27E−00	2.25E+04	5.64E+01
48.115	7.10E−01	2.20E−00	70000.0	4.12E+02	1.42E−01	9.31E−02	1.65E+03	4.12E−00
49.113M	2.90E−01	7.30E−02	70000.0	5.59E−00	1.93E−03	1.26E−03	2.23E+01	4.47E−02
49.114M	9.70E−01	2.40E+01	70000.0	6.15E+03	2.12E−00	1.38E−00	2.46E+04	4.92E+01
49.115M	2.60E−01	1.90E−01	70000.0	1.30E+01	4.51E−03	2.94E−03	5.22E+01	1.04E−01
49.115	1.70E−01	4.80E+01	70000.0	2.15E+03	7.46E−01	4.86E−01	8.62E+03	1.72E+01
50.113	3.20E−01	2.70E+01	70000.0	2.55E+03	8.84E−01	5.77E−01	9.13E+03	4.56E+02
50.125	9.40E−01	7.50E−00	70000.0	2.08E+03	7.21E−01	4.70E−01	7.45E+03	3.72E+02
51.122	8.20E−01	2.60E−00	70000.0	6.08E+02	2.10E−01	1.37E−01	2.25E+03	6.76E+01
51.124	1.60E−00	2.30E+01	70000.0	1.05E+04	3.63E−00	2.37E−00	3.89E+04	1.16E+03
51.125	4.30E−01	3.60E+01	70000.0	3.65E+03	1.26E−00	8.23E−01	1.35E+04	4.05E+02
52.125M	1.50E−01	1.20E+01	70000.0	7.23E+02	2.50E−01	1.63E−01	1.90E+03	4.75E+02
52.127M	3.20E−01	1.30E+01	70000.0	1.67E+03	5.78E−01	3.77E−01	4.39E+03	1.09E+03
52.127	2.40E−01	3.80E−01	70000.0	3.66E+01	1.26E−02	8.26E−03	9.64E+01	2.41E+01
52.129M	1.10E−00	1.00E+01	70000.0	4.41E+03	1.52E−00	9.97E−01	1.16E+04	2.90E+03

Table 3.2–1 (continued)
DOSE CONVERSION FACTORS

Identification (Z.A.)	Energy (MeV)	T(1/2) Days	Weight in Grams	Inh. Dose (rem/Ci)	Dose/Conc rem/Ci-s/m³ Hi-Rate	Dose/Conc rem/Ci-s/m³ Lo-Rate	Organ Dose (rem/Ci)	Ing. Dose (rem/Ci)
TOTAL BODY (continued)								
52.129	9.80E−01	5.10E−02	70000.0	2.00E+01	6.94E−03	4.53E−03	5.28E+01	1.32E+01
52.131M	1.60E−00	1.15E−00	70000.0	7.39E+02	2.55E−01	1.66E−01	1.94E+03	4.86E+02
52.132	1.90E−00	2.60E−00	70000.0	1.98E+03	6.86E−01	4.47E−01	5.22E+03	1.30E+03
53.126	2.30E−01	1.21E+01	70000.0	2.20E+03	7.63E−01	4.97E−01	2.94E+03	2.94E+03
53.129	8.90E−02	1.38E+02	70000.0	9.73E+03	3.36E−00	2.19E−00	1.29E+04	1.29E+04
53.131	4.40E−01	7.60E−00	70000.0	2.65E+03	9.17E−01	5.98E−01	3.53E+03	3.53E+03
53.132	1.70E−00	9.70E−02	70000.0	1.30E+02	4.52E−02	2.95E−02	1.74E+02	1.74E+02
53.133	8.40E−00	8.70E−01	70000.0	5.79E+03	2.00E−00	1.30E−00	7.72E+03	7.72E+03
53.134	1.50E−00	3.60E−02	70000.0	4.28E+01	1.48E−02	9.66E−03	5.70E+01	5.70E+01
53.135	1.30E+02	8.00E−01	70000.0	8.24E+04	2.85E+01	1.86E+01	1.09E+05	1.09E+05
55.131	2.90E−02	8.75E−00	70000.0	2.01E+02	6.96E−02	4.54E−02	2.68E+02	2.68E+02
55.134M	1.90E−01	1.30E−01	70000.0	1.95E+02	6.77E−02	4.41E−02	2.61E+01	2.61E+01
55.134	1.10E−00	6.50E+01	70000.0	5.66E+05	1.96E+02	1.27E+02	7.55E+04	7.55E+04
55.135	6.60E−02	7.00E+01	70000.0	3.66E+04	1.26E+01	8.26E−00	4.88E+03	4.88E+03
55.136	6.50E−01	1.10E−00	70000.0	5.66E+03	1.96E−00	1.27E−00	7.55E+02	7.55E+02
55.137	5.90E−01	7.00E+01	70000.0	3.27E+05	1.13E+02	7.38E+01	4.36E+04	4.36E+04
56.131	3.80E−01	9.80E−00	70000.0	1.10E+03	3.81E−01	2.48E−01	3.93E+03	1.96E+02
56.140	2.30E−00	1.07E+01	70000.0	7.28E+03	2.52E−00	1.64E−00	2.60E+04	1.30E+03
57.140	1.90E−00	1.68E−00	70000.0	8.43E+02	2.91E−01	1.90E−01	3.37E+03	3.37E−01
58.141	2.10E−01	3.00E+01	70000.0	1.66E+03	5.76E−01	3.75E−01	6.66E+03	6.66E−01
58.143	9.70E−01	1.33E−00	70000.0	3.40E+02	1.17E−01	7.69E−02	1.36E+03	1.36E−01
58.144	1.30E−00	1.91E+02	70000.0	6.56E+04	2.27E+01	1.48E+01	2.62E+05	2.62E+01
59.142	8.50E−01	8.00E−01	70000.0	1.79E+02	6.21E−02	4.05E−02	7.18E+02	7.18E−02
59.143	3.20E−01	1.35E+01	70000.0	1.14E+03	3.94E−01	2.57E−01	4.56E+03	4.56E−01
60.144	2.00E+01	6.56E+02	70000.0	3.46E+06	1.19E+03	7.82E+02	1.38E+07	1.38E+03
60.147	4.00E−01	1.11E+01	70000.0	1.17E+03	4.05E−01	2.64E−01	4.69E+03	4.69E−01
60.149	1.10E−00	8.30E−02	70000.0	2.41E+01	8.34E−03	5.44E−03	9.65E+01	9.65E−03
61.147	6.90E−02	3.83E+02	70000.0	6.98E+03	2.41E−00	1.57E−00	2.79E+04	2.79E−00
61.149	5.40E−01	2.20E−00	70000.0	3.13E+02	1.08E−01	7.08E−02	1.25E+03	1.25E−01
62.147	2.30E+02	6.56E+02	70000.0	3.98E+07	1.37E+04	8.99E+03	1.59E+08	1.59E+04
62.151	4.20E−02	6.45E+02	70000.0	7.15E+03	2.47E−00	1.61E−00	2.86E+04	2.86E−00
62.153	3.00E−01	1.95E−00	70000.0	1.54E+02	5.34E−02	3.48E−02	6.18E+02	6.18E−02
63.152	8.80E−01	3.80E−01	70000.0	8.83E+01	3.05E−02	1.99E−02	3.53E+02	3.53E−02
63.152	6.60E−01	5.59E+02	70000.0	9.75E+04	3.37E+01	2.20E+01	3.90E+05	3.90E+01
63.154	1.30E−00	5.72E+02	70000.0	1.96E+05	6.79E+01	4.43E+01	7.86E+05	7.86E+01
63.155	1.60E−00	3.14E+02	70000.0	1.32E+05	4.59E+01	2.99E+01	5.31E+05	5.31E+01
64.153	1.70E−01	1.65E+02	70000.0	7.41E+03	2.56E−00	1.67E−00	2.96E+04	2.96E−00
64.159	3.60E−01	7.50E−01	70000.0	7.13E+01	2.46E−02	1.61E−02	2.85E+02	2.85E−02
65.160	8.50E−01	6.60E+01	70000.0	1.48E+04	5.12E−00	3.34E−00	5.93E+04	5.93E−00
66.165	5.10E−01	9.70E−02	70000.0	1.30E+01	4.52E−03	2.95E−03	5.22E+01	5.22E−03
66.166	7.90E−01	3.40E−00	70000.0	7.09E+02	2.45E−01	1.60E−01	2.83E+03	2.83E−01
67.166	7.00E−01	1.10E−00	70000.0	2.03E+02	7.04E−02	4.59E−02	8.14E+02	8.14E−02
68.169	3.70E−01	9.30E−00	70000.0	9.09E+02	3.14E−01	2.05E−01	3.63E+03	3.63E−01
68.171	6.50E−01	3.10E−01	70000.0	5.32E+01	1.84E−02	1.20E−02	2.13E+02	2.13E−02
69.170	3.40E−01	1.07E+02	70000.0	9.61E+03	3.32E−00	2.16E−00	3.84E+04	3.84E−00
69.171	3.00E−02	3.42E+02	70000.0	2.71E+03	9.38E−01	6.11E−01	1.08E+04	1.08E−00
70.175	1.60E−01	4.10E−00	70000.0	1.73E+02	5.99E−02	3.91E−02	6.93E+02	6.93E−02
71.177	1.70E−01	6.70E−00	70000.0	3.01E+02	1.04E−01	6.79E−02	1.20E+03	1.20E−01

Table 3.2–1 (continued)
DOSE CONVERSION FACTORS

Identification (Z.A.)	Energy (MeV)	T(1/2) Days	Weight in Grams	Inh. Dose (rem/Ci)	Dose/Conc rem/Ci-s/m³ Hi-Rate	Dose/Conc rem/Ci-s/m³ Lo-Rate	Organ Dose (rem/Ci)	Ing. Dose (rem/Ci)
TOTAL BODY (continued)								
72.181	5.00E−01	4.30E+01	70000.0	5.68E+03	1.96E−00	1.28E−00	2.27E+04	2.27E−00
73.182	1.10E−00	7.60E+01	70000.0	2.20E+04	7.64E−00	4.98E−00	8.83E+04	8.83E−00
74.181	2.00E−01	1.00E−00	70000.0	6.34E+01	2.19E−02	1.43E−02	2.11E+02	2.11E+01
74.185	1.40E−01	1.00E−00	70000.0	4.44E+01	1.53E−02	1.00E−02	1.48E+02	1.48E+01
74.187	6.80E−01	5.00E−01	70000.0	1.07E+02	3.73E−02	2.43E−02	3.59E+02	3.59E+01
75.183	2.40E−01	6.40E−00	70000.0	8.11E+02	2.80E−01	1.83E−01	1.62E+03	8.11E+02
75.186	3.80E−01	2.50E−00	70000.0	5.02E+02	1.73E−01	1.13E−01	1.00E+03	5.02E+02
75.187	1.20E−02	7.00E−00	70000.0	4.44E+01	1.53E−02	1.00E−02	8.88E+01	4.44E+01
75.188	9.40E−01	6.40E−01	70000.0	3.17E+02	1.10E−01	7.17E−02	6.35E+02	3.17E+02
76.185	5.10E−01	2.00E−00	70000.0	3.23E+02	1.11E−01	7.30E−02	1.07E+03	1.07E+02
76.191M	6.00E−02	4.50E−01	70000.0	8.56E−00	2.96E−03	1.93E−03	2.85E+01	2.85E−00
76.191	1.60E−01	1.80E−00	70000.0	9.13E+01	3.15E−02	2.06E−02	3.04E+02	3.04E+01
76.193	3.80E−01	8.00E−01	70000.0	9.64E+01	3.33E−02	2.17E−02	3.21E+02	3.21E+01
77.190	3.70E−01	7.50E−00	70000.0	8.80E+02	3.04E−01	1.98E−01	2.93E+03	2.93E+02
77.192	1.10E−00	1.58E+01	70000.0	5.51E+03	1.90E−00	1.24E−00	1.83E+04	1.83E+03
77.194	8.10E−01	7.60E−01	70000.0	1.95E+02	6.75E−02	4.40E−02	6.50E+02	6.50E+01
78.191	7.00E−01	2.70E−00	70000.0	5.99E+02	2.07E−01	1.35E−01	1.99E+03	1.99E+02
78.193M	7.50E−02	3.00E−00	70000.0	7.13E+02	2.46E−02	1.61E−02	2.37E+02	2.37E+01
78.193	4.30E−02	2.40E+01	70000.0	3.27E+02	1.13E−01	7.38E−02	1.09E+03	1.09E+02
78.197M	5.50E−01	5.60E−02	70000.0	9.76E−00	3.37E−03	2.20E−03	3.25E+01	3.25E−00
78.197	2.60E−01	7.30E−01	70000.0	6.01E+01	2.08E−02	1.35E−02	2.00E+02	2.00E+01
79.196	4.60E−01	5.40E−00	70000.0	7.87E+02	2.72E−01	1.77E−01	2.62E+03	2.62E+02
79.198	5.80E−01	2.60E−00	70000.0	4.78E+02	1.65E−01	1.07E−01	1.59E+03	1.59E+02
79.199	1.80E−01	3.10E−00	70000.0	1.76E+02	6.12E−02	3.99E−02	5.89E+02	5.89E+01
80.197M	3.00E−01	9.10E−01	70000.0	1.81E+02	6.28E−02	4.10E−02	2.88E+02	2.16E+02
80.197	9.70E−02	2.10E−00	70000.0	1.35E+02	4.69E−02	3.06E−02	2.15E+02	1.61E+02
80.203	2.50E−01	8.20E−00	70000.0	1.36E+03	4.72E−01	3.08E−01	2.16E+03	1.62E+03
81.200	4.00E−01	9.20E−01	70000.0	1.86E+02	6.45E−02	4.21E−02	3.89E+02	1.75E+02
81.201	1.70E−01	1.90E−00	70000.0	1.63E+02	5.67E−02	3.69E−02	3.41E+02	1.53E+02
81.202	3.80E−01	3.50E−00	70000.0	6.74E+02	2.33E−01	1.52E−01	1.40E+03	6.32E+02
81.204	2.50E−01	5.00E−00	70000.0	6.34E+02	2.19E−01	1.43E−01	1.32E+03	5.94E+02
82.203	2.20E−01	2.17E−00	70000.0	1.46E+02	5.06E−02	3.30E−02	5.04E+02	4.03E+01
82.210	5.20E−00	1.20E+03	70000.0	1.91E+06	6.61E+02	4.31E+02	6.59E+06	5.27E+05
82.212	8.20E+01	4.40E−01	70000.0	1.10E+04	3.82E−00	2.49E−00	3.81E+04	3.05E+03
83.206	1.80E−00	2.80E−00	70000.0	1.38E+03	4.79E−01	3.12E−01	5.32E+03	5.32E+01
83.207	1.00E−00	5.00E−00	70000.0	1.37E+03	4.75E−01	3.10E−01	5.28E+03	5.28E+01
83.210	1.00E+01	2.50E−00	70000.0	6.87E+03	2.37E−00	1.55E−00	2.64E+04	2.64E+02
83.212	8.30E+01	4.20E−02	70000.0	9.58E+02	3.31E−01	2.16E−01	3.68E+03	3.68E+01
84.210	5.50E+01	2.50E+01	70000.0	4.07E+05	1.40E+02	9.18E+01	1.45E+06	8.72E+04
85.211	6.10E+01	3.00E−01	70000.0	1.67E+04	5.78E−00	3.78E−00	2.23E+04	2.23E+04
88.223	2.80E+02	5.90E−00	70000.0	6.98E+05	2.41E+02	1.57E+02	1.74E+06	5.23E+05
88.224	2.80E+02	2.30E−00	70000.0	2.72E+05	9.42E+01	6.14E+01	6.80E+05	2.04E+05
88.226	1.10E+02	9.00E+02	70000.0	4.18E+07	1.44E+04	9.44E+03	1.04E+08	3.13E+07
88.228	2.30E+02	2.30E+02	70000.0	2.23E+07	7.73E+03	5.04E+03	5.59E+07	1.67E+07
89.227	2.00E+02	6.00E+03	70000.0	2.78E+08	9.62E+04	6.28E+04	1.11E+09	1.11E+05
89.228	2.30E+02	2.60E−01	70000.0	1.58E+04	5.46E−00	3.56E−00	6.32E+04	6.32E−00
90.227	2.00E+02	1.84E+01	70000.0	9.72E+05	3.36E+02	2.19E+02	3.89E+06	3.89E+02
90.228	2.30E+02	6.91E+02	70000.0	4.20E+07	1.45E+04	9.47E+03	1.68E+08	1.68E+04

Table 3.2–1 (continued)
DOSE CONVERSION FACTORS

Identification (Z.A.)	Energy (MeV)	T(1/2) Days	Weight in Grams	Inh. Dose (rem/Ci)	Dose/Conc rem/Ci-s/m³ Hi-Rate	Dose/Conc rem/Ci-s/m³ Lo-Rate	Organ Dose (rem/Ci)	Ing. Dose (rem/Ci)
TOTAL BODY (continued)								
90.230	4.80E+01	5.70E+04	70000.0	1.43E+08	4.96E+04	3.23E+04	5.73E+08	5.73E+04
90.231	1.80E−01	1.07E−00	70000.0	5.09E+01	1.76E−02	1.14E−02	2.03E+02	2.03E−02
90.232	6.20E+01	5.70E+04	70000.0	1.85E+08	6.40E+04	4.18E+04	7.40E+08	7.40E+04
90.234	9.10E−01	2.41E+01	70000.0	5.79E+03	2.00E−00	1.30E−00	2.31E+04	2.31E−00
91.230	2.90E+02	1.77E+01	70000.0	2.70E+05	9.38E+01	6.12E+01	1.08E+06	1.08E+02
91.231	1.40E+02	4.10E+04	70000.0	4.01E+08	1.38E+05	9.05E+04	1.60E+09	1.60E+05
91.233	3.20E−01	2.74E+01	70000.0	2.31E+03	8.01E−01	5.22E−01	9.26E+03	9.26E−01
92.230	3.50E+02	1.47E−00	70000.0	1.35E+05	4.70E+01	3.06E+01	5.43E+05	5.43E+01
92.232	2.80E+02	1.00E+01	70000.0	7.40E+05	2.56E+02	1.67E+02	2.96E+06	2.96E+02
92.233	5.00E+01	1.00E+02	70000.0	1.32E+06	4.57E+02	2.98E+02	5.28E+06	5.28E+02
92.234	4.90E+01	1.00E+02	70000.0	1.29E+06	4.48E+02	2.92E+02	5.18E+06	5.18E+02
92.235	4.60E+01	1.00E+02	70000.0	1.21E+06	4.20E+02	2.74E+02	4.86E+06	4.86E+02
92.236	4.70E+01	1.00E+02	70000.0	1.24E+06	4.29E+02	2.80E+02	4.96E+06	4.96E+02
92.238	4.30E+01	1.00E+02	70000.0	1.13E+06	3.93E+02	2.56E+02	4.54E+06	4.54E+02
92.240	1.30E−00	5.85E−01	70000.0	2.09E+02	7.23E−02	4.71E−02	8.03E+02	8.03E−00
93.237	4.90E+01	3.90E+04	70000.0	1.39E+08	4.82E+04	3.14E+04	5.57E+08	5.57E+04
93.239	2.90E−01	2.33E−00	70000.0	1.78E+02	6.17E−02	4.03E−02	7.14E+02	7.14E−02
94.238	5.70E+01	2.20E+04	70000.0	1.44E+08	4.99E+04	3.26E+04	5.78E+08	5.78E+04
94.239	5.30E+01	6.40E+04	70000.0	1.60E+08	5.54E+04	3.61E+04	6.40E+08	6.40E+04
94.240	5.30E+01	6.30E+04	70000.0	1.59E+08	5.53E+04	3.60E+04	6.39E+08	6.39E+04
94.241	2.30E−00	4.50E+03	70000.0	2.56E+06	8.88E+02	5.79E+02	1.02E+07	1.02E+03
94.242	5.10E+01	6.50E+04	70000.0	1.54E+08	5.34E+04	3.48E+04	6.17E+08	6.17E+04
94.243	3.70E−01	2.08E−01	70000.0	2.03E+01	7.03E−03	4.59E−03	8.13E+01	2.44E−03
94.244	4.80E+01	6.50E+04	70000.0	1.80E+08	6.12E+04	3.99E+04	7.09E+08	2.12E+04
95.241	5.70E+01	1.80E+04	70000.0	1.36E+08	4.72E+04	3.08E+04	5.46E+08	5.46E+04
95.242M	7.30E+01	1.50E+04	70000.0	1.38E+08	4.77E+04	3.12E+04	5.52E+08	5.52E+04
95.242	8.00E+01	6.67E−01	70000.0	1.18E+04	4.09E−00	2.67E−00	4.74E+04	4.74E−00
95.243	5.40E+01	2.00E+04	70000.0	1.33E+08	4.61E+04	3.01E+04	5.33E+08	5.33E+04
95.244	4.40E+01	1.81E−02	70000.0	2.10E+02	7.28E−02	4.74E−02	8.41E+02	8.41E−02
96.242	8.00E+01	1.61E+02	70000.0	3.40E+06	1.17E+03	7.68E+02	1.36E+07	1.36E+03
96.243	6.00E+01	8.40E+03	70000.0	1.03E+08	3.57E+04	2.33E+04	4.13E+08	4.13E+04
96.244	6.00E+01	5.20E+03	70000.0	7.51E+07	2.59E+04	1.69E+04	3.00E+08	3.00E+04
96.245	5.60E+01	2.40E+04	70000.0	1.45E+08	5.01E+04	3.27E+04	5.80E+08	5.80E+04
96.246	5.60E+01	2.40E+04	70000.0	1.45E+08	5.01E+04	3.27E+04	5.80E+08	5.80E+04
96.247	5.50E+01	2.40E+04	70000.0	1.42E+08	4.92E+04	3.21E+04	5.69E+08	5.69E+04
96.248	5.20E+01	2.40E+04	70000.0	1.17E+09	4.06E+05	2.64E+05	4.68E+09	4.68E+05
96.249	5.20E−00	4.40E−02	70000.0	6.04E+01	2.09E−02	1.36E−02	2.41E+02	2.41E−02
97.249	3.80E−00	2.89E+02	70000.0	2.90E+05	1.00E+02	6.54E+01	1.16E+06	3.48E+01
97.250	5.70E+01	1.34E−01	70000.0	2.01E+03	6.98E−01	4.55E−01	8.07E+03	2.42E−01
98.249	6.00E+01	4.70E+04	70000.0	1.75E+08	6.06E+04	3.95E+04	7.01E+08	2.10E+04
98.250	6.20E+01	3.50E+03	70000.0	5.57E+07	1.92E+04	1.25E+04	2.23E+08	6.69E+03
98.251	5.90E+01	5.30E+04	70000.0	1.74E+08	6.04E+04	3.94E+04	6.99E+08	2.09E+04
98.252	2.10E+02	7.93E+02	70000.0	4.40E+07	1.52E+04	9.93E+03	1.76E+08	5.28E+03
98.253	7.30E+01	1.80E+01	70000.0	3.47E+05	1.20E+02	7.83E+01	1.38E+06	4.16E+01
98.254	3.80E+03	5.60E+01	70000.0	5.62E+07	1.94E+04	1.26E+04	2.24E+08	6.74E+03
99.253	7.30E+01	2.00E+01	70000.0	3.85E+05	1.33E+02	8.70E+01	1.54E+06	4.63E+01
99.254	1.30E+02	1.60E−00	70000.0	5.49E+04	1.90E+01	1.24E+01	2.19E+05	6.59E−00
99.254	1.20E+02	4.76E+02	70000.0	1.50E+07	5.22E+03	3.40E+03	6.03E+07	1.81E+03

Table 3.2–1 (continued)
DOSE CONVERSION FACTORS

Identification (Z.A.)	Energy (MeV)	T(1/2) Days	Weight in Grams	Inh. Dose (rem/Ci)	Dose/Conc rem/Ci-s/m³ Hi-Rate	Dose/Conc rem/Ci-s/m³ Lo-Rate	Organ Dose (rem/Ci)	Ing. Dose (rem/Ci)
TOTAL BODY (continued)								
99.255	7.50E+01	3.00E+01	70000.0	5.94E+05	2.05E+02	1.34E+02	2.37E+06	7.13E+01
100.254	1.30E+02	1.35E−01	70000.0	4.63E+03	1.60E−00	1.04E−00	1.85E+04	5.56E−01
100.255	7.50E+01	8.96E−01	70000.0	1.77E+04	6.14E−00	4.00E−00	7.10E+04	2.13E−00
100.256	3.60E+03	1.11E−01	70000.0	1.05E+05	3.65E+01	2.38E+01	4.22E+05	1.26E+01

2. $\text{Dose} = 1.48 \times 10^{6} \dfrac{\text{rem}}{\text{Ci, inhaled}} \times 10^{-6}\ \text{Ci} = 1.48\ \text{rem} \approx 1.5\ \text{rem.}$

Sample Calculation 2. If the air concentration is 10^{-6} μCi/cc of I-135 and the individual will be exposed for 8 hr, the estimation of dose may be made as follows:

1. If the exposure is occupational, assume the high breathing rate factor (Ci-sec/m³).

2. If the thyroid is the organ of interest, the conversion factor 4.28 E + 01 or 4.28×10^{1} rem/ci-sec/m³.

3. 10^{-6} μCi/cc = 10^{-6} Ci/m³.

4. $\text{Dose} = 10^{-6}\ \text{Ci/m}^3 \times 8\ \text{hr} \times 3{,}600 \dfrac{\text{sec}}{\text{hour}} \times \dfrac{4.28 \times 10^{1}\ \text{rem}}{\text{Ci-sec/m}^3} \approx 1.2\ \text{rem.}$

Sample Calculation 3. If 1 Ci of Sr-89 were released from a stack and the dose at 16 km (10 miles) were desired, it could be calculated as follows:

1. Relative axial concentration at 10 miles (Pasquill, Class F – strong inversion from Figure 3.2–1) is $2 \times 10^{-5}\text{m}^{-2}$. This must be divided by the wind speed (4 m/sec) to obtain relative dilution of 5×10^{-6} sec/m³.

2. Bone dose conversion factor for Sr-89 = 9.42 E + 01 or 9.42×10^{1} rem/Ci-sec/m³ (lower breathing rate).

3. $\text{Dose} = 1\ \text{Ci} \times 5 \times 10^{-6} \dfrac{\text{sec}}{\text{M}^3} \times 9.42 \times 10^{1} \dfrac{\text{rem}}{\text{Ci-sec/m}^3} = 4.7 \times 10^{-4}\ \text{rem.}$

Alternate Data

Since it may be desirable to use parameters other than those used in these conversion factors, a nomogram has been prepared and included as Figure 3.2–2 so that any or all of the parameters may be varied to obtain different conversion factors. This nomogram also is based on Equation 3.2–1 with the assumption that $e^{-1.26 \times 10^4 T_E}$ is near zero. The nomogram may be used as follows:

1. *rem/Ci-sec/m³*

a. Draw a line from the time-integrated concentration through the assumed breathing rate to derive curies inhaled.

b. Draw a line from curies inhaled through f_a to derive curies in organ.

c. Draw a line from curies in organ through the effective energy, mark the point, and draw a line through the effective half-life to the next column.

d. Draw a line from this column through the organ weight to derive dose in rem.

2. *rem/curie inhaled*: Done in the same fashion as 1 above except that Step a is eliminated; begin at Column 3. To determine rem/curie ingested, substitute f_w for f_a.

3. *rem/curie in organ*: Same as 1 above; eliminate Steps a and b and begin at Column 5.

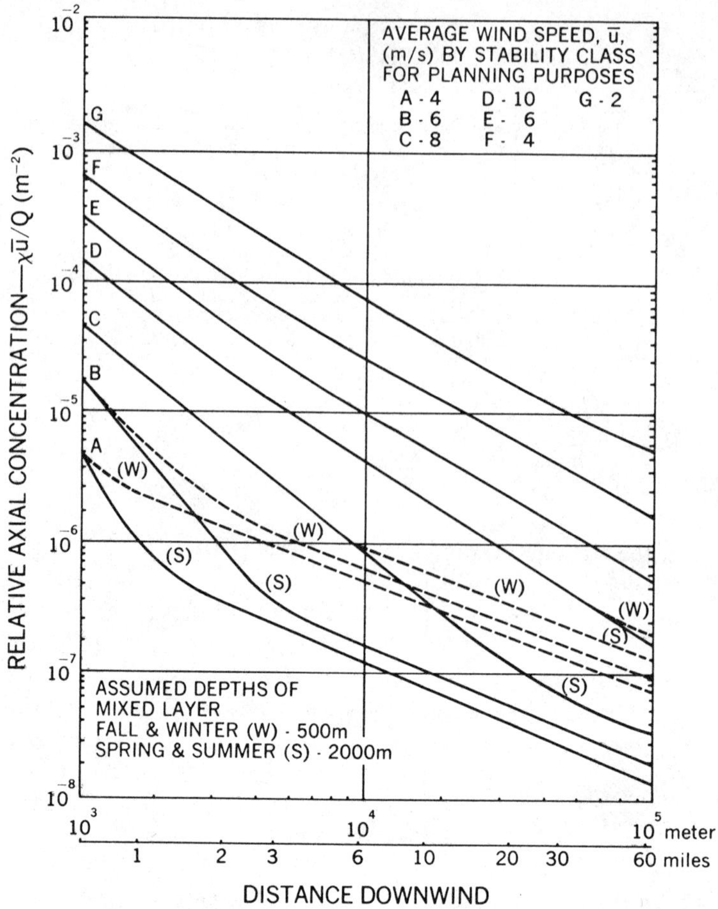

FIGURE 3.2–1. Average relative axial concentration by stability. (From Markee, E. H., Jr., *NRTS Meteorological Information Bulletin,* Number 2, April 1966.

For the short-lived isotopes, the use of a single exponential model may grossly overestimate the actual dose. The conversion factors for these isotopes should be corrected by $T_R/(T_R + T_U)$ where T_R is the radiological half-life and T_U is the half-time for uptake (0.25 day for iodine). The corrected iodine factors are shown in Table 3.2–2. It can be seen that many of these isotopes would not constitute a significant hazard if mixed fission products were released. In addition, for most of the organs a few isotopes constitute 80 to 98^+% of the exposure. Therefore, for estimating dose from mixed fission products it is not necessary to make a calculation for the contribution of every dose.

Legal Notice

This report was prepared as an account of Government-sponsored work. Neither the United States, nor the Commission, nor any person acting on behalf of the Commission (A) makes any warranty or representation, express or implied, with respect to the accuracy, completeness, or usefulness of the information contained in this report, or that the use of any information, apparatus, method, or process disclosed in this report may not infringe privately owned rights, or (B) assumes any liabilities with respect to the use of, or for damages resulting from the use of, any information, apparatus, method, or process disclosed in this report.

As used in the above, "person acting on behalf of the Commission" includes any employee or contractor of the Commission, or employee of such contractor, to the extent that such employee provides access to any information pursuant to his employment or contract with the Commission, or his employment with such contractor.

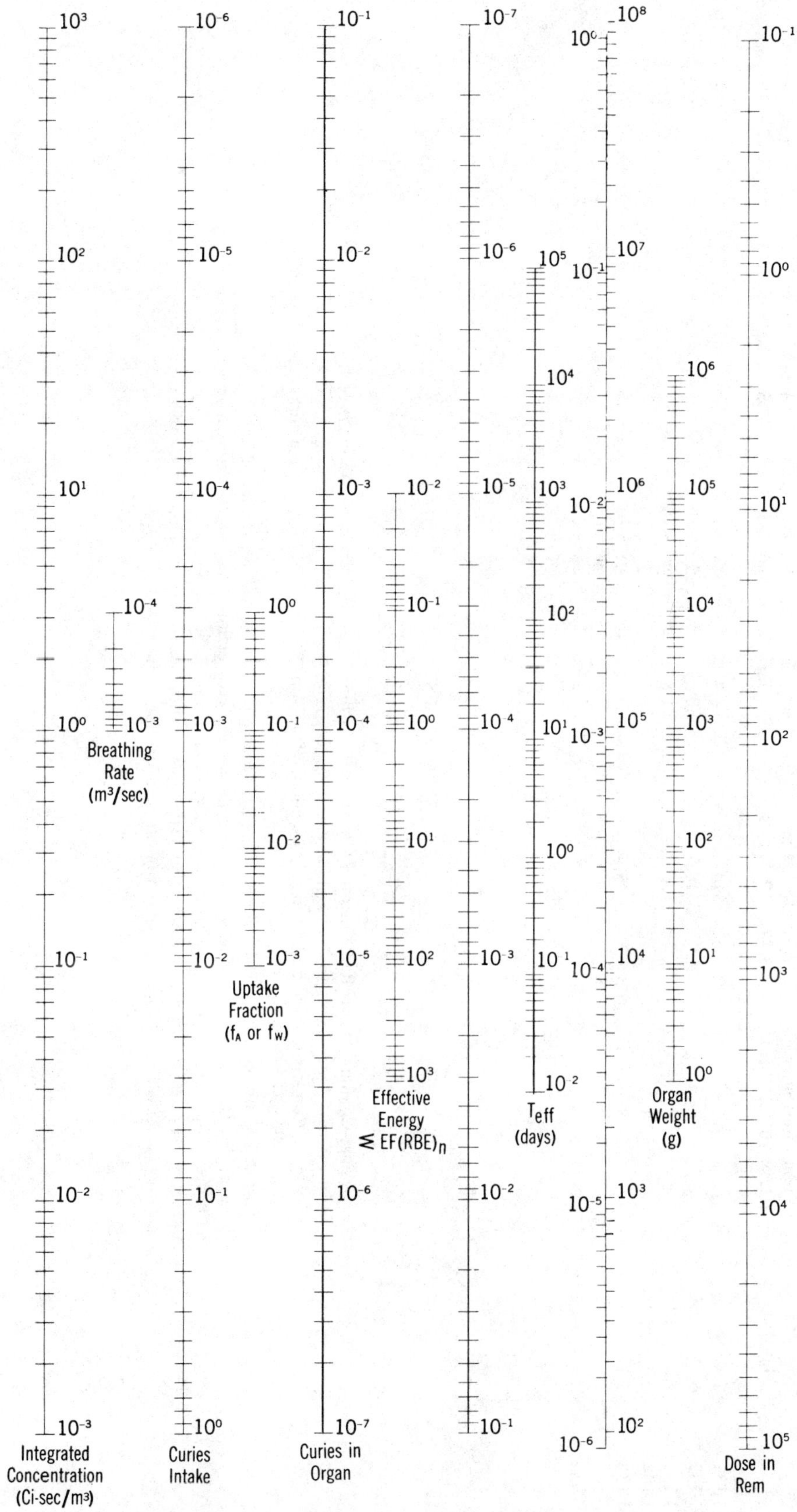

FIGURE 3.2–2. Nomogram for infinity dose calculations.

Table 3.2–2
THYROID CONVERSION FACTORS

	ICRP II				Corrected for uptake decay		
	rem/Ci inhaled	rem/Ci-sec/m^3 (2.32×10^{-4} m^3/sec)	rem/Ci-sec/m^3 (3.47×10^{-4} m^3/sec)	rem/Ci organ	rem/Ci inhaled	rem/Ci-sec/m^3 (2.32×10^{-4} m^3/sec)	rem/Ci-sec/m^3 (3.47×10^{-4} m^3/sec)
I-131	1.48×10^6	343	514	6.3	1.44×10^6	330	500
I-132	5.35×10^4	12.4	18.5	0.23	1.5×10^4	3.5	5.2
I-133	4×10^5	92.8	139	1.8	3.1×10^5	72	110
I-134	2.5×10^4	5.8	8.7	0.11	3.3×10^3	0.75	1.1
I-135	1.24×10^5	28.8	43.0	0.54	6.6×10^4	15	23

Note: $T_R (T_U + T_R)$ where $T_U = 0.25$ day

REFERENCES

1. International Commission on Radiological Protection, *Radiation Protection,* Report of Committee II on Permissible Dose for Internal Radiation, ICRP-2,.Pergamon Press, New York, 1959.
2. International Commission on Radiological Protection, *Radiation Protection,* ICRP-6, Pergamon Press, New York, amended 1959, revised 1962.
3. **Vennart, J. and Minski, M.,** Radiation doses from administered radionuclides, *J. Br. Radiol.,* 35, 372, 1963.
4. **Loevinger, R., Holt, J. G., and Hine, G. J.,** Internally administered radioisotopes, in *Radiation Dosimetry,* Hine, G. J. and Brownell, G. L., Eds., Academic Press, New York, 1956, chap. 17.
5. **Markee, E. H., Jr.,** A simplified method of estimating environmental hazards from accidental airborne release of radioactive material, *NRTS Meteorological Information Bulletin,* Number 2, April 1966.

3.3 LIMITS FOR RADIOACTIVE SURFACE CONTAMINATION

G. D. Schmidt
U.S. Public Health Service

The control of surface contamination has been a common procedure of radiation control programs since the early days of the atomic industry. These procedures include both the direct measurement of total surface contamination using portable survey meters, and smear or wipe tests for sampling the "removable" surface contamination. Smears have proven effective in detecting and measuring the extent of radioisotope spills and contamination tracked into a clean area. The smear samples are usually counted in sensitive laboratory instruments and provide measurement of "removable" contamination, which may be related to the extent of possible internal exposure (due to ingestion, inhalation, wound contamination, and so forth).

Standards for the permissible amounts of radioactive surface contamination have been established by governmental authority in the United States only for radioactive material shipments and by the Department of Defense. However, surface contamination limits have been adopted by most nuclear organizations on an individual basis. These limits have not necessarily been derived on the basis of the associated personnel hazards, but rather may be minimum levels obtainable without economic hardship. The general philosophy adopted in the United States has been to limit surface contamination to "the lowest practicable level." In some cases, the values adopted appear to be extremely conservative; however, the facility has been able to live with these values. Surface contamination limits have also been set at a low level because of a desire to limit (prevent) the contamination of sensitive areas and materials (such as in low level counting rooms and cross contamination of experiments).

In order to summarize the available information and guidance on contamination levels, a review was undertaken in 1965 of adopted guides and the data relating to personnel hazards from surface contamination. A compilation was made of the surface contamination limits used in the United States and in other countries, based on literature and a questionnaire sent to various organizations. The limits for use in occupational radiation areas and release to noncontrolled areas (unrestricted or general public use) are summarized in Table 3.3–1.

The contamination categories used in Table 3.3–1 involve an arbitrary assignment of the actual data, which included limits for more than 30 specific categories. Generally similar values are used for many of these categories, and it is felt that only the following categories merit specific guides:

1. *Occupational: Basic Guide* – The "basic guide" is applicable for use in a radiation area where control over the entry and activity of individuals is exercised for the purposes of radiation protection. The basic guide is applicable for floors, benches, tools, and equipment. A similar guide is also generally used for protective clothing, as seen in Table 3.3–1.
2. *Occupational: Clean Area* – The "clean area" guide is for use in areas where radioactive materials are generally not utilized (or used only in low levels), but the area is subject to contamination from radiation areas. These may include nonradiation laboratories, service areas, and areas sensitive to radioactive contamination.
3. *Noncontrolled: Skin and Personal Clothing* – This guide applies to personal clothing, shoes, hands, and skin of the whole body.
4. *Noncontrolled: Release of Materials* – This guide applies to equipment, tools, vehicles, and facilities for use by the general public without any radiation protection precautions.

It is observed that the contamination limits for the same category used by different organizations vary by factors of up to 1,000. There are basically two different limits used: (1) that recommended by Dunster[1,2] and Barnes[3] and used by most foreign organizations, and (2) the "lowest practicable limit," generally used in the United States. The basic guide recommended by Dunster for readily removable contamination is 22,000 dpm/100 cm^2 for alpha activity.

The literature contains only a few cases where work in contaminated areas has resulted in a detectable body burden. Eisenbud, Blatz, and Barry[4] reported that significant radium burdens

Table 3.3–1
SUMMARY OF CONTAMINATION GUIDES, RECOMMENDATIONS, AND 1965 FACILITY SURVEY[a]

Application	Alpha (dpm/100 cm²)		Beta/gamma	
	Total	Removable	Total (mR/hr)	Removable (dpm/100 cm²)
Occupational				
Basic guides	200–22,000	N.D.[b]–22,000	0.1–1.0	N.D.–220,000
U.S. practice	300–2,000	N.D.–540	0.1–1.0	N.D.–2,200
Clean areas	N.D.–2,200	N.D.–2,200	N.D.–0.5	N.D.–660,000
U.S. practice	N.D.–1,000	N.D.–54	N.D.–0.5	N.D.–220
Protective clothing	N.D.–22,000	N.D.–22,000	0.1–2.5	80–220,000
U.S. practice	N.D.–5,000	N.D.–300	0.1–2.5	80–1,000
Noncontrolled				
Skin and clothing	N.D.–2,200	N.D.–2,200	0.05–2.0	N.D.–22,000
U.S. practice	N.D.–1,500	N.D.–5	0.05–2.0	N.D.–80
Material				
U.S. Practice	N.D.–5,000	N.D.–500	N.D.–0.3	N.D.–200

[a]The ranges of contamination limits presented have been compiled from the responses of various organizations in the U.S. and other countries as well as published guides and recommendations.
[b]N.D.: nondetectable.

From Schmidt, G. D., in *CRC Handbook of Laboratory Safety,* 2nd ed., Steere, N.V., Ed., Chemical Rubber Co., Cleveland, 1971, 478.

were observed only when the contamination was greater than about 100,000 dpm alpha activity per 100 cm^2 of surface. Schultz and Becher[5] showed a correlation between the amount of surface contamination, the airborne concentrations, and the urinary excretion of uranium. A contamination level of 40,000 dpm/100 cm^2 was correlated with the maximum permissible air concentration for uranium. The body burden data on three families who had lived in a radium-contaminated residence were reported by Evans.[6] Radium was processed in the basement until 1941. Surveys of the residence in 1964 showed extensive alpha contamination levels in the range of 10,000 to 100,000 dpm/100 cm^2 with peak levels greater than 1,000,000 dpm/100 cm^2. The results of radium body burden measurements were negative with the exception of those persons who participated in the radium processing. The negative findings included individuals who lived in the house as young children.

The results obtained from a survey in the United Kingdom of 23 luminising establishments, which provided 75 workers for body radioactivity measurements, are summarized and discussed by Duggan and Godfrey.[7] They found that widespread contamination of about 6,600 dpm ^{226}Ra/100 cm^2 (loose) and 66,000 dpm ^{226}Ra/100 cm^2 (total) in allegedly controlled areas is associated with undesirably high radium body burdens in the workers (i.e., of the order of the maximum permissible body burden (MPBB) of 0.1 μCi ^{226}Ra). Duggan and Godfrey recommend levels lower than those of the U.K. Ministry of Health, based upon operational experience. However, the authors do report that radiation hygiene was generally not of the highest standard.

A study of contamination in aircraft instrument repair facilities, which in some cases were involved in the stripping of radium luminous dial instruments, was reported by the Bureau of Radiological Health and Eberline Instrument Corporation. Radium body burdens of 40% of the MPBB were observed in two individuals from one facility which had alpha surface contamination levels of 1,650,000 dpm/100 cm^2 maximum total, 50,000 dpm/100 cm^2 average total, and 2,700 dpm/100 cm^2 maximum removable.

Since 1965[a] there have appeared in the literature additional recommended guides and adopted

[a]Recommendations concerning radioactive surface contamination, prior to 1965, were made by Saenger,[9] General Dynamics,[10] Los Alamos Scientific Laboratory,[11] American Standards Association,[12] National Committee on Radiation Protection,[13] and the International Atomic Energy Agency.[14]

levels for surface contamination, which are summarized in Table 3.3–2. Included in Table 3.3–2 are the values adopted in 1964 by the U.K. Ministry of Health[15] which are based on the recommendation of Dunster.[2] Similar values were later adopted by the U.K. Department of Employment and Productivity.[16] The values adopted by the U.K. Ministry of Health allow significantly higher levels of contamination than any of the other referenced guides. The values adopted by the Department of Defense in DSAM 4145.8[17] are quite typical of U.S. practice. The U.S. Department of Transportation,[18] however, has set limits for contamination of packages offered for transport intermediate between U.S. practice and the U.K. Ministry of Health levels.

Based upon the adopted radioactive surface contamination guides and the operational experience relating internal exposure to actual surface contamination, it appears that the limits used in the United States are conservative. Therefore, the suggested radioactive surface contamination guides in Table 3.3–3 appear reasonable and should not represent a health hazard. The recommendations, however, are subject to the following conditions and interpretations:

1. These limits are to be used as guides, and in practice professional judgment should be used by the radiological physicist to determine the acceptability of the actual contamination.

2. The values are for the most hazardous radionuclides, and a factor-of-10 increase in the levels may be allowed for other radionuclides. The most hazardous radionuclides (high radiotoxicity) are generally taken as those in Class 1 of the IAEA classification[14] which includes, for instance, ^{90}Sr, ^{226}Ra, ^{239}Pu, and ^{241}Am.

3. Although it is felt that the recommended values should not result in a health hazard, good radiation protection practice dictates that a reasonable effort be made to keep actual contamination levels below these values.

4. Compliance with contamination guides should not be used as evidence that exposure of persons to internal sources of radiation is within the prescribed standards. Biological sampling or whole-body counting should be used to ascertain internal exposures.

5. For release of material to the general public:

a. A reasonable effort should be made to minimize the contamination (i.e., the application of additional decontamination procedures have little effect on the contamination levels).

b. Surfaces of premises or equipment likely to be contaminated that are inaccessible for measurement shall be presumed to be contaminated in excess of the above limits and should not be released.

6. Total contamination, as used here, refers to that contamination measured with an appropriate survey instrument probe in direct contact with the surface. Removable contamination refers to activity removed from the surface by wiping the surface with a filter paper or other similar material moistened with a detergent solution (or appropriate solvent); the wipe is dried and counted, and the activity removed per 100 cm^2 of surface calculated.

Table 3.3–2
REFERENCED CONTAMINATION GUIDES

Facility or reference	Alpha (dpm/100 cm²)		Beta/gamma (mR/hr and dpm/100 cm²)		Application
	Total	Removable	Total	Removable	
DOT, 49 CFR 173[18] [a]		220		2,200	Package
		220	<0.5	2,200	Vehicles
PHS No. 999-RH-36[8]	25,000 (maximum) 5,000 (average)	500			Radium dial stripping
DSAM 4145.8[17]	1,000	200	2.0	200	Radiation laboratory
	1,000	1,000	2.0	2.0	Contamination clothing
	600	30	0.25	100	Nonradiation laboratory
	200	200	0.2	0.2	Personal clothing
	200	0	0.06	0	Skin-body
U.K. Ministry of Health[15] [b]		22,000		220,000	Radiation area
		2,200		2,200	Nonradiation area (body, personal clothing)
Duggan and Godfrey[7] [c]	6,600	660			Radium dial painting

[a] For natural uranium, natural thorium, and depleted uranium use values higher by a factor of 10.
[b] The guide is the maximum permissible levels of contamination not known to be fixed to the surface.
[c] Value is for dpm $^{226}Ra/100\ cm^2$; note that one might expect to observe 2–3 alpha daughters associated with radium contamination.

From Schmidt, G. D., in *CRC Handbook of Laboratory Safety,* 2nd ed., Steere, N. V., Ed., Chemical Rubber Co., Cleveland, 1971, 479.

Table 3.3–3
SUGGESTED RADIOACTIVE SURFACE CONTAMINATION GUIDES

Application	Alpha (dpm/100 cm^2)		Beta/gamma	
	Total	Removable	Total (mR/hr)	Removable (dpm/100 cm^2)
Occupational				
Basic guide	25,000 (maximum) 5,000 (average)	500	1.0	5,000
Clean area	1,000	100	0.5	1,000
Noncontrolled				
Skin, personal clothing	500	N.D.[a]	0.1	N.D.
Release of material	2,500 (maximum) 500 (average)	100	0.2	1,000

[a]N.D.: nondetectable.

From Schmidt, G. D., in *CRC Handbook of Laboratory Safety,* 2nd ed., Steere, N. V., Ed., Chemical Rubber Co., Cleveland, 1971, 480.

REFERENCES

1. **Dunster, H. J.,** Contamination of surfaces by radioactive materials: the derivation of maximum permissible levels, *Atomics,* 6(8), 233, 1955.
2. **Dunster, H. J.,** Surface contamination measurements as an index of control of radioactive materials, *Health Phys.,* 8, 353, 1962.
3. **Barnes, D. E.,** Basic Criteria in the Control of Air and Surface Contamination, Health Physics in Nuclear Installations; Report of Symposium Organized at the Danish Atomic Energy Center of Riso, May 1959.
4. **Eisenbud, M., Blatz, H., and Barry, E. V.,** How important is surface contamination? *Nucleonics,*12(8), 12, 1954.
5. **Schultz, N. B. and Becher, A. F.,** Correlation of uranium alpha surface contamination, airborne concentrations, and urinary excretion rates, *Health Phys.,* 9, 901, 1963.
6. **Evans, R. D.,** Long-term effects of residence in a highly contaminated house, in *Radium and Mesothorium Poisoning and Dosimetry and Instrumentation Techniques in Applied Radioactivity,* M.I.T. 952, M.I.T. Press, Cambridge, Mass., 1964, 15.
7. **Duggan, M. and Godfrey, B.,** Some factors contributing to the internal radiation hazard in the radium luminising industry, *Health Phys.,* 13, 613, 1967.
8. Evaluations of Radium Contamination in Aircraft Instrument Repair Facilities, Public Health Service Publ. No. 999-Rh-36, Bureau of Radiological Health, Public Health Service, U.S. Dept. of Health, Education, and Welfare, Rockville, Md., and Eberline Instrument Corporation, Santa Fe, N. M.
9. **Saenger, E. L.,** *Medical Aspects of Radiation Accidents: A Handbook for Physicians, Health Physicists and Industrial Hygienists,* U.S. Atomic Energy Commission, Washington, D.C., February 1963.
10. *Health Physics Handbook,* OSP-279, Health Physics Office Dept. 3, General Dynamics Corporation, Fort Worth, Tex., April 1963.
11. **Dummer, J. E., Jr., Ed.,** *General Handbook for Radiation Monitoring,* LA-1835, U.S. Atomic Energy Commission, Superintendent of Documents, U.S. Government Printing Office, Washington, D.C., November 1958.
12. *Radiation Protection in Nuclear Fuel Fabrication Plants,* N 7.2, American Standards Association Inc., New York, July 8, 1963.
13. National Committee on Radiation Protection, *Safe Handling of Radioactive Materials,* National Bureau of Standards Handbook 92, U.S. Department of Commerce, Superintendent of Documents, U.S. Government Printing Office, Washington, D.C., March 1964.
14. *Safe Handling of Radioisotopes,* Safety Series No. 1, International Atomic Energy Agency, Vienna, March 1962.

15. United Kingdom Ministry of Health, Code of Practice for the Protection of Persons Against Ionizing Radiations Arising from Medical and Dental Use, 1964.
16. United Kingdom Department of Employment and Productivity, *Ionizing Radiations: Precautions for Industrial Users,* Safety, Health, and Welfare New Series No. 13, HMSO, 1969.
17. Defense Supply Agency, Department of Defense, Radioactive Commodities in the DoD Supply Systems, DSAM 4145.8, Superintendent of Documents, U.S. Government Printing Office, Washington, D.C., December 1967.
18. Code of Federal Regulations, Title 49, Chapter 1–Department of Transportation, Part 173 Shippers, Subsection 173.393 (h) General packaging requirements, *Fed. Register,* 33 (194), Part II, Superintendent of Documents, U.S. Government Printing Office, Washington, D.C., October 4, 1968.

3.4 DETERMINING INDUSTRIAL HYGIENE REQUIREMENTS FOR INSTALLATIONS USING RADIOACTIVE MATERIALS[a]

Allen Brodsky
Mercy Hospital
Duquesne University
Pittsburg, Pennsylvania

INTRODUCTION

The purpose of this section is to present some guidelines that have proved useful in evaluating radiation safety requirements for various operations over a wide range of types, quantities, and forms of radioactive material. These guidelines are illustrated by the development of a table that first arranges the commonly used radionuclides in eight groups corresponding to the relative magnitudes of their maximum radiotoxicities. Then the table presents curie quantities above which the need for certain safeguards should be examined. For reasons given below, the eight groups are chosen to correspond to the eight orders of magnitude over which the estimated maximum doses per curie range when the radioactive material is delivered in a single intake by inhalation. However, safeguards against external exposure are based on the external dose rate for each radionuclide.

Although the use of single intakes by inhalation may seem to apply only to accidental situations, it also turns out to yield about the same ordering of radionuclides that one would select on the basis of permissible concentrations for continuous exposure under routine operations. Also, since the calculations of dose by inhalation take into account an appreciable fraction of material assumed to pass through the gastrointestinal tract within the first few days after exposure, the relative grouping of radionuclides is also consistent with the relative ingestion toxicity, within one order of magnitude.

Some words of caution should be inserted at this point. Since the generalized calculations in this section cannot take into account additional safety considerations, or operating conditions that may be designed into a particular facility, the guides given herein should not ordinarily be used in themselves as the sole criteria for selecting safeguards. The numbers given in the table are chosen to represent (in general) the magnitudes at which each safeguard should be considered if materials are to be handled in the most dispersible form, or if processes are varied or unpredictable in regard to the unwanted release of material from process locations. The values in the table may be scaled upward, for example, for situations in which the radioactive material is normally diluted with other materials that would prevent an intake of enough radionuclide to produce serious exposures, or for processes normally enclosed in a manner known to prevent the escape of significant quantities. Some suggested scaling factors are presented with the table of safeguards to take into account the degree of hazard contributed by the nature of the operation. On the other hand, in unusual circumstances where, for example, good ventilation control is absent and the process is known to disperse the radioactive material toward the operator, operation in chemical hoods might be required at even lower curie levels than those indicated in the table. Principles of ventilation design to deal with these lower levels of radioactivity would be the same as those used in general industrial hygiene practice.

The methods of hazard estimation and the numbers given in this section are intended primarily for use as guides by health physicists or industrial hygienists experienced in radiation control, and require addition of the other considerations that apply to specific operations, particular materials, or the possible chemical reactions, in order to avoid the application of larger and more expensive factors of safety than are necessary for the protection of health. Further, although the methods described in this section have been based on the general limits of radiation exposure recommended by the National Committee for Radiation Protection and Measurement (NCRP) and the Federal Radiation Council (FRC), the use of these methods does not automatically guarantee that all other applicable safety or regulatory requirements will be met. The industrial hygienist or health physicist must balance the combination of safeguards in order to (1) protect

[a]This section is reprinted from *Am. Ind. Hyg. Assoc. J.*, 26, 294, 1965. With permission.

health, (2) meet additional regulatory requirements, and (3) ensure that the first two objectives are met with minimum cost to the employer.

CLASSIFICATION OF RADIONUCLIDES INTO GROUPS

A number of different classifications have been suggested for purposes of estimating hazards of given radionuclides.[1] Many of these classifications recognize the differences in relative toxicity per curie of the various radionuclides, but divide them arbitrarily into only three or four groups for purposes of selecting safeguards. Since the maximum dose per curie inhaled varies over about nine orders of magnitude or more for the radioactive materials of interest, the author has found that the use of only three groups may result in applying similar safeguards to two radionuclides that differ by a factor of approximately 1,000 in relative radiotoxicity. This means that either the occasional introduction of unnecessary factors of safety of about 1,000 would be encouraged by such guidelines, or else dangerous reductions in the factor of safety may occur. Although the uncertainties[2] in present methods of dose estimation are recognized, the author believes that relative maximum radiotoxicities per curie are generally reliable to within a factor of 10 for most radionuclides of concern.[3] Also, the author has generally found that dose estimates obtained by using present ICRP-NCRP methods[2] tend to be on the safe side when compared with information given in the literature. A considerable part of the wide variation among radionuclides in dose per curie inhaled can be seen to result from the wide range in types and energies of radiation emitted, the wide range in physical half-lives, and the wide ranges in linear energy transfer per unit path through tissue. The uncertainties in these physical parameters are generally small compared to uncertainties in biological factors such as fractional uptake in the critical organ, the biological rate of elimination, or the relative biological effectiveness of the energy deposited per gram of tissue. The uncertainty of these biological factors may range up to factors of 10 to 100.[3] Thus, the radionuclides will be arranged in order of dose per curie (values in Table 3.4–1 will be given in the inverse units of curies per 15 rem for convenience in showing the minimum activity that must be inhaled to give a significant dose). On the assumption that the relative ordering of the maximum dose per curie inhaled (under the worst chemical and physical conditions) is reliable on the average to within a factor of 10, the radionuclides are then arranged in groups that cover only a factor of 10 in relative radiotoxicity. Exceptions to this grouping are made in the case of natural uranium and natural thorium, which have such low specific activities that even in the pure state many milligrams of material would have to be inhaled to produce an appreciable dose (see Table 3.4–1).

The total dose to the critical organ in 50 years per microcurie intake, assuming instantaneous uptake in the critical organ from a single exposure of duration to be short compared to the effective half-life of the radionuclide in the body, was calculated from the equation

$$\text{Dose (rem/}\mu\text{Ci)} = \int_{t=0}^{t=50\times 365} I_0 e^{-0.693t/T_{1/2}}$$

$$= 1.44 I_0 T[1 - \exp(1.265 \times 10^4/T)] \qquad (3.4\text{–}1)$$

where I_0 = initial dose rate to the critical organ per microcurie intake, or $f_a R/q f_2$ (with the standard symbols from Reference 2); f_2 = the fraction of radionuclide in the organ of reference divided by that in the total body (Table 12 of Reference 2); q = body burden listed beside the corresponding critical organ in Table 1 of Reference 2; T = effective half-life in the body in days, from Table 12 of Reference 2; and R = permissible dose rate for continuous exposure in rem per day for the body organ concerned, obtained from Reference 2 as 0.1/7 rem/day for irradiation of the whole body, 0.08 rem/day for bone, 0.1/7 rem/day for the gonads, 0.6/7 rem/day for the thyroid and skin, and 0.3/7 rem/day for other parts of the body. The expression in brackets in Equation 3.4–1 is essentially 1, except for the few bone-seeking radionuclides that have effective half-lives that are not short compared to 50 years. For strontium-90, with an effective half-life of 6.4 × 10^3 days, the factor in brackets becomes 0.861. For purposes of this section, the same relative dose from daughter products is assumed for single intake or for continuous exposure from radionuclides that have radioactive daughters building up in the body. The contribution from daughters is thus taken into account by using the total body

Table 3.4–1
RADIOTOXICITY VS. LEVELS ABOVE WHICH VARIOUS SAFEGUARDS MAY BE REQUIRED

RADIONUCLIDE AND GROUP		PHYSICAL PROPERTIES			RADIOBIOLOGICAL PROPERTIES — SOLUBLE					RADIOBIOLOGICAL PROPERTIES — INSOLUBLE		RELATIVE RADIOTOXICITY	
	Radionuclide Group	Half-Life (Physical) (days)	Specific Activity (curies per gram)	External Gamma Dose Rate (R/hour at one meter per curie)	Critical organ (receiving highest proportion of permissible dose from solubilized activity [after inhalation; other than GI tract]) A	Body burden (for permissible continuous dose rate in critical organ) (microcuries) B	Effective half-life in critical organ (days) C	Permissible concentration in water (for continuous use by employees 40 hrs per week) (microcuries per milliliter) D	Permissible concentration in air (for continuous occupation exposure 40 hrs per week) (microcuries per cc) E	Permissible concentration in water (for continuous use by employees 40 hrs per week) (microcuries per ml.) F	Permissible concentration in air (for continuous occupational exposure 40 hrs per week) (microcuries per cc) G	Single inhalation in curies to give 15 REM to critical organ (curies per 15 REM)	Single inhalation in curies to give 15 REM to lung (" insol materials) (curies per 15 REM)
I	H-3	4.5×10^3	9.78×10^3	<0.0002	Total water	10^3	12	1×10^{-1}	5×10^{-6}	1×10^{-1}	5×10^{-6}	6.15×10^{-2}	—
	C-14	2.0×10^6	4.61	<0.01	Fat (as CO_2)	300	12	2×10^{-2}	4×10^{-6}	—	5×10^{-5} (as CO_2)	2.88×10^{-2}	—
	Tc-99m	0.25		0.063					4×10^{-5}				
II	Br-82	1.5	1.06×10^6	—	Total body	10	1.3	8×10^{-3}	1×10^{-6}	1×10^{-3}	2×10^{-7}	7.47×10^{-3}	—
	Cr-51	27.8	9.2×10^4	—	Total body	800	26.6	5×10^{-2}	1×10^{-5}	5×10^{-2}	2×10^{-6}	8.84×10^{-2}	5.3×10^{-3}
	Fe-55	1.1×10^3	2.51×10^3	—	Spleen	10^3	388	2×10^{-2}	9×10^{-7}	7×10^{-2}	1×10^{-6}	2.17×10^{-3}	2.3×10^{-3}
III	S-35	87.1	4.28×10^4	<0.01	Testes	90	76.5	2×10^{-3}	3×10^{-7}	8×10^{-3}	3×10^{-7}	7.23×10^{-4}	6.9×10^{-4}
	Au-198	2.7	2.44×10^5	0.25	GI(LLI)	—	26 (Total body)	2×10^{-3}	3×10^{-7}	1×10^{-3}	2×10^{-7}	7.25×10^{-4}	5.3×10^{-4}
	Ca-47	4.9	5.9×10^5	—	Bone	5	4.9	1×10^{-3}	2×10^{-7}	1×10^{-3}	2×10^{-7}	2.59×10^{-4}	4.6×10^{-4}
	I-132	0.097	1.05×10^7	—	Thyroid	0.3	0.097	2×10^{-3}	2×10^{-7}	5×10^{-3}	9×10^{-7}	4.5×10^{-4}	—
	Ce-141	32	2.8×10^4	—	Bone	30	31	3×10^{-3}	4×10^{-7}	3×10^{-3}	2×10^{-7}	7.06×10^{-4}	4.2×10^{-4}
	Mixed fission products a	a	$<4\times10^{11}$	—	Bone, Lung							1.7×10^{-4}	
	Sr-85	65	2.37×10^4	—	Total body	60	64.7	3×10^{-3}	2×10^{-7}	5×10^{-3}	1×10^{-7}	2.0×10^{-3}	2.7×10^{-4}
	La-140	1.68	5.61×10^5	0.95	GI(LLI)	9	1.68	7×10^{-4}	2×10^{-7}	7×10^{-4}	1×10^{-7}	4.2×10^{-4}	2.6×10^{-4}
	Nb-95	35	3.93×10^4	—	Total body	40	33.5	3×10^{-3}	5×10^{-7}	3×10^{-3}	1×10^{-7}	3.6×10^{-3}	2.3×10^{-4}
	Zn-65	245	8.2×10^3	—	Prostate	60 (Total body)	13 (Prostate) 194 (Total body)	3×10^{-3}	1×10^{-7}	5×10^{-3}	6×10^{-8}	2.6×10^{-4}	1.5×10^{-4}
	Co-58	72	3.13×10^4	—	Total body	30	8.4	4×10^{-3}	8×10^{-7}	3×10^{-3}	5×10^{-8}	8.4×10^{-3}	1.3×10^{-4}
	Fe-59	45.1	4.92×10^4	0.65	Spleen	20	41.9	2×10^{-3}	1×10^{-7}	2×10^{-3}	5×10^{-8}	3.0×10^{-4}	1.3×10^{-4}
IV	Hf-181	46	1.62×10^4	—	Spleen	4	41	2×10^{-3}	4×10^{-8}	2×10^{-3}	7×10^{-8}	9.94×10^{-5}	1.92×10^{-4}
	Pm-147	920	9.25×10^2	<0.01	Bone	60	570	6×10^{-3}	6×10^{-8}	6×10^{-3}	1×10^{-7}	8.9×10^{-5}	2.3×10^{-4}
	P-32	14.3	2.85×10^5	<0.01	Bone	6	14.1	5×10^{-4}	7×10^{-8}	7×10^{-4}	8×10^{-8}	8.7×10^{-5}	2.1×10^{-4}
	Ba-140	12.8	7.3×10^4	1.54	Bone	4	10.7	8×10^{-4}	1×10^{-7}	7×10^{-4}	4×10^{-8}	1.4×10^{-4}	8.6×10^{-5}
	Th-234	24.1	2.32×10^4	—	Bone	4	24.1	5×10^{-4}	6×10^{-8}	5×10^{-4}	3×10^{-8}	8.5×10^{-5}	7.3×10^{-5}
	Kr-85	3.9×10^3	3.96×10	0.0019	Total body	—	—		$1\times10^{-5}_{(sub)}$	—	—	6.9×10^{-2}	5.8×10^{-5}
	Ir-192	74.5	9.16×10^3	0.51	Kidney	6	30	1×10^{-3}	1×10^{-7}	1×10^{-3}	3×10^{-8}	3.2×10^{-4}	6.9×10^{-5}
	Cl-36	1.2×10^8	3.21×10^{-2}	—	Total body	80	29	2×10^{-3}	4×10^{-7}	2×10^{-3}	2×10^{-8}	2.7×10^{-3}	5.3×10^{-5}
	Y-91	58	2.50×10^4	—	Bone	5	58	8×10^{-4}	4×10^{-8}	8×10^{-4}	3×10^{-8}	5.0×10^{-5}	7.3×10^{-5}
	Ta-182	112	6.2×10^3	—	Liver	7	88	1×10^{-3}	4×10^{-8}	1×10^{-3}	2×10^{-8}	1.1×10^{-4}	5.0×10^{-5}
	Ca-45	164	1.77×10^4	<0.01	Bone	30	162	3×10^{-4}	3×10^{-8}	5×10^{-3}	1×10^{-7}	4.3×10^{-5}	2.6×10^{-4}
	Sr-89	50.5	2.77×10^4	—	Bone	4	50.4	3×10^{-4}	3×10^{-8}	8×10^{-4}	4×10^{-8}	4.0×10^{-5}	8.5×10^{-5}
	Cs-137	1.1×10^4	9.85×10	0.36	Total body	30	70	4×10^{-4}	6×10^{-8}	1×10^{-3}	1×10^{-8}	2.6×10^{-4}	3.0×10^{-5}
	Co-60	1.9×10^3	1.14×10^3	1.32	Total body	10	9.5	1×10^{-3}	3×10^{-7}	1×10^{-3}	9×10^{-9}	2.6×10^{-3}	2.2×10^{-5}

Table 3.4–1 (continued)
RADIOTOXICITY VS. LEVELS ABOVE WHICH VARIOUS SAFEGUARDS MAY BE REQUIRED

RADIONUCLIDE AND GROUP		PHYSICAL PROPERTIES			RADIOBIOLOGIC PROPERTIES							RELATIVE RADIOTOXICITY	
					SOLUBLE					INSOLUBLE			
	Radionuclide Group	Half-Life (Physical) (days)	Specific Activity (curies per gram)	External Gamma Dose Rate (R/hour at one meter per curie)	Critical organ (receiving highest proportion of permissible dose from solubilized activity [after inhalation; other than GI tract]) A	Body burden (for permissible continuous dose rate in critical organ) (microcuries) B	Effective half-life in critical organ (days) C	Permissible concentration in water (for continuous use by employees 40 hrs per week) (microcuries per milliliter) D	Permissible concentration in air (for continuous occupation exposure 40 hrs per week) (microcuries per cc) E	Permissible concentration in water (for continuous use by employees 40 hrs per week) (microcuries per ml.) F	Permissible concentration in air (for continuous occupational exposure 40 hrs per week) (microcuries per cc) G	Single inhalation in curies to give 15 REM to critical organ (curies per 15 REM)	Single inhalation in curies to give 15 REM to lung (insol materials) (curies per 15 REM)
IV (*cont.*)	Ce-144	290	3.18×10^{3}	0.20	Bone	5	243	3×10^{-4}	1×10^{-8}	3×10^{-4}	6×10^{-9}	1.4×10^{-5}	1.5×10^{-5}
	I-126	13.3	7.8×10^{4}	—	Thyroid	1	12.1	5×10^{-5}	8×10^{-9}	3×10^{-3}	3×10^{-7}	1.4×10^{-5}	7.3×10^{-4}
	Eu-154	5.8×10^{3}	1.45×10^{2}	—	Kidney	5	1.18×10^{3}	6×10^{-4}	4×10^{-9}	6×10^{-4}	7×10^{-9}	1.3×10^{-5}	1.6×10^{-5}
	I-131	8	1.24×10^{5}	0.25	Thyroid	0.7	7.6	6×10^{-5}	9×10^{-9}	2×10^{-3}	3×10^{-7}	1.2×10^{-5}	7.3×10^{-4}
	Tm-170	127	6.08×10^{3}	0.004	Bone	9	113	1×10^{-3}	4×10^{-8}	1×10^{-3}	3×10^{-8}	3.8×10^{-5}	7.5×10^{-5}
	I-125								5×10^{-9}				
V	I-129	6.3×10^{9}	1.62×10^{-4}	—	Thyroid	3	138	1×10^{-5}	2×10^{-9}	6×10^{-3}	7×10^{-8}	2.3×10^{-6}	1.6×10^{-4}
	Tc-99	7.7×10^{7}	1.71×10^{-2}	—	GI(LLI) Kidney	— 10	—	1×10^{-2}	2×10^{-6}	5×10^{-3}	6×10^{-8}	9.1×10^{-6}	—
VI	Ra-223	11.7	5.0×10^{4}	—	Bone	0.05	11.7	2×10^{-5}	2×10^{-9}	1×10^{-4}	2×10^{-10}	3.9×10^{-6}	5.3×10^{-7}
	Po-210	138.4	4.5×10^{3}	<0.00005	Spleen	0.03	42	2×10^{-5}	5×10^{-10}	8×10^{-4}	2×10^{-10}	1.3×10^{-6}	5.0×10^{-7}
	Th-227	18.4	3.17×10^{4}	—	Bone	0.02	18.4	—	—	—	—	5.5×10^{-7}	4.6×10^{-7}
	Sr-90	1×10^{4}	1.44×10^{2}	<0.01	Bone	2	6.4×10^{3}	1×10^{-5} b	1×10^{-9} b	1×10^{-3}	5×10^{-9}	3.9×10^{-7}	1.3×10^{-5}
	Pb-210	7.1×10^{3}	8.8×10	—	Kidney	0.04	494	4×10^{-6}	1×10^{-10}	5×10^{-3}	2×10^{-10}	3.2×10^{-7}	5.3×10^{-7}
	Cm-242	162.5	3.34×10^{3}	<0.01	Liver	0.05	154.3	7×10^{-4}	1×10^{-10}	7×10^{-4}	2×10^{-10}	3.0×10^{-7}	4.6×10^{-7}
	U-233	5.9×10^{7}	0.01 (with 80 ppm U^{232})	0.0002 (with 20 ppm U^{232})	Bone	0.05	300	9×10^{-4}	5×10^{-10}	9×10^{-4}	1×10^{-10}	7.0×10^{-7}	2.7×10^{-7}
	U-235 (+1% U-234)	2.6×10^{11}	2.15×10^{-6}	<0.002	{Kidney Bone}	0.03 0.06	15 300	8×10^{-4} —	5×10^{-10} —	8×10^{-4} —	1×10^{-10} —	1.1×10^{-6}	2.6×10^{-7}
	U-238 and U–nat'l	1.6×10^{12}	3.34×10^{-7}	<0.002	Kidney	0.005	15	1×10^{-3}	7×10^{-11}	1×10^{-3}	1×10^{-10}	1.9×10^{-7}	3.0×10^{-7}
	Th-232 and Th–nat'l	5.1×10^{12}	1.11×10^{-7}	<0.0002	Bone	0.01	7.3×10^{4}	5×10^{-5} b	3×10^{-11} b	1×10^{-3}	3×10^{-11}	2.25×10^{-9}	2.6×10^{-8}
VII	Sm-147	4.8×10^{13}	1.95×10^{-8}	—	Bone	0.1	1.5×10^{3}	2×10^{-3}	7×10^{-11}	2×10^{-3}	3×10^{-10}	7.7×10^{-8}	6.9×10^{-7}
	Nd-144	7.3×10^{17}	4.97×10^{-15}	—	Bone	0.1	1.5×10^{3}	2×10^{-3}	8×10^{-11}	2×10^{-3}	3×10^{-10}	7.7×10^{-8}	7.3×10^{-7}
	Ra-226	5.9×10^{5}	1.00	—	Bone	0.1	1.6×10^{4}	4×10^{-7}	3×10^{-11}	9×10^{-4}	5×10^{-11} b	4.9×10^{-8}	1.5×10^{-8}
	Cm-244	6.7×10^{3}	8.2×10	—	Bone	0.1	2.1×10^{3}	2×10^{-4}	9×10^{-12}	8×10^{-4}	1×10^{-10}	1.1×10^{-8}	2.3×10^{-7}
VIII	Am-243	2.9×10^{6}	1.85×10^{-1}	—	Bone	0.05	7.1×10^{4}	1×10^{-4}	6×10^{-12}	8×10^{-4}	1×10^{-10}	7.6×10^{-9}	2.7×10^{-7}
	Am-241	1.7×10^{5}	3.21	0.039	Bone	0.05	5.1×10^{4}	1×10^{-4}	6×10^{-12}	8×10^{-4}	1×10^{-10}	6.6×10^{-9}	2.7×10^{-7}
	Np-237	8×10^{8}	6.9×10^{-4}	—	Bone	0.06	7.3×10^{4}	9×10^{-5}	4×10^{-12}	9×10^{-4}	1×10^{-10}	5.2×10^{-9}	2.7×10^{-7}
	Ac-227	8×10^{3}	7.2×10	—	Bone	0.03	7.2×10^{3}	6×10^{-5}	2×10^{-12}	9×10^{-3}	3×10^{-11}	3.0×10^{-9}	6.9×10^{-8}
	Th-230	2.9×10^{7}	1.97×10^{-2}	0.009	Bone	0.05	7.3×10^{4}	5×10^{-5}	2×10^{-12}	9×10^{-4}	1×10^{-11}	2.8×10^{-9}	2.3×10^{-8}
	Pu-242	1.4×10^{8}	3.9×10^{-3}	—	Bone	0.05	7.3×10^{4}	1×10^{-4}	2×10^{-12}	9×10^{-4}	4×10^{-11}	2.5×10^{-9}	8.5×10^{-8}
	Pu-238	3.3×10^{4}	2.68×10	<0.02	Bone	0.04	2.3×10^{4}	1×10^{-4}	2×10^{-12}	8×10^{-4}	3×10^{-11}	2.2×10^{-3}	7.3×10^{-8}
	Pu-240	2.4×10^{6}	2.27×10^{-1}	<0.001	Bone	0.04	7.1×10^{4}	1×10^{-4}	2×10^{-12}	8×10^{-4}	4×10^{-11}	2.0×10^{-9}	8.5×10^{-8}
	Pu-239	1.9×10^{6}	6.17×10^{-2}	<0.00001	Bone	0.04	7.2×10^{4}	1×10^{-4}	2×10^{-12}	8×10^{-4}	4×10^{-11}	2.0×10^{-9}	8.5×10^{-8}

Table 3.4–1 (continued)
RADIOTOXICITY VS. LEVELS ABOVE WHICH VARIOUS SAFEGUARDS MAY BE REQUIRED

RADIONUCLIDE AND GROUP			FACILITIES AND EQUIPMENT							SITE	PROCEDURES							
	Radionuclide Group	Chemical hood required (curies)	Glovebox required (curies)	Glovebox inside hot cell or cave (Based on gamma dose rates) (curies)	1 Absolute filter (in the exhaust from active atmosphere) (curies)	2 Absolute filters in series (in the active exhaust) (curies)	Continuous general air sampler with alarm (in work areas) (curies)	Continuous exhaust stack monitor and alarm (to protect public) (curies)	Building containment or controlled leak rate (to protect public) (curies)	Radius of low population zone X (in m.) dist. beyond which cloud dose is less than 15 Rem for Q curies released, where Q = fC, for C curies in process	Personnel monitoring and/or appropriate shielding vs. external gamma radiation (curies)	Occasional excretion radio-assay spot checks of operating personnel (curies)	Routine excretion assay of all operating personnel (curies)	Emergency dosimeters worn to measure high external doses (curies)	Routine environmental monitoring of site and community (curies)	Pre-planned written emergency procedures and drills (curies)	Written routine operating procedures (curies)	Written pre-operational analysis of maximum credible accidents (including doses to people) (curies)
I	H-3, C-14, Tc-99m	1	10	—	10	10^4	10^4	10^5	10^6	$X=0.47Q^{2/3}$	—	10	10^2	—	1,000	10^4	10^6	10
II	Br-82, Cr-51, Fe-55	0.1	1	5	1	10^3	10^3	10^4	10^5	$2.2Q^{2/3}$	0.5	1	10	50	100	10^3	10^5	1
III	S-35	10^{-2}	0.1	100	0.1	10^2	10^2	10^3	10^4	$10Q^{2/3}$	10	0.1	1	1,000	10	10^2	10^4	0.1
	Au-198			4							0.4			40				
	Ca-47																	
	I-132																	
	Ce-141																	
	Mixed fission products[c]			1							0.1			10				
	Sr-85																	
	La-140			1							—			—				
	Nb-95																	
	Zn-65																	
	Co-58																	
	Fe-59	10^{-2}	0.1	2	0.1	10^2	10^2	10^3	10^4	$10Q^{2/3}$	0.2	0.1	1	20	10	10^2	10^4	0.1
IV	Hf-181	10^{-3}	10^{-2}	100	10^{-2}	10	10	10^2	10^3	$47Q^{2/3}$	10	10^{-2}	0.1	1,000	1	10	10^3	10^{-2}
	Pm-147			100							10			1,000				
	P-32			0.5							0.05			5				
	Ba-140																	
	Th-234			500							50			5,000				
	Kr-85																	
	Ir-192			2							0.02			20				
	Cl-36																	
	Y-91																	
	Ta-182			—							—			—				
	Ca-45			100							10			1,000				
	Sr-89																	
	Cs-137			3							0.3			30				
	Co-60	10^{-3}	10^{-2}	1	10^{-2}	10	10	10^2	10^3	$47Q^{2/3}$	0.1	10^{-2}	0.1	10	1	10	10^3	10^{-2}

Table 3.4–1 (continued)
RADIOTOXICITY VS. LEVELS ABOVE WHICH VARIOUS SAFEGUARDS MAY BE REQUIRED

RADIONUCLIDE AND GROUP		FACILITIES AND EQUIPMENT								SITE	PROCEDURES							
Radionuclide Group		Chemical hood required (curies)	Glovebox required (curies)	Glovebox inside hot cell or cave (Based on gamma dose rates) (curies)	1 Absolute filter (in the exhaust from active atmosphere) (curies)	2 Absolute filters in series (in the active exhaust) (curies)	Continuous general air sampler with alarm (in work areas) (curies)	Continuous exhaust stack monitor and alarm (to protect public) (curies)	Building containment or controlled leak rate (to protect public) (curies)	Radius of low population zone X (in m.) dist. beyond which cloud dose is less than 15 REM for Q curies released, where Q = fC, for C curies in process	Personnel monitoring and/or appropriate shielding vs. external gamma radiation (curies)	Occasional excretion radio-assay spot checks of operating personnel (curies)	Routine excretion assay of all operating personnel (curies)	Emergency dosimeters worn to measure high external doses (curies)	Routine environmental monitoring of site and community (curies)	Pre-planned written emergency procedures and drills (curies)	Written routine operating procedures (curies)	Written pre-operational analysis of maximum credible accidents (including doses to people) (curies)
IV (*cont.*)	Ce-144	10^{-3}	10^{-2}	5	10^{-2}	10	10	10^2	10^3	$47Q^{2/3}$	0.05	10^{-2}	0.1	5	1	10	10^3	10^{-2}
	I-126	↕	↕		↕	↕	↕	↕	↕	↕		↕	↕		↕	↕	↕	↕
	Eu-154	↕	↕		↕	↕	↕	↕	↕	↕		↕	↕		↕	↕	↕	↕
	I-131	↕	↕	4	↕	↕	↕	↕	↕	↕	0.4	↕	↕	40	↕	↕	↕	↕
	Tm-170	10^{-3}	10^{-2}	250	10^{-2}	10	10	10^2	10^3	$47Q^{2/3}$	25	10^{-2}	0.1	2,500	1	10	10^3	10^{-2}
V	I-129, Tc-99	10^{-4}	10^{-3}	—	10^{-3}	1	1	10	10^2	$220Q^{2/3}$	100	10^{-3}	10^{-2}	10^4	0.1	1	10^2	10^{-3}
VI	Ra-223	10^{-5}	10^{-4}		10^{-4}	0.1	0.1	1	10	$1,000Q^{2/3}$		10^{-4}	10^{-3}		10^{-2}	0.1	10	10^{-4}
	Po-210	↕	↕	20,000	↕	↕	↕	↕	↕	↕	10	↕	↕	1,000	↕	↕	↕	↕
	Th-227	↕	↕		↕	↕	↕	↕	↕	↕		↕	↕		↕	↕	↕	↕
	Sr-90	↕	↕	100	↕	↕	↕	↕	↕	↕	10	↕	↕	1,000	↕	↕	↕	↕
	Pb-210	↕	↕		↕	↕	↕	↕	↕	↕		↕	↕		↕	↕	↕	↕
	Cm-242	↕	↕	100	↕	↕	↕	↕	↕	↕	1	↕	↕	100	↕	↕	↕	↕
	U-233	↕	↕	5,000 (500 kg)	↕	↕	↕	↕	↕	↕	50	↕	↕	5,000	↕	↕	↕	↕
	U-235 (+% U-234)	↕	↕		↕	↕	↕	↕	↕	↕		↕	↕		↕	↕	↕	↕
	U-238 and U-nat'l	↕	↕		↕	↕	↕	↕	↕	↕		↕	↕		↕	↕	↕	↕
	Th-232 and Th-nat'l	10^{-5}	10^{-4}		10^{-4}	0.1	0.1	1	10	$1,000Q^{2/3}$		10^{-4}	10^{-3}		10^{-2}	0.1	10	10^{-4}
VI	Sm-147, Nd-144, Ra-226, Cm-244	10^{-6}	10^{-5}	—	10^{-5}	10^{-2}	10^{-2}	0.1	1	$4,700Q^{2/3}$	—	10^{-5}	10^{-4}	—	10^{-3}	10^{-2}	1	10^{-5}
VIII	Am-243	10^{-7}	10^{-6}		10^{-6}	10^{-3}	10^{-3}	10^{-2}	0.1	$22,000Q^{2/3}$		10^{-6}	10^{-5}		10^{-4}	10^{-3}	0.1	10^{-6}
	Am-241	↕	↕	25	↕	↕	↕	↕	↕	↕	0.25	↕	↕	25	↕	↕	↕	↕
	Np-237	↕	↕		↕	↕	↕	↕	↕	↕		↕	↕		↕	↕	↕	↕
	Ac-227	↕	↕		↕	↕	↕	↕	↕	↕		↕	↕		↕	↕	↕	↕
	Th-230	↕	↕	100	↕	↕	↕	↕	↕	↕	1	↕	↕	100	↕	↕	↕	↕
	Pu-242	↕	↕	—	↕	↕	↕	↕	↕	↕	—	↕	↕	—	↕	↕	↕	↕
	Pu-238	↕	↕	50	↕	↕	↕	↕	↕	↕	0.5	↕	↕	50	↕	↕	↕	↕
	Pu-240	↕	↕	1,000	↕	↕	↕	↕	↕	↕	10	↕	↕	1,000	↕	↕	↕	↕
	Pu-239	10^{-7}	10^{-6}	(Based on criticality)	10^{-6}	10^{-3}	10^{-3}	10^{-2}	0.1	$22,000Q^{2/3}$	(Based on criticality)	10^{-6}	10^{-5}	(Based on criticality)	10^{-4}	10^{-3}	0.1	10^{-6}

Table 3.4–1 (continued)
RADIOTOXICITY VS. LEVELS ABOVE WHICH VARIOUS SAFEGUARDS MAY BE REQUIRED

[a]For mixed gross fission products of 0- to 4-year reactor operations, the overall relative hazard per curie changes relatively slowly with decay time for decay times shorter than 30 days, although for very short decay times (less than 1 day) the thyroid dose predominates (see Brodsky, A., Criteria for acute exposure to mixed fission product aerosols, *Health Phys.*, 11, 1017, 1965).

[b]Permissible concentrations are taken from the Code of Federal Regulations, Title 10, Part 20, August 1966, since licensees of the U.S. Nuclear Regulatory Commission must comply with pertinent requirements of these regulations. Other values under "Radiobiologic Properties" are taken from the report of Subcommittee 11, International Commission on Radiological Protection (ICRP), *Health Phys.*, Vol. 3, 1960. In general, the permissible concentrations in the Federal Regulations are based on MPC's of the ICRP. However, for reasons presented in the regulations there are minor differences (less than about an order of magnitude) for the isotopes ^{90}Sr and ^{232}Th.

[c]Taken for gross fission product mixtures, separated from reactor fuel less than 30 days after reactor shutdown, as calculated in Brodsky, A., Criteria for acute exposure to mixed fission product aerosols, *Health Phys.*, 11, 1017, 1965.

NOTE: The curie levels for protection against inhalation and contamination are intended for dry, dusty materials that may be easily dispersed in concentrated form. For liquids, or where active material will be diluted with other materials, the above safeguard levels may be raised by factors of ten or more. For simple storage of stock solutions, or where operations are conducted only in such a way that no materials of specific activity greater than 0.1 mCi/g can be dispersed as an aerosol, the above levels concerned with protecting against intake of loose radioactive materials may be multiplied by 100 or more, depending on the nature of the material and the particular combination of safeguards selected (see text).

The reader should be sure to check values in the table against the most recent recommendations of the NCRP and any local or federal regulations that may pertain to his operation.

burdens, q, to give dose rates, R. These body burdens were calculated by the ICRP Committee, taking into account daughter products that build up in the body.

Since for insoluble materials an average of 12½% of the material inhaled may remain in the lung, with a half-life of 120 days, the lung must also be taken into consideration as a possible critical organ. Single intake doses based on the lung dose were calculated from the equation.

Dose to the lung per microcurie (μCi) inhaled (insoluble)

$$= \frac{\text{Permissible dose rate per week}}{(\text{MPC}_{\text{air}} \text{ based on continuous exposure to lung})} \times 1.4 \times 10^{8} \text{ cc/week}$$

$$= 2.14 \times 10^{-9}/\text{MPC}_{\text{air}} \text{ based on lung} \qquad (3.4\text{–}2)$$

since the equilibrium dose rate from continuous exposure is the same as the average dose rate from a series of single intakes of the same total quantity of radioactivity per week, for effective half-lives short compared to 50 years.

The smallest value in curies per 15 rem was selected as a basis for ordering the radionuclides in order to derive initial criteria that would be safe even for processes in which the radionuclides might occur in chemical forms that could give the highest potential internal exposures. This method provides a safe starting point from which other correction factors may be applied where they are determinable in specific circumstances. The final relative ordering of radionuclides for purposes of selecting safeguards against internal exposure is given in Table 3.4–1.

METEOROLOGICAL CALCULATIONS

In Figure 3.4–1, the minimum curie quantities that would need to be released at zero point in order to produce an estimated dose of 15 rem to an adult standing X m downwind are plotted for some of the radionuclides frequently in use. The ordinates at a distance of 10 m show that the radionuclides do conveniently group themselves into eight groups corresponding to eight orders of magnitude of relative maximum hazard of various radionuclides under different proximities of the potential point of release and likely points of exposure. Although the meteorological calculations as well as the single intake doses used to obtain Figure 3.4–1 would yield considerable factors of safety under most actual situations, the use of even raw guides such as Figure 3.4–1 when additional information about materials and procedures cannot be specified in advance may often result in less safety overdesign than the use of "professional judgment" alone without any consideration of relative radiotoxicity. Figures 3.4–2 and 3.4–3 have also been used in estimating exposures and exposure durations from released materials. The Appendix of this section presents a discussion of the calculations used in obtaining Figures 3.4–1 through 3.4–3, and the bases for estimating the fractions of released material that may be inhaled or deposited under various circumstances.

DERIVATION OF SAFEGUARD GUIDE LEVELS

In the introduction, arguments were presented for beginning the estimation of hazards and safeguard requirements by a consideration of the relative internal or external doses per curie of the radionuclides involved under the various possible modes of exposure. This part will outline the derivation of the base-line levels in Table 3.4–1 above which the need for various safeguards should be examined. These levels were derived with the aid of the basic information on relative radiotoxicity and the methods of estimating general hazard potential discussed in earlier sections. The industrial hygienist will ordinarily need to consider additional factors such as specific activity, chemical dilution and form, volatility, types of unit operation, probability of accidents indicated by process history, or other aspects of the particular industrial or laboratory operations in order to determine which combination of safeguards is actually required at what curie level.

Radionuclide Properties and Radiotoxicity Groups

Some of the properties of the radionuclides encountered in industry and their relative radiotoxicities are listed in the first half of Table 3.4–1. These data are presented here not only for convenient reference, but also because the

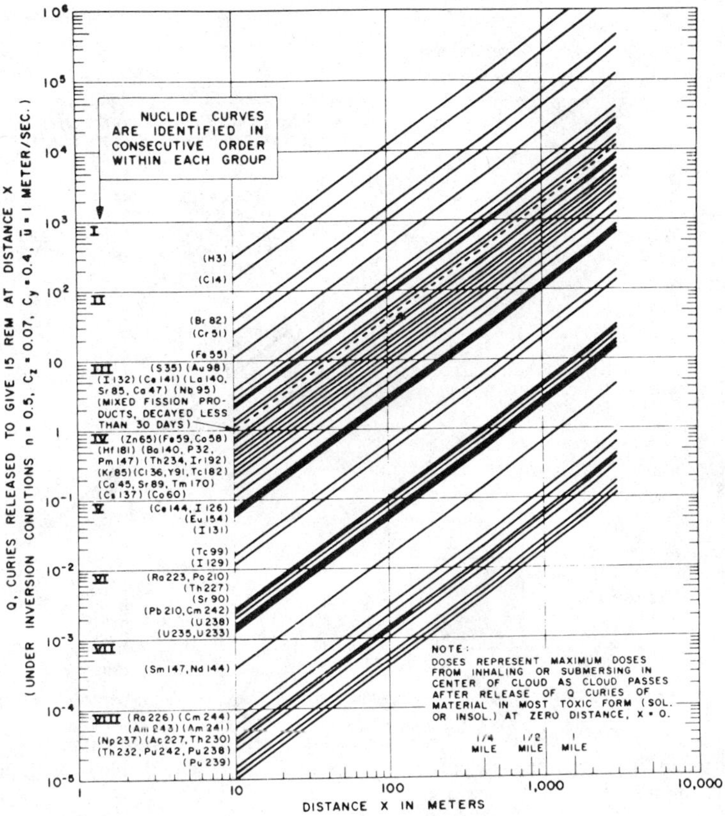

FIGURE 3.4–1. Curies released to give 15 rem at distance X for various nuclides.

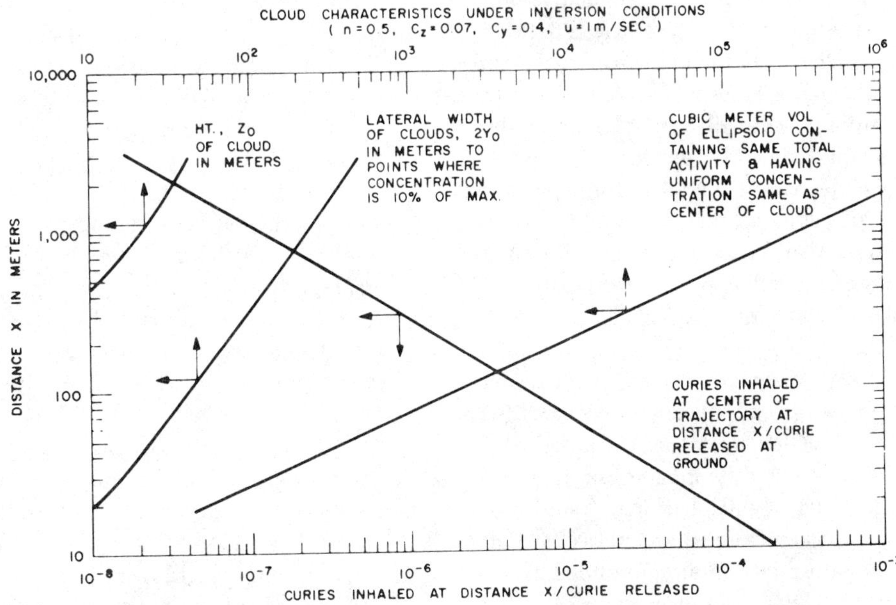

FIGURE 3.4–2. Cloud characteristics under inversion conditions.

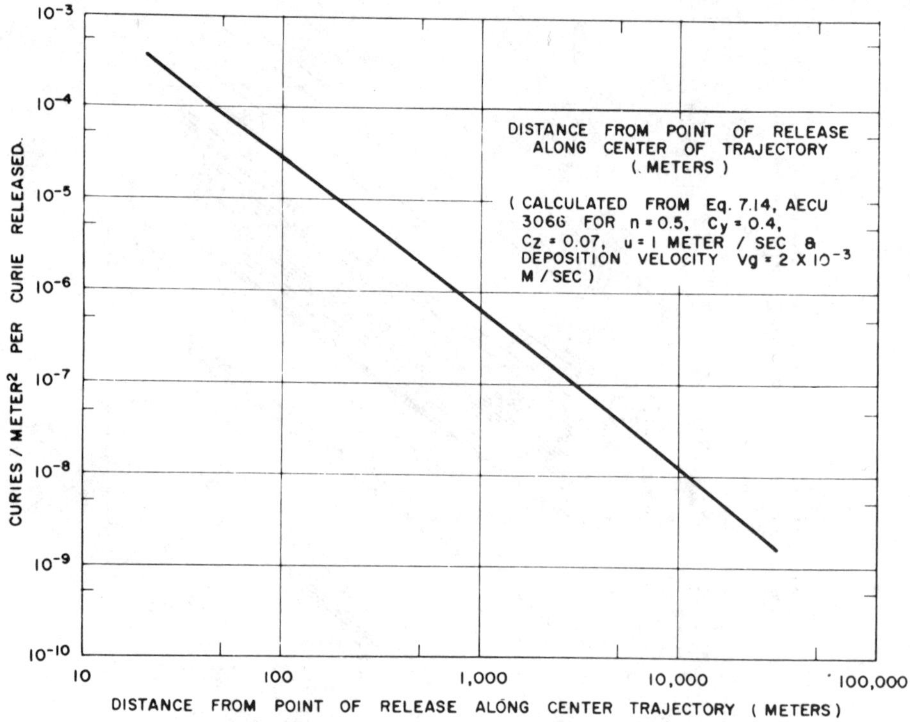

FIGURE 3.4–3. Typical deposition of cloud activity plotted vs. distance, with activity deposit in curies per square meter per curie released given as the probable upper limit along the center of the trajectory under inversion conditions.

repeated use of such a table with the magnitudes of important variables revealed helps to generate additional perspective regarding the relative importance of various factors in hazard evaluation. In Table 3.4–1 other radiobiologic properties such as MPC's are also listed for reference and to illustrate that the radionuclides will generally order themselves into approximately the same groupings if the permissible concentrations for continuous inhalation are used as the criterion for relative radiotoxicity. Also, examination of the relationships between permissible intakes for ingestion and inhalation will show that the inhalation hazard is generally the limiting consideration in the following selection of safeguard levels. The maximum radiation doses from inhaled substances were obtained from ICRP data[2] as described above. Specific activities and gamma dose rates were either calculated from half-lives and decay schemes[2,5] or taken from compilations.[6,7] Values in the gamma dose rate column preceded by < are generally calculated upper limits to the bremsstrahlung dose rate, with self-absorption of the X-radiation and absorption by any container neglected.

In a previous paper,[4] Figure 3.4–1 was plotted from the single intake doses calculated by Fairbairn and Dunning,[7] whose values were obtained in a similar fashion to those given here. Fairbairn and Dunning gave values generally within a factor of two of those given in Table 3.4–1, except for ^{3}H, ^{14}C, ^{32}P, and ^{230}Th. In this section the effective lung dose from inhaled insoluble particles containing ^{3}H or ^{14}C is assumed to be much less than the dose calculated from a uniform distribution of absorbed energy throughout lung tissue. The maximum ^{3}H beta-ray range is only about 6 μm in tissue,[8] and the average range would be about 1 μm.[9] Absorption with the insoluble particle or phagocytic cell and the mucous linings of the tracheobronchial and alveolar surfaces would be expected to reduce sharply the dose to sensitive tissue.[10] However, since the mucoid film on the alveolar surfaces is only about 0.2-μm thick, these assumptions may not be applicable in the case of fine particulates that remain dispersed throughout the lung for long effective half-lives.

Estimation of Safeguard Requirements

Chemical Hood – A value of 0.1 μCi (10^{-7} Ci) was selected for the most toxic group as the lowest quantity requiring that operations potentially capable of dispersing the radioactive material be conducted within a hood. This is the value given in Reference 11 as the minimum significant quantity for dusty operations, and it is the value in Reference 11, Table 2, requiring chemical hoods for radioisotope laboratory operations (10 μCi), multiplied by the factor of 0.01 for dry and dusty operations. Also, it is about 1/30 of the approximate level at which some laboratories have found it generally desirable to require drybox operations.

For the most hazardous nuclides in Table 3.4–1, 0.1 μCi is only about two-and-one-half times the maximum permissible body burden.[2] In Group VIII of Table 3.4–1, all the important nuclides have long, effective half-lives in the body. Thus, it seems reasonable to attempt to minimize the probability of inhaling a significant fraction of the permissible body burden.

Since both the permissible continuous intake values[2,12] and the single intake values in Table 3.4–1 range over about eight orders of magnitude, the values requiring hood operation have been scaled up a factor of ten each for Groups I to VII.

For some of the uranium and thorium nuclides and their natural mixtures, the extremely low specific activity even in the carrier-free state almost guarantees a reduction in the probability of intake[1] by inhalation. Therefore, these nuclides have been placed in higher categories than would be indicated purely on the basis of dose per curie. The low specific activity also limits the external dose rates from these elements in bulk form even in multicurie quantities, and this further limits the external radiation safeguards needed for large quantities. Such an a priori consideration of specific activity[1] was not considered valid for any of the other nuclides in Table 3.4–1.

Glovebox – The level where gloveboxes may begin to be desirable was set at 1 μCi for Group VIII based primarily on experience[13] with ^{239}Pu. This level of 1 μCi comes to 10 times the quantity where operations within a chemical hood may be required, and is about 25 times the permissible body burden for the most hazardous nuclides. Scaling upward was done on the same basis as for chemical hoods.

Glovebox inside hot cell or cave (see also Personnel monitoring and/or shielding, below) – The point at which operations should be conducted within a hot cell or cave, as well as a glovebox (for loose material), was based primarily on the external gamma, X-ray, or bremsstrahlung dose rate from the quantity of material as a point source. Since this external dose rate varies widely even within the same internal toxicity category, a separate number has been listed where applicable in Table 3.4–1, appropriate to the individual nuclide. This level is based on the quantity that could give approximately 1 R/hr at 1 m under some conditions of geometry. Of course, for materials such as natural uranium and natural thorium, the specific activity is so low in any case that such high external radiation levels are not possible (as a result of self-absorption of the radiation within the material).

Although a hot cell or cave, if provided, would give a certain additional amount of containment against dispersal, this was not factored into consideration in other parts of Table 3.4–1 since the degree of containment is too uncertain and dependent on other facility design parameters. This is one of the many other factors that must remain subject to the judgment of the reviewing industrial hygienist or health physicist in deciding what combination of safeguards and what overall degree of containment must finally be imposed on the particular operations to be conducted.

Of course, for the well-designed glovebox[13] we may assume for our purposes that less than about 10^{-8} of the material within the box will escape from the box on the average as a result of ordinary operations, including small glove leaks. The reviewing health physicist or industrial hygienist must ensure that proper glovebox design and ventilation are provided and that proper procedures for changing gloves, removing materials or equipment, and other activities will be carried out to meet the expected containment requirements.

One absolute filter (in exhaust from active atmosphere) – The micrograms per square meter times factors of 10^{-6} to 10^{-4} gives generally the air concentrations in micrograms per cubic meter above a contaminated ground area where ordinary human activities are conducted.[6,14] Recent work by Jones and Pond[15] and by Stewart[16] has also produced resuspension factors in this range, and

Stewart recommends the factor of 10^{-6} as an appropriate average value for use in hazard evaluation both in the laboratory and in the field. Assuming that in a glovebox with finite floor area and controlled air flow a factor of 10^{-6} may be applied, then we might expect that

$$\text{Air concentration } (\mu\text{Ci/ml}) = 10^{-12} \times \mu\text{Ci/m}^2 \qquad (3.4\text{–}3)$$

Thus, the number of microcuries at which no absolute filter should be required was taken as the quantity that could, if spread over 1 m^2 of the glovebox surface, produce an air concentration averaging about the permissible concentration for occupational exposure.[2] This would require a filtered air exhaust whenever the exhaust air leaving the box may be expected to exceed occupational MPC, which is thus nominally taken as a level of operations at which $10^{12} \times MPC_{air}$ μCi or more are handled. Additional dilution as the exhaust air flows to the stack, and atmospheric dilution, would thereby ensure that concentrations in controlled areas would be well below maximum permissible levels. Of course, appropriate monitoring would be required, as shown below, to ensure that build-up of significant contamination in the surrounding environment does not occur, and that 10 CFR 20 requirements are met.[17]

The above resuspension factor of 10^{-6} has been compared with the value of 35 $\mu g/m^2$ for permissible ^{239}Pu contamination in controlled areas.[6] It is found that 35 $\mu g/m^2$ divided by 10^{12} gives the occupational MPC of ^{239}Pu in air, so these figures seem to be in good agreement.

For Groups I to VII, the values have been scaled upward by factors of ten each to take into account the increasing MPC's of these groups.

Two absolute filters in series (in exhaust from active atmosphere) – This level was set as 1,000 times the quantity requiring one absolute filter in the exhaust stream. Opinions gathered by the author indicate that a reduction in air concentrations by a factor of 1,000 may be assumed for each filter over long operating times when the filter is properly tested and installed.[18]

Additional filters beyond two may be required for facilities handling very large quantities of material in one batch.[18] The attenuation factors for each filter beyond the second would have to be based on further information on the specific absolute filters to be used and the particle size characteristics of the material handled.

Continuous room air sampler with alarm – To arrive at a reasonable level at which we should consider requiring a continuous air sampler with alarm, in addition to the routine air sampling and breathing zone sampling program that may be required, the suggestion by the NCRP in *Handbook 69*[12] that an individual could be exposed if necessary to 1,200 times the MPC of a nuclide for 1 hr, if previous records and monitoring results showed that this would not cause the individual to exceed the quarterly dose limit, was considered. It was also considered that it would be easy to construct a device to detect 1,000 MPC (40-hr occupational) of any radionuclide within 1 hr.

Then, the resuspension factor above was used to calculate the number of microcuries spread per square meter that would produce an air concentration of 1,000 MPC. This gave, for ^{239}Pu, a quantity of

$$10^{12} \times 10^3 \times 2 \times 10^{-12} = 2{,}000\ \mu\text{Ci/m}^2 \qquad (3.4\text{–}4)$$

which was rounded off to 1 mCi, and it was assumed that this quantity of material spread over any area of less than 1 m^2 would not be likely to produce the same concentration as this quantity per square meter spread over an infinite plane surface.[16]

It should be noted that this requirement would not take the place of routine monitoring and other safeguards, but would be required primarily for the purpose of detecting unexpected incidents in time to prevent individuals from receiving biologically serious overexposures.

Continuous exhaust stack monitor and alarm – This item would also be required when appropriate to prevent serious results from unexpected incidents and would not take the place of routine effluent sampling and monitoring, although the same instrumentation could be arranged to serve both purposes in some respects.

The level at which consideration should be given to requiring a stack monitor and alarm, in addition to a routine sampling and monitoring program, was established on the basis that less than 1/10,000 of the material released if a filter should fail would be inhaled by an individual standing at 15 m from the stack opening (see Figure 3.4–2). Multiplying the level of 1 μCi, at which an absolute filter in the exhaust might be required, by 10,000 gives a level of 10 mCi as the minimum amount of ^{239}Pu spread on a 1-m^2 drybox floor that could produce a concentration of the order of MPC at 15 m from the stack if the filter should fail. Also, if the entire 10 mCi of ^{239}Pu should be released up the stack in a single incident, the inhaled dose of ^{239}Pu to an individual 15 m from the stack could be of the order of 1 μCi, which would deliver 7,100 rem to the bone over 50 years; however, the individual would be unlikely to inhale this much material unless directly downwind from the stack and at the same height as the center of the cloud of activity. The maximum dose would be 710 rem at 66 m, or 71 rem at 280 m. These numbers also apply only to extremely unfavorable meteorological conditions, as discussed previously.

The other groups of nuclides were scaled up according to the maximum dose per curie, as before.

Building containment (airtight) – If it is assumed that there is a 1% release from a drybox and that 1% of that released from the drybox escapes an ordinary building, then $10^4 \times 10^{-5}$ = 0.1 Ci of ^{239}Pu could give a total bone dose of 15 rem to a person standing 10m away (Figure 3.4–1). Thus, above this level of operation the advisability of requiring sealed entrances and other protective features should be considered, depending on the type of operations, the site, and other factors. As before, the quantities in the less toxic groups were sealed up according to the relative magnitude of the possible dose per curie (adjusted for specific activity in the case of natural uranium and natural thorium).

Site radius of low population zone – The proposed radius for the low population zone for each group is given by the equation

$$X = A_i Q^{2/3} \qquad (3.4\text{–}5)$$

where X is the distance from the point of release in meters, Q is the quantity of the material released = fC, C is the number of curies in process within one building at one time, and f is the assumed fraction of material in process that is released (for example, 1 for noble gases, 0.5 for halogens and other gases, and 0.01 for nuclides incorporated in materials in solid or liquid form under operating conditions.)[19] For Groups I through VIII, A_i becomes 0.47, 2.2, 10, 47, 220, 1,000, 4,700, and 22,000, respectively. This corresponds to the equation giving the distance at which Q Ci of the most hazardous radionuclide in the group will give a dose of less than 15 rem to a person standing directly downwind from this point of release under inversion conditions: n = 0.5, C_z = 0.07, C_y = 0.4; and wind speed u = 1 m/sec. The equations are consistent with the diffusion equation specified in 10 CFR 100 and include the assumption that all the radioactive material remains in the cloud as it travels downwind.[19] The value of 15 rem was selected as the 1-year occupational dose limit used by the NCRP-ICRP Committee in its calculations, based on individual body organs other than the skin and thyroid or whole body.[2] It was used for all the nuclides in the table, since this permissible level is applicable to most of them, and differences in permissible levels for other nuclides are smaller than the uncertainties in dose estimation.

The dose of 15 rem may be compared to the whole-body dose of 25 rem and the thyroid dose of 300 rem used in 10 CFR 100 for defining the low population zone. It is noted that for Groups II through VIII the internal dose to the critical organ is the limiting consideration, except for ^{85}Kr, which is limited by the external beta dose. A skin dose of 30 rather than 15 rem would not change the group to which ^{85}Kr belongs. Previously,[4] the 15-rem dose for mixed gross fission products of 180-day irradiation time was obtained by assuming, as in Reference 20, that a dose of 25 rem to the whole body is equivalent to an exposure to 10 Ci/m^3 for 1 sec. Further estimates[23] for various reactor operation and decay times indicate that 1 Ci-sec/m^3 produces a combined internal irradiation from various fission product nuclides that would be approximately equivalent to a 25-R external gamma exposure, within a factor of two for 0 to 3 years of operation and decay times of less than 30 days. An exposure of 1 Ci-sec/m^3 would actually deliver an estimated combination of 0.28 R to the whole body, 36 rem to the lungs over several months, about 40 to 50 rem to the bones within 50 years, about 7 rem to the gastrointestinal tract within 2 days, about 10 rem to the thyroid within several weeks, and an undetermined amount of radiation to the blood and other organs.[23] Although the relative effectiveness of such a combination of doses is not well known, it would not be expected to differ from the gross effect of a 25-R whole-body exposure by more than an order of magnitude. Since this equivalent dose includes an external dose of 0.28 R, the use of this equivalent automatically limits the external gamma radiation levels when used to establish safeguards against the release of radioactive material. Since pressure-resistant containment structures are not usually required in nuclear material licensing, the "worst case" was used in developing these criteria by assuming that all the material to be released is done so within a short span of time. This makes the dose from cloud passage the important consideration, since there is no fixed gamma radiation source[20] for any appreciable period of time.

Personnel monitoring and/or shielding – The level at which we might wish to require personnel monitoring and/or shielding, rather than rely on the judgment of supervisors alone to control exposure times and other factors, was based on the quantity of material that could produce a dose rate of 10 mR/hr at 1 m. Lower or higher levels than this might be more appropriate, depending on the experience of operating personnel, the type of operation, and other variables. The level selected here is intended to avoid a dogmatic requirement for personnel monitoring at dose rates low enough that well-trained personnel could maintain appropriate exposure limits by administrative procedures. However, the author actually prefers the philosophy that either shielding or personnel monitoring be provided whenever recordable occupational exposures may be received, unless economic considerations are prohibitive and compliance with permissible exposure limits can be otherwise assured.

Occasional bioassay spot checks for operating personnel – These levels were the same as those requiring glovebox operation, for similar reasons.

Continued routine bioassay – The level selected at which consideration should be given to requiring routine bioassay (unless containment is considered absolutely foolproof) is set at ten times the level for occasional bioassay checks. It is also several hundred thousand times the quantity that could give an internal dose of 15 rem. For ^{239}Pu, the quantity becomes 10 μCi, which is the quantity that, if released in a cloud, could give a dose of about 15 rem to the bone of a person standing 10 m away, under the inversion parameters given in Figure 3.4–1.

Bioassay sampling frequency – The bioassay sampling frequency depends on the degree of hazard, since in no case would it be desirable to wait so long between samples that serious internal exposures could go undetected. However, an upper limit on the length of time between samples should be set also on the basis that sufficient material will be excreted at the time(s) of sampling so that any preceding intermittent or continuous exposures averaging above MPC can be detected. Thus, the maximum length of time between bioassay samples should be set on the basis that a single intake of the material at the beginning of the sampling period that would lead to an average exposure exceeding MPC should be detectable at the end of the sampling period. To establish this time, both the number of curies to deliver the permissible average dose and the effective biological half-life of the radionuclide in the body must be considered. Suggested sampling intervals[4] have been removed from Table 3.4–1, since further information on elimination of inhaled substances is needed.[2,3,10]

A minimum length of time between samples may be based on the upper limit of the dose that could be received in the interval, with consideration given to the other safeguards and emergency procedures that would be established in cases where high potential exposures are possible. Also, some consideration should be given to making a more accurate measurement of repeated acute doses from fractions of the radionuclide that are eliminated with shorter biological half-lives than of those governing calculation of maximum permissible continuous intake,[2] in situations where these fractions could possibly contribute the major part of the cumulative dose over a long period of time.

Thus, in general, the maximum time interval is limited by the ability to detect the quantity excreted at the end of the interval as the result of an intake at the beginning of the interval that would deliver an average dose during the interval exceeding permissible levels. The minimum time interval would be set, in the most hazardous operations, so that an individual would not be likely (in the event of safeguard failures) to inhale a serious dose of material before the condition is detected.

Emergency dosimeter program – Where the radionuclide emits gamma- or X-rays of sufficient penetrating power to be measured by the ordinary personnel dosimeter, the quantity of nuclide that can produce up to 10 R/hr at 1 m has been chosen as the level at which the dosimetry should be capable of measuring up to lethal doses of radiation. Special consideration will have to be given to neutron-emitting materials, and to situations where criticality is possible.

Environmental sampling and monitoring – A lost batch of 100 μCi is enough to contaminate about 3,000,000 ft^2 (or 0.1 mi^2) to the minimum level of 60 dpm/ft^2 detected at Los Alamos after 9 years of operating[24] releasing about 1 Ci of ^{239}Pu. The other groups were scaled up by factors of ten as before, 100 μCi taken as the level at which environmental monitoring be considered for ^{239}Pu. This approach is generally sufficiently conservative from ecological as well as immediate inhalation or ingestion considerations as a result of the usually small fraction of material that would be dispersed either routinely or accidentally.

Of course, the frequency and extent of the environmental monitoring program would depend on many factors such as the maximum possible quantity released, the stack height, the half-life of the radionuclide, and the sensitivity of analytical techniques.

Preplanned emergency procedures – This is set at the same level as that where a continuous air monitor and alarm are required. The two safeguards are obviously related – when the alarm sounds, personnel should be prepared to act appropriately.

Written routine operating instructions – Written operating instructions for personnel should be required where the potential magnitude of the hazard is sufficient to require sealed building entrances and exitways, unless facilities within the building are obviously so foolproof that no individual inside or outside the building could receive serious exposure even in the event of the maximum credible accident. Thus, the values are the same as those listed under building containment.

Written preoperational analysis of maximum credible accidents – The level at which a written analysis of the maximum credible accident is required is taken to be the same as that at which a glovebox is required, since this is the approximate level at which serious overexposures in the event of an accident begin to have a significant probability if appropriate safeguards fail or are not provided. In all applications, of course, the applicant should analyze the potential hazards and indicate safeguards provided to minimize them. The extent and detail with which the additional evaluation of "maximum credible accidents" should be given would of course depend on the number and the degree of hazard of the possible types of accidents.

DISCUSSION

This section has attempted to illustrate methods of estimating potential hazards and safeguard requirements from the types and quantities of radioactive material to be handled. No recommendations on specific equipment or industrial hygiene procedures have been included. There are already many good references[7,11,13,14,18,25-29] on appropriate procedures or proper design considerations that can be applied once the need for certain types of procedures or facilities is established.

The guidelines given in Table 3.4-1 refer only to those safeguards that are subject to quantitative assessment within an order of magnitude or two. Thus, the table is only an aid to, and cannot be a substitute for, the professional judgment of an industrial hygienist experienced in health physics and in balancing the relative proportions of various safeguards designed into a specific industrial or laboratory facility. Furthermore, an examination of the items included in Table 3.4–1 will show that the safeguards listed are neither mutually exclusive nor independent of one another. In many instances, combinations of several of the protective measures in various relative proportions can best be used to achieve a desired level of protection.

The unit for quantity of radioactive material in Table 3.4–1 has been the curie rather than the gram, since measurements of radioactivity and quantities shipped to users are usually reported in curies or in other units proportional to the curie. Except for natural uranium (mainly ^{238}U) or natural thorium (^{232}Th), no a priori allowance has been made for the specific activity of the material in listing the relative radiotoxicity, as others have done in estimating the relative inherent hazard of the radionuclides.[1,7,11,28,30] For the purposes of this section, we have considered that the probability of intake is not likely to be related to specific activity for any radionuclides in which inhaled quantities of only 10 mg or less can produce internal doses of 15 rem or more. In most industrial or laboratory operations, the specific activity of the radioactive material in process and the probabilities or potentials for intake will depend not on the specific activity of the pure nuclide, but on variables specific to the individual operation, such as the quantities and kinds of chemical reagents with which the radionuclide is mixed, the potential mechanisms of dispersal, and local ventilation characteristics. For all the radionuclides listed in Table 3.4–1 except natural uranium and natural thorium, the amount of inhaled nuclide needed to give a serious internal dose is far less than 10 mg and is usually in the submicrogram range (see columns "Specific Activity" and "Relative Radiotoxicity" in Table 3.4–1). These mass quantities are so small that regardless of their exact magnitude they could become attached to a single dust particle.[10,26]. Thus, the specific radioactivity of a nuclide cannot be included in setting generalized guidelines such as those in Table 3.4–1, but it must be considered by the industrial hygienist along with all the other specific operational characteristics of an industrial process in selecting the optimum and most economic combination of industrial hygiene facilities and procedures.

APPENDIX

The values in Figure 3.4–1 were obtained by calculating the integrated time-concentration exposures in units of curie-seconds per cubic meter at various distances, X, along the center-line trajectory by using a modified form of Sutton's diffusion equation[31]

$$(\text{Ci-sec/m}^3) = \frac{2Q_0}{u\pi C_y C_z X^{(2-n)}} \exp\left(\frac{-4VX^{n/2}}{nu\pi^{1/2}C_z}\right) \qquad \text{(A.3.4–1)}$$

where Q_0 is the number of curies released as a point source at $X = 0$. Conservative parameters representing inversion weather conditions were selected consistent with those given in proposed 10 CFR 100[19]: $n = 0.5$, $C_z = 0.07$, $C_y = 0.40$, and $u = 1$ m/sec wind velocity. The particle settling velocity, V, was taken to be 2×10^{-3} m/sec, corresponding to a particle about 4 to 5 μm in diameter,[31] of density 2.5 g/cm^3. Figure 3.4–1 shows that the exponential settling term is not an important factor in determining cloud concentrations at distances less than 2 mi for particles of sizes smaller than about 4 μm. Dense particles of plutonium would have a higher settling velocity for a given particle size, but the probable smaller size of PuO fumes escaping in an accidental release would compensate for this. Thus, the settling velocity of 2×10^{-3} m/sec is believed to be a reasonably conservative choice for maximizing cloud concentrations[16,31] (see also Reference 16, in which a similar choice has been made).

After the chosen parameters are inserted, Equation A.3.4–1 reduces to

$$(\text{Ci-sec/m}^3) = (22.7\, Q_0/X^{1.5}) \exp(-0.129\, X^{0.25}). \qquad \text{(A.3.4–2)}$$

By the use of Equation A.3.4–2, the probable limit to the number of curies inhaled at distance X downwind per curie released at ground level zero point was plotted in Figure 3.4–2. This curve was used to obtain the curves for each nuclide in Figure 3.4–1 and is also useful in general hazard evaluation work for other nuclides or other modes of exposure than those assumed for Figure 3.4–1.

For convenience in estimating the total area or number of people that may be affected by an accidental release of radioactivity, the cloud width out to the points where the concentration is 10% of that in the center of a Gaussian cloud, the cloud height to 10% concentration, and the equivalent cloud volume are also plotted in Figure 3.4–2 for the case of a cloud at ambient temperatures with the center line of the trajectory remaining at ground level. The cloud volume is calculated as the volume of an ellipsoid that would give the same inhaled dose, if it traveled along the same trajectory as the diffusing cloud at a wind speed of 1 m/sec, to a person breathing at the center of the trajectory, under the conditions that the ellipsoid contained the same total activity as the cloud and had a uniform concentration of activity equal to the concentration at the center of the Gaussian cloud. Thus, the curve of cloud volume vs. distance gives an easy and conservative estimate of volumes over which the activity may be assumed to be dispersed at various distances and, together with wind speed, may be used to estimate the maximum exposure to individuals at various distances as well as the average exposure and the numbers of persons exposed. The cloud width and height were calculated from the equations for $2y$ and z_0 in Reference 31. The cloud volume was calculated from the equation

$$V = (\pi^{1/2}\, X^{(2-n)/2})^3 C_y^2 C_z \qquad \text{(A.3.4–3)}$$

Ground contamination levels outside of nuclear facilities, for either accidental or routine release of radioactive material, were estimated from the equation[32]

$$S = \frac{2Q_0 V}{u\pi C_y C_z X^{(2-n)}} \exp\left(\frac{-4VX^{1\ 1/2}}{nu\pi^{1/2}C_z}\right) \qquad \text{(A.3.4–4)}$$

where S is the curies per square meter deposited along the center of the trajectory at distance X meters from the release point of Q_0 curies at ground level, V is the average deposition velocity[16] of the particles in meters per second, u is the wind speed in meters per second, and the other symbols are diffusion parameters that characterize weather conditions, as in Equation A.3.4–1. In Figure 3.4–3 the number of curies deposited per square meter per curie released at zero point is plotted as a function of the distance X in meters along the center of the cloud trajectory, for inversion conditions[19] in which n = 0.5, C_y = 0.4, and C_z = 0.07, and the wind speed, u, is only 1 m/sec. Figure 3.4–3 tends to overestimate contamination levels in curies per square meter, but the cloud dimensions from similar meteorological conditions in Figure 3.4–2 would tend to be correspondingly diminished for any given curie level released. Thus, the area of most immediate concern and the likely upper limits of dose rate or ingestion rate tend to be emphasized by Figures 3.4–2 and 3.4–3.

The above methods of estimating dispersion and deposition of radioactive materials released outside nuclear facilities have been used to determine base-line protection factors provided by atmospheric dilution in the selection of appropriate safeguard levels in Table 3.4–1. Although the values in Figures 3.4–1 through 3.4–3 tend to overestimate population exposures, they often suffice in evaluating most facility requirements to determine whether various safeguards are unnecessary or required, since the relative hazard potentials often fall either well below or well above the orders of magnitude over which the meteorological factors alone may vary. However, in situations where the balance between safety and economic considerations becomes marginal, the industrial hygienist may be obligated to consider the effects of more representative meteorological conditions, particularly when dealing with methods for controlling long-term, relatively low-level population exposures from routine plant operations. Various methods[31,33–40] have been developed for estimating meteorological dilution factors under different weather conditions, for different source configurations other than a point release, and for clouds of material released above ambient temperature or from elevated stacks. Consideration of these other conditions will generally lead to estimated population exposures lower than those that would be obtained from Figures 3.4–1 through 3.4–3. Even the single change to neutral or average meteorological conditions[33,36] would give maximum cloud concentrations one tenth or less of those implied in Figures 3.4–1 through 3.4–3. When a detailed picture of likely average or single-incident diffusion patterns is desired for emergency planning or hazard evaluation, an experience meteorologist should be consulted.

For estimating inhalation exposures from accidents within buildings, the Fickian equations for molecular diffusion would in most cases[37] not be applicable, since there would be interference from normal room air currents, exhaust ventilation, "splash" of particles from operations such as cutting or grinding, or explosions and fires that are often associated with accidental releases.[25,26] However, estimates of the upper limits of possible inhalation exposure are often useful. For example, when immediate alarm and evacuation procedures are in effect, an individual more than a few feet from the point of release would generally not remain for more than a minute in the vicinity of the accident. For an unsuspected instantaneous point release, the maximum quantity of activity inhaled can be estimated by assuming that the released material is immediately dispersed in a hemisphere of radius r feet equal to the distance between the point of release and the breathing zone. If the worker is assumed to remain at a fixed position, then the hemisphere may be assumed to be ventilated by at least a rate $A = 15\pi r^2$ ft^3/min, where a minimum convection velocity of 15 ft/min is adopted as recommended by Hemeon.[25] The concentration of a contaminant introduced at time t = 0 into an air space of volume V cubic feet is

$$X = X_0 e^{-At/V} \qquad \text{(A.3.4–5)}$$

where X is the concentration at time t minutes, X_0 is the initial concentration of the material dispersed instantaneously throughout volume V cubic feet, and A is the rate of ventilation of volume V

in cubic feet per minute. Thus, by assuming a maximum breathing rate of 3.5 ft^3/min (100 l/min) the number of curies, C, inhaled by a worker remaining r feet from a point release would be less than

$$C = 3.5 \int_{t=0}^{\infty} (C_0/V)e^{-22.5t/r}\,dt$$

$$= 0.074\ C_0/r^2 \qquad \text{(A.3.4–6)}$$

where C_0 is the number of curies released instantaneously (less than 1 sec), $V = (2/3)\pi r^3$ is the volume of the hemisphere up to the breathing zone, and r is the distance in feet from the point of release to the breathing zone. It is noted that the mean duration of the activity in such a hemisphere, on the assumption of further dilution by natural convection only, would be only $r/22.5$, which is a small fraction of a minute for distances of a few feet. Also, Equation A.3.4–6 shows that someone more than 3 ft from the point of release would probably inhale less than 1% of the escaping material. Since the usual air velocities outdoors would be greater than 150 ft/min, an individual more than 3 ft from an outdoor release would probably inhale less than 1/1,000 of the total quantity released (see also Figure 3.4–2).

For the continuous release of C_0 Ci/min, the build-up equations given by Hemeon[25] can be integrated in similar fashion to give the inhaled dose. Similar calculations would apply for a worker moving in a large ventilated room in which the radioactive material is introduced, except that the volume, V, would be that of the room, and the ventilation rate, Q, in the derivation of Equation A.3.4–6 would be the general ventilation rate of the room.

REFERENCES

1. **Morgan, K. Z., Snyder, W. S., and Ford, M. R.,** Relative hazard of the various radioactive materials, *Health Phys.*, 10, 171, 1964.
2. *Report of Committee II on Permissible Dose for Internal Radiation,* Pergamon Press, London, 1959; *Health Phys.*, 3, 1, 1960.
3. **Snyder, W. S.,** Range of uncertainty of MPC values, in *Hearings before the Joint Committee on Atomic Energy,* U.S. Government Printing Office, Washington, D.C., May 1960, 338. (This would still be about true as of 1975).
4. **Brodsky, A.,** Determining Industrial Hygiene Requirements for Installations Using Radioactive Materials, presented at the 1964 Conference of the American Industrial Hygiene Association, Philadelphia, April 29, 1964.
5. **Strominger, D., Hollander, J. M., and Seaborg, G. T.,** *Table of Isotopes,* UCRL-1928 (2nd rev.); *Rev. Mod. Phys.*, 30(2), Part II, 1958.
6. **Brodsky, A. and Beard, G. V.,** *A Compendium of Information for Use in Controlling Radiation Emergencies,* U.S. AEC Publ. TID-8206 (rev.) 52–100, September 1960.
7. **Fairbairn, A. and Dunning, N. J.,** The classification of radioisotopes for packaging, in *Regulations for the Safe Transport of Radioactive Materials – Notes on Certain Aspects of the Regulations,* Safety Series No. 7, International Atomic Energy Agency, Vienna, 1961, 25.
8. **Slack, L. and Way, K.,** *Radiations from Radioactive Atoms in Frequent Use,* U.S. AEC rep. M-6965, National Research Council, February 1959.
9. **Cronkite, E. P., Fliedner, T. M., Kilman, S. A., and Rubini, J. R.,** *Tritium in the Physical and Biological Sciences,* Vol. 2, International Atomic Energy Agency, Vienna, 1962, 191.
10. **Hatch, T. F. and Gross, P.,** *Pulmonary Deposition and Retention of Inhaled Aerosols,* AIHA-AEC Monograph, Academic Press, New York, 1964, 9–18, 74–78, 132–135.
11. *Safe Handling of Radioisotopes,* Safety Series No. 1, International Atomic Energy Agency, Vienna, 1958, 99.
12. Maximum permissible body burdens and maximum permissible concentrations of radionuclides in air and in water for occupational exposure, recommendations of the National Committee on Radiation Protection, in *NBS Handbook 69,* National Bureau of Standards, Washington, D.C., June 5, 1959.
13. **Garden, N. B.,** *Report on Glove Boxes and Containment Enclosures,* U.S. AEC Publ. TID-16020, June 20, 1962 (available from Office of Technical Services, Department of Commerce, Washington, D.C.).
14. **Bradley, F. J.,** Nuclear plant engineering and maintenance, in *Production Handbook,* 2nd ed., Carson, G. B., Ed., Ronald Press, New York, 1958, 24.

15. **Jones, I. S. and Pond, S. F.,** *Some Experiments to Determine the Resuspension Factor of Plutonium from Various Surfaces,* AERE-R4635, Health Physics and Medical Division, UKAEA Research Group, Atomic Energy Research Establishment, Harwell, Eng., May 1964; see also Fairbairn, A., The derivation of maximum permissible levels of radioactive surface contamination of transport containers and vehicles, in *Regulations for the Safe Transport of Radioactive Materials – Notes on Certain Aspects of the Regulations,* Safety Series No. 7, International Atomic Energy Agency, Vienna, 1961, 79.
16. **Stewart, K.,** The Resuspension of Particulate Material from Surfaces, paper presented at the International Symposium on Surface Contamination, Gatlinburg, Tenn., June 1964.
17. *Standards for Protection against Radiation,* Title 10, Code of Federal Regulations, Part 20, effective January 1961, U.S. Atomic Energy Commission, Washington, D.C.
18. *High Efficiency Particulate Air Filter Units,* U.S. AEC Publ., August 1961 (available from Office of Technical Services, Department of Commerce, Washington, D.C.).
19. *Reactor Site Criteria,* U.S. Atomic Energy Commission Regulation 10 CFR Part 100, 27, Federal Regulation 309, April 12, 1962.
20. **Cowan, F. P. and Kuper, J. B. H.,** Exposure criteria for evaluating the public consequences of catastrophic accidents in large nuclear plants, *Health Phys.,* 1, 76, 1958; see also Beck, C. K. et al., *Theoretical Possibilities and Consequences of Major Accidents in Large Nuclear Power Plants,* U.S. AEC rep. WASH-740, March 1957.
21. **Brodsky, A.,** Acceptable Emergency Doses from Fission Product Clouds, presented at the Fall Symposium of the Baltimore-Washington Chapter of the Health Physics Society, Bethesda, Md., October 12, 1963 (available from Graduate School of Public Health, University of Pittsburgh).
22. **Suzuki, M.,** A biological consideration on variability of doses received by exposure to radioactive aerosol released in a hypothetical big reactor accident and evaluation of exposure hazard, in *Reactor Safety and Hazard Evaluation, II,* Symposium in Vienna, May 14–18, 1962, International Atomic Energy Agency, Vienna, August 1962.
23. **Brodsky, A.,** Criteria for acute exposure to mixed fission product aerosols, *Health Phys.,* 11, 1017, 1965.
24. **Jordan, H. S. and Black, R. E.,** Evaluation of the air pollution problems resulting from discharge of a radioactive effluent, *Am. Ind. Hyg. Assoc. J.,* 19, 20, 1958.
25. **Hemeon, W. C. L.,** *Plant and Process Ventilation,* 2nd ed., Industrial Press, New York, 1963, 217.
26. **Drinker, P. and Hatch, T.,** *Industrial Dust,* McGraw-Hill, New York, 1954, 32, 243.
27. **Blatz, H.,** *Radiation Hygiene Handbook,* McGraw-Hill, New York, 1959.
28. *Safe Handling of Radioisotopes – Health Physics Addendum,* Safety Series No. 2, International Atomic Energy Agency, Vienna, 1960; see also *Medical Addendum,* Safety Series No. 3, International Atomic Energy Agency, Vienna, 1960.
29. Safety standard for non-medical X-ray and sealed gamma-ray sources, in *NBS Handbook 93,* National Bureau of Standards, Washington, D.C., January 3, 1964; see also Control and removal of radioactive contamination in laboratories, in *NBS Handbook 48,* 1951, and other handbooks of this series.
30. **Duhamel, F. and Lavie, J. M.,** La limitation des quantités de substances radioactives manipulés en transportes, *Health Phys.,* 10, 453, 1964.
31. **Wexler, H. et al.,** *Meteorology and Atomic Energy,* AECU 3066, U.S. Weather Bureau-U.S. AEC Document, July 1955 (available from U.S. Government Printing Office, Washington, D.C.).
32. **Wexler, H. et al.,** *Meteorology and Atomic Energy,* AECU 3066, U.S. Weather Burea-U.S. AEC Document, July 1955, 93.
33. **Pack, D. H.,** Meteorological aspects of nuclear emergencies, in *Reactor Safety and Hazards Evaluation,* Division of Radiological Health, Public Health Service, R. A. Taft Sanitary Engineering Center, Cincinnati, February 1963.
34. **Gifford, F. G.,** Use of routine meteorological observations for estimating atmospheric dispersion, *Nucl. Saf.,* 2(4), 47, 1961.
35. **Cramer, H. E.,** Engineering estimates of atmospheric dispersal capacity, *Am. Ind. Hyg. Assoc. J.,* 20, 183, 1959.
36. **Pasquill, F.,** *Atmospheric Diffusion,* D. Van Nostrand, London, 1962, 192–198, 204–213, 263–270.
37. **Sutton, O. G.,** *Micrometeorology,* McGraw-Hill, New York, 1953, 133–140, 273–288.
38. **Smith, M. E. and Singer, I. A.,** Diffusion and deposition in relation to reactor safety problems, *Am. Ind. Hyg. Assoc. Q.,* 18, 4, 319, 1957.
39. **Healy, J. W.,** Calculations of Environmental Consequences of Reactor Accidents, Report HW-54128, General Electric Co., Hanford, Washington, December 11, 1957.
40. **Fitzgerald, J. J., Hurwitz, H., Jr., and Tonks, L.,** Method for Evaluating Radiation Hazards from a Nuclear Incident, Report KAPL-1045, Knolls Atomic Power Laboratory, Schenectady, N.Y., March 26, 1954.

3.5 GENERAL NUCLEAR PROPERTIES

Table 3.5–1
NUCLEAR REACTIONS

Fission Reactions

Energy Released E (MeV), Prompt Neutrons ν, Ratio Values of Delayed to Prompt Neutrons β, per Thermal Fission

Isotope	Total energy of light fragments	Total energy of heavy fragments	Total energy of gamma rays	Total energy of fission neutron	Total energy of beta rays	Total energy	ν	β
U^{233}	97	66	14	5	9	191	2.51	0.0026
U^{235}	98	67	15	4.9	9	194	2.47	0.0064
Pu^{239}	100	72	14	5.8	9	201	2.91	0.0021

Breeding Processes

$$U^{238} + n = U^{239} + \gamma;\ U^{239} \xrightarrow{\text{23 min.}} Np^{239} + \beta^-;\ Np^{239} \xrightarrow{\text{2.3 days}} Pu^{239} + \beta^-$$

$$Th^{232} + n = Th^{233} + \gamma;\ Th^{233} \xrightarrow{\text{23 min.}} Pa^{233} + \beta^-;\ Pa^{233} \xrightarrow{\text{27.4 days}} U^{233} + \beta^-$$

Thermonuclear Reactions

$$D + D \begin{cases} \nearrow T + P + 4.0\ \text{MeV} \\ \searrow He^3 + n + 3.2\ \text{MeV} \end{cases}$$

$$D + T \rightarrow He^4 + n + 17.6\ \text{MeV}$$

From Katcoff, S., *Nucleonics,* Brookhaven National Laboratory, Upton, L.I., N.Y., November 1960. With permission.

Table 3.5–2
FISSILE AND FERTILE RADIOACTIVE MATERIALS

	Uranium 233	Uranium 235	Uranium 238	Plutonium 239	Thorium 232	Natural uranium (>99% U^{238})
Energy, MeV	4.8	4.4 and 4.6		5.1	4.0	4.2
Half-life, millions of years	0.162	713		0.024	13,900	4,510
Neutron binding energy, MeV	6.7	6.4		6.4	5.1	4.8
Critical energy for fission, MeV[a]	5.5	5.8		5.5	5.9	5.9
Fission cross section, barns[b]	527.	577.	–	742.		4.2
Capture cross section, barns[b]	54.	106.	2.71	287.		3.5
Total absorption, barns[b]	581.	683.	2.71	1,029.		7.7
Absorbed neutrons for fission, %	90.5	84.5	–	72.		

[a]Minimum energies of photons to cause fission.
[b]Data are for 2,200 m/sec thermal neutrons.

From Bolz, R. E. and Tuve, G. L., Eds., *Handbook of Tables for Applied Engineering Science,* 2nd ed., CRC Press, Cleveland, 1973, 411.

Table 3.5–3
NEUTRON SOURCES

Alpha Sources

Alpha source	*Alpha energies, MeV*	*Half-life*
Radium 226	7.683, 5.996, 5.305 . . . ., 4.59	1620 years
Actinium 227	7.36, 6.617, 6.273 . . . ., 4.942	22 years
Thorium 228	8.780, 6.775, 6.272 . . . ., 5.338	1.91 years
Uranium 232	8.780, 6.775, 6.272 . . . ., 5.261	74 years
Polonium 210	5.305	138.4 days
Radium D (Pb^{210})	5.305	19.4 years
Plutonium 238	5.495, 5.452	86.4 years
Plutonium 239	5.147, 5.134, 5.096	24360 years
Americium 241	5.534, 5.50, 5.477, 5.435	458 years
Americium 242	6.110, 6.066	100 years
Curium 242	6.110, 6.066	162.5 days

Neutron Yield vs. Alpha Energy

Alpha energy, MeV	*Neutron yield from beryllium target, neutrons per million alphas*
4.0	24
5.0	54
6.0	105
7.0	185

Neutron Yields of Various Targets

The following table shows yields from different target materials when bombarded with the alphas from polonium 210, which has an alpha energy of 5.30 MeV.

Target	*Yield, neutrons per million alphas*
Lithium	2.7
Beryllium	77
Boron	22
Carbon	0.1
Fluorine	12

Practical Yields from Alpha-neutron Sources

Source	*Yield, n/sec/curie*
Ra–Be	$1.0–1.5 \times 10^7$
RaD–Be	2.5×10^6
Po–Be	2.5×10^6
Ac–Be	2.0×10^7
Pu^{239}–Be	2.2×10^6
RdTh–Be	2.0×10^7

Gamma Outputs of Alpha-neutron Sources

Source	*Gamma output, mrhm/curie*	*Gamma output, mrhm/10^6 n/sec*
Ra–Be	974	54
RaD–Be	33	13.3
Po–Be	0.1	0.04
Ac–Be	145	5.5
Pu^{239}–Be	11	4.9
RdTh–Be	944	34

Properties of Target Materials

The following table gives the reaction energy and the approximate neutron yield for a number of target materials when the projectile particles are the 5.30 MeV alphas of polonium 210.

Also given is the Coulomb repulsion energy or electrostatic barrier: the alpha particle and the target nucleus are both positively charged; as a result, only the alpha particles that have sufficient energy to overcome the electrostatic barrier will enter the nucleus and initiate the reaction.

Target	*Isotopic abundance*	*Reaction energy, MeV*	*Coulomb repulsion, MeV*	*Yield, n/sec/curie*
Lithium 6	7.5%			1.0×10^5 (Lithium 6 and 7)
Lithium 7	92.5%	−2.79	1.64 (Lithium 6 and 7)	
Beryllium 9	100 %	5.74	2.10	2.85×10^6
Boron 10	18.8%	1.37	2.57	8.1×10^5 (Boron 10 and 11)
Boron 11	81.2%	0.27	2.52	
Carbon 12	98.9%	−8.40	2.97	3.7×10^3 (Carbon 12 and 13)
Carbon 13	1.1%	2.36	2.92	
Fluorine 19	100 %	−1.95	4.05	4.44×10^5

From: Neutron Sources and Their Characteristics, Tech. Bull. NS-2, Commercial Products, Atomic Energy of Canada, Ltd., Ottawa. With permission.

Table 3.5–4
TOTAL CHAIN YIELD FROM THERMAL NEUTRON FISSIONS IN U^{235}[a]

The total integrated chain yield is equivalent to the total fission yield of nuclei having a particular mass. The fission product given uniquely characterizes the total chain yield for the respective mass number.

Mass No.	*Fission product*	*% yield*	*Mass No.*	*Fission product*	*% yield*
72	Zn^{72} (49 hr)	0.000016	125	Sb^{125} (2.0 yr)	0.021
73	Ga^{73} (5.0 hr)	0.00011	126	Sb^{126} (9 hr)	0.05[b]
77	As^{77} (38.7 hr)	0.0083	127	Sb^{127} (91 hr)	0.13[c]
78	As^{78} (91 min)	0.021	128	Sn^{128} (57 min)	0.37
79	As^{79} (9.0 min)	0.056	129	I^{129} (1.7×10^7 yr)	0.9
81	Se^{81} (17.6 min)	0.14	130	Sb^{130} (10 min)	2.0
83	Kr^{83} (stable)	0.544	131	Xe^{131} (stable)	2.93, 2.88[b]
84	Kr^{84} (stable)	1.00	132	Xe^{132} (stable)	4.38, 4.31[b]
85	Rb^{85} (stable)	1.30	133	Cs^{133} (stable)	6.59, 6.49[b]
86	Kr^{86} (stable)	2.02	134	Xe^{134} (stable)	8.06, 7.9[b]
87	Rb^{87} (6×10^{10} yr)	2.49	135	Cs^{135} (2.6×10^6 yr)	6.41, 6.31[b]
88	Sr^{88} (stable)	3.57[c]	136	Xe^{136} (stable)	6.46, 6.36[b]
89	Sr^{89} (51 days)	4.79	137	Cs^{137} (29 yr)	6.15, 6.05[b]
90	Sr^{90} (28 yr)	5.77[c]	138	Ba^{138} (stable)	5.74
91	Zr^{91} (stable)	5.84	139	Ba^{139} (84 min)	6.55[c]
92	Zr^{92} (stable)	6.03	140	Ce^{140} (stable)	6.44[c,d]
93	Zr^{93} (1.1×10^6 yr)	6.45	141	Ce^{141} (33 days)	~6.0
94	Zr^{94} (stable)	6.40	142	Ce^{142} (stable)	5.95
95	Mo^{95} (stable)	6.27	143	Nd^{143} (stable)	5.98[c]
96	Zr^{96} (stable)	6.33	144	Nd^{144} (5×10^{15} yr)	5.67[c]
97	Mo^{97} (stable)	6.09	145	Nd^{145} (stable)	3.95[c]
98	Mo^{98} (stable)	5.78	146	Nd^{146} (stable)	3.07[c]
99	Mo^{99} (66 hr)	6.06[c]	147	Sm^{147} (1.3×10^{11} yr)	2.38
100	Mo^{100} (stable)	6.30	148	Nd^{148} (stable)	1.70[c]
101	Ru^{101} (stable)	5.0	149	Sm^{149} (stable)	1.13
102	Ru^{102} (stable)	4.1	150	Nd^{150} (stable)	0.67[c]
103	Ru^{103} (39.7 days)	3.0	151	Sm^{151} (80 yr)	0.45
104	Ru^{104} (stable)	1.8	152	Sm^{152} (stable)	0.285
105	Ru^{105} (4.45 hr)	0.90[c]	153	Sm^{153} (47 hr)	0.15[c]
106	Ru^{106} (1.01 yr)	0.38	154	Sm^{154} (stable)	0.077
107	Rh^{107} (22 min)	0.19	155	Sm^{155} (24 min)	0.033
109	Pd^{109} (13.4 hr)	0.030	156	Eu^{156} (15.4 days)	0.014[c]
111	Ag^{111} (7.6 days)	0.019	157	Eu^{157} (15.4 hr)	0.0078
112	Pd^{112} (21 hr)	0.010[c]	158	Eu^{158} (60 min)	0.002
115	Cd^{115} (53 hr) + Cd^{115}[e] (43 days)	0.011	159	Gd^{159} (18 hr)	0.00107[b]
117	Cd^{117}[e] (3.0 hr)	0.011	161	Tb^{161} (6.9 days)	0.000076
121	Sn^{121} (27.5 hr)	0.015			

[a] See original source for extensive references.
[b] Average of values in references cited in original source.
[c] Based on a yield of $(6.15 + 5.94)/2 = 6.05$ for Cs^{137} (Steinberg, E. P., personal communication).
[d] Measured absolute yield is 6.32%. The number 6.44% is used to normalize other yields.
[e] Metastable.

Table 3.5–4 (continued)
TOTAL CHAIN YIELD FROM THERMAL NEUTRON FISSIONS IN U^{235}[a]

REFERENCES

Petruska, J. A., Thode, H. G., and Tomlinson, R. H., The absolute fission yields of twenty-eight mass chains in the thermal neutron fission of U^{235}, *Can. J. Phys.,* 33(11), 693, 1955.

Steinberg, E. P. and Glendenin, L. E., Survey of radiochemical studies of the fission process, in *Proc. 1955 Geneva Conf.,* 7, Paper No. 614, 3.

From *Reactor Physics Constants,* 2nd ed., ANL-5800, Argonne National Laboratory, U.S. Atomic Energy Commission, July 1963.

Table 3.5–5
DELAYED NEUTRON FRACTION (β) FOR FAST AND THERMAL FISSION

Following is a summary of the more useful values of the delayed neutron fraction, i.e., the relative number of delayed neutrons per fission, for the various isotopes.

Isotope	Fast fission	Thermal fission
U^{233}	0.0027 ± 0.0002	0.00264 ± 0.0002
U^{235}	0.0065 ± 0.0003	0.0065 ± 0.0003
U^{238}	0.0157 ± 0.0012	
Pu^{239}	0.0021 ± 0.0002	0.0021 ± 0.0002
Pu^{240}	0.0026 ± 0.0003	
Th^{232}	0.022	

From *Reactor Physics Constants,* 2nd ed., ANL-5800, Argonne National Laboratory, U.S. Atomic Energy Commission, July 1963.

Table 3.5–6

CUMULATIVE YIELDS OF VARIOUS FISSION PRODUCTS FROM THERMAL NEUTRON FISSIONS IN U^{233}, U^{235}, AND Pu^{239}

Data for U^{235} in Table 3.5–4 also apply but are not repeated here. Cumulative yields given below do not necessarily represent total chain yields, since a later member of the chain may also be formed directly from fission. In any disagreement between Table 3.5–4 and this table, the former values are considered more reliable.

Mass No.	Fission product	U^{233} % yield	U^{235} % yield	Pu^{239}, % yield a,b,c	Mass No.	Fission product	U^{233} % yield	U^{235} % yield	Pu^{239}, % yield a,b,c
72	Zn^{72} (49 hr)			0.00012	98	Mo^{98} (stable)	5.18		
77	Ge^{77} (12 hr)	0.010	0.0031		99	Mo^{99} (66 hr)	4.8		5.9
77	As^{77} (38.7 hr)	0.019			100	Mo^{100} (stable)	4.40		
78	Ge^{78} (86 min)		0.020		101	Mo^{101} (14.6 min)		~5.6	
81	Se^{81d} (57 min)		0.0084		101	Ru^{101} (stable)	3.00		
83	Se^{83} (25 min)		0.22		102	Mo^{102} (11.5 min)		4.3	
83	Br^{83} (2.4 hr)	0.79	0.51	0.085	102	Ru^{102} (stable)	2.37		
83	Kr^{83} (stable)	1.14			103	Ru^{103} (39.7 days)	1.6		5.8
84	Br^{84} (31.8 min)		0.90		104	Ru^{104} (stable)	0.96		
84	Kr^{84} (stable)	1.90			105	Rh^{105} (35.3 hr)			3.9
85	Kr^{85} (10.3 yr)	0.56	0.293		106	Ru^{106} (1.01 yr)	0.28		5.0
86	Kr^{86} (stable)	3.18			109	Pd^{109} (13.4 hr)	0.040	0.030	1.5
87	Br^{87} (55 sec)		3.1		111	Ag^{111} (7.6 days)	0.025		0.27
89	Sr^{89} (51 days)	6.5		1.9	112	Pd^{112} (21 hr)	0.016		0.10
90	Sr^{90} (28 yr)		5.8		115	Ag^{115} (21 min)		0.0077	
91	Sr^{91} (9.7 hr)		5.8	2.4	115	Cd^{115d} (43 days)	0.001	0.0007	0.003
91	Sr^{91} (2.7 hr)		5.3		115	Cd^{115} (53 hr)	0.019	0.0097	0.038
91	Y^{91} (58 days)		5.35	3.0	121	Sn^{121} (27.5 hr)	0.018		0.044
91	Zr^{91} (stable)	6.53			123	Sn^{123} (136 days)		0.0013	
92	Zr^{92} (stable)	6.70			125	Sn^{125} (9.6 days)	0.050	0.013	0.072
93	Zr^{93} (1.1×10^6 yr)	7.10			127	Sb^{127} (91 hr)			0.39
94	Y^{94} (16.5 min)		5.4		127	Te^{127d} (105 days)		0.035	
94	Zr^{94} (stable)	6.82			129	Te^{129d} (37 days)		0.35	
95	Zr^{95} (65 days)	5.9	6.2	5.9	131	Sb^{131} (23 min)		2.6	
95	Mo^{95} (stable)	6.10			131	Te^{131d} (30 hr)		0.44	
96	Zr^{96} (stable)	5.60			131	I^{131} (8.05 days)	2.7	3.1^{a}	3.8
97	Zr^{97} (17 hr)		5.9	5.7	131	Xe^{131} (stable)	3.74		2.87
97	Mo^{97} (stable)	5.35			132	Te^{132} (77 hr)		4.7	5.2

Table 3.5–6 (continued)
CUMULATIVE YIELDS OF VARIOUS FISSION PRODUCTS FROM THERMAL NEUTRON FISSIONS IN U^{233}, U^{235}, AND Pu^{239}

Mass No.	Fission product	U^{233}	U^{235}	Pu^{239},	Mass No.	Fission product	U^{233}	U^{235}	Pu^{239},
		% yield	% yield	% yield [a,b,c]			% yield	% yield	% yield [a,b,c]
132	Xe^{132} (stable)	5.10		4.02	143	Nd^{143} (stable)	5.2		6.31
133	Sb^{133} (4.1 min)		4.0		144	Ce^{144} (285 days)	4.1	6.0	5.28
133	Te^{133d} (63 min)		4.9		144	Nd^{144} (stable)	4.0		5.29
133	I^{133} (21 hr)		6.9	5.3	145	Nd^{145} (stable)	3.0		4.24
133	Xe^{133} (5.27 days)		6.62	5.27	146	Nd^{146} (stable)	2.3		3.53
133	Cs^{133} (stable)	6.18		5.27	147	Nd^{147} (11 days)		2.7	
134	Te^{134} (44 min)		6.9[a]		147	Sm^{147} (stable)	1.71		2.92
134	Xe^{134} (stable)	6.54		5.69	148	Nd^{148} (stable)	1.15		2.28
135	I^{135} (6.7 hr)	5.1	6.1[a]	5.8	149	Pm^{149} (5.6 hr)		1.4	
135	Cs^{135} (2.6×10^6 yr)	>4.9		5.53	149	Sm^{149} (stable)	0.61		1.89
136	I^{136} (86 sec)	1.7	3.1	2.1	150	Nd^{150} (stable)	0.48		1.38
136	Xe^{136} (stable)	<8.9		5.06	151	Sm^{151} (80 yr)	0.27		1.17
137	Cs^{137} (29 yr)	7.16	6.15	5.24	152	Sm^{152} (stable)	0.17		0.83
138	Cs^{138} (32 min)		5.74		153	Sm^{153} (47 hr)	0.095		0.41
139	Ba^{139} (84 min)			5.7	154	Sm^{154} (stable)	0.037		0.32
140	Ba^{140} (12.8 days)	6.0	6.44[e]	5.68	155	Sm^{155} (24 min)			0.22
140	Ce^{140} (stable)	5.6		5.68	155	Eu^{155} (1.9 yr)		0.03	
141	Ce^{141} (33 days)			5.2	156	Sm^{156} (10 hr)		0.013	
142	Ce^{142} (stable)	5.6		6.69	156	Eu^{156} (15.4 hr)			0.12
143	Ce^{143} (33 hr)		5.7	5.4					

[a]The Cs, Nd, and Sm yields were measured by Wiles, Petruska, and Tomlinson,[1] and corrected to 5.6% yield for Ba^{140}. For some of the Ce and Nd isotopes, the ratios of yields per Reference 2 were used.

[b]Most values are from References 3, corrected to a Ba^{140} yield of $1.06 \times 5.32\% = 5.68\%$.

[c]The xenon isotope yields were determined by Fleming and Thode.[4] These were normalized to corrected Cs yields at mass 133.

[d]Metastable.

[e]Measured value is 6.32%. The factor 1.06 is introduced to bring the total yield for light and heavy groups to 100% each.

REFERENCES

1. **Wiles, D. M., Petruska, J. A., and Tomlinson, R. H.,** Some cumulative yields of isotopes formed in the thermal neutron fission of Pu^{239}, *Can. J. Chem.*, 34(3), 227, 1956.
2. **Krizhansky, L. M., et al.,** Rare-earth isotope yields in the fission of Pu^{239} by pile neutrons, *Sov. J. At. Energy*, 2(3), 334, 1957.
3. **Steinberg, E. P. and Freedman, M. S.,** *Summary of Results of Fission-Yield Experiments,* Book 3, National Nuclear Energy Series, McGraw-Hill, New York, 1951, 1378.
4. **Fleming, W. H. and Thode, H. G.,** The relative yields of the isotopes of xenon in plutonium fission, *Can. J. Chem.*, 34(3), 193, 1956.

From *Reactor Physics Constants,* 2nd ed., ANL-5800, Argonne National Laboratory, U.S. Atomic Energy Commission, July 1963.

Table 3.5–7
CUMULATIVE PERCENTAGE YIELDS FROM FISSION SPECTRUM NEUTRON-INDUCED FISSIONS IN Pu^{239}, U^{238}, AND Th^{232}

Mass No.	*Fission product*	Pu^{239}[a]	U^{238}[b]	Th^{232}[c]	*Mass No.*	*Fission product*	Pu^{239}[a]	U^{238}[b]	Th^{232}[c]
72	Zn^{72} (49 hr)			0.00033	111	Ag^{111} (7.6 days)		0.073	0.052
73	Ga^{73} (5.0 hr)			0.00045	112	Pd^{112} (21 hr)	0.14	0.046	0.057
77	Ge^{77} (12 hr)			0.009	115	Cd^{115}[e] (43 days)		0.003	0.003
77	As^{77} (39 hr)		0.0038	0.020	115	Cd^{115} (53 hr)	0.069	0.037	0.072
83	Br^{83} (2.4 hr)			1.9	127	Sb^{127} (93 hr)		0.12	
83	Kr^{83} (stable)		0.40	1.99	131	I^{131} (8.05 days)			1.2
84	Kr^{84} (stable)		0.85	3.65	131	Xe^{131} (stable)		3.2	1.62
85	Kr^{85} (10.3 yr)		0.153	0.87	132	Te^{132} (77 hr)		4.7	2.4
86	Kr^{86} (stable)		1.38	6.0	132	Xe^{132} (stable)		4.7	2.87
89	Sr^{89} (51 days)		2.9	6.7	133	Cs^{133} (stable)		5.5 (8.08)[d]	
90	Sr^{90} (28 yr)	2.2	3.2	6.8	134	Xe^{134} (stable)		6.6	5.38
91	Sr^{91} (9.7 hr)			7.2	135	Cs^{135} (2.6×10^6 yr)		6.0[d]	
95	Zr^{95} (65 days)		5.7		136	Xe^{136} (stable)		5.9	5.65
97	Zr^{97} (17 hr)	5.2		5.2	137	Cs^{137} (29 yr)	6.6	6.2 (7.11)[d]	6.3
99	Mo^{99} (67 hr)	5.9	6.3	2.7	140	Ba^{140} (12.8 days)	5.0	5.7	6.2
103	Ru^{103} (40 days)		6.6	0.16	141	Ce^{141} (33 days)			9.0
105	Rh^{105} (35 hr)			0.07	144	Ce^{144} (290 days)		4.9	7.1
106	Ru^{106} (1.0 yr)		2.7	0.042	153	Sm^{153} (47 hr)	0.48		
109	Pd^{109} (13.4 hr)	1.9	0.32	0.055	156	Eu^{156} (15.4 days)		0.066	

[a]Values from Shuey, R. L., Instrumentation for Energy Determination of High-energy Particles, UCRL-793, July 1950.
[b]Most values are averages of data in Keller, R. N., Steinberg, E. P., and Glendenin, L. E., *Phys. Rev.*, 94(4), 969, 1954, and Engelkemier, D. W. et al., *Fission Yields in* U^{238}, Book 3, National Nuclear Energy Series, McGraw-Hill, New York, 1951, 1375, normalized to absolute of Mo^{99} in Terrell, J., et al., internal memorandum, Los Alamos Scientific Laboratory, LADC-1463, 1953.
[c]Most values from Turkevich, A. and Niday, J. B., *Phys. Rev.*, 84(1), 52, 1951. Kr and Xe yields measured by T. J. Kennett and H. G. Thode, and reported by private communication to S. Katcoff.
[d]Cs^{133} and Cs^{135} yields derived from ratios to Cs^{137} measured by R. H. Tomlinson and reported by private communication to S. Katcoff.
[e]Metastable.

From *Reactor Physics Constants,* 2nd ed., ANL-5800, Argonne National Laboratory, U.S. Atomic Energy Commission, July 1963.

Table 3.5–8
DELAYED NEUTRON YIELD FROM THERMAL NEUTRON-INDUCED FISSION IN U^{233}, U^{235}, AND Pu^{239}

Isotope	*Delayed neutrons fission*	*Group index, i*	*Half-life, T_i*	*Decay constant, λ_i*	*Relative abundance, a_i*	*Absolute group yield, %*
U^{233}	0.0066 ± 0.0003	1	55.00 ± 0.54	0.0126 ± 0.0002	0.086 ± 0.003	0.057 ± 0.003
		2	20.57 ± 0.38	0.0337 ± 0.0006	0.299 ± 0.004	0.197 ± 0.009
		3	5.00 ± 0.21	0.139 ± 0.006	0.252 ± 0.040	0.166 ± 0.027
		4	2.13 ± 0.20	0.325 ± 0.030	0.278 ± 0.020	0.184 ± 0.016
		5	0.615 ± 0.242	1.13 ± 0.40	0.051 ± 0.024	0.034 ± 0.016
		6	0.277 ± 0.047	2.50 ± 0.42	0.034 ± 0.014	0.022 ± 0.009
U^{235}	0.0158 ± 0.0005	1	55.72 ± 1.28	0.0124 ± 0.0003	0.033 ± 0.003	0.052 ± 0.005
		2	22.72 ± 0.71	0.0305 ± 0.0010	0.219 ± 0.009	0.346 ± 0.018
		3	6.22 ± 0.23	0.111 ± 0.004	0.196 ± 0.022	0.310 ± 0.036
		4	2.30 ± 0.09	0.301 ± 0.012	0.395 ± 0.011	0.624 ± 0.026
		5	0.61 ± 0.083	1.13 ± 0.15	0.115 ± 0.009	0.182 ± 0.015
		6	0.23 ± 0.025	3.00 ± 0.33	0.042 ± 0.008	0.066 ± 0.008
Pu^{239}	0.0061 ± 0.0003	1	54.28 ± 2.34	0.0128 ± 0.0005	0.035 ± 0.009	0.021 ± 0.006
		2	23.04 ± 1.67	0.0301 ± 0.0022	0.298 ± 0.035	0.182 ± 0.023
		3	5.60 ± 0.40	0.124 ± 0.009	0.211 ± 0.048	0.129 ± 0.030
		4	2.13 ± 0.24	0.325 ± 0.036	0.326 ± 0.033	0.199 ± 0.022
		5	0.618 ± 0.213	1.12 ± 0.39	0.086 ± 0.029	0.052 ± 0.018
		6	0.257 ± 0.045	2.69 ± 0.47	0.044 ± 0.016	0.027 ± 0.010

REFERENCE

Keepin, G. R. et al., Delayed neutrons from fissionable isotopes of uranium, plutonium, and thorium, *Phys. Rev.*, 107(4), 1044, 1957.

From *Reactor Physics Constants,* 2nd ed., ANL-5800, Argonne National Laboratory, U.S. Atomic Energy Commission, July 1963.

Table 3.5–9

CROSS SECTIONS FOR NATURALLY OCCURRING ELEMENTS

Thermal neutron cross sections for a most probable velocity of 2,200 m/sec (0.0253 eV or a wavelength of 1.80 Å) are given in terms of absorption and scattering (subscripts a and s). The total cross section (subscript t) determines the diminution of a neutron beam as it traverses a sample. The microscopic cross section applied to a single nucleus; the macroscopic cross section, $\Sigma = N\sigma$, is equivalent to the cross section per cm^3 for the target material, which contains N nuclei per cm^3. Because σ cm^2 is the effective area per single nucleus, i.e., the "cross section" per single nucleus, the dimensions of Σ are of reciprocal length. One barn is a unit of 10^{-24} cm^2 per nucleus.

For density in kg/m^3, multiply the value in g/cm^3 by 1,000. For density in lb/ft^3, multiply the value in g/cm^3 by 62.42.

Symbols:

$1 - \bar{\mu}_0$ = transport scattering factor
ξ = logarithmic energy loss

Atomic No.	Element or compound	Atomic or mol. wt.	Density, g/cm^3	Nuclei per unit vol. $\times 10^{-24}$	$1-\bar{\mu}_0$	ξ	Microscopic cross section, barn			Macroscopic cross section, cm^{-1}		
							σ_a	σ_s	σ_t	Σ_a	Σ_s	Σ_t
1	H	1.008	8.9[a]	5.3[a]	0.3386	1.000	0.33	38	38	1.7[a]	0.002	0.002
	H_2O	18.016	1	0.0335[b]	0.676	0.948	0.66	103	103	0.022	3.45	3.45
	D_2O	20.030	1.10	0.0331[b]	0.884	0.570	0.001	13.6	13.6	3.3[a]	0.449	0.449
2	He	4.003	17.8[a]	2.6[a]	0.8334	0.425	0.007	0.8	0.807	0.02[a]	2.1[a]	2.1[a]
3	Li	6.940	0.534	0.0463	0.9047	0.268	71	1.4	72.4	3.29	0.065	3.35
4	Be	9.013	1.85	0.1236	0.9259	0.209	0.010	7.0	7.01	124[a]	0.865	0.865
	BeO	25.02	3.025	0.0728[b]	0.939	0.173	0.010	6.8	6.8	73[a]	0.501	0.501
5	B	10.82	2.45	0.1364	0.9394	0.171	755	4	759	103	0.346	104
6	C	12.011	1.60	0.0803	0.9444	0.158	0.004	4.8	4.80	32[a]	0.385	0.385
7	N	14.008	0.0013	5.3[a]	0.9524	0.136	1.88	10	11.9	9.9[a]	50[a]	60[a]
8	O	16.000	0.0014	5.3[a]	0.9583	0.120	20[a]	4.2	4.2	0.000	21[a]	21[a]
9	F	19.00	0.0017	5.3[a]	0.9649	0.102	0.001	3.9	3.90	0.01[a]	20[a]	20[a]
10	Ne	20.183	0.0009	2.6[a]	0.9667	0.0968	<2.8	2.4	5.2	7.3[a]	6.2[a]	13.5[a]
11	Na	22.991	0.971	0.0254	0.9710	0.0845	0.525	4	4.53	0.013	0.102	0.115
12	Mg	24.32	1.74	0.0431	0.9722	0.0811	0.069	3.6	3.67	0.003	0.155	0.158
13	Al	26.98	2.699	0.0602	0.9754	0.0723	0.241	1.4	1.64	0.015	0.084	0.099
14	Si	28.09	2.42	0.0522	0.9762	0.0698	0.16	1.7	1.86	0.008	0.089	0.097
15	P	30.975	1.82	0.0354	0.9785	0.0632	0.20	5	5.20	0.007	0.177	0.184
16	S	32.066	2.07	0.0389	0.9792	0.0612	0.52	1.1	1.62	0.020	0.043	0.063
17	Cl	35.457	0.0032	5.3[a]	0.9810	0.0561	33.8	16	49.8	0.002	80[a]	0.003
18	Ar	39.944	0.0018	2.6[a]	0.9833	0.0492	0.66	1.5	2.16	1.7[a]	3.9	5.6[a]

Table 3.5–9 (continued)
CROSS SECTIONS FOR NATURALLY OCCURRING ELEMENTS

Atomic No.	Element or compound	Atomic or mol. wt.	Density, g/cm^3	Nuclei per unit vol. $\times 10^{-24}$	$1-\bar{\mu}_0$	ξ	Microscopic cross section, barn			Macroscopic cross section, cm^{-1}		
							σ_a	σ_s	σ_t	Σ_a	Σ_s	Σ_t
19	K	39.100	0.87	0.0134	0.9829	0.0504	2.07	1.5	3.57	0.028	0.020	0.048
20	Ca	40.08	1.55	0.0233	0.9833	0.0492	0.44	3.0	3.44	0.010	0.070	0.080
21	Sc	44.96	2.5	0.0335	0.9852	0.0438	24	24	48	0.804	0.804	1.61
22	Ti	47.90	4.5	0.0566	0.9861	0.0411	5.8	4	9.8	0.328	0.226	0.555
23	V	50.95	5.96	0.0704	0.9869	0.0387	5	5	10.0	0.352	0.352	0.704
24	Cr	52.01	7.1	0.0822	0.9872	0.0385	3.1	3	6.1	0.255	0.247	0.501
25	Mn	54.94	7.2	0.0789	0.9878	0.0359	13.2	2.3	15.5	1.04	0.181	1.22
26	Fe	55.85	7.86	0.0848	0.9881	0.0353	2.62	11	13.6	0.222	0.933	1.15
27	Co	58.94	8.9	0.0910	0.9887	0.0335	38	7	45	3.46	0.637	4.10
28	Ni	58.71	8.90	0.0913	0.9887	0.0335	4.6	17.5	22.1	0.420	1.60	2.02
29	Cu	63.54	8.94	0.0848	0.9896	0.0309	3.85	7.2	11.05	0.326	0.611	0.937
30	Zn	65.38	7.14	0.0658	0.9897	0.0304	1.10	3.6	4.70	0.072	0.237	0.309
31	Ga	69.72	5.91	0.0511	0.9925	0.0283	2.80	4	6.80	0.143	0.204	0.347
32	Ge	72.60	5.36	0.0445	0.9909	0.0271	2.45	3	5.45	0.109	0.134	0.243
33	As	74.91	5.73	0.0461	0.9911	0.0264	4.3	6	10.3	0.198	0.277	0.475
34	Se	78.96	4.8	0.0366	0.9916	0.0251	12.3	11	23.3	0.450	0.403	0.853
35	Br	79.916	3.12	0.0235	0.9917	0.0247	6.7	6	12.7	0.157	0.141	0.298
36	Kr	83.80	0.0037	2.6[a]	0.9921	0.0236	31	7.2	38.2	81[a]	19[a]	99[a]
37	Rb	85.48	1.53	0.0108	0.9922	0.0233	0.73	12	12.7	0.008	0.130	0.138
38	Sr	87.63	2.54	0.0175	0.9925	0.0226	1.21	10	11.2	0.021	0.175	0.195
39	Yt	88.92	5.51	0.0373	0.9925	0.0223	1.31	3	4.3	0.049	0.112	1.160
40	Zr	91.22	6.4	0.0423	0.9927	0.0218	0.185	8	8.2	0.008	0.338	0.347
41	Nb	92.91	8.4	0.0545	0.9928	0.0214	1.16	5	6.16	0.063	0.273	0.336
42	Mo	95.95	10.2	0.0640	0.9931	0.0207	2.70	7	9.70	0.173	0.448	0.621
43	Tc	98	—	—	0.9932	0.0203	22	—	—	—	—	—
44	Ru	101.1	12.2	0.0727	0.9934	0.0197	2.56	6	8.56	0.186	0.436	0.622
45	Rh	102.91	12.5	0.0732	0.9935	0.0193	149	5	154	10.9	0.366	11.3
46	Pd	106.4	12.16	0.0689	0.9937	0.0187	8	3.6	11.6	0.551	0.248	0.799
47	Ag	107.88	10.5	0.0586	0.9938	0.0184	63	6	69	3.69	0.352	4.04
48	Cd	112.41	8.65	0.0464	0.9940	0.0178	2,450	7	2,457	114	0.325	114

Table 3.5–9 (continued)
CROSS SECTIONS FOR NATURALLY OCCURRING ELEMENTS

Atomic No.	Element or compound	Atomic or mol. wt.	Density, g/cm^3	Nuclei per unit vol. $\times 10^{-24}$	$1-\bar{\mu}_0$	ξ	Microscopic cross section, barn			Macroscopic cross section, cm^{-1}		
							σ_a	σ_s	σ_t	Σ_a	Σ_s	Σ_t
49	In	114.82	7.28	0.0382	0.9942	0.0173	191	2.2	193	7.30	0.084	7.37
50	Sn	118.70	6.5	0.0330	0.9944	0.0167	0.625	4	4.6	0.021	0.132	0.152
51	Sb	121.76	6.69	0.0331	0.9945	0.0163	5.7	4.3	10.0	0.189	0.142	0.331
52	Te	127.61	6.24	0.0295	0.9948	0.0155	4.7	5	9.7	0.139	0.148	0.286
53	I	126.91	4.93	0.0234	0.9948	0.0157	7.0	3.6	10.6	0.164	0.084	0.248
54	Xe	131.30	0.0059	2.7[a]	0.9949	0.0152	35	4.3	39.3	95[a]	12[a]	0.001
55	Cs	132.91	1.873	0.0085	0.9950	0.0150	28	20	48	0.238	0.170	0.408
56	Ba	137.36	3.5	0.0154	0.9951	0.0145	1.2	8	9.2	0.018	0.123	0.142
57	La	138.92	6.19	0.0268	0.9952	0.0143	8.9	15	24	0.239	0.403	0.642
58	Ce	140.13	6.78	0.0292	0.9952	0.0142	0.73	9	9.7	0.021	0.263	0.283
59	Pr	140.92	6.78	0.0290	0.9953	0.0141	11.3	4	15.3	0.328	0.116	0.444
60	Nd	144.27	6.95	0.0290	0.9954	0.0138	46	16	62	1.33	0.464	1.79
61	Pm	145	—	—	0.9954	0.0137	60	—	—	—	—	—
62	Sm	150.35	7.7	0.0309	0.9956	0.0133	5,600	5	5,605	173	0.155	173
	Sm_2O_3	348.70	7.43	0.0128[b]	0.974	0.076	16,500	22.6	16,500	211	0.289	211
63	Eu	152	5.22	0.0207	0.9956	0.0131	4,300	8	4,308	89.0	0.166	89.2
	Eu_2O_3	352.00	7.42	0.0127[b]	0.978	0.063	8,740	30.2	8,770	111	0.383	111
64	Gd	157.26	7.95	0.0305	0.9958	0.0127	46,000	—	—	1,403	—	—
65	Tb	158.93	8.33	0.0316	0.9958	0.0125	46	—	—	1.45	—	—
66	Dy	162.51	8.56	0.0317	0.9959	0.0122	950	100	1,050	30.1	3.17	33.3
	Dy_2O_3	372.92	7.81	0.0126[b]	0.993	0.019	2,200	214	2,414	27.7	2.7	30.4
67	Ho	164.94	8.76	0.0320	0.9960	0.0121	65	—	—	2.08	—	—
68	Er	167.27	9.16	0.0330	0.9960	0.0119	173	15	188	5.71	0.495	6.20
69	Tm	168.94	9.35	0.0333	0.9961	0.0118	127	7	134	4.23	0.233	4.46
70	Yb	173.04	7.01	0.0244	0.9961	0.0115	37	12	49	0.903	0.293	1.20
71	Lu	174.99	9.74	0.0335	0.9962	0.0114	112	—	—	3.75	—	—
72	Hf	178.5	13.3	0.0449	0.9963	0.0112	105	8	113	4.71	0.359	5.07
73	Ta	180.95	16.6	0.0553	0.9963	0.0110	21	5	26	1.16	0.277	1.44
74	W	183.86	19.3	0.0632	0.9964	0.0108	19.2	5	24.2	1.21	0.316	1.53
75	Re	186.22	20.53	0.0664	0.9964	0.0107	86	14	100	5.71	0.930	6.64
76	Os	190.2	22.48	0.0712	0.9965	0.0105	15.3	11	26.3	1.09	0.783	1.87

Table 3.5–9 (continued)
CROSS SECTIONS FOR NATURALLY OCCURRING ELEMENTS

Atomic No.	Element or compound	Atomic or mol. wt.	Density, g/cm^3	Nuclei per unit vol. $\times 10^{-24}$	$1-\bar{\mu}_0$	ξ	Microscopic cross section, barn			Macroscopic cross section, cm^{-1}		
							σ_a	σ_s	σ_t	Σ_a	Σ_s	Σ_t
77	Ir	192.2	22.42	0.0703	0.9965	0.0104	440	—	—	30.9	—	—
78	Pt	195.09	21.37	0.0660	0.9966	0.0102	8.8	10	18.8	0.581	0.660	1.24
79	Au	197	19.32	0.0591	0.9966	0.0101	98.8	9.3	107.3	5.79	0.550	6.34
80	Hg	200.61	13.55	0.0407	0.9967	0.0099	380	20	400	15.5	0.814	16.3
81	Tl	204.39	11.85	0.0349	0.9967	0.0098	3.4	14	17.4	0.119	0.489	0.607
82	Pb	207.21	11.35	0.0330	0.9968	0.0096	0.170	11	11.2	0.006	0.363	0.369
83	Bi	209	9.747	0.0281	0.9968	0.0095	0.034	9	9	0.001	0.253	0.256
84	Po	210	9.24	0.0265	0.9968	0.0095	—	—	—	—	—	—
85	At	211	—	—	0.9968	0.0094	—	—	—	—	—	—
86	Rn	222	0.0097	2.6[a]	0.9970	0.0090	0.7	—	—	—	—	—
87	Fr	223	—	—	0.9980	0.0089	—	—	—	—	—	—
88	Ra	226.05	5	0.0133	0.9971	0.0088	20	—	—	0.266	—	—
89	Ac	227	—	—	0.9971	0.0088	510	—	—	—	—	—
90	Th	232.05	11.3	0.0293	0.9971	0.0086	7.56	12.6	20.2	0.222	0.369	0.592
91	Pa	231	15.4	0.0402	0.9971	0.0086	200	—	—	8.04	—	—
92	U	238.07	18.9	0.04783	0.9972	0.0084	7.68	8.3	16.0	0.367	0.397	0.765
	UO_2	270.07	10	0.0223[b]	0.9887	0.036	7.6	16.7	24.3	0.169	0.372	0.542
93	Np	237	—	—	0.9972	0.0084	170	—	—	—	—	—
94	Pu	239	19.74	0.0498	0.9972	0.0083	1,026	9.6	1,036	51.1	0.478	51.6
95	Am	242	—	—	0.9973	0.0082	8.000	—	—	—	—	—
96	Cm	245	—	—	0.9973	0.0081	—	—	—	—	—	—
97	Bk	249	—	—	0.9973	0.0081	500	—	—	—	—	—
98	Cf	249	—	—	0.9973	0.0079	900	—	—	—	—	—
99	E	253	—	—	0.9974	0.0079	160	—	—	—	—	—
100	Fm	256	—	—	0.9974	0.0078	—	—	—	—	—	—
101	Mv	260	—	—	0.9974	0.0077	—	—	—	—	—	—

[a] Value has been multiplied by 10^5.
[b] Molecules/cm^3.

Table 3.5–9 (continued)
CROSS SECTIONS FOR NATURALLY OCCURRING ELEMENTS

REFERENCES

Hughes, D. J. and Schwartz, R. B., *Neutron Cross Sections,* 2nd ed., BNL-325, Brookhaven National Laboratory, Upton, L.I., N.Y., 1958 (see also Supplement I to second edition, 1960).

For a table of scattering cross sections giving data for each isotope of the element, see Gray, D. E., Ed., *American Institute of Physics Handbook,* 2nd ed., McGraw-Hill, New York, 1963, 8-148.

From *Reactor Physics Constants,* 2nd ed., ANL-5800, Argonne National Laboratory, U.S. Atomic Energy Commission, July 1963.

Table 3.5–10
CHARACTERISTIC DECAY OF A RADIOISOTOPE

At the end of each half-life interval, one-half of the starting material will remain; at the end of two half-lives, one-fourth; and at the end of four half-lives, one-sixteenth.

Half-lives	*F* [a]	*Half-lives*	*F* [a]	*Half-lives*	*F* [a]	*Half-lives*	*F* [a]
0.00	1.000	0.70	0.616	1.65	0.319	3.20	0.109
0.02	0.986	0.75	0.595	1.70	0.308	3.30	0.102
0.04	0.973	0.80	0.574	1.75	0.297	3.40	0.095
0.06	0.959	0.85	0.555	1.80	0.287	3.50	0.088
0.08	0.946	0.90	0.535	1.85	0.277	3.60	0.083
0.10	0.933	0.95	0.518	1.90	0.268	3.70	0.077
0.12	0.920	1.00	0.500	1.95	0.259	3.80	0.072
0.14	0.908	1.05	0.483	2.00	0.250	3.90	0.067
0.16	0.895	1.10	0.467	2.10	0.233	4.00	0.063
0.18	0.883	1.15	0.451	2.20	0.218	4.10	0.058
0.20	0.871	1.20	0.435	2.30	0.203	4.20	0.054
0.25	0.841	1.25	0.421	2.40	0.189	4.30	0.051
0.30	0.812	1.30	0.406	2.50	0.177	4.40	0.047
0.35	0.785	1.35	0.393	2.60	0.165	4.50	0.044
0.40	0.758	1.40	0.379	2.70	0.154	4.60	0.041
0.45	0.732	1.45	0.367	2.80	0.144	4.70	0.039
0.50	0.707	1.50	0.354	2.90	0.134	4.80	0.036
0.55	0.683	1.55	0.342	3.00	0.125	4.90	0.034
0.60	0.660	1.60	0.330	3.10	0.117	5.00	0.031
0.65	0.638						

[a] F = fraction remaining.

From Wang, Y., Ed., *CRC Handbook of Radioactive Nuclides,* Chemical Rubber Co., Cleveland, 1969, 4.

Table 3.5–11
RADIOACTIVE ISOTOPES

The engineering uses of radioactive isotopes are limited; for more extensive data see the Wang, Y., Ed., *CRC Handbook of Radioactive Nuclides,* Chemical Rubber Co., Cleveland, 1969.

The atomic weight of an atom depends on the number of neutrons in the nucleus as well as the number of protons (and electrons) indicated by the atomic number. For each atomic number, which defines the element, there can be several atomic weights, depending on the number of neutrons. Atoms that differ only in the number of neutrons are known as isotopes or nuclides. While there are more than 20 anisotopic elements (one atomic weight), many others have 2 to 10 isotopes. When there are too many or too few neutrons, the atom becomes unstable or radioactive, and a statistical "decay" occurs, involving the release of radiations of one kind or another. Isotopes may be produced artificially by fission or more likely by bombardment, as in a reactor or a cyclotron.

Isotope differences other than mass are very small and not easy to detect. In a few cases the physical or chemical properties of the isotopes are sufficiently different to permit separation by some common process. This is true with isotopes of boron, carbon, hydrogen, lithium, nitrogen, and oxygen. Separation methods include distillation, electrolysis, electromigration, and chemical exchange.

Mass Number and Half-life; Commercially Available Isotopes

Abbreviations:

d = days
h = hours
y = years

Element and mass No.		*Half-life*
Hydrogen	3	12.3 y
Beryllium	7	53 d
	10	2.7×10^6 y
Carbon	14	5.73×10^3 y
Fluorine	18	1.8 h
Sodium	22	2.58 y
	24	15.0 h
Magnesium	28	21.3 h
Aluminum	26	7.4×10^5 y
Silicon	31	2.62 h
Phosphorus	32	14.3 d
	33	25 d
Sulfur	35	86.7 d
Chlorine	36	3.0×10^5 y
	38	37.3 m
Argon	37	35.1 d
Potassium	42	12.4 h
	43	22 h
Calcium	45	165 d
	47	4.7 d

Element and mass No.		*Half-life*
Scandium	43	3.9 h
	44m	2.4 d
	44	4.0 h
	46	84 d
	47	3.4 d
Titanium	44	48 y
Vanadium	48	16.1 d
	49	330 d
Chromium	51	27.8 d
Manganese	52	5.7 d
	53	2.0×10^6 y
	54	303 d
	56	2.6 h
Iron	52	8.3 h
	55	2.7 y
	59	45 d
Cobalt	56	77.3 d
	57	267 d
	58	71 d
	60	5.26 y
Nickel	63	92 y
	65	2.56 h

Element and mass No.		*Half-life*
Copper	61	3.3 h
	64	12.9 h
	67	61 h
Zinc	65	245 d
	69m	14 h
Gallium	66	9.5 h
	67	78 h
	72	14.1 h
Germanium	68	282 d
	71	11 d
	77	11 h
Arsenic	74	18 d
	76	26.5 h
	77	39 h
Selenium	75	120 d
Bromine	77	58 h
	82	35.3 h
Krypton	79	34.9 h
	83m	1.86 h
	85m	4.4 h
	85	10.76 y

Table 3.5–11 (continued)
RADIOACTIVE ISOTOPES

Mass Number and Half-life; Commercially Available Isotopes (continued)

Element and mass No.		*Half-life*
Rubidium	83	83 d
	84	33 d
	86	18.7 d
Strontium	85	64 d
	87m	2.8 h
	89	50.4 d
	90	28 y
Yttrium	87	80 h
	88	108 d
	90	64.2 h
	91	59 d
Zirconium	95	65 d
	97	17 h
Niobium	95	35 d
Molybdenum	99	66 h
Technetium	99m	6.0 h
	99	2.1×10^5 y
Ruthenium	97	2.9 d
	103	40 d
	106	1.0 y
Rhodium	102m	2.9 y
	105	36 h
Palladium	103	17 d
	109	13.5 h
Silver	105	40 d
	110m	260 d
	111	7.5 d
Cadmium	109	1.3 y
	115m	43 d
	115	2.3 d
Indium	111	2.8 d
	114m	50 d
Tin	113	118 d
	119m	250 d
	121	27 h
Antimony	122	2.8 d
	124	60 d
	125	2.7 y
Tellurium	125m	58 d
	127m	105 d
	127	9.3 h
	129m	34 d
	129	1.1 h
	132	3.2 d
Iodine	123	13 h
	124	4.2 d
	125	60.2 d
	126	13.2 d
	129	1.6×10^7 y
	130	12.5 h
	131	8.05 d
	132	2.3 h
	133	21 h
Xenon	131m	12 d
	133	5.3 d
Cesium	131	9.7 d
	132	6.6 d
	134	2.1 y
	137	30 y
Barium	131	11.6 d
	133	7.2 y
	140	12.8 d
Lanthanum	140	40 h
Cerium	139	140 d
	141	32.5 d
	143	1.4 d
	144	285 d
Praseodymium	142	19.2 h
	143	13.7 d
Neodymium	147	11.1 d
	149	1.8 h
Promethium	147	2.7 y
	149	2.2 d
	151	1.2 d
Samarium	153	2 d
Europium	152m	9.3 h
	152	12.4 y
	154	16 y
	155	1.8 y
Gadolinium	153	240 d
	159	18 h
Terbium	160	72 d
	161	6.9 d
Dysprosium	165	2.3 h
Holmium	166m	1.2×10^3 y
Erbium	169	9.4 d
	171	7.5 h
Thulium	170	130 d
Ytterbium	169	32 d
	175	4.2 d
Lutetium	177	6.8 d
Hafnium	175	70 d
	181	45 d
Tantalum	182	115 d
Tungsten	181	130 d
	185	74 d
	187	24 h
Rhenium	183	70 d
	186	3.8 d
	188	17 h
Osmium	185	94 d
	191m	14 h
	191	15 d
	193	32 h
Iridium	192	74 d
	194	19 h
Platinum	193m	4.4 d
	197	20 h
Gold	195	183 d
	198	2.7 d
	199	3.15 d
Mercury	197m	24 h
	197	65 h
	203	47 d
Thallium	202	12 d
	204	3.8 y
Lead	210	22 y
Bismuth	206	6.24 d
	207	30 y
	210m	2.6×10^6 y
Polonium	208	2.9 y
	210	138 d
Radon	222	3.8 d
Radium	224	3.64 d
	226	1.62×10^3 y
	228	5.7 y
Actinium	227	21.8 y
Thorium	228	1.91 y
	230	7.6×10^4 y
Protactinium	231	3.2×10^4 y
	233	27.4 d
	234	6.7 h

TABLE 3.5–11 (continued)
RADIOACTIVE ISOTOPES

Mass Number and Half-life; Commercially Available Isotopes (continued)

Element and mass No.		*Half-life*	*Element and mass No.*		*Half-life*	*Element and mass No.*		*Half-life*
Uranium	232	72 y	**Plutonium**	237	45.6 d	**Americium**	241	458 y
	233	1.6×10^5 y		239	2.4×10^4 y	**Curium**	242	163 d
				240	6.7×10^3 y		244	18 y
Neptunium	237	2.1×10^6 y						

REFERENCES

Wang, Y., Ed., *CRC Handbook of Radioactive Nuclides,* Chemical Rubber Co., Cleveland, 1969.

Weast, R. C., Ed., *CRC Handbook of Chemistry and Physics,* 56th ed., CRC Press, Cleveland, 1975.

Data from Baker, P. S., in *CRC Handbook of Radioactive Nuclides,* Wang, Y., Ed., Chemical Rubber Co., Cleveland, 1969.

Table 3.5–12
RADIOISOTOPES FOR INDUSTRIAL USE

Typical Applications and Isotopes in Common Use

Isotopes	*Applications*	*Isotopes*	*Applications*
CHEMICAL INDUSTRY		ELECTRICAL INDUSTRY	
S^{35}, H^{3}	Efficiency of separation	Kr^{85}	Leak testing
Au^{198}, I^{131}, Na^{24}, Mn^{56}, Br^{82}, Cr^{51}	Thoroughness of mixing	Hg^{197}	Mercury-switch studies
I^{131}, Br^{82}, Na^{24}, H^{3}	Leak location	H^{3}, Kr^{85}, Pm^{147}	Luminous dials
Co^{60}, Cs^{137}	Gaging of liquid or solid levels	Co^{60}, Ni^{63}	Pre-ionization of gases in electronic tubes
Rb^{86}	Study of process-stream flow patterns	Sr^{90}	Power for navigational lights
Au^{198}	Location of pipe obstructions	METALS INDUSTRY	
Sb^{124}	Study of mass balances in refinery stream	Fe^{59}	Tracing blast furnace operations
C^{14}, H^{3}	Study of reaction mechanisms	H^{3}	Study of hydrogen embrittlement
Xe^{133}, Kr^{85}	Measurement of gas-flow velocities	Fe^{59}, Cu^{64}, Zn^{65}, Cr^{51}, Ni^{63}	Study of piston-ring and bearing wear
$ZrNb^{95}$, Co^{60}	Catalyst flow studies	Co^{60}	Control of discharge in coke ovens
C^{14}	Study of carbon deposits in fuel research	Sr^{90}, Kr^{85}, Tm^{170}, Eu^{155}, Ce^{144}, Cs^{137}	Thickness gaging
P^{32}, Co^{58}, Co^{60}, C^{14}	Drug-metabolism studies	Co^{60}	Measuring wear of firebrick linings
S^{35}	Study of vulcanizing process and tire wear	Co^{60}, Ir^{192}, Cs^{137}, Sm^{145}, Gd^{153}, Eu^{155}, Ce^{144}, Tm^{170}, Ta^{182}, Yb^{169}	Detecting thickness variation and defects in castings; weld inspection
C^{14}	Study of frictional forces in rubber	TRANSPORTATION INDUSTRY	
Zn^{65}	Evaluation of plastic blood bags	Fe^{55}, Zn^{65}	Measurement of wear in pistons and bearings
Ca^{45}, Na^{24}, Cl^{38}	Study of diffusion in glass	Xe^{133}	Studies of sediment and sand movement
Kr^{85}, Xe^{133}	Determination of air pollution from refinery	Au^{198}	Evaluation of rail life
Sr^{90}, Kr^{85}	Control of rubber thickness on tire ply	Co^{60}, Ir^{192}, Cs^{137}	Gaging of automobile sheet steel
Cs^{137}	Control of rock wool production	H^{3}, Kr^{85}, Pm^{147}	Luminous locks and dials
Co^{60}, Cs^{137}	Initiation of chemical reactions; effecting of polymerization		
Sr^{90}, Kr^{85}	Elimination of static		
Co^{60}, Cs^{137}	Sterilization of medical supplies		
Sr^{90}, Kr^{85}	Thickness gaging of paper and plastics		

From Wang, Y., Ed., *CRC Handbook of Radioactive Nuclides,* Chemical Rubber Co., Cleveland, 1969.

Table 3.5–13
STABILITY OF RADIOACTIVE COMPOUNDS

Most users of compounds labeled with radioisotopes recognize that such compounds decompose on storage and that the decomposition is accelerated by self irradiation. The degree of the decomposition in relation to the storage conditions of the compound and the measures that can be taken to control and minimize the rate of self radiolysis are not always so well known. Information on this subject is largely empirical. Not only are a large number of labeled compounds – particularly organic compounds – extensively used as tracers, but many applications demand a very high purity. Fractions of a percent of radiochemical impurity can sometimes lead to incorrect deductions from a tracer investigation, and under these conditions the problem of decomposition by self irradiation becomes a very serious one.

Compounds labeled with the pure beta-emitting radioisotopes (C^{14}, tritium, S^{35}, P^{32}, and Cl^{36}) are most commonly used in tracer investigations. Compounds labeled with the gamma-emitting radioisotopes such as I^{125}, I^{131}, Co^{57}, Co^{58}, and Se^{75} have special application in medicine. Some properties of these radionuclides are shown in Table A.

Table A. Physical Properties of Some Radionuclides

Radionuclide	*Half-life*	*Beta energy, MeV*		*Specific activity, mCi/mA*		*Daughter nuclide (stable)*
		Max.	*Mean*	*Maximum*	*Common values for compounds*	
H^3	12.26 years	0.018	0.0057	2.9×10^4	10^2–10^4	He^3
C^{14}	5700 years	0.159	0.050	64	1–10^2	N^{14}
S^{35}	87.2 days	0.167	0.049	1.5×10^6	1–10^2	Cl^{35}
Cl^{36}	3.03×10^5 years	0.714	0.3	1.2	10^{-3}–10^{-1}	Ar^{36}
P^{32}	14.3 days	1.71	0.69	9.3×10^6	10–10^2	S^{32}
I^{131}	8.04 days	0.81	0.19	1.7×10^7	10^2–10^4	Xe^{131}
I^{125}	60 days	Electron capture		2.2×10^6	10^2–10^4	Te^{125}
Co^{57}	270 days	Electron capture		4.9×10^5	10^3–10^5	Fe^{57}
Co^{58}	71 days	Electron capture $+\beta^+$		1.9×10^6	10^3–10^5	Fe^{58}
Se^{75}	121 days	Electron capture		1.1×10^6	10–10^3	As^{75}

Decomposition depends in part on the amount of energy absorbed by the compound during its useful life, so that, for a given amount of activity, the radiation energy emitted should be a guide to the seriousness of the problem. The problem of decomposition by self irradiation might be expected to increase in magnitude as the series of pure beta emitters in Table A is descended, but, in fact, almost the reverse is true. This occurs largely for three reasons:

1. The fraction of energy absorbed is much less than unity for the more energetic beta emitters such as P^{32}; on the other hand, almost complete total absorption of the beta energy occurs with tritium compounds. Gamma energy is, in general, little absorbed by the compound itself or by its immediate environs.

2. The decomposition also depends on the specific activity of the compound; as can be seen from Table A, the specific activities of tritiated compounds in current use are usually much higher than those for compounds labeled with other pure beta-emitting radionuclides.

3. The absorbed energy decreases exponentially with time; this is an important factor for compounds labeled with radionuclides having a short half-life such as I^{131} or P^{32}.

The reason why labeled compounds decompose

Table 3.5–13 (continued)
STABILITY OF RADIOACTIVE COMPOUNDS

is not difficult to understand: the radiation energy will be commonly absorbed by the compound itself or by its environs. If the former occurs, then the excited molecules may break up in some manner; if the latter occurs, the radiation energy can produce free radicals and other reactive species, which may then cause destruction of the molecules of the labeled compound.

The modes by which decomposition of labeled compounds can arise have been classified as shown in Table B.

Table B. Modes of Decomposition of Labeled Compounds

Mode of decomposition	*Cause*	*Method for control*
Primary (internal)	Natural isotopic decay	None, for a given specific activity [a]
Primary (external)	Direct interaction of the radioactive emission (alpha, beta, or gamma) with molecules of the compound	Dispersal of the labeled molecules
Secondary	Interactions of excited products with molecules of the compound	Dispersal of active molecules; cooling to low temperatures; scavenging of free radicals
Chemical	Thermodynamic instability of compounds; poor choice of environment	Cooling to low temperatures; removal of harmful agents

[a]Note that dilution with the inactive form of the compound subseqnent to preparation is not beneficial in this case.

Primary decomposition is the production of an impurity due to the disintegration of the unstable nucleus. Secondary decomposition is commonly the most damaging and the most difficult to control. Chemical decomposition (by oxidation, hydrolysis, etc.) is even more likely to occur with radioactive chemicals. It may also be necessary to guard against photochemical or microbiological decomposition of the compound.

Tables reporting the various kinds of decomposition for a great variety of labeled compounds are available. Over 30 pages of such tables are given in the CRC Handbook of Radioactive Nuclides; this source also cites a large number of references.

While it is true that no measurable decomposition has occurred in many labeled compounds that have been stored for months or even years, there are other compounds showing major decomposition in a matter of days. It is thus very important to take proper account of the condition and the stability of any compounds being used.

Adapted from Wang, Y., Ed., *CRC Handbook of Radioactive Nuclides,* Chemical Rubber Co., Cleveland, 1969.

Table 3.5–14
RADIOISOTOPE FUEL DATA

This table represents radioisotopic fuel data only and does not include the capsule or containment materials that would reduce the overall heat source, specific power, and the power density, depending on the application and the heat source design.

For data on the thermophysical, mechanical, chemical, biological, shielding, and electrical properties of these fuels, consult the listed references. Material compatibility data is also cited in these references.

Radioisotope	*Weight of active isotope in compound*	*Half-life*	*Melting point, °C*	*Specific power, watts/g*[a]	*Power density, watts/cm³*
Co^{60}					
Metal	Pure Co^{60} (theoretical—not available)		1,495	17.7	156
Co-Ni[b]	17.5% Co^{60}	5.26 yr	1,487	3.1	27
Co-Ni[c]	35% Co^{60}		1,481	6.2	55
Sr^{90}					
Metal	55.0% Sr^{90}	28.0 yr	772	0.5	1.28
$SrTiO_3$	24.5% Sr^{90}		1,910	0.23	1.17
SrO	44.0% Sr^{90}		2,457	0.42	1.94
SrF_2	36.0% Sr^{90}		1,463	0.34	1.44
Sr_2TiO_4	31.5% Sr^{90}		1,860	0.30	1.48
Cs^{137}					
CsCl	28.9% Cs^{137}	30.0 yr	645	0.12	0.37
Cs_2SO_4	26.9% Cs^{137}		1,019	0.11	0.46
Ce^{144}					
Ce_2O_3	10.8% Ce^{144}		2,190	2.76	19
Ce_2O_2S	10.3% Ce^{144}	285 days	2,000	2.64	15.8
Ce_2S_3	9.4% Ce^{144}		1,890	2.41	12.5
CeF_3	9.0% Ce^{144}		1,437	2.31	14.3
Pm^{147}					
Metal	95% Pm^{147}	2.62 yr	865	0.31	2.30
Pm_2O_3	78% Pm^{147}		2,130	0.27	1.87
Tm^{170}					
Metal	16.3% Tm^{170}		1,545	2.24	20.5
	2.9% Tm^{171}	128 days			
Tm_2O_3	14.3% Tm^{170}		2,375	2.24	19.7
	2.5% Tm^{171}	(Tm^{171}— 700 days)			
Po^{210}					
Metal	95% Po^{210}	138 days	254	144	1,324
GdPo[d]			1,675 (GdPo)	1.6	16.6
GdPo[e]			1,675 (GdPo)	7.5	77.2
Pu^{238}					
Metal[f]	80% Pu^{238}	87 yr	575–615	0.45	6.8
PuO_2[g]	71% Pu^{238}		2,150	0.40	2.7
PuC	74.6% Pu^{238}		1,654	0.42	5.7
PuN	74.6% Pu^{238}		2,570	0.42	6.2
PuZr[h]	73% Pu^{238}		730	0.41	5.6
PuZr[i]	77% Pu^{238}		615	0.44	6.5
Cm^{242}					
Cm_2O_3–AmO_2	35.7% Cm^{242}	163 days	2,000	42.8	500
Cm^{244}					

Table 3.5–14 (continued)
RADIOISOTOPE FUEL DATA

Radioisotope	*Weight of active isotope in compound*	*Half-life*	*Melting point, °C*	*Specific power, watts/g*[a]	*Power density, watts/cm³*
Metal	95.5% Cm^{244}	18.1 yr	1,340	2.67	36
Cm_2O_3	86.9% Cm^{244}		1,950	2.42	26.1
Cm_2O_2S	84.4% Cm^{244}		2,000	2.35	23.3
CmF_3	77.5% Cm^{244}		1,406	2.15	21.1

[a]For horsepower per pound of fuel, multiply these values of watts per gram by 0.61.
[b]200 Ci/g.
[c]400 Ci/g.
[d]98% Ta matrix.
[e]91% Ta matrix.
[f]Production grade.
[g]Microspheres; packing fraction 65%.
[h]20% Zr alloy.
[i]10% Zr alloy.

REFERENCES

Fulham, H. T. and Van Tuyl, H. H., Promethium Isotopic Power Data Sheets, BNWL-363, Battelle/Northwest, Richland, Wash., February 1967.[j]

Mound Laboratory, Plutonium-238 and Polonium-210 Data Sheets, MLM-1441, Monsanto Research Corp., Miamisburg, Oh., September 29, 1967.[j]

Pacific Northwest Laboratory, Quarterly Report, BNWL-680, Division of Isotope Development Programs, Battelle/Northwest, Richland, Wash., October-December 1967.[j]

Rimshaw, S. J. and Ketchen, E. E., Cerium-144 Data Sheets, ORNL-4185, Oak Ridge National Laboratory, Oak Ridge, Tenn., November 1967.[j]

Rimshaw, S. J. and Ketchen, E. E., Cesium-137 Data Sheets, ORNL-4186, Oak Ridge National Laboratory, Oak Ridge, Tenn., December 1967.[j]

Rimshaw, S. J. and Ketchen, E. E., Curium Data Sheets, ORNL-4186, Oak Ridge National Laboratory, Oak Ridge, Tenn., December 1967.[j]

Rinshaw, S. J. and Ketchen, E. E., Strontium-90 Data Sheets, ORNL-4188, Oak Ridge National Laboratory, Oak Ridge, Tenn., December 1967.[j]

Smith, P. K., Keski, J. R., and Angerman, C. L., Properties of Thulium Metal and Oxide, DP-1114, Savannah River Laboratory, Aiken, S. C., June 1967.[j]

Winbley, W. C., Jr., Properties of Co^{60} and Cobalt Metal Fuel Forms, DP-1051 (Rev. 1), Savannah River Laboratory, E. I. du Pont de Nemours & Co., for U.S. Atomic Energy Commission, October 1966.

[j]Available from Clearinghouse for Federal Scientific and Technical Information, National Bureau of Standards, Springfield, VA 22151.

From Carpenter, R. T., in *Handbook of Tables for Applied Engineering Science,* 2nd ed., Bolz, R. E. and Tuve, G. L., Eds., CRC Press, Cleveland, 1973, 430.

3.6 REACTOR MATERIALS

Table 3.6–1
REACTOR MATERIALS

Each engineering material used in the construction of a nuclear reactor may serve one or more functions. In most cases the physical, mechanical, and thermal properties of the material, as well as the nuclear properties, are of some importance. In the following list each function is briefly defined, and some of the important properties are noted.

Fuel – The fuel elements are uranium, plutonium, and thorium. The fissionable isotopes are U^{233}, U^{235}, and Pu^{239}. Although uranium has 14 isotopes, all radioactive, natural uranium contains 99.28% of U^{238} and only 0.7% of U^{235}. The most common fuel is natural uranium, "enriched" with added U^{235}. Isotopes that can be converted into fissile fuels are called "fertile" materials (see Table 3.5–2). These include U^{238}, which forms U^{239} by capture of a neutron and undergoes transition to Pu^{239} by beta decompositions, and thorium 232, which is converted to Th^{233} by neutron fluxes and forms U^{233} by beta decay.

Reactor-fuel elements should be strong and corrosion resistant and have high thermal conductivity. A minimum of fission products should enter the coolant. Alloying, cladding, or plating are required to attain these properties. Stainless steel, aluminum, titanium, molybdenum, beryllium, and zirconium are used, in addition to graphite and some ceramic materials.

General physical and mechanical properties of some of these materials are given in Table A.

Table 3.6–1 (continued)
REACTOR MATERIALS

Table A. Reactor Fuels and Associated Materials

Typical Properties

For specific heat in J/kg·K, multiply tabular values by 4187. For thermal conductivity in Btu/hr·ft·°F, multiply by 57.8. For N/m^2 multiply psi by 6895.

Material[a]	Specific gravity	Specific heat, 25°C	Thermal conductivity, $\frac{watts}{cm°C}$	Melting point, °C	Coefficient of linear expansion/°C ($\times 10^6$)	Tensile strength, psi	Modulus of elasticity, psi (millions)	Absorption cross section, σ_a, barns
Uranium	19.	0.03	0.25	1,132	46.	60,000.	23.	7.6
UO_2	10.96		0.08	2,600	10.	–	–	7.6
Plutonium	19.8	0.03	0.08	640.	54.	–	14.	–
Thorium	11.7	0.03	0.4	1,750.	11.5	32,000	9.	7.5
Beryllium	1.85	0.44	2.18	1,285	12.	30,000	42.	0.01
Zirconium	6.5	0.07	0.21	1,852	5.8	30,000	13.7	0.18
Cadmium	8.65	0.05	0.93	321	32.	10,300	8.	2,450.
Aluminum	2.7	0.21	2.37	660.	26.	13,000	10.	0.24
Stainless steel (347)	8.0	0.12	0.2	1,400.	16.	75,000	29.	3.0
Titanium	4.5	0.125	0.2	1,670.	8.5	9,000	16.	5.8
Graphite	2.2	0.17	0.24	>3,500	3.	100	0.7	0.004

[a] Some metals become radioactive and dangerous to handle after exposure to the neutron flux of a reactor, but these would not be used in the structure of a thermal reactor. Tungsten and cobalt are the worst in this respect, but tantalum, chromium, and manganese may also attain a high induced radioactivity, and copper a considerably lower one.

Table 3.6–1 (continued)
REACTOR MATERIALS

Moderator – For thermal reactors it is necessary to reduce the kinetic energy of the "fast" neutrons by successive scattering collisions. Materials effective for this "slowing down" of neutrons to the thermal range include graphite, water, heavy water, beryllium and beryllium oxide, and hydrocarbons (see Tables 3.6–5 and 3.6–7; also Tables 3.6–18 through 3.6–20. A moderator should be resistant to corrosion and stable at high temperatures and in high radiation flux. A moderator may serve as a fuel diluent or fuel container; in the latter case mechanical strength is required.

Reflector – Moderating materials that also scatter neutrons back into the core to reduce "leakage" are termed reflectors. A "bare reactor" is one without such reflectors. Savings in fuel and smaller critical size are attained with reflectors.

Control material – Control rods or other arrangements for varying the reactor output are of three classes: (1) *shim rods* for coarse control during startup, (2) *regulating rods* for small but quick control, and (3) *safety rods* or "scram" control for emergencies. The earlier methods used materials with large capture cross sections to absorb neutrons (cadmium or boron), but it is more economical to employ uranium, cobalt, or other material that produces a secondary product such as a desired isotope. Other control methods involve the positioning of the fuel, moderator, or reflector elements. Typical properties of the more effective isotopes of the high absorption materials are given in Table B.

Table B. Properties of Neutron Absorbers (Poisons)

Isotope	Thermal microscopic cross sections, σ, barns	Major resonance	
		Energy, eV	σ_a, barns
Cadmium 113	20,000	0.18	7,200
Boron 10	3,850	None	
Samarium 149	41,000	0.096	16,000
Samarium 152	220	8.2	15,000
Gadolinium 155	56,000	2.6	1,400
Gadolinium 157	240,000	17.	1,000
Europium 151	7,800	0.46	11,000
Europium 153	450	2.46	3,000
Silver 107	31	16.6	630
Silver 109	87	5.1	12,500

Coolant – In some reactors it may be necessary to dissipate almost the entire heat output to a heat sink; in reactors for power generation, the primary coolant becomes the heat source for the power cycle, and two or more coolant circuits are involved. Practical coolants are water, liquid metals, and various gases. Heavy water is superior but costly. Organic liquids tend to leave surface deposits in the fuel section. High heat capacity and high thermal conductivity are desired, with low vapor pressure and stability under high temperature and radiation. For specific properities of materials in Table C, see Table 3.6–8.

Table 3.6–1 (continued)
REACTOR MATERIALS

Table C. Reactor Coolants

Liquids and liquid metals		Gases and vapors	
Diphenyl	NaK	Air	Mercury
Dowtherm®[a]	Potassium	Carbon dioxide	Neon
Gallium	Rubidium	Helium	Nitrogen
Heavy water	Sodium	Hydrogen	Steam
Lithium	Tin		
Mercury	Water		

[a]Dowtherm Heat Transfer Agent, Dow Chemical Co., Midland, Mich.

Shielding material – Design of shields for personnel protection depends on objectives, space and weight requirements, and types of radiation. A comparison of the common shielding materials is presented in Table 3.6–15 and Table 3.6–11 gives further data.

From Bolz, R. E. and Tuve, G. L., Eds., *Handbook of Tables for Applied Engineering Science,* 2nd ed., CRC Press, Cleveland, 1973, 432.

Table 3.6–2
ENERGY LIBERATED BY FISSION

The following values are approximations for the energy quantities involved in fission of any one of the three common nuclear fuels – uranium 233 or 235 or plutonium 239 – since there is a relatively small difference between these weights.

One pound of fissile material will liberate approximately 10 million kwh. The fission of 1 g/day is roughly equal to 1 mw. The "burnup" of 1,000 kg (2,200 lb) of uranium would deliver 100 mw for about 25 years.

While almost all of the fission energy ultimately appears as heat within the matter surrounding the reaction, only about 80% of the fission energy is immediately converted to heat from the kinetic energy of the fission fragments. About 95% of the energy eventually appears as heat, but the 5% represented by the neutrinos essentially escapes from the reactor.

Percentage Distribution of Fission Energy

Kinetic energy of fission fragments	82.5%
Instantaneous gamma-ray energy	3.5%
Gamma-rays from fission products	3.0%
Beta particles from fission products	3.5%
Kinetic energy of fission neutrons	2.5%
Neutrinos	5.0%

From Bolz, R. E. and Tuve, G. L., Eds., *Handbook of Tables for Applied Engineering Science,* 2nd ed., CRC Press, Cleveland, 1973, 433.

Table 3.6–3
DECAY OF REACTOR FISSION PRODUCTS

Days from Shutdown, After Infinite Operation; Various Levels of Effective Energy (EE)

	Approximate decay, MeV/W–s						
Time, days	**Group 1, EE = 0.4 MeV**	**Group 2, EE = 0.8 MeV**	**Group 3, EE = 1.3 MeV**	**Group 4, EE = 1.7 MeV**	**Group 5, EE = 2.2 MeV**	**Group 6, EE = 2.5 MeV**	**Group 7, EE = 2.8 MeV**
0.1	1.5×10^9	1×10^{10}	1.1×10^9	4.3×10^9	1.2×10^9	3.4×10^8	1.7×10^9
0.2	1.5×10^9	9.5×10^9	6.8×10^8	3.5×10^9	7.4×10^8	3×10^8	9×10^8
0.4	1.45×10^9	8.8×10^9	3.7×10^8	3.1×10^9	3×10^8	2.9×10^8	2.8×10^8
0.7	1.35×10^9	8.0×10^9	2×10^8	3×10^9	1.15×10^8	2.8×10^8	4.7×10^7
1.0	1.3×10^9	7.5×10^9	1.2×10^8	3×10^9	8×10^7	2.7×10^8	
2.0	1.1×10^9	6.4×10^9	2.3×10^7	3×10^9	5.6×10^7	2.6×10^8	
4.0	8×10^8	5.4×10^9		2.9×10^9	4×10^7	2.4×10^8	
7.0	6×10^8	4.7×10^9		2.6×10^9	3×10^7	2.1×10^8	
10.	4.8×10^8	4×10^9		2.2×10^9	2.3×10^7	1.7×10^8	
20.	3×10^8	3.2×10^9		1.3×10^9	1.7×10^7	1.0×10^8	
40.	1.3×10^8	2.4×10^9		4×10^8	1.55×10^7	4×10^7	
70.	5×10^7	1.8×10^9		1×10^8	1.4×10^7	1.3×10^7	
100	2.4×10^7	1.5×10^9		3×10^7	1.3×10^7		

Note: These data for infinite-power operation may be used for estimating by calculation the sources resulting from finite-time operation before shutdown.

Data from Rockwell, T., III, Ed., *Reactor Shielding Design Manual,* USAEC Report No. TID-7004, U.S. Atomic Energy Commission, March 1956.

Table 3.6–4
HEAT PRODUCTION AFTER REACTOR SHUTDOWN

Uranium Fuel Afterheat

Time after shutdown		Duration of full-power operation				
		1 hr	1 day	1 week	1 month	1 year
Seconds	Other units	Percentage of full power after shutdown				
3×10^2	5 min	1.0				
6×10^2	10 min	0.8	1.4			
1.8×10^3	30 min	0.39	1.1	1.3		
3.6×10^3	1 hr	0.24	0.85	1.1		
7.2×10^3	2 hr	0.12	0.61	0.85	1.00	
2.88×10^4	8 hr	0.03	0.28	0.50	0.63	
8.64×10^4	1 day		0.09	0.27	0.41	0.58
2.6×10^5	3 days		0.03	0.13	0.28	0.42
6.1×10^5	1 week		0.010	0.08	0.18	0.31
1.21×10^6	2 weeks			0.03	0.10	0.22
2.6×10^6	30 days			0.015	0.06	0.16
7.8×10^6	90 days				0.02	0.08
3.1×10^7	1 year					0.015

Based on *Heat Generation in Irradiated Uranium,* ANL-4790, Argonne National Laboratory, U.S. Atomic Energy Commission.

Table 3.6–5
MODERATOR MATERIALS

Nuclear Characteristics of Potential Moderators

For density in lb/ft^3, multiply the value in g/cm^3 by 62.42.

Element or compound	*A, atomic or molecular weight*	*Density*[a] *g/cm^3*	*N,*[b] *$\times 10^{-24}$*	*Scattering cross section σ_s epithermal, barns/atom*	*Absorption cross section σ_a, 0.025 eV, barns/atom or molecule*
Hydrogen, H	1.008	0.0090	0.0054	20.3	0.33
Deuterium, D	2.02	0.0180	0.0054	3.3	0.00046
Helium, He	4.00	0.0180	0.0027	0.8	—
Beryllium, Be	9.01	1.85	0.124	6.1	0.009
Carbon (graphite), C	12.0	1.60	0.080	4.7	0.0045
Oxygen, O	16.0	0.014	0.0054	3.8	0.0002
Sodium, Na	23.0	0.97	0.0254	3.0	0.49
Magnesium, Mg	24.3	1.74	0.043	3.4	0.059
Aluminum, Al	27.0	2.70	0.060	1.35	0.215
Beryllia, BeO	25.0	3.025	0.073	9.9	0.009
Beryllium carbide, Be_2C	30.0	2.4	0.048	16.9	0.023
Beryllium fluoride, BeF_2	47.0	1.986	0.025	15.4	0.029
Light water, H_2O	18.0	1.00	0.033	44.4	0.66
Heavy water, D_2O	20.0	1.10	0.033	10.5	0.0011
Sodium hydroxide, NaOH	40.0	2.1	0.032	27.1	0.82
Zirconium hydride, ZrH_2	93.2	5.61	0.036[d]	48.6	0.84
Biphenyl, $C_{12}H_{10}$	154.2	0.87	0.0034	259.4	2.54
Polystyrene, $(CH)_n$	13.0	1.07	0.038[e]	25.0[g]	0.335[g]
Paraffin, $(CH_2)_n$	14.0	0.9	0.030[f]	45.3[h]	0.665[h]

Element or compound	*Macroscopic absorption cross section, $\Sigma_a = N\sigma_a$ cm^{-1}*	*Macroscopic scattering cross section, $\Sigma_s = N\sigma_a$ cm^{-1}*	*Logarithmic mean energy loss/ collision,*[c] *ξ*	*Slowing-down power, $\xi\Sigma_s$*	*Moderating ratio, $\sigma_s\xi/\sigma_a$*
Hydrogen, H	0.0018	0.11	1.000	0.11	61
Deuterium, D	2.5×10^{-6}	0.018	0.725	0.013	5,200
Helium, He	—	0.0022	0.425	0.0009	∞
Beryllium, Be	0.0011	0.76	0.206	0.16	145
Carbon (graphite), C	0.00036	0.38	0.158	0.060	165
Oxygen, O	1.1×10^{-6}	0.021	0.12	0.0025	230
Sodium, Na	0.012	0.076	0.083	0.0063	0.53
Magnesium, Mg	0.0025	0.15	0.073	0.011	4.4
Aluminum, Al	0.013	0.081	0.071	0.0058	0.45
Beryllia, BeO	0.00066	0.72	0.173	0.12	190
Beryllium carbide, Be_2C	0.0011	0.81	0.193	0.16	145
Beryllium fluoride, BeF_2	0.00074	0.39	0.151	0.058	84
Light water, H_2O	0.022	1.47	0.925	1.36	62
Heavy water, D_2O	36×10^{-6}	0.35	0.504	0.18	5,000
Sodium hydroxide, NaOH	0.026	0.87	0.77	0.67	26
Zirconium hydride, ZrH_2	0.030	1.75	0.84	1.47	49
Biphenyl, $C_{12}H_{10}$	0.00862	0.880	0.812	0.715	83
Polystyrene, $(CH)_n$	0.013	0.95	0.842	0.80	62
Paraffin, $(CH_2)_n$	0.020	1.36	0.913	1.24	62

[a]For the gases H_2, D_2, He, O_2 an arbitrary density 100 times the density of NTP is assumed; i.e., a pressure on the order of 1,500 psi.

Table 3.6–5 (continued)
MODERATOR MATERIALS

Nuclear Characteristics of Potential Moderators (continued)

[b]Number of atoms or mole per cm^3
[c]$\xi = 1 - [(A-1)^2/2A] \ln [(A+1)/(A-1)]$.
[d]Below 800°C.
[e]Number of (CH) units/cm^3.
[f]Number of (CH_2) units/cm^3.
[g]Per (CH) unit.
[h]Per (CH_2) unit.

From Tipton, C. R., Jr., Ed., *Reactor Handbook,* 2nd ed., Vol. 1: Materials, Wiley Interscience, New York, 1960. With permission.

Table 3.6–6
GRAPHITE FOR REACTORS

Commercial graphite for reactors is a mixture of crystalline graphite and cross-linking intercrystalline carbon. The physical properties that are measured are the result of contributions from both sources. Graphite for nuclear reactors is a material for neutron moderators, reflectors, thermal columns, and exponential piles. Its desirable properties for these uses include low neutron-capture cross section, high-temperature stability, and machinability.

Table A lists common properties of two typical nuclear graphite materials that differ primarily in purity. Many special grades are also available to meet such requirements as high density, low porosity, and high permeability.

Graphite is available as a matrix material containing dispersions of uranium and/or thorium for high-temperature fuel elements or boron for control-rod and neutron-shielding purposes.

The comparative position of graphite as a neutron moderating and reflecting element is indicated in Table B. *Moderator figure-of-merit* is the ratio of the life of a fast neutron before absorption to the time required to slow it down to an energy of 1 eV. *Reflector figure-of-merit* is the ratio of the probability of thermal neutron scattering back into a reactor core to the probability of absorption.

Reactor graphite is made from petroleum coke, which is calcined, crushed, and screened, then mixed with a coal-tar pitch binder and fired at high temperature. Size of grains depends on the source of the raw material as well as on the processing. Physical properties and crystal orientation will be somewhat different for an extruded product than for a molded product. Impurities and trace elements will vary with the raw materials; the principal impurities are iron, silicon, calcium, and aluminum, each of these typically less than 500 ppm; other metals are typically less than 70 ppm.

Additional factors that affect the properties of the final product are size of the finished piece, the particle-size distribution in the original mix, the maximum processing temperature, and the temperature at which the properties are being measured. The tensile strength of graphite is about one half its flexural strength, while the crushing strength is about twice the flexural strength. Strength increases with temperature, in the usual range, as do the specific heat and the coefficient of thermal expansion. Thermal conductivity decreases with increase in temperature.

Table 3.6–6 (continued)
GRAPHITE FOR REACTORS

Table A. Typical Properties of Nuclear Graphite

Property	Grade A	Grade B
Slow neutron absorption, cross section per carbon atom × 10^{-27}, mb[a]	3.95	4.5
Total ash content, percent	0.01	0.06
Boron content, ppm	0.2	0.3
Specific resistance, microhm-meter		
Longitudinal	5.40	6.50
Transverse	11.75	11.80
Thermal conductivity, Btu·ft/hr·ft²·°F		
Longitudinal	141	116
Transverse	62	69
Coefficient of thermal expansion, 10^{-6}/°F		
Longitudinal	0.2	0.6
Transverse	1.9	1.7
Bulk density, g/cm³	1.73	1.71
Tensile strength, psi, longitudinal	1,400	1,300
Flexural strength, psi, longitudinal	3,400	2,700
Compressive strength, psi, longitudinal	5,000	5,000
Elastic modulus, 10^6 psi		
Longitudinal	2.3	1.8
Transverse	0.8	1.0

[a]mb = millibarns.

Courtesy of Carbon Products Division, Union Carbide Corporation, New York.

Table B. Comparative Moderating and Reflector Characteristics

	H_2O	D_2O	Be	BeO	Graphite
Moderator figure-of-merit	209	17,500	456	604	695
Reflector figure-of-merit	172	15,300	788	774	1,430

Adapted from *The Industrial Graphite Engineering Handbook,* Carbon Products Division, Union Carbide Corporation, New York, 6.03.

Table 3.6–7
NUCLEAR PROPERTIES OF WATER AND HEAVY WATER

Quantity	*Light water*	*Heavy water*
Abundance, %	99.9849 to 99.9861	0.0139 to 0.0151
Molecular weight	18.016	20.028
Molecular density at 20°C, cm^{-3}	0.334×10^{24}	0.0332×10^{24}
Neutron macroscopic absorption cross section at 2,200 m/sec, cm^{-1}	0.0220	0.000040
Thermal neutron diffusion area, cm^2	8.12	—
For 0.16% H_2O content		13,500
Corrected for H_2O absorption		25,000
Thermal neutron macroscopic transport cross section, cm^{-1}	2.10	0.395
Average cosine of neutron scattering angle	0.34	0.15
Epithermal neutron macroscopic scattering cross section	1.49	0.349
Epithermal neutron macroscopic slowing-down cross section, cm^{-1}	1.38	0.178
Fission neutron age, cm^2		
To indium	30.4	100
To thermal	31.4	125

From Tipton, C. R., Jr., Ed., *Reactor Handbook*, 2nd ed., Vol. 1: Materials, Wiley Interscience, New York, 1960. With permission.

Table 3.6–8
ACTIVATION DATA FOR SOME REACTOR COOLANTS

Target isotope	Isotopic abundance, %	Activation cross section, barn	Radioactive product of reaction	Half-life	Energy of radiation, MeV (γ)	Gammas per disintegration of active atom
Na^{23}	100	0.53[a]	Na^{24}	14.9 hr	2.76; 1.38	1; 1
K^{41}	6.8	1.15[a]	K^{42}	12.4 hr	1.51	0.25
O^{18}	0.204	0.00021[a]	O^{19}	29.4 sec	1.6	0.7
O^{16}	99.8	$\sim 0.019 \times 10^{-3}$[b]	N^{16}	7.4 sec	6.13; 7.10	0.76; 0.06
O^{17}	0.039	$\sim 0.0052 \times 10^{-3}$[b]	N^{17}	4.1 sec	1 (neutron)	1 (neutron)

[a] Neutron activation cross section at 2,200 m/sec.

[b] Fast (n,p.) reactions; cross sections averaged over fission spectrum. Data are from Roys, P. A. and Shure, K., Production cross section of N^{16} and N^{17}, in *Abstracts, 1958 Ann. Meeting Am. Nucl. Soc.*, Paper No. 3-8, Los Angeles, June 1958. A previous determination in Henderson, W. J. and Tunnicliffe, P. R., The production of N^{16} and N^{17} in the cooling water of the NRX reactor, *Nucl. Sci. Eng.*, 3(2), 145, 1958, gives cross sections of 0.0185×10^{-3} barns for N^{16} and 0.0093×10^{-3} barns for N^{17}.

REFERENCE

Rockwell, T., III, Ed., *Reactor Shielding Design Manual,* USAEC Report No. TID-7004, U.S. Atomic Energy Commission, March 1956, Table 4.1.

From *Reactor Physics Constants,* 2nd ed., ANL-5800, Argonne National Laboratory, U.S. Atomic Energy Commission, July 1963.

Table 3.6–9
REPROCESSING URANIUM FUEL

How Reprocessing Plants Fit Into the Uranium Fuel Cycle

Fuel reprocessing technology is being spurred by recent AEC guidelines pertaining to the deposition of high-level radioactive wastes and the discharge of radioactive effluents from plants. These guidelines direct that all high-level wastes must be solidified within 5 years of production and buried within 10 years; they propose that radiation-exposure levels for individuals remain under 5 mrem/year incremental dosage from plant effluents.

When defined as a series of processing steps, the uranium fuel cycle consists of four vital recurring operations (mining is performed just once): fuel fabrication, burning, reprocessing, and enrichment. A fissionable U^{235} atom is likely to undergo these steps several times before being "burned."

In commercial power reactors the fuel consists almost entirely of uranium in metallic or oxide form, of which 3 to 4% is U^{235}, with the remainder being nonfissionable U^{238}. In the reactor only a fraction of the already dilute U^{235} is burned by the time thermal efficiency drops off, and the fuel assembly has to be replaced. Consequently, the "spent" fuel contains most of the valuable uranium, as well as plutonium and neptunium – fission products that are worth recovering.

The bulk of the remaining fission products pose an environmental hazard and an economic handicap. As worthless radioactive materials they must be sealed off from the biosphere for hundreds of years; as nuclear "poisons" they impede fission efficiency and cannot be allowed to build up in the fuel cycle.

The function of reprocessing plants is to conserve valuable uranium, plutonium, and neptunium, while converting worthless fission products into safe, disposable forms. Uranium is usually produced at reprocessing plants in the form of uranyl nitrate, which must be fluorinated to UF_6 prior to enrichment. A trend is developing for reprocessors to include fluorination as one of their services. Previously, this has been a separate mini-step in the fuel cycle.

During enrichment, the concentration of U^{235} is brought back to the 3 to 4% level required for power generation; then the fuel goes back to the fabricator who prepares new fuel rods.

From *Chem. Eng.*, 78, 19, 1971. With permission.

Table 3.6–10
RADIATION EXPOSURE AND SHIELDING

Data on Hazards from Alpha, Beta, and Gamma Radiation

The following miscellaneous items of information on penetration and shielding are useful in evaluating certain radiation hazards and in planning personnel protection.

Gamma rays – As an approximation for the dose rate at a distance of 1 ft from a point source of gamma radiation, the value in rems per hour is six times the product of total gamma energy emitted per disintegration of the parent, in MeV, times number of curies of the parent nuclide (accuracy ± 20 % from 0.07 to 4 MeV.)

The attenuation of dose rate with distance, from 100 Ci of cobalt 60, is given in the following table. Account has been taken of air absorption and build-up factor as well as inverse-square attenuation.

Table A. Dose Rate from 100 Ci of Co^{60}

Dose rate, rems/hour	Distance, feet	Dose rate, rems/hour	Distance, feet
1,500	1	0.0075	400
15	10	0.0012	800
0.6	50	0.0006	1,000
0.15	100		

Warning note: Hazardous beta rays may also be present. The above data provide the dose rates from gamma rays only.

Table B gives relative shield thickness for gamma radiation. The build-up factor to correct exponential to broad-beam radiation has been applied.

Table B. Shield Thickness vs. Gamma-dose Transmission

	Shield thickness, inches		
Broad-beam transmission	**Concrete, 147 lb/ft³**	**Iron**	**Lead**
Radium (11 Principal Gammas, 0.24–2.20 MeV)			
0.1	10	3.1	1.6
0.01	19	6.2	3.5
0.001	28	9.1	5.5
0.0001	38	12.0	7.8
0.00001	47	15.3	10.2
Cobalt 60 (1.33 + 1.17 MeV per Disintegration)			
0.1	11	3.2	1.7
0.01	19	6.0	3.3
0.001	27	8.8	4.8
0.0001	35	11.4	6.5
0.00001	43	14.6	8.1

Table 3.6–10 (continued)
RADIATION EXPOSURE AND SHIELDING

Data on Hazards from Alpha, Beta, and Gamma Radiation (continued)

Table B. Shield Thickness vs. Gamma-dose Transmission

Broad-beam transmission	Shield thickness, inches Concrete, 147 lb/ft^3		Iron		Lead
Cesium 137 (0.66 MeV)					
0.1	8.5		2.6		0.85
0.01	15		4.7		1.7
0.001	22		6.8		2.5
0.0001	28		8.9		3.4
0.00001	34		11.0		4.2
Iridium 192 (Gammas from 0.13–0.87 MeV, Averaging 0.3 MeV)					
0.1	7				0.48
0.01	13				1.1
0.001	18.3				1.9
0.0001	24				2.6
0.00001	30				3.5
Gold 198 (0.41-, 0.68-, and 1.1- MeV Gammas)					
0.1	6.6				0.35
0.01	12.0				0.83
0.001	17.4				1.7
0.0001	22.6				2.8
0.00001	28.0				4.3
Iodine 131 (0.08–0.723 MeV, Predominantly 0.36 MeV)					
0.1	6				
0.01	12				
0.001	18				
Barium 140 + Lanthanum 140 (0.030–2.5 MeV, Averaging about 1.6 MeV)					
0.1	25	9.8	3.4	1.6	0.87
0.01	44	18	6.4	3.4	2.0
0.001	64	27	9.2	5.2	3.0
0.0001	81	35	11.8	7.1	4.1
0.00001	104	44	14.8	9.0	5.2
0.000001		51			6.3

[a]After several mean free paths, every 10 in. of concrete reduces the radiation by another factor of ten.

From Brodsky, A. and Beard, G. V., *A Compendium of Information for Use in Controlling Radiation Emergencies,* TID – 8206 (rev.), U.S. Atomic Energy Commission, 1960.

Table 3.6–10 (continued)
RADIATION EXPOSURE AND SHIELDING

Alpha and beta particles – The energy required to just penetrate the 0.07-mm protective layer of skin is 7.500 keV for an alpha particle but only 70 keV for a beta particle. The alpha activity of several common sources is shown in Table C.

Table C. Alpha Activity in Particles per Minute per Microgram

Plutonium	140,000.
Neptunium	1,519.
Natural uranium	1.5
Uranium 238	0.741
Thorium 232	0.247

The range of beta particles in air is about 12 ft/MeV. The air dose in rads per hour at 1 ft from a beta point source is about 200 times its value in curies (absorption neglected).

Beta-ray surface dose rates for several materials are shown in Table D.

Table D. Beta-ray Surface Dose Rates

Material	mrad/hr	Material	mrad/hr
Thorium, 4–5 years after separation	40	Uranium slug, natural	233
Tuballoy, D-38	200	UO_2, brown oxide	207
Oralloy		UF_4, green salt	179
40%	180	$UO_2(NO_3)_2$–$6H_2O$	111
93%	140	UO_3, orange oxide	204
Plutonium 239		U_3O_8, black oxide	203
Nickel coated	360	UO_2F_2	176
Uncoated	440	$Na_2U_2O_7$	167
Uranium 233			
1-month U^{232} build-up	7,000		
1-year U^{232} build-up	58,000		

From Brodsky, A. and Beard, G. V., *A Compendium of Information for Use in Controlling Radiation Emergencies,* TID-8206 (rev.), U.S. Atomic Energy Commission, 1960.

Data for Table 3.6–10 from Wang, Y., Ed., *CRC Handbook of Radioactive Nuclides,* Chemical Rubber Co., Cleveland, 1969.

Table 3.6–11
PROPERTIES OF SHIELDING MATERIALS

Gamma-ray Mass-absorption Coefficients for Various Materials Used in Shielding

For density in lb/ft³, multiply g/cm³ by 62.42

Material	*Density, ρ, g/cm³*	*Mass-absorption coefficient, μ, cm⁻¹*			*Material*	*Density, ρ, g/cm³*	*Mass-absorption coefficient, μ, cm⁻¹*		
		1 MeV	*3 MeV*	*6 MeV*			*1 MeV*	*3 MeV*	*6 MeV*
Air	0.001294	0.0000766	0.0000430	0.0000304	**Flesh**[b]	1	0.0699	0.0393	0.0274
Aluminum	2.7	0.166	0.0953	0.0718	Fuel oil				
Ammonia (liquid)	0.771	0.0612	0.0322	0.0221	(medium)	0.89	0.0716	0.0350	0.0239
Beryllium	1.85	0.104	0.0579	0.0392	Gasoline	0.739	0.0537	0.0299	0.0203
Beryllium carbide	1.9	0.112	0.0627	0.0429	Glass				
Beryllium oxide					Borosilicate	2.23	0.141	0.0805	0.0591
(hot-pressed blocks)	2.3	0.140	0.0789	0.0552	Lead (Hi-D)	6.4	0.439	0.257	0.257
Bismuth	9.80	0.700	0.409	0.440	Plate (avg)	2.4	0.152	0.0862	0.0629
Boral	2.53	0.153	0.0865	0.0678	Iron	7.86	0.470	0.282	0.240
Boron (amorphous)	2.45	0.144	0.0791	0.0679	Lead	11.34	0.797	0.468	0.505
Boron carbide					Lithium hydride				
(hot pressed)	2.5	0.150	0.0825	0.0675	(pressed powder)	0.70	0.0444	0.0239	0.0172
Bricks					Lucite® (polymethyl				
Fire clay	2.05	0.129	0.0738	0.0543	methacrylate)	1.19	0.0816	0.0457	0.0317
Kaolin	2.1	0.132	0.0750	0.0552	Paraffin	0.89	0.0646	0.0360	0.0246
Silica	1.78	0.113	0.0646	0.0473	Rocks				
Carbon	2.25	0.143	0.0801	0.0554	Granite	2.45	0.155	0.0887	0.0654
Clay	2.2	0.130	0.0801	0.0590	Limestone	2.91	0.187	0.109	0.0824
Cements					Sandstone	2.40	0.152	0.0871	0.0641
Colemanite borated	1.95	0.128	0.0725	0.0528	Rubber				
Plain (1 Portland					Butadiene				
cement: 3 sand					copolymer	0.915	0.0662	0.0370	0.0254
mixture)	2.07	0.133	0.0760	0.0559	Natural	0.92	0.0652	0.0364	0.0248
Concretes					Neoprene	1.23	0.0813	0.0462	0.0333
Barytes	3.5	0.213	0.127	0.110	Sand	2.2	0.140	0.0825	0.0587
Barytes-boron frits	3.25	0.199	0.119	0.101	Type 347				
Barytes-limonite	3.25	0.200	0.119	0.0991	stainless steel	7.8	0.462	0.279	0.236
Barytes-lumnite-					Steel (1%C)	7.83	460	0.276	0.234
colemanite	3.1	0.189	0.112	0.0939	Uranium	18.7	1.46	0.813	0.881
Iron-Portland[a]	6.0	0.364	0.215	0.181	Uranium hydride	11.5	0.903	0.504	0.542
MO (ORNL mixture)	5.8	0.374	0.222	0.184	Water	1.0	0.0706	0.0396	0.0277
Portland (1 cement:					Wood				
2 sand: 4 gravel					Ash	0.51	0.0345	0.0193	0.0134
mixture)	2.2	0.141	0.0805	0.0592	Oak	0.77	0.0521	0.0293	0.0203
	2.4	0.154	0.0878	0.0646	White pine	0.67	0.0452	0.0253	0.0175

[a]Elemental composition, wt %: hydrogen, 1.0; oxygen, 52.9; silicon, 33.7; aluminum, 3.4; iron, 1.4; calcium, 4.4; magnesium, 0.2; carbon, 0.1; sodium, 1.6; potassium, 1.3.

[b]Composition, wt%: oxygen, 65.99; carbon, 18.27; hydrogen, 10.15; nitrogen, 3.05; calcium, 1.52; phosphorus, 1.02.

From Tipton, C. R., Jr., Ed., *Reactor Handbook,* 2nd ed., Vol. 1: Materials, Wiley Interscience, New York, 1960. With permission.

Table 3.6–12
GAMMA-RAY SHIELD DESIGN[a]

Effects of Ionizing Radiation

The effects of radiation on human beings and the criteria for protection are usually discussed in terms of low-LET radiation, i.e., X-rays and gamma rays. Gamma rays of cobalt 60 and 200 to 250 kV X-rays have been used as the reference radiation. To obtain the dose equivalent in rems, in which all dose-limiting regulations are stated, the energy absorption in rads must be multiplied by QF, a quality factor that takes the relative biological effectiveness into account. The QF for all beta, gamma, and electron or positron radiation and X-rays is recommended as unity by the NCRP; for neutrons, protons, alpha particles, and fission fragments it is greater than unity and may be as high as 20.[b]

Shield Design Procedure

The design of gamma shields is straightforward. Given the amount of activity to be shielded, the source geometry (so that the amount of self absorption can be estimated), and the energy of the gammas emitted, the shield thickness required to yield a given radiation dose rate at the shield surface can be calculated from the fundamental data for mass attenuation coefficients given in Tables 3.6–11 and 3.6–18. For thick shields a build-up factor should be applied, i.e.,

$$D = \frac{SBe^{-upt}}{K4\pi r^2}$$

where D = dose rate, rem/hr, S = source strength, photon/s, B = build-up factor, μ = mass attenuation coefficient, cm^2/g, ρ = shield material density, g/cm^3, t = shield thickness, cm, K = photons/s to give 1 rem/hr, and r = shield radius for an equivalent point source, cm.

Shields for seven typical radioactive isotopes can be designed, using Table 3.6–10, if the shielding material is concrete, iron, or lead. This table includes the build-up factor, i.e., the build up of lower energy gammas as a consequence of degradation of the original, higher energy, gamma rays.

[a]For data on the effects of radiation on engineering materials, see Table 3.6–16.
[b]See also Basic Radiation Protection Criteria, NCRP Report No. 39, National Council on Radiation Protection and Measurements, January 1971, 80.

From Bolz, R. E. and Tuve, G. L., Eds., *Handbook of Tables for Applied Engineering Science,* 2nd ed., CRC Press, Cleveland, 1973, 442.

Table 3.6–13
NEUTRON ATTENUATION IN WATER

This table gives fast neutron dose in water for a number of monoenergetic sources and a fission source. Calculations have been made by the moments method (From NDA 15C-42 and NDA 15C-60).

Source	Distance, r, in cm			
	30	60	90	120
	$4\pi r^2 D(r)$ Mrep cm²/hr per N/s			
2 MeV	8×10^{-5}	1×10^{-7}		
4 MeV	8.4×10^{-4}	1×10^{-5}	7×10^{-8}	
6 MeV	2.5×10^{-3}	1.3×10^{-4}	5.5×10^{-6}	1.7×10^{-7}
8 MeV	3.6×10^{-3}	3×10^{-4}	2.1×10^{-5}	1.3×10^{-6}
10 MeV	3.9×10^{-3}	4.2×10^{-4}	4×10^{-5}	3.5×10^{-6}
14 MeV	4.4×10^{-3}	6.3×10^{-4}	7.5×10^{-5}	8×10^{-6}
Fission	2.7×10^{-4}	1×10^{-5}	4×10^{-7}	

REFERENCE

Attenuation in Water of Radiation from the Bulk Shielding Reactor, ORNL-2518, Oak Ridge National Laboratory, July 1958.

From Goldstein, H., *The Attenuation of Gamma Rays and Neutrons in Reactor Shields,* U.S. Atomic Energy Commission, May 1957.

Table 3.6–14
FISSION REACTOR SHIELDING

Effects of Shield Thickness on Dose Rates

Shield Design Procedure

The design of even idealized reactor shields is rendered extremely complex by the generation of gamma rays by the inelastic scattering and capture of neutrons. The neutron dose at the outer surface of a shield of a given thickness is relatively insensitive to the composition of the shield if the materials are close to their theoretical density and the shield is free of voids; however, the gamma dose varies widely with the composition. Layers of heavy material for attenuating gamma rays should be placed far enough out in the shield so that they will not be a serious source of secondary gamma rays.

Experimental Data on Typical Shielding

Extensive experiments at the Bulk Shielding Facility of the U.S. Atomic Energy Commission have yielded useful data, and reports from this source should be consulted. While many shields of complex make-up have been tested, it is possible to give useful estimates from the basic data on two common shielding materials – concrete and water. Roughly, 25.4 cm (10 in.) of concrete or 40.7 cm (16 in.) of water reduces the *gamma* radiation by a factor of ten. The attenuation per unit thickness for concrete will, of course, depend on the mix, but it remains fairly constant across a thick shield.

One set of test results at the Bulk Shielding Facility was reported by J. L. Meem and H. E. Hungerford in AEC Report No. ORNL 1147, April 1952. In these tests observations were made at distances of 20 to 120 cm from the surface of the reactor core. The core was equivalent to a nearly right circular cylinder of 33-cm radius, and the results were normalized to a uniform spherical source.

For a shield of *concrete*, an attenuation of 10/1 was produced for each increment of about 25 cm (10 in.) of concrete for *gamma* radiation and about 20 cm (7.9 in.) of concrete for *neutron* radiation, regardless of thickness.

For shielding by *water*, an attenuation from 10 rep/hr · W near the reactor wall to 0.001 rep/hr · W required 174 cm of water for gamma radiation but less than 60 cm of water for neutron radiation. In other terms, the mean thickness in this range for an attenuation of 10/1 by water shielding was slightly over 43 cm (17 in.) for gamma rays, but less than 15 cm (5.9 in.) for neutron radiation. In all of these tests, the outer layers of shield were less effective (per unit thickness) than those near the radiation source.

Other tests were made using a compound shield of water and 3 in. of iron near the core, plus nine 1-in. thicknesses of lead spaced from 5 to 75 cm from the reactor core surface. For this shield the gamma dose, in rep/hr · W, at 200 cm from the surface of the core was 3×10^{-7} rep/hr · W. The positions of the lead layers in several lead-water shields tested were close to that for minimum weight of an idealized, nearly spherical shield around a reactor having a core diameter of about 1 m. By adding 0.4% boron in the form of boric acid to the water in a lead-water shield, the generation of hydrogen-capture gammas in the water was suppressed sufficiently to reduce the gamma dose rate by a factor of over four.

Similar data are given in other tables in this section. The general conclusion must be drawn that simple ratios for shield effectiveness, concrete vs. water or neutron vs. gamma radiation, should be used only for rough estimates.

From Bolz, R. E. and Tuve, G. L., Eds., *Handbook of Tables for Applied Engineering Science,* 2nd ed., CRC Press, Cleveland, 1973, 443.

Table 3.6–15
COMPARATIVE VALUES OF SHIELDING MATERIALS

Equivalent Thickness for Various Shielding Materials

Material	Specific gravity	Relaxation length, inches[a] Gamma rays 8 MeV	Relaxation length, inches[a] Gamma rays 4 MeV	Relaxation length, inches[a] Fast neutrons
Lead	11.3	0.75	0.95	3.5
Iron	7.8	1.73	1.45	2.4
Iron concrete	4.3	3.94	3.25	2.5
Aluminum	2.7	6.7	5.1	3.9
Concrete	2.3	7.1	5.5	4.7
Beryllium oxide	2.3	10.	7.1	3.5
Beryllium	1.85	12.	8.	3.6
Graphite	1.7	10.	7.5	3.6
Water	1.00	16.	12.	4.

[a]Relaxation length is related to exponential decay and is defined as the thickness required to attenuate the radiation by a factor of $1/e = 1/2.72 = 0.368$.

Note: the similarity of shielding materials for fast neutrons as compared with the wide differences when gamma rays are involved.

From Bolz, R. E. and Tuve, G. L., Eds., *Handbook of Tables for Applied Engineering Science,* 2nd ed., CRC Press, Cleveland, 1973, 444.

Table 3.6–16
RADIATION EFFECTS ON ENGINEERING MATERIALS

Dose Limits and Property Changes for Ionizing Radiation

Radiation Types and Measuring Units

Effects of radiation on materials depend on the type of radiation involved, its energy, and the duration of exposure. Changes in the material properties such as strength of a solid or viscosity of a liquid are dependent on the total radiation received and the type of radiation, i.e., gamma rays, neutrons, or other. Quantitative description of the radiation is actually related to the method for its measurement. By definition, dosimetry is concerned with the amount of radiation absorbed by the material in the test sample, but the measuring device uses a different material; hence, the measurement is indirect. For these reasons several units have been used for expressing the radiation dose, and direct conversions by numerical factors are not always applicable. The unit used in this table is the *rad*(rd), defined basically as 100 erg/g of the material under test.

Classes of Materials

Coverage in the following table is limited to several classes of common engineering materials, largely organic. A great many test results are available covering specific materials and various test conditions (see the references).

Metals and Alloys – Metals are only very slightly affected by gamma radiation, but they are damaged by energetic particles, which generate solid-state defects. When the fluence[a] is less than about 10^{17}, the radiation effects include (1) increases in strength, hardness, and electrical resistivity and (2) reductions in ductility and toughness. Metals in nuclear reactors are exposed to both high neutron and gamma dosage.

Ceramics – Gamma radiation can cause color changes in ceramics. Neutron radiation may cause the ceramic to expand, and the thermal conductivity is lowered.

[a]Refers to neutrons having more than 1 MeV of energy.

Typical Effects of Radiation on Materials

Class of material	Limiting dose, rd: Negligible effect	Limiting dose, rd: Major (50%) effect	Typical effect of radiation
Dielectrics			
Inorganic	10^8	10^{10}	Increased loss factor
Organic	10^7	10^9	Increased loss factor
Elastomers	10^6	10^8	Reduced strength, elasticity, durability; compression set
Fibers and textiles	10^6	10^8	Loss in tenacity and strength
Fuels (liquid)	10^8	10^9	Lowered volatility; increased gum
Glasses	10^5	10^7	Coloration; opaqueness
Lubricants (petroleum)	10^7	10^9	Increased viscosity; dark color; gas formation; lowered oxidation resistance
Paints and coatings	10^8	10^9	Cracking, blistering, peeling, porosity
Plastics[b]	10^7	10^9	Reduced strength; swelling; brittleness; gas evolution
Wood products	10^6	10^8	Loss of strength; "dry rot"

Table 3.6–16 (continued)
RADIATION EFFECTS ON ENGINEERING MATERIALS

Dose Limits and Property Changes for Ionizing Radiation (continued)

[b]Some plastics are more radiation-resistant than others, e.g., PVC and polystyrene.

Note: Most organic materials are damaged by radiation dosage of 1 Mrd or more, regardless of the source or kind of radiation.

Damage to semiconductors depends on nature and use of the semiconductor device.

Certain common materials and many assembled components are more sensitive to radiation than the above table would indicate, especially in prolonged exposure. Examples are foods, Teflon®, neoprene, explosives, capacitors (paper), rectifiers, and dry batteries.

Any application for which the exposure approaches the limits listed should be investigated with a goal of reducing the radiation or using less sensitive materials.

Effects of temperature are not considered in the above table.

REFERENCES

Bolt, R. O. and Carroll, J. G., Eds., *Radiation Effects on Organic Materials,* Academic Press, New York, 1963.

Kircher, J. F. and Bowman, R. E., *Effects of Radiation on Materials and Composites,* Reinhold, New York, 1964.

Rockwell, T., III, Ed., *Reactor Shielding Design Manual,* USAEC Report No. TID-7004, U.S. Atomic Energy Commission, March 1956.

From Bolz, R. E. and Tuve, G. L., Eds., *Handbook of Tables for Applied Engineering Science,* 2nd ed., CRC Press, Cleveland, 1973, 445.

Table 3.6–17
GAMMA-RAY ACTIVITY DUE TO THERMAL NEUTRON CAPTURE

Following are activation data for isotopes produced by capture of thermal neutrons. Gamma-ray activity data are given for both primary possible secondary products of neutron capture.

Radio-active isotope	Half-life	E_γ, MeV	Yield-% per disintegration	Parent isotope	Isotopic-% or half-life	Activation cross section, barn
B^{12}	0.027 sec	~4.5	~4	B^{11}	81.2	<50 mb
N^{16}	7.4 sec	6.1	55	N^{15}	0.37	24±8 μb
		7.1	20			
O^{19}	29.4 sec	1.4	70	O^{18}	0.20	0.21±0.04 mb
F^{20}	10.7 sec	1.63	100	F^{19}	100	9±2 mb
Ne^{23}	40.2 sec	0.44	29	Ne^{22}	8.8	36±15 mb
		1.65	1			
Na^{24}	15.1 hr	2.76	99.96	Na^{23}	100	0.53±0.02
		1.38	99.96			
		4.14	0.04			
Mg^{27}	9.45 min	1.015	29	Mg^{26}	11.3	26±2 mb
		0.181	1			
		0.834	70			
Al^{28}	2.3 min	1.78	100	Al^{27}	100	0.21±0.02
Si^{31}	2.65 hr	1.26	0.07	Si^{30}	3.12	0.11±0.01
S^{37}	5.04 min	3.12	90	S^{36}	0.016	0.14±0.04
Cl^{38*}	1.0 sec	0.66	100	Cl^{37}	24.5	~5 mb
Cl^{38}	37.5 min	2.15	47	Cl^{37}	24.5	0.56±0.12
		1.65	31	Cl^{38*}	1 sec	
A^{41}	1.83 hr	1.37	99.3	A^{40}	99.6	0.53±0.02
K^{42}	12.46 hr	1.53	18	K^{41}	6.91	1.15±0.11
Ca^{47}	4.9 days	1.31	71	Ca^{46}	0.0033	0.25±0.10
		0.82	5			
		0.49	5			
Ca^{49}	8.8 min	3.10	89.78	Ca^{48}	0.185	1.1±0.1
		4.05	9.88			
Sc^{46*}	19.5 sec	0.14	100	Sc^{45}	100	10±4
Sc^{46}	85 days	1.12	99.9	Sc^{45}	100	12±4
		0.89	100	Sc^{46*}	19.5 sec	

Radio-active isotope	Half-life	E_γ, MeV	Yield-% per disintegration	Parent isotope	Isotopic-% or half-life	Activation cross section, barn
Sc^{47}	3.43 days	0.16	74	Ca^{47}	4.9 days	
Ti^{51}	5.79 min	0.605	1.4	Ti^{50}	5.25	0.14±0.03
		0.928	4.2			
		0.323	95.8			
V^{52}	3.76 min	1.44	100	V^{51}	99.8	4.5±0.9
Cr^{51}	27.8 days	0.32	10	Cr^{50}	4.31	13.5±1.4
Mn^{56}	2.58 hr	2.98	0.4	Mn^{55}	100	13.4±0.3
		2.13	15			
		2.65	1.8			
		1.81	24			
		0.845	99			
Fe^{59}	45.1 days	0.191	3	Fe^{58}	0.31	0.9±0.2
		1.29	43			
		1.10	57			
Co^{60*}	10.5 min	0.058	99.7	Co^{59}	100	16±3
Co^{60}	5.3 yr	1.17	99.9	Co^{59}	100	20±3
		1.33	100	Co^{60*}	10.5 min (99.7%)	
Co^{61}	1.65 hr	0.07	100	Co^{60}	5.28 yr	6±2
Ni^{65}	2.56 hr	0.37	4.1	Ni^{64}	1.16	1.6±0.2
		1.49	24.9			
		1.12	18.1			
Cu^{64}	12.87 hr	1.34	0.4	Cu^{63}	69	3.9±0.8
Cu^{66}	5.15 min	1.05	9	Cu^{65}	31	1.8±0.4
Zn^{65}	245 days	1.11	45.5	Zn^{64}	48.9	0.5±0.1
Zn^{69*}	13.8 hr	0.439	100	Zn^{68}	18.6	97±10 mb
As^{77}	38.8 hr	0.524	0.52	Ge^{77*}	52 sec (85.8%)	

Table 3.6–17 (continued)
GAMMA-RAY ACTIVITY DUE TO THERMAL NEUTRON CAPTURE

Radio-active isotope	Half-life	E_γ, MeV	Yield-% per disinte-gration	Parent isotope	Isotopic-% or half-life	Activation cross section, barn
As^{77}	38.8 hr	0.246	1.51	Ge^{77}	12 hr	
Se^{75}	119.9 days	0.121[a]	24.7	Se^{74}	0.87	26 ± 6
		0.265	58.7			
		0.280	17.8			
		0.136	59.9			
		0.402	17.8			
Se^{77*}	17.5 sec	0.162	100	Se^{76}	9	7 ± 3
Se^{79*}	3.9 min	0.096	100	Se^{78}	23.5	0.4
Zn^{71*}	3 hr	0.38	100	Zn^{70}	0.62	?
		0.49	100			
		0.61	100			
Zn^{71}	2.2 min	0.51	100	Zn^{70}	0.62	85 ± 20 mb
Ga^{70}	21 min	1.04	0.8	Ga^{69}	60.2	1.4 ± 0.3
Ga^{72}	14.2 hr	0.69[a]	28.2	Ga^{71}	39.8	5.0 ± 1.0
		0.89	105			
		1.86	30.3			
		2.51	53.1			
Ge^{75*}	48 sec	0.139	100	Ge^{74}	36.7	?
Ge^{75}	82 min	0.63[a]	0.80	Ge^{74}	36.7	0.45 ± 0.08
		0.264	11.15	Ge^{75*}	48 sec	
Ge^{77*}	52 sec	0.159	14.2	Ge^{76}	7.7	30 ± 20 mb
Ge^{77}	12 hr	0.56[a]	21.6	Ge^{76}	7.7	0.2 ± 0.1
		0.79	4.3	Ge^{77*}	52 sec (14.2%)	
		1.36	4			
		2.0	1			
As^{76}	26.8 hr	2.05	1	As^{75}	100	4.2 ± 0.8
		1.40	2			
		1.20	9.22			
		0.64	8.78			
		0.55	39.78			

Radio-active isotope	Half-life	E_γ, MeV	Yield-% per disinte-gration	Parent isotope	Isotopic-% or half-life	Activation cross section, barn
Se^{81*}	56.8 min	0.103	100	Se^{80}	49.8	30 ± 10 mb
Se^{83}	25 min	0.176	100	Se^{82}	9.2	4 ± 2 mb
		0.950	100			
Br^{83}	2.33 hr	0.046	20	Se^{83*}	67 sec	
				Se^{83}	25 min	
Kr^{83*}	1.88 hr	0.032	100	Br^{83}	2.33 hr	
		0.009	100			
Br^{80*}	4.5 hr	0.048	100	Br^{79}	50.54	2.9 ± 0.5
		0.036	100			
Br^{80}	18.5 min	0.62	14	Br^{79}	50.54	8.5 ± 1.4
				Br^{80*}	4.5 hr	
Br^{82}	35.9 hr	0.688[a]	152.4	Br^{81}	49.5	3.5 ± 0.5
		0.817	108			
		1.469	69.8			
Kr^{79}	34.5 hr	0.217[a]	34.4	Kr^{78}	0.35	2.0 ± 0.5
		0.398	12.7			
		0.833	5.84			
Kr^{85*}	4.4 hr	0.305	22	Kr^{84}	57	0.10 ± 0.03
		0.150	78			
Kr^{87}	78 min	2.57	21.9	Kr^{86}	17.4	60 ± 20 mb
		2.05	3.57			
		0.85	13.57			
		0.403	86.9			
Rb^{86*}	1.02 min	0.56	100	Rb^{85}	72.15	?
Rb^{86}	18.6 days	1.08	10	Rb^{85}	72.15	0.72 ± 0.15
				Rb^{86*}	1.02 min	
Rb^{88}	17.8 min	2.68	6.1	Rb^{87}	27.85	0.12 ± 0.03
		0.91	13.6			
		1.85	24			
		4.89	Small			

Table 3.6–17 (continued) GAMMA-RAY ACTIVITY DUE TO THERMAL NEUTRON CAPTURE

Radio-active isotope	Half-life	E_γ, MeV	Yield-% per disintegration	Parent isotope	Isotopic-% or half-life	Activation cross section, barn
Sr^{85*}	70 min	0.233	1.3	Sr^{84}	0.55	?
		0.007	84.7			
		0.225	84.7			
		0.150	14			
Sr^{85}	64 days	0.513	100	Sr^{84}	0.55	1.0±0.3
				Sr^{85*}	70 min	
					(86%)	
Sr^{87*}	2.8 hr	0.388	100	Sr^{86}	9.87	1.3±0.4
Zr^{95}	63.3 days	0.754	54	Zr^{94}	17.4	0.09±0.03
		0.722	43			
Zr^{97}	17 hr	1.15	80	Zr^{96}	2.8	0.10±0.05
		1.72	20			
		0.58	80			
Nb^{94*}	6.6 min	0.0414	99.9	Nb^{93}	100	1.0±0.5
		0.90	~0.1			
Nb^{95*}	84 hr	0.235	100	Zr^{95}	63.3 days	
					(3%)	
Nb^{95}	35 days	0.770	99	Nb^{95*}	84 hr	
				Zr^{95}	63.3 days	
					(97%)	
Nb^{97*}	60 sec	0.747	100	Zr^{97}	17 hr	
Nb^{97}	74 min	0.665	100	Nb^{97*}	60 sec	
Mo^{93*}	6.8 hr	0.26	100	Mo^{92}	15.86	<6 mb
		0.68	100			
		1.48	100			
Mo^{99}	67 hr	0.780[a]	8	Mo^{98}	23.75	0.45±0.10
		0.740	6			
		0.377	1			
		0.041	5			
		0.181	6			
Mo^{101}	14.6 min	0.515[a]	96	Mo^{100}	9.62	0.20±0.05

Radio-active isotope	Half-life	E_γ, MeV	Yield-% per disintegration	Parent isotope	Isotopic-% or half-life	Activation cross section, barn
Mo^{101}	14.6 min	0.083	93.5			
		0.896	3.5			
		2.08	8.5			
Tc^{99*}	6.04 hr	0.142	1	Mo^{99}	67 hr	
		0.002	99		(94%)	
		0.140	99			
Tc^{101}	14 min	0.307[a]	87	Mo^{101}	14.6 min	
		0.939	3			
		0.545	10.4			
		0.130	3.5			
Ru^{97}	2.8 days	0.57	1	Ru^{96}	5.5	10±4 mb
		0.325	3.5			
		0.216	95.5			
Tc^{97*}	92 days	0.099	100	Ru^{97}	2.8 days	
		0.090	100			
Ru^{103}	39.7 days	0.610	6.45	Ru^{102}	31.5	1.2±0.3
		0.495	90.5			
		0.055	1.09			
Ru^{105}	4.5 hr	0.730	100	Ru^{104}	18.7	0.7±0.2
Rh^{103*}	54 min	0.040	100	Ru^{103}	41 days	
				Pd^{103}	17 days	
					(94%)	
Rh^{105*}	40 sec	0.130	100	Ru^{105}	4.5 hr	
Rh^{105}	36.5 hr	0.310	30	Rh^{105*}	40 sec	
Rh^{104*}	4.4 min	0.051	100	Rh^{103}	100	12±2
		1.53	0.0065			
Rh^{104}	42 sec	1.24	0.11	Rh^{104*}	4.4 min	
		0.556	1.98	Rh^{103}	100	140±30
Pd^{103}	17 days	0.498	4	Pd^{102}	0.96	4.8±1.5
		0.362	2.84			
		0.298	2.05			

Table 3.6–17 (continued)
GAMMA-RAY ACTIVITY DUE TO THERMAL NEUTRON CAPTURE

Radio-active isotope	Half-life	E_γ, MeV	Yield-% per disintegration	Parent isotope	Isotopic-% or half-life	Activation cross section, barn
Pd^{109*}	4.8 min	0.17	100	Pd^{108}	26.71	?
Ag^{109*}	39.2 sec	0.0875	100	Pd^{109}	13.6 hr	
				Cd^{109}	1.30 yr	
Pd^{111}	22 min	0.73	<1	Pd^{110}	11.81	0.3±0.1
		0.65	<1			
		0.56	<1			
Ag^{108}	2.3 min	0.60	0.22	Ag^{107}	51.35	44±9
		0.62	0.8			
		0.43	0.28			
Ag^{110*}	270 days	0.883^a	203	Ag^{109}	48.65	2.8±0.5
		0.945	48			
		1.382	40			
		1.519	8			
Ag^{111}	7.5 days	0.340	8	Pd^{111}	22 min	
		0.243	1			
Cd^{107}	6.7 hr	0.846	<1	Cd^{106}	1.21	1±0.5
Ag^{107*}	44.3 sec	0.094	100	Cd^{107}	6.7 hr	
Cd^{111*}	48.6 min	0.148	100	Cd^{110}	12.4	0.2±0.1
		0.247	100			
Cd^{113*}	5.1 yr	0.265	0.1	Cd^{112}	24.1	30±15 mb
Cd^{115*}	43 days	1.30	1	Cd^{114}	28.86	0.14±0.03
		0.935	2			
Cd^{115}	53 hr	0.033	1	Cd^{114}	28.86	1.1±0.3
		0.523	23.8			
		0.49	12.4			
		0.26	2.32			
In^{115*}	4.5 hr	0.335	95	Cd^{115}	53.0 hr	
Cd^{117*}	2.9 hr	0.84	61	Cd^{116}	7.58	1.5±0.3
		1.27	30			
		1.55	9			
		0.425	61			

Radio-active isotope	Half-life	E_γ, MeV	Yield-% per disintegration	Parent isotope	Isotopic-% or half-life	Activation cross section, barn
Cd^{117*}	2.9 hr	0.281	91			
In^{117*}	1.9 hr	0.312	22	Cd^{117*}	2.9 hr	
		0.160	23	Cd^{117}	50 min	
In^{117}	66 min	0.562	91	In^{117*}	1.9 hr	
		0.712	9		(22%)	
		0.160	91			
Cd^{117}	50 min	0.425	100	Cd^{116}	7.58	?
		0.281	100			
Sn^{117*}	14 days	0.562	100	In^{117}	66 min	
		0.160	100			
In^{114*}	50 days	0.192	96.5	In^{113}	4.2	56±12
		0.722	3.5			
		0.556	3.5			
In^{114}	72 sec	1.300	0.09	In^{113}	4.2	2±0.6
In^{116*}	54.2 min	0.137^a	3	In^{115}	95.8	145±15
		1.49	21			
		0.40	25			
		1.27	129			
		2.09	25			
Sn^{113}	112 days	0.400	100	Sn^{112}	0.95	1.3±0.3
In^{113*}	1.75 hr	0.258	11	Sn^{113}	112 days	
		0.393	89			
		0.135	11			
Sn^{117*}	14 days	0.162	99	Sn^{116}	14.24	6±2 mb
		0.320	1			
		0.159	99			
Sn^{119*}	275 days	0.065	100	Sn^{118}	24.01	10±6 mb
		0.024	100			
Sn^{123}	40 min	0.153	100	Sn^{122}	4.71	0.16±0.04
Sn^{125}	9.5 min	1.39	1.9	Sn^{124}	5.98	0.2±0.1
		0.326	99.7			

Table 3.6–17 (continued)
GAMMA-RAY ACTIVITY DUE TO THERMAL NEUTRON CAPTURE

Radio-active isotope	Half-life	E_γ, MeV	Yield-% per disintegration	Parent isotope	Isotopic-% or half-life	Activation cross section, barn
Sb^{125}	2.4 yr	0.637^a	22	Sn^{125}	9.5 min	
		0.175	11	Sn^{125}	9.5 days	
		0.465	60			
		0.0354	11			
Te^{125}	58 days	0.110	100	Sb^{125}	2.4 yr	
		0.0354	100			
Sn^{125}	9.5 days	1.97	1.2	Sn^{124}	5.98	4±2 mb
		0.81	1			
		1.07	3.75			
Sb^{122*}	3.5 min	0.0753	100	Sb^{121}	57.25	?
		0.0607	100			
Sb^{122}	2.8 days	1.137	0.73	Sb^{121}	57.25	6.8±1.5
		0.686	3.4	Sb^{122*}	3.5 min	
		1.258	0.66			
		0.566	66.3			
Sb^{124*}	21 min	0.0185	100	Sb^{123}	42.75	30±15 mb
Sb^{124*}	1.3 min	0.012	100	Sb^{123}	42.75	30±15 mb
Sb^{124}	60.9 days	1.37^a	3.75	Sb^{123}	42.75	2.5±0.5
		2.09	49.2	Sb^{124*}	21 min	
		0.967	2.16	Sb^{124*}	1.3 min	
		0.644	108.6			
		0.725	16			
Te^{121*}	154 days	0.0818	100	Te^{120}	0.089	?
		0.214	100			
Te^{121}	17 days	0.575	87	Te^{121*}	154 days	
		0.506	13			
Te^{123*}	104 days	0.0887	100	Te^{122}	2.46	1.1±0.5
		0.159	100			
Te^{125*}	58 days	0.110	100	Te^{124}	4.61	5±3
		0.0354	100			
Te^{127*}	105 days	0.089	98	Te^{126}	18.71	90±20 mb
		0.0585	1.5			

Radio-active isotope	Half-life	E_γ, MeV	Yield-% per disintegration	Parent isotope	Isotopic-% or half-life	Activation cross section, barn
Te^{127}	9.35 hr	0.418	Small	Te^{126}	18.71	0.8±0.2
		0.370	Small	Te^{127*}	105 days	
		0.170	Small		(98%)	
Te^{129*}	33 days	0.106	100	Te^{128}	31.8	15±5
Te^{129}	74 min	1.12	10.4	Te^{128}	31.8	0.13±0.03
		0.21	1.7	Te^{129*}	33 days	
		0.72	2			
		0.475	17.1			
		0.027	98			
Te^{131*}	30 hr	0.180^a	21.7	Te^{130}	34.5	<8 mb
		0.239	70			
		1.12	90			
		0.099	84			
		0.446	50			
Te^{131}	24.8 min	0.773	5	Te^{130}	34.5	0.22±0.05
		0.446	45	Te^{131*}	30 hr	
		0.147	60		(21.7%)	
		0.099	40			
		0.051	40			
I^{128}	24.98 min	0.540	1.8	I^{127}	100	5.5±0.5
		0.455	17.3			
		0.990	0.2			
Xe^{125}	18 hr	0.243	95	Xe^{124}	0.096	?
		0.056	5			
		0.187	5			
I^{125}	60 days	0.035	100	Xe^{125}	18 hr	
Xe^{127}	36.4 days	0.368	10	Xe^{126}	0.09	?
		0.170	85			
		0.200	85			
		0.145	2			
		0.056	4			
Xe^{129*}	8 days	0.0400	100	Xe^{128}	1.919	?

Table 3.6–17 (continued)
GAMMA-RAY ACTIVITY DUE TO THERMAL NEUTRON CAPTURE

Radio-active isotope	Half-life	E_γ, MeV	Yield-% per disintegration	Parent isotope	Isotopic-% or half-life	Activation cross section, barn
Xe^{129*}	8 days	0.196	100			
Xe^{133*}	2.3 days	0.232	100	Xe^{132}	26.89	?
Xe^{133}	5.27 days	0.081	100	Xe^{132}	26.89	0.2±0.1
				Xe^{133*}	2.3 days	
Xe^{135*}	15.6 min	0.52	100	Xe^{134}	10.44	?
Xe^{135}	9.2 hr	0.370	2	Xe^{134}	10.44	0.2±0.1
		0.620	3	Xe^{135*}	15.6 min	
		0.250	97			
Cs^{134*}	3.2 hr	0.137	0.8	Cs^{133}	100	17±4 mb
		0.0105	98.2			
		0.127	98.2			
Cs^{134}	2.3 yr	0.200[a]	13	Cs^{133}	100	26±5
		0.801	210	Cs^{134*}	3.2 hr	
		1.17	4.02			
		1.37	6.70			
Ba^{131}	11.6 days	1.03[a]	2.9	Ba^{130}	0.101	10.1±1.0
		0.82	5.6			
		0.495	79.7			
		0.245	17.2			
		0.122	61			
Ba^{133*}	38.8 hr	0.276	100	Ba^{132}	0.097	?
		0.012	100			
Ba^{133}	7.2 yr	0.360	96.5	Ba^{132}	0.097	7±2
		0.070	3.5	Ba^{133*}	38.8 hr	
		0.292	3.5			
		0.081	100			
Ba^{135*}	28.7 hr	0.268	100	Ba^{134}	2.42	?
Ba^{137*}	2.60 min	0.661	100	Ba^{136}	7.81	?
Ba^{139}	85 min	1.43	19	Ba^{138}	71.66	0.5±0.1
		0.163	85			
La^{140}	40.2 hr	2.57[a]	20	La^{139}	99.9	8.4±1.7
La^{140}	40.2 hr	0.920	40			
		0.486	73			
		1.60	76			
		0.130	26			
Ce^{137*}	35 hr	0.255	100	Ce^{136}	0.193	0.6±0.2
Ce^{137}	8.7 hr	0.445	3	Ce^{136}	0.193	6.3±1.5
				Ce^{137*}	35 hr	
Ce^{139*}	55 sec	0.740	100	Ce^{138}	0.250	7±5 mb[b]
Ce^{139}	140 days	0.1665	100	Ce^{138}	0.25	0.6±0.3
				Ce^{139*}	55 sec	
Ce^{141}	32 days	0.1449	75	Ce^{140}	88.48	0.31±0.10
Ce^{143}	32 hr	1.10[a]	6	Ce^{142}	11.07	0.95±0.05
		0.861	17			
		0.351	24			
		0.294	21			
		0.057	36			
Pr^{142}	19.3 hr	1.61	2.8	Pr^{141}	100	10±3
Nd^{147}	11.06 days	0.69	3	Nd^{146}	17.26	1.8±0.6
		0.318	20			
		0.165	3			
		0.532	25			
		0.092	60			
Nd^{149}	2.0 hr	0.65[a]	15	Nd^{148}	5.74	3.7±1.2
		0.124	81			
		0.285	68			
Pm^{149}	52 hr	0.285	100	Nd^{149}	2 hr	
		1	100			
Nd^{151}	12 min	0.73[a]	40	Nd^{150}	5.63	?
		0.085	60			
		0.117	120			
		0.421	60			
		1.14	100			

Table 3.6–17 (continued)
GAMMA-RAY ACTIVITY DUE TO THERMAL NEUTRON CAPTURE

Radio-active isotope	Half-life	E_γ, MeV	Yield-% per disintegration	Parent isotope	Isotopic-% or half-life	Activation cross section, barn
Pm^{151}	27.5 hr	0.715[a]	10	Nd^{151}	12 min	
		0.340	42			
		0.177	115			
		0.275	70			
		0.069	105			
Sm^{145}	400 days	0.0613	100	Sm^{144}	3.16	<2
Sm^{153}	47 hr	0.17	25	Sm^{152}	26.63	140±40
		0.07	15			
		0.1	53			
Sm^{155}	24 min	1.05	100	Sm^{154}	22.53	5.5±1.1
		0.246	100			
Eu^{155}	1.7 yr	0.102	50	Sm^{155}	24 min	
		0.084	27			
		0.018	27			
Eu^{152}	9.2 hr	1.39[a]	1.54	Eu^{151}	47.8	1400±300[b]
		0.983	21.4			
		0.344	2.8			
		0.122	15.2			
Gd^{153}	236 days	0.1	100	Gd^{152}	0.2	<125
Gd^{159}	18 hr	0.364	10	Gd^{158}	24.87	4±2
		0.23	6			
		0.136	4			
		0.079	2			
		0.056	12			
Gd^{161}	3.63 min	0.360	10	Gd^{160}	21.90	0.8±0.3
		0.102	90			
		0.316	90			
		0.060	10			
Tb^{161}	7 days	0.106[a]	2	Gd^{161}	3.63 min	
		0.0573	18			
		0.0277	27			
		0.0783	5			

Radio-active isotope	Half-life	E_γ, MeV	Yield-% per disintegration	Parent isotope	Isotopic-% or half-life	Activation cross section, barn
Tb^{160}	73 days	0.093[a]	80	Tb^{159}	100	>22
		0.976	63			
		1.45	26			
		0.466	39			
		0.297	63			
Dy^{165*}	1.25 min	0.108	90	Dy^{164}	28.18	510±20
		0.515	6			
		0.361	4			
Dy^{165}	139 min	0.279[a]	1	Dy^{164}	28.18	2100±300[b]
		0.361	40	Dy^{165*}	1.25 min (90%)	
		1.02	8			
		0.71	2			
		0.094	10			
Ho^{166}	27.2 hr	1.378	1	Ho^{165}	100	60±12
		0.080	48			
Er^{169}	9.4 days	0.0084	15	Er^{168}	27.1	2±0.4
Er^{171}	7.5 hr	0.308[a]	95	Er^{170}	14.9	9±2
		0.013	3			
		0.126	40			
		0.118	55			
		0.005	65			
Tm^{170}	127 days	0.084	24	Tm^{169}	100	118±30
Yb^{169}	32 days	0.0084[a]	60	Yb^{168}	0.14	11000±3000[b]
		0.094	60			
		0.118	60			
		0.20	50			
		0.31	40			
Yb^{175}	4.1 days	0.396[a]	10	Yb^{174}	31.84	60±40
		0.144	5			
		0.251	2			
		0.138	3			
		0.114	13			

Table 3.6–17 (continued)
GAMMA-RAY ACTIVITY DUE TO THERMAL NEUTRON CAPTURE

Radio-active isotope	*Half-life*	E_γ, *MeV*	*Yield-% per disinte-gration*	*Parent isotope*	*Isotopic-% or half-life*	*Activation cross section, barn*
Yb^{177}	1.8 hr	1.24[a]	4	Yb^{176}	12.73	5.5±1
		0.140	1			
		0.147	9			
		0.118	3			
Lu^{176*}	3.7 hr	0.089	90	Lu^{175}	97.4	35±15
Lu^{177}	6.7 days	0.321	2	Lu^{176}	2.6	4000±800
		0.208	2	Yb^{177}	1.8 hr	
		0.072	3			
		0.250	3			
		0.113	5			
Hf^{175}	70 days	0.430	1.5	Hf^{174}	0.19	?
		0.089	13.7			
		0.342	97.5			
Hf^{180*}	5.5 hr	0.444	80	Hf^{179}	13.75	?
		0.333	20			
		0.216	20			
		0.093	80			
		0.0576	100			
Hf^{181}	44.6 days	0.482[a]	89.3	Hf^{180}	35.25	10±3
		0.137	110			
		0.0039	1			
		0.346	13.2			
Ta^{182*}	16.5 min	0.180	100	Ta^{181}	100	30±10 mb
Ta^{182}	115 days	0.229[a]	45	Ta^{181}	100	19±7
		0.100	65	Ta^{182*}	16.5 min	
		0.084	55			
		1.289	94			
W^{181}	140 days	0.152	0.176	W^{180}	0.135	10±10
		0.136	0.114			
W^{185*}	1.7 min	1.30	100	W^{184}	30.6	2.1±0.6
		1.65	100			
W^{187}	24 hr	0.686	12	W^{186}	28.4	34±7

Radio-active isotope	*Half-life*	E_γ, *MeV*	*Yield-% per disinte-gration*	*Parent isotope*	*Isotopic-% or half-life*	*Activation cross section, barn*
W^{187}	24 hr	0.480	8			
		0.206	5			
		0.072	3			
		0.132	3			
Re^{186}	91 hr	0.768	0.05	Re^{185}	37.07	100±20
		0.123	2			
		0.137	22			
Re^{188*}	18.7 min	0.060	100	Re^{187}	62.93	?
		0.105	100			
Re^{188}	17 hr	1.96[a]	0.20	Re^{187}	62.93	75±15
		0.633	1	Re^{188*}	18.17 min	
		0.155	9			
Os^{185}	95 days	0.879[a]	15	Os^{184}	0.018	<200
		0.233	2			
		0.160	2			
		0.646	85			
Os^{191}	14 hr	0.074	100	Os^{190}	26.4	?
Ir^{191*}	4.9 sec	0.042	100	Os^{191}	16 days	
		0.129	100			
Os^{193}	31.5 hr	0.387[a]	5	Os^{192}	41	1.6±0.4
		0.460	5			
		0.281	12			
		0.139	7			
		0.073	36			
Ir^{192*}	1.45 min	0.056	100	Ir^{191}	38.5	260±100
Ir^{192}	74.5 days	0.485[a]	208	Ir^{191}	38.5	700±200
		0.613	23	Ir^{192*}	1.45 min	
		1.06	0.04			
Ir^{194}	19 hr	2.05[a]	0.14	Ir^{193}	61.5	130±30
		1.66	8.9			
		1.18	9.4			
		0.643	18.5			

Table 3.6–17 (continued)
GAMMA-RAY ACTIVITY DUE TO THERMAL NEUTRON CAPTURE

Radio-active isotope	Half-life	E_γ, MeV	Yield-% per disintegration	Parent isotope	Isotopic-% or half-life	Activation cross section, barn
Ir^{194}	19 hr	0.328	19.3			
Pt^{193}*	3.5 days	0.130	100	Pt^{192}	0.78	90±40
Pt^{195}*	~6 days	0.130	60	Pt^{194}	32.8	?
		0.031	40			
		0.099	40			
Pt^{197}	18 hr	0.279	0.9	Pt^{196}	25.4	0.80±0.10
		0.191	10.6			
		0.077	99.1			
Pt^{199}	31 min	0.96[a]	3	Pt^{198}	7.2	3.9±0.8
		0.54	41			
		0.197	81			
		0.246	42			
		0.316	41			
Au^{199}	3.15 days	0.209	3.7	Pt^{199}	31 min	
		0.050	20.6			
		0.159	89.9			
Au^{198}	2.7 days	1.089	0.16	Au^{197}	100	96±10
		0.4118	99.018			
Hg^{197}*	24 hr	0.164	96.6	Hg^{196}	0.146	?
		0.133	96.6			
Hg^{197}	65 hr	0.191	1.2	Hg^{197}*	24 hr (96.6%)	
		0.077	100			
Au^{197}*	7.4 sec	0.407	2	Hg^{197}*	24 hr (3.4%)	
		0.130	1.4			
		0.277	1.4			
Hg^{203}	45.4 days	0.279	100	Hg^{202}	29.8	3.8±0.8
Hg^{205}	5.5 min	0.203	Small	Hg^{204}	6.85	0.43±0.10
Tl^{204}	3 yr	0.38	0.008	Tl^{203}	29.5	8±3
Th^{233}	23.5 min	0.662	0.05	Th^{232}	100	7.34±0.15
		0.448	0.10			
		0.350	0.004			
		0.172	0.03			
		0.098	0.25			
Pa^{233}	26.95 days	0.417[a]	20	Th^{233}	23.5 min	
		0.104	35			
		0.313	80			
		0.058	111			
		0.028	39			
U^{239}	23.5 min	0.075	100	U^{238}	99.274	2.76±0.09
Np^{239}	2.33 days	0.067[a]	62	U^{239}	23.5 min	
		0.285	87			
		0.105	30			
		0.228	4			

[a] Effective value.
[b] Reactor neutrons.
*Metastable.

REFERENCES

Hughes, D. J. and Schwartz, R. B., *Neutron Cross Sections,* BNL-325, Brookhaven National Laboratory, Upton, L.I., N.Y., July 1958.
The Reactor Handbook, Vol. 1: Physics, AECD-3645, U.S. Atomic Energy Commission, 1955. 158.
Sullivan, W. H., Trilinear Chart of Nuclides, 2nd rev., U.S. Atomic Energy Commission, January 1957.

From *Reactor Physics Constants,* 2nd ed., ANL-5800, Argonne National Laboratory, U.S. Atomic Energy Commission, July 1963.

Table 3.6–18
TOTAL MASS ATTENUATION COEFFICIENTS

In cm^2/g

The following table gives the total gamma-ray attenuation coefficient, μ/ρ, for some common materials. The product of these numbers and the density of the material gives the familiar cross sections, μ, Table 3.6–19.

Material	*Gamma-ray energy, MeV*						
	0.1	*0.2*	*0.5*	*1.0*	*2*	*5*	*10.0*
H	.295	.243	.173	.126	.0876	.0502	.0321
Be	.132	.109	.0773	.0565	.0394	.0234	.0161
C	.149	.122	.0870	.0636	.0444	.0270	.0194
N	.150	.123	.0869	.0636	.0445	.0273	.0200
O	.151	.123	.0870	.0636	.0445	.0276	.0206
Na	.151	.118	.0833	.0608	.0427	.0274	.0215
Mg	.160	.122	.0860	.0627	.0442	.0286	.0228
Al	.161	.120	.0840	.0614	.0432	.0282	.0229
Si	.172	.125	.0869	.0635	.0447	.0296	.0243
P	.174	.122	.0846	.0617	.0436	.0290	.0242
S	.188	.127	.0874	.0635	.0448	.0302	.0255
A	.188	.117	.0790	.0573	.0407	.0279	.0241
K	.215	.127	.0852	.0618	.0438	.0305	.0267
Ca	.238	.132	.0876	.0634	.0451	.0316	.0280
Fe	.344	.138	.0828	.0595	.0424	.0313	.0294
Cu	.427	.147	.0820	.0585	.0418	.0316	.0305
Mo	1.03	.225	.0851	.0575	.0414	.0344	.0359
Sn	1.58	.303	.0886	.0568	.0408	.0355	.0383
I	1.83	.339	.0913	.0571	.0409	.0361	.0394
W	4.21	.708	.125	.0640	.0437	.0409	.0465
Pt	4.75	.795	.135	.0659	.0445	.0418	.0477
Tl	5.16	.866	.143	.0675	.0452	.0423	.0484
Pb	5.29	.896	.145	.0684	.0457	.0426	.0489
U	1.06	1.17	.176	.0757	.0484	.0446	.0511
Air	.151	.123	.0868	.0655	.0445	.0274	.0202
NaI	1.57	.305	.0901	.0577	.0412	.0347	.0366
H_2O	.167	.136	.0966	.0706	.0493	.0301	.0219
Concrete[a]	.169	.124	.0870	.0635	.0445	.0287	.0229
Tissue	.163	.132	.0936	.1683	.0478	.0292	.0212

[a]Type 04.

From *Reactor Physics Constants,* 2nd ed., ANL-5800, Argonne National Laboratory, U.S. Atomic Energy Commission, July 1963.

Table 3.6–19
TOTAL GAMMA-RAY ATTENUATION CROSS SECTIONS

In cm^{-1}

Material	*Density, g/cm³*	*Gamma-ray energy, MeV*						
		0.1	*0.2*	*0.5*	*1.0*	*2*	*5*	*10.0*
Be	1.85	.244	.202	.1430	.1045	.0729	.0433	.0298
C	2.25	.335	.275	.1958	.1431	.0999	.0608	.0437
Na	.9712	.147	.115	.0809	.0590	.0415	.0266	.0209
Mg	1.741	.279	.212	.1497	.1092	.0770	.0498	.0397
Al	2.70	.435	.324	.2268	.1658	.1166	.0761	.0618
Si	2.42	.416	.303	.2103	.1537	.1082	.0716	.0588
P	1.83	.318	.223	.1548	.1129	.0798	.0531	.0443
S	2.07	.389	.263	.1809	.1314	.0927	.0625	.0328
K	0.87	.187	.110	.0741	.0538	.0381	.0265	.0232
Ca	1.55	.369	.205	.1358	.0983	.0699	.0490	.0434
Fe	7.86	2.704	1.085	.6508	.4677	.3333	.2460	.2311
Cu	8.933	3.814	1.313	.7325	.5226	.3734	.2823	.2725
Mo	9.01	9.280	2.027	.7668	.5181	.3730	.3099	.3190
Sn	7.298	11.53	2.211	.6466	.4145	.2978	.2591	.2795
I	4.94	9.040	1.675	.4510	.2821	.2020	.1783	.1946
W	19.3	81.25	13.66	2.413	1.235	.8434	.7894	.8975
Pt	21.37	101.51	16.99	2.885	1.408	.9510	.8933	1.019
Tl	11.86	61.20	10.27	1.696	.8005	.5361	.5017	.5740
Pb	11.34	59.99	10.16	1.644	.7757	.5182	.4831	.5545
U	18.7	19.82	21.88	3.291	1.416	.9051	.8340	.9556
NaI	3.667	5.757	1.118	.3304	.2116	.1511	.1272	.1342
H_2O	1.00	.167	.136	.0966	.0706	.0493	.0301	.0219
Concrete[a]	2.35	.397	.291	.2045	.1492	.1046	.0674	0538

[a]Type 04.

From *Reactor Physics Constants,* 2nd ed., ANL-5800, Argonne National Laboratory, U.S. Atomic Energy Commission, July 1963.

Table 3.6–20
REMOVAL CROSS SECTIONS FOR VARIOUS MATERIALS

The removal cross section is a measure of the ability of a material to remove fast neutrons for shielding attenuation. It is most often applied to a wall of solid material between the fission source and a layer of water or hydrogenous material. The solid wall reduces the neutron energy to such an extent that it will be thermalized and captured in the water.

Symbols:

σ_R = microscopic removal cross section, barns per atom
Σ_R = macroscopic removal cross section per centimeter

Material	σ_R, *barn*	N_0 *at 20°C, atom/cm³*	Σ_R, *cm⁻¹*	*Material*	σ_R, *barn*	N_0 *at 20°C, atom/cm³*	Σ_R, *cm⁻¹*
Hydrogen	1.00±0.05	—	—	Lead	3.53±0.30	0.0330	0.116
Deuterium	0.92±0.10[a]	—	—	Bismuth	3.49±0.35	0.0282	0.098
Lithium	1.01±0.04	0.0460×10^{24}	0.046	Uranium	3.6±0.4	0.0473	0.17
Beryllium	1.07±0.06	0.120	0.128	Boric oxide (B_2O_3)	4.30±0.41	—	—
Boron	0.97±0.10	0.139	0.135	Boron carbide (B_4C)	5.1±0.4	—	—
Carbon (graphite)	0.72±0.05	0.113	0.081	Fluorothene (C_2F_3Cl)	6.66±0.8	—	—
Oxygen	0.92±0.05	—	—	Heavy water (D_2O)	2.76±0.11	—	—
Fluorine	1.29±0.06	—	—	Lithium fluoride (LiF)	2.43±0.34	—	—
Aluminum	1.31±0.05	0.0603	0.079	Oil (CH_2)	2.84±0.11	—	—
Chlorine	1.2±0.8	—	—	Paraffin ($C_{30}H_{62}$)	80.5±5.2	—	—
Iron	1.98±0.08	0.0848	0.168	Perfluoroheptane (C_7F_{16})	26.3±0.8	—	—
Nickel	1.89±0.10	0.0913	0.173				
Copper	2.04±0.11	0.0846	0.173				
Zirconium	2.36±0.12	0.0423	0.10				
Tungsten	3.13±0.25	0.0631	0.198				

[a]Calculated: $\sigma_R(D_2O)$ = 2.76 b.

REFERENCES

Chapman, G. T. and Storrs, C. L., *Effective Neutron Removal Cross Sections for Shielding,* AECD-3978 (ORNL-1843), U.S. Atomic Energy Commission, September 19, 1955.

From *Reactor Physics Constants,* 2nd ed., ANL-5800, Argonne National Laboratory, U.S. Atomic Energy Commission, July 1963.

Table 3.6–21
CRITICAL MASS AND ITS PARAMETERS

Critical mass, the minimum quantity of fissile material capable of sustaining a fission chain, varies greatly with reactor design (as in the 1- to 100-kg range); thus, no simple formula involving the parameters is possible. A number of theoretical models have been set up, and the calculation procedure described, but the ultimate answers are experimental.[a]

Among the factors or conditions determining criticality are

1. The fuel and fuel enrichment
2. Absorption and leakage of neutrons
3. Size and shape of the system.

More specifically, the parameters are set up in terms of more narrowly defined factors, such as the following:

p = *resonance escape probability,* the fraction of source neutrons that escape capture while being slowed down to a particular energy level (the term "resonance" refers to the resonance region of the absorber)

f = *thermal utilization,* the ratio of thermal neutrons absorbed in the fuel to the total thermal neutrons absorbed

n = *liberation ratio,* the number of neutrons liberated per neutron absorbed in the fuel

e = *fast-fission factor,* the ratio of fast neutrons slowing down, to those produced by thermal-neutron fissions

$k\infty$ = *infinite multiplication factor,* ratio of neutrons from fission to neutrons absorbed in the preceding generation, in a system of infinite size

Values of *n*

	Natural uranium	U^{233}	U^{235}	Pu^{239}
Fast neutrons	1.09	2.60	2.18	2.74
Thermal neutrons	1.33	2.27	2.06	2.10

The following table illustrates the wide variation in critical mass for cylindrical and slab cores, unreflected or water reflected, contained in either stainless steel or aluminum. It will be noted that the critical mass varies in the range of 1 to 46 kg.

[a]For extensive tables of critical mass data, see Sections 3 and 4 of the original source of this table.

Table 3.6–21 (continued)
CRITICAL MASS AND ITS PARAMETERS

Cylindrical and Slab Cores Containing UO_2F_2-H_2O Solutions, Various Concentrations; Uranium about 93% Enriched in U^{235}

Core diameter, cm	H/U^{235} atom ratio	U^{235} concentration, g/cm³ sol.	Critical core height, cm	Critical mass, kg U^{235}
ALUMINUM-WALLED CYLINDERS AND SLABS				
Water-reflected				
15.2	27.1	0.8288	89.3	13.47
	43.2	0.537	70.1	6.87
	58.8	0.415	71.8	5.44
16.5	26.2	0.827	44.5	7.91
	44.3	0.5376	38.7	4.45
	78.7	0.315	42.6	2.87
	119.0	0.212	52.6	2.39
20.3	29.9	0.759	20.7	5.09
	49.5	0.488	18.8	2.97
	78.7	0.315	19.4	1.98
	192.0	0.134	28.1	1.22
	290.0	0.0881	40.1	1.15
25.4	27.1	0.8288	12.4	5.2
	43.2	0.537	12.5	3.40
	51.5	0.470	11.4	2.72
	127	0.199	14.4	1.45
	328.7	0.0787 sol.	22.4	0.893
	499	0.0522	35.2	0.0930
38.1	27.1	0.8288	7.7	7.3
	52.9	0.459	7.90	4.14
	221	0.116	11.30	1.49
	499	0.0522	16.90	1.01
	755	0.0343	27.10	1.02
	999	0.0260	44.30	1.31
76.2	27.1	0.8288	5.0 ± 1	18.9 ± 3.8
	44.3	0.5376	4.8 ± 1	11.8 ± 2.5
	72.4	0.3423	5.5 ± 1	8.6 ± 1.6
50.8 × 50.8 sq.	27.1	0.8288	6.3 ± 1	13.5 ± 2.2
	44.3	0.5376	4.3 ± 1	6.0 ± 1.4
	72.4	0.3423	6.2 ± 1	5.5 ± 0.9
Bare				
22.3	44.3	0.5376	219	45.8
	50.1	0.480	202.2	37.8
	55.4	0.437	171.6	29.2
	60.8	0.402	162.5	25.4
	66.1	0.373	159.8	23.2
	71.5	0.350	163.2	22.2
25.4	27.1	0.8288	38.9	16.4
	44.3	0.5376	35.1	9.6
	50.1	0.480	34.8	8.40
	52.9	0.459	34.0	7.90
	55.4	0.437	34.3	7.60
	60.8	0.402	34.1	6.96
	66.1	0.373	34.1	6.45
	73.4	0.3370	33.7	5.8
	83.1	0.300	34.4	5.22
	169	0.151	41.2	3.15
	328	0.0785	147.8	5.83
	331	0.0779	170.1	6.72
Bare (Continued)				
30.5	44.3	0.5376	23.2	9.1
	50.1	0.480	22.6	7.92
	55.4	0.437	22.7	7.25
	60.8	0.402	22.7	6.67
	331	0.0779	32.8	1.86
38.1	27.1	0.8288	18.5	17.5
	44.3	0.5376	17.9	11.0
	50.1	0.480	17.9	9.79
	74.6	0.3314	16.8	6.4
	169.0	0.151	18.5	3.18
	328.7	0.0787	21.7	1.95
	331	0.0779	22.9	2.03
	499	0.0522	27.4	1.63
	755	0.0343	43.6	1.70
50.8	27.1	0.8288	15.8	26.5
	44.3	0.5376	15.0	16.3
	50.1	0.480	15.4	14.9
	60.8	0.402	15.3	12.5
	73.4	0.3370	15.2	10.4
	325	0.0791	18.7	2.97
76.2	44.3	0.5376	13.7	33.6
	50.1	0.480	13.8	30.2
	72.4	0.3423	13.9 ± 0.5	21.6 ± 0.7
	331	0.0779	16.3	5.79
50.8 × 50.8 sq.	27.1	0.8288	15 ± 1	32 ± 2
	72.4	0.3423	14.3	12.6
	331	0.0779	17.9	3.60
Partially Water-Reflected[b]				
15.2	44.3	0.5376	75.0	7.34
19.1	44.3	0.5376	25.7	3.93
20.3	44.3	0.5376	23.6	4.12
	51.5	0.470	23.8	3.64
	72.4	0.3423	23.3	2.59
25.4	43.2	0.537	17.3	4.71
	72.4	0.3423	16.7	2.90
38.1	74.6	0.3314	12.0	4.5
50.8	72.4	0.3423	10.6	7.3
76.2	72.4	0.3423	9.2	14.4
76.2 × 152.4[c]	57.0	0.4240	8.4	41.5

Section 4

Biomedical Materials

4.1 ENGINEERING APPROACHES TO LIMB PROSTHETICS AND ORTHOTICS

A. B. Wilson, Jr.
Committee on Prosthetics Research and Development
National Research Council
E. F. Murphy
Research Center for Prosthetics
Veterans Administration

THE GOALS

This section is directed to a variety of audiences for diverse purposes. Primarily it is addressed to professors of bioengineering and of medical specialties like orthopedic and general surgery or physical medicine, all concerned in varying ways with improving prosthetics or orthotics services to amputees or brace wearers. To a secondary extent, through these professors, the review is addressed to postdoctoral and graduate students, to residents and interns, and to research fellows. Likewise, in this era of increasing awareness on the part of engineering, aerospace, and defense industries of the potentialities for using their talents and special knowledge to attack human problems through rehabilitation engineering, there is a need to point out crucial problems, priorities, relevant areas, and sources of information.

Thus, a skilled person in other areas who is a relative newcomer in prosthetics and orthotics may be able to select those areas where he can best make major contributions, building efficiently upon the prior efforts summarized by Wilson,[1,2] and making a flying start in a relatively new field. To aid all these groups, some typical indexes, major journals, and important reference collections are noted in Appendix A. In an interdisciplinary field like this, the usual single specialty index or library may leave large gaps. A list of foreign language journals dealing with prosthetics and orthotics is given in Appendix B.

Another purpose of the present review, though, is to assist the relatively sophisticated workers already in this field in selecting from among the rapidly growing masses of literature a relatively manageable number of key articles, useful references, and clues as to major problems. Appendix C is based upon analyses of major problems, by an interdisciplinary group of experts convened during 1970[3] by the Subcommittee on Design and Development of the Committee on Prosthetics Research and Development, National Research Council, a group which advises several government agencies. (In 1975 a new edition was being prepared; it may be published by the time this volume is ready for printing.) These suggested problems should help to frame thesis topics, sensible proposals to funding agencies, and whole career areas for well-trained young investigators.

It will be obvious that problems abound in all areas, so the prospective investigator, with any expert guidance and counsel available, needs self-criticism and self-analysis – leading to legitimate self-confidence – in deciding where his own interests, talents, enthusiasms, and available resources may best make contributions and gain proper recognition. Selections of a problem area and a specific task are best done by the equivalent of vocational guidance.

Examples of Superficial and Real Problems

There are fashions in research, though the cycle is slower than in clothing. For decades there was enthusiasm for knee locks in above-knee prostheses leading to numerous devices, patents, and projects. Later understanding of biomechanical principles of fit and alignment, helping thousands of amputees, has greatly reduced the demand for knee locks. The Mauch S-N-S, developed and evaluated with great effort, appears a highly satisfactory answer to automatic knee control but is still somewhat short of voluntary control.

Some other areas for research and development have been widely publicized for well over a decade, though relatively few patients have actually benefited. Myoelectric control of external power, Figure 4.1–1, using the tiny electrical signals of muscles (or usually of the remnants of amputated muscles) in the forearm to control the switching of external power to an electrically driven artificial hand, has been an extremely dramatic topic since the extensive early publicity on the Russian hand

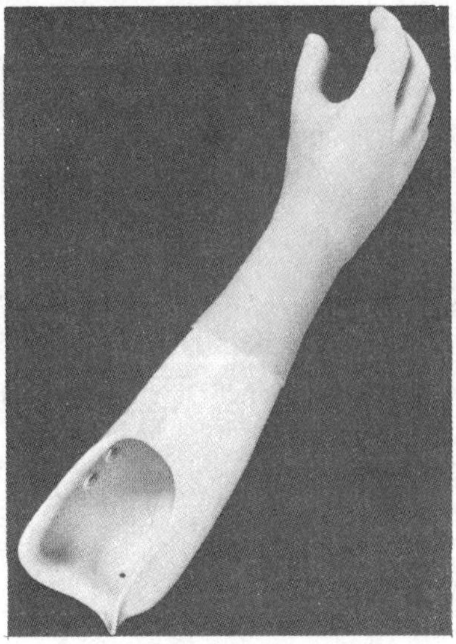

FIGURE 4.1–1. Self-contained below-elbow prosthesis powered by electricity and controlled by myoelectric currents, developed at Northwestern University.[9] Note the electrodes installed on the interior surface of the socket. (From Childress and Billock, VA-sponsored contract.)

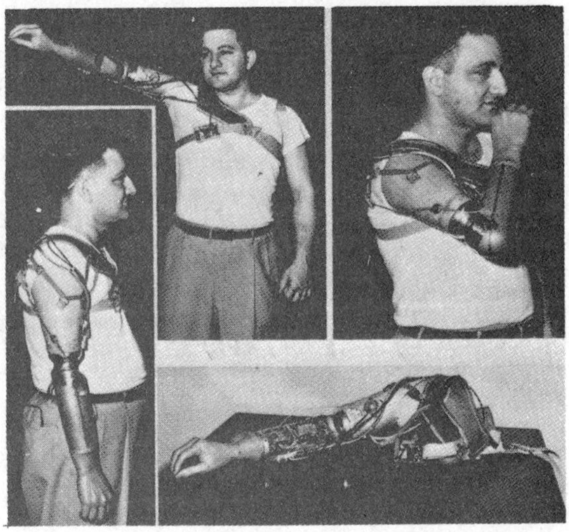

FIGURE 4.1–2. The Alderson above-elbow arm, Model IV (circa 1954.) (From *Human Limbs and their Substitutes,* Advisory Committee on Artificial Limbs, National Academy of Sciences-National Research Council, McGraw-Hill, New York, 1954, 393. With permission.)

about 1957 and the public demonstration in 1960 and paper by Kobrinskii et al.[4] (The usual newspaper articles, of course, understandably overlooked a long history of externally powered hands dating from German work,[5,5a] during World Wars I and II, the series of externally powered hands, wrists, elbow locks, and other components developed in the United States by Alderson[6] (Figure 4.1–2) after World War II, the analyses of switches vs. myoelectric signals – with techniques available in 1950-52 – made at New York University for Alderson, and the pioneering demonstration of the possibilities of control of an artificial hook from myoelectric signals by Battye, Nightingale, and Whillis[7] in England.) Some articles overlooked the distinctions between levels of amputation, the use of remnants of an amputated muscle normally related to a function vs. retraining of more proximal and essentially unrelated muscles, control by on-off "switch" vs. proportional (even if nonlinear) measures of muscle tension, or problems of sensory feedbacks of position, force, and velocity in relation to muscle shortening, contraction (joint displacement, if available), or other kinesthetic cues. As Basmajian[8] emphasizes, honesty and candor are needed, and important areas of research are often tackled only after some years of more empirical development.

Nevertheless, myoelectric control or external power, or both, have served as bait attracting numerous talented, eager workers into prosthetics and orthotics. For a decade and a half there have been numerous seminar papers, theses, reports, papers at bioengineering congresses, and technical articles. It is perhaps dramatically useful – or possibly lamentably wasteful – that most of the very few workers who have persisted in this complex field of prosthetics and orthotics have eventually faced not only questions of microvolts or even of basic research on muscle but the very real, serious, and potentially fruitful problems of prosthetics and orthotics. These latter include, as we shall see, not only fitting, suspension, harnessing, and control of devices but also laboratory and clinical evaluation, quality control in manufacture, deployment with all its demands for training and maintenance, and other aspects of a total system. Multiple government agencies at several levels, universities, rehabilitation centers, private manufacturers, commercial and/or institutional prosthetics facilities, and several medical and paramedical specialties all are involved. These broader systems questions, the economic and psychological aspects, and especially the biomechanical concepts of fitting and control are finally crucial to success of a novel method.[9]

Probably it is too much to expect a single enthusiastic graduate student, working on a narrow aspect of electrodes or of signals, to perceive the whole problem early – or even late – in his studies, but might not a dedicated university or rehabilitation center, over a period of years, expand its horizons and points of view from the highly specialized toward the global or even the universal?

Perspectives and Purposes

Obviously, then, this review is written with certain points of view and of bias. Based on a quarter century of effort at a national level, a frequent restudy of classic sources from many countries and multiple disciplines as well as persistent checking of obscure references, participation in numerous conferences, and authorship – or complicity, one may even say – in a substantial number of the standard documents and texts of two decades, the present authors can scarcely plead ignorance, innocence, or lack of involvement. Nevertheless, this relationship has allowed some perspective, possibly mellowing of prejudices, and even some attempts at philosophy.

The current review, in an attempt to help bioengineers and physicians to leap quickly to serious and fruitful problems, provides a very quick survey of standard literature, then focuses on recent articles, books, and key journals. The review is predominantly of American literature, or at least that in the English language, though clearly there is an extensive, valuable, and steadily growing literature in many foreign languages.

Perhaps the two key themes are interdisciplinary work, based on humility and mutual respect, and systematic transition (with frequent recycling) from basic research through development, evaluation growing to clinical trials, education of all professions concerned, and finally service to the actual individual patient – all based on themes of persistence and continuity. As Talley[10] showed, and Gershenson and Holsberg[10a] have reiterated, this program not only achieves tangible results benefiting thousands of patients but also dramatically reduces patient-care costs for the sponsor, more than paying for the research program.

Some Standard References

Although there had been occasional intensive efforts in other eras and countries,[11,12] engineering had very little influence on the development and application of artificial limbs and orthopedic braces in the United States until the government in 1945 began support of research and development in this area.[13] Complementary efforts are now being carried out under various sponsors in many parts of the world.[14] The International Society for Prosthetics and Orthotics[a] has recently been formed to take over and expand the roles in research and education of the International Committee on Prosthetics and Orthotics of the International Society for Rehabilitation of the Disabled. As a result of these programs, not only devices and fitting techniques but also overall systems and detailed procedures for handling amputees and many other types of orthopedic disability have been changed radically.

In the course of the last quarter century of government-sponsored research, development, and education programs, a number of pertinent technical publications have emerged. *Human Limbs and Their Substitutes,*[13] originally published by McGraw-Hill in 1954 (and reissued with an updated bibliography by Hafner Publishing Company in 1968), was prepared by the Advisory Committee on Artificial Limbs, National Research Council (operating wing of the National Academy of Sciences) to make available results of research during the first few years of the program. Much of the material in *Human Limbs and Their Substitutes* still is quite useful. Although outdated in parts, this book belongs on the shelf of many serious students of prosthetics history and of every student of biomechanics because of the basic data on locomotion, arm and hand motions and forces, phantom pain, etc., and the scholarly analyses of applications to clinical problems even in the face of inadequate data. A sequel to this volume is needed.

The special problems arising in the management of the child amputee are covered comprehensively in *The Limb-deficient Child,*[15] a 390-page volume prepared by the Child Amputee Prosthetics Project, University of California, Los Angeles. Though published 8 years ago, and though a good deal of

[a]Represented in the United States by the U.S. National Committee for the International Society for Prosthetics and Orthotics, 1440 N St., NW, Washington, D.C. 20005.

progress has been made in some hardware development, the basic rules and philosophy set forth are quite valid. *The Limb-deficient Child* should be the bible for every clinic responsible for child amputees.

Since 1954 the Committee on Prosthetics Research and Development or CPRD[a] (successor to the Advisory Committee on Artificial Limbs), alone at first and in recent years in cooperation with the Committee on Prosthetic-Orthotic Education or CPOE[a] (both in the National Research Council), published the journal *Artificial Limbs* primarily for the purpose of making available to clinical personnel results of research. In addition to major papers with extensive references in each issue, there were occasional annotated bibliographies. The News and Notes in each issue summarized meetings of CPRD and its panels, prosthetics education activities, and other news. *Artificial Limbs* ceased publication with Volume 16, Number 1 in Spring 1972.

Because of repeated demand for certain classical articles that appeared in *Artificial Limbs* through the years, the Committee arranged for publication by Krieger in 1970 of a volume entitled *Selected Articles from Artificial Limbs.*[16] This volume includes those articles felt to contain basic information useful to the serious student.

CPOE has published annotated bibliographies on various topics and lists of visual aids. Some were published in *Artificial Limbs* but others have been duplicated for distribution by CPOE.

CPRD and its subgroups routinely hold interdisciplinary specialized conferences. Some of the meetings held by panels are reported in multilithed documents[17] distributed to the attendees, committee members, sponsors, and a relatively few others. These reports serve as historic records of progress in the field and as major sources of rapid communication not only among the participants but also among other workers here and abroad. These references also remain available in very limited quantities to future investigators, enabling them to benefit from these contemporary records of step-by-step development and initial evaluations. CPRD also occasionally prepares major monographs concentrating on research aspects, e.g., of external power, geriatric amputees, and sensory aids. Based on an earlier CPRD report, CPOE[18] recently published a monograph emphasizing the clinical aspects of management of the geriatric amputee, particularly the desirability and possibility of conserving the knee joint by a successful below-knee amputation. The document summarizes medical care, amputation surgery, rigid postoperative dressing, prostheses, and rehabilitation. The prognosis is highly individualistic, being dependent on economics and motivation as well as on medical and prosthetics aspects, but generally patients with below-knee amputations appear to have a lower mortality rate, and such a large number of elderly amputees survive for long periods that a rehabilitation program is definitely indicated.

A major text concentrating on amputations for peripheral vascular disease is that by Warren and Record.[19] It represents collaboration of a general surgeon with special interest in the problem (and leadership of a cooperative study by a group of Veterans Administration surgeons) with an orthopedic surgeon active in amputation surgery since World War II who has served as chief of prosthetics clinic teams almost since their origin. While some of the statements may appear somewhat superficial or dogmatic, they are based upon exceptional opportunities for observation.

Mital and Pierce[20] have published a concise text on amputations and prostheses. It is intended primarily for residents, but it should also be very useful for the experienced general surgeon now renewing his interest and anxious to learn of modern methods and devices.

Weiss,[21] of Poland, a dynamic promoter of the concept of immediate postsurgical prosthetics fitting of amputees, has published a book through the U. S. Government Printing Office. In addition to his concern for psychological values of early standing and ambulation of amputees, Dr. Weiss is especially interested in electromyographic study of the roles of muscles concerned with standing, balancing, and walking in both normal and amputated persons. One of his major goals in immediate postoperative fitting and very early ambulation has been to provide correspondingly early stimulation of the central nervous system and opportunities for reeducation of the muscles remaining available to the amputee. There appear to be numerous opportunities for other research groups to provide independent tests of these concepts and to enlarge the understanding of

[a]2101 Constitution Avenue, Washington, D.C. 20418.

subconscious neuromuscular feedback systems. More research is also needed (perhaps along the lines suggested by the Panel on Sensory and Neuromuscular Implications of the CPRD Conference on the Geriatric Amputee, 1961[22]) on means by which the amputee can best learn to substitute for the more direct sensory inputs and spinal reflexes in the normal a new rapid interpretation of the changing patterns of pressure between the prosthesis and his skin, so as to make subconscious corrective actions in order to counteract imbalances or stumbling which would be trivial to the normal individual but so often serious to the amputee. Recently, Frankel and his colleagues at Case Western Reserve University have made some preliminary studies in this area, but further theoretical studies and clinical observations are indicated.

Also a product of the research program is the *Bulletin of Prosthetics Research* published twice a year by the Veterans Administration and available through the Government Printing Office. Interested parties, by contacting the Research Center for Prosthetics[a] of the Veterans Administration, may have their names placed on a mailing list to receive notification and price of each issue as it becomes available. As the name implies, the primary purpose of this journal is to report results of research. However, it usually carries a number of articles of interest to clinicians in many fields. *Bulletin of Prosthetics Research* BPR 10-21, Spring 1974, contains a consolidated index of the first 10 years, BPR 10-1 through 10-20.

The American Orthotic and Prosthetic Association[b] publishes quarterly *Orthotics and Prosthetics*, primarily to keep prosthetists and orthotists informed of the latest developments, though much is also useful to physicians, therapists, and others. The *ISPO Bulletin* is now published quarterly by the International Society for Prosthetics and Orthotics.[c] This journal, which has been available in several languages, is designed to provide useful information to all members of the rehabilitation team.

The American Academy of Orthopaedic Surgeons, with assistance from the Department of the Army and the Veterans Administration, published the *Orthopaedic Appliances Atlas* primarily as a reference for clinical personnel. Volume 1, published in 1952,[23] is devoted to "braces, splints, and shoe alterations" while Volume 2, published in 1960,[24] provides a comprehensive coverage of artificial limbs. Although obsolete in many parts, these volumes also contain much information valuable to the student and not to be found elsewhere. A completely new book, probably to be entitled *Atlas of Orthoetics,* is currently being prepared by the AAOS to replace Volume 1, and should be published early in 1976. Although a number of texts and newsletters on management of the amputee (regardless of age) have been published since Volume 2 of the "Atlas" appeared, none is as comprehensive as Volume 2, which, of course, needs to be brought up to date.

Prosthetic and Orthotic Practice[25] is a collection of articles based on papers given at an international conference on prosthetics and orthotics held in Dundee, Scotland, in June 1969. (The International Society for Prosthetics and Orthotics and the World Veterans Federation conducted a world congress in Vienna in 1973, and the ISPO held its first international congress in Montreux, Switzerland, in 1974.) These volumes reflect not only the current practices throughout much of the world but also recent research and evaluation philosophy and results as well.

Biomedical Engineering Systems, edited by Clynes and Milsum,[26] contains a number of chapters valuable directly or indirectly. The authors of this review have a chapter, mainly on limb prosthetics, with a shorter discussion on orthotics. Those interested in the growing possibilities of stimulation of paralyzed muscle will benefit from the chapter by Greatbatch on physiological stimulators. Though primarily concentrating on cardiac pacemakers, he also discusses bladder stimulators, implantable materials, electrodes and their polarization problems, and implantable power supplies. Konikoff also provides a chapter on energy sources for implanted devices. Ko's chapter on biotelemetry has obvious implications for research on myoelectric control (and conversely on muscle stimulation). He provides illustrative circuit dia-

[a]252 Seventh Ave., New York, N.Y. 10001.
[b]1440 N St., NW, Washington, D.C. 20005.
[c]P.O. Box 42, DK 2900 Hellerup, Denmark.

grams and very brief comments on materials, sealing, and power supply.

Numerous organizations have occasional papers, sessions, or entire symposia related to prosthetics, orthotics, or both. The Annual Conferences on Engineering in Medicine and Biology[a] are now sponsored by an Alliance for Engineering in Medicine and Biology which includes a number of medical and surgical societies as well as the electronic and other engineering societies who have operated these successful conferences for over 20 years. The Institute of Electrical and Electronic Engineers, the American Society of Mechanical Engineers, and occasionally other engineering groups hear papers at sessions organized by subgroups interested in biomedical engineering, human factors, or biomechanics. Papers are usually printed in a convention record, as preprints, or as monographs.

Orthotics etcetera[27] consists of 28 articles (by 26 authors) intended to be especially helpful to the young physician. Most of the articles are concerned with views on patient care. The preface discusses the origins of terms such as prosthetics and orthotics, and the final chapter on orthotic eponyms defines or provides references to a great variety of devices.

Technical Aids

Most texts and journals noted here are primarily concerned with artifical limbs and braces, or prosthetic and orthotic devices custom-fitted for wear by the individual patient, but there is also a considerable need for those technical aids which are used by, but not worn by, the patient, e.g., feeders such as seen in Figure 4.1–3. For certain regulatory purposes, the Veterans Administration uses "prosthetic device" as a broad term embracing many types of aids. Some enthusiasts for "technical aids," conversely, use that as the general or all-embracing term to include prosthetic and orthotic devices, wheelchairs, crutches, and canes as subsets, and perhaps architectural modifications, automotive adaptive equipment, or special vehicles as further subgroups. Other writers consider prostheses, orthoses, and technical aids (or adaptive devices) as three separate categories.

The International Committee on Technical Aids of the International Society for Rehabilitation of the Disabled and the Handicap Institute of Sweden[b] have cooperated in publishing periodically loose-leaves (texts in English and other languages with ample illustrations) on technical aids from many countries. These groups also have published a variety of reports of conferences on transportation and other problems of the disabled. Mrs. Charlot Rosenberg[28] completed just before her death a slender paperback book of simple aids mainly for cerebral palsy cases. Many of the aids described can be constructed with hand tools.

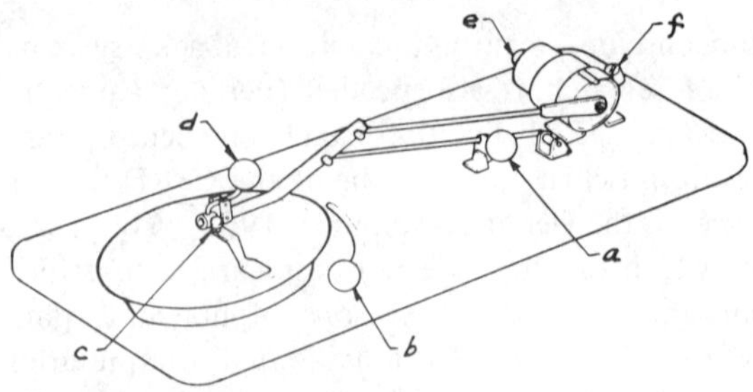

FIGURE 4.1–3. A device designed at the Ontario Crippled Children's Centre to permit children with certain types of cerebral palsy to eat independently. (From *Artif. Limbs*, 13(2), 65, 1969.)

Staff members of the Institute of Rehabilitation Medicine, New York University Medical Center, have published a variety of monographs, texts, and bibliographies. Lowman and Klinger[29] and their associates have published a relatively expensive but detailed and valuable book on a great number of adaptive devices or technical aids, largely available by mail order from sources listed yet mainly completely unknown to most physicians, practically all patients, and well-meaning private inventors who attempt to help disabled friends.

Numerous groups have worked on wheelchairs. Peizer and Wright[30] have summarized the results of 5 years of evaluation by the Bioengineering Research Service of the Veterans Administration Prosthetics Center of a great number of wheelchairs, curb or stair climbers, and accessories. These efforts included mechanical tests in the laboratory, limited trials on volunteers, and broader clinical trials of some devices with the cooperation of several VA hospitals. The paper proposes standards for conventional wheelchairs and criteria for powered wheelchairs. Lipskin[31] is

[a]ACEMB, Suite 1350, 5454 Wisconsin Avenue, Chevy Chase, Md. 20015.
[b]Handikappinstitutet, Fack, 161 03 Bromma 3, Sweden.

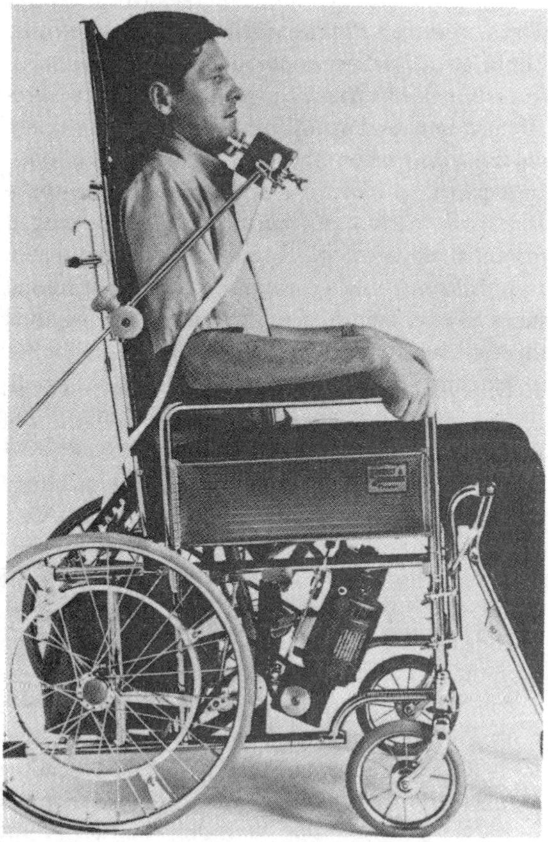

FIGURE 4.1–4. Everest and Jennings Chin Control for electrically powered wheelchair. (From *Bull. Pros. Res.*, 10-17, 221, 1972.)

actively evaluating in the laboratory and in VA hospitals more elaborate control systems, e.g., Figure 4.1–4, for electrically powered wheelchairs used by high-level quadriplegics and other patients. The most difficult cases have such severe involvement of arms (as well as legs) that radial pegs on the driving rim of a manually propelled chair are out of the question and the hand-operated tillers or joy-stick switches often used to control electrically driven chairs for lower-level quadriplegics are inadequate. Chin-operated switches, tongue switches, a sonic control detecting humming of notes, a photodector sensing the difference between iris and cornea as the eye is moved, and a pneumatic method controlled by blowing and sucking are among the systems available for using

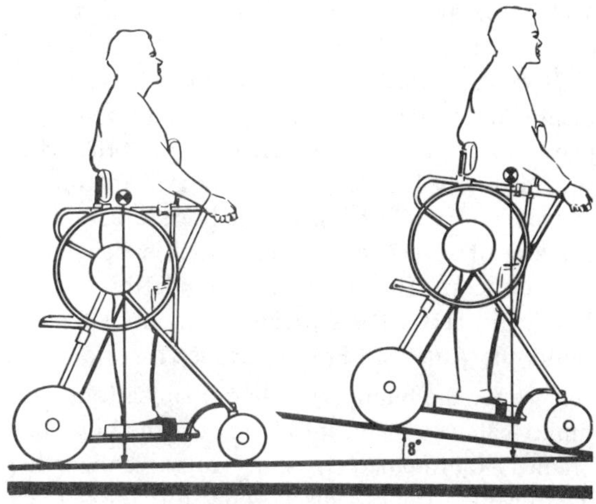

FIGURE 4.1–5. Shift for c.g. projection when Stand-Alone is on (a) level plane and (b) 8° incline. Hand wheels drive rear wheels by chains. Adjustable pads hold paralyzed user in standing position. (From *Bull. Pros. Res.*, 10-2, 7, 1962.)

remaining abilities that are under evaluation, but probably in need of further development and reevaluation. The progress reports of VAPC in each issue of the *Bulletin of Prosthetics Research* also typically discuss wheelchairs.

Several devices, supplementing wheelchairs, are available to provide motion in other than the sitting position, to increase head height and arm reach, or to shift pressures from the ischial tuberosities and thighs to other parts of the body. Peizer and Bernstock[32] reported good biomechanical evaluation and generally favorable response to the Stand-Alone (Figure 4.1–5) therapeutic aid during a field test involving 32 subjects. Eagleson et al.[33] reported on a somewhat curved "surfboard" placed obliquely over a wheelchair frame so the user, in the prone position and partially "standing" against slanted footrests with his head higher than usual, could propel himself.

At least two magazines attempt to bring information on new technical aids, transportation methods, and other matters directly to severely disabled patients. The *Rehabilitation Gazette* (formerly *Toomey J. Gazette*)[a] is the publication, of a group of "iron-lung" polio patients, other quadriplegics from high spinal cord injuries, and their friends. Limited funds typically permit only one issue per year. *Accent on Living,*[b] a quarterly, appears to cover a wider range of disabilities.

[a]P.O. Box 149, Chagrin Falls, Oh. 44022.
[b]P.O. Box 726, Gillum Road and High Drive, Bloomington, Ill. 61701.

These journals also provide a heartwarming source of inspiration.

It would seem highly desirable that engineering faculty and students, before attempting to invent a page-turner, a novel wheelchair, or other aid, should review the designs already available in the above references, other available literature, and the catalogs of several companies which specialize in such devices. (The American Foundation for the Blind, the Howe Press of Perkins School for the Blind, the American Printing House for the Blind, Science for the Blind, and other organizations have similar catalogs of technical aids for the blind like brailled thermometers, slide rules, and other instruments, games, maps, etc. for the totally blind, and various forms of magnifiers for the partially sighted. There are also at least half a dozen designs, in varying stages of commercialization, of closed-circuit television cameras and monitors to allow independent adjustment of magnification, brightness, contrast, and usually reversal of black and white.)

FUNDAMENTAL STUDIES

Locomotion Studies

Several groups in the United States and Canada are equipped to carry out advanced studies of locomotion building upon the classic studies at the University of California (Figure 4.1–6). These groups have met repeatedly under the auspices of the Committee on Prosthetics Research and Development to formulate methods so that the data collected by any one group can be used easily by any other group.[34] In recent years a well-equipped locomotion laboratory has been established at the Biomechanical Research and Development Unit, Roehampton, London, England. Lamoreux[34a] has provided an extensive survey of kinematic aspects, especially computer techniques, with an extensive bibliography.

Meanwhile, a number of papers have been published that provide further understanding of various aspects of human locomotion. To supplement a substantial series of papers by Murray and various colleagues on locomotion of normal men of various age groups and of men with specific impairments, Murray, Gore, and Clarkson[35] analyze walking without support by patients with hip pain. Lurching of the head and trunk was common, and 19 of the 26 abducted the arm on the painful side, presumably to shift center of gravity and thus reduce compressive muscular forces. A cane or walking stick on the sound side would also have been effective in reducing these forces in accordance with well-known biomechanical analyses, and many of the patients normally did use a cane when walking long distances.

A paper by Murray, Kory, and Sepic[36] describes walking patterns of normal women, developed from a study involving 30 subjects of various ages, each wearing high-, medium-, and low-heeled shoes. The findings should be useful to physical therapists responsible for gait training.

Seireg, Murray, and Scholz[37] described an instrumented cane, or walking stick, definition of axes, and sample data, and soon afterwards[38] presented data on 53 men with various unilateral disabilities walking with a single cane. There were wide variations in forces and timing between subjects, e.g., 20 patients with bone or joint disabilities applied greater forces for longer periods, while 20 hemiplegics tended to exert one or more short, sharp peak forces during a long-sustained contact with the floor. A given patient walked with relatively reproducible loading patterns during successive steps. The low bending moments showed that the patients tended to apply loads close to the cane shaft. In addition to the obvious use of the cane under nearly vertical and axial loading as a prop, there were other uses for pulling, pushing, or restraining forces and for sensory feedback. (One may imagine that the series of sharp peak forces in the hemiplegics was at least partially the result of balancing tremor and spasticity but perhaps also partially caused by detecting and correcting through the cane incipient loss of balance which was not readily detected and corrected by the hemiplegic leg.)

Robinson[39] has described a cane with a load cell. He found striking differences in loading between contralateral and ipsilateral canes, of which the patient was unaware, with unilateral hip involvement. Plank,[40] working with Frankel and Burstein, has also analyzed the forces on a crutch.

Kennaway,[41] in a unique paper, discusses the factors affecting friction between crutch tips and hazardous walking surfaces. He concludes that high frictional resistance on well-lubricated surfaces is likely to be obtained with high hysteretic rubber, breaking up of the rubber surface by as many edges as practical, soft (50 or less on the Shore-A scale) material, and natural rubber or

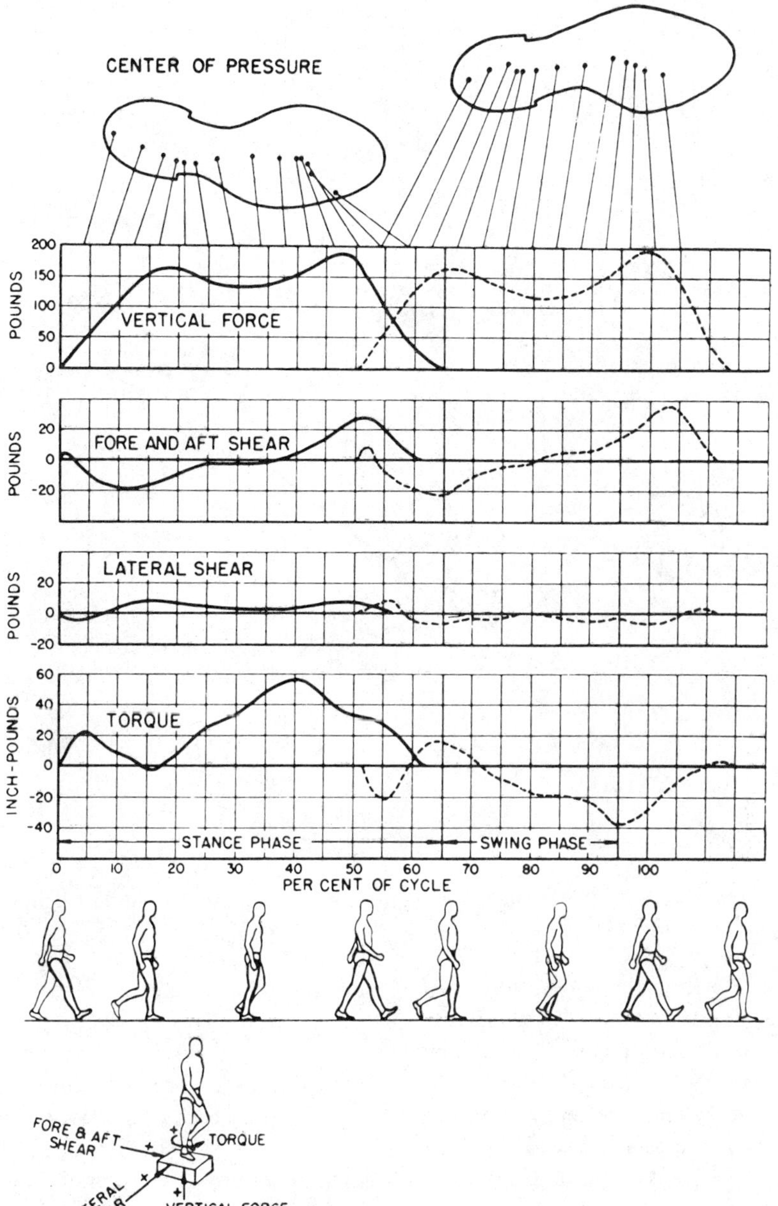

FIGURE 4.1–6. Typical results of force-plate studies for a normal subject during level walking, illustrative of the information obtained at the Biomechanics Laboratory, University of California (Berkely and San Francisco), in their study of human locomotion.[13] (From *Human Limbs and their Substitutes,* Advisory Committee on Artificial Limbs, National Academy of Sciences-National Research Council, McGraw-Hill, New York, 1954, 453. With permission.)

oil-extended natural or styrene-butadiene compounds. He considers that use of crutch or cane tips is so variable that clinical trial is preferable to simulated service testing. Nevertheless, the correlation of force data from instrumented crutches and canes and from forceplates and combination with the concepts outlined by Kennaway should assist not only the interpretation of clinical trials but also the further refinement of these often-neglected but crucial accessories.

Shambes and Waterland[42] describe the body-sway characteristics during stance of a quadrimembral amputee with bare stumps of partial feet

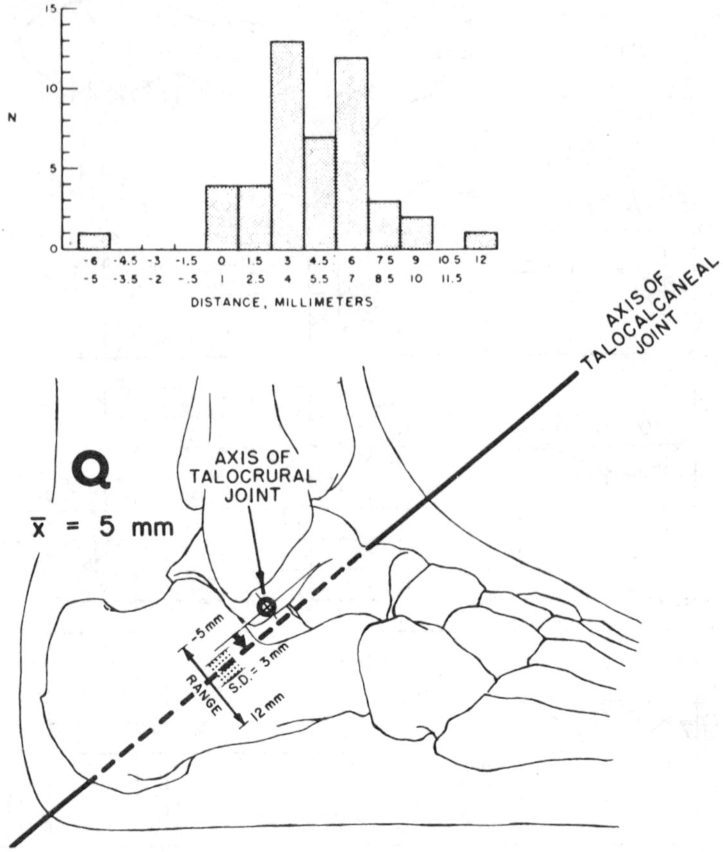

FIGURE 4.1–7. Typical data on anthropometry of the foot from Isman and Inman.[43] (From *Bull. Pros. Res.*, 10-11, 127, 1969.)

and with well-worn orthopedic shoes, which are interesting but apparently not broadly significant. (They submitted a further paper[42a] analyzing body sway with better-fitting shoes; the patient apparently still can balance better with direct sensory feedback from the bare stumps.)

Individual joints as well as total locomotion patterns have been studied. Isman and Inman[43] emphasize the wide variation in the degree of joint motions of the foot-ankle complex among individuals (Figure 4.1–7).

Morrison has provided two papers[44,45] describing forces acting about the knee joint. This information should be useful to those engaged in reconstructive surgery of the knee joint, the design of parts for knee-joint replacement, and the study of lubrication of the knee joint.

Johnston and Smidt,[46] as part of a comprehensive symposium on the currently exciting topic of surgical replacement of the total hip joint (both the head of the femur and the acetabulum), describe measurements of hip motion by electrogoniometers. This objective measurement at the University of Iowa of substantial numbers of patients with diseased hips, before and after treatment by each of a variety of widely advocated methods, is a valuable attempt to reduce subjective opinion and to guide surgical judgment.

Chapman and Kurokawa,[47] in reporting on studies involving transverse rotations of the human trunk during locomotion, show that rotation of the pelvis and the upper part of the torso is not quite 180° out of phase, as has been reported previously. Also, no significant differences were found between male and female in this respect.

Energy consumption during locomotion has been measured by a variety of techniques in a number of laboratories, notably by Ralston at the University of California. Corcoran and Brengelmann[48] have measured oxygen uptake by having the subject (normal or handicapped) walk next to a velocity-controlled cart carrying the relatively bulky respirometer. They agreed with Ralston and others that the comfortable speed of walking selected by each subject is at or near his minimal oxygen consumption per meter walked. Stoboy et al.[49] have measured energy expenditure during wheelchair propelling.

Needed at this time is a compilation in one volume of all that is known about human locomotion. Such a synthesis has been a long-term goal of Inman and his colleagues at the University of California.

Pressure Studies

It has only been in recent years that research groups have directed attention to the effect of pressure on human tissues, primarily because means of measuring forces and pressures between a mechanical device and the human body without significantly disturbing the original condition have not been available. To explore measurement techniques that might be useful, the Committee on Prosthetics Research and Development in 1968 held a conference on the subject.[50] Cochran[51] reviewed not only the principles but also a wide variety of methods for measuring forces and pressures involved in clinical efforts to treat abnormalities of the musculoskeletal system by corrective and supportive forces. He supplies an extensive bibliography.

Appoldt, Bennett, and Contini[52] studied the influence of the thickness of the sensing element on data obtained. To no one's surprise, an element even 1/16-in. thick protruding into the socket and thus into the stump produced an increase in the pressure produced when used to measure pressure between an above-knee stump and a total-contact socket. The effect varied considerably, according to the authors, because of the different "densities" or stiffnesses of tissues in various parts of the stump, so the error caused by a gauge of finite thickness at a bony area with only a thin layer of soft tissue (where pressure is of the greatest clinical interest) would be particularly serious. The same group reported on a related study involving measurement of tangential (shear and frictional) unit forces in above-knee sockets.[53] From data collected with a sensor of their own design (Figure 4.1–8) on two subjects, the authors conclude that, in a well-fitted socket, tangential unit forces are of little significance. In general, tangential unit forces were found to be small, ranging from 0 to less than 4 psi.

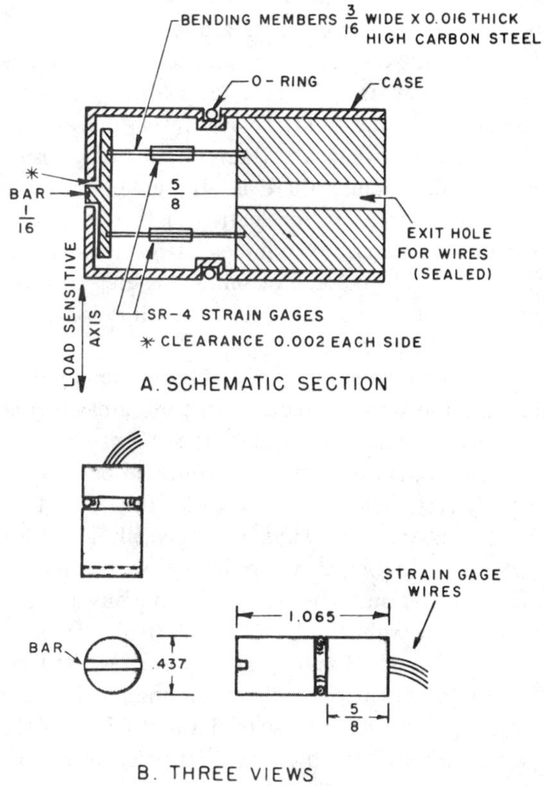

FIGURE 4.1–8. Tangential pressure transducer developed at New York University.[53] (From *Bull. Pros. Res.*, 10-13, 73, 1970.)

In an attempt to collect more meaningful data than that obtained by a single transducer, Correll developed a five-point array which, with the aid of a computer, is intended to yield a pattern of pressures over the area.[50] The technique was used by Sonck, Cockrell, and Koepke[54] in an attempt to analyze the merits of silicone gel as a liner for sockets for below-knee prostheses. Data collected on 23 patients showed that the sockets lined with silicone gel produced lower loads in general when compared to sockets with no liner or with a liner of Kem-Blo® sponge rubber. However, the experiment is not completely convincing, since there is no reference to quality of fit of the sockets nor to reproducibility of the data. Further, there was no control series, in that most patients were provided with only one prosthesis.

A very simple approach to pressure measurement that may have wide clinical application is the pressure-sensitive membrane developed by Brand at the U.S. Public Health Service Hospital in Carville, La.[55,56] with the cooperation of South-west Research Institute. The membrane, consisting of microcapsules containing a dye suspended in a 0.050-in.-thick sheet of open-celled polyurethane foam, may be used flat or made into slipper socks, gloves, or stump socks. Dye is released from the microcapsules when a certain amount of pressure is applied. Because there is also an effect from mechanical fatigue, the patient is instructed to perform a prescribed number of motions (steps, grips, etc.). The colored membrane is then compared with a standard color intensity index to estimate the pressures.

Both Brand and many others have attempted to provide footwear to reduce risks of damage from excessive pressure to feet made insensitive and often deformed or partially amputated because of leprosy (Hansen's disease), diabetes, or other conditions. The problems are especially grave for very poor patients doing heavy work in hot climates. Various types of clogs, generally padded with foam rubber, have been tried. Tuck,[57] Jopling,[58] and Mondl et al.[59] have reported on use of Plastazote, a resilient thermoplastic foam which when hot can be shaped directly against the patient's body. Its thermal diffusivity is so low that the patient is not burned.

Another simple but useful approach in measuring the effects of pressure, a matrix of switches in a plastic envelope which can be inflated to open the contacts, is given by Mooney et al.[60] in a comparative study of seat cushions and their relationships to decubitus ulcers. Unfortunately, none of the available cushions tested was safe for prolonged sitting by paralyzed, insensitive patients.

Bush,[61] in a paper receiving the Baruch award, reviewed methods of pressure measurement, then tested normal individuals sitting in wheelchairs. Like his predecessors, he showed that allowing the feet to hang free would reduce pressure on the ischial tuberosities, but he did not seem concerned about the potentially high pressure at the fulcrum between the anterior edge of the chair seat and the lower surfaces of the thigh with consequent restriction of return circulation from the distal portions.

The project at Seattle under Burgess on immediate postoperative fitting of prostheses and the Veterans Administration Prosthetics Center have collaborated in efforts to measure pressure within the rigid dressing and forces transmitted to the temporary prosthesis. Figure 4.1–9 shows an instrumented temporary prosthesis.

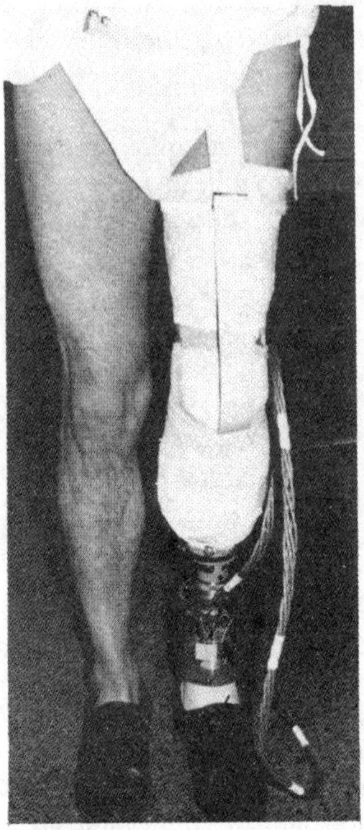

FIGURE 4.1–9. Artificial leg fitted with instruments to measure the forces and pressures taken by the prosthesis and by certain areas on the stump during postoperative care. (Courtesy Veterans Administration Prosthetics Center.)

To date, no truly satisfactory methods are available for measuring pressure between devices and the human body and thus for studying in the laboratory and especially clinically the quantitative effects on human tissues of these pressures, times of exposure and relief, and moisture levels. When these problems in engineering and in biology – more complex than mechanical fatigue and creep of metals – are solved, great advances in rehabilitation and orthopedic surgery can be expected. Without awaiting full understanding, though, bioengineers and clinicians can begin to use – and improve upon – the limited tools, data, and tentative concepts.

UPPER-EXTREMITY PROSTHETICS

During the past decade and a half, an extraordinary amount of effort, especially in Europe, has been devoted to the application of external power to upper-extremity prosthetics, particularly with myoelectric control. This emphasis no doubt arises because most engineers can be of help in solving some of the problems of the physically disabled. As we noted in the introduction, enthusiasm for external power seems to be the "bait" which attracts engineers; eventually, most grow to appreciate the importance of fitting of sockets and harnessing which was noted in 1961[62] and 1966.[63]

The Yugoslav Committee for Electronics and Automation (ETAN) has sponsored symposia in 1962, 1966, 1969, and 1972 on the use of external power in prosthetics and orthotics. These symposia have attracted an ever-increasing number of participants. The proceedings[64] of the third international symposium contain reports on nearly all projects in relevant topics throughout the world. The report is divided into sections as follows:

- General Considerations
- Hand Prostheses
- Manipulators
- Legged Locomotion Studies
- Functional Stimulation
- Components

Some of the papers are excellent. As a whole, the part on upper limbs indicates that the hardware, or component, designers are still more advanced than those who are concerned about the control of the devices. It was demonstrated during evaluation of the Alderson arm in the early Fifties[65] that the control of externally powered upper-extremity prostheses must be, for the most part, carried out subconsciously if the patients are to accept them. (The patients could perform mental arithmetic calculations or operate the Alderson arm but not both tasks simultaneously.) This theory is accepted by many, but, unfortunately, less experienced investigators spend a lot of time developing interesting hardware without giving adequate consideration to the control problem.

Electrical signals from the contraction of muscles (myoelectric signals, often loosely termed electromyographic or EMG) as a source of control have intrigued investigators in many countries for about two decades. Electric artificial hands for below-elbow amputees are now available commercially from sources in Russia, Austria, Germany, and Italy for those who feel that the

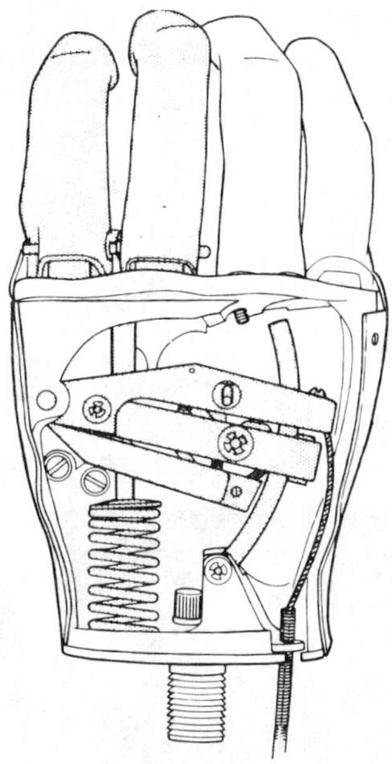

FIGURE 4.1–10. The voluntary-closing hand developed by the Army Prosthetics Research Laboratory.[13] The hand is covered with a thin, lifelike, plastic glove. (From American Academy of Orthopaedic Surgeons, *Orthopaedic Appliances Atlas, Vol 2 – Artificial Limbs,* J. W. Edwards, Ann Arbor, Mich., 1960, 32. With permission.)

advantages offered are worth the additional costs compared with the harness-operated APRL hand or hook (Figures 4.1–10 and 4.1–11). The type of harness shown in Figure 4.1–11 provides rapid operation of and substantial sensory feedback from terminal devices of either voluntary-closing (as shown) or voluntary-opening, spring-closing types.

Evaluation is a major interdisciplinary problem, too often overlooked. The Yugoslav report has a series of chapters on various forms of evaluation of the Belgrade, Italian, and Austrian electric hands, and of a multifunction control. The volume also includes the history of development and very limited trials on one patient of a Japanese hand and the dynamic assessment of above-knee prostheses (mainly offering increased stance-phase knee stability). Evaluation should occupy a greater portion of future efforts, budgets, and publications.

To provide guidance to developers, CPRD sponsored three workshops on external power[17],

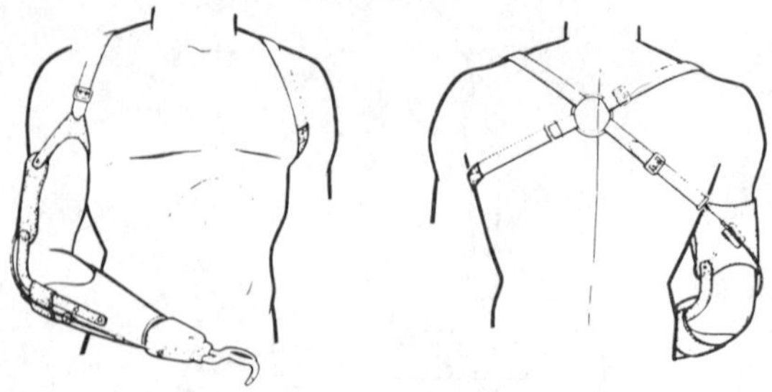

FIGURE 4.1–11. Typical harness pattern for body-operated below-elbow prosthesis. (From *Artif. Limbs,* 7(2), 29, 1963.)

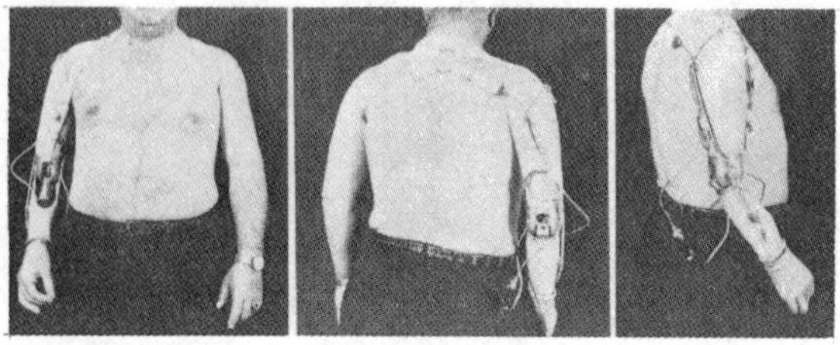

FIGURE 4.1–12. Above-elbow amputee with prosthesis using the electric elbow developed at Rancho Los Amigos Hospital. This is the only electric elbow readily available commercially in the United States today. (From *Artif. Limbs,* 15(1), 75, 1971.)

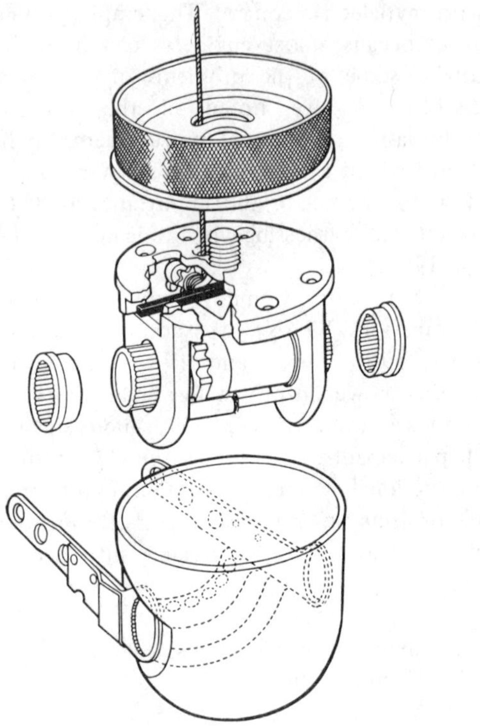

FIGURE 4.1–13. The artificial elbow with an alternating lock operated by shoulder depression, developed in the late Forties by Northrop Aviation, is still used almost universally by above-elbow amputees. (From American Academy of Orthopaedic Surgeons, *Orthopaedic Appliances Atlas, Vol. 2 – Artificial Limbs,* J. W. Edwards, Ann Arbor, Mich., 1960, 50. With permission.)

[66,67] during 1968-70. In addition, it sponsored the evaluation of all externally powered elbow units then available,[68] e.g., Figure 4.1–12. At each of the meetings efforts were made to develop improved criteria for design and control of externally powered artificial limbs. So far, despite considerable publicity, no externally powered elbow unit has been found to be generally useful by typical unilateral above-elbow amputees in preference to conventional harness-operated locks (Figure 4.1–13) unless they are handicapped by problems in addition to the amputation, such as a fused shoulder joint. Much can still be done by combinations of external power and harness (Figure 4.1–14). A CPRD workshop in 1971[68a] encouraged endoskeletal upper-limb prostheses with improved cosmetic covers, recommended expanded investigation of upper-limb sockets and harnesses, and reviewed the status of numerous externally powered and harness-powered devices.

The Veterans Administration (with the cooperation of two Army hospitals) is completing the report on a clinical application study of external power for young veterans, including power-driven elbows for above-elbow amputees. These VAPC elbows, based on the harmonic-drive principle, were designed to fit within the dimensions of the conventional mechanical elbow lock.

The VA clinical application study also includes the system that seems to offer the most to the below-elbow patient, the self-contained, self-suspended electrical prosthesis noted above, Figure 4.1–1, developed by Childress and Billock[9] at Northwestern University, including modification of the Münster fitting technique developed by Hepp and Kuhn. One of the valuable results of the work at Northwestern University has been this recent emphasis on development of improved designs for sockets in an effort to eliminate harnessing. (The designs, which take into consideration anatomical and physiological principles, promise improvement for some types of lower-limb amputees as well as those with amputations through the upper limb.) Control of the present unit uses electromyographic signals from remnants of forearm muscles beginning immediately after

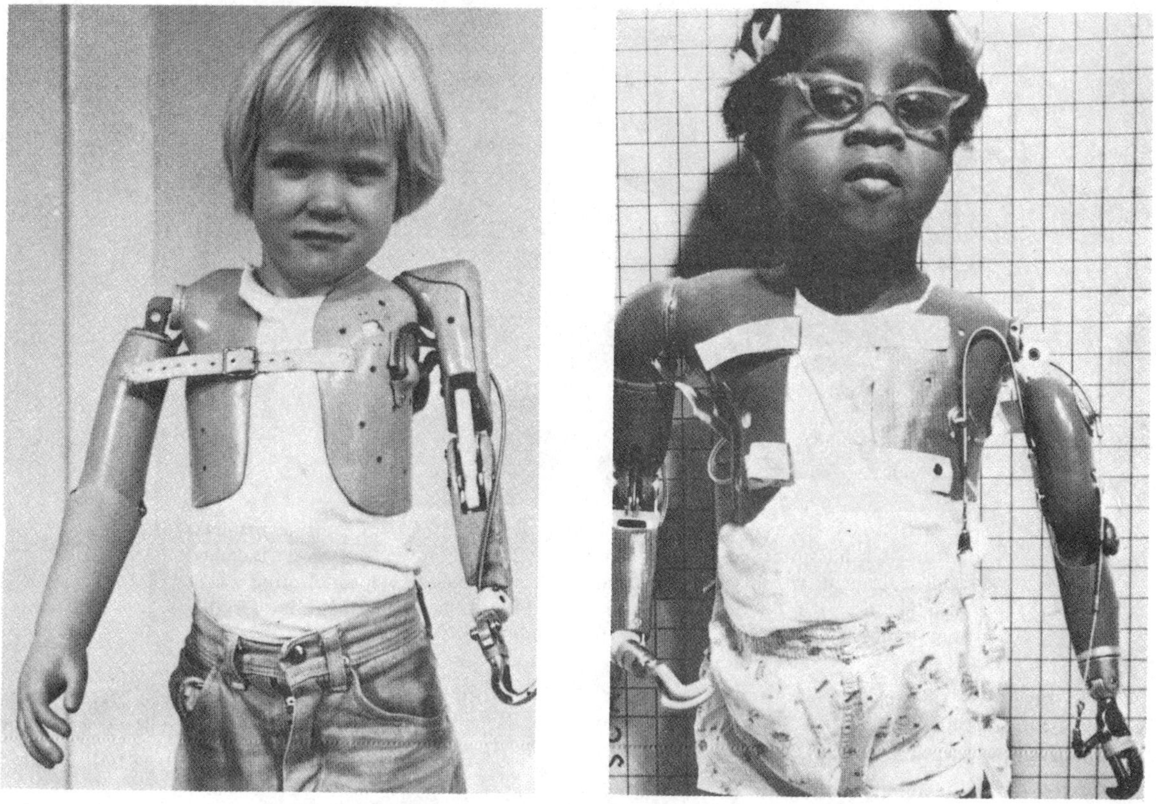

FIGURE 4.1–14. Two children with bilateral losses fitted experimentally with a combination of externally powered and body-powered devices. (From Maternal and Child Health Research, HEW-sponsored work.)

amputation surgery (Figure 4.1–15). The batteries, control unit, and actuator are enclosed in the permanent arm and hand. Unfortunately, an externally powered hook is not available for evaluation, though several groups are working on the problem.

In their excellent article Childress and Billock[9] point out that elimination of harness and all other paraphernalia external to the prosthesis makes such units much more desirable than previously available externally powered units. The indications for prescription of the Northwestern system using a hand are given as follows:

1. The amputee desires this type of system and can afford it.
2. He is engaged in light work (e.g., student, salesman, etc.).
3. He is a unilateral below-elbow amputee.
4. He has suitable stump musculature.
5. Appearance is important to him.

In the same issue of the *Bulletin of Prosthetics Research* Peizer, Wright, Pirrello, and Mason[69] attempt to make a case for the use of mechanically operated electric switches mounted in the harness (possibly lighter and more comfortable than usual) to control electrical components for above-elbow amputees. Switches presumably should be simpler, less expensive, and possibly more reliable than myoelectric control. In contrast, there is some evidence that myoelectric control may be superior in terms of "naturalness" (if appropriate muscles remain available), minimal need for excursion and force, and possibilities for eliminating harness completely. (The possibility of combining the simplicity and economy of switches and some of the advantages of myoelectric control by placing microswitches in the socket wall to be operated by muscle bulges has received little attention. The usual fear of inadvertent operation from displacement of the socket on the stump by external loading may now be partially overcome by better fitting of the socket brim as well as total contact of the distal end.)

The most promising sophisticated control

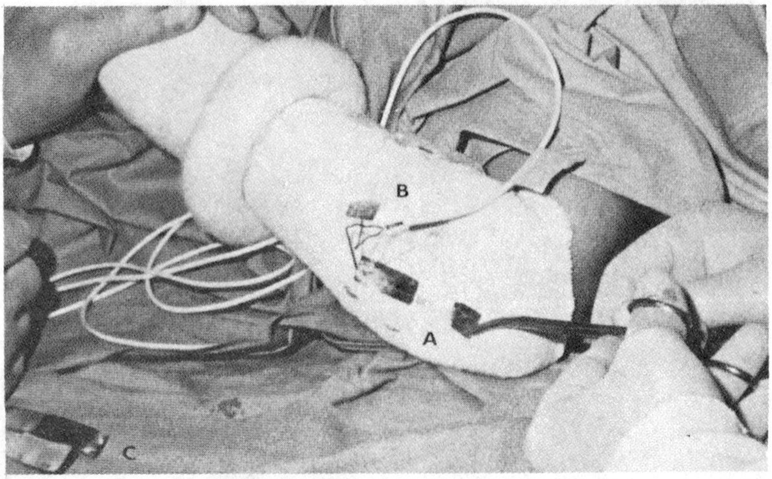

FIGURE 4.1–15. Attachment of electrodes for myoelectrically controlled artificial hand immediately after surgery. At A the ground electrode is being pulled through two slits in the tubular bandage. The two extensor electrodes can be seen near B after they have been taped down. Two foil electrodes for the flexor muscles appear near C. (From *Artif. Limbs,* 13(2), 60, 1969.)

system presently seems to be the myoelectric system using pattern-recognition techniques.[70,71] Under development at Moss Rehabilitation Center, the system collects signals from nine or more sites about the shoulder and processes them in such a way as to give coordinated and simultaneous motion about the elbow axis, the forearm axis, and the humeral axis. Though a computer is used in research, simple resistance weighting circuits are anticipated for routine use in wearable, portable prostheses.

LOWER-EXTREMITY PROSTHETICS

Improvements in the design of components and in the methods of fitting and alignment of artificial legs for all levels of amputation have been introduced steadily through the years. Some are reflected in patents, others in standard texts[24] and in the materials used at the university-level prosthetics education programs at New York University, Northwestern University, and the University of California at Los Angeles, and in reports of international meetings (for example, Reference 25). Other accounts are widely scattered in medical, engineering, and prosthetics literature. A review of the more recent advances in above-knee prosthetics was given by Wilson[72] in 1968 and in below-knee prosthetics, with the many variants on the PTB, by the same author[73] in 1969. LeBlanc[74] provides an overview of recent developments in Syme prostheses, particularly variants on an elastic liner to provide suspension above the bulbous end of the stump yet allow easy donning.

One of the most important trends in lower-extremity prosthetics has been toward more distal, or lower, levels of amputation, particularly to conserve the knee joint whenever feasible even when amputation is required because of vascular disease (Figure 4.1–16). For many years most general and vascular surgeons felt that amputations for vascular disease typically, or even always, should be performed at the junction of the middle and lower thirds of the thigh or even higher. It was true that such amputations generally *healed* reasonably rapidly, but very often the amputee was rehabilitated only very slowly or not at all. Often the elderly patient was left on crutches or perhaps in a wheelchair, many times with instructions to apply elastic bandages routinely until the stump shrank but with very little supervision to ensure adequate bandaging, prevent hip flexion contracture from prolonged sitting, or maintain the strength of the remaining muscles needed for the relatively difficult task of controlling the free knee joint of an above-knee prosthesis. Sometimes there was no real plan to provide a prosthesis (because of lack of interest, supervision, or financial resources), or the decision was deferred so long that flabby stump, severe hip contrac-

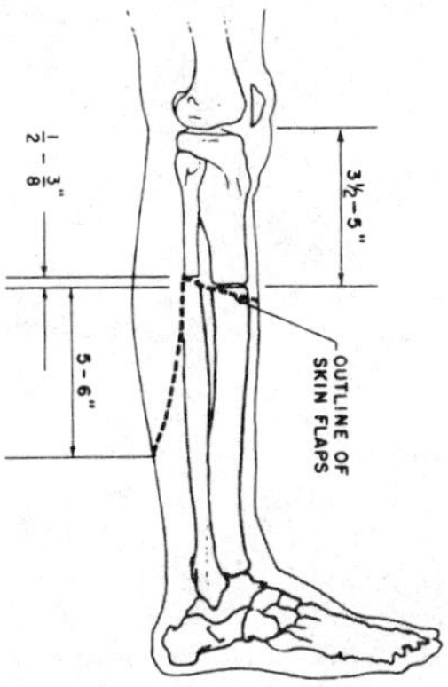

FIGURE 4.1–16. Suggested outline of flaps for a below-knee amputation when circulation is deficient. (From *Inter-Clinic Inf. Bull.*, 8(4), 4, 1969.)

ture, feeble muscles, and low motivation made the chances for prosthetic rehabilitation very poor.

Nevertheless, a few widely scattered voices urged more frequent use of below-knee amputation (even if longer time and more meticulous care were needed to assure healing of the skin below the knee). Pedersen,[75] for example, alone or with colleagues through numerous writings, lectures, and instructional courses, preached this message. There were striking regional differences in ratio of below-knee to above-knee cases because of interest and conviction, without the high incidence of reamputation of below knee to above knee which the pessimists assumed to justify routine primary above knee amputation. Glattly,[76] in reporting a study of amputees reaching a prosthetics facility, or limbshop, for the first time for fitting, noted two major nearby cities with reversal of ratios of below knee/above knee: approximately 20/80% vs. 80/20%, for no obvious geographic, climatic, economic, or real medical reasons.

The patellar-tendon-bearing below-knee prosthesis (Figure 4.1–17) introduced by the University of California, Berkeley, about 1958 has been improved by the introduction of new methods of suspension just above the femoral condyles, and the air-cushion socket[77] (Figure 4.1–18) developed by the originators of the "PTB" provides for a better distribution of weight-bearing loads over the distal areas of the stump in certain instances. CPRD conducted a conference on PTB variants[73] which reviewed and offered a classification scheme for these new approaches.

Swing phase of walking with an above-knee

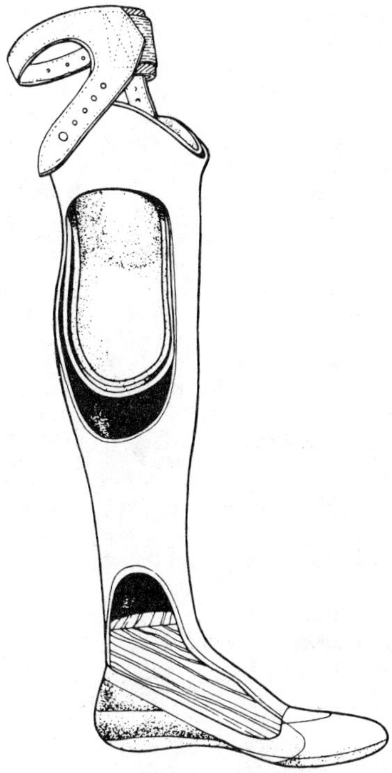

FIGURE 4.1–17. The classical patellar-tendon-bearing prosthesis for below-knee amputees, designed by the Biomechanics Laboratory, University of California (Berkeley and San Francisco). (From *Artif. Limbs,* 13(2), 2, 1969.)

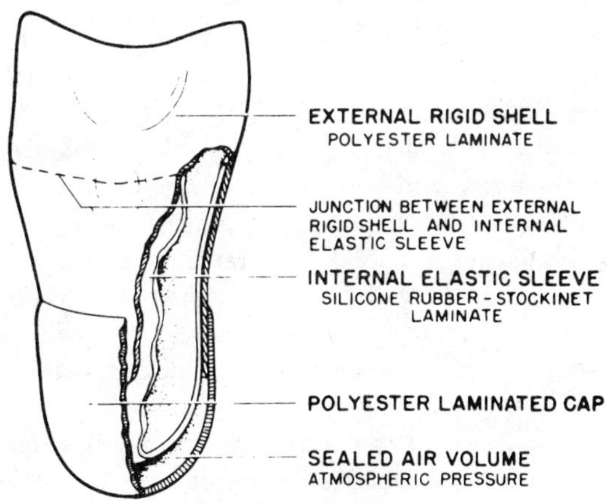

FIGURE 4.1–18. The air-cushion type of patellar-tendon-bearing socket, designed to provide optimum load-bearing pressures on the distal portion of the stump. (From *Artif. Limbs,* 14(1), 24, 1970.)

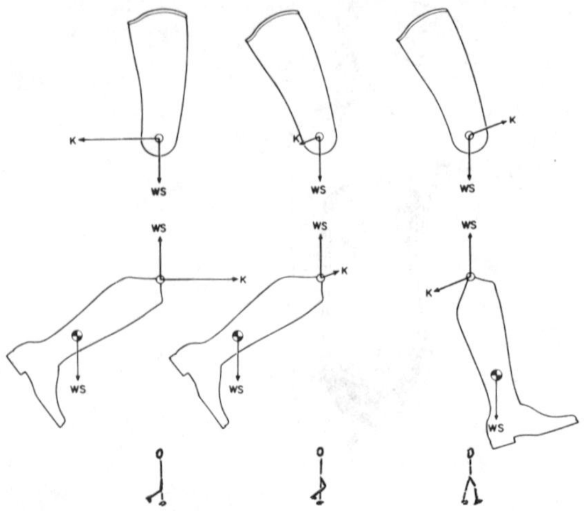

FIGURE 4.1–19. Forces developed during rotation of the shank about the knee joint during forward swing of the thigh. (From American Academy of Orthopaedic Surgeons, *Orthopaedic Appliances Atlas, Vol 2 – Artificial Limbs,* J. W. Edwards, Ann Arbor, Mich., 1960, 172. With permission.)

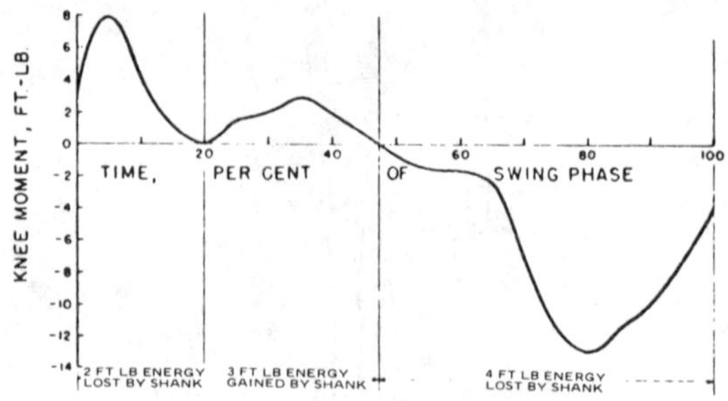

FIGURE 4.1–20. Knee-moment pattern required for natural swing of an artificial leg. This curve was developed by the University of California Biomechanics Laboratory based on data collected during the study of human locomotion. (From *Artif. Limbs,* 12(2), 10, 1968.)

prosthesis (Figures 4.1–19 and 4.1–20), is important to permit graceful gait and modest energy consumption at a variety of walking speeds and cadences. The various hydraulic and pneumatic swing-phase knee control units for above-knee prostheses, e.g., Figures 4.1–21 and 4.1–22, developed during the past 20 years,[78,79] are being well accepted in spite of higher cost. Wallach and Saibel[80] have developed performance curves for tall, medium, and short men using Dr. Murray's data. The DuPaCo Hermes® swing-phase unit is mentioned, although no comparison is made between the theoretical and actual performance. The swing-and-stance unit (Figure 4.1–23) introduced by Mauch in 1968,[81] after prolonged development and evaluation, is being received quite well. Note that these knees absorb and dissipate energy as brakes, at rates increasing with speed of walking. For level walking, at least by unilateral amputees, there seems no need to introduce external power. For climbing stairs or hills, especially by the bilateral above-knee amputee, any attempts at external power would introduce formidable problems of energy storage and of control. The hip-disarticulation case (or the hemipelvectomy) can be fitted successfully with a prosthesis with movable joints (Figure 4.1–24). Several improvements have been made in the socket designs.

The use of temporary prostheses for lower-limb amputees, long advocated by a few but criticized because of being too often characterized by poor fitting and alignment, is now widespread, largely as a result of the introduction of early fitting[82] and immediate postsurgical fitting procedures, e.g., Figures 4.1–25 and 4.1–26.[83,84] More than a hundred articles and brochures have appeared on immediate postsurgical fitting, almost all of them favorable in principle. Burgess and other authors stress the need for a competent, interdisciplinary team and for availability of service at all hours, implying the desirability of an amputation and prosthetic treatment center. Baumgartner[85] of Zurich, Switzerland, for instance, stresses that good results of immediate postoperative fitting of the patient amputated for vascular diseases are possible but are bound to a number of conditions of the patient, the doctor, and the availability of the team. Because his own clinic is only occasionally able to observe all these conditions, it prefers (evidently reluctantly) to use early fitting.

The most authoritative and up-to-date instructions on immediate postsurgical fitting are given by Burgess et al.[84] Although it is highly recommended that clinical personnel undertaking this work be trained formally, this booklet contains sufficient information to enable competent clinic teams to carry out these procedures successfully. The article on early fitting by Goldner et al.[82] is not nearly so specific, but, if good prosthetic practices are followed, successful results can be obtained.

A number of designs of temporary prostheses (Figure 4.1–27 and 4.1–28 have been developed in an attempt to permit economical application of

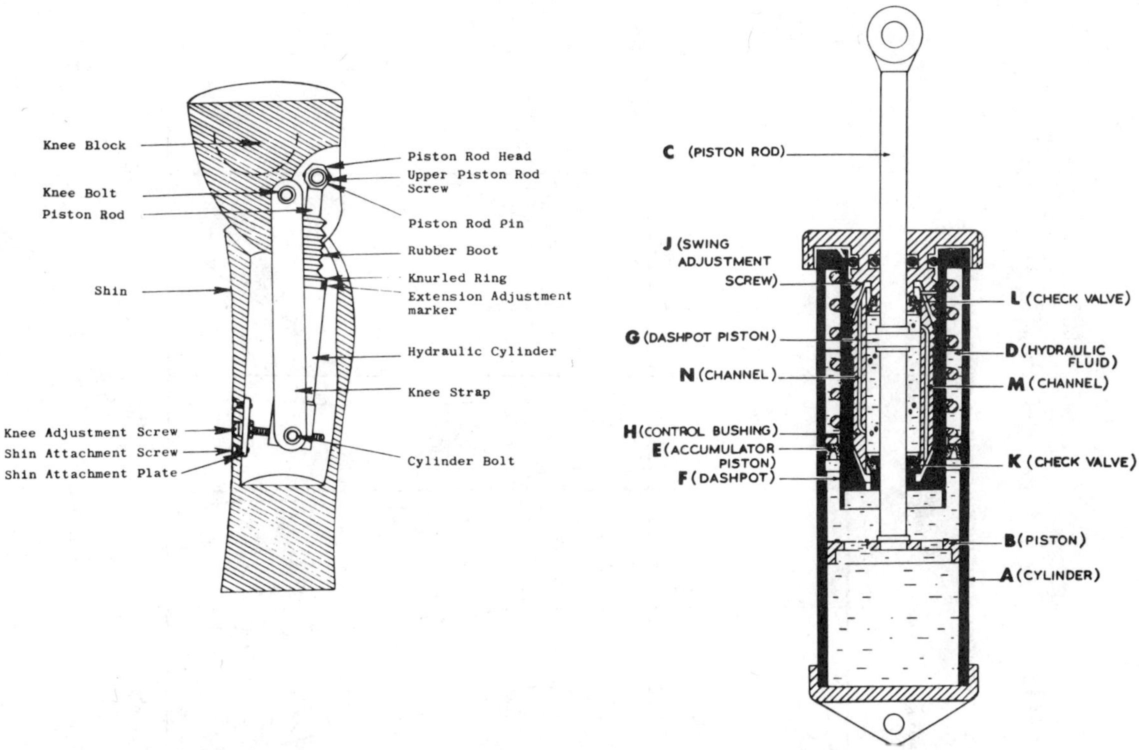

FIGURE 4.1–21. Schematic diagram of the Henschke-Mauch Model B knee unit for control of the shank of an above-knee prosthesis during the swing phase of walking. (From *Artif. Limbs*, 14(2), 8, 1970.)

early fitting, immediate postsurgical fitting, or both types. All of these employ endoskeletal aluminum tubing to carry axial, bending, and torsional loads and all provide for adjustment of alignment. Most are acceptable from the standpoints of function and strength and are marginally acceptable in appearance, at least while used in a hospital.

As yet, no fully satisfactory yet economical cosmetic finishing technique has been developed. Gardner[86] gives a relatively superficial description of one device of his own design, but glosses over some real problems such as man-hours and elapsed time required, and durability. The Committee on Prosthetics Research and Development sponsored a conference on the subject in March 1971, and a report was issued.[86a] There was general agreement among the conferees on the feasibility of prefabricated modular components, on rapid adjustments for size and alignment, and on the desirability of a relatively soft, fleshlike texture and a dull finish (Figure 4.1–29), in contrast to the hard surface and glossy finish of conventional wood, metal, or plastic-laminate exoskeletal prostheses.

The modifier "modular" has been used to describe adjustable pylon-type (endoskeletal prostheses using certain prefabricated components, with some parts standard, some selected from a few stock sizes, and some cut to size to fit the individual patient. These prostheses are of a type that permit rapid assembly and disassembly and perhaps interchangeability without necessarily consisting of measurement modules in the sense used in architecture. The goal is an adjustable prosthesis that is inexpensive yet durable enough so that it can be used not only for the "temporary" (a few weeks) prosthesis but also for the "interim" (weeks or months), and preferably for all subsequent "definitive" prostheses with greater emphasis on cosmetic factors and fatigue strength but less need for readjustment of alignment.

Though there has been since World War I an old concept of modular prostheses, with standard interchangeable parts (including gauging) and tubular structures cut to the necessary length, there always have been practical difficulties. Such structures in principle allowed rapid selection of parts, assembly, and choice of alignment to suit individual needs, utilizing or even extending the possibilities of "adjustable legs" such as the

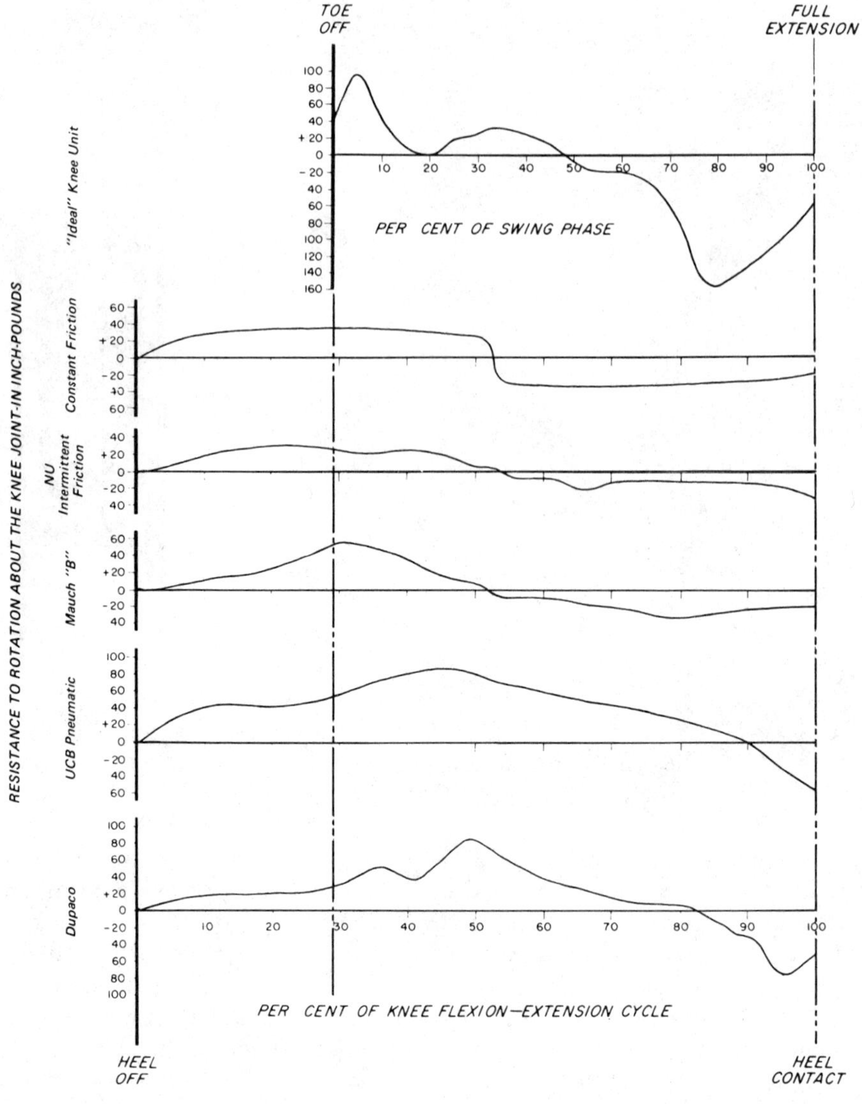

FIGURE 4.1-22. Knee-moment patterns of various swing-phase units for above-knee prostheses. Data were taken at the Veterans Administration Prosthetics Center. The knee units were adjusted for intermediate resistance and were subjected to 43 cycles per minute. (From *Artif. Limbs,* 12(2), 12, 1968.)

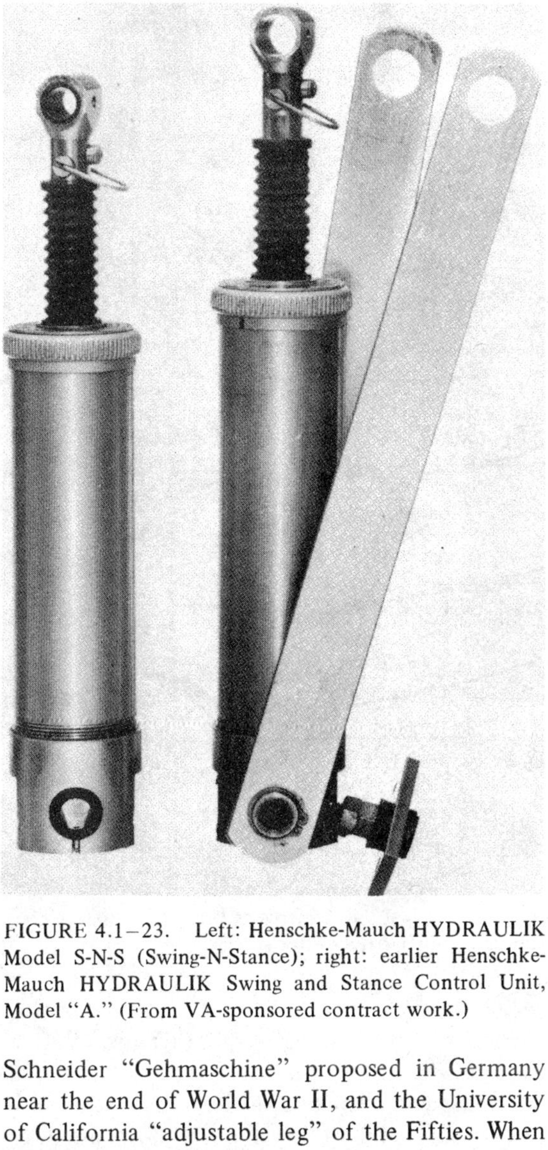

FIGURE 4.1–23. Left: Henschke-Mauch HYDRAULIK Model S-N-S (Swing-N-Stance); right: earlier Henschke-Mauch HYDRAULIK Swing and Stance Control Unit, Model "A." (From VA-sponsored contract work.)

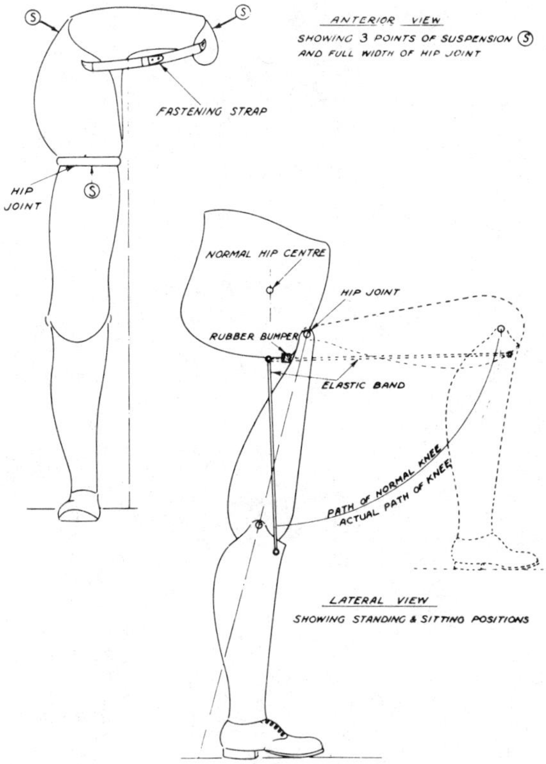

FIGURE 4.1–24. Schematic drawing showing the mechanics of the Canadian-type prosthesis for the hip-disarticulation case. The Canadian type is the most functional prosthesis yet designed for the hip-disarticulation amputee. (From American Academy of Orthopaedic Surgeons, *Orthopaedic Appliances Atlas, Vol. 2 – Artificial Limbs,* J. W. Edwards, Ann Arbor, Mich., 1960, 231. With permission.)

Schneider "Gehmaschine" proposed in Germany near the end of World War II, and the University of California "adjustable leg" of the Fifties. When these adjustable devices were first proposed, conventional trial-and-error alignment often was sadly lacking in precision, forcing the amputee to limp badly (or to use a cane) in an attempt to compensate.

Biomechanical analyses[87] did, indeed, explain why a limp shifting the center of gravity toward the above-knee prosthesis, carrying of a load like a briefcase on the disabled side, or use of a cane in the hand on the unaffected side could reduce gluteus medius forces and thus pressures between the prosthetic socket and the lateral distal aspect of the above-knee stump. Correspondingly, the pressure on the proximal medial aspect could be reduced. Likewise, insightful anatomical plus physiological and biomechanical analyses[88] explained the roles of anatomical alignment to match the remnant of the anteriorly bowed femur, of vigorous hip extension, and of the couples from pressures between the stump and the thigh socket (or by analogy, presumably the cuff in the case of the UCLA flail-leg brace described below) in stabilizing the artificial knee joint against buckling. While some of these concepts evidently were appreciated by Schede,[89] even at the time of World War I, they tended repeatedly to disappear from practice.

Though the growing understanding of biomechanical and anatomical principles allowed better preliminary casting, e.g., Figure 4.1–30, and more accurate "bench" alignment, reducing the gross errors so common in the past, it inspired and guided ever finer adjustment or "tuning" of alignment during the trial of the new prosthesis. Thus, because the newer adjustable "modular"

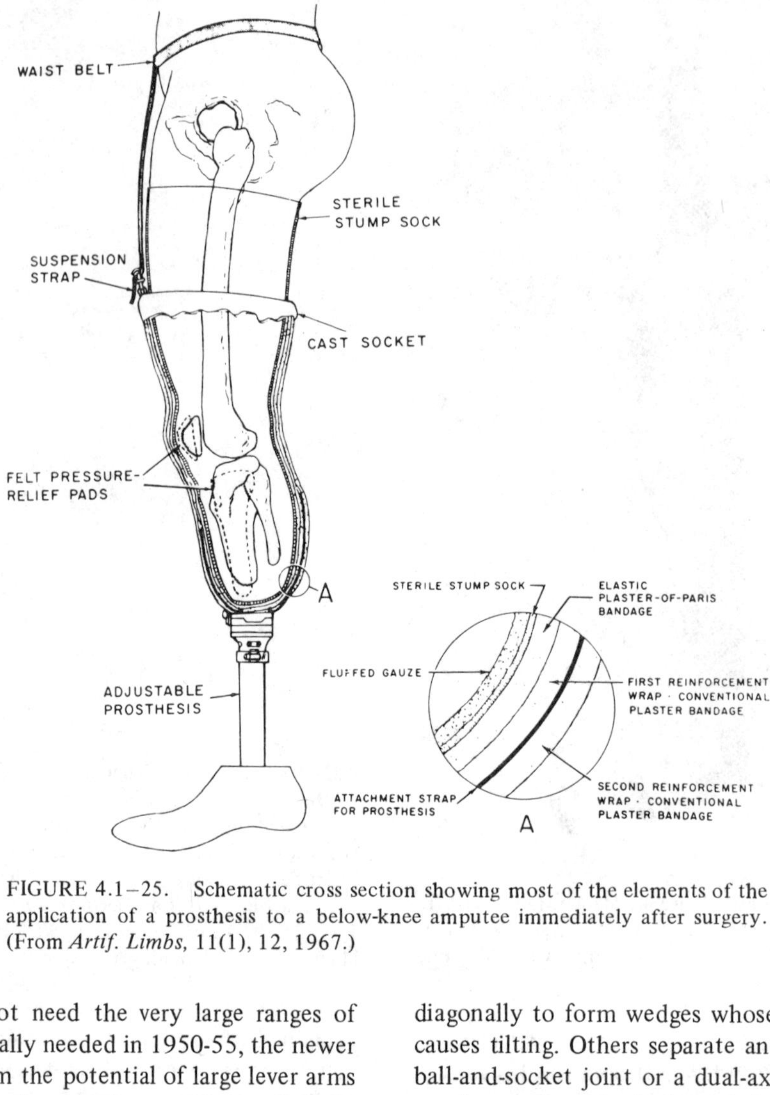

FIGURE 4.1–25. Schematic cross section showing most of the elements of the application of a prosthesis to a below-knee amputee immediately after surgery. (From *Artif. Limbs,* 11(1), 12, 1967.)

prostheses do not need the very large ranges of adjustment typically needed in 1950-55, the newer designs, free from the potential of large lever arms after correcting the relative positions of body weight and point of support, can be lighter and simpler. The greater understanding of biomechanics of "bench" alignment accompanied the presumption that the adjustable element could be replaced by an element cut to the new shape. This idea is balanced by greater urgency to refine (and perhaps, over months, repeatedly to readjust) the alignments in all planes, yet to provide secure locking of all adjustments between deliberate adjustments.

Various recent designs, therefore, furnish an adequate variety of alignment devices in the modular prostheses.[72,73] Some provide ball-and-socket joints at top and bottom of a structural tube so that compensatory angular motions create translation. Other designs use a cylinder cut diagonally to form wedges whose relative rotation causes tilting. Others separate angular motion of a ball-and-socket joint or a dual-axis joint (generally at the same end of the tube) from translatory adjustments. In addition to the considerations of strength and the maintenance of the desired position, there are implications for criteria for rather flexible cosmetic covers. Some of these problems are implied, if not always clearly defined, in CPRD conference minutes.

Adjustment for longitudinal growth of children has been approached in various ways. Some have believed a vigorous child would wear out his prosthesis and require a replacement before growth alone would require change, much as some children wear out shoes. Other clinicians (perhaps viewing different ages, levels of handicap, or level of activity) believe provision for frequent adjustment for growing children is necessary. For these needs, the modular concept, especially with adjust-

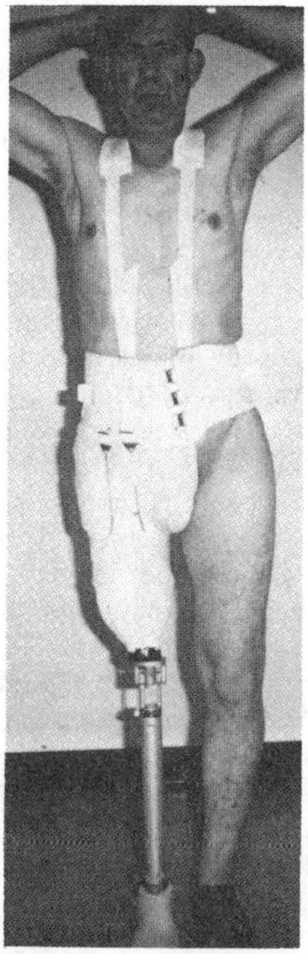

FIGURE 4.1–26. Above-knee amputee provided with prosthesis immediately after surgery. (From Burgess et al., VA-sponsored contract work.)

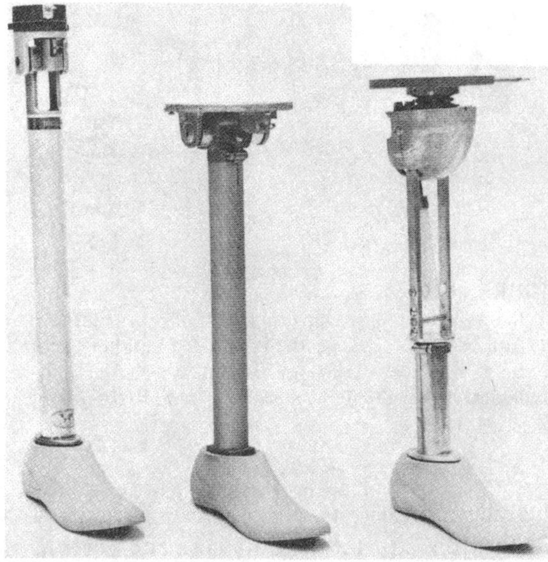

FIGURE 4.1–28. Pylon-type prostheses for above-knee amputees that will accept various knee-control devices. Left, United States Manufacturing Co. unit; center, Hosmer Corporation unit; right, Veterans Administration Prosthetics Center Multiplex unit. (From *Artif. Limbs,* 12(2), 9, 1968.)

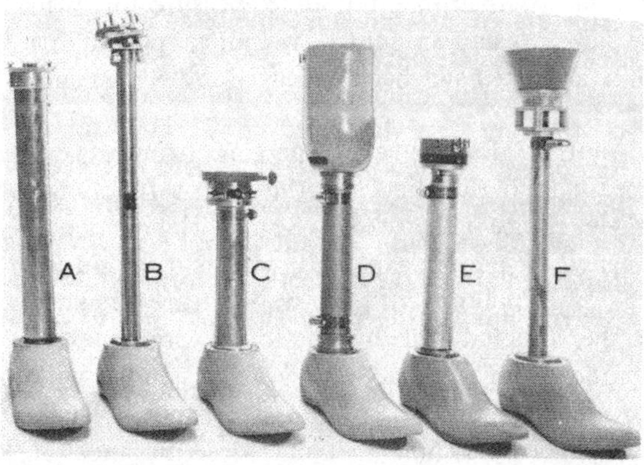

FIGURE 4.1–27. Below knee pylon-type prostheses that can be used for fitting immediately after surgery. A, Hosmer postoperative pylon; B, Northwestern pylon (Hosmer); C, Veterans Administration Prosthetics Center "standard" pylon; D, Canadian "instant" prosthesis (Hosmer); E, United States Manufacturing Co. pylon; F, Finnie-Jig (Arthur Finnieston Co.). Metal straps for attachment to a plaster-of-Paris socket are available, but not shown. (Courtesy of Veterans Administration Prosthetics Center).

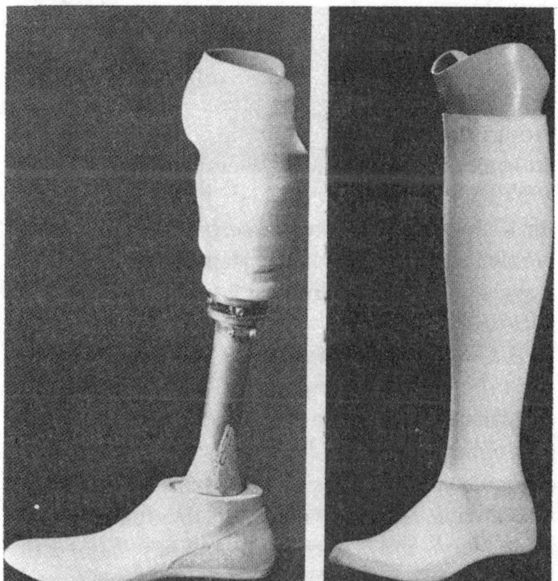

FIGURE 4.1–29. Left, an adjustable below-knee pylon-type prosthesis; right, same prosthesis with carved foam filler to provide support for an elastomeric cover for a natural appearance. (From *Artif. Limbs,* 14(1), 20, 1970.)

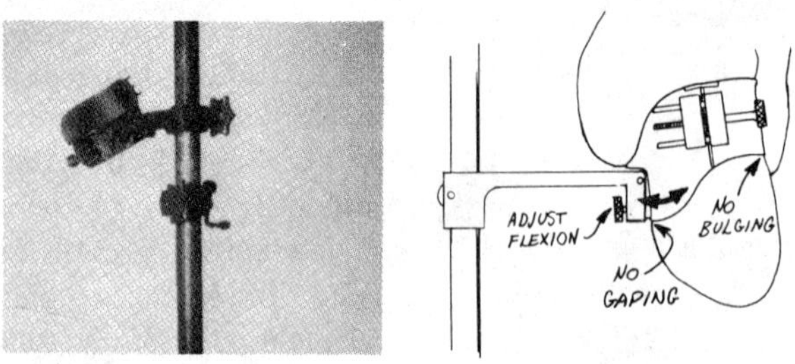

FIGURE 4.1–30. A tool developed by the Biomechanics Laboratory, University of California (Berkeley and San Francisco), to aid the prosthetist in taking a cast of an above-knee stump for use in fabrication of a total-contact, quadrilateral socket. (From *Artif. Limbs,* 12(12), 4, 1968.)

able alignment left in place and opportunity to lengthen the structure and its cosmetic covering, would seem desirable. An economical but better method for giving good appearance and resiliency is badly needed in order to take advantage of the many benefits offered by the modular concept.

Circumferential and diametric measures are also important for both children and adults. One problem in lower-extremity prosthetics that has never been approached to any significant degree is an adjustable socket to maintain biomechanically correct fit by conforming to temporary swelling or shrinking and to slower permanent changes (e.g., atrophy of unused muscles). Perhaps the reason for this deficiency is our lack of detailed knowledge concerning the effects of pressure on various areas of a given stump, especially in the presence of diabetes, degenerative diseases, or growth of children. In 1971, CPRD conducted a small conference to explore certain aspects of this problem. Hopefully, this conference set priorities and encouraged further research.

LOWER-EXTREMITY ORTHOTICS

Although the first 10 to 15 years of the government-sponsored research program were devoted almost exclusively to artificial limbs, it was always felt that much of the fundamental information and some of the lessons learned in limb prosthetics would be applicable in the solution of problems involving orthopedic bracing, or orthotics. Such has been the case, and during the past 15 years considerable effort has been directed toward lower-extremity orthotics. The Committee on Prosthetics Research and Development has sponsored eight workshops directed solely to this area, and has recently engaged in the evaluation of more than a dozen devices. All concerned are aware that the number of patients needing orthotic devices is larger than the number of amputees requiring prostheses, but that problems of orthotic devices are even more complex in some respects than those involving prostheses. Braces are even more limited in bulk because they must be worn outside the body. Muscles may be flaccid or spastic. Sensation may be normal, abnormal, or missing. Bracing is indicated for a wide range of conditions.

An excellent review of principles of lower-extremity bracing was edited by Perry and Hislop,[90] with authors from the staff of Rancho Los Amigos Hospital, Downey, Calif., a center for treatment of numerous chronic diseases. Simplified diagrams are used to illustrate normal anatomy and physiology and then to point out the effects of various diseases or injuries.

One of the first devices to be developed as a direct adaptation from the artificial limb program was the patellar-tendon-bearing below-knee brace (Figure 4.1–31),[91] developed in the Veterans Administration Prosthetics Center to relieve the shank or foot and ankle of some of the weight-bearing function. The PTB brace has proven to be useful in the treatment of a variety of conditions.[92] Instructions for fabrication are available.[93] Instrumentation to measure forces in the side-bars and amount of weight supported in a bilateral case have been reported by Staros and Peizer.[94]

Another development emerging from experience in limb prosthetics is the UCLA functional long-leg brace (Figure 4.1–32).[95] This device is designed to permit a patient with one flaccid leg (e.g., from poliomyelitis) to walk with a free knee on the affected side by employing principles of alignment and ankle action learned in studies with above-knee prostheses.

A paper by Lehmann et al.[96] describes most of the devices under test by CPRD, e.g., Figures 4.1–33 through 4.1–35, as well as some of the newer devices that have been accepted widely in recent years, and contains a bibliography. Little is given in reference to indications for prescription, no doubt because insufficient experience has been had as yet for detailed instructions to be developed. Another paper in the same issue by Lehmann and his colleagues[97] evaluates the

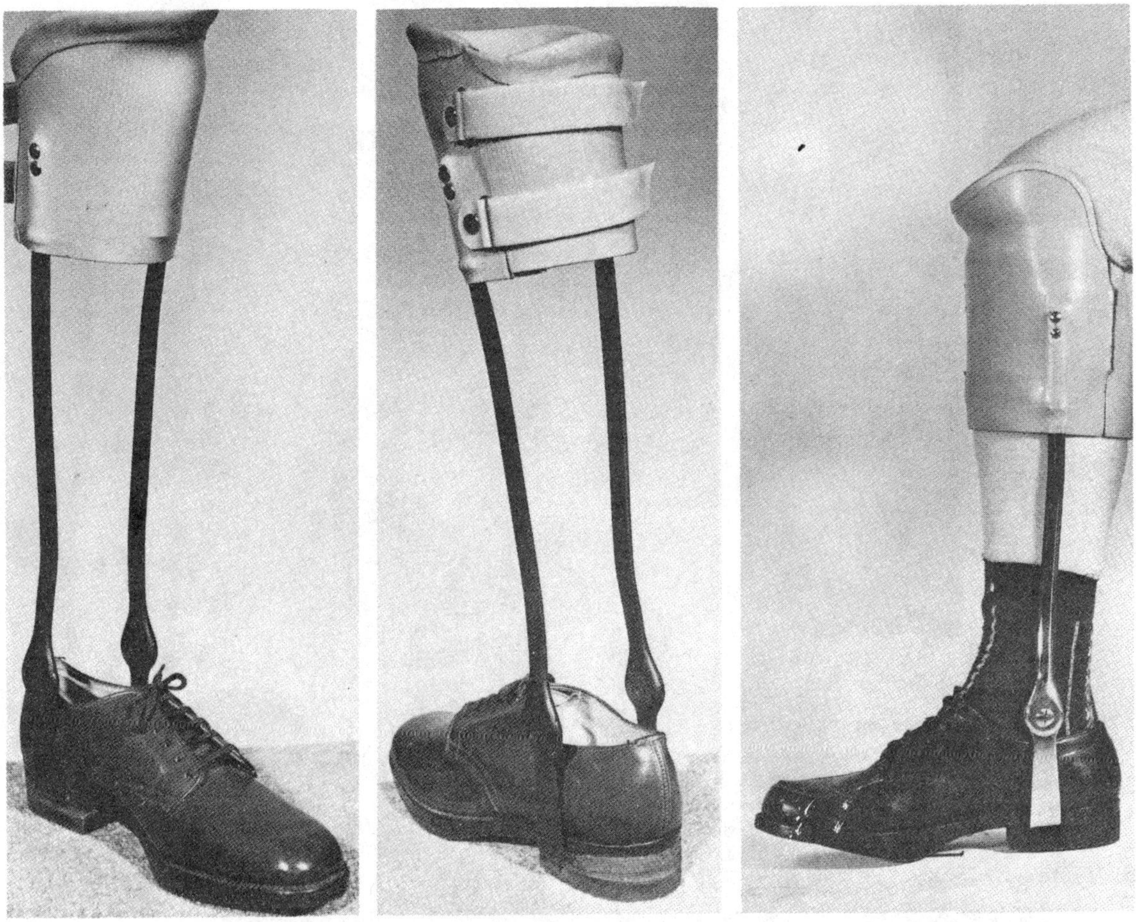

FIGURE 4.1–31. The Veterans Administration Prosthetics Center's patellar-tendon-bearing brace. This device reduces the weight-bearing function of the shank sufficiently to relieve pain in many instances. (From *Artif. Limbs,* 15(1), 48, 1971.)

relative effectiveness of several types of braces intended to remove body weight from the leg; it points out the importance of training the patient to use muscles at the thigh cuff.

The best article to appear so far on the philosophy of orthotics is by Henderson and Lamoreux.[98] They point out that any functional effect of an orthotic device is always the direct result of the forces and moments which it applies or which are applied through it to the human body. They go on from there to emphasize the need for a thorough analysis of the patient's functional deficiencies and the development of a prescription to assist in overcoming the deficiencies without restricting residual functions.

The American Academy of Orthopaedic Surgeons has developed a technical analysis form[99] to assist the clinician in evaluation of the patient, and the functional nomenclature currently being developed by the Committee on Prosthetic-Orthotic Education should prove to be very helpful in developing prescriptions as recommended by Henderson and Lamoreux. These efforts, especially after they are published in the forthcoming *Atlas of Orthotics*, should have profound impact both on research and development and on clinical practices.

In 1958, the University of California at San Francisco, as a pilot program for development of orthopedic braces, set out to design a brace for the foot and ankle that would be an analog of the ankle and subtalar joints and thus allow independent use of springs or other means to replace weakened muscles in performing certain functions without inducing relative motion between the brace and the human body. So far, this dual-axis brace is primarily useful in control of swing phase, where forces are modest. In the course of this

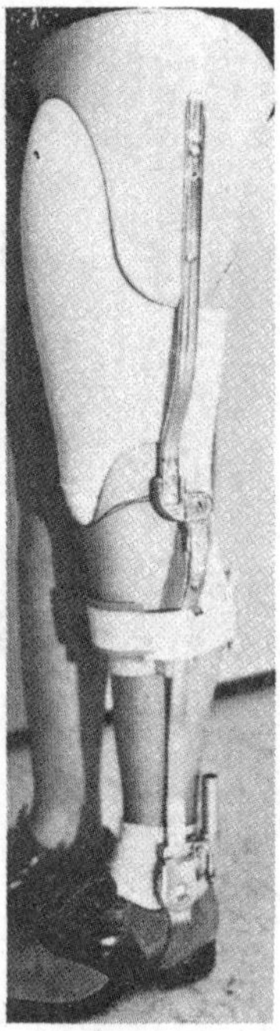

FIGURE 4.1–32. The functional long-leg brace developed at the University of California, Los Angeles, to permit a patient with one flaccid, totally paralyzed lower limb to walk without the need for a "locked knee joint." (From VA- and HEW-sponsored work.)

FIGURE 4.1–33. Orthosis developed at the Army Medical Biomechanical Research Laboratory for "drop-foot" condition. Posterior rod is a composite of fiberglass and epoxy resin. (From *Orth. Pros.*, 25(1), 8, 1971. With permission.)

project a molded plastic "shoe insert" was developed to provide better control of the foot in stance phase, and indeed it has been found to be quite useful independently of the dual-axis brace. The biomechanical considerations are covered quite lucidly by Inman;[100] the engineering design considerations are detailed very competently by Lamoreux[101] and Campbell et al.[102] Henderson and Campbell[103] give quite adequate instructions for making and fitting these devices. Although these papers give brief reviews of clinical experience, advantages, and limitations at the developer's laboratory, clear indications for use of the dual-axis brace are not given; probably they must be developed from wider clinical application studies.

Reimann and Lyquist[104] have developed a splint to be used in treatment for clubfoot in children based on the principles of the University of California dual-axis brace. By permitting motion of the ankle, subtalar, and midtarsal joints only about normal axes, the developers feel, this device encourages normal development of the deranged foot structure. Initial trials in Scandinavia are encouraging, and clinical trials on a broad basis should be initiated in the United States and elswhere. As with most new developments in prosthetics and orthotics, evaluation is a difficult, time-consuming task.

Another outgrowth of experience in limb prosthetics is the use of prothetics and orthotics principles in the treatment of fractures of the long bones of the limbs, especially the lower. Ambulatory weight-bearing treatment of certain types, using plaster casts, had been advocated by Dehne et al.[105] and Brown and Urban.[106] CPRD sponsored a small conference[107] at Duke University to survey the concept and a later meeting at Denver[108] to bring together those working in the field. Probably understandably, neither meeting resulted in agreement on details, though there were general agreements on goals and basic principles.

Sarmiento has used some of the principles developed for fitting below-knee amputees and in the PTB weight-bearing brace to guide the application of a plaster cast or molded plastic brace so that a patient with a fractured tibia can be made

FIGURE 4.1–34. Two-bar type of orthosis developed by the Army Medical Biomechanical Research Laboratory for "drop-foot" condition. The medial and lateral rods are a composite material of fiberglass and epoxy resin. (From *Orth. Pros.*, 25(1), 6, 1971. With permission.)

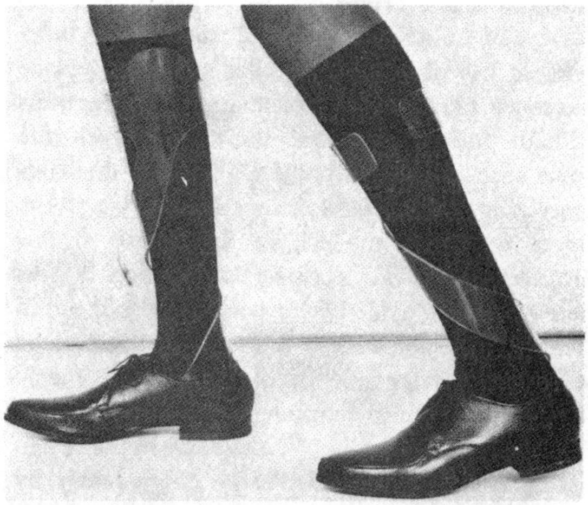

FIGURE 4.1–35. The spiral orthosis developed at New York University. This device has the potential of controlling the foot as well as the ankle.

ambulatory much sooner than is normally the case. In a recent article[109] he reports on experience with 135 cases. Two methods of making the brace are described: one uses plaster of Paris, the other is a later development using perforated synthetic balata (Figures 4.1–36 and 4.1–37) which is lighter and cooler. The data given show that successful healing can be facilitated while maintaining freedom of the knee and ankle. Further studies have been made of the role of the interosseus membrane.

Mooney has applied principles developed in above-knee prosthetics to treatment of certain types of femoral fractures. His excellent report to his sponsor,[110] the Social and Rehabilitation Service of the Department of Health, Education,

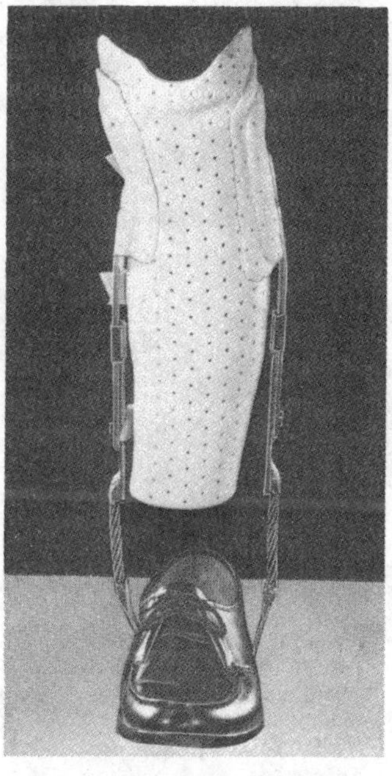

FIGURE 4.1–36. Brace developed at the University of Miami for treatment of fractured tibiae. The cable "ankle joints" are attached to fittings on the shoes that allow the patient to remove the shoe easily. (In a later design, flexible plastic strips attached to a shoe insert served as "ankle joints," maintained vertical position of the brace on the shank, and allowed removal of the shoe.)

FIGURE 4.1–37. Patient at the University of Miami with brace for treatment of fractured femur on the right, and prothesis fitted to below-knee stump immediately after surgery on the left.

and Welfare, shows that the typical patient treated in this manner, which permits early ambulation, completed treatment an average of 10 weeks (14.5 vs. 24.7) before the usual patient treated with conventional methods. This report also gives a good history of the research effort, including measurements of pressures and forces and studies of materials as well as clinical trials, pointing out approaches that were unsuccessful. Unfortunately, the report does not include instructions for fabrication and application that are sufficiently detailed for others to follow.

Improved medical procedures have been responsible for the increased survival rate of children with cerebral palsy and spina bifida (myelomeningocele). The most comprehensive publication concerning recent research, development, and application of devices for these two conditions is a report of a workshop sponsored by the Committee on Prosthetics Research and Development in 1969.[111] Unfortunately, this report contains very few references. However, it does contain much original material, along with fairly clear articles on present treatment programs in various places in the United States and Canada. Of especial interest are the reports from the Ontario Crippled Children's Centre (Toronto) on the engineering approach being taken there to provide mechanical devices as part of a total treatment program for these severely disabled children. Work in this area surely needs to be intensified. Attention is called to the thorough and fascinating annual reports of this center.

Another poorly understood disease of childhood is Legg-Calvé-Perthes syndrome in which the hip joint begins to become malformed. The more conventional method of treatment has been to relieve the affected side of all weight-bearing responsibility. A more recent theory is that when the thigh is maintained in a position to 45° of abduction and weight bearing is permitted, new growth of bone will form in a manner of preserve hip-joint function. Apparent success has been achieved using plaster-of-Paris casts but at great cost and inconvenience. Bobechko et al.[112] have developed a clever, inexpensive brace (Figure 4.1–38 and 4.1–39) that permits ambulation and sitting yet is easy to keep clean. The roles of bone growth and calcification in response to stress (assumed in response to fractures of normal bone and to the distortion of malnutrition) seem uncertain if not brushed aside. Presumably abnormal bone requires different thresholds, at least.

The use of electricity to cause paralyzed muscle to contract in a useful way was pioneered by

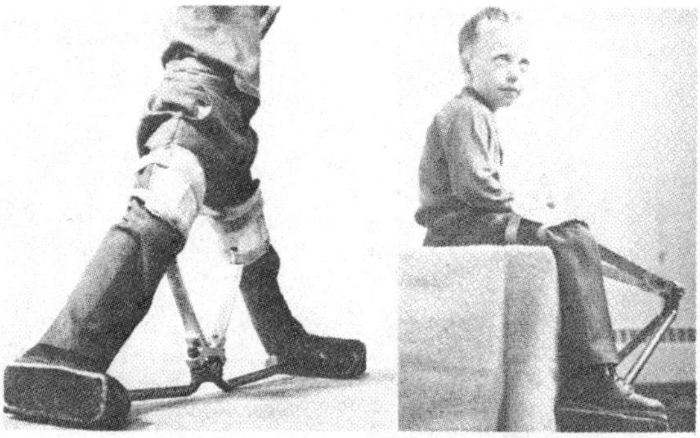

FIGURE 4.1–38. An orthosis developed at the Ontario Crippled Children's Centre (Toronto) for treatment of child with Legg-Calvé-Perthes disease. Unique design permits knee flexion yet maintains head of femur in the acetabulum. (From *Artif. Limbs,* 12(2), 38, 1968.)

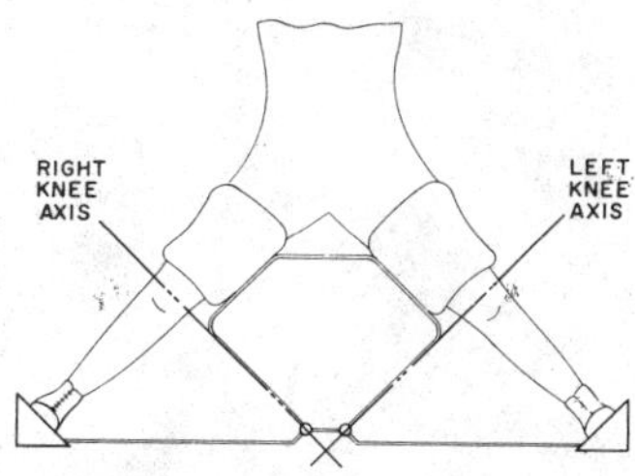

FIGURE 4.1–39. Schematic diagram for the Ontario Crippled Children's Centre brace for treatment of children with Legg-Calvé-Perthes disease. (From *Artif. Limbs,* 12(2), 39, 1968.)

Liberson[113] in the Fifties. Although at least two companies have made stimulator units available commercially for application to individuals who would ordinarily need a "drop-foot" brace, little use has been made of this concept. Three institutions are now deeply involved in this work at the present time: the University of Ljubljana in Yugoslavia, Case Western Reserve University, and Rancho Los Amigos Hospital. The group at Ljubljana has produced a unit using surface electrodes[64] that has been evaluated by CPRD.[64a] All three groups also are concerned with systems that use implanted electrodes in order to eliminate the problem of finding the "trigger points" on the skin over the "motor points" for best stimulation of the muscle, with negligible (or least) discomfort each time the "brace" is applied. An evaluation coordinated by CPRD indicates favorable results for very carefully selected patients in the care of expert teams.

UPPER-EXTREMITY ORTHOTICS

Upper-extremity orthotics covers the design, fabrication, and application of devices ranging from simple hand splints for maintaining bones and muscles in position for healing or for relief of pain to elaborate externally powered exoskeletal devices for restoration of function for completely flail arms. The need for simple devices in the course of therapy has always been evident, but the need for the more complex devices has been brought about originally by successful treatment of respiratory polio and in recent years largely by the improved medical management that has resulted in keeping alive patients with severe spinal-cord injuries.

Bunnell[114] pioneered in the design and use of simple therapeutic splints, and many of his designs are in use today. The Georgia Warm Springs Foundation developed a number of assistive devices in an effort to meet the needs of poliomyelitis patients.[23] The use of wrist extension to provide hand prehension was introduced by Bisgrove,[115] used extensively at Rancho Los Amigos[116] in the Fifties, and refined by Engen (Figure 4.1–40)[117-119] during the Sixties. Though simple in concept, full potential of these devices can be achieved only when applied by trained teams.

A current research project under McDowell at the Veterans Administration Hospital, Richmond, Va., is studying application of sterilizable unpadded hand and forearm splinting over the postoperative dressing immediately after tendon or joint operations to maintain position and allow the desired range of controlled motion.

In application of prefabricated splints such as the Engen designs anthropometric data on hands are desirable. Birdsell[120] studied hand sizes of adult males in preparation for construction of the APRL artificial hand. Fletcher and Leonard,[121] based on the interpretation by DeFries[122] of Birdsell's data, plus additional measurements on adult females and children, described the design of five sizes of artificial hands. Recently, Garrett[123] has published size and gripping-force data on the adult human hand.

Bennett[124] and his colleagues at Warm Springs pioneered the development of simple devices to support most or all of the weight of the arms of

FIGURE 4.1–40. Two views of the reciprocal-type Engen plastic hand orthosis. Extension of wrist results in palmar prehension. The effective length of the rod in the four-bar linkage is adjustable. (From *Artif. Limbs,* 13(1), 16, 1969.)

FIGURE 4.1–41. Child using the Ontario Crippled Children's Centre feeder. (From *Artif. Limbs,* 13(2), 66, 1969.)

severely paralyzed patients so that feeble muscles could produce purposeful motions. Thus, the patient was enabled to eat, for example, food on a high "lapboard" on the arms of a wheelchair. These so-called "feeders" were further developed at several centers. Kay and Appoldt[125] reported on an analysis by New York University of the features of existing designs in preparation for a CPRD-sponsored conference.[126] These devices have proven to be quite useful to certain types of patients.

During the past 15 years, considerable attention has been devoted to providing patients who have severely paralyzed arms with powered assistive devices, but, for the most part, the results have been discouraging.[127,128] Some highly motivated patients have found externally powered devices useful, and new powered components and simpler orthoses,[128a] as well as new control systems,[128b,128c] are being evaluated in both institutional and home use (Engen[129] has sometimes successfully combined power with the feeder principle). More often than not the average patient finds that in the noninstitutional setting the result obtained is not worth the concentration required. It seems reasonable to assume that often members of the family or others are quite willing to assist the patient in carrying out functions that might be done with the device.

The primary reason for lack of unqualified success of the various externally powered devices, no doubt, is the method of control. To date, robotlike devices controlled by stored programs have not been proven practical or acceptable to patients. The only voluntary-control systems that have been developed are sequential, requiring considerable concentration and not a little skill. As in the case of externally powered devices for high-level arm amputees, some means of coordinated motion that can be carried out subconsciously must be found if these types of patients are to achieve much useful function with mechanical devices.

A different feeding device developed at the Ontario Crippled Children's Centre, primarily for cerebral palsy patients, Figure 4.1–41, is described in detail by Rife and Kennedy.[130] This device is merely one example of the possibilities for a wide variety of technical aids, many suited to a few specific cases, but candidates for evaluation on broader scales with emphasis on identification of prescription criteria.

Currently, relatively little attention is being paid to the problems of patients with paralyzed or partially paralyzed upper limbs. In 1970 CPRD held a workshop on the subject,[131] emphasizing the use of wrist flexion (available in many quadriplegics) to produce prehension, but proposing future workshops on arthritic splints and on splints for burns. Later in the year a group sponsored by CPRD[3] considered the needs and possibilities. The observations and recommendations of these groups are as follows:

Quadriplegia

A. Known treatment procedures and devices for quadriplegic patients should be subjected to comparative analyses to determine relative merits and prescription criteria for each. Systems to be studied should include those developed at Rancho Los Amigos Hospital, the Texas Institute for Rehabilitation and Research, The Rehabilitation Institute of Chicago, and the Institute of Rehabilitation Medicine, New York University. A great deal of work has been done during the last few years by the Veterans Administration, particularly through the VA Prosthetics Center with the cooperation of a number of VA Spinal Cord Injury Centers.

B. Fundamental studies of the dynamics and kinematics of hand, wrist, elbow, and shoulder motion should be undertaken immediately. Such studies should include wrist flexion and extension, and hand-deviation positions in relation to the performance of various common tasks and the frequency of use of these positions.

C. With regard to the total rehabilitation engineering effort directed toward the quadriplegic patient (and perhaps other types of patients (Figure 4.1–41) with upper-extremity disabilities as well), it is noted that braces or assistive devices may be of two broad general types:

1. Those that are attached to, or worn by, the patient
2. Those that are not fitted to the patient but rather are attached to a wheelchair, table, desk, or other stable base.

While even less is known about the role and value of the latter group of appliances (or manipulators), their potential appears worthy of assessment.

D. The possible value of computer-aided decision-making assistance for the severely handicapped patient should also be investigated in the spinal-cord-injury centers.

Hemiplegia

At the present time no specific recommendations are offered concerning static splinting for the prevention of hand deformities or contractures in hemiplegia, except that studies in this area should be carried out.

No satisfactory functional braces for the hemiplegic are currently available, partly because the spasticity present is a bar to effective bracing. It is recommended, therefore, that the investigation of methods to relieve spasticity should be intensified. Techniques investigated should include neurophysiological stimulation with mechanical as well as electrical inputs. A coordinated effort involving orthopedic surgery and neurosurgery to study neuromuscular control mediated by both central and peripheral factors is recommended.

The effect of loss of sensory inputs on the functions of the hand and arm with available motor power is noted. It is recommended that at an early date a laboratory or center with appropriate psychological and neurophysiological capability study the significance of the loss of specific sensory modalities on functional capacity.

Rheumatoid Arthritis

Past and current orthotic management of the rheumatoid hand has proven to be ineffective for the most part, although some progress has been made in surgical management. Therefore, it is recommended that investigation of new orthotic management methods be initiated, and that these investigations include such areas as plastic supportive gloves and internal injection of plastics for joint stabilization.

Peripheral Nerve and Thermal Injuries

It is recognized that upper-extremity disabilities arising from peripheral nerve injuries present a wide range of complex problems. At this time the committee is unable to make specific recommendations in this area. Data arising from studies previously recommended should throw further light on these problems and provide insights concerning required research.

Training of Occupational Therapists in Orthotics

It is apparent that upper-extremity splinting and bracing are being done by occupational therapists as well as by orthotists and physicians. It is recommended that a survey of training programs for occupational therapists in this area, as well as literature available for this purpose, be made in order to determine the possible need for intensified or expanded educational efforts.

SPINAL ORTHOTICS

No doubt because of the difficulties encountered in measuring motion and forces about the spine and torso, biomechanics of the spine is by no means fully understood. Often the cause of pathological conditions of the spine is not known, and consequently treatment methods are often empirical and not very effective.

A survey made by Perry[132] of more than 3,100 orthopedic surgeons on the use of external support in the treatment of low back pain indicates that a myriad of designs is available, that preferences for some devices are regional, and that treatment methods are based more on experience and intuition than on a sound knowledge of biomechanics. A survey involving other problems related to the spine would no doubt produce the same kind of answer.

Research workers at the University of California, San Francisco, have been the leaders in recent studies of the biomechanics of the spine. Their reports form a basis for future studies.[133-135] The UC group continues to stress the significance of rotation about the longitudinal axis of the spine.

In recent years investigators working under Hirsch in Sweden have been studying motion in the spine. Two examples of this work are given in papers by Lysell[136] and White.[137]

An outstanding contribution to treatment of idiopathic scoliosis in children has been the so-called Milwaukee brace, developed by Blount and co-workers in the late Forties. In the hands of skilled clinic teams, the Milwaukee brace has proven to be quite effective. One of the chief objections has been the tendency in some cases for dental and facial deformities to develop. Orthodontists in this country (e.g., see Reference 138) and abroad have attempted to combat the problem. In an attempt to determine optimum fitting and treatment procedures, several groups have tried to measure the forces actually applied by the Milwaukee brace (presumably varying with time as a result of muscular effort or relaxation), the latest being Galante et al.[139] Only tentative conclusions have been reached in this long-range study. Two of the principal factors in successful use of the brace appear to be understanding and cooperation of the patient (usually a teenage girl) and the family in a prolonged program of exercise as well as wearing of the brace.

The use of air-filled structures for support of the spine and torso has been proposed by several groups in recent years. A device known as Minerva Pneumatica[140] has been developed and offered commercially by a group in Italy. The Minerva Pneumatica, designed to support the cervical spine, consists of anterior and posterior portions, each containing a bladder that can be filled with air to provide the support required. The developers claim that the device, which is manufactured in three sizes, has been especially useful in providing support to victims of trauma during rescue operations and that many ambulances in Europe have been equipped with it. They also recommend its use after certain types of surgery to maintain position until healing is effected, and in some cases for long-term use.

In order to encourage initiation of research and to review previous work in spinal orthotics, CPRD sponsored a workshop in 1969. The report of the workshop[141] should be of use to those embarking in the field. During 1970 CPRD sponsored still another meeting,[3] one of its purposes being to set forth areas in which work is believed to be needed. The observations and recommendations made were as follows:

Standardization of Nomenclature

Of fundamental importance to cooperative research efforts and eventually to teaching in spinal orthotics is standardization of the nomenclature. It is recommended that the New York University scheme of categorizing and naming braces, including cervical, thoracic, and lumbar orthoses, be adopted as standard.

Serious consideration should be given to the prescription of orthotic appliances based upon functional analysis as proposed by the American Academy of Orthopaedic Surgeons.[99]

External Appliances

Cervical Appliances

A detailed analysis should be made of the functions and limitations of standard available cervical appliances, for example, the so-called soft and hard collars, four-poster cervical brace, and cervical brace with trunk fixation. Specifically, resistance to, and limitation of, motions about the various axes should be studied by techniques such as carefully controlled cineradiography; also, measurement of unit contact pressures and net forces at the skin-appliance interface (mandibular and occipital regions) should be made. Both normal volunteers and patients with pathological conditions of the cervical spine should be studied. Evaluation of such factors as comfort, durability, cost, and ease of fitting should be included. Based upon the results of these studies, it is hoped that improvement in the design of cervical appliances can be undertaken subsequently.

The Abdominal Pressure Spinal Support

The presently available abdominal pressure spinal support (University of California-San Francisco Royalite® jacket)[141] should continue under investigation for function and possible improvement of materials and fitting techniques.

New materials with graduated flexibility (especially those allowing flexible margins for greater comfort) should be evaluated by fabricating comparable appliances over the same plaster model used for the basic Royalite jacket. Polyethylene, polycarbonate (Lexan®, a transparent material), and perhaps other materials should be compared with Royalite. The Engen technique[142] for corrugating a thermoplastic material for stiffness (by placing cords on the model before draping the warm, soft material) should be considered.

Measurements should be taken to record allowed motions, pressure in the abdominal cavity, venous flow, and vital capacity for each of the test braces for comparative evaluation.

Improved fixation to the pelvis for the basic Royalite jacket should be investigated, perhaps utilizing techniques of the Milwaukee scoliosis brace.

Instrumentation for Research Studies

Instrumentation and techniques for studying spinal orthotics should be improved.

A surgically implantable transducer for measuring relative motions between vertebrae such as linear displacements between the spinous processes should be developed. The transducer should exert negligible force on the spinous processes and be self contained if possible (i.e., information would be retrieved by telemetry or by measurements of inductance or capacitance through the skin interface). Such a transducer would reduce or eliminate radiation exposure now needed by X-ray studies and in most, if not all, cases would eliminate the need for pins projecting through the skin; these are troublesome and in most instances inconvenient.

Pending availability of such a transducer, X-ray techniques for measuring motion of vertebrae should be standardized. Implanted radiopaque markers on selected vertebrae, standardized positions and loadings, and defined tube and screen positions should be considered.

Miniaturized pressure gauges suitable for implanting in the nucleus pulposus of the disc should be considered, subjected to trial on large animals, and, if appropriate, tested on humans.

Pneumatic Devices

Pneumatic or inflatable bracing should be developed within a feasibility program.

For the abdominal-pressure (Royalite) brace, the compression of the abdomen might be provided by an inflatable bag inside an essentially flat abdominal wall.

For cerebral palsy, scoliosis bracing, pneumatic (or hydraulic) bracing or pressure pads, with selective weaves to limit distortion when inflated, should be developed. Such inflated structures would replace metal bars in stiffening corsetlike structures; others would serve as pads to distribute appropriate forces over the body.

The concept of pulsating or alternating pressure between adjacent pads to exert intermittently high forces while allowing periodic release of pressure and blood flow to proceed should be pursued. An appliance embodying this concept might first be tried on a case of cerebral palsy scoliosis where corrective forces have been traditionally limited to low levels by skin tolerance.

An organization with a substantial case load of cerebral palsy cases, versatile fabrication facilities, and interdisciplinary personnel including physicians, orthotists, and engineers would be appropriate for experiments with these pneumatic appliances for CP patients. Development of experimental models would probably require at least 2 years.

Pneumatic devices for muscular dystrophy cases should be developed to permit the patient to vary pressure in a given pad by control valves regulating flow of air or liquid between pads so as to allow alternating or shifting pressures and slight shifts of positions. External automatic or patient-regulated pulsating pads within rigid outer abdominal or thoracic shells might also function as respirators.

Electrical Stimulation

The possibilities of electrical stimulation of muscles on the convex side (or perhaps of electrical inhibition of muscles on the concave side) of spinal curvatures, perhaps controlled by microswitches sensitive to sagging into the curved position, should be explored. Such work might be initiated very modestly with a commercially available stimulator, but serious investigation on a substantial scale would probably require much more fundamental knowledge of muscle properties, muscle stimulation, and biomechanics of spinal curvatures and rotations.

In view of the obvious need for development in the next few years of bracing in the cervical and lumbar region as well as the relatively satisfactory (although far from perfect) appliances for the thoracolumbar region (including scoliosis), no attempt is made here to suggest development or modification of such appliances. Likewise, at this time, no specific recommendation is made for analysis of beds, mattresses, chairs, or cushions, important as these obviously are in spinal posture, pressure distributions, and pain.

Internal Splinting Devices

Severe spinal deformities are best managed by some form of internal splinting or corrective device in conjuction with external support during maturation of fusion in the involved area. Current devices are only partially successful in achieving these objectives. It is anticipated that, with recent knowledge of the mechanical properties of the vertebral column, improvements in current designs and new concepts of correction and stabilization will be possible. Experimentation with such systems as multipoint fixations, dual-cable, or multi-cable designs for simultaneous correction of curvature and rotation deformities should be encouraged.

In order to define the status of present instrumentation and to stimulate new approaches to the design and development of improved splinting and fixation devices, a workshop is recommended. Participants should include orthopedic surgeons knowledgeable in the use and problems of such surgical implants, and engineers knowledgeable in biomechanics of the spine and implantation materials.

Fundamental Studies

Experimental studies have produced mechanical data for the isolated spinal column by measurements of stiffness for all its modes of deformation. It is recommended that these studies be expanded to measure the same mechanical constants of the spine with the surrounding musculature and attached structures intact so that this external support (i.e., the support provided by the trunk musculature) may be evaluated in mechanical terms. Data of the spinal column in vivo should then be used to analyze those forces developed within this column by the application of back braces or other devices (internal or external) so that improved designs may proceed on a rational basis. The data from these studies should also be applied to the design and development of a disc prosthesis to replace or reinforce a worn or damaged disc so that normal mechanical function would be approached. This work is currently under way at the University of California but should be encouraged at one or more additional centers.

Instrumentation for these measurements is presently unavailable. It is anticipated that development of such instrumentation will be extremely difficult and time consuming. Nevertheless, the value of such data warrants the investigation necessary. A time of no less than 3 years appears reasonable for this study and will require a cooperative approach between bioengineering and medical personnel.

The use of both external and internal support appliances or devices is based on the assumption that restriction of motion is necessary for the alleviation of painful symptoms. This is undoubtedly true in most cases and to at least a certain degree in all cases.

However, a glaring basic defect or hiatus in our knowledge hampers a completely rational approach to management of back pain. Specifically, we do not know the etiologic mechanism of pain in the large majority of so-called "low back" patients. Presumably, it is related to disc degeneration, but whether mechanical, chemical, or other unknown factors are directly responsible for the pain remains unknown. Elucidation of the factor or factors responsible is a prerequisite to more rational and significiantly improved treatment of "low back pain."

APPENDIX A

An initial reading list of periodicals related to the many fields covered seems appropriate. As in other interdisciplinary fields, the literature is widely scattered.

Periodicals

Bulletin of Prosthetics Research, Prosthetic and Sensory Aids Service, Veterans Administration, 252 Seventh Ave., New York, N.Y., 10001.

Dow Corning Center for Aid to Medical Research Bulletin, Midland, Mich. 48640.

Human Factors, Journal of Human Factors Society, Johns Hopkins Press, Baltimore, Md. 21218. (Nonmember subscriptions should be sent to Johns Hopkins Press.)

Rehabilitation Literature, National Society for Crippled Children and Adults, Inc., 2023 West Ogden Ave., Chicago, Ill. 60612.

IEEE Transactions on Bio-Medical Electronics, The Institute of Electrical and Electronics Engineers, 345 E. 47th St., New York, N.Y. 10017.

IEEE Transactions on Human Factors in Electronics, The Institute of Electrical and Electronics Engineers, 345 E. 47th St., New York, N.Y. 10017.

ISPO Bulletin, International Society for Prosthetics and Orthotics, P. O. Box 42, DK 2900 Hellerup, Denmark.

Transactions of the American Society for Artificial Internal Organs, Department of Medicine, Georgetown University Hospital, Washington, D.C. 20007.

Indexes

Engineering Index Monthly Bulletin, Engineering Index, Inc., 345 E. 47th St., New York, N.Y. 10017.

Index Medicus, National Library of Medicine, U.S. Department of Health, Education, and Welfare, Public Health Service. (Subscriptions should be addressed to the Superintendent of Documents, U.S. Government Printing Office, Washington, D.C. 20402.)

Reference Collections

Easter Seal Library of the National Society for Crippled Children and Adults, Inc., 2023 West Ogden Ave., Chicago, Ill. 60612.

John Crerar Library (medical and engineering), Illinois Institute of Technology, Chicago, Ill. 60616.

National Library of Medicine, Bethesda, Md. 20014.

Research Center for Prosthetics (reference collection), Veterans Administration, 252 Seventh Ave., New York, N.Y. 10001.

APPENDIX B

Orthopadie Technik
Wilhelmstr. 42
62 Wiesbaden, West Germany

Medizinische Technik
Medicus Verlag GmbH
Klingsorstr. 21
1000 Berlin 41, West Germany

Orthopaedics, Traumatology and Prosthetics
Editorial Office
11 Olminskovo St.
Khorkov 461, U.S.S.R.

A.P.O Revue: Annals of the Swiss Association for Prostheses and Orthotics
Editions Prolux SA
10, Grand-Pont
1003 Lausanne, Switzerland

APPENDIX C[a]

Synthesis

Research and education in limb prosthetics and orthotics in the United States are making progress at a steady pace in an orderly way with rather well-defined programs in fundamental studies, design and development, evaluation, and education. Amputees and artificial limbs have been studied the longest and represent the most advances, yet further improvements can be predicted. Lower- and upper-extremity orthotics have recently come to receive the attention they deserve, and progress is being made in these areas. Spinal orthotics and problems of the back have long been neglected, and work in sensory aids for the blind and deaf-blind has suffered because of a lack of fiscal support.

The Education Program is a productive one, but a reappraisal of its goals should probably be made in view of the changing methods of patient management.

The method of funding research by relatively small grants has been effective, but it appears that the time has come when support of special centers devoted to integrated rehabilitation engineering programs seems desirable.

Given below is a synthesis of the detailed recommendations made by various study groups sponsored by CPRD.

Research and Development

Lower-extremity Prosthetics

1. Continuation of promotion and teaching of immediate postsurgical fitting and early fitting.
2. Continuation of work on forming sockets directly over the stump.
3. Refinement and standardization of modular prostheses with emphasis on cosmesis.
4. Development of adjustable above-knee sockets.
5. Refinement of suspension methods for above-knee prostheses for geriatric patients.
6. Development of swing- and stance-phase control devices for hip-disarticulation prostheses.
7. Acceleration of final development of knee-disarticulation prostheses using four-bar linkage and swing-phase control units.

[a]This appendix is based upon analyses of major problems during 1970. A new edition was in preparation during 1975, and may be published by the time Volume III of the *CRC Handbook of Materials Science* is ready for press.

Lower-extremity Prosthetics (continued)

8. Determination of feasibility of voluntary control of knee units for above-knee amputees.
9. Acceleration of development of Mauch hydraulic foot-ankle unit.
10. Continuation of development of ankle-rotation mechanisms.
11. Development of improved procedures for measuring distribution of pressures between stump and socket. Both research and clinical instruments are needed.

Upper-extremity Prosthetics

1. A complete restudy of body-powered prostheses. Studies involving new methods of fitting, harnessing, and modular prostheses should be integrated and completed. Of especial interest are the socket designs developed at Northwestern University for the self-contained prostheses.
2. Improvement in cosmetic appearance of artificial hooks, hands, and arms.
3. A survey of the upper-extremity amputee population to determine the degree of usefulness provided by present prostheses.
4. Increased emphasis on development of controls and sensory feedback mechanisms for externally powered prostheses.
5. Continuation of development of externally powered components at about the same level of effort, recognizing that control of these devices is critical to their success. Electricity is considered the best power source at this time.

Lower-extremity Orthotics

1. Development of modular, adjustable, temporary orthoses for clinical use.
2. Development of electromyographic signals to control orthotic and natural joints.
3. Development of simple methods for modifying standard shoes to accommodate foot deformities.
4. Development of knee units that will permit better control of the unstable knee.
5. Development of a method to provide assistance to hip motion.
6. Development of a push-off device.
7. Continuation of study of neuromuscular electrical stimulation as a substitute for or supplement to external bracing, and for control of spasticity.
8. Activation of studies to delineate the anatomical and physiological problems in weight-bearing braces.
9. Expansion of the development of orthotic devices for children with spina bifida and for children with cerebral palsy.
10. Continuation of studies on orthotic devices for management of fractures.
11. Establishment of a Panel of Orthotic Aids to encourage further development of such devices as wheelchairs, crutches, and walkers.

Upper-extremity Orthotics

1. A survey to determine:
   a. The number of patients with various types of dysfunction of the upper extremity.
   b. Rehabilitation potential of each type of dysfunction.
   c. The methods of treatment available for each type of patient.
2. A comparative study of present treatment procedures and devices available for quadriplegic patients.
3. Development of a method to relieve spasticity in hemiplegic patients.
4. Development of improved splints for the rheumatoid arthritic hand to be used with and without surgery, drugs, or both.
5. Study of combinations of surgery, splinting, and functional electrical stimulation.
6. A survey of what is currently being taught to occupational therapists in the field of upper-extremity orthotics, especially in arthritis.

Spinal Orthotics

1. Acceptance of the scheme of categorizing and naming spinal braces developed in recent years by New York University Prosthetic and Orthotic Studies and dissemination thereof through publications of the American Academy of Orthopaedic Surgeons and catalogs of manufacturers.
2. An analysis of the functions provided by the standard available cervical appliances.
3. An analysis of the functions provided by the standard lumbosacral and thoracolumbar braces and corsets.
4. Continuation of further development and investigation of the effectiveness of the Abdominal Pressure Spinal Support (Royalite) Jacket developed at the University of California, San Francisco.
5. Development of an implantable transducer for measurement of relative motions between vertebrae.
6. Development of an implantable transducer for measuring disc pressure.
7. Investigation of the usefulness of inflated structures in the design of spinal orthoses.
8. Investigation of the possibilities of electrical stimulation of appropriate muscles to provide a balanced state.
9. Development of improved methods of internal splinting of the spinal column.
10. Development of a prosthetic intervertebral disc.
11. Determination of the factor or factors responsible for "low-back pain."

Sensory Aids

1. Although support for research and development in sensory aids for the blind and deaf has been meager, a number of promising devices are emerging, therefore, initiation of a formal evaluation program is strongly recommended.
2. A conference on standards, specifications, and use of the typhlocane is needed.
3. A conference on problems of individuals with severely impaired (low) vision is indicated.
4. A conference on reading machines with emphasis on evaluation techniques is needed.
5. A conference on Braille output devices is needed.

Fundamental Studies

1. A thorough study of the feasibility of functional electrical stimulation of the neuromuscular system.
2. Further study of the biomechanics of the spine.
3. Continuing studies of normal and pathological human gait, including forces imposed on joints of the lower limb.
4. Initiation of an integrated program for the development of internal prostheses, especially hip, knee elbow, and finger joints.
5. Acceleration of work in skeletal attachment of external prostheses.
6. Studies to determine the properties of biological materials, biological structures, and the effect of pressure on human tissues.
7. Restoration of studies of biomechanics of the upper extremity.
8. Investigation of the formation and treatment of neuromas.
9. Physiological studies of muscles involving force, length, time, and myoelectric signals.
10. The development of procedures to insure the safety and efficacy of internal structural prostheses.
11. Studies of the interaction between neuromuscular control systems and internal structural prostheses.

Surgery

1. Study of the response of human tissues to pressure, especially with regard to circulation and wound healing.
2. Study of the formation and presentation of neuroma.
3. Evaluation of osteoplasty techniques as practiced in amputation surgery by Dederich, Murdoch, Deffer, and others.
4. Evaluation of implant procedures designed to increase weight-bearing ability of stumps.

Surgery (continued)

5. Further study of bone overgrowth in amputation stumps.
6. Evaluation of muscle fixation techniques in amputation stumps.
7. Determination of the feasibility of a substitute for the fatty tissues that often protect bony prominences.
8. Revival of studies of the phantom-limb sensation.
9. Continuation of development of improved surgical techniques for management of children with congenital limb deficiencies.
10. Expansion of present efforts in developing methods for attaching limb prostheses directly to the long bones.
11. Reevaluation of the concept of muscle attachment to external prostheses primarily for improved control.
12. Reevaluation of methods of amputation through the foot.
13. Development of new implantable devices for replacement of most of the joints of the extremities.
14. Development of an implantable artificial muscle.
15. Development of methods to control spasticity.
16. Further study of fracture bracing, especially for the femur.
17. Development of better mechanical devices for use in bone-lengthening procedures.
18. Reassessment of techniques for epiphyseal stimulation.
19. Development of better methods of replacing large bone deficits.
20. Continuation of studies involving limb replantation but at a very low priority.

Evaluation

1. Continuation of the present evaluation program is recommended.
2. Expansion to permit cooperation with countries abroad should result in better exchange of information and thus permit the United States to introduce useful foreign improvements at an earlier date than is now the case.

Children's Problems

1. The formation of four to six specialized centers in the United States to provide prosthetics management for severely disabled child amputees.
2. The development of a Cooperative Clinical Program in orthotics for children parallel to the Cooperative Amputee Clinic Program.
3. The development of locomotion aids in addition to braces.
4. The development of communication aids for children with speech impairments.

Methodology – Rehabilitation Engineering Centers

1. The establishment of multidisciplinary centers to carry out integrated programs of research, development, evaluation, and education in rehabilitation engineering.
2. Subjects that might serve as central themes for study are
   a. Functional Electrical Stimulation
   b. Internal Structural Prostheses
   c. Upper-extremity Paralysis
   d. Paraplegia
   e. Hemiplegia
   f. The Effect of Pressure on Human Tissues
   g. Control of Neuromusculoskeletal Systems and Their Substitutes
   h. Problems of the Spine
   i. Sensory Aids for the Blind and Deaf-blind
   j. Surgical Procedures
   k. Lower-extremity Prosthetics

REFERENCES

1. **Wilson, A. B., Jr.,** Limb prosthetics – 1970, *Artif. Limbs,* 14, 1, 1970.
2. **Wilson, A. B., Jr.,** The prosthetics and orthotics program, *Artif. Limbs,* 14, 1, 1970.
3. Committee on Prosthetics Research and Development, *Rehabilitation Engineering – A Plan for Continued Progress,* National Academy of Sciences, Washington, D. C., 1971.
4. **Kobrinskii, A. E., Bolkhovitin, S. V., Voskoboinikova, L. M., Ioffe, D. M., Polyan, E. P., Popov, B. P., Slavutskii, Ya. L., Sysin, A. Ya., and Yakobson, Ya. S.,** Problems of bioelectric control, in *Automatic and Remote Control: Proceedings of the First International Congress of the International Federation of Automatic Control,* Vol. 2, Moscow, 1960, 619.
5. **Schlesinger, G.,** Der mechanische Aufbau der Künstlichen Glieder, in *Ersatzglieder und Arbeitshilfen,* Borchardt, M., et al., Eds., J. Springer, Berlin, 1919.

5a. **Reiter, R.,** Eine neue Electrokuntshand, *Grenzgeb. Med.,* 4, 133, 1948.

6. **Alderson, S. W.,** The electric arm, in *Human Limbs and Their Substitutes,* Klopsteg, P. E. and Wilson, P. D., McGraw-Hill, New York, 1954, 368 (reprinted by Hafner, New York, 1968).
7. **Battye, C. K., Nightingale, A., and Whillis, J.,** The use of myoelectric currents in the operation of prostheses, *J. Bone Jt. Surg.,* 37-B, 506, 1955.
8. **Basmajian, J. V.,** Human motors and their substitutes: some desperate needs, *Bull. N. Y. Acad. Med.,* 47, 671, 1971.
9. **Childress, D. S. and Billock, J. N.,** Self-containment and self-suspension of externally powered prostheses for the forearm, *Bull. Pros. Res.,* BPR 10-14, 4, 1970.
10. **Talley, W. H.,** Prosthetics research – a cost reduction program: an editorial, *Bull. Pros. Res.,* BPR 10-10, 1, 1968.

10a. **Gershenson, M. and Holsberg, W. G.,** *Bull. Pros. Res.,* BPR 10-21, Spring 1974.

11. **Borchardt, M. et al., Eds.,** *Ersatzglieder und Arbeitshilfen,* J. Springer, Berlin, 1919.
12. **Little, E. M.,** *Artificial Limbs and Amputation Stumps,* H. K. Lewis and Co., London, and Blakiston, Philadelphia, 1922.
13. **Klopsteg, P. E. and Wilson, P. D.,** *Human Limbs and Their Substitutes,* McGraw-Hill, New York, 1954 (reissued with selected bibliography by Hafner, New York, 1968).
14. Committee on Prosthetics Research and Development, *Research in Limb Prosthetics and Orthotics – A Report of an International Conference, April 28–May 2, 1969,* National Academy of Sciences, Washington, D.C., 1969.
15. **Blakeslee, B., Ed.,** *The Limb-deficient Child,* University of California Press, Berkeley, 1963.
16. Committee on Prosthetics Research and Development, *Selected Articles from Artificial Limbs, January 1954 – Spring 1966,* Robert E. Krieger, Huntington, N.Y., 1970.
17. Committee on Prosthetics Research and Development, *Report of Eighth Workshop Panel on Upper-extremity Prosthetics, March 31–April 2, 1970,* National Academy of Sciences, Washington, D.C., 1970.
18. Committee on Prosthetic-Orthotic Education, *The Geriatric Amputee: Principles of Management,* National Academy of Sciences, Washington, D.C., 1971.
19. **Warren, R. and Record, E. E.,** *Lower Extremity Amputation for Arterial Insufficiency,* Little, Brown, Boston, 1967.
20. **Mital, M. A. and Pierce, D. S.,** *Amputees and Their Prostheses,* Little, Brown, Boston, 1971.
21. **Weiss, M., et al.,** *Myoplastic Amputation, Immediate Prosthesis and Early Amputation,* U.S. Government Printing Office, Washington, D.C., 1971.
22. Committee on Prosthetics Research and Development, *The Geriatric Amputee,* National Academy of Sciences-National Research Council Publication 919, Washington, D.C., 1961.
23. American Academy of Orthopaedic Surgeons, *Orthopaedic Appliances Atlas, Vol. 1 – Braces, Splints and Shoe Alterations,* J. W. Edwards, Ann Arbor, Mich., 1952.
24. American Academy of Orthopaedic Surgeons, *Orthopaedic Appliances Atlas, Vol. 2 – Artificial Limbs,* J. W. Edwards, Ann Arbor, Mich., 1960.
25. **Murdoch, G., Ed.,** *Prosthetic and Orthotic Practice,* Edward Arnold, London, 1970.
26. **Clynes, M. and Milsum, J. H., Eds.,** *Biomedical Engineering Systems,* McGraw-Hill, New York, 1970.
27. **Licht, S., Ed.,** *Orthotics etcetera,* Elizabeth Licht, New Haven, Conn., 1966.
28. **Rosenberg, C.,** *Assistive Devices for the Handicapped,* American Rehabilitation Foundation, Minneapolis, and National Medical Audiovisual Center of the National Library of Medicine, Atlanta, 1968.
29. **Lowman, E. W., and Klinger, J. L.,** *Aids to Independent Living,* McGraw-Hill, New York, 1969, 9.
30. **Peizer, E. and Wright, D. W.,** Five years of wheelchair evaluation, *Bull. Pros. Res.,* BPR 10-11, 9, 1969.
31. **Lipskin, R.,** An evaluation program for powered wheelchair control systems, *Bull. Pros. Res.,* BPR 10-14, 121, 1970.
32. **Peizer, E. and Bernstock, W. M.,** Bioengineering evaluation and field test of the Stand-Alone therapeutic aid, *Bull. Pros. Res.,* BPR 10-2, 1, 1964.
33. **Eagleson, H. M., Jr., Craig, R. E., Best, E. E., and Knudson, A. B. C.,** A specially adapted mobile standing device (surfboard) for spinal cord injury patients, in *Proc. 15th Annual Clinical Spinal Cord Injury Conference,* November 7–9, 1966, Veterans Administration, Washington, D.C., undated.
34. Committee on Prosthetics Research and Development, *Report of Twenty-second Meeting, May 17, 1971,* National Academy of Sciences, Washington, D.C., 1971.

34a. **Lamoreux, L. W.,** Kinematic measurements in the study of human walking, *Bull. Pros. Res.,* BPR 10-15, 3, 1971.

35. **Murray, M. P., Gore, D. R., and Clarkson, B. H.,** Walking patterns of patients with unilateral hip pain due to osteo-arthritis and avascular necrosis, *J. Bone Jt. Surg.,* 53-A, 259, 1971.

36. **Murray, M. P., Kory, C., and Sepic, S. B.,** Walking patterns of normal women, *Arch. Phys. Med.,* 51, 637, 1970.

37. **Seireg, A. H., Murray, M. P., and Scholz, R. C.,** Method for recording the time, magnitude, and orientation of forces applied to walking sticks, *Am. J. Phys. Med.,* 48, 307, 1968.

38. **Murray, M. P., Seireg, A. H., and Scholz, R. C.,** Survey of the time, magnitude, and orientation of forces applied to walking sticks by disabled men, *Am. J. Phys. Med.,* 48, 1, 1969.

39. **Robinson, H. S.,** Cane for measurement and recording of stress, *Arch. Phys. Med.,* 50, 457, 1969.

40. **Plank, M. P.,** Analysis of Forces Exerted on the Crutch during the Swing-through Gait, Master's thesis, Division of Orthopaedic Surgery, Case Western Reserve University, Cleveland, 1967.

41. **Kennaway, A.,** On the reduction of slip of rubber crutch-tips on wet pavement, snow, and ice, *Bull. Pros. Res.,* BPR 10-14, 130, 1970.

42. **Shambes, G. M. and Waterland, J. C.,** Stance characteristics of a quadrilateral amputee, *Bull. Pros. Res.,* BPR 10-13, 173, 1970.

42a. **Shambes, G. M. and Waterland, J. C.,** Stance characteristics of a quadrilateral amputee – addendum, *Bull. Pros. Res.,* BPR 10-15, 102, 1971.

43. **Isman, R. E. and Inman, V. T.,** Anthropometric studies of the human foot and ankle, *Bull. Pros. Res.,* BPR 10-11, 97, 1969.

44. **Morrison, J. B.,** Function of the knee joint in various activities, *BioMed. Eng.,* 4, 573, 1969.

45. **Morrison, J. B.,** The mechanics of the knee joint in relation to normal walking, *J. Biomech.,* 3, 51, 1970.

46. **Johnston, R. C. and Smidt, G. L.,** Hip motion measurements for selected activities of daily living, *Clin. Orthop.,* 72, 205, 1970.

47. **Chapman, M. W. and Kurokawa, K. M.,** Some observations on the transverse rotations of the human trunk during locomotion, *Bull. Pros. Res.,* BPR 10-11, 38, 1969.

48. **Corcoran, P. J. and Brengelmann, G. L.,** Oxygen uptake in normal handicapped subjects, in relation to speed of walking beside velocity-controlled cart, *Arch Phys. Med.,* 51, 78, 1970.

49. **Stoboy, H., Wilson-Rich, B., and Lee, M.,** Workload and energy expenditure during wheelchair propelling, *Paraplegia,* 8, 223, 1971.

50. Committee on Prosthetics Research and Development, Pressure and Force Measurement – A Report of a Workshop, May 27-28, 1968, National Academy of Sciences, Washington, D.C., 1968.

51. **Cochran, G. V. B.,** The clinical measurement and control of corrective and supportive forces, *Clin. Orthop.,* 75, 209, 1971.

52. **Appoldt, F. A., Bennett, L., and Contini, R.,** Socket pressure as a function of pressure transducer protrusion, *Bull. Pros. Res.,* BPR 10-11, 236, 1969.

53. **Appoldt, F. A., Bennett, L., and Contini, R.,** Tangential pressure measurements in above-knee suction sockets, *Bull. Pros. Res.,* BPR 10-13, 70, 1970.

54. **Sonck, W. A., Cockrell, J. L. and Koepke, G. H.,** Effect of liner materials on interface pressures in below-knee prostheses, *Arch. Phys. Med.,* 51, 666, 1970.

55. **Brand, P. W. and Ebner, J. D.,** A pain substitute pressure assessment in the insensitive limb, *Am. J. Occup. Ther.,* 23, 479, 1969.

56. **Brand, P. W., Salim, T. D., and Burke, J. F.,** Sensory Denervation – a Study of its Causes and its Prevention is Leprosy and of the Management of Insensitive Limbs, Final Report SRS Project No. RC-40-M, U.S. Public Health Service Hospital, Carville, La., Aug. 1, 1966 – June 30, 1970.

57. **Tuck, W. H.,** The use of Plastazote to accommodate deformities in Hansen's disease, *Lep. Rev.,* 40, 171, 1969.

58. **Jopling, W. H.,** Observations of the use of Plastozote insoles in England, *Lep. Rev.,* 40, 175, 1969.

59. **Mondl, A. M., Gardiner, J., and Bisset, J.,** The use of Plastazote in footwear for leprosy patients, *Lep. Rev.,* 40, 177, 1969.

60. **Mooney, V., Einbund, M. J., Rogers, J. E., and Stauffer, E. S.,** Comparison of pressure distribution qualities in seat cushions, *Bull. Pros. Res.,* BPR 10-15, 129, 1971.

61. **Bush, C. A.,** Study of pressures on skin under ischial tuberosities and thighs during sitting, *Arch. Phys. Med.,* 50, 207, 1969.

62. Committee on Prosthetics Research and Development, The Application of External Power in Prosthetics and Orthotics, National Academy of Sciences – National Research Council Publication 874, Washington, D.C., 1961.

63. Committee on Prosthetics Research and Development, The Control of External Power and Upper-extremity Rehabilitation, National Academy of Sciences Publication 1352, Washington, D.C., 1966.

64. **Gavrilović, M. M. and Wilson, A. B., Jr., Eds.,** *Advances in External Control of Human Extremities,* Proceedings of the Fourth International Symposium on External Control of Human Extremities, Yugoslav Committee for Electronics and Automation, Belgrade.

64a. Committee on Prosthetics Research and Development, *Clinical Evaluation of the Ljubljana Functional Electrical Peroneal Brace,* Report E-7, National Academy of Sciences, Washington, D.C., 1973.

65. **Gottlieb, M., Santschi, W., and Lyman, J.,** *Evaluation of the Model IV-E2 Electric Arm,* Spec. Tech. Rep. No. 19, Department of Engineering, University of California (Los Angeles), November 1953.

66. Committee on Prosthetics Research and Development, *Report of Sixth Workshop Panel on Upper-extremity Prosthetics, October 21-23, 1968,* National Academy of Sciences, Washington, D.C., 1969.
67. Committee on Prosthetics Research and Development, *Report of Seventh Workshop Panel on Upper-extremity Prosthetics, July 30-31, 1969,* National Academy of Sciences, Washington, D.C., 1969.
68. **LeBlanc, M. A.,** Clinical evaluation of externally powered prosthetic elbows, *Artif. Limbs,* 15, 70, 1971.
68a. Committee on Prosthetics Research and Development, *Report of Ninth Workshop Panel on Upper-limb Prosthetics, October 25-27, 1971,* National Academy of Sciences, Washington, D.C., 1972.
69. **Peizer, E., Wright, D. W., Pirrello, T., Jr., and Mason, C. P.,** Current indications for upper-extremity powered components, *Bull. Pros. Res.,* BPR 10-14, 22, 1970.
70. **Finley, F. R.,** Electromyographic patterns of multiple muscle sources, in *The Control of External Power in Upper-extremity Rehabilitation,* National Academy of Sciences Publication 1352, Washington, D.C., 1966, 61.
71. **Wirta, R. W., and Taylor, D. R., Jr.,** Development of a multiple-axis myoelectrically controlled prosthetic arm, in *Advances in External Control of Human Extremities,* Yugoslav Committee for Electronics and Automation (ETAN), Belgrade, 1970, 245.
72. **Wilson, A. B., Jr.,** Recent advances in above-knee prosthetics, *Artif. Limbs,* 12, 1, 1968.
73. **Wilson, A. B., Jr.,** Recent advancements in below-knee prosthetics, *Artif. Limbs,* 13, 1, 1969.
74. **LeBlanc, M. A.,** Elastic-liner type of Syme prosthesis: basic procedure and variations, *Artif. Limbs,* 15, 22, 1971.
75. **Pedersen, H. E., LaMont, R. L., and Ramsey, R. H.,** Below-knee amputation for gangrene, *South. Med. J.,* 57, 820, 1964 (reprinted in *Orthop. Pros. Appl. J.,* 18, 281, 1964.)
76. **Glattly, H. W.,** A statistical study of 12,000 new amputees, *South. Med. J.,* 47, 1373, 1964.
77. **Wilson, L. A., Lyquist, E., and Radcliffe, C. W.,** Air-cushion socket for patellar-tendon-bearing below-knee prosthesis, *Bull. Pros. Res.,* BPR 10-10, 5, 1968.
78. **Murphy, E. F.,** The swing phase of walking with above-knee prostheses, *Bull. Pros. Res.,* BPR 10-1, 5, 1964.
79. **Staros, A. and Murphy, E. F.,** Properties of fluid flow applied to above-knee prostheses, *Bull. Pros. Res.,* 10-1, 40, 1964.
80. **Wallach, J. and Saibel, E.,** Control mechanism performance criteria for an above-knee leg prothesis, *J. Biomech.,* 3, 87, 1970.
81. **Mauch, H. A.,** Stance control for above-knee artificial legs – design considerations in the S-N-S knee, *Bull. Pros. Res.,* BPR 10-10, 61, 1968.
82. **Goldner, J. L., Clippinger, F. W., and Titus, B. R.,** Use of temporary plaster or plastic pylons preparatory to fitting a permanent above-knee or below-knee prosthesis, *Bull. Pros. Res.,* BPR 10-13, 87, 1970.
83. **Burgess, E., Traub, J., and Wilson, A. B., Jr.,** Immediate Postsurgical Prosthetics in the Management of Lower-extremity Amputees, TR 10-5, Veterans Administration, Washington, D.C., 1967.
84. **Burgess, E. M., Romano, R. L., and Zettl, J. H.,** The Management of Lower-extremity Amputations, TR 10-6, Veterans Administration, Washington, D.C., 1969.
85. **Baumgartner, R.,** The early fitting of the vascular amputee, *A. P. O. Rev.,* 1, X (October), 13, 1970.
86. **Gardner, H. F.,** Endoskeletal structures for lower-extremity prostheses, *Bull. Pros. Res.,* BPR 10-13, 113, 1970.
86a. Committee on Prosthetic Research and Development, *Cosmesis and Modular Limb Prostheses,* National Academy of Sciences, Washington, D.C., 1971.
87. **Radcliffe, C. W.,** Functional considerations in the fitting of above-knee prostheses, in *Selected Articles from Artificial Limbs,* Robert E. Krieger, Huntington, N.Y., 1970, 5.
88. **Brunnstrom, S.,** Anatomical and physiological considerations in the clinical application of lower-extremity prosthetics, in *Orthopaedic Appliances Atlas, Vol. 2 – Artificial Limbs,* American Academy of Orthopaedic Surgeons, J. W. Edwards, Ann Arbor, Mich., 1960, 359.
89. **Schede, F.,** Theoretische Grundlagen für den Bau von Kunstbeinen; inbesondere für den Oberschenkelamputierten, *Z. Orthop. Chirurgie,* Beilageheft (Suppl. to) Vol. 39, Ferdinand Enke, Stuttgart, 1919.
90. **Perry, J. And Hislop, H. J.,** *Principles of Lower-extremity Bracing,* American Physical Therapy Association, New York, 1967.
91. **McIlmurray, W. and Greenbaum, W.,** A below-knee weight bearing brace, *Orthop. Pros. Appl. J.,* 12, 81, 1958.
92. **Kay, H. W.,** The Veterans Administration Prosthetics Center Patellar-Tendon-Bearing Brace, Report E-2, Committee on Prosthetics Research and Development, National Academy of Sciences, Washington, D.C., 1970.
93. Veterans Administration Prosthetics Center, A Manual for Fabrication and Fitting of the Below-Knee Weight-Bearing Brace (PTB), The Center, New York, 1967.
94. **Staros, A. and Peizer, E.,** Application of the Veterans Administration Prosthetics Center below-knee, weight-bearing brace to a bilateral case, *Artif. Limbs,* 9, 35, 1965.
95. **Anderson, M. H. and Bray, J. J.,** The U.C.L.A. functional long leg brace, *Clin. Orthop.,* 37, 98, 1964.
96. **Lehmann, J. F., DeLateur, B. J., Warren, C. G., and Simons, B. C.,** Trends in lower extremity bracing, *Arch. Phys. Med.,* 51, 338, 1970.
97. **Lehmann, J. F., Warren, C. G., DeLateur, B. J., Simons, B. C., and Kirkpatrick, G. S.,** Biomechanical evaluation of axial loading in ischial weight-bearing braces of various designs, *Arch. Phys. Med.,* 51, 331, 1970.
98. **Henderson, W. H. and Lamoreux, L. W.,** The orthotic prescription derived from a concept of basic orthotic functions, *Bull. Pros. Res.,* BPR 10-11, 89, 1969.

99. **McCollough, N. C., Fryer, C. M., and Glancy, J.,** A new approach to patient analysis for orthotic prescription – Part 1: the lower extremity, *Artif. Limbs,* 14, 68, 1970.
100. **Inman, V. T.,** UC-BL dual-axis ankle-control system and UC-BL shoe insert: biomechanical considerations, *Bull. Pros. Res.,* BPR 10-11, 130, 1969.
101. **Lamoreux, L. W.,** UC-BL dual-axis ankle-control system: engineering design, *Bull. Pros. Res.,* BPR 10-11, 146, 1969.
102. **Campbell, J. W., Henderson, W. H., and Patrick, D. E.,** UC-BL dual-axis ankle-control system: casting, alignment, fabrication, and fitting, *Bull. Pros. Res.,* BPR 10-11, 184, 1969.
103. **Henderson, W. H. and Campbell, J. W.,** UC-BL shoe insert, casting, and fabrication, *Bull. Pros. Res.,* BPR 10-11, 215, 1969.
104. **Reimann, I. and Lyquist, E.,** Dynamic splint in the treatment of club foot, *Acta Orthop. Scand.,* 40, 817, 1970.
105. **Dehne, E., Deffer, P. A., Hall, R. M., Brown, P. W., and Johnson, E. V.,** The natural history of the fractured tibia, in *The Surgical Clinics of North America,* Vol. 46, W. B. Saunders, Philadelphia, 1961, 6.
106. **Brown, P. W. and Urban, J. G.,** Early weight-bearing treatment of open fractures of the tibia, *J. Bone J. Surg.,* 51-A, 59, 1969.
107. Committee on Prosthetics Research and Development, Fracture Bracing, A Report of a Workshop Feb. 28-Mar. 1, 1969, National Academy of Sciences, Washington, D.C., 1969.
108. Committee on Prosthetics Research and Development, Cast-Bracing of Fractures, A Report on a Workshop Jan. 27-28, 1971, National Academy of Sciences, Washington, D.C., 1971.
109. **Sarmiento, A.,** A functional below-the-knee brace for tibial fractures, *J. Bone J. Surg.,* 52-A, 295, 1970.
110. **Mooney, V. and Harvey, J. P., Jr.,** Application of Lower Extremity Orthotics to Weight-Bearing Relief, Final Report to Social and Rehabilitation Service, Department of Health, Education, and Welfare, Project No. RD-2580 by Rancho Los Amigos and University of Southern California, Los Angeles, 1970.
111. Committee on Prosthetics Research and Development, Bracing of Children with Paraplegia Resulting from Spina Bifida and Cerebral Palsy: Report of a Workshop, Oct. 2-4, 1969, National Academy of Sciences, Washington, D.C., 1970.
112. **Bobechko, W. P., McLaurin, C. A., and Motloch, W. M.,** Toronto orthosis for Legg-Perthes disease, *Artif. Limbs,* 12, 36, 1968.
113. **Liberson, W. T., Holmquest, H. J., Scot, D., and Dow, M.,** Functional electrotherapy stimulation of the peroneal nerve synchronized with the swing phase of the gait of hemiplegic patients, *Arch. Phys. Med.,* 42, 101, 1961.
114. **Bunnell, S.,** *Surgery of the Hand,* 3rd ed., J. B. Lippincott, Philadelphia, 1956.
115. **Bisgrove, J. G.,** A new functional dynamic wrist extension-finger flexion hand splint, a preliminary report, *J. Assoc. Phys. Ment. Rehab.,* 8, 162, 1954.
116. **Anderson, M. H.,** *Functional Bracing of the Upper Extremities,* Charles C Thomas, Springfield, Ill., 1958.
117. **Engen, T. J.,** Development of upper extremity orthotics, part I, *Orthot. Pros.,* 24, 12, 1970.
118. **Engen, T. J.,** Development of upper extremity orthotics, part II, *Orthot. Pros.,* 24, 1, 1970.
119. **Kay, H. W.,** Clinical evaluation of the Engen plastic hand orthosis, *Artif. Limbs,* 13, 13, 1969.
120. **Birdsell, J. B.,** A Survey to Size the 4-B Prosthetic Hand, Spec. Tech. Rep. No. 16, Department of Engineering, University of California (Los Angeles), 1950.
121. **Fletcher, M. J. and Leonard, F.,** The principles of artificial-hand design, *Artif. Limbs,* 2, 78, 1955.
122. **DeFries, M. G.,** Sizing of Cosmetic Hands to Fit the Child and Adult Amputee Population, Tech. Rep. No. 5441, Army Prosthetics Research Laboratory, Washington, D.C., 1954.
123. **Garrett, J. W.,** The adult human hand: some anthropometric and biomechanical considerations, *Hum. Factors,* 13, 117, 1971.
124. **Bennett, R. L.,** The evolution of the Georgia Warm Springs Foundation feeder, *Artif. Limbs,* 10, 5, 1966.
125. **Kay, H. W. and Appoldt, N. V.,** Preliminary design analysis of linkage feeders, *Artif. Limbs,* 10, 10, 1966.
126. **Kay, H. W.,** Conclusions of a conference on linkage feeders, *Artif. Limbs,* 10, 20, 1966.
127. **Karchak, A., Jr., Allen, J. R., and Nickel, V. L.,** The application of external power and control to orthotic systems, in *Advances in External Control of Human Extremities,* Yugoslav Committee for Electronics and Automation (ETAN), Belgrade, 1970, 107.
128. Case Western Reserve University, Final Report, Biomedical Research Program on Cybernetic Systems for the Disabled, EDC Report No. 4-70-29, Social and Rehabilitation Service Grant No. RD-1814-M, undated.
128a. **Seamone, W. and Schmeisser, G., Jr.,** Interdisciplinary development and evaluation of externally powered upper-limb prosthesis (program report, other VA research programs), *Bull. Pros. Res.,* BPR 10-20, 321, 1973.
128b. **Conry, J. and Hassard, G. H.,** Wink switch activation of powered flexor hinge splint, *Bull. Pros. Res.,* BPR 10-18, 84, 1972.
128c. **Hoshall, C. H.,** Displacement sensors and their application to control of synthetically powered prostheses and orthoses, *Bull. Pros. Res.,* BPR 10-20, 4, 1973.
129. **Engen, T. and Spencer, W. A.,** Development of Externally Powered Upper Extremity Orthotics. Final Report to the Social Rehabilitation Service, Department of Health, Education, and Welfare, under Research Grant RD-1564, 1969.
130. **Rife, S. S. and Kennedy, E.,** A feeding device, *Artif. Limbs,* 13, 64, 1969.
131. Committee on Prosthetics Research and Development, Report of First Workshop Panel on Upper-extremity Orthotics, July 8-10, 1970, National Academy of Sciences, Washington, D.C., 1971.

132. **Perry, J.,** The use of external support in the treatment of low back pain, *Artif. Limbs,* 14, 49, 1970.
133. **Morris, J. M., Lucas, D. B., and Bresler, B.,** Role of the trunk in stability of the spine, *J. Bone J. Surg.*, 43-A, 327, 1961.
134. **Gregersen, G. G. and Lucas, D. B.,** An *in vivo* study of the axial rotation of the human thoracolumbar spine, *J. Bone J. Surg.*, 49-A, 247, 1967.
135. **Lumsden, R. M., II, and Morris, J. M.,** An *in vivo* study of axial rotation and immobilization at the lumbrosacral joint, *J. Bone J. Surg.*, 50-A, 1591, 1968.
136. **Lysell, E.,** Motion in the cervical spine – an experimental study on autopsy specimens, *Acta Orthop. Scand.*, (Suppl.), No. 123, 1969.
137. **White, A. A., III,** Analysis of the mechanics of the thoracic spine in man – an experimental study of autopsy specimens, *Acta Orthop. Scand.* (Suppl.), No. 127, 1969.
138. **Luedtke, G. L.,** Management of the dentition of patients under treatment for scoliosis using the Milwaukee brace, *Am. J. Orthod.*, 57, 607, 1970.
139. **Galante, J., Schultz, A., Dewald, R. L., and Ray, R. D.,** Forces acting in the Milwaukee brace on patients undergoing treatment for idiopathic scoliosis, *J. Bone J. Surg.*, 52-A, 498, 1970.
140. **Pavetto, G. C. and Zumaglini, C.,** The Minerva Pneumatica – clinical report on its use over a period of five years, *Sci. Tecnica,* N 122, 1970.
141. Committee on Prosthetics Research and Development, Spinal Orthotics – A Report of a Workshop, Mar. 28-29, 1969, National Academy of Sciences, Washington, D.C., 1970.
142. **Engen, T. J.,** unpublished manual.

4.2 POLYMERS AS SURGICAL IMPLANTS

R. I. Leininger
Battelle/Columbus Laboratories

INTRODUCTION

This review is concerned with the polymers used in surgical implants, the requirements for such uses, and the past, present, and, hopefully, future research approaches for the development of needed materials. It is not concerned with the multitudinous devices used or proposed except for illustration. Once a suitable material is available, engineering design and production are the problems. Obviously there must be interaction between the materials scientist and the engineer if the needed implant is to be realized.

Foreign-body implants are not new; they probably date from the beginnings of man himself. Unfortunately, these first implants resulted from having to "suffer the slings and arrows of outrageous fortune" and were hardly therapeutic. As man progressed to higher levels of civilization, morals, and ethics, crude stone arrowheads were replaced by implants of various metals propelled by devices more sophisticated than bows. The astonishing thing is that if no vital organs were disrupted or no fatal infections or poisons accompanied the implants, the recipient usually survived. This implant method unfortunately still survives. Numerous veterans of past wars and numerous civilians carry unwanted metallic memorabilia of their mishaps. The point is that while the body can tolerate implants, for surgical implants both efficacy and safety are overriding considerations. Minimum interference with the biological system by the implant is desired.

Tens of thousands of people are benefiting from surgical implants of many types. Thousands of artificial heart valves and heart pacemakers are implanted each year and more than 50,000 women have implanted breast prostheses. More than 5,000 patients with little or no kidney function survive with the aid of artificial kidneys, which, although external to the body, can be considered as implants because they are extensions of the circulatory system.

Polymers have long been used as surgical implants because of the wide variety of mechanical properties they offer, their general inertness to conventional environments, and their ready formability. Thus, some polymers are available with properties closely approximating those of soft tissue; others are suitable for hip sockets where load bearing is necessary. As can be imagined, polymers were first chosen from commercially available materials, and even now many polymeric surgical implants differ from their commercial counterparts only in care in handling and packaging. Development of polymers specifically suited to implant usage has received major emphasis only recently. In 1964, programs were begun in the U.S. by the National Institutes of Health on the development of materials suitable for an artificial heart or assist heart and for artificial kidneys. These material development programs, still being carried on by the National Heart and Lung Institute and the National Institute of Arthritis and Metabolic Diseases, have produced a number of polymers and modified polymers as candidates for specific applications. Interest by the scientific community has been high, as shown by the number of meetings, symposia, and short courses being offered – in one recent year 14 national meetings on biomaterials were held. A *Handbook of Biomedical Plastics* describes plastics used for different devices through mid-1969. An earlier monograph, Use of Artificial Materials in Surgery, was published in 1966. A number of such references are given in the bibliography.

In choosing or developing polymers for surgical implants, it is recognized that certain requirements must be met. These requirements include:

1. Compatibility
2. Nontoxicity (including noncarcinogenicity and nonteratogenicity)
3. Requisite physical properties (including permeability)
4. Stability in the body (maintenance of properties)
5. Sterilizability
6. Formability

Of the above requirements, physical properties

and formability present comparatively few problems to the polymer chemist except perhaps in combination with other specific requirements. The other requirements present definite challenges, as seen in the following sections of this review.

General Considerations of Compatibility

A compatible implant would have no effect on the adjacent tissue; that is, the nearby cells would show no abnormalities, no variant type cells would appear, there would be no inflammatory reactions, and of course there would be no cell necrosis. In short, the histology of the surrounding tissues would be altogether normal. The healing-in process would closely resemble the healing following a sham operation in which no polymer was implanted. In the special case of blood compatibility, this no-damage requirement extends to the coagulation system, proteins, formed elements, and other components of the blood.

In practice, no polymeric material (or metal or ceramic) has been found to be perfectly compatible. The body recognizes the implant as foreign and isolates it by encapsulation. Commonly, the thickness of the encapsulating membrane is often taken as a measure of the compatibility toward tissue. The thinner the membrane, the better the compatibility. Unfortunately, the methods of determination of compatibility are not standardized as to shape and size of the polymer sample, animal used, location within the animal, or the quantification of compatibility. Progress is being made by such groups as the American Society for Testing Materials Subcommitte F-4 on Surgical Implants and by similar groups in Australia and Europe. Further hope for standardization on a wide scale is the recent interest by the International Standards Organization in surgical implants. Still further impetus will result from the probable passage of legislation giving the U.S. Food and Drug Administration regulatory authority over medical devices and thereby over materials.

For convenience, studies on tissue and blood compatibilities are treated separately.

COMPATIBILITY

Tissue Compatibility

Tissue compatibilities of polymeric materials of a wide variety have been studied, and several polymers are commonly used in clinical practice. In judging the compatibility of a polymer, difficulty arises because of the additives often used to modify the properties or to stabilize the polymer. Adverse tissue reactions encountered can usually be traced to deliberate or adventitious additives. Experience has shown that stable "pure" polymers, with few exceptions, are acceptable to the surrounding tissue. It should be noted that the polymers most widely used in contact with soft tissue (silicone rubber, polytetrafluoroethylene, polyolefins, and polyethyleneglycol terephthalate) are used in a relatively pure form. Compounded polymers, notably polyvinyl chloride, are used for temporary, relatively short-term implantations, e.g., as catheters. Because of the general tissue compatibility of uncompounded polymers, effort toward the development of new structures for this implant use has not been necessary. Special grades of some materials are available; for example, a "medical grade" of silicone rubber, Silastic®, is prepared with attention to the purity of starting materials, choice of catalyst, and care during processing and packaging. Likewise, plasticized polyvinyl chloride is compounded with plasticizers and stabilizers approved for medical uses.

In reviewing the literature dealing with tissue compatibility of polymers, one is presented with the difficulty that not only have different methods been used but also the polymers being evaluated are usually identified only as to generic type. In addition, polymers studied in a given investigation are usually ranked on a comparative scale of tissue reactivity and not on an absolute basis. Because of this, no attempt is made here to include even a majority of papers on tissue compatibility; instead representative papers are included and some emphasis is given to methods that show promise of quantifying polymer-tissue interactions.

Four points will be noted in any study of the literature on the compatibility of polymeric implants.

1. There is no clear-cut distinction between imcompatibility and toxicity.
2. It is apparent that so long as a polymer is essentially stable in the body and does not degrade, incidences of incompatibility are usually caused by additives used with the polymers.
3. With few exceptions, polymers have not been designed expressly for implant use. This is because of the general body acceptability of

polymers and wide choice of properties given by available materials.

4. Emphasis has been on the preparation of polymers so that possible irritants were avoided and, of course, on care in handling to avoid contamination.

Polymers either of relatively pure materials or those in compounded forms have been used for implants for more than 30 years. The choice of polymers has been based on availability, stability, nonreactivity, and mechanical properties that appeared suited to the application. In the early 1940s, use of polymeric implants began on a wider scale and increased so rapidly that in 1947 Ingraham et al.[1] reviewed plastic materials in surgery. Since that time, the use of such implants has increased tremendously and has been the subject of further reviews and books.

In 1939, Dennis[2] used polymethyl methacrylate tubes 1 cm O.D. × 0.7 cm I.D. to allow drainage of osteomyelitic cavities transcutaneously. Tubular and sponge rubber were first implanted into thigh muscles of dogs and suppuration occurred. Lucite® was implanted similarly and found to cause no gross or microscopic evidence of inflammatory reaction after implantation for 4 weeks. There was a fibrous capsule (1.5 mm) surrounding the implant. Clinically the drains were left in place from 1½ to over 30 months without compatibility problems.

In 1941, Aries[3] compared nylon and silk in various locations in dogs and found that the nylon caused less tissue reaction than silk. It caused no adhesions and was encapsulated by the body. At about the same time, Narat[4] compared Vinyon® (copolymer of vinyl chloride and vinyl acetate), nylon, and catgut with silk in clinical applications. Again the tissue reaction of the synthetic material was found smaller than with the natural materials.

A series of polymers which were studied by Blum[5] in orthopedic applications included polymethylmethacrylate, cellulose acetate, nylon, urea formaldehyde, and phenol formaldehyde. These were compared with protein plastics prepared by reaction of formaldehyde with casein, red blood cells, and whole blood. The first group of polymers was found stable for periods up to 9 months, as bone implants in dogs. Polymethylmethacrylate showed the least tissue reaction. The protein plastics were found to be absorbed and replaced by new bone. Blum's work is especially significant in that the materials were also evaluated by tissue-culture techniques using embryonic tissue. It was found that polymethylmethacrylate, nylon, and phenol formaldehyde resin permitted normal growth, whereas urea formaldehyde resin and cellulose acetate inhibited cell growth. Presumably, unreacted formaldehyde in the resin and plasticizers used in the cellulose acetate were responsible for the inhibition. In any event, the tissue-culture studies correlated with the tissue response as seen in vivo.

Polyethylene[6] was considered as an implant directly into the cerebral cortex and as a replacement of an area of excised dura mater of brain in cats, dogs, monkeys, and rabbits. Segments of polyethylene tubing or film were implanted directly into the brain, and after periods of up to 150 days there was no evidence of gross reaction around the implant. There was no adhesion of the tissue to the implant. More importantly, histologic sections through the area of implantation (up to 10 days) showed no more tissue reaction than that found adjacent to a control stab wound. Sections through the sites after 10 days showed a thin capsule formation that increased to 6 to 8 cells in thickness in 90 days. This was found comparable to the capsule formed in response to tantalum, Vitallium®, and polymethylmethacrylate. Two polyethylene films were used as replacement of the dura mater in cats. In some cases, the bone was replaced and in others removed. With one polyethylene film considerable reaction including formation of hematomas was found, but with the second film there was no early reaction and the film was gradually encapsulated by a thin membrane. The authors considered the first film to contain additives such as antioxidants, whereas the second film was more nearly pure.

In 1946, Poppe and Oliveira[7] used films as possible wrappings for large blood vessels in the treatment of aneurysms or to produce occlusions. A noncoated cellophane was found to produce little reaction, whereas polyethylene was found to produce a marked fibrosis reaction. The type of polyethylene is not specified and it is probable that the sample contained additives. It is also possible that it was coated cellophane. A purified cellulose acetate film was found to produce little tissue reaction. This result can be compared with results obtained by Blum[5] who found cellulose acetate (presumably plasticized) to be tissue reactive and to inhibit cell growth. It should be noted that these writers were seeking a material that would cause fibrosis so as to reduce aneurysms or obliterate a patent ductus. Further work on polyethylene to excite a tissue response was done by Yeager and Cowley.[8] They found pure polyethylene to be well tolerated and to give minimal tissue response. Another polyethylene stimulated fibrous tissue production, and it was found that this film had been solvent cast and contained a low percentage (probably less than 1%) of dicetyl phosphate. The authors warned that the clinical application of polyethylene necessitated knowledge of the purity of the material.

In 1949, LeVeen and Barberio[9] considered the effect of the physical form of the implanted polymer. By then, a number of polymers had been studied and found to incite minimal tissue reactions. The authors considered that this could have been the result of the limited surface area of the bulk or film implants. Accordingly, they implanted powders or shavings of nylon-66, celluloid, polymethylmethacrylate, Teflon®, and cellophane. The finely divided materials were implanted through a needle into the peritoneal cavity of guinea pigs with the following results. Celluloid was found to produce an intense fibrinoplastic reaction and an inflammatory reaction subsided within 70 days, but adhesions were formed. This intense reaction was probably caused by the plasticizer, usually camphor, generally used with cellulose nitrate in making celluloid, even though the authors considered the celluloid to be pure. The polymethylmethacrylate caused an early reaction similar to that from celluloid, but not as intense. After 70 days, the same dense fibrotic changes were seen. The reaction to nylon was identical to that of the polymethylmethacrylate, as was the reaction caused by cellophane. Teflon, on the other hand, caused no inflammatory reactions and there was no gross evidence of tissue reaction after 70 days. There was no fibroplastic reaction and the individual particles could be easily removed from the filmy capsules. The authors concluded that it is necessary to use large surface areas to determine the true tissue reactivity of polymers. They felt that plastics produced tissue response but, even so, might still be useful, and recommended that the surface should be well polished so as to reduce surface area. The nonreactivity of the Teflon was ascribed to assumed nonadsorption of proteins on the surface.

In 1957, Harrison et al.[10] compared tissue reactions of polymers usually used at that time as implants. These included Dacron®, Ivalon®, nylon, Orlon®, and Teflon. The materials were used in the form of fabrics, except Ivalon, which was used as a compressed sponge. The materials were implanted in the thoracic and abdominal aortas of dogs, and the surrounding tissue was studied at intervals up to a year. With nylon there was initial edema with sufficient fluid in some cases to necessitate drainage. This subsided after about 2 weeks. There was also an initial acute inflammatory reaction that subsided over a period of 6 months. Fibrosis, evident on the 7th day, proceeded to completely encapsulate the implant. After encapsulation in about 5 months only occasional chronic inflammatory cells were seen. Dacron caused less edema, but the cell reaction was similar to that on nylon. Orlon also caused less edema and the acute and chronic inflammatory reaction and evidence of fibrosis were less than with nylon and Dacron. The Ivalon behaved similarly to Dacron and nylon. Teflon, however, caused no edema. Although an acute foreign-body reaction was observed, it was considerably less than that with the other materials. Similarly, the capsule formed with Teflon was thinner than that formed about the other materials. It should also be noted that more intense reaction surrounded the silk sutures used than the synthetic materials. The authors concluded that, although there is a tissue response, the synthetic materials are suitable and Teflon is the most unreactive material.

Polyvinylformal (Ivalon) was used extensively as an implant material for some time.[11-16] The polyvinylformal was used as a sponge and prepared by the reaction of formaldehyde with polyvinylalcohol. It was used to fill defects, reinforce aneurysms, repair septal defects, and for mammary augmentation. Because it was in sponge form, tissue infiltrated so that the implant increased greatly in hardness. There were instances of dilation and rupture of the material when used as arterial grafts. The material was found to become friable and deteriorate in the body. For these reasons it is no longer used as an implant material.[17]

Usher and Wallace[18] studied the compatibility of nylon (DuPont), Dacron (DuPont), Teflon (DuPont), and Marlex® (Phillips). These authors followed the lead of LeVeen and Barberio[9] in using the materials in a finely divided form. Marlex, a high density polyethylene, was used in the form of "tiny" pellets and the others were used in the form of undyed, shredded yarn. Ten grams of each material were placed in the peritoneal cavity of a dog. Implants were followed for only 7 days. Findings were similar to those of LeVeen and Barberio in that nylon caused the production of ascitic fluid and there were adhesions (for example of bowel-to-bowel) along with inflammatory reactions. Orlon produced much the same reaction but a smaller amount of free fluid. Dacron caused less adhesions but there was still a foreign-body reaction and a moderate amount of fluid. Teflon, as found by LeVeen and Barberio,[9] caused only a few filmy adhesions which were easily separated. There was no free fluid and the inflammation was reduced. Marlex produced no adhesions, although there was some free fluid. There was considerably less gross reaction than with nylon, Orlon, Dacron, or Teflon. Some inflammatory reaction was present.

In 1959, a group headed by Lieb[19,20] evaluated a number of plastics as small strips subcutaneously and intragluteally for periods up to 10 months. A second group of materials was implanted in the anterior chamber of rabbits and in one dog for periods up to 10 months. For polymethylmethacrylate, celluloid, Dacron, nylon, Orlon, polyethylene, Ivalon, nylon-6, Teflon, Vinyon, and Vinyon-N®, the results, in general, followed those of earlier workers. Polymethylmethacrylate, nylon-6, and Teflon appeared to cause less tissue reaction than the other materials. It is of interest that nylon-6 appeared to be less reactive than nylon-66; in fact, the capsule formed around the nylon-6 was comparable to that of the Teflon. As expected, the celluloid gave an intense reaction, but here again it subsided after about 2 months with a continuation of chronic inflammation.

These investigators also evaluated the compatibility of several plastics in the anterior chambers of rabbits. Samples, in the form of 5- × 2- × 1-mm strips with smooth round edges, were chosen on the basis of their optical properties. With CR-39® (polyester of diallyl glycol carbonate), hyperemia was present in the first 2 weeks of implantation and pigment precipitated on the anterior and posterior surfaces. No remarkable cellular reaction was observed but in most of the eyes the strips were surrounded by a fibrogenous membrane. Gafite® (polymethyl-α-chloroacrylate) gave a reaction similar to that of CR-39 with the addition of a mild nonspecific acute inflammation. Lexan® caused prolonged irritation with an inflammatory exudate. Because this was an early sample, it was considered that the reactions might be caused by foreign materials that would not be present in later commercial materials. Two samples of polymethylmethacrylate were evaluated: Lucite and Plexiglas®. A vey small tissue reaction was found with Lucite, but with the Plexiglas there were acute and chronic inflammatory reactions. This again emphasizes that while the base polymer may show good compatibility, additives present may give rise to tissue reactions. Silicone B695-106-1 (Dow-Corning) was found to have excellent compatibility. It is of interest that in the eyes of 3 of the 54 animals, the plastic was covered by a single layer of endothelial cells 4 months after implantation. The authors noted that the formation of this tissue layer may have accounted for the extreme chemical inertness. Styrex 767®, styrene-acrylonitrile, was found to produce an acute inflammatory reaction and later a chronic inflammation. Styron-666® (polystyrene) was found to be well tolerated by the eye tissue with only little postoperative reaction. The increased reactivity of the acrylonitrile-containing material may be explained by later results[21] which showed that similar materials had extractable components. It will be remembered that polyacrylonitrile fibers (presumably pure) have been found compatible.[10]

Boretos et al.[22] compared the tissue reactivity of a polyether segmented polyurethane (Lycra®) with that of a polyester polyurethane (Estane® 5701, B.F. Goodrich Co.). The implants were 25-mm I.D. by 39-mm O.D. by 0.75-mm-thick rings implanted subcutaneously in dogs. Significant adverse tissue reaction was noted for the polyester urethanes, whereas the polyether urethanes showed no noticeable fibrous reaction.

Polyvinylchloride (PVC) presents a special case because it is usually used modified with plasticizers and other additives including stabilizers so that it could be expected to give varying tissue responses. During the 1960s interest rose in the prolonged use of endotracheal and nasotracheal tubes as an alternative to tracheostomy. It was found that the use of some compounded PVC tubes led to white plaquelike tissue reactions at the glottis. Because of this, Guess and Stetson in 1968[23,24] investigated the tissue compatibility and toxicity of tubing materials both by implanting and by tissue-culture technique (their technique is described later in this review). It was found that those samples showing toxic reactions by tissue culture and causing necrosis when they were implanted intramuscularly contained organotin compounds as stabilizers; those samples without organotin stabilizers did not show such reactions.

Guess and Haberman[25] investigated the additives used in compounding polyvinylchloride in more detail. Starting with base PVC, they found that two samples gave a definite reaction as implants, while a third did not. Reasoning that the reaction was caused by some extractable component, they extracted one of the PVC resins with alcohol. The extracted material was shown to cause visible necrosis when injected into the muscle of the rabbit, and the PVC resin residue now revealed only the common slight foreign-body reaction. The extracted material was not identified. The material could have been residual catalyst, catalyst reaction products, or suspending agent. It should be noted that as far back as 1953, Scales[26] found that uncompounded, low-molecular-weight PVC as multifilament gave no inflammatory reaction or increased vascularity in 289 days' implantation in a guinea pig.

The authors then extended their work to include plasticizers and stabilizers that might possibly be used with vinyl plastics. They used saturated saline solutions of plasticizers intradermally or intraperitoneally in rabbits and mice, respectively. The pure materials and saturated saline solutions also were used in tissue culture. Among the 44 different plasticizers evaluated, 10 of the 17 phthalate plasticizers showed no toxicity by the method of measurement used, while 7 showed some degree of toxicity either in the undiluted form or the saturated aqueous solution. The five sebacates evaluated showed almost no toxicity in the systems. Among the stabilizers investigated, all of the organotin compounds showed some degree of toxicity. Note that the five lead stabilizers evaluated showed a surprising lack of toxicity. As might be expected, the saturated saline solutions of the compounds tested often showed less reaction than the pure compounds.

A number of polymers were evaluated either as intramuscular implants or by the effect of alcohol extracts on cell cultures and as intradural injections. As is seen in Table 4.2–1, all of the bulk polymers were found to be compatible as intramuscular implants. Alcoholic extracts of the ABS and cellulose triacetate were found to have an effect on the cell culture. Note that some extracts showed a positive effect when injected intradermally, which was not correlated with the amount of extractant, or the intramuscular or cell-culture results. The extracts of Teflon, polypropylene, and polycarbonate had a definite effect in the intradermal test, whereas the two nylons with a much higher content of extractables showed no such effect. The amount of alcohol extractables in this series did not affect the tissue compatibility.

It has been known that phthalate plasticizers are extracted by biological fluids.[27] While physio-

Table 4.2–1
TOXICITY PROFILE OF SEVERAL PLASTIC TYPES

		Cell culture toxicity		Rabbit toxicity	
Plastic	**% Extract alcohol**	**Plastic**	**Extract**	**Plastic, i.m.**	**Extract, i.d.**
Teflon®	0.25	–	–	–	++++
Nylon 66	1.3	–	–	–	–
Nylon 6	1.3	–	–	–	–
Polypropylene	2.0	–	–	–	++
Polycarbonate	0.25	–	–	–	++
Acrylonitrilbutadiene styrene	6.1	–	+	–	–
Polyphenylene oxide	10.8	–	–	–	++
Cellulose triacetate	15.9	–	+	–	+++
Polyethylene	0.36	–	–	–	–
Teflon FEP	0.02	–	–	–	–

From Guess, W. L. and Haberman, S., *J. Biomed. Mater. Res.*, 2, 313, copyright © 1968, John Wiley & Sons, Inc. With permission.

logical saline solutions were not able to extract diethylhexylphthalate from PVC tubing even after 6 hr, 4% bovine serum albumin and whole blood both were able to extract appreciable quantities of the plasticizer. The extracted plasticizer was almost entirely in the plasma fraction and was specifically associated with a lipoprotein fraction. That the extraction of plasticizer can proceed to extreme degrees was shown by Drenick and Lipset in 1971[28] who found that a vinyl nasogastric tube, which had apparently been in place for several weeks, had kinked and become so rigid that it was withdrawn only with great difficulty.

Silicone rubber implants have been and are being widely used because of their good tissue compatibility and mechanical properties. One of the first studies of the compatibility of silicone rubbers was by Marzoni et al. in 1959.[29] Sponges of three densities (not given) were implanted in place of rib sections next to bones and subcutaneously in dogs for periods up to a year. The lowest density sponges showed encapsulation by moderately fibrous tissue, and there was some tissue invasion but much less than that seen with Ivalon. The more dense sponge showed almost no tissue reaction and only a very thin encapsulation. Implants of these materials in rib beds showed that regeneration of the bone had occurred under the implant. The reactions of the implants on the bone and on soft tissue were identical. It was concluded that the two sponges of higher density produced strikingly less tissue reaction and fluid accumulation than other materials used as implants, and that clinical applications were justified. Because silicone polymers have been and are being very widely used as implants, the literature is extensive. References 30 through 35 are but a small portion of the published information.

The use of lower-molecular-weight silicone polymers as fluids has been reviewed by Blocksma.[36] The fluid, implanted by injection of very small drops, is recommended for the correction of facial hemiatropy, depressed facial scars, and other facial conditions. Compatibility of the fluid is good. Presently the use of the fluid is restricted to selected surgeons. Use for mammary augmentation is illegal in the U.S. Relatively recently, a new class of polymers, hydrogels, has been proposed for implant and ophthalmological uses.[37] These are polymers of 2-hydroxyethylmethacrylate or glycerolmethacrylate. Both polymers are cross-linked by copolymerization with small amounts of a difunctional methacrylate. Because they are hydrophilic they swell in water, but then do not dissolve because of the cross-linking.

Preparation of the polymers is usually by polymerization of the monomer or monomers in a water solution containing ammonium persulfate as a catalyst.[38] The concentration of the monomer in the solution determines the water content of the finished polymer. After equilibration in physiologic saline, polymers containing 70, 75, and 80% water were implanted subcutaneously and intraperitoneally into rats. All samples were tolerated well in both locations with no reaction other than the usual encapsulation which began after about 24 hr. Within 10 days all implanted samples had adhered firmly to the underlying material and surrounding tissue. Vascularization and ingrowth of cellular elements from the surroundings led to the fixation of the implant after about 2 to 3 weeks. It was proposed that such plastics might be used to fill void spaces after lobectomies and for cosmetic or reconstructive surgery. Šprincl et al.[39] studied the effect of water content of polyglycolmethacrylate gels on the healing-in process. The polymers were prepared containing 50, 60, 70, 80, and 90% water by volume in the usual manner in a water solution containing ammonium persulfate as the initiator. With 2% of glycoldimethyacrylate as a cross-linking agent, gels containing up to about 45% water in the initial mixture are homogeneous and optically transparent, while gels containing greater amounts of water are heterogeneous and opaque. These polymers were implanted into rats subcutaneously as disks 15 mm in diameter and 5 mm thick. The samples were not equilibrated in saline before implantation. Macroscopically the test implants were well tolerated by the organism and did not provoke any unfavorable reaction. With samples containing 50 to 60% water, a cellular capsule was formed within 30 days. By 90 days the capsule was formed of collagenous, fibrous tissue and capillaries were found sporadically in the external layer of the capsule. With the sample containing 70% water, in 30 days the capsule consisted of fine collagen fibers, fibroblasts, and fibrocytes. With the hydrogel containing 80% water, penetration by capillaries had occurred again by 30 days. This penetration was deeper than for the 70% sample, and in 90 days bands containing numerous capillaries and collagenous fibrous tissue penetrated deeply into the implant. The capsule for the sample containing 90% water was not completely formed within 30 days, and the external layer of the implant was practically replaced by a continuous and considerably cellular zone of newly formed capillaries, fibrocytes, and fibroblasts. By 90 days the newly formed capillaries occupied a larger part of the implant but never penetrated it throughout. Thus, it can be considered that the inhomogeneous gels containing 70% or more water contained void spaces or pores large enough to permit the ingrowth of tissue and capillaries. The authors conclude that gels containing 70% or less water are the most suitable for implantations in plastic surgery because the surface was not damaged by the tissue. They suggest that when a softer prosthesis is required (containing a larger amount of water) the surface should be protected by a layer of less-porous polymer.

Refojo[40,41] discusses the swelling characteristics of another hydrogel, polyglycerolmethacrylate, and the ophthalmological applications of this gel. In this application the material would be implanted in the dehydrated state and allowed to swell to fill the body cavity, for example, the vitreous cavity in the eye. Such implants in organs or tissues could be introduced through a smaller incision than that required for a preformed gel or the usual plastic implant.

Modification of this type of hydrogel can affect the tissue compatibility. Barvič and co-workers[42] prepared copolymers of glycolmethacrylate with dimethyaminoethylmethacrylate and methacrylic acid, and terpolymers containing both dimethylaminoethylmethacrylate and and methacrylic acid. With gels containing 20% water and equilibrated with saline before implantation in rats, it was found that the copolymers containing up to 20% methacrylic acid and the interpolymers were well tolerated by the tissue, comparable to the tolerance of homopolymer itself. The copolymers of the dimethylaminoethylmethacrylate, however, were tolerated only at a low content of comonomer, 3 to 5%. Samples with 15 to 20% of the dimethyl amino compound showed a pronounced reaction starting with the first day. The reaction increased until about day 10 and subsided in 3 weeks; after 24 to 30 days the macroscopic picture was within normal limits. The implant capsule, however, persisted in a very pronounced form and after a longer time usually resulted in a neoplastic reaction.

Another type of hydrogel is the ionic complex prepared from polyvinylbenzyltrimethylammoniumchloride and poly(sodium)styrenesulfonate.[43] These can be prepared as neutral polymers or with an excess of either the anionic or cationic component. In subcutaneous implants in rats it was found that 3 of 16 rats implanted with the cationic complex developed sarcomas after a 12-month period. No evidence of tumors was found in the rats implanted with either the anionic or neutral complex. The possible significance of the effect of the excess cationic groups on this polymer and the neoplastic effect of the dimethylaminomethacrylate in the methacrylate hydrogels is discussed later.

One of the few polymers prepared expressly for implant use was a 90/7.5/2.5 polymer of butylacrylate, methylacrylate, and methacrylamide.[44] This was prepared as a foamed material which would have mechanical properties similar to those of arteries. After a subcutaneous implant in foam form for 4 months, necrosis and foreign-body-type giant cells were absent. There was scanty marginal reaction to the foam as compared to that for film. There was tissue ingrowth into the pores of the foam (pore size from 30 to 600 μm with a medium of perhaps 250 μm).

Measurement of Tissue Compatibility

In the foregoing general review of tissue compatibility of a number of polymers, compatibility was measured qualitatively by either macroscopic or microscopic examination of the surrounding tissues and the implanted polymer. In several investigations, efforts were made to quantify the tissue-implant interaction. Sewell et al.[45] in 1955 devised a system in which the number of different types of cells, the overall cellular concentration, and the diameter of the inflammatory response were considered. Each factor was numerically weighted and the summation of these weighted factors was taken as a measurement of the host response to the implant. Table 4.2–2 shows the assessment of tissue compatibility by his system. This system was devised to compare the tissue reaction of sutures of ovine catgut in three species. This method was used by Postlethwait,[46] who compared the tissue compatibility of polyglycolic acid sutures with those of catgut, silk, and Dacron. Table 4.2–3 shows the data obtained using Sewell and co-workers' scheme at intervals up to 8 months. It should be noted that the polyglycolic

Table 4.2–2
ASSESSMENT OF TISSUE COMPATIBILITY

Total points	Grade of tissue reaction	Name assigned to the grade
0–16	1	Very slight
17–32	2	Slight
33–48	3	Slight to moderate
49–64	4	Moderate
65–80	5	Moderate to marked
81–96	6	Marked
97–112	7	Marked to extensive
Greater than 112	8	Extensive

From *Surg. Gynecol. Obstet.,* 100, 485, 1955. By permission of *Surgery, Gynecology, and Obstetrics.*

acid and catgut sutures are absorbable and were 85 to 100% absorbed in the first 4 months of implantation. This no doubt explains the sharp increase found in compatibility of these sutures after 4 months' implantation. It is interesting to note that silk and Dacron had essentially the same tissue compatibility in this work, although previous observers, using more qualitative methods, have considered Dacron to be less tissue reactive than silk.

Salthouse and Williams[47] evaluated a series of sutures by measuring the enzyme activity in the surrounding tissue. Table 4.2–4 summarizes the results obtained using this method. The column headed "tissue reaction" gives overall reactions as measured by the method of Sewell discussed previously. The data show that acid hydrolytic and proteolytic enzymes constitute the principal type of activity present at the sites of tissue reaction to the sutures. Such activity, higher with collagenous sutures, may be responsible for the absorption of these sutures. Salthouse and Willigan[48] extended this method to nine samples of plastics. Table 4.2–5 shows the results after 7 and 14 days of implantation of different tin-stabilized polyvinylchloride compounds intramuscularly in rats, compared to a U.S.P. negative (high-density polyethylene) and a sham wound. High enzyme activity is seen with these plasticized compounds stabilized with tin compounds. Table 4.2–6 shows similar data for unmodified polymers. The enzyme activity induced by these polymers was much less than that induced by the compounded polyvinylchlorides. The polyethylene induced the greatest reaction and can be compared to the U.S.P. negative sample used in the previous table. The polypropylene sample was surgical suture. The author suggested other enzyme activity demonstrable by histochemical procedures also would be of value in interpreting the mechanism and cellular dynamics of implant-tissue interactions.

The change in the ratio of γ-globulin to albumin in the serum surrounding an implant was used as a measure of the tissue compatibility by Atsumi and co-workers.[49] Figures 4.2–1 and 4.2–2 show the change in this ratio for several materials implanted subcutaneously in mice. The usual maximum tissue response is shown, and it seems to occur 1 to 2 months after implantation; after that period the reaction diminishes. This corresponds to histological observations in general. In these data, silicone rubber also shows a low tissue reaction. The influence of the additives is

Table 4.2–3
TISSUE COMPATIBILITY OF SUTURES MEASURED BY THE SEWELL METHOD[45]

Interval	PGA O[a] reaction grade	PGA 3-0 reaction grade	Chromic O reaction grade	Chromic 3-0 reaction grade	Silk reaction grade	Dacron® reaction grade
3 days	35.6	35.7	36.9	34.1	41.1	36.5
7 days	43.2	42.0	37.9	38.9	43.4	40.9
14 days	31.9	29.1	43.0	42.2	31.9	31.8
28 days	29.4	26.5	41.8	42.0	26.2	25.9
42 days	24.2	24.7	42.2	44.6	28.2	25.2
2 months	25.1	21.3	48.4	46.6	24.0	22.4
4 months	10.2	10.5	26.4	21.3	18.5	17.5
6 months	9.0	10.0	21.1	12.1	13.1	15.2
8 months	–	10.0	16.6	18.8	19.2	16.7

[a]PGA: polyglycolic acid

From *Arch. Surg. (Chicago),* 101, 489, 1970. With permission. Copyright 1970, American Medical Association.

Table 4.2–4
SUMMARY OF TISSUE REACTION AND ENZYME HISTOCHEMISTRY

Suture size 3-0	Tissue reaction	Alkaline phosphatase	Acid phosphatase	Aminopeptidase	Proteas
Ethilon nylon	Minimal	+; present in occasional neutrophil only	Neg.	Neg.	NT
Mersilene® polyester fiber	Slight	+ in some monocytic cells adjacent to suture	+ in fibroblasts and macrophages	Neg.	NT
Silk	Marked	± in occasional neutrophil only (2+ within strand at 18 days)	+ in occasional macrophage; 3+ in giant cells (2+ in fibroblasts after 18 days)	± (no increase at 18 days)	NT
Catgut, plain	Moderate	± an occasional neutrophil	2+ in some fibroblasts and all macrophages. Heaviest activity adjacent to suture	+	2+
Catgut, chromic	Moderate	+ in occasional neutrophil (no change at 21 days)	2+ in fibroblasts and macrophages (no change after 21 days)	+ (3+ after 21 days)	2+
Collagen, plain	Moderate	± in occasional neutrophil (no change at 11, 18, or 21 days)	2+ in fibroblasts and macrophages. Heaviest activity adjacent to suture (+ at 11, 18, and 21 days)	+ (2+ after 11 days, 3+ after 18 and 21 days)	2+
Collagen, chromic	Moderate	± in occasional neutrophil	2+ in fibroblasts and microphages	+	NT
Cotton	Extensive	2+; confined to cells within suture strands	3+ in all cells of tissue reaction	NT	NT

Table 4.2–5
TISSUE COMPATIBILITY, TIN-STABILIZED COMPOUNDS

Sample No.	Zone of inhibition, mm, succinic dehydrogenase[a]	Acid phosphatase, pH 5.2	Aminopeptidase, pH 6.8	Aminopeptidase, pH 5.2 (Cathepsin B)	Adenosine triphosphatase, pH 7.0
7 days postimplant					
1 U.S.P. negative	0.15	+	±	–	+B
2 PVC A	0.68	++A	+A	±A	++B
3 PVC B	0.85	+++ A	++A	+AC	++B
4 PVC C	0.48	++A	+A	±A	++B
5 PVC D	0.76	+++A	+A	+AC	++B
6 Sham	0.02	±	–	–	–
14 days postimplant					
1 U.S.P. negative	0.12	±	±	–	+B
2 PVC A	0.76	++A	+A	+AC	+++B
3 PVC B	0.96	++A	+A	+AC	+++B
4 PVC C	0.64	+A	+A	±AC	+++B
5 PVC D	1.05	+++AC	++A	++AC	+++B
6 Sham	0.00	–	–	–	–

[a]Measured from polymer surface to enzyme active muscle (mean of five readings)

Notes:

A = activity present in giant cells
B = activity confined to blood vessel endothelium
C = extracellular activity in addition to intracellular
– = no activity
± = slight activity
+ = moderate activity
++ = abundant activity
+++ = intense activity

(From Salthouse, T. N. and Willigan, D. A., *J. Biomed. Mater. Res.*, 6, 105, copyright © 1972, John Wiley & Sons, Inc. With permission.)

shown by the difference between the medical-grade silicone rubber and another silicone rubber cured with benzoylperoxide. It is also shown by the difference between a natural rubber conventionally cured with sulfur and a purified rubber vulcanized with a peroxide.

Tissue culture has been used by a number of workers to measure the compatibility and toxicity of implant materials. In 1945, Blum[5] stated, "The most sensitive index of tissue response to foreign bodies is the effect produced by the substance on explants of embryonic tissues." Tables 4.2–7 and 4.2–8 show the results obtained by this method on a series of protein and other type plastics. The protein plastics were condensation products of formaldehyde with the proteins named. The results of the tissue growth work agreed well with the implantation findings except in the case of the ureaformaldehyde resin which was toxic to the growing cells, although well tolerated as an implant. The toxicity of the cellulose acetate was probably due to a plasticizer. Rosenbluth et al.[50] used a tissue-culture method to screen a number of plastic items used in medical practice and compared the results with tissue culture to those obtained by intramuscular implantation. The method involved growing a monolayer of mouse cells (strain L929) on the bottom of a glass petri dish in Eagle's medium. Plastic test samples were placed in the petri dishes and incubated. Death, deterioration, or any other observable changes in the cells in relation to control cells were noted as positive or toxic responses. Most positive responses were easily observable in 24 hr, and no additional evidences of toxicity could be attributed to the test samples after 48 hr. For photographic records the cells were stained with a crystal violet solution that stains the living cells and does not stain

Table 4.2–6
TISSUE COMPATIBILITY, UNMODIFIED COMPOUNDS

Sample No.	Zone of inhibition, mm, succinic dehydrogenase[a]	Acid phosphatase, pH 5.2	Aminopeptidase, pH 6.8	Aminopeptidase, pH 5.2 (Cathepsin B)	Adenosine triphosphatase, pH 7.0
7 days postimplant					
7 Silicone rubber	0.20	+A	+A	Trace	+B
8 PTFE	0.18	±	Trace	–	–
9 Polyethylene	0.26	++A	++A	+AC	+B
10 Polypropylene	0.04	±	Trace	–	–
11 Sham	0.03	Trace	–	–	–
14 days postimplant					
7 Silicone rubber	0.28	+	+	Trace	+B
8 PTFE	0.20	+	+	–	–
9 Polyethylene	0.30	++	++	+C	+B
10 Polypropylene	0.06	Trace	±	–	–
11 Sham	0.00	–	–	–	–

[a]Measured from polymer surface to enzyme active muscle (mean of five readings)

Notes:

A	=	Activity present in giant cells
B	=	Activity confined to blood vessel endothelium
C	=	Extracellular activity in addition to intracellular
–	=	No activity
Trace	=	Trace of activity
±	=	Slight activity
+	=	Moderate activity
++	=	Abundant activity
+++	=	Intense activity

From Salthouse, T. N. and Willigan, D. A., *J. Biomed. Mater. Res.*, 6, 105, copyright © 1972, John Wiley & Sons, Inc. With permission.

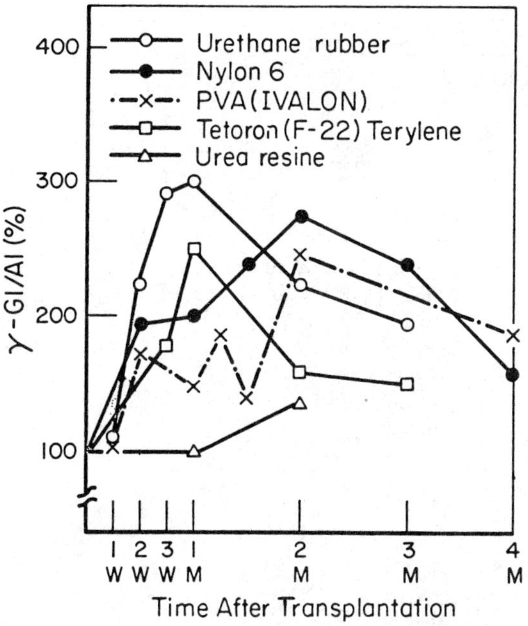

FIGURE 4.2–1. Variation of γ-globulin/albumin after transplantation of various materials. (From *Trans. Am. Soc. Artif. Intern. Organs,* 9, 324, 1963. With permission.)

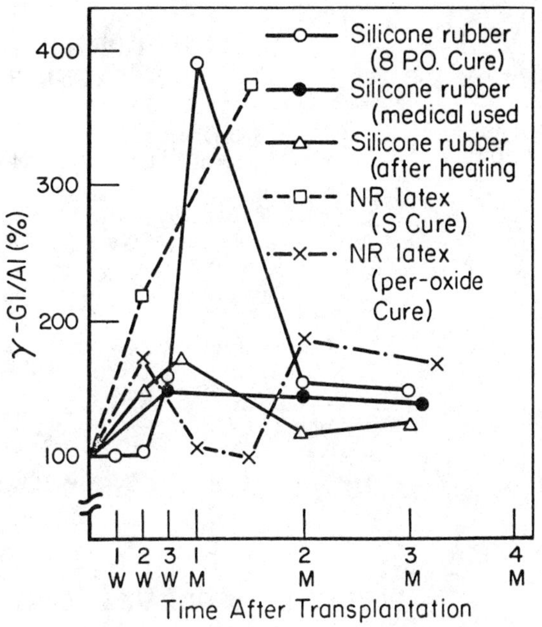

FIGURE 4.2–2. Variation of γ-globulin/albumin after transplantation of silicone rubber and natural rubber. (From *Trans. Am. Soc. Artif. Intern. Organs,* 9, 324, 1963. With permission.)

attenuated or dead cells. Table 4.2–9 shows the comparison of tissue-culture results with those obtained by intramuscular implantation. The plastics used were obtained from medical devices such as tubing, and their composition was not specified. It is seen that the majority of the results obtained by tissue culture were in agreement with those obtained by implantation. In some cases materials that inhibited cell growth were judged nontoxic in the implantation tests. No samples showed a toxic reaction intramuscularly and a nontoxic reaction in tissue culture. These data indicate that the tissue-culture technique proposed here is more sensitive than the implantation method. However, intramuscular implants were evaluated on a microscopic basis. Table 4.2–10 shows that prior extraction of the materials with ethyl alcohol usually decreased the toxicity as shown by tissue culture and macroscopic examinations of the implant sites. It is seen that the results obtained by histopathological examination of tissue from the implantation site agree with the tissue-culture results. This tissue-culture method was modified to cover the layer of monocells in a culture plate with a thin layer of agar before placing the implant. Homsy and co-workers[52,53] used a tissue culture system somewhat similar to that used above. The cells used were from pooled, newborn mouse hearts and were anchored to the culture dish with a thin plasma chick embryo clot which immobilized the tissue cells and provided a satisfactory substrate for cell migration. The results with various materials are shown in Table 4.2–11.

In 1968, Johnsson and Hegyeli[54] suggested using tissue-culture techniques for evaluating plastics-tissue interactions, on the basis that the physiological conditions encountered in the body could be simulated and controlled by this method. Furthermore, they proposed that the degradation of materials could be followed by radioactive tagging of the prosthetic material with subsequent examination of the cells and the medium for migration of radioactivity. Johnsson and Hegyeli[54] have studied cell-polymer interaction of materials proposed for use in the artificial heart program. In the tissue-culture phase of this study, human amnion and L929 mouse fibroblast cells were used. The results were compared with cell growth of control cultures on glass coverslips (Table 4.2–12, adapted from Tables 4.2–6 through 4.2–9,[55] shows the results using unmodified polymers). Silicone rubber, polypropylene, and Teflon showed essentially normal growth

Table 4.2–7
EFFECT OF NONABSORBABLE PLASTIC ON EMBRYONIC EXPLANTS OF TISSUE

Substance (3 plates of each)	24-hr results	48-hr results
Urea (and formaldehyde)	No growth	Dead
Phenol (and formaldehyde)	Fair growth	Growing well. Bits of plastic incorporated in tissue
Cellulose acetate	2 no growth, 1 fair growth	2 dead, 1 poor, slight growth
Methylmethacrylate	Fair growth	Growing well
Nylon	Fair growth	Growing well
Steel	Fair growth	Growing badly
Control	Fair growth	Normal

From *Br. J. Surg.*, 33, 246, 1945. With permission.

Table 4.2–8
EFFECT OF ABSORBABLE PLASTIC ON EMBRYONIC EXPLANTS OF TISSUE

Substance (3 plates of each)	24-hr results	48-hr results
Fibrin plastic	Good growth	Growing well. Bits of plastic incorporated in tissue
Casein plastic	Fair growth	Growing well. Bits of plastic incorporated in tissue
Fibrinogen plastic	Fair growth	Growing well. Bits of plastic incorporated in tissue

From *Br. J. Surg.*, 33, 245, 1945. With permission.

Table 4.2–9
COMPARISON OF TOXICITY SCREENING RESULTS BY TISSUE CULTURE METHOD OPPOSED TO INTRAMUSCULAR IMPLANTATION METHOD

Positive T.C., positive I.M.	Negative T.C., negative I.M.	Positive T.C., negative I.M.	Negative T.C., positive I.M.
36	56	20	0

From *J. Pharm. Sci.*, 54, 156, 1965. Reproduced with permission of the copyright owner.

compared to the control surface. Hydrin® rubber (Goodyear) inhibited cell growth. The work carried on by Pennington et al.[56] was concerned primarily with evaluation of experimental polymers and polymers modified by experimental treatments for nonthrombogenicity.

Block copolymers that had been hydroxylated, carboxylated, or sulfochlorinated showed toxic responses in tissue-culture tests at Batelle/Columbus. As other polymers with such functional groups did not show similar responses, these responses were probably due to the presence of extractable materials.

Powell et al.[57] used implantation and a tissue-culture method as described by Rosenbluth et al.[50,51] to evaluate a number of materials shown in Table 4.2–13. Again it is shown that the tissue-culture method is more sensitive than implantation when each is graded by macroscopic methods.

Homsy[52] has advocated measuring the extractables by a "pseudoextracellular fluid" as an

Table 4.2–10
COMPARISON OF TOXICITY RESULTS BETWEEN SELECTED SAMPLES BEFORE AND AFTER HARSH ETHANOL EXTRACTION

Plastic code	Tissue culture	Intramuscular implantation	Histopathological examination of tissue from intramuscular implantation site
X-37	+[a]	+	+
X-37 HEE[c]	–[b]	–	
X-74	+	+	+
X-74 HEE	–	–	
X-152	+	+	+
X-152 Hee	–	–	–
X-154	–	–	–
X-154 HEE	–	–	–
X-168	+	+	+
X-168 HEE	+	–	+
X-169	–	–	–
X-169 HEE	–	–	–
X-la[d]	–	–	–
X-175[d]	–	–	–
X-55[e]	+	+	+

[a]Positive or toxic reaction.
[b]Negative or nontoxic reaction.
[c]HEE: plastic sample after it was extracted with ethyl alcohol by refluxing for 24 hr.
[d]X-la and X-175 were used as control negative plastic samples. These samples have been used repeatedly and have demonstrated that they produce no toxic effect.
[e]Sample X-55 represents a positive control.

From *J. Pharm. Sci.*, 54, 156, 1965. Reproduced with permission of the copyright owner.

indication of tissue compatibility of a polymer. The fluid proposed is shown in Table 4.2–14 and is similar to the salt content of the blood. Candidate materials in an amount sufficient to have about 400 cm^2 of surface area are exposed to 250 cc of this fluid for 63 hr at 15°C, 30 psia. The aqueous fluid is then extracted in carbon tetrachloride and the moles per million equivalent of normal hexanol are determined by infrared analysis. Results obtained on a number of polymers were shown in Table 4.2–11. It is seen that some of those polymers with lower amounts of extractables are more inert in the tissue-culture test. There are exceptions, however; for example, vinylidene fluoride (No. 15 in the table) has extractables of only 3 mpm, but has a tissue-culture response of 2 to 3. Polymers having a moderate response to tissue-culture tests (+2) had extractables from below the limit of detection up to 198 mpm equivalent. It is apparent that the compounds extractable from the different polymers vary widely in toxicity in a tissue-culture test. On the other hand, it is also possible that biological fluids may extract materials not extractable by salt solutions. An example of this was cited earlier[26,27] in which it was found that plasticizers could be extracted from compounded polyvinylchloride by plasma or whole blood, whereas they were not extractable by saline solutions. Extraction tests have also been proposed by Autian.[58] Solvents in this case are pure water and 95% alcohol. With water the test material is autoclaved for an hour and the resulting eluate is examined by both biological and physicochemical methods to detect whether constituents have been extracted. Biological tests that may be used are tissue-culture tests, intradermal, intravenous, and intraperitoneal injections for toxicity, and chick-embryo tests for

Table 4.2–11
BIOCOMPATIBILITY SCREENING RESULTS,[a] RESIN EXPOSED TO PECF, 63 HR, 115°C, 30 psia

Polymer	Tissue culture response	Eluate expressed as MPM[b] equivalent of *n*-hexanol
1. Chopped graphite fiber	+1	n.d.[c]
2. Silastic® (Dow-Corning 372 nonreinforced)	+1	5
372 reinforced fabric	+1	5
3. Polyethylene (U. of Texas)[d] (SG: 0.96)	+1	17
4. Vitreous carbon frit	+1–+2	n.d.
5. Polytetrafluoroethylene (PTFE) bleached fiber	+1–+2	n.d.
6. Fluorinated ethylene propylene (FEP T-160)	+1–+2	n.d.
7. Polyphenylene oxide (Grade 731)	+1–+2	27
8. Polyethylene (SG:0.96)	+2	n.d.
9. Acrylic molding power (V-415)	+2	n.d.
10. Polyphenylene oxide (Grade 534-801)	+2	17
11. Polyethylene (SG: 0.925)	+2	17
12. Fluorinated ethylene propylene (FEP T-100)	+2	23
13. Ionomer (1550)	+2	142
14. Polypropylene (Grade 114, Food Grade)	+2	198
15. Vinylidene fluoride (Grade 200)	+2–+3	3
16. Nylon (Grade 101)	+2–+3	14
17. Ionomer (AD 8043)	+2–+3	30
18. Cellulose propionate	+3	81.7
19. Polystyrene (HH401)(Grade 300)	+3	168
20. Nylon (Grade 38)	+4	12
21. Polyvinylchloride (Grade 5430)	+3–+4	277
22. Polyurethane (Grade 58093)	+4	89
23. Polyurethane (Grade 16139)	+4	328
24. Polyvinylchloride (U. of Texas)[e]	+4	514
25. ABS (Grade X7-1000)	+4	516

[a]Scale
+1 – Some vacuolization and growth inhibition but nominally as control cultures.
+2 – Moderate vacuolization, morphological changes, and growth inhibition.
+3 – Severe growth inhibition and vacuolization.
+4 – Total growth inhibition.
[b]Moles per million.
[c]Not detectable.
[d]University of Texas, Austin, Drug/Plastic Research Laboratory, negative standard.
[e]University of Texas, Austin, Drug/Plastic Research Laboratory, intensely toxic standard.

From Homsy, C. A., *J. Biomed. Mater. Res.*, 4, 341, copyright © 1970, John Wiley & Sons, Inc. With permission.

Table 4.2–12
IN VITRO CELL BIOCOMPATIBILITY STUDIES, PHOTOGRAPHIC DOCUMENTATION

Sterilized prosthetic material	Type of cells used	Presence of cell		Rate of growth		Morphology of cells as compared to controls		Evidence of cytotoxicity	
		24 hr	48 hr	24 hr	48 hr	24 hr	48 hr	24 hr	48 hr
Control cultures grown on glass coverslips	Wish human amnion	Yes	–	5–12[a]	–		–		–
Silicone rubber Not heparinized	Wish human amnion	Yes	–	8–20	–	Normal	–	No	–
Heparinized lot No. 1 (styrene grafting method)	Wish human amnion	Yes	–	6–11	–	Cells flattened	–	No	–
Control cultures grown on glass coverslips	L929	Yes	Yes	3–5[b]	5–6			No	No
Silicon rubber									
Not heparinized	L929	Yes	Yes	3	5	Normal	Normal	No	No
Heparinized lot No. 2 (styrene grafting method)	L929	Yes	Yes	2–3	3	Cells flattened	Cells abnormal	No	?
Polypropylene									
Not heparinized	L929	Yes	Yes	6[b]	7	Normal	Normal	No	No
Heparinized	L929	Yes	Yes	4	5–7	Normal	Normal	No	No
Teflon®									
Not heparinized	L929	?	Yes	2	4	Rounded up	Normal	?	No
Heparinized	L929	?	Yes	2	3	Rounded up	Normal	?	No
Hydrin rubber									
Not heparinized	L929	No	No	Nil	Nil	–	–	–	–
Heparinized	L929	No	No	Nil	Nil	–	–	–	–

[a] 330X magnification, cells per microscope field.
[b] 500X magnification, cells per microscope field.

From Johnsson, R. I. and Hegyeli, A. F., *J. Biomed. Mater. Res.*, 3, 115, copyright © 1969, John Wiley & Sons, Inc. With permission.

Table 4.2–13
SOLID MATERIALS TESTED FOR IRRITANCY

Name	i.m. rabbits,[a] 1-week implantation	Tissue culture[b]
Vinyl tubing (C-1)	0	N
Plastic tubing (Y-12)	0	N
Delrin® (Y-26)	0	N
Vinyl tubing (Y-46)	3	P
Vinyl material (Y-78) containing plasticizer and dibutyltin diisoocytyl-thioglycolate (2.4%)	3	P
Nylon-6,6 (Y-93)	0	N
Polycarbonate (Y-94)	0	N
Gum rubber (Y-104)	0	P
Shur-Weld (Y-115)[c]	0	N
Bonfil® (Y-117)[c]	0	N
Syntex® F (Y-118)[c]	0	N
Zoe B & T (Y-119)[c]	0	P
Tenacin® (Y-121)[c]	0	N
Cranioplastic kit (Y-123)[c]	0	N
Polyethylene tubing (Y-124)	0	N
Oriented polystyrene film (Y-133)	0	N
Intra-Cardiac Path-Teflon® (Y-138)	0	N
Polybutylene film (Y-147)	0	N

[a]Response graded from 0 (nonreactive) to 3 (most reactive).
[b]P – positive (toxic); N – negative (nontoxic).
[c]Solid dental material formed by combining liquid and powder as directed (place in contact with cells 24 hr after mixing).

From Powell, D., Lawrence, W. H., Turner, J., and Autian, J., *J. Biomed. Mater. Res.*, 4, 583, copyright © 1970, John Wiley & Sons, Inc. With permission.

Table 4.2–14
PSEUDOEXTRACELLULAR FLUID (PECF) ($NaHCO_3$, K_2HPO_4, NaCl, KCl)[a]

Ion	Concentration (mEq/liter)	
	Physiological	PECF
NA^+	145	145
K^+	5	5
CL^-	113	118
HCO^-	30	30
$HPO_4^=$	2	2

[a]Reagent grade chemicals.

From Homsy, C. A., *J. Biomed. Mater. Res.*, 4, 341, copyright © 1970, John Wiley & Sons, Inc. With permission.

teratogenic effects. In alcohol extraction the material is extracted in a Soxhlet for 24 hr with 95% alcohol. The alcohol is evaporated and any remaining residue is reconstituted with saline or an oil prior to the above biological tests. Extraction tests are specified in the U.S. Pharmacopeia for biological tests of plastic containers. The extractants specified are sodium chloride solution, a 1 in 20 solution of alcohol in sodium chloride solution, polyethylene glycol 400, and vegetable oil (freshly refined sesame oil or cottonseed oil). The extractants are evaluated biologically by injection into mice or rabbits. Implantations in rabbits are also specified. In both the extractant-toxicity tests and the implantation tests, evaluation is made after 72 hr.

Structural and Physical Factors

While much work has been done relating the effects of nonpolymeric additives on the compatibility of polymers with tissue, essentially nothing has been done on the possible relation of polymer structure to tissue compatibility. Little and Parkhouse[59] tried to relate the degree of crystallinity of a polymer and the size of the filler used to the compatibility. They measured degree of crystallinity and filler particle size by X-ray diffraction using a fiber type camera and copper γ-radiation. The degree of crystallinity was ranked by sharpness of the diffraction pattern, and filler size was estimated to be 0.1 μm or above when diffraction rings had a speckled appearance. Polymers studied included polyethylenes of varying density, polyvinylchloride, polypropylene, polytetrafluoroethylene, nylon, Dacron, polystyrene, hydrocarbon polymers, Mylar®, Delrin®, regenerated cellulose, cellulose acetate, and silicone rubber. The authors concluded that polymers such as high-density polyethylene and polyvinylchloride (undescribed) and those with fillers with particle sizes above 0.1 μm by their measurement would give rise to fibroblast reactions. The other polymers studied gave more diffuse X-ray diffraction patterns and did not excite a fibroblast reaction unless they contained larger particle size crystalline fillers. Their finding that high-density polyethylene is more tissue reactive than medium- or low-density polyethylene is questionable in view of the results of other investigators; their finding that polyvinylchloride is more crystalline than polymers such as low-density polyethylene, nylon, and Teflon is contrary to accepted evidence. It is of course possible that fillers of large particle size might give rise to increased tissue reaction through mechanical effects of surface roughness. It might be noted that the only polymer that is normally filled when used as an implant is silicone rubber in which the filler is a silica with a particle size of about 30 μm. Little and Parkhouse noted that the chemical nature of the filler is also important; iron oxide as a filler in silicone rubber excited a greater reaction than the other fillers used.

There seems to be no question that mechanical trauma may be associated with an implant, and this may lead to apparent changes in the host response. Ocumpaugh and Lee[60] subcutaneously implanted dumbell-shaped bars of etched and unetched Teflon, annealed Teflon, and polyurethane in dogs. The reaction of tissue to all of these samples was very similar. There was, however, more advanced capsular development at sharp angular surfaces, e.g., at the sharp angled end of the dumbbell implant. In addition, in those areas there was more persistent formation of ground substance containing lipoproteins and acid mucopolysaccharides. Such compounds are usually formed in response to trauma. Thickening of capsules and corners of implants has also been noted with metal implants.[61]

Conclusions

From the evidence now at hand it may be concluded that polymers which are reasonably stable in the body may be implanted without inciting more than slight and tolerable tissue reactions. Where greater reactions have been caused, it is usually apparent that they result from lower-molecular-weight, extractable materials. There is general agreement that high-density polyethylene, Teflon, and silicone rubber incite less reaction than other polymers. It is not known if these polymers are inherently less tissue reactive than other polymers or whether they contain less-reactive extractables or less extractable material. Until rigorously purified polymers are evaluated as to their tissue reactivity, this question cannot be answered.

Two polymers apparently free of additives have been found to cause neoplastic reactions; these are cationic polymers containing dimethylamino ethylmethacrylate[42] and vinyl benzyltrimethylammonium chloride,[43] respectively. It is not known if this is characteristic of this type polymer. Positively charged polymers such as protamine and polylysine which are soluble have not been shown to cause such effects.

Blood Compatibility

The requirements of compatibility of a material with blood are the same as for materials in contact with soft or hard tissue – no damage immediate or later to the biological environment and no reciprocal damage to the material. With blood, this requirement is much more stringent because incompatibility may be manifested immediately and drastically. With tissue implants, some degree of incompatibility can be tolerated; inflammatory reactions, unless so great as to constitute toxic reactions resulting in necrosis, will not impair the function of the tissue implant; as time goes on, capsule formation walls off the implant and isolates it from the tissue. For a blood-contacting material, the first evidence of incompatibility may be the initiation of coagulation which may proceed to the point where function of the device is compromised. Blood flow may cease or thrombi may be carried downstream to enter some vital organ.

Induction of coagulation is only one facet of incompatibility. There also must be no disturbance of any enzyme system, alteration of proteins, or disruption or modification of function of the formed blood elements. This latter would include such adverse reactions as lysis of red blood cells, impairment of the phagocytic action of the white blood cells, or initiation of the platelet-release reaction. Because thrombogenesis is the most obvious consequence of incompatibility, it has received major attention. It has long been known that foreign materials induce coagulation. Hewson[62] observed that blood would clot quickly in a bowl even though it had remained fluid for hours in an isolated venous segment. This promotion of coagulation has been found to be a general property of foreign materials. The coagulation process in this case is initiated either by the activation by the foreign surface of Factor XII (Hageman factor) or by disruption of a formed element, e.g., platelet. Blood coagulation is a multistep process involving the sequential activation of a series of clotting factors, culminating in the polymerization of fibrin to the cross-linked polymer which together with entrapped formed elements comprises the clot. Figure 4.2–3[63] depicts the generally accepted coagulation scheme. Because this is an enzyme process, there is amplification at each stage so that activation of only a small amount of the first enzyme, Factor XII, results in production of a large amount of final clot. Note that absence or inactivation of any of the clotting factors will interrupt the clotting sequence. For example, hemophiliacs lack Factor VIII, and Hageman patients lack Factor XII.

Coagulation can also result from the release of tissue thromboplastin into the blood from the site of an injured blood vessel wall or from clotting factors released from injured platelets. These latter mechanisms are the natural body defenses against blood loss after a breach in the vessel wall.

Materials suitable for use in contact with blood are necessary for both long- and short-term applications. The former include catheters which are placed within the bloodstream for monitoring, diagnostic purposes, or for hyperalimentation, as well as more dramatic applications for heart-lung machines, oxygenators, artificial kidneys, and assist hearts. Long-term devices requiring blood-compatible materials are needed for repair or replacement of diseased or damaged arteries. Artificial hearts, of course, must be made from blood-compatible materials.

Blood-incompatible materials have been practical for long-term use in the vascular system – fabrics, woven or knitted, in the form of tubes for arterial replacements. There is some immediate coagulation but the clot is eventually replaced by living tissue on the walls of the tube which serves as a neointima. Formation of a neointima may be compared to the encapsulation of a soft-tissue implant in that it shields the prosthesis from contact with the blood. The source of the cells forming the neointima may arise from the adjacent portions of the natural artery, cells deposited from the bloodstream, or from ingrowth from the outside of the prosthesis through the porous walls. The fiber used in making such prostheses apparently has little effect upon the formation of the neointima; Dacron, Teflon, Orlon, rayon, and other polymers, as well as glass and metal fibers, have supported such tissue growth. In a similar fashion, felts and velours have served as supports for such tissue growth. Because this portion of the review is concerned with research directed toward blood-compatible materials, the use of such fibrous materials is not covered. They have been thoroughly reviewed in Wesolosky's book listed in the Background Bibliography and in journal

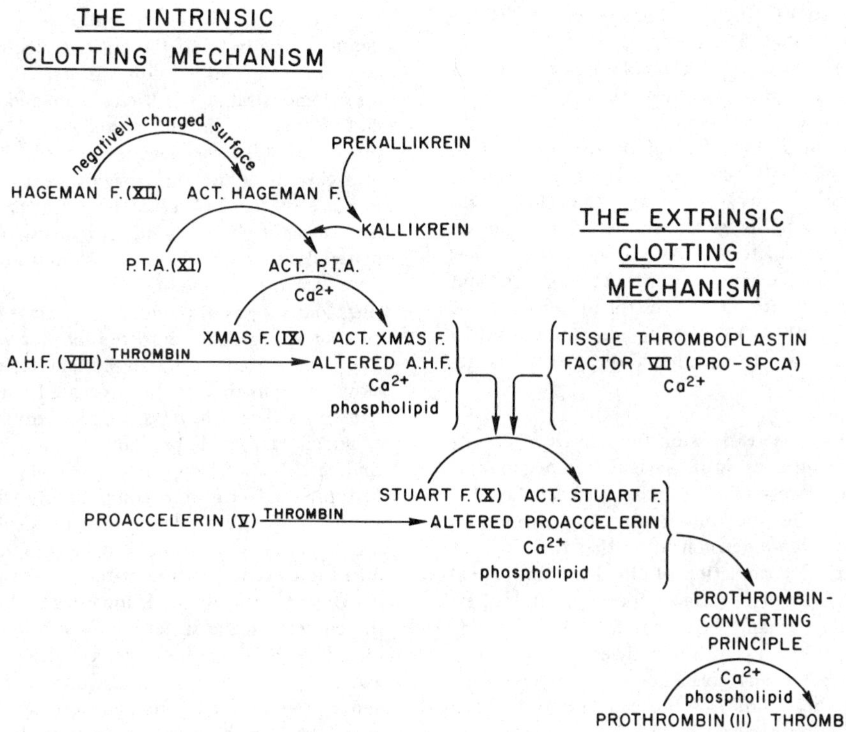

FIGURE 4.2–3. Sequence of steps leading to the coagulation of blood. (From Ratnoff, O. D., in *Cecil-Loeb Textbook of Medicine,* 13th ed., Beeson, P. B. and McDermott, W., Eds., W. B. Saunders, Philadelphia, to be published. With permission.)

articles such as those by Harrison[64-66] in 1958-59 and Shumacker and Muhm[67] in 1969.

Thrombogenic materials have also been used as materials of construction of heart valves with reasonable success. In these applications, perhaps because of the high velocity of blood flow past the devices, polymers and metals that are thrombogenic can be used. However, improved materials are needed because the formation of emboli is not uncommon with present materials even with the use of systemic anticoagulants. In addition, adverse effects on the platelets are a usual consequence of artificial valve implantations.

The study of thrombogenicity of polymers dates from at least 1931 when Lampert[68] found that a coating of paraffin glass led to an increased clotting time compared to that of bare glass. In 1941, Lozner and Taylor[69] determined that the coagulation time of collodium (nitrocellulose) was extended by a factor of about six over that of glass and that paraffin and Lusteroid extended the time by a factor of about five. Attempts were made to relate the thrombogenicity to some characteristic or property or structure of the polymer and reviews were published in 1964[70] and 1965.[71]

Speculation has been made as to the factors controlling the thrombogenicity of foreign surfaces. The following excerpt from a literature survey in 1964[70] summarizes progress up to that point.

> The effect of foreign surfaces on blood coagulation has been studied for many years. The studies, for the main part, have been empirical and have not considered the specific mechanisms involved. Historically, the following factors have been considered as being important in the interaction between blood and a foreign surface: roughness, wettability, and electrical chemical nature of the surface. In this section each of these factors is briefly discussed.
>
> **Surface Roughness**
>
> The surface roughness of a prosthetic material has long been considered as an important cause of thrombosis formation.[72] It was shown that a rough surface destroyed platelets and caused more rapid clot formation.[73] Duc[74] and Frei and Fleisch[75,76] found that surface roughness caused damage to platelets, erythrocytes, and leukocytes. *In vivo* studies[77] in which plastic patches were suspended in the middle of the bloodstream between the right atrium and right ventricle showed that rough surfaces always produced excessive clot formation. Smooth surfaces remained free from clots as long as contact with the thrombus formed along the suture line was avoided.

Thus it would appear that a partial solution to the problem can be found by simply using very smooth surfaces. The problem is not this simple, however. It was found that using highly polished surfaces of glass, methylmethacrylate, polyethylene, and stainless steel caused only a slight prolongation of clotting times.[78] *In vivo* tests of smoothly lined vascular prostheses of polyurethane,[79] nylon, polyethylene, and methylmethacrylate[80] showed that early thromboses occurred with these materials. As stated earlier, it has been shown that a certain amount of toughness and porosity is necessary for the successful performance of vascular fabric grafts where some degree of initial coagulation is necessary to seal the fabric against leakage.

Surface Wettability

Because of early observations on the relative effects of glass and paraffin surfaces on the coagulation of blood, it was felt that the wettability of the surface played a decisive role in the mechanism of clot formation. Lampert[68] formulated a generalization that the effect of a surface in delaying coagulation of blood is proportional to the capacity of that surface for repelling water. Donovan and Zimmermann[81] attributed the long clotting times with polyethylene to its nonwettability. These assumptions are somewhat inaccurate since the intima itself is hydrophilic. Studies of the capillary rise[82] and the contact angle of bubbles with capillary endothelium[83] have indicated that the wettability of the intima is more similar to that of glass than to hydrophobic materials such as paraffin. In addition, the wettability of various materials as indicated by their contact angles with water could not be directly related to their clotting times with blood.[84] In addition, collodion, which is comparatively hydrophilic, is about as effective as paraffin as a coating for glass. The introduction of sulfonic acid groups to the relatively hydrophobic surface of polystyrene (thus making it hydrophilic) has been reported to lengthen clotting time.[85] In this case, however, the electrical properties and chemical nature of the surface are also drastically changed. Finally, *in vivo* tests have indicated that vascular grafts composed of water-repellent materials were occluded by thrombi after their implantation in animals.[86] Thus, it seems evident that the wettability of the surface is not an important factor in clot formation.

Electrical Nature of Surfaces

The nature of the electrical charge on the surface of materials may be an important factor in determining the stability of blood in contact with these materials. There exists in blood a complex mixture of proteins, lipids, and cells, all of which carry a negative electrical charge.[86] These substances will either be attracted or repelled by a surface charge. The electrical phenomena of normal and injured blood vessels have been investigated by Sawyer and Pate[87] who found that the intima has a negative potential of 1 to 5 mv with respect to the adventia. When the blood vessel is sufficiently damaged the charge changes from negative to positive and does not return to negative until healing is advanced. Addition of heparin to the blood increased the magnitude of the negative charge on the intima.[88] Sawyer and Pate[87] found that clots could be produced at the positive electrode by causing a very small current to flow through blood vessels. It was also demonstrated by Richardson and Schwartz[89] that clot formation in an injured vessel could be totally prevented if the vessel were negatively charged by passage of a current of the same magnitude as the injury current. Because of these observed electrical effects and because the vessel itself is conductive, many workers felt that materials used for prostheses should also be conductive. As a test of this hypothesis, Gott et al.[90] inserted steel rings and a series of polycarbonate rings coated with silver powder and graphite into the inferior vena cava of dogs. The rings coated with graphite had superior resistance to clots as compared to the uncoated rings. However, the steel rings caused heavy thrombus formation and were all occluded within 1 hr. This is in accord with earlier findings.[74] Thus, conductivity of the prosthesis is not the determining factor in its compatibility with blood.

The profound effect of the electrical potential of the intimal surface on thrombosis has prompted investigation into the electrical nature of the surfaces of materials used in contact with blood. Hubbard and Lucas[91] estimated the charges on the surfaces of a wide variety of materials, including glass, silica, clays, cellulose, assorted plastics, and tissue, by determining the uneven distribution of Br^- and $Ag(NH_3)_2^+$ ions on their surfaces. The conclusions from the work appear to have been drawn too hastily in that the authors state that any negatively charged surface will initiate the clotting of blood.

The magnitude of the charge on the surface in contact with a liquid (the zeta potential) may be determined by measurement of the streaming potential of the liquid passing over the surface.[92,93] Early zeta-potential measurements of glass showed that it has a relatively high negative potential of around −30 mv. The potential of paraffin was zero.[94] Later measurement of paraffin showed a zeta potential of −17.2 mv in contact with Ringer's solution. Work on glass and silicone-coated glass[95] and on various plastics[93,94] has shown that glass has a somewhat higher potential (−20 to −30 mv) than do most plastics (−15 to −20 mv). When the zeta potentials of glass and plastics were measured in plasma or in various protein fractions they fell almost immediately to zero, indicating that the surfaces were becoming coated with constituents from the blood. Mirkovitch et al.[96,97] were unable to establish a direct relationship between zeta potentials and clotting activities of the materials. From these facts, it is evident that the other factors besides the surface charge influence the clot-forming mechanism of blood. However, it is recognized that the surface charge plays a part in adsorption onto the surface, and this adsorption may well influence the coagulation process.

Chemical Nature of the Surface

The chemical nature of the surface exposed to blood is closely related to the electrical nature because the type of functional groups present will determine the type and magnitude of the charge carried on the surface. Studies of the chemical nature of the intima surface have established the presence of mucopolysaccharides, particularly chondroitin sulfate and heparin sulfate.[89] The anticoagulant activity of heparin is ascribed to the presence of sulfated amino groups. Removal of these groups from the heparin molecule by basic hydrolysis results in loss of its anticoagulant activity.

The importance of the nature of surface groups has been dramatically illustrated by Gott.[100] Plastic rings treated successively with graphite, Zephiran® (a quaternary ammonium compound containing a long alkyl chain), the heparin displayed remarkable antithrombogenic activity both *in vivo* and *in vitro*. Implanted rings with this multiple coating showed no clotting after 14 days; the treated test tubes used in *in vitro* tests did not clot after 10 hours. Implanted plastic valves treated in the same way showed little or no clot formation after 1 year.

Another example of the effect of surface heparinization was presented by Chvapil and Krajieek.[101] Treatment of collagen tubes with a heparin solution increased the clotting time from 1 minute to over 10 minutes. Similar results were obtained by sulfonating the surface of polystyrene.[85]

The effect of surface chemistry has been demonstrated using glass itself.[102] Treatment of glass tubes with chromosulfuric acid for 36 hours almost doubled the cephaline clotting times. The original clotting activity could be restored by treating the acid-washed glass with a sodium hydroxide solution. The above data show that very fruitful results have already been obtained by altering the surface chemistry of plastics, and they indicate further investigation in this area.

The work by Whiffin and Gott[103] with the graphite-benzalkonium chloride-heparin treated polymer surfaces resulted in the first nonthrombogenic or thromboresistant materials.[a] It should be noted, however, that other aspects of blood compatibility of these materials were not known. Because test implant samples had a very small surface-area-to-blood-volume ratio, untoward effects on the formed elements or blood proteins might not be seen because of the natural body defenses against altered blood components. There was and still is uncertainty over the interaction (or noninteraction) of these surfaces with blood, resulting in observed nonthrombogenicity. Because it was shown that the heparin was eluted by the blood,[104] some thought that the nonthrombogenicity was a result of systemic heparinization by the eluted heparin or that the eluted heparin was at least sufficient to heparinize the boundary layer of the blood in contact with the treated surface. The treatment did have the drawback that is was not suitable to polymers such as silicone rubber and Teflon because of adhesion problems. Also, because of the dependence on adhesion of the graphite layer to the base polymer, use with substrates that would be either flexed or stretched was not very practical. Despite these drawbacks, development of the GBH treatment was the first major step toward obtaining nonthrombogenic surfaces.

The first concerted effort toward nonthrombogenic surfaces came with the initiation of the Artificial Heart Program of the National Heart Institute under the leadership of the late Dr. Frank Hastings. It is continuing under successor organizations in the NHLI.

The most widely used in vivo method to determine the thrombogenicity of polymers has been the vena caval Gott ring.[105] The ring used is 9 mm long with an O.D. of 8 mm and an I.D. of 7 mm with streamlined leading and trailing edges. It is implanted in the inferior vena cava and examined for thrombus formation after 2 hr in the acute study and after 2 weeks in the chronic test. This is a very severe test, as all unaltered polymers ordinarily will severely thrombose and be occluded within 2 hr. The venous system was chosen because such rings will remain open in the more swiftly flowing arterial system.

A number of other methods of in vivo, in vitro, and ex vivo testing have been used. In vitro tests are usually some modification of the Lee-White test[106] in which the coagulation time is measured in a simple tube at 37°C. This test grossly measures the integrity of the entire clotting system as shown in Figure 4.2–3 plus any effects of contact of the blood with injured tissue and contact with the needle, syringe, or tubings during withdrawal. With possible thromboresistant materials, it is hoped that activation of Factor XII is the primary variable measured. Among modifications to this test have been the substitution of films for tubes[107,108] and the exclusion of air. The effect of the latter was shown by Spaet and co-workers.[109] In this work, blood was drawn from animals and humans directly into lengths of "plastic" tubing, displacing isotonic saline so that the blood did not come into contact with air. Clotting times of 1½ to 3 hr were obtained in contrast to approximately ½ hr with an air interface.

Leonard and co-workers[110] have used an ex

[a]Materials which do not initiate coagulation or the formation of thrombi have been described as nonthrombogenic, thromboresistant, antithrombogenic, and athrombogenic. Nonthrombogenic and thromboresistant are used in this review.

Table 4.2–15
COMPARISON OF THROMBUS FORMATION BY DIFFERENT MATERIALS

Coating	Substrate	Source	Mean thrombus weight deviation from polystyrene controls, %
TDMAC/heparin	Polystyrene	Battelle	–75.4 ± 8.1
TDMAC/heparin	Polypropylene	Battelle	–60.7 ± 11.6
GBH	Polypropylene	U. of California	–40.7 ± 10.0
Parylene "N"	Polypropylene	Union Carbide	–17.9 ± 16.3
Parylene "C"	Polypropylene	Union Carbide	–12.1 ± 18.2
Collagen	Polypropylene	Cornell	+ 6.3 ± 17.6
Collagen	Polystyrene	Ethicon	–15.1 ± 21.2
Polypropylene	–	Tenite 4230®	+ 7.7 ± 8.3[a]
Polystyrene (control)	–	Lustrex GP	0

[a]Not statistically significant.

From Leonard, E. F., Koffsky, R. M., Carter, J. H., Litwak, R. S., Aledort, L., and Aservatham, J., in *Proceedings 5th Annual Contractors' Conference of the Artificial Kidney Program of the NAIMD,* Krueger, K. K., Ed., National Institute of Arthritis, Metabolism, and Digestive Diseases, National Institutes of Health, Bethesda, Md., 1972.

vivo device for testing the thrombogenicity of different materials. The material in the form of a "captured vortex device" is placed in a carotid-jugular shunt and the thrombus formation measured after a given time. As the name of the device implies, a captured vortex is obtained which is highly thrombogenic. Results are shown in Table 4.2–15.

Approaches to Thromboresistant Polymers

It will be seen that the approaches are essentially empirical and that, while definite progress has been made in developing suitable materials and in measuring their interactions with blood and blood components, the bases for and the details of blood-surface interactions remain obscure.

Gott et al.[111] have classified thromboresistant surfaces into the following three categories:

1. Heparinized surfaces
2. Surfaces with anionic radicals or imposed negative electrical charges
3. Surfaces of relatively inert materials.

This classification is used here although it may be too general in that diverse materials are included under each classification.

A fourth class, "biologic surfaces," has been added to include those materials modified by the addition of a biologic material other than heparin. The third class is somewhat misnamed because, as will be seen, none of the materials studied are completely inert with respect to blood. At the least, adsorption of proteins occurs and formed elements from the blood adhere in some degree to all surfaces. By analogy with polymers in contact with tissues (where even the most compatible materials incited the formation of a filmy capsule), it is quite possible that materials completely inert toward blood will not be found.

Heparinized Materials

As was previously shown, heparinization of materials began with the GBH process which involved the physical bonding of graphite to a polymer surface followed by the physical bonding of a benzylkoniumchloride to the graphite which then served to ionically bond the heparin to the surface. Heparin, the structure of which is shown in Figure 4.2–4, is highly negatively charged because of the sulfate groups. Because of the negative groups it will complex strongly with quaternary salts, and it is this complex formation which binds the heparin to the GBH surfaces. In 1964 the group at Battelle/Columbus modified a number of polymers by covalently attaching quaternary groups to the surfaces.[112-117] These surfaces would then bind heparin in the same manner that a graphite-benzylkoniumchloride surface would be in the Gott process. The advantage was that the method would permit heparinization of polymers that could not be treated by the GBH process. Quaternary groups were added to polymer surfaces by a variety of methods. One of

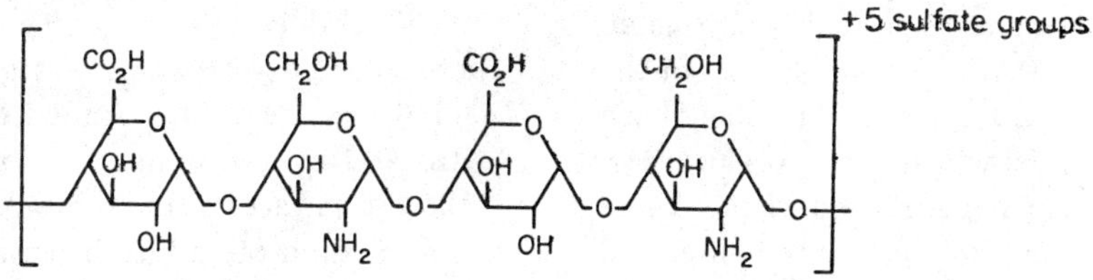

FIGURE 4.2–4. Structure of heparin.

the most general methods was the grafting of styrene to polymer surfaces followed by the chloromethylation of the grafted styrene and conversion to the quaternary salt by treatment with amines such as dimethyl aniline. Another method utilized comonomers such as vinylpyridine or dimethylaminoethyl methacrylate which could later be quaternized and serve as binding sites for heparin. A simpler heparinization procedure was later devised by that group in which the polymer surface was exposed to a solution of a quaternary salt such as tridodecylmethylammonium chloride (TDMAC).[118] The hydrocarbon portion of such a quaternary salt is soluble in polymers, and the salt remains locked in position upon removal of the solvent. The quaternary groups then serve as complexing sites when the quaternized surface is contacted with a heparin solution. The important advantage of this two-step method of heparinization over the chloromethylation of polymers is that it can be used for devices containing more than one material, whereas the treatment conditions for the chloromethylation method had to be tailored to individual polymers. Later, the method was further simplified by the prior formation of an organic soluble complex of heparin with a long-chain tertiary amine[119,120] such as TDMAC or tricaprylmethylammonium chloride (TCMAC).

A solution of this complex usually can be used to treat preformed devices by simple immersion of the device in the solution followed by drying. With materials such as Silastic and vinyl compounds, immersion for a short time is sufficient for treatment. With polymers such as Teflon and high-density polyethylene and polypropylene, treatment at reflex temperatures for perhaps an hour is necessary because of the lower solubility of the solvents and the complex in these polymers. There is initial elution of heparin from surfaces treated by either the one- or two-step process. There is also initial interaction of these surfaces with blood components as shown by hemolysis and platelet disappearance. In vitro measurements

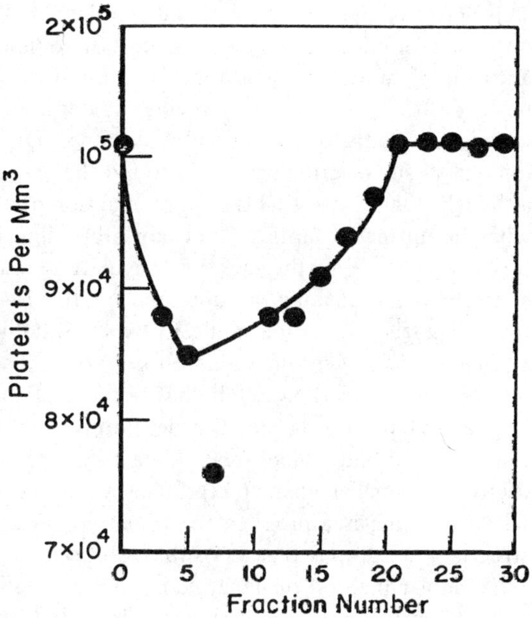

FIGURE 4.2–5. Platelet interaction: heparin-TDMAC complex treated silicone rubber. (Unpublished data of G. A. Grode and J. P. Crowley.)

show these effects to be confined to the first portions of blood contacting the treated surfaces. Figure 4.2–5 shows a typical pattern of surface interaction with platelets. In this experiment whole ACD blood was passed over a column containing about 220 cm^2 of treated Silastic. As can be seen, there is a definite interaction which diappears after contact with about 100 cc of blood.

Clinical results using devices treated with the one-step TDMAC process were summarized at the 1972 meeting of the American Society of Artificial Organs.[120,121] Very promising results were reported in a number of applications. These included the use of treated shunts for by-passing blood during the repair of aortic aneurysms and for carotid artery arteriectomies, the use of treated catheters for monitoring and diagnostic purposes, or for hyperalimentation. Furthermore, dogs were maintained for 48 hr[122] on partial by-pass using a

Lande-Edwards oxygenator that had been treated with the one-step TCMAC heparin process. In the oxygenator work, the treated surface had been reacted with gluteraldehyde in an effort to cross-link and stabilize the heparin, and then treated with a 25% saline solution to remove unreacted complex.

In a different application of this treatment, Dacron fabric grafts were treated with the two-step TDMAC procedure[123] and implanted in both the inferior vena cava and the abdominal aorta. Implantation was in dogs for 2 months. In both locations the grafts remained patent and there were relatively thin neointimal linings. Only minimal tissue reaction was seen around the grafts, although there was slightly more reaction than with the untreated fabrics. Transfabric bleeding, a problem because of the lack of coagulation, was controlled by packing around the graft. Two treated grafts were used clinically when suitable saphenous veins were unavailable for grafting. The results were good. It was felt that because of the lack of initial clotting and the development of a thinner neointima, such treated grafts would be useful for smaller diameter grafts. Initial trans-fabric bleeding is a problem for which composite structures are proposed as a possible solution.

A similar method of heparinizing polyethylene and polypropylene surfaces was described by Eriksson et al.[124,125] In this method the polymer is exposed to a solution of a cationic surfactant such as cetyltrimethylamınonium bromide at a temperature close to the softening temperature of the polymer. The surfactant hydrocarbon chain is adsorbed into the surface zone of the polymer, and, when the temperature is lowered, the penetrating molecules become "frozen" in the plastic surface. Later treatment with heparin results in heparin being complexed to the surface of the polymer. Here also there is some loss of heparin upon contact with blood. To decrease this loss from these surfaces, Lagergren and Eriksson in 1971[126] cross-linked the surface heparin by treating with a 1% gluteraldehyde solution for 10 min at 50 to 60°C. Using dogs, noncross-linked heparinized arterial-venous shunts were found to lose about 75% of the initial amount of heparin in 3 hr. With cross-linked heparinized shunts there was no detectable loss of heparin in 12 hr.

Another procedure for electrostatically bonding heparin to a surface, in this case cellophane for artificial kidney membranes, was developed by Merrill and co-workers in 1966.[127-129] In this procedure ethyleneimine was grafted to cellulose and the grafted polymer used to bind heparin. In vitro and in vivo evaluations of such heparinized cellulose surfaces were promising.

A method of surface heparinization of silicone rubbers was developed at Carnegie-Mellon University by Merker et al.[130,131] In this method gamma-aminopropyltriethoxysilane is reacted with the silica filler of the silicone rubber to provide sites for the electrostatic bonding of heparin.

These methods of heparinizing polymer surfaces depend on the electrostatic bonding of heparin sometimes aided by post cross-linking. This method has the drawback that the heparin is elutable to some degree by blood. Treatment of such heparinized surfaces with aldehydes as done by Lagergren and Eriksson[126] and later by Rea et al.[122] cross-link the heparin and greatly reduce, if not eliminate, the elution of the heparin.

Covalent bonding of heparin has been studied because, presumably, covalently bonded heparin would be much more resistant to removal. Heparin (see Figure 4.2–4) contains hydroxyl, carboxyl, and sulfate-amino groups which presumably could be used for covalent bonding. Problems arise in that heparin is soluble only in water or formamide. Halpern and Shibakawa[132] coupled heparin dissolved in formamide to polyisocyanatostyrene. In vitro thromboresistance was promising. The group at Battelle/Columbus[114,128,129] also studied the covalent bonding of heparin primarily through the use of a cyanuric chloride/heparin adduct. This adduct was prepared in an aqueous environment at 4°C with sodium bicarbonate as the acid acceptor. The adduct was then reacted with amino groups on polymer surfaces as, for example, with silicone rubber which had been treated with gamma-aminopropyltriethoxysilane.

An alternative approach was to react cyanuric chloride (or diacid chlorides or diisocyanates) with the heparin portion of the TDMAC-heparin complex. The organic solubility of the complex permitted reaction of the heparin with the coupling agents. The adducts could then be reacted in a nonaqueous solution with functional groups on polymer surfaces to covalently bond the adduct to the surface. Removal of the TDMAC portion of the adduct by extraction with concentrated saline solution would leave heparin covalently bonded to this polymer surface. Substantial amounts of heparin were covalently bonded to surfaces by

Table 4.2–16
CASTING SYRUP FOR CROSS-LINKED HEPARIN-PVAC FILMS

Material	wt %
Polyvinyl alcohol (Du Pont Elvanol® 73-125G)	6.9
Formaldehyde	6.6
Glutaraldehyde	0.25
Glycerol	4.0
$MgCl_2 \cdot 6H_2O$	5.0
Sodium heparin	0.1

From Merrill, E. W., Salzman, E. W., Wong, P. S. L., and Damus, P., in *Proceedings 5th Annual Contractors' Conference of the Artificial Kidney Program of the NIAMD,* Krueger, K. K., Ed., National Institute of Arthritis, Metabolism, and Digestive Diseases, National Institutes of Health, Bethesda, Md., 1972.

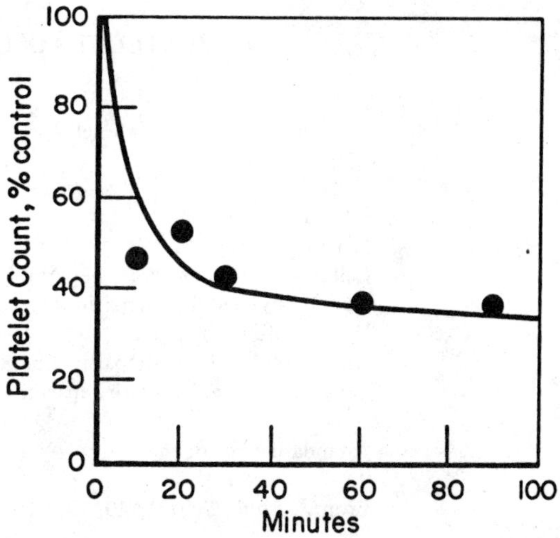

FIGURE 4.2–6. Change in platelet count with time in heparinized platelet-rich plasma rotated in heparinized cellophane tubes. (From Salzman, E. W. et al., *J. Biomed. Mater. Res.*, 3, 69, copyright © 1969, John Wiley & Sons, Inc. With permission.)

both the aqueous and nonaqueous technique, but the surfaces were not as thromboresistant as those prepared by the one-step TDMAC treatment.

Another method of covalently bonding heparin has been studied by Merrill and co-workers.[133-136] In this method heparin is cross-linked with polyvinyl alcohol in solution by reaction with aldehydes. Thromboresistance is good and the heparin is resistant to elution. Films of the hydrated polyvinyl alcohol-heparin gel are weak but they can be reinforced by nonwoven nylon. Reasonably tough materials can be prepared by the following modification of the original casting syrup (Table 4.2–16);[135] this solution is dehydrated and then heated to 80°C for 100 min and rehydrated to give a material which might be useful for tubing and other devices. These cross-linked heparin-polyvinyl alcohol materials do show good thromboresistance. Heparinized surfaces do adsorb proteins and platelets. For example, Table 4.2–17[137] shows the drop in platelet count after exposure of several surfaces to platelet-rich plasma. Note that prior exposure of the surface to albumen reduced the platelet adsorption. From work reported by Salzman et al.,[138] the number of platelets adsorbed on heparinized surfaces reaches a maximum as shown in Figure 4.2–6. It is possible that the use of heparinized devices clinically might be limited by the surface area because of this platelet adsorption.

Polymers containing dissolved heparin have also been used by Hufnagel and co-workers[139,140] and by Salyer et al.[141-143] Hufnagel et al. dissolved heparin in silicone rubber and found in vivo thromboresistance. Salyer et al. added bulk heparin to an epoxy polymer in a quantity up to seven parts per hundred. Thromboresistance was increased and the authors felt that the epoxy and the heparin were chemically bonded because of difficulty of leaching heparin from the polymer.

Surfaces with Anionic Radicals or Imposed Negative Electrical Charges

Through analogy with the blood vessel wall it has been postulated that surfaces of high negativity would be thromboresistant. Because of this, groups have prepared copolymers of varying anionic nature. Byck[144] and a group at the Union Carbide Corporation prepared copolymers of ethylene and acrylic acid and vinyl acetate and their ionomers. The series of materials had no significant thromboresistance and exhibited a fairly uniform behavior toward blood. The only significant exception to this uniformity was the heavy platelet adherence to the dimethylethanolamine salt of the ethylene/acrylic acid copolymer.

A study at Battelle/Columbus[118] was concerned with the effect of surface chemistry on interaction with blood components. The functional groups studied were hydroxyl, sulfate,

Table 4.2–17
PLATELET ADSORPTION BY SURFACES

Material	% change in platelet count after 20 min	SE[a]
Soda glass	12.8	3.2
Siliconized glass	16.5	1.5
Heparin hydrogel-G	29.8	1.1
Cellulose acetate butyrate heparin	31.5	2.4
Aminoethyl cellulose-heparin	37.0	3.3
Hep gel, albumin coated	20.5	1.5
Aminoethyl cellulose-heparin, albumin coated	18.0	3.4
Cellulose acetate butyrate heparin, albumin coated	18.5	3.5

[a]Standard error of the mean.

From *J. Appl. Physiol.*, 29, 723, 1970. With permission.

sulfate ester, sulfhydryl, carboxylic acid, amino, sulfamido, and nitro. These groups were studied as attachments to the surface of cross-linked polystyrene beads. Plasma, platelet-rich plasma platelet-poor plasma, or whole blood was passed over a column containing the chemically modified polystyrene beads at 37°C. Compared to those for the unmodified polystyrene, the following results were found:

1. Hydroxyl – approximately the same as unmodified polystyrene.
2. Sulfate-ester – lower platelet adsorption, elongation of clotting time.
3. Sulfhydryl – very little interaction as measured by platelet adsorption, platelet release reaction or change in clotting time.
4. Aliphatic carboxyl – platelet adsorption with some platelet release reaction; clotting times are increased.
5. Aromatic carboxylic acid – platelet absorption but no apparent platelet release reaction; clotting time is prolonged. (The surface is similar to the aliphatic carboxylic acid but with a lower platelet interaction.)
6. Polyglutamic acid – similar to the aliphatic and aromatic acid.
7. Aromatic amine – marked interaction with the platelets as shown by disappearance of virtually all the platelets during passage of the first 20 fractions; no evidence of platelet release reaction was seen. This reaction was compared with that of an aliphatic amine modified silicone rubber surface. In this case there was also heavy platelet absorption which was accompanied by an increase in acid phosphatase, indicating platelet release reaction.
8. Nitrate – surface adsorbed very few platelets and appeared virtually inert; some lengthening of clotting time.

Of these varied surfaces, those containing sulfhydryl or nitro groups appeared to be essentially innocuous. The most damaging surface is that containing the amine groups. Ziegler and Miller[145] modified the surface of polytetrafluoroethylene (PTFE) by grafting or by chemical treatment. Table 4.2–18 shows that highly sulfonated PTFE had much higher in vivo thrombogenicity. The authors found that the sulfonated material adsorbed significantly more fibrinogen and concluded that the increased nonthrombogenicity was a result of the higher protein adsorption. Salyer[142] prepared copolymers of ethylene/acrylic acid and ethylene/vinyl alcohol/methacrylic acid with varying numbers of acid groups. No significant variation in in vitro clotting times was found between pure polyethylene and polyethylene copolymers containing as high as 55% acidic monomers. Note that, inasmuch as these polymers must have varied greatly in wettability, this factor considered by itself also did not appreciably affect the in vitro clotting behavior.

However, a group of anionic polymers does possess appreciable thromboresistance. The Amicon Corporation[146-148] has been concerned with the development of thromboresistant polyelectrolyte complexes based on polyvinylbenzyltrimethylammonium chloride and polystyrenesul-

Table 4.2–18
RELATIONSHIP BETWEEN THROMBOGENICITY OF SURFACES AND FIBRINOGEN ON SURFACES

Solvent	Fibrinogen on surface, μg/cm² Picked up[a]	Adsorbed[b]	Nonthrombogenic character[c]
Glass	3.14, 2.37	0.119	Very poor, minutes[d]
Polytetrafluoroethylene (PTFE)	0.28	0.156	Fair, days to weeks
Etched PTFE	1.02, 2.58	0.465	Poor to good
Hydroxyl-coated PTFE	2.18	0.471	Poor, days to weeks
Carboxyl-coated PTFE (Van de Fraaff)	2.11, 2.58	0.308, 0.362	Poor
PTFE coated with sulfated hydroxyl	2.16, 2.02	0.460, 0.326	Fair, weeks
Ethylenesulfonic coated PTFE	1.61	0.542	Fair, weeks
Lightly sulfonated PTFE	1.15	0.472	Fair to good, weeks to months
Highly sulfonated PTFE	1.31, 2.89	0.625, 0.647	Very good, months

[a]This figure includes undenatured fibrinogen adsorbed from the underlying solution which contained 0.05 to 0.10 mg fibrinogen per 100 ml.
[b]From solution containing 0.33 mg fibrinogen in 100 ml water.
[c]Cannulae, having the surfaces under test, were implanted in the blood vessels of dogs and the length of time the dogs lived was used as a measure of the thrombogenicity.
[d]Based on in vitro tests.

From Ziegler, T. F. and Miller, M. L., *J. Biomed. Mater. Res.*, 4, 259, copyright © 1970, John Wiley & Sons, Inc. With permission.

fonate. Of the possible ratios of the two ionic polymers making up the complex, the one with 0.5 mEq anionic excess of polystyrenesulfonate showed the best thromboresistance. The neutral complex showed less thromboresistance; the cationic complex containing 1.3 mEq excess and the moderately anionic containing 0.86 mEq excess were still lower in thromboresistance. The polymers are mechanically weak because of the high water content so that methods have been developed to graft them to substrates such as natural or butadiene rubber or to polyurethanes. Both high-energy radiation and ceric-ion-grafting methods have been used. Composites prepared from impregnated fabrics and velours have also been used at higher strength materials. It is noteworthy to observe that not only are the neutral and cationic polymers thrombogenic but more highly anionic polymers are also less thromboresistant than the one containing 0.5 mEq excess of anionic polymer.

A polymer containing 90 parts butylacrylate, 7.5 parts methylmethacrylate, and 2.5 methacrylamide has been prepared by emulsion polymerization, using a sodium alkyl aryl sulfonate as the emulsifying agent.[149] The terpolymer latex was compounded with 37 pph of polyethylmethacrylate latex (prepared as was the terpolymer) and 1.765 pph of formaldehyde. The polyethylmethacrylate serves as a filler and the formaldehyde as a cross-linking agent by reacting with the amide groups in the polymer. The latex particles were negatively charged and the terpolymer was found to have a zeta potential of −66.1 to −68.1 mV in Ringer's solution. The terpolymer was found to be nonthrombogenic by the Gott implant test.

Carboxymethylcellulose has been found[150,151] to have a moderate degree of thromboresistance. In order to obtain a moldable material, the carboxymethylcellulose was acetylated. Careful hydrolysis, removing acetyl groups from the surface, restored the original carboxymethylcellulose. Although the thromboresistance was insufficient to warrant continued work, this is another instance of modifying the surface of a bulk polymer with desirable mechanical properties to improve its thromboresistance.

In addition to the thesis that surfaces with negatively charged groups would possess some degree of thromboresistance, there had been the finding that a negative galvanic charge applied to a steel insert implanted in a canine femoral artery did not lead to clotting. Thus, it was reasonable to investigate the thromboresistance of electrets with a negative charge (an electret is an insulator having persistent internal electrical polarization, often considered to be an electrical analog of a permanent magnet). Negatively charged electrets have been found to have significant thromboresistance.[152,153]

Inert Surfaces

"Inert" is probably a misnomer for this group of materials because no surface has been found completely inert to blood – at the least, proteins from the blood are adsorbed onto the surface as are formed elements such as platelets. The group does differ from the other groups in that it contains neither highly negative groups nor imposed electrical charges. Structures and compositions vary widely, from hydrogels at one extreme to hydrophobic, low-surface-energy materials at the other.

Hydrogels of both hydroxyethylmethacrylate and acrylamide have been found to be thromboresistant. Levowitz and co-workers[154] used them as caval implants in dogs in much the same manner as in the Gott test. In their series, 11 of 15 dogs lived more than 1 week after implantation, with times running up to 30 weeks. Some thrombus was usually found, but the implants, being neither streamlined nor smooth, made the test more severe. There does not seem to be further work on this system although Yohji and Nosé[108] mention Hydron® being only slightly better than Silastic as a prototype thrombogenic material. Polyacrylamide hydrogels[155,156] have shown significant nonthrombogenic properties, but the gels have been too weak for practical use. While reinforcement with fabric was insufficient, methods have been developed to graft a thin layer of the acrylamide hydrogel to polyurethane by the use of ceric-IV ion catalyzed technique. Grafting to the surfaces of hydroxyl-containing polymers was also accomplished.

Hoffman and a group at the University of Washington[157] have been grafting hydrophilic methacrylates to silicone rubber and polyurethane substrates using colbalt-60 radiation to provide sites for the bonding of biological molecules such as proteins, enzymes, and mucopolysaccharides. Albumin, thrombin, plasmin, and heparin have been bonded to the grafted hydrogels and, in some cases, have been shown to retain their biological function. Schmer,[158] also at the University of Washington, in a related study, has bound heparin to Agarose® by various methods – some include an intermediate "arm" between the support and the heparin. Heparin bound via the intermediate arm exhibited significantly increased biological activity, as indicated by increased binding of heparin cofactor. The interaction of these surfaces with blood has not been reported. Peterlin and Ikenberry[159] also took the approach of grafting hydrophilic monomers. They prepared grafts of dimethylaminoethyl methacrylate, 2-sulfoethyl methacrylate, acrylic acid, and 4-vinyl pyridine to polyethylene. It is of interest to note that grafted hydrogels have shown essentially the same protein adsorption as nonhydrogel surfaces.[160]

Fluorosilicone polymers have been studied by Musolf and Metevia.[161] These polymers are high-molecular-weight hydroxyl end-blocked trifluoropropylmethylsiloxane polymers containing small amounts of methylvinylsiloxane. They are cured using 2, 4-dichlorobenzoyl peroxide, as is the usual silicone rubber, polydimethylsiloxane. The fluorine-containing silicone rubbers have significantly improved thromboresistance compared to that of the usual silicone rubber. It may or may not be significant that blood wets the fluorinated polymers more readily than the nonfluorinated rubbers.

Segmented or block polyurethanes have shown considerable promise as nonthrombogenic materials. Boretos and Pierce[162] found a Spandex® type (Lycra® – DuPont) to be nonthrombogenic in an acute caval ring test. This is of interest because it is the only instance uncovered wherein a commercially available material was found to be thromboresistant without modification. Considerable work has been done on the synthesis of new polyurethane materials. A group at Stanford Research Institute[163] has prepared a number of polyether urethane elastomers based on polytetramethylene glycol or polypropylene glycol and methylenebis(4-phenylisocyanate) and a variety of diamines. These polymers range from tough, hard materials to soft rubbers, depending upon the composition. Good resistance has been shown by the Gott test. A similar segmented polyurethane has been prepared by a group at the University of Utah.[164] The polymer used contained polypropylene glycol of molecular weight 1025 and was coupled with ethylenediamine. It too showed promising thromboresistance in the chronic ring test. Another polyurethane that has received considerable study is a block copolymer of a polyurethane and polysiloxane[165–169] that contains a minor portion of polydimethylsiloxane segments and a major portion of polyurethane segments. The polyurethane segments may be either polyethers or polyesters. In one example in the patent the polyurethane segment was prepared from a polypropylene glycol of molecular weight 450 and 2,4-toluenediisocyanate. Approximately 400 parts of the polyurethane were reacted with 45 parts of an acetate end-blocked polydimethylsiloxane. Such polymers have shown good thromboresistance and have been used clinically in intraaortic balloon pumps (IABP). Leinbach et al.[170] reported on the use of such pumps prepared from the block copolymer in 70 cardiac patients. All patients suffering from shock and myocardial infarction (MI) were heparinized and most of them also received low-molecular-weight dextran (Rheomacrodex®, mol wt = 40,000, 10% solution in saline at 10 ml/hr). Postoperative cases were given only dextran because heparinization was not feasible with the condition of thc patients involved. The following is a direct quotation from the findings in this work.

1. To assess the hematologic effects of IABP assistance in man, the changes which were observed in the major hematologic variables during the balloon augmentation of 65 cases were analyzed. In 44 cases, the patients had cardiogenic shock secondary to MI and, in 21 cases, the patients were IABP assisted after open heart surgery and cardiopulmonary bypass. All MI patients were heparinized but the postoperative cases were given only dextran.

2. The results obtained in the IABP-assisted groups of patients were compared, respectively, to those obtained (a) in a group of patients who also had MI which was less severe and did not require circulatory assistance, and (b) in a group of patients who underwent open heart surgery with cardiopulmonary bypass, but were able to regain cardiac function without assistance.

3. The IABP treatment of the above-specified types of patients was found to produce no discernible effect on leukocyte counts or morphology.

4. In IABP-assisted patients, erythrocyte-related hematologic variables may be at somewhat lower levels than corresponding controls, a difference which can be accounted for by (a) the greater volume of blood withdrawn by the frequent sampling of these patients, (b) the greater use of non-cellular plasma expanders, and (c) a possible peripheral sequestering effect of counterpulsation. As yet, there is no definite evidence of balloon-induced hemolysis.

5. In IABP-treated patients, fribrinogen levels were found to increase, in general, from normal to above normal levels which indicates no activation of the clotting mechanism.

6. Platelet levels fall in IABP-treated patients with MI but, as a group, the depression is relatively not larger than that of the control. In postoperative cases, the platelet recovery following the depressive effect of cardiac surgery may be slowed. This effect is clinically manageable and does not negate the clinical use of IABP's without heparin in that critical period.

Further work in animals and in a few selected patients without the use of heparin was done with this intraaortic balloon prepared from the Avcothane-51 polyurethane-siloxane block copolymer by DeLaria and co-workers.[171] It was found that this copolymer could be used safely in calves in the absence of heparin for a period of 8 hr and that the use of low-molecular-weight dextran did not appear to enhance the thromboresistance for this system. At autopsy no thrombi or emboli were found in any of the 12 calves. Occasional small fibrin flecks (less that 2 mm) were seen on several balloons in what appeared to be small surface imperfections. However, some thrombus was observed rountinely on the catheters which were not coated with the elastomer. Three patients whose conditions precluded the use of heparin were pumped; one patient survived and was discharged from the hospital. Pumping times ranged from 24 to 72 hr. Immediately following withdrawal, the balloons were examined and then rapidly fixed in isotonic buffered 1% gluteraldehyde solution. Gross evidence of balloon thrombi was not detected nor were there any ultramicroscopic surface distinctions between the balloons and those balloons previously used in patients with continuous systemic heparinization. Coating the catheter with thromboresistant material appears to be warranted in an attempt to further decrease the likelihood of thromboembolic complications.

Bishop, Porter, and Eisenhut[172] prepared three-segment block copolymers for potential use in cardiovascular devices. Components of these block copolymers were primarily polyvinylcyclohexane and ethylene/propylene copolymer. Materials exhibited only limited thromboresistance. In an effort to improve the thromboresistance, these materials were modified so as to contain hydroxyl, sodium carboxylate, or sodium sulfonate functional groups. None of these modifications showed more than moderate thromboresistance in the in vivo tests at Johns Hopkins.

Blends of polymers and compounded polymers have also been evaluated as to thromboresistance. For example, Serres[173] found coatings of a mixture of graphite, polyurethane, and polyvinylchloride to give thromboresistance to a shunt. The coating solution used on polyvinylchloride tubing consisted of 150 g of polyurethane and 60 g of polyvinylchloride dissolved in 6 l of tetrahydrofuran to which was added 600 ml of colloidal graphite. Tubing to be coated was immersed in this solution and allowed to air dry for 24 hr. The treated tubing was used along with a pulsatile pump lined with Dacron velour for a left heart by-pass procedure. No clots were found after 4-hr use with dogs, and the system was used successfully to resect thoracic aneurysms in four patients. It will be remembered that graphite coatings had been found earlier to effect some improvement in thromboresistance. Monsanto[141-143] has found that the addition of polyoxypropylene/polyoxyethyleneglycol to epoxy resins improved the thromboresistance so that the material was thrombus free in the chronic Gott implant test. On the other hand, Atsumi et al.[49] purified natural rubber and cured it by irradiation or with peroxides. This purified rubber was found to have much greater in vitro thromboresistance than the rubber cured by the usual means. This finding is analogous to that found with "pure" polymers in tissue compatibility.

Taylor and co-workers[174] prepared a series of polyester-polyurethanes containing increasing amounts of acetylene black as fillers. The sample containing about 17% acetylene black was found to have the best thromboresistance as shown by arterial grafts in dogs. The potency rate was 81% over a 21-day implantation period. The authors concluded that a successful arterial graft should have a resistivity approximately that of the natural artery or less and in addition should have a negative zeta potential. Samples with a lower acetylene black content had negative zeta potentials but their resistivities were higher, whereas samples with higher acetylene black contents had lower resistivities and positive zeta potentials.

Biologic Surfaces

Effort has been made on the modification of the surfaces through the use of biological materials. One of the more interesting of these approaches is that of Kusserow and co-workers[175] who added urokinase to a heparinized surface. This was done on the premise that even if thrombi were initiated on the heparinized surface, the urokinase would activate plasminogen which would then act to lyse the thrombus formed. Heparin and urokinase were physically adsorbed to graphite or graphite-benzylkoniumchloride surfaces. In vitro experiments showed that the urokinase adsorbed on a graphite surface, preserving its plasminogen-activating capacity, resulted in the lysis of clots. There is no doubt that this research direction will be continued with the investigation of chemical bonding of the enzymes to the surface for greater permanence. Nosé and a group at the Cleveland Clinic[176,177] have been working with "biolized" materials for cardiovascular prostheses. In general, these materials are prepared from natural rubber to which has been added albumin, but for some, gelatin and heparin were added. Irrespective of the addition, the modified rubber was treated with gluteraldehyde or formaldehyde. The genesis of this treatment was (1) the protein adsorption on polymers exposed to blood and (2) the observation that the tendency toward clot formation on a cardiovascular prosthesis diminished with successive implantations. The sample of rubber containing 14% dog albumin treated with 4% formaldehyde for 20 hr showed superior thromboresistance to silicone rubber and Hydron. A sample containing both gelatin and heparin cross-linked with gluteraldehyde showed even greater thromboresistance.

Several groups[178,179] have pursued somewhat the same direction by coating surfaces with albumin with resultant increase in thromboresistance and decrease in the adsorption of platelets.

Factors Affecting Thrombogenicity of Polymers

This discussion so far has been limited to the thrombogenic aspects of polymers. It has been seen that a number of different polymers such as various types of polyurethanes, neutral and anionic hydrogels, fluorinated silicone polymers, as well as polymers having heparinized surfaces possess significant degrees of thromboresistance as compared with the usual unmodified polymer or glass. It is remarkable that no common structural feature appears to be related to thromboresistance. All, with the possible exception of the neutral hydrogels, have a negative zeta potential in respect to saline, but this is a feature of many other materials which are highly thrombogenic. The surface energies or wettabilities of these polymers range from very low to very high again as do thrombogenic materials. Bulk electrical conductivities are low except for the carbon-filled polyurethane, but low conductivity is a feature of polymers in general.[a] The polymers all adsorb proteins from the blood, including the thrombogenic polymers. It is possible that thromboresistant polymers show an adsorption pattern toward blood such that the resultant protein surface is in itself nonthrombogenic. This might be analogous to the capsule formed about a tissue implant which isolates it from the biological environment. Some evidence supports this in that GBH coated caval rings are known to lose heparin and remain open and thrombus free after the heparin content has decreased. Gott et al.[180] implanted a ring that had been graphite coated and treated with benzylkoniumchloride into a dog which was systemically heparinized. The ring remained open because of the systemic heparinization and was later implanted into another dog which was not heparinized. That the ring in the second dog remained clot free was attributed to adsorption of heparin from the bloodstream of the first dog into the ring, thus making it nonthrombogenic. An alternative explanation is that the ring had become passivated or nonthrombogenic through adsorption of proteins. This possibility could be tested by implanting normally thrombogenic materials into well-heparinized animals and, after a period of time, eliminating the anticoagulant and determining whether the implants remained thrombus free. This does not occur in artificial heart valves; they are known to produce emboli despite anticoagulation of the patient. Here, continual abrasion of portions of the valve may prevent build-up of a passivated surface.

In 1967[117] the Battelle group showed that their heparinized surfaces did not activate Factor XII. It was later found by Merrill et al.[137] that the heparin-hydrogel surfaces similarly did not activate Factor XII. It is not known whether other surfaces which have shown appreciable thromboresistance similarly fail to activate Factor XII. The nonactivation of Hageman factor by the heparinized surfaces is probably not a result of the negative charges. Nossel and co-workers[181] found that positively charged molecules such as protamine and polylysine inhibited Hageman factor activation either in solutions or when they were adsorbed on surfaces, whereas heparin in solution did not inhibit Factor XII activation. It was also found[117] that a polymer surface containing quaternary ammonium groups did not activate Factor XII. This cationic surface, however, was thrombogenic, probably as a result of the interaction between the positively charged surface and the formed elements. For example, there was pronounced lysis of red cells. Disruption of the formed elements could release clotting factors that would initiate coagulation even though the Hageman factor had not been activated.

Coagulation can also be initiated via the extrinsic system through initiation of the platelet-release action. The role of platelets in blood-surface interactions has been extensively reviewed by Salzman.[182] The adsorption or deposition of platelets on the foreign surface is preceded by the formation of a protein layer.[183,184] Lyman et al.[185] have found a dependence of platelet adhesion on the critical surface tension of hydrophobic polymers. The number of platelets adsorbed increased as the critical surface tension increased. Friedman et al.[186] measured the platelet adhesion to a series of polymers including hydrophobic and hydrophilic with critical surface tensions ranging from 10 dyn/cm for polymethacrylic ester of perfluorinated octanol to 47 dyn/cm for a polyamide. No influence of the critical surface tension on the number of platelets adsorbed was found; rather, the number of platelets that adhered to the surface depended primarily on the blood flow rate and the time of exposure. The apparent discrepancy between these two studies is a result of the short time (1 min) used in the Lyman measurements as compared with the longer times (up to 60 min) in the Friedman study. Maximum platelet deposition occurred within 3 to 10 min in the Friedman study. It may be concluded that while the surface energy of the substrate may have an influence on the initial rate of deposition, the final number of platelets deposited on a surface is independent of the critical surface tension. This independence is probably a result of the intervening protein layer deposited on the surface prior to platelet deposition. Weiss and Blemenson[187] found that cell adhesion in serum-free media was greatest on high-energy surfaces and least on low-energy surfaces. However, these differences essentially

[a]Actually, any polymer in contact with blood will have an appreciable surface conductivity because of the protein layer deposited from the ionic plasma.

Table 4.2–19
CHANGE IN CRITICAL SURFACE TENSION DURING IMPLANTATION

Material	Critical surface tension, dyn/cm Initial	After implantation
Epoxy-pluronic	45	26
Acrylic terpolymer	30–40	25
Graphite	50	33
Albuminized surface	30	33
Polyurethane electret	48	31
Carboxycellulose	43	42
TDMAC-heparin	40	39

Adapted from Gott and Baier.[189]

disappeared when serum was added to the growth medium. The fate of the platelets adsorbed on a surface is important because if the platelets undergo the platelet-release action, clotting factors will be available to initiate coagulation. It was found by Merrill et al.[137] that platelets adsorbed in vivo on a heparinpolyvinyl alcohol hydrogel largely disappeared in 24 hr, leaving a thin layer of adsorbed material (perhaps protein), and then remained free of cells for prolonged periods. Platelets adsorbed to surfaces heparinized by the chloromethylation technique were studied for evidence of the platelet-release reaction, but no evidence of release reaction was found.[117]

Baier and colleagues[188–190] have studied the surfaces of a number of materials both before and after implantation. The surfaces were characterized by contact-angle measurements to determine critical surface tension, by multiple attenuated internal reflectance infrared to determine surface composition and presence of protein layer, and by the "Teflon" test to determine the presence of desorbable surface active agents. Implantation was for 2 hr. Table 4.2–19 shows the initial and after-implantation values of the critical surface tension for several materials.[189]

All surfaces were found to adsorb proteins from the blood during implantation, and it is seen that the adsorption decreased the differences in the critical surface tensions of the materials used. It was also found that the epoxy-pluronic, acrylic terpolymer, TDMAC-heparin, and albuminated materials had desorbable surface active materials. Baier et al.[190] draw the conclusion "that three different mechanisms could tentatively be correlated with thromboresistance of various materials: (1) intrinsically low-energy surfaces apparently resisted thrombus formation because they adsorbed a passivating layer of compatible protein, (2) intrinsically high-energy surfaces also rapidly adsorbed thin protein films from blood that converted their surfaces to a thromboresistant low-energy character, (3) materials with desorbable surface-active agents (surfactants, heparin, albumin, and the like) slowly and continuously leached them into adjacent liquid phases and thus shed thrombus as it developed. In all three instances, a proteinaceous layer deposited on implanted materials rapidly and spontaneously, and it was this layer that determined subsequent substrate reactivity to blood. Thus, ultimate thromboresistance or thrombogenicity of candidate biomaterials is intimately related to the nature of the 'conditioning' layer of protein initially adsorbed."

Nyilas[166] considers that the heat adsorption of proteins is of prime importance because the heat liberated during adsorption can denature Factor XII and, thus, initiate the coagulation sequence. Accordingly, he concludes that surfaces containing chemical groups that can be hydrogen-bond donors, acceptors, or both are the most detrimental; they are followed by chemical groups containing highly polarized covalent bonds that can induce dipoles in the adsorbing molecules. Extending this reasoning, he considers the thromboresistance of the polyurethane-polysiloxane block copolymer to result from:

1. Reduction of potential hydrogen bonding sites either by blocking the urethane groups or by rendering them inaccessible by steric hindrance that results from the copolymerization.

2. Combination of two different types of polymers introducing into the surface a variety of different groups that are only capable of van der Waals' bonding (as a result of the different nature of these groups, it can be postulated that the field of van der Waals' forces in the near vicinity of the surface will be nonuniform, i.e., it will vary locally from one type of surface group to another).

While supportive experimental evidence is not available, the thromboresistance of the highly polar, heparinized surfaces is in direct opposition to this theory.

Surface Reactions Other than Thrombogenesis

Reactions other than thrombogenesis of blood to foreign surfaces are important. Serious deleterious effects on the blood have been noticed during the use of assist hearts or total replacement hearts. Interpretation of the data is difficult because the devices are usually made from more than one material and inclusion of a pump may give rise to mechanical trauma to the blood.

For example, the group at the University of Utah[191] has found that when total artificial hearts are implanted in calves, the limitations to long-term implantation have been the thromboembolic complications arising presumably from the artificial surfaces and perhaps from the abnormal flow patterns inherent in the devices. The thrombotic tendencies associated with these prostheses have been considered as causing the associated disseminated intravascular coagulation but evidence for this is incomplete. On the other hand, LaFarge and co-workers[192] have been able to implant assist pumps for periods of observation as long as 203 days without detecting blood or organ abnormalities. The materials of construction of the prostheses were similar, and observed differences in results may arise from the fact that in the Utah study the natural heart was removed so that the total blood flow resulted from the implanted artificial heart. In the second study, pumps accepted blood from the normal ventricle. Therefore, the hemodynamics were different and it is conceivable that there is some enzyme or other biochemical function of the intact heart for which compensation must be made if the heart is removed.

Kusserow et al.[193,194] have implanted aortic grafts above the renals so that effects are limited to one material with no pumping. The abdominal aorta is occluded so that the blood flow goes through the kidneys, which act as filters and indicators of any emboli produced by the interaction of blood with the surface. Aortic ring implants of various fluorosilicone and dimethyl silicone polymers, epoxy and modified epoxy rings, and elastomeric materials such as saturated hydrocarbons and carboxylated and sulfochlorinated block polymers were implanted. The results indicated that all rings gave evidence either of some degree of surface thrombosis or of renal embolization with infarction. The saturated-hydrocarbon-block copolymer rings appeared to give the more promising overall results of the surfaces used. There has also been evidence that the phagocytic activity of the leukocytes is impaired by contact with the artificial surfaces.

The effects of surfaces on plasma proteins and plasma enzymes have been measured in vitro by a group at Cordis Laboratories.[195-197] In the test, 35 to 40 ml of plasma was allowed to remain in contact with 8 ml of finely divided polymer (minus 20 plus 40 mesh) for 24 and 48 hr at 37°C. Changes in plasma proteins and plasma enzymes were determined. Results are as follows:

Silicone rubber was found to have insignificant effects on both the enzymes and the plasma proteins. Silicone rubber heparinized by an aminoorganosilane coupling agent caused no significant changes in the concentrations of the 17 proteins or in the enzyme activities.

Segmented polyurethane (Lycra) did not significantly alter any of the plasma proteins; there was a possible decrease in isocitric and lactic dehydrogenase activity and some slight prolongation of the clotting time of the exposed plasma. With Lycra, heparinized by the two-step TDMAC process, the concentrations of the alpha and beta lipoproteins dropped significantly.

Silicone rubber heparinized by the two-step method – As with the treated segmented polyurethane the beta and alpha-1 lipoproteins strikingly decreased, often becoming undetectable after 48 hr. Enzyme assays revealed little change except for a slight possible decrease in cholinesterase activity. The plasma samples became turbid and this turbidity was cholesterol from denatured alpha and beta lipoproteins. It was found that these effects were caused by excess TDMAC on the silicone polymer surface.

Silicone rubber heparinized by the one-step TDMAC process did not produce any of the adverse effects noted previously in an unpublished study by Cordis Laboratories of the interaction between the one-step process heparinized polymers and plasma.[198] No striking changes were noted in the levels of the 21 plasma proteins quantitated by immunodiffusion, nor were significant differences seen in the levels of the enzymes assayed. The plasma cholesterol concentrations were essentially unchanged. That the clotting times were indefinitely prolonged suggests that appreciable heparin leached.

Silicone rubber covalently heparinized gave results essentially similar to those for the one-step heparinization.

Segmented polyurethane elastomer and fluorinated silicone rubber elastomer and saturated hydrocarbon block copolymers showed no marked differences in any of the plasma enzyme activities or in the plasma protein concentrations.

Ioplex® 101 (0.5 mEq anionic excess charge, 58% water – Amicon Corporation) displayed a trend toward lowered activity of acid and alkaline phosphatases. Of the proteins, fibrinogen showed a consistent reduction in level.

Epoxy polymer with 5% pluronic F-68 (Monsanto Research Corporation) maintained normal enzyme activity and plasma protein concentrations.

PPA-sialic acid – cholesterol coated, acrylamide hydrogel, ethylene vinyl acetate, polyester polyurethane created no striking changes in either the proteins or the enzymes.

Teflon showed increases in the activities of both the lactic and isocitric dehydrogenases.

Researchers at the University of North Carolina[199] also studied the interaction of surfaces with plasma proteins. In their work, the plasma was exposed to glass, silicone-coated glass, Teflon, polyethylene, Cuprothane®, and a methacrylate hydrogel. The proteins adsorbing to the surfaces were recovered by elution with Triton® and examined by disc electrophoresis. The 9 to 12 proteins recovered from each of the test surfaces included albumen, fibrinogen, several alpha and beta globulins, and gamma globulin. There appeared to be no preferential adsorption.

Effects on Formed Elements

Effects on formed elements are of obvious importance, and the interaction of platelets with surfaces has been discussed previously. Hemolysis is a frequent consequence of the use of heart-lung machines, assist hearts, and heart valves. The relative contributions of the pump or valve trauma and the effects of the surface are not known. Dyck and Winters[21] measured the hemolytic effects of commercial plastics. They found the depression in surface tension of physiological saline solution exposed to the plastic surfaces correlated with the hemolytic behavior of the plastics in contact with blood. Table 4.2–20 shows the degree of hemolysis caused by the polymers in contact with blood for 48 hr compared with the surface tension of physiologic saline after a corresponding exposure to the polymers. It is seen that polymers with low amounts of surface-active materials extractable by the saline (such as Silastic, polyfluorocarbons, polypropylene, and others) are relatively nonhemolytic. The authors propose that such an extractability test be used to predict the hemolytic effect of polymers.

Table 4.2–20
EFFECT OF SURFACE TENSION DECREASE ON EXTRACTABLES

Material	Decrease in surface tension, dyn/cm²	Hemolysis in 24 hr, mg/ml
Silastic® 372	3.4	0
Estane® 58091	35.3	55
Estane 58092	0.3	3
PTFE	–5.5	0
Tyril®	32.3	59
Lexan®	–5.4	2
Zytel® 61	12.0	0
Zytel 2345	–0.4	0
Epalene E-12	22.2	12
Moplen®	–0.8	0
Blank (glass)	–	0

Adapted from Dyck, M. F. and Winters, P. R., *J. Biomed. Mater. Res.*, 5, 207, copyright © 1971, John Wiley & Sons, Inc. With permission.

Summary of Blood-polymer Interactions

Table 4.2–21 shows the present state of the knowledge of blood-polymer interactions. Despite the incompleteness of this knowledge at the time of this compilation (mid-1972), materials are available with sufficient thromboresistance to be used for days or weeks. Their long-term thromboresistance has not yet been determined with certainty. From in vitro work it appears that there are polymer surfaces compatible with the blood proteins and the blood enzyme systems. Less information is available on the interactions between polymer surfaces and blood-formed elements. Interactions are shown by the adsorption of formed elements to the surfaces, but the extent of the damage, if any, to the formed element is not yet clear. The effects so far seen are obscured by the possibility of hydrodynamic

Table 4.2–21
POLYMER/SURFACE INTERACTION WITH BLOOD COMPONENTS

Protein adsorption –	General, regardless of surface energy or water content of the surface. Differences in adsorption patterns not known but probable. Denaturation of proteins apparently uncommon.
Platelet adsorption –	General, greater on cationic surfaces.
Platelet-release reaction –	Known to occur on some cationic surfaces.
Hemolysis –	Definite in some cases where extractable surfactants are present. Low with pure neutral or annionic polymers.
Hageman factor activation –	Not activated by heparinized or quaternized surfaces. Effect of other polymers not determined.
Other clotting factors –	Slightly or not affected.
Phagocytic activity of leukocytes –	Some evidence of decreased activity.

damage to the formed elements resulting from pumping or from implanted valves. It appears certain that the surface-formed element interaction is mediated by proteins adsorbed on the polymer surface from the blood, but there is essentially no knowledge as to the structure of the deposited protein film or its composition. Until more fundamental information is available regarding the effects of this protein layer on formed elements, the development of truly blood-compatible polymers will continue on an empirical basis.

EFFECT OF THE BIOLOGIC ENVIRONMENT UPON POLYMERS

A 1964 literature survey for information regarding the change in plastics properties caused by implantation indicated that comparatively little work has been done.[200] Unfortunately, this situation still prevails although certain polymers such as silicone rubbers, polyethylene, Dacron, Teflon, and polymethylmethacrylate have been implanted for many years without evidence of gross failure.

Harrison[64] studied fabric and grafts of nylon, Dacron, Orlon, and Teflon in dogs to replace portions of the ascending thoracic aorta. Table 4.2–22 shows the loss of strength over a period encompassing 2 to 3 years after implanting. Nylon was seen to have lost about 80% of its tensile strength in 3 years, while Dacron lost about 11% in about 2 years. Presumably the nylon used was nylon-66, although it was not specified. Harrison and Adler[201] continued this study with essentially the same results. Further work[202,203] has been reported on the change in properties of implanted nylon; these results are shown in Tables 4.2–23 and 4.2–24.

Table 4.2–22
CHANGES IN TENSILE STRENGTH OF PLASTICS AFTER IMPLANTATION

Material	Days implanted	Loss of tensile strength, %
Nylon	1,073	80.7
Dacron®	780	11.4
Orlon®	735	23.8
	670	1.0
Teflon®	677	5.3
	675	7.0

Modified from Harrison, J. H., *Am. J. Surg.*, 95, 3, 1958.

In contrast to these changes found with nylon, a group headed by Hardy[44] studied the change in properties of acrylate-amide films implanted subcutaneously in dogs for up to 4 months and found essentially no change in the stress-strain curves.

Mirkovitch et al.[204] implanted a polyester-polyurethane intramuscularly in dogs in the form of strips 9 cm by 0.2 mm. Table 4.2–25 shows the extreme loss in tensile strength of this material in the first 8 months. Within 6 months the materials had disintegrated to the point where tensile strength could not be measured. On the other

Table 4.2–23
COMPARISON OF INITIAL AND FINAL STRENGTHS OF NYLON NO. 5 GAGE SUTURES REMOVED FROM HUMAN TISSUE

Case	Initial strength, lb	Straight pull breaking load number of tests	Final strength, lb	Percentage loss	Time buried, years
1	6.5 (est.)	2	4.0	38	11
2	6.5 (est.)	4	5.2	21	11
3	6.5 (est.)	4	5.9	9	6
4	6.9	9	6.9	23	4
5	6.6	9	6.4	3	2½

From Maloney, G. E., *Br. J. Surg.*, 48, 528, 1960. With permission.

Table 4.2–24
COMPARISON IN INITIAL AND FINAL STRENGTHS OF NYLON NO. 7 GAGE SUTURES REMOVED FROM HUMAN TISSUES

Case	Initial strength, lb	Straight pull breaking load number of tests	Final strength, lb	Percentage loss	Time buried, years
1	11.5 (est.)	12	8.1	30	11
2	11.5 (est.)	14	8.0	30	11
3	11.4 (est.)	5	9.8	14	10
4	11.3	4	10.2	10	4
5	11.4 (est.)	7	9.0	21	3
6	11.5	6	9.6	16	2½
7	11.8	11	10.5	11	2½
8	11.5	4	12.2	–	2½
9	12.5	8	10.2	18	1

From Maloney, G. E., *Br. J. Surg.*, 48, 528, 1960. With permission.

hand, a polyether-polyurethane[162] showed no significant degradation of tensile properties in 28 days in blood, although all samples were uniformly altered in color. From private communications from other workers in the field, it seems that the polyether-based polyurethanes are much less susceptible to degradation by the body than are those polyurethanes based on polyesters, although no long-term studies seem to have been done.

Leininger et al.[200] studied the changes in properties of several plastics in the form of intramuscular films in dogs for periods up to 17 months. Again, nylon was found to lose an appreciable amount of strength, whereas Mylar and Silastic were essentially unchanged. Teflon and polyethylene both showed changes.

Shumacker and Muhn[67] extensively surveyed arterial suture techniques and grafts, past, present, and future. They too reached the conclusion that nylon had been demonstrated to lose a substantial portion of its original strength relatively soon after implantation and that such materials as Dacron and Teflon retained tensile strength quite well. They also concluded that prosthetic grafts of Teflon and especially of Dacron are proving very satisfactory, from the standpoint of both implantation and long and continued function. They do question, however, whether the ideal prosthetic graft has yet been devised, and note that a question remains as to the suitability over periods of time measured in decades. Since this question is not answered, they believe that satisfactory alternative procedures utilizing autogenous tissue are often applicable.

Polyethylene has been followed for as long as 21 years as human implants, e.g., ear, nose, and

Table 4.2–25
CHANGE IN PROPERTIES OF POLYURETHANE

Time, months	Tensile strength, psi
0	8,150
8	1,846
16	Disintegrated

From Mirkovitch, V., Akutsu, T., and Kolff, W. J., *Trans. Am. Soc. Artif. Intern. Organs,* 8, 79, 1962. With permission.

Table 4.2–26
BREAKDOWN OF LABELED POLYMERS

Polymer	Weeks for C^{14} to appear in urine
Polystyrene ($-C_6H_5CH-C^{14}H_2-$)	21
Polyethylene ($-CH_2-C^{14}H_2-$)	26
Polymethylmethyacrylate ($-CH_2C(CH_3)COOC^{14}H_3-$)	54

From Oppenheimer, B. S., Oppenheimer, E. T., Danishefsky, I., Stout, A. P., and Eirich, F., *Cancer Res.,* 15, 333, 1955. With permission.

skull.[205] Although no actual measurements of mechanical properties were made, the authors concluded that " time has shown polyethylene to be extremely stable in the living human host." They also found the foreign body-host reaction to be minimal.

Although it appears that several polymers are not degraded appreciably in the body it is equally clear that the natural body defenses are effective against a number of polymers. Oppenheimer et al.[206] implanted radioactively tagged polystyrene, polyethylene, and polymethylmethacrylate in rats and measured the time for radioactivity to appear in the urine. Table 4.2–26 shows the results. Even a presumably inert polymer such as polyethylene breaks down in the body. Presumably this is by enzymatic oxidative chain scission. Some years ago, Epstein[207] of Battlelle was able to produce changes in polyethylene films in a matter of weeks by exposing them to bacterial cultures, producing relatively large amounts of oxidative enzymes. The changes (stress-strain curve) resembled the changes found during implantation of polyethylene films in dogs. It appears quite possible that naturally occurring enzymes such as amidase and esterase have the ability to break down synthetic polymers in the body. Braley[34] has speculated that the good stability of the polydimethylsiloxanes in the body is because of the O-Si-O bonds which are not found in nature and perhaps not susceptible to enzymatic breakdown.

There are, however, applications for polymers within the body in which controlled degradation is desired. One of these is as surgical adhesives or as hemostatic agents in which it is desired that the material disappear over a period of time sufficient for knitting of the tissues to be reestablished. At that point, the presence of the foreign material is of no further purpose. There has been considerable work on two systems: (1) the cyanoacrylates and (2) a gelatin-resorcinol-formaldehyde combination. Both of these materials are biodegradable. They are not treated in this review because of the appearance of comprehensive reviews on the subjects.[208,209]

Recently a novel application of the gelatin-resorcinol-formaldehyde (GRF) has been proposed by Grode and co-workers.[210] They employed the GRF tissue adhesive to block the fallopian tube of rabbits, and it was shown that the blockage can be easily and reproducibly assured. Preliminary evidence exists that the blockage may continue concurrent with the ingrowth of tissue, causing permanent tubal occlusion. Research on this method is continuing because of its promise as an easy, safe, and reliable method of outpatient female sterilization.

Degradable polymers are desired as sutures so that later removal is not necessary. Naturally occurring materials, e.g., catgut, have been used for this purpose for many years, but they have the disadvantage of nonuniformity because they are natural materials and their stability is controlled by the degree of tanning of the suture. More recently considerable effort has been devoted to polymers of lactic and glycolic (hydroxyacetic) acid. Some of the first papers in this field were by Dardik et al.[211] and Postlethwait[46] in 1970 who studied polyglycolic acid as a suture material. The sutures were used for femoral and carotid artery anastomoses. There were no disruptions, leaks, or false aneurysms in either group. Glycolic acid could be identified microscopically in the arterial

Table 4.2–27
IN VITRO LIPID ABSORPTION

Material	Weight gain, %	Volume increase, %
Silicone control	0.70	0.46
Fluorosilicone	0.64	0.17
Nitrile	2.92	4.72
Viton®	0.99	1.19
Urethane	4.51	6.56
Chlorobutyl	2.63	3.95
Ethylene-propylene rubber	6.50	10.13
Neoprene	6.10	11.75
Butyl	2.93	12.17

From Carmen, R. and Kahn, P., *J. Biomed. Mater. Res.*, 2, 457, copyright © 1968, John Wiley & Sons, Inc. With permission.

wall tissue up to 40 days, and fragments could be identified up to 90 days after surgery. In specimens studied 325 to 460 days after surgical operation no suture material could be identified, but silk-suture lines were marked by varying degrees of inflammatory responses. The PGA anastomoses evoked a much milder early inflammatory response which subsided, and in some specimens it was virtually impossible to identify a PGA anastomotic site either grossly or microscopically.

Polylactic acid polymers have been reviewed by Kulkarni.[212] This material has been proposed for use not only as a suture, but also as a form of rod and film. The rods, for example, might be used as reinforcement in orthopedic applications wherein the polymer degrades and disappears as it is replaced by bone.

Obviously, any degradable polymer used in the body must degrade into products that are either nontoxic or whose toxicity is so low that no harm will result. Degradation products from both polylactic acid and polyglycolic acid are normal constituents of the metabolic processes in the body and, as such, they are not toxic.

In addition to degradation of the polymer implant by the biologic environment, changes in properties may occur through absorption of components from the tissues or blood. For example, many artificial heart valves are of a check-valve design containing a poppet ball of silicone rubber. In some cases sufficient lipids may be absorbed to lead to failures of the balls. Carmen and Kahn[213] studied the lipid content of variant balls and found that it varied from 0.1 to 5.5 wt % and that the high lipid pickup was accompanied by volume increases up to 4%. At least half of the absorbed material was free lipids including fatty acids, neutral fats, steroids, steroid esters, and phospholipids. The remainder was a complex mixture of relative polar materials and possibly included monoglycerides, glycolipids, lipoproteins, etc. The lipid absorption was simulated in vitro and surprisingly, the lipid pickup increased with additional postcuring of the silicone rubber sample. Added filler, however, decreased the lipid pickup. Table 4.2–27 gives the comparative lipid increase and volume increase of a number of polymers. It should be noted that they reported that approximately only 2 to 3% of the silicone rubber balls from aortic ball valve prostheses returned for examination after implantation showed variations from normal. Most variation in the balls has been found in valves in the aortic position, but some have been found in the mitral position.[214] Because of this change in properties of the silicone balls, some researchers are using hollow metal balls, e.g., titanium.

SUMMARY AND CONCLUSIONS

The literature on polymers as surgical implants has been reviewed primarily from the standpoint of compatibility of polymers with tissue and with blood. Previous work indicates that in general pure polymers are compatible with tissue, although apparently measurable differences do exist. Apparent incompatibility can generally be traced to the presence of extractable materials that in themselves give rise to toxic reactions. In only two instances are the polymers indicated to be inherently carcinogenic. In both, the polymers are catonic, one because of an excess of dimethylaminoethylmethacrylate and the other because of an excess of the comonomer vinylbenzyltrimethylammonium chloride. As these two materials have been found to be tumorogenic only in rats and in preliminary experiments, the results are therefore open to question.

For polymers to be used on contact with blood, the situation is different in that all but a few polymers have been found to cause blood coagulation. There seems to be agreement that polymers containing surface-bound heparin show a high degree of thromboresistance. Certain polyether-based polyurethanes and polyurethane-polydimethylsiloxane block copolymers and water-gel polymers likewise show some degree of thromboresistance. The water-gel polymers include those based on polyacrylamide and the ionic polymers known as Ioplex. Two possible causes of thrombogenicity of polymers are recognized: the activation of Factor XII (Hageman factor) and the disruption of blood formed elements (especially platelets) by the foreign surfaces. The extent of damage for foreign surfaces to formed elements is uncertain because of the difficulty in separating purely surface effects from hemodynamic effects.

Available evidence indicates that, in general, pure polymers cause little if any damage to soluble components of blood such as the proteins, enzymes, and clotting factors with possible exception of the Hageman factor.

It is interesting to note that cationic polymers show a greater deposition of platelets and cause the platelet release reaction, although, in general, all polymers regardless of wettability, water content, or degree of negative charge adsorb about the same number of platelets and about the same amount of protein from the blood plasma.

Although there is evidence that the enzyme systems in blood attack polymers with the possible exception of polydimethylsiloxane, the rate of attack is low enough that polymers such as polyethylene, Teflon, polymethylmethacrylate, polypropylene, and polyethyleneglycolterephthalate are sufficiently stable toward blood to be used safely over a number of years. There is no evidence that long-term use of polymers such as polydimethylsiloxane, polyethylene, or polyethyleneglycolterephthalate leads to tumor formation in humans.

BACKGROUND BIBLIOGRAPHY

Articles

Bruck, S. D., Biomaterials in medical devices, *Trans. Am. Soc. Artif. Intern. Organs,* 18, 1, 1972.

Bruck, S. D., Macromolecular aspects of biocompatible materials, *J. Biomed. Mater. Res.*, 6, 173, 1972.

Bruck, S. D., Polymeric materials: current status of biocompatibility, *Biomaterials, Medical Devices and Artificial Organs,* in press.

Hastings, G. W., Macromolecular chemistry and medicine, *Ange. Chem. Int. Ed. Eng.*, 9(5), 332, 1970.

Kammermeyer, K., Biomaterials – developments and applications, *Chem. Technol.*, 1, 719, 1971.

Rainer, G. W., Goals for standardization and legislation in the medical device domain, *J. Assoc. Adv. Med. Instrum.*, 6(2), 105, 1972.

Rendell-Baker, L. and Miller, M. J., AAMI/FDA 1972 National Conference on Medical Device Standards, *J. Assoc. Adv. Medical Instrum.*, 6(2), 126, 1972.

Sanders, H. J., Artificial organs, *Chem. Eng. News,* Part 1, April 5, 1971, 32; Part 2, April 12, 1971, 68.

Salzman, E. W., Nonthrombogenic surfaces: critical review, *Blood,* 38, 4, 509, 1971.

Wesolowski, S. A., Martinex, A., and McMahon, J. D., Use of artificial materials in surgery, *Current Problems in Surgery,* Monthly Clinical Monographs, Year Book Medical Publishers, Chicago, December, 1966.

Books

Manly, R. S., Ed., *Adhesion in Biological Systems,* Academic Press, New York, 1970.

Hegyeli, R. J., Ed., Artificial Heart Program Conference, Proceedings June 9-13, 1969, Washington, D.C., U.S. Department of Health, Education and Welfare, Public Health Service, National Institutes of Health, Bethesda, Md.

Bement, A. L., Jr., Ed., *Biomaterials Bioengineering Applied to Materials for Hard and Soft Tissue Replacement,* published for Battelle/Seattle Research Center by University of Washington Press, Seattle, 1971.

Conference on Moving Blood, Federation Proceedings, San Diego, Calif., January 13-15, 1971, sponsored by the National Heart and Lung Institute, National Institutes of Health, Bethesda, Md. September-October, 1971.

Wesolowski, S. A., in *Evaluation of Tissue and Prosthetic Vascular Grafts,* De Bakey, M. E., Ed., Charles C Thomas, Springfield, Ill., 1962.

Lee, H. and Neville, K., Eds., *Handbook of Biomedical Plastics*, Pasadena Technology Press, Pasadena, Calif., 1971.

Gould, R. F., Ed., *Interaction of Liquids as Solid Substrates,* Advances in Chemistry Series, based on symposia sponsored by the Division of Organic Coatings and Plastics Chemistry of the American Chemical Society.

Gregor, H. P., Ed., *Medical Applications of Plastics,* Biomedical Materials Symposium No. 1, Interscience, New York, 1971.

Lefaux, R., Ed., *Practical Toxicology of Plastics,* Iliffe Books, London, 1968 (English translation).

Technical Reports and Contracts, The Medical Devices Applications Program of the NHLI, reports available from the National Technical Information Service, Springfield, Va.

Matsumoto, T., *Tissue Adhesives in Surgery,* Medical Examination Publishing Co., Flushing, N. Y.

REFERENCES

1. **Ingraham, F. D., Alexander, E., Jr., and Matson, D. D.,** Synthetic plastic materials in surgery, *N. Engl. J. Med.,* 236, 362, 1947.
2. **Dennis, C.,** Prolonged dependent drainage with "Lucite" drains, *Surgery,* 13(6), 900, 1943.
3. **Aries, L. J.,** Experimental studies with synthetic fiber (nylon) as a buried suture, *Surgery,* 9, 51, 1941.
4. **Narat, J. K.,** Use of synthetic, nonabsorbable suture material in surgery, *Surg. Gynecol. Obstet.,* 73, 819, 1941.
5. **Blum, G.,** Experimental observations of the use of absorbable and nonabsorbable plastics in bone surgery, *Proc. R. Soc. Med.,* 38, 169, 1945; or see **Blaine, G.,** Experimental observations on the use of absorbable and nonabsorbable plastics in bone surgery, *Br. J. Surg.,* 33, 245, 1945.
6. **Ingraham, F. D., Alexander, E., Jr., and Matson, D. D.,** Polyethylene, a new synthetic plastic for use in surgery, *J.A.M.A.,* 135, 82, 1947.
7. **Poppe, J. K. and de Oliveira, H. R.,** Treatment of syphilitic aneurysms by cellophane wrapping, *J. Thorac. Surg.,* 15, 186, 1946.
8. **Yeager, G. H. and Cowley, R. A.,** Studies on the use of polythene as a fibrous tissue stimulant, *Ann. Surg.,* 128, 509, 1948.
9. **LeVeen, H. H. and Barberio, J. R.,** Tissue reaction to plastics used in surgery with special reference to Teflon, *Ann. Surg.,* 129, 74, 1949.
10. **Harrison, J. H., Swanson, D. S., and Lincoln, A. F.,** A comparison of the tissue reactions to plastic materials, *A. M. A. Arch. Surg.,* 74, 139, 1957.
11. **Chen, R. W., Musser, A. W., and Postlethwait, R. W.,** Alterations of and tissue reaction to polyvinyl alcohol sponge implants, *Surgery,* 66, 889, 1969.
12. **Gale, J. W., Curreri, A. R., Young, W. P., and Dickie, H. A.,** Plastic sponge prosthesis following resection in pulmonary tuberculosis, *J. Thorac. Surg.,* 24, 587, 1952.
13. **Grindlay, J. H. and Waugh, J. M.,** Plastic sponge which acts as a framwork for living tissue. Experimental studies and preliminary report of use to reinforce abdominal aneurysms, *Arch. Surg.,* 63, 288, 1951.
14. **Harrison, J. H.,** Ivalon sponge (polyvinyl alcohol) as a blood vessel substitute (failure in experimental animals), *Surgery,* 41, 729, 1957.
15. **Schwartz, A. W. and Erich, J. B.,** Experimental study of polyvinyl-formal (Ivalon) sponge as a substitute for tissue, *Plast. Reconstr. Surg.,* 25, 1, 1960.
16. **Shumway, N. W., Gliedman, M. L., and Lewis, F. J.,** An experimental study of the use of polyvinyl sponge for aortic grafts, *Surg. Gynecol. Obstet.,* 100, 703, 1955.
17. **Brown, J. B., Ohlwiler, D. A., and Fryer, M. P.,** Investigations of and use of dimethyl siloxanes, halogenated carbons and polyvinyl alcohol as subcutaneous prostheses, *Ann. Surg.,* 152, 534, 1960.
18. **Usher, F. C. and Wallace, S. A.,** Tissue reaction to plastics, *A.M.A. Arch. Surg.,* 79, 997, 1958.
19. **Lieb, W. A., Geeraets, W. J., Guerry, D., III, and Dickerson, T.,** Tissue tolerance of plastic resins, *Eye, Ear, Nose, Throat Mon.,* 38, 210, 1959.
20. **Lieb, W. A., Geeraets, W. J., Guerry, D., III, and Dickerson, T.,** Tissue tolerance of plastic resins, *Eye, Ear, Nose, Throat Mon.,* 38, 303, 1959.
21. **Dyck, M. F. and Winters, P. R.,** Lytic effects of plastic surfaces on erythrocytes, *J. Biomed. Mater. Res.,* 5, 207, 1971.
22. **Boretos, J. W., Detmer, D. E., and Donachy, J. H.,** Segmented polyurethane: a polyether polymer. II. Two years experience, *J. Biomed. Mater. Res.,* 5, 373, 1971.
23. **Guess, W. L. and Stetson, J. B.,** Tissue reactions to organotin-stabilized polyvinyl chloride (PVC) catheters, *J.A.M.A.,* 204, 118, 1968.
24. **Stetson, J. B.,** Prolonged endotracheal intubation, *Br. J. Anaesth.,* 40, 712, 1968.
25. **Guess, W. L. and Haberman, S.,** Toxicity profiles of vinyl and polyolefinic plastics and their additives, *J. Biomed. Mater. Res.,* 2, 213, 1968.
26. **Scales, J. T.,** Tissue reaction to synthetic materials, *Proc. R. Soc. Med.,* 46, 647, 1953.
27. **Jaeger, R. J. and Rubin, R. J.,** Plasticizers from plastic devices: extraction, metabolism, and accumulation by biological systems, *Science,* 170, 460, 1970.
28. **Drenick, E. J. and Lipset, M.,** Difficulty with removal of plastic nasogastric tube, *J.A.M.A.,* 218, 1573, 1971.
29. **Marzoni, F. A., Upchurch, S. E., and Lambert, C. J.,** An experimental study on silicone as a soft tissue substitute, *Plast. Reconstr. Surg.,* 24, 600, 1959.
30. **Speirs, A. C. and Blocksma, R.,** New implantable silicone rubbers, *Plast. Reconstr. Surg.,* 31, 166, 1963.
31. **Blocksma, R. and Braley, S.,** The silicones in plastic surgery, *Plast. Reconstr. Surg.,* 35, 366, 1965.
32. **Bader, K. and Curtin, J. W.,** Clinical survey of silicone underlays and pulleys in tendon surgery in hands, *Plast. Reconstr. Surg.,* 47, 576, 1971.
33. **Braley, S.,** The chemistry and properties of the medical-grade silicones, *J. Macromol. Sci. Chem.,* A43, 529, 1970.
34. **Braley, S.,** Acceptable plastic implants, in *Modern Trends in Biomechanics,* Vol. 1, Simpson, D. C., Ed., Butterworths, London, 1970, chap. 2.
35. **Spina, V., Kamakura, L., and Psillakis, J. M.,** Total reconstruction of the ear in congenital microtia, *Plast. Reconstr. Surg.,* 48, 349, 1971.

36. **Blocksma, R.**, Experience with dimethylpolysiloxane fluid in soft tissue augmentation, *Plast. Reconstr. Surg.*, 48, 564, 1971.
37. **Wichterle, O. and Lim, D.**, Hydrophilic gels for biological use, *Nature*, 185, 117, 1960, and U.S. patent 2,976,576, March 28, 1961.
38. **Barvič, M., Kliment, K., and Zavadil, M.**, Biologic properties and possible uses of polymer-like sponges, *J. Biomed. Mater. Res.*, 1, 313, 1967.
39. **Šprincl, L., Kopeček, J., and Lim, D.**, Effect of porosity of heterogeneous poly(glycol monomethacrylate) gels on the healing-in of test implants, *J. Biomed. Mater. Res.*, 5, 447, 1971.
40. **Refojo, M. F.**, Polymers in ophthalmic surgery, *J. Biomed. Mater. Res.*, 5, 113, 1971.
41. **Refojo, M. F.**, The swelling implant, in *Medical Applications of Plastics*, Gregor, H. P., Ed., Interscience, New York, 1971, 179.
42. **Barvič, M.**, Tolerance of modified poly(glycol methacrylates) by the organism, *J. Biomed. Mater. Res.*, 5, 225, 1971.
43. **Marshall, D. W., Cross, R. A., and Bixler, H. J.**, An evaluation of polyelectrolyte complexes as biomedical materials, *J. Biomed. Mater. Res.*, 4, 357, 1970.
44. **Hardy, R. W.**, An acrylic-amide foam arterial prosthesis, *J. Thorac. Cardiovasc. Surg.*, 38, 652, 1959.
45. **Sewell, W. R., Wiland, J., and Craver, B. N.**, A new method of comparing sutures of ovine catgut with sutures of bovine catgut in three species, *Surg. Gynecol. Obstet.*, 100, 483, 1955.
46. **Postlethwait, R. W.**, Polyglycolic acid surgical suture, *Arch. Surg.*, 101, 489, 1970.
47. **Salthouse, T. N. and Williams, J. A.**, Histochemical observations of enzyme activity at suture implant sites, *J. Surg. Res.*, 9, 481, 1969.
48. **Salthouse, T. N. and Willigan, D. A.**, An enzyme histochemical approach to the evaluation of polymers for tissue compatibility, *J. Biomed. Mater. Res.*, 6, 105, 1972.
49. **Atsumi, K., Sakurai, Y., Atsumi, E., Narausawa, S., Kunisawa, S., Okikura, M., and Kimoto, S.**, Application of specially cross-linked natural rubber for artificial internal organs, *Trans. Am. Soc. Artif. Intern. Organs*, 9, 324, 1963.
50. **Rosenbluth, S. A., Weddington, G. R., Guess, W. L., and Autian, J.**, Tissue culture method for screening toxicity of plastic materials to be used in medical practice, *J. Pharm. Sci.*, 54, 156, 1965.
51. **Guess, W. L. and Rosenbluth, S. A.**, Agar diffusion method for toxicity screening of plastics on cultured cell monolayers, *J. Pharm. Sci.*, 54, 1545, 1965.
52. **Homsy, C. A.**, Bio-compatibility in selection of materials for implantation, *J. Biomed. Mater. Res.*, 4, 341, 1970.
53. **Homsy, C. A., Ansevin, K. D., O'Bannon, W., Thompson, S. A., Hodge, R., and Estrella, M. E.**, *J. Macromol. Sci. Chem.*, A4(3), 615, 1970.
54. **Johnsson, R. I. and Hegyeli, A. F.**, Tissue culture techniques for screening of prosthetic materials, *Ann. N.Y. Acad. Sci.*, 146, 66, 1968.
55. **Johnsson, R. I. and Hegyeli, A. F.**, Interaction of blood and tissue cells with foreign surfaces, *J. Biomed. Mater. Res.*, 3, 115, 1969.
56. **Pennington, C. J., Boatman, J. B., Peters, A. C., and Peterson, L. L.**, PH-43-67-1404, PB 201 433, PB 201 434, PB 201 435, Biological Evaluation of Prosthetic Materials, April 30, 1971 (order documents by PB number from the National Technical Information Service, Springfield, Va.).
57. **Powell, D., Lawrence, W. H., Turner, J., and Autian, J.**, Development of a toxicity evaluation program for dental materials and products, *J. Biomed. Mater. Res.*, 4, 583, 1970.
58. **Autian, J.**, Toxicologic aspects of implants, *J. Biomed. Mater. Res.*, 1, 433, 1967.
59. **Little, K. and Parkhouse, J.**, Tissue reactions to polymers, *Lancet*, 2, 857, 1962.
60. **Ocumpaugh, D. E. and Lee, H. L.**, Foreign body reactions to plastic implants, *J. Macromol. Sci. Chem.*, A4(3), 595, 1970.
61. **Wood, N. K., Kaminski, E. J., and Oglesby, J.**, The significance of implant shape in experimental testing of biological materials: disc vs. rod, *J. Biomed. Mater. Res.*, 4, 1, 1970.
62. **Gulliver, G.**, *Works of William Hewson*, Sydenham Society, London, 1846. (Cited in Salzman, E. W., Nonthrombogenic surfaces: critical review, *Blood*, 38, 4, 509, 1971.)
63. **Ratnoff, O. D.**, Coagulation defects, in *Cecil-Loeb Textbook of Medicine*, 13th ed., Beeson, P. B. and McDermott, W., Eds., W. B. Saunders, Philadelphia, to be published.
64. **Harrison, J. H.**, Synthetic materials as vascular prostheses. I. A comparative study on small vessels of nylon, Dacron, Orlon, Ivalon sponge and Teflon, *Am. J. Surg.*, 95, 3, 1958.
65. **Harrison, J. H.**, Synthetic materials as vascular prostheses. II. A comparative study of nylon, Dacron, Orlon, Ivalon sponge and Teflon in large blood vessels with tensile strength studies, *Am. J. Surg.*, 95, 16, 1958.
66. **Harrison, J. H.**, Synthetic materials as vascular prostheses. III. Long term studies on grafts of nylon, Dacron, Orlon and Teflon replacing large blood vessels, *Surg. Gynecol. Obstet.*, 108, 433, 1959.
67. **Shumacker, H. B., Jr. and Muhm, H. U.**, Arterial suture techniques and grafts: past, present, and future, *Surgery*, 66, 419, 1969.
68. **Lampert, H.**, *Die physikalische Seite des Blutgerunnings Problems*, George Thieme, Leipzig, 1931.
69. **Lozner, E. L. and Taylor, F. H. L.**, *J. Clin. Invest.*, 21, 241, 1942.
70. **Falb, R. D. and Leininger, R. I.**, Literature survey on blood-plastic interaction, PB 169 094, *Technical Reports from Contracts of the Medical Devices Applications Program*, NHLI, October 20, 1964.

71. **Leininger, R. I.,** Surface effects in blood-plastic compatibility, in *Biophysical Mechanisms in Vascular Homeostasis and Intravascular Thrombosis,* Sawyer, P. N., Ed., Appleton-Century-Crofts, New York, 1965.
72. **Moolten, S. E., Vroman, L., Vroman, G. M. S., and Goodman, B.,** Role of blood platelets in thromboembolism, *Arch. Intern. Med.,* 84, 667, 1949.
73. **Silberberg, M.,** The causes and mechanism of thrombosis, *Physiol. Rev.,* 18, 197, 1938.
74. **Duc, G.,** Influences sur le sang de matériaux utilisables en circulation extracorporelle, *Vox Sang.,* 7, 63, 1962.
75. **Frei, P. C. and Fleisch, A.,** Procédé nouveau permettant de diminuer l'hémolyse du sang conservé, *Helv. Physiol. Acta,* 19, C19, 1961.
76. **Frei, P. C. and Fleisch, A.,** De la nocivité de quelques matériaux employés dans la construction des organes artificiels et de leur nettoyage par des enzymes protéolytiques, *Vox Sang.,* 6, 489, 1961.
77. **Mirkovitch, V., Akutsu, T., and Kolff, W. J.,** Intracardiac thrombosis on plastic in relation to construction of artificial valves, *J. Appl. Physiol.,* 16, 381, 1961.
78. **Rose, J. and Broida, H.,** Effect of plastic and steel surfaces on clotting time of human blood, *Proc. Soc. Exp. Biol. Med.,* 86, 384, 1954.
79. **Dreyer, B., Akutsu, T., and Kolff, W. J.,** Aortic grafts of polyurethane in dogs, *J. Appl. Physiol.,* 15, 18, 1960.
80. **Hufnagel, C. A.,** The use of rigid and flexible plastic prostheses for artificial replacement, *Surgery,* 37, 165, 1955.
81. **Donovan, T. J. and Zimmermann, B.,** The effect of artificial surfaces on blood coagulability, with special reference to polyethylene, *Blood,* 4, 1310, 1946.
82. **Harvey, E.,** *Med. Phys.,* 2, 138, 1950.
83. **Copley, A. I., Glouer, F. A., and Scott-Blair, G. W.,** Wettability of fibrinized surfaces and of living vascular endothelium by blood, *Biorheology,* 2, 29, 1964.
84. **Ross, J., Jr., Greenfield, L., Bowman, R., and Morrow, A. G.,** in *Prosthetic Valves for Cardiac Surgery,* Sawyer, P. N., Ed., Charles C Thomas, Springfield, Ill., 1961, 212.
85. **Lovelock, J. E. and Porterfield, J. S.,** Blood coagulation: its prolongation in vessels with negatively charged surfaces, *Nature,* 167, 39, 1951.
86. **Mirkovitch, V.,** Bioelectric phenomena, thrombosis and plastics: a review of current knowledge, *Cleveland Clin. Q.,* 30, 241, 1963.
87. **Sawyer, P. N. and Pate, J. W.,** Bio-electric phenomena as an etiologic factor in intravascular thrombosis, *Am. J. Physiol.,* 175, 103, 1953.
88. **Schwartz, S. I.,** Prevention and production of thrombosis by alterations in electric environment, *Surg. Gynecol. Obstet.,* 108, 533, 1959.
89. **Richardson, J. W. and Schwartz, S. I.,** Prevention of thrombosis with the use of a negative electric current, *Surgery,* 52, 636, 1962.
90. **Gott, V. L., Koepke, D. E., Dagget, R. L., Zarnstorff, W., and Young, W. P.,** The coating of intravascular plastic prostheses with colloidal graphite, *Surgery,* 50, 382, 1961.
91. **Hubbard, D. and Lucas, G. L.,** Ionic charges of glass surfaces and other materials, and their possible role in the coagulation of blood, *J. Appl. Physiol.,* 15, 265, 1960.
92. **Glasstone, S. and Lewis, D.,** *The Elements of Physical Chemistry,* D. Van Nostrand Co., Princeton, N. J., 1958, 566.
93. **Davies, J. T. and Rideal, E. K.,** *Interfacial Phenomena,* 2nd ed., Academic Press, New York, 1963, 118.
94. **Gortner, R. A. and Briggs, D. R.,** *Proc. Soc. Exp. Biol. Med.,* 25, 820, 1928.
95. **Horan, F. E., Hirsch, F. G., Wood, L. A., and Wright, I. S.,** Surface effects on blood-clotting components as determined by zeta potentials, *J. Clin. Invest.,* 29, 202, 1950.
96. **Beck, R. E., Mirkovitch, V., Andrus, P. G., and Leininger, R. I.,** Apparatus for determining zeta potentials from streaming potentials, *J. Appl. Physiol.,* 18, 1263, 1963.
97. **Mirkovitch, V., Beck, R. E., Andrus, P. G., and Leininger, R. I.,** The zeta potential and blood compatibility characteristics of some selected solids, *J. Surg. Res.,* 4, 395, 1964.
98. **Gore, I. and Larkey, B. J.,** Functional activity of aortic mucopolysaccharides, *J. Lab. Clin. Med.,* 56, 839, 1960.
99. **Ohta, G., Sasaki, H., Fugitsugu, M., Tanishima, K., and Watanabe, S.,** *Proc. Soc. Exp. Biol. Med.,* 109, 298, 1962.
100. **Gott, V. L., Whiffen, J. D., and Dutton, R. C.,** Heparin bonding on colloidal graphite surfaces, *Science,* 142, 1297, 1963.
101. **Chvapil, M. and Krajicek, M.,** Principle and construction of a highly porous collagen-fabric vascular graft, *J. Surg. Res.,* 3, 358, 1963.
102. **Wasler, B. A.,** *Scand. J. Clin. Lab. Invest.,* 2, Suppl. 37, 1959.
103. **Whiffen, J. D. and Gott, V. L.,** Bacteriological radioactive benzalkonium-heparin surface, *J. Surg. Res.,* 5, 51, 1964.
104. **Gott, V. L., Whiffen, J. D., and Valiathan, M.,** Graphite-benzalkonium-heparin coatings on plastics and metals, *Ann. N.Y. Acad. Sci.,* 146, 21, 1968.
105. **Gott, V. L., Ameli, M. M., Whiffen, J. D., Leininger, R. I., and Falb, R. D.,** Newer thromboresistant surfaces: evaluation with a new *in-vivo* technique, *Surg. Clin. North Am.,* 47, 1443, 1967.
106. **Lee, R. I. and White, P. D. A.,** A clinical study of the coagulation time of blood, *Am. J. Med. Sci.,* 145, 495, 1913.
107. **Weing, W. J.,** Device for measuring blood compatibility, *J. Biomed. Mater. Res.,* 2, 179, 1968.
108. **Yohji, Y. and Nosé, Y.,** A new method for evaluation of antithrombogenicity of materials, *J. Biomed. Mater. Res.,* 6, 165, 1972.

109. **Spaet, T., Cintron, J., and Kropatkin, B. S.,** A technique for determining whole blood clotting times in plastic tubes, *J. Lab. Clin. Med.,* 54, 467, 1959.
110. **Leonard, E. F., Koffsky, R. M., Carter, J. H., Litwak, R. S., Aledort, L., and Aservatham, J.,** in *Proceedings 5th Annual Contractors Conference of the Artificial Kidney Program of the NAIMD,* Krueger, K. K., Ed., National Institute of Arthritis, Metabolism, and Digestive Diseases, National Institutes of Health, Bethesda, Md., 1972.
111. **Gott, V. L., Ramos, M. D., Najjar, F. B., Allen, J. L., and Becker, K. E.,** The *in vivo* screening of potential thromboresistant materials, in *Proceedings Artificial Heart Program Conference,* National Heart Institute, Bethesda, Md., 1969, 181.
112. **Falb, R. D., Grode, G. A., Epstein, M. M., Brand, B. D., and Leininger, R. I.,** Development of blood-compatible polymeric materials, PB 168 861, *Technical Reports from Contracts of the Medical Devices Applications Program,* National Heart and Lung Institute, Bethesda, Md., June 29, 1965.[a]
113. **Leininger, R. I., Cooper, C. W., Epstein, M. M., Falb, R. D., and Grode, G. A.,** Nonthrombogenic plastic surfaces, *Science,* 152, 1625, 1966.
114. **Falb, R. D., Grode, G. A., Luttinger, M., Epstein, M. M., Drake, B., and Leininger, R. I.,** Development of blood-compatible polymeric materials, PB 173 053, June 22, 1966.
115. **Leininger, R. I., Epstein, M. M., Falb, R. D., and Grode, G. A.,** Preparation of nonthrombogenic plastic surfaces, *Trans. Am. Soc. Artif. Intern. Organs,* 12, 151, 1966.
116. **Falb, R. D., Takahashi, M. T., Grode, G. A., and Leininger, R. I.,** Studies on the stability and protein adsorption characteristics of heparinized polymer surfaces by radioisotope labeling techniques, *J. Biomed. Mater. Res.,* 1, 239, 1967.
117. **Falb, R. D., Grode, G. A., Grotta, H. M., Wright, R. A., Poirier, R. H., Takahashi, M. T., and Leininger, R. I.,** Development of blood-compatible materials, PB 175 668, March 30, 1967; PB 183 317, October 1967.
118. **Grode, G. A., Anderson, S., and Falb, R. D.,** Development of blood compatible materials, PB 188, 108, October, 1968; PB 188, 111, October, 1969.
119. **Grode, G. A., Crowley, J. P., Payne, S. T., and Falb, R. D.,** PB 195, 727, October, 1970; PB 205, 475, November, 1971.
120. **Grode, G. A., Falb, R. D., and Crowley, J. P.,** Biocompatible materials for use in the vascular system, *Biomedical Materials Symposium,* No. 3, Interscience, New York, 1972, 77.
121. **Leininger, R. I., Crowley, J. P., Falb, R. D., and Grode, G. A.,** Three years' experience *in vivo* and *in vitro* with surfaces and devices treated by the heparin complex method, *Trans. Am. Soc. Artif. Intern. Organs,* 17, 312, 1972.
122. **Rea, W. J., Whitley, D., and Eberle, J. W.,** Long-term membrane oxygenation without systemic heparinization, *Trans. Am. Soc. Artif. Intern. Organs,* 18, 316, 1972.
123. **Najjar, F. B. and Gott, V. L.,** The use of small-diameter Dacron grafts with wall bonded heparin for venous and arterial replacement: canine studies and preliminary clinical experience, *Surgery,* 68, 1053, 1970.
124. **Eriksson, J. D., Gillberg, G., and Lagergren, H.,** A new method for preparing nonthrombogenic plastic surfaces, *J. Biomed. Mater. Res.,* 1, 301, 1967.
125. **Lagergren, H., Johansson, L., and Eriksson, J. C.,** Die einwirkung elektrisch gelandener Flachen auf die Blutkoagulation, *Thoraxchirurgie,* 12, 172, 1964.
126. **Lagergren, H. and Eriksson, J. C.,** Plastics with a stable surface monolayer of cross-linked heparin: preparation and evaluation, *Trans. Am. Soc. Artif. Intern. Organs,* 17, 10, 1971.
127. **Merrill, E. W., Salzman, E. W., Lipps, B. J., Jr., Gilliland, E. R., Austen, W. G., and Joison, J.,** Antithrombogenic cellulose membranes for blood dialysis, *Trans. Am. Soc. Artif. Intern. Organs,* 12, 139, 1966.
128. **Britton, R. A., Merrill, E. W., Gilliand, E. R., Salzman, E. W., Austen, W. G., and Kemp, D. S.,** Antithrombogenic cellulose film, *J. Biomed. Mater. Res.,* 2, 429, 1968.
129. **Salzman, E. W., Austen, W. G., Lipps, B. J., Merrill, E. W., and Joison, J.,** A new antithrombotic surface: development and *in vitro* and *in vivo* characteristics, *Surgery,* 61, 1, 1967.
130. **Merker, R. L., Elyash, L. J., Mayhew, S. H., and Wang, J. Y. C.,** The heparinization of silicone rubber using aminoorganosilane coupling agents, in *Artificial Heart Program Conference Proceedings,* Hegyeli, R. J., Ed., National Heart Institute Artificial Heart Program, Washington, D.C., 1969, 29.
131. **Merker, R. L., Elyash, L. J., and Mayhew, S. H.,** Studies relative to materials suitable for use in artificial hearts, prepared for *Artificial Heart Program,* PB 199 909, March 1, 1971.
132. **Halpern, B. D. and Shibakawa, R.,** Heparin covalently bonded to polymer surface, in *Interactions of Liquids at Solid Substrates,* Gould, R. F., Ed., American Chemical Society Publications, Washington, D.C., 1968.
133. **Merrill, E. W., Wong, P. S. L., Salzman, E. W., and Gusten, W. G.,** Pilot model of an artificial kidney: biomaterials, *3rd Annual Contractors' Conference of the Artificial Kidney Program of the NIAMD,* 1970.
134. **Merrill, E. W., Salzman, E. W., Wong, P. S. L., Smith, K. A., Colton, C. K., Bellantoni, E., and Damus, P.,** Nonthrombogenic surfaces and artificial kidney design, in *4th Annual Contractors' Conference of the Artificial Kidney Program of the NIAMD,* National Institute of Arthritis, Metabolism, and Digestive Diseases, National Institutes of Health, Bethesda, Md., 1971.

[a]All PB reports are from work sponsored by the National Heart and Lung Institute, National Institutes of Health, and are obtainable from the National Technical Information Service of the U.S. Department of Commerce, 5285 Port Royal Road, Springfield, Va., 22151.

135. **Merrill, E. W., Salzman, E. W., Wong, P. S. L., and Damus, P.,** Nonthrombogenic surfaces and artificial kidney design, in *Proceedings 5th Annual Contractors' Conference of the Artificial Kidney Program of the NIAMD,* Krueger, K. K., Ed., National Institute of Arthritis, Metabolism, and Digestive Diseases, National Institutes of Health, Bethesda, Md., 1971.
136. **Wong, P. S. L., Merrill, E. W., and Salzman, E. W.,** A nonthrombogenic hydrogel cast from solution, *Fed. Proc.,* 28, 441, 1969.
137. **Merrill, E. W., Salzman, E. W., Wong, P. S. L., and Ashford, T. P.,** Polyvinyl alcohol-heparin hydrogel "G," *J. Appl. Physiol.,* 29, 723, 1970.
138. **Salzman, E. W., Merrill, E. W., Binder, A., Wolf, C. F., Ashford, T. P., and Austen, W. G.,** Protein-platelet interaction on heparinized surfaces, *J. Biomed. Mater. Res.,* 3, 69, 1969.
139. **Hufnagel, C. A., Conrad, P. W., Gillespie, J. F., Pifarre, R., and Ilano, A.,** A new method for the prevention of thrombus formation in valvular prostheses, in *American College of Cardiology Abstracts,* 1963.
140. **Hufnagel, C. A. and Conrad, P. W.,** Comparative study of some prosthetic valves for aortic and mitral replacement, *Surgery,* 57, 205, 1965.
141. **Salyer, I. O. and Weesner, W. E.,** Materials and components for circulatory assist devices, in *Artificial Heart Program Conference Proceedings,* U. S. Department of Health, Education and Welfare, Washington, D.C., 1969, 59.
142. **Salyer, I. O.,** Materials and components for circulatory assist devices, PB 188 873, Annual Report, November 1969, and PB 197 357, Annual Report, December, 1970.
143. **Salyer, I. O., Blardinelli, A. J., Ball, G. L., Weesner, W. E., Gott, V. L., Ramos, M. D., and Furuse, A.,** in *Biomedical Materials Symposium No. 1: Medical Applications of Plastics,* Interscience, New York, 1971, 105.
144. **Byck, J. S., Miller, W. A., Spivack, M. A., Gonsior, L. J., Williams, V. Z., Jr., Barth, B. P., Bassett, H. D., Dearing, R. K., Gaasch, J. F., Schober, D. L., Stewart, D. D., and Tittmann, F. R.,** Polymeric materials for circulatory assist devices, in PB 201 935, March 1971.
145. **Ziegler, T. F. and Miller, M. L.,** Adsorption of fibrinogen on modified polytetrafluoroethylene surfaces, *J. Biomed. Mater. Res.,* 4, 259, 1970.
146. **Bixler, H. J., Cross, R. A., and Marshall, D. W.,** Polyelectrolyte complexes as antithrombogenic materials, in *Artificial Heart Program Conference Proceedings,* U.S. Department of Health, Education and Welfare, Washington, D.C., 1969, 79.
147. **Marshall, D. W.,** Development of polyelectrolyte complexes as thromboresistant materials for use in components of artificial hearts, in PB 198 403, February 1971.
148. **Nelson, L. M., Cross, R. A., Vogel, M. A., Gott, V. L., and Fadali, A. M.,** Synthetic thromboresistant surfaces from sulfonated polyelectrolyte complexes, *Surgery,* 67, 826, 1970.
149. **Leonard, F., Nielson, C. A., Fadali, A. M., and Gott, V. L.,** Thromboresistant polymers by emulsion polymerization with anionic surfactants. I., *J. Biomed. Mater. Res.,* 3, 455, 1969.
150. **Memchin, R. G.,** Evaluation of biopolymers as blood compatible materials, in PB 188 874, Annual Report, January 1970.
151. **Memchin, R. G.,** Evaluation of biopolymers as blood compatible materials, in PB 197 355, Annual Report, December 1970.
152. **Murphy, P., Lacroix, A., and Merchant, S.,** Studies relative to materials suitable for use in artificial hearts, in *Artificial Heart Program Conference Proceedings,* U.S. Department of Health, Education, and Welfare, Washington, D.C., 1969, 99.
153. **Murphy, P., Lacroix, A., and Merchant, S.,** Studies relative to materials suitable for use in artificial hearts, in PB 189 246, December 1969.
154. **Levowitz, B. S., La Guerre, J. N., Calem, W. S., Gould, F. E., Scherrer, J., and Schoenfeld, H.,** Biologic compatibility and applications of hydron, *Trans. Am. Soc. Artif. Intern. Organs,* 14, 82, 1968.
155. **Halpern, B. D., Cheng, H., Kuo, S., and Greenberg, H.,** Hydrogels as nonthrombogenic surfaces, in *Artificial Heart Program Conference Proceedings,* U.S. Department of Health, Education, and Welfare, Washington, D.C., 1969, 87.
156. **Halpern, B. D., McGonigal, P. J., Greenberg, H., and Heldon, A.,** Non-clotting plastic surfaces, in PB 190 665, February, 1970 and PB 200 987, January 1971.
157. **Hoffman, A. S., Schmer, G., Harris, C., and Kraft, W. G.,** Covalent binding of biomolecules to radiation-grafted hydrogels on inert polymer surfaces, *Trans. Am. Soc. Artif. Intern. Organs,* 18, 10, 1972.
158. **Schmer, G.,** The biological activity of covalently immobilized heparin, *Trans. Am. Soc. Artif. Intern. Organs,* 18, 321, 1972.
159. **Peterlin, A.,** Improved nonthrombogenic materials, in *Proceedings 5th Annual Contractors' Conference of the Artificial Kidney Program of the NIAMD,* Krueger, K. K., Ed., National Institute of Arthritis, Metabolism, and Digestive Diseases, National Institutes of Health, Bethesda, Md., 1972, 107.
160. **Hoffman, A. S.,** Private communication, 1972.
161. **Musolf, M. C. and Metevia, V. L.,** Studies relative to materials suitable for use in artificial hearts, in PB 190 666, October 1969; Musolf, M. C. and Metevia, V. L., Development of blood compatible silicone elastomers III, in PB 203 591, April 1971.
162. **Boretos, J. W. and Pierce, W. S.,** Segmented polyurethane: a new elastomer for biomedical applications, *Science,* 158, 1481, 1967.

163. **Brash, J. L., Fritzinger, B. K., and Loo, B. H.,** Development of materials for heart-assist devices, in PB 197 352, October 1970 and PB 210 665, April 1972.
164. **Lyman, D. J., Kwan-Gett, C., Zwart, H. H. J., Bland, A., Eastwood, N., Kawai, J., and Kolff, W. J.,** The development and implantation of a polyurethane hemispherical artificial heart, *Trans. Am. Soc. Artif. Intern. Organs,* 17, 456, 1971.
165. **Nyilas, E.,** Development of blood-compatible elastomers, U.S. patent 3,562,352, 1971.
166. **Nyilas, E.,** Development of blood-compatible elastomers: theory, practice, and *in vivo* performance, *Proc. 23rd Annual Conf. Eng. Med. Biol.,* 12, 147, 1970.
167. **Nyilas, E.,** Development of blood-compatible elastomers – performance of avcothane-51 blood contact surfaces and correlations with theory, Preprint of Paper No. 35b, *Symposium on Biomaterials in the Cardiovascular System,* 70th National Meeting of the American Institute of Chemical Engineers, Atlantic City, N. J., August 29 – September 1, 1971.
168. **Nyilas, E.,** Development of blood-compatible elastomers. II. Performance of avcothane blood contact surfaces in experimental animal implantation, *J. Biomed. Mater. Res.,* 3, 97, 1972.
169. **Nyilas, E., Leinback, R. C., Caulfield, J. B., Buckley, M. J., and Austen, W. G.,** Development of blood-compatible elastomers. III. Hemotologic effects of avcothane intra-aortic balloon pumps in cardiac patients, *J. Biomed. Mater. Res.,* 3, 129, 1972.
170. **Leinbach, R. C., Nyilas, E., Caulfield, J. B., Buckley, M. J., and Austen, W. G.,** Evaluation of hematologic effects of intra-aortic balloon assistance in man, *Tarns. Am. Soc. Artif. Intern. Organs,* 18, 493, 1972.
171. **DeLaria, G. A., Nyilas, E., and Bernstein, E. F.,** Intra-aortic balloon pumping without heparin, *Trans. Am. Soc. Artif. Intern. Organs,* 18, 501, 1972.
172. **Bishop, E. T., Porter, L. M., and Eisenhut, W. O.,** Development of segmented elastomers for use in blood pumps and oxygenators, in PB 200 989, January 1971.
173. **Serres, E. J.,** Open chest pulsatile left heart bypass without coagulation, *Arch. Surg.,* 101, 18, 1970.
174. **Taylor, B. C., Sharp, W. V., Wright, J. I., Ewing, K. L., and Wilson, C. L.,** The importance of zeta potential, ultrastructure, and electrical conductivity to the *in vivo* performance of polyurethane-carbon black vascular prostheses, *Trans. Am. Soc. Artif. Intern. Organs,* 17, 22, 1971.
175. **Kusserow, B. K., Larrow, R., and Nichols, J.,** The urokinase-heparin bonded synthetic surface, an approach to the creation of a prosthetic surface possessing composite antithrombogenic and thrombolytic properties, *Trans. Am. Soc. Artif. Intern. Organs,* 17, 1, 1971.
176. **Imai, Y., Tajima, K., and Nosé, Y.,** Biolized materials for cardiovascular prosthesis, *Trans. Am. Soc. Artif. Intern. Organs,* 17, 6, 1971.
177. **Nosé, Y., Tajima, K., Imai, Y., Klain, M., Mrava, G., Schriber, K., Urbanek, K., and Ogawa, H.,** Artificial heart constructed with biological material, *Trans. Am. Soc. Artif. Intern. Organs,* 17, 482, 1971.
178. **Lyman, D. J., Klein, K. G., Brash, J. J., Fritzinger, B. D., Andrade, J. D., and Bonomo, F. S.,** *Thromb. Diath. Haemorrh.,* Suppl. 42, 109, 1971.
179. University of Denver, Denver Res. Inst., Coagulation-resistant coatings, in PB 185 709, Annual Report, April 1969.
180. **Gott, V. L., Whiffen, J. D., Dutton, R. C., Leininger, R. I., and Young, W. P.,** Discussion: biophysical studies on various graphite-benzalkonium-heparin surfaces, in *Biophysical Mechanisms in Vascular Homeostasis and Intravascular Thrombosis,* Sawyer, P. N., Ed., Appelton-Century-Crofts, New York, 1965, 304.
181. **Nossel, H. L., Rubing, H., Drillings, M., and Hsieh, R.,** Inhibition of Hageman factor activation, *J. Clin. Invest.,* 46, 1172, 1968.
182. **Salzman, E. W.,** Role of platelets in blood-surface interactions, in *Federation Proceedings of Conference on Mechanical Surface and Gas Layer Effects on Moving Blood,* Federation of American Societies for Experimental Biology, Bethesda, Md., 1971, 1503.
183. **Petschek, H., Adamis, D., and Kantrowitz, A. R.,** Stagnation flow thrombus formation, *Trans. Am. Soc. Artif. Intern. Organs,* 14, 256, 1968.
184. **Dutton, R. C.,** Microstructure of initial thrombus formation on foreign materials, *J. Biomed. Mater. Res.,* 3, 13, 1969.
185. **Lyman, D. J., Klein, K. G., Brash, J. L., and Fritzinger, B. K.,** *Thromb. Diath. Haemorrh.,* 23, 120, 1970.
186. **Friedman, L. I., Liem, H., Grabowski, E. F., Leonard, E. F., and McCord, C. W.,** Inconsequentiality of surface properties for initial platelet adhesion, in *Trans. Am. Soc. Artif. Intern. Organs,* 14, 63, 1970.
187. **Weiss, L. and Blemenson, L. E.,** Dynamic adhesion and separtaion of cells *in vitro.* II. Interaction of cells with hydrophilis and hydrophobic surfaces, *J. Cell. Physiol.,* 79, 23, 1967.
188. **Baier, R. E., Gott, V. L., and Feruse, A.,** Surface chemical evaluation of thromboresistant materials before and after venous implantation, *Trans. Am. Soc. Artif. Intern. Organs,* 16, 50, 1970.
189. **Gott, V. L. and Baier, R. E.,** Twelve month progress report on contract to study materials compatible with blood, in PB 197 355, April 30, 1970.
190. **Baier, R. E., Loeb, G. I., and Wallace, G. T.,** Role of an artificial boundary in modifying blood proteins, in *Federation Proceedings of the Conference on Mechanical Surface and Gas Layer Effects on Moving Blood,* Federation of American Societies for Experimental Biology, Bethesda, Md., 1971, 1523.
191. **Hershgold, E. J., Kwan-Gett, C. W., Kawai, J., and Rowley, K.,** Hemostasis, coagulation and the total artificial heart, *Trans. Am. Soc. Artif. Intern. Organs,* 18, 181, 1972.

192. **LaFarge, C. G., Carr, J. G., Coleman, S. J., and Bernhard, W. F.,** Hemodynamic studies during prolonged mechanical circulatory support, *Trans. Am. Soc. Artif. Intern. Organs,* 18, 186, 1972.
193. **Kusserow, B. K., Larrow, R., and Nichols, J.,** Analysis and measurement of the effects of materials on blood leukocytes, erythrocytes and platelets, in PB 195 452, Annual Report, October 1969.
194. **Anon.,** Annual Report of the Medical Devices Applications Branch, National Heart and Lung Institute, July 1, 1971 through June 30, 1972, MD-50.
195. **Halbert, S. P., Anken, M., and Ushakoff, A. E.,** Studies on the compatibility of various plastics with the proteins in human plasma, *Proceedings of the Artificial Heart Program Conference,* 1969, 223 (available from Superintendent of Documents, U.S. Government Printing Office, Washington, D.C.).
196. **Halbert, S. P., Anken, M., and Ushakoff, A. E.,** Compatibility of blood with materials useful in the fabrication of artificial organs, in PB 197 251, Annual Report, June 1970.
197. **Halbert, S. P., Ushakoff, A. E., and Anken, M.,** Compatibility of various plastics with human plasma proteins, *J. Biomed. Mater. Res.,* 4, 549, 1970.
198. **Anken, M.,** private communication to G.A. Grode, December 30, 1970.
199. **Mason, R. G., Rodman, N. F., Scarborough, D. E., and Idenberry, L. D.,** Blood compatibility of some polymeric and nonpolymeric materials, in *Proceedings of the Artificial Heart Program Conference,* 1969, 193 (available from Superintendent of Documents, U.S. Government Printing Office, Washington, D.C.).
200. **Leininger, R. I., Mirkovitch, V., Peters, A., and Hawks, W. A.,** changes in properties of plastics during implantation, *Trans. Am. Soc. Artif. Intern. Organs,* 10, 320, 1964.
201. **Harrison, J. H. and Adler, R. H.,** Nylon as a vascular prosthesis in experimental animals with tensile strength studies, *Surg. Gynecol. Obstet.,* 103, 613, 1956.
202. **Maloney, G. E.,** The effect of human tissues on the tensile strength of implanted nylon sutures, *Br. J. Surg.,* 48, 528, 1960.
203. **Szilagyi, D. E.,** Long-term evaluation of plastic arterial substitutes; an experimental study, *Surgery,* 55(1), 165, 1964.
204. **Mirkovitch, V., Akutsu, T., and Kolff, W. J.,** Polyurethane aortas in dogs. Three-year results, *Trans. Am. Soc. Artif. Intern. Organs,* 8, 79, 1962.
205. **Rubin, L. R., Bromberg, B. E., and Walden, R. H.,** Long-term human reaction to synthetic plastics, *Surg. Gynecol. Obstet.,* 132, 603, 1971.
206. **Oppenheimer, B. S., Oppenheimer, E. T., Danishefsky, I., Stout, A. P., and Eirich, F.,** Further studies of polymers as carcinogenic agents in animals, *Cancer Res.,* 15, 333, 1955.
207. **Epstein, M. E.,** unpublished results.
208. **Matsumoto, T.,** *Tissue Adhesives in Surgery,* Medical Examination Publishing Co., Flushing, N.Y., January 1972.
209. **Manly, R. S., Ed.,** *Adhesion in Biological Systems,* Academic Press, New York, 1970.
210. **Grode, G. A., Pavkov, K. L., and Falb, R. D.,** Feasibility study on the use of a tissue adhesive for the nonsurgical blocking of fallopian tubes, Phase I: evaluation of a tissue adhesive, *Fertil. Steril.,* 20, 552, 1971.
211. **Dardik, H., Dardik, I., Katz, A. R., Smith, R. B., Schwibner, B. H., and Laufman, H.,** A new absorbable synthetic suture in growing and adult primary vascular anastomoses: morphologic study, *Surgery,* 68(6), 112, 1970.
212. **Kulkarni, R. K., Moore, E. G., Hegyeli, A. F., and Leonard, F.,** Biodegradable poly(lactic acid) polymers, *J. Biomed. Mater. Res.,* 5, 169, 1971.
213. **Carmen, R. and Kahn, P.,** In vitro testing of silicone rubber heart-valve poppets for lipid absorption, *J. Biomed. Mater. Res.,* 2, 457, 1968.
214. **Leatherman, L. L., McConn, R. G., and Cooley, D. A.,** Malfunction of mitral ball-valve prostheses due to swollen poppet, *J. Thorac. Cardiovasc. Surg.,* 57, 160, 1969.

Section 5

Graphitic Materials

5.1 PROPERTIES OF CARBON AND GRAPHITE

Manufactured carbon and graphite should not be viewed as single specific materials, but rather as families of materials. Each member of the family is essentially pure carbon, but each varies from the other in such characteristics as orientation of the crystallites, the size and number of pore spaces, grain size, degree of crystallization, and apparent density.

In the carbon industry the term *carbon* is used to refer to materials in which the small crystallites have low orientation. The term *graphite* is used to refer to material that has a highly ordered structure. Specific grades, having controlled characteristics, are produced by selecting raw materials and by varying processing techniques. A broad familiarity with the properties and characteristics of the various grades is important when selecting materials for a specific application.

Information regarding characteristics and recommended applications for various types of industrial graphite is available from Carbon Products Division, Union Carbide Corporation, New York.

Table 5.1–1
TYPICAL PROPERTIES OF CARBON AND GRAPHITE AT ROOM TEMPERATURE

Type of product	*Bulk density*	*Porosity, percent*	*Strength, psi*		*Elastic modulus, 10^6 psi*	*Specific resistance, 10^{-5} ohm-in.*	*Thermal conductivity, Btu·ft / hr·ft²·deg F*	*Coefficient of thermal expansion, 10^{-7}/deg F*
			Compressive	*Flexural*				
CARBON								
Porous carbon, grade 45	1.04	47	900	500	0.4	700	1	16
Carbon furnace lining (24″ × 30″ cross section)	1.60	21	2 500	600	0.8	160	9	13
Carbon refractory brick	1.63	17	4 400	1 200	1.2	195	16	18
Carbon chemical brick	1.56	22	8 800	2 600	1.9	160	4	13
Carbon pipe	1.55	22	9 000	2 600	1.9	150	3	13
Carbon electrodes, 8-in. diam. and equivalent rectangular	1.57	21	2 400	1 100	1.2	110	9	13
Carbon electrodes, 17- to 45-in. diam. and equivalent rectangular	1.60	21	1 700	400	0.7	170	9	13
GRAPHITE								
Porous graphite, grade 45	1.04	53	500	300	0.3	130	45	11
Nuclear graphite *see* Table 3.6–7								
Fine-grain, premium graphite	1.73	23	8 300	4 000	1.5	43	68	13
High-density, premium graphite	1.84	19	8 400	3 700	1.7	47	63	11
Recrystallized graphite	1.95	13	7 200	5 400	2.7	28	104	3
Graphite brick	1.56	31	3 100	1 650	1.4	34	86	10
Graphite pipe	1.67	26	5 000	2 800	1.7	34	86	10
Medium-grain, dense graphite, cylinders, and plates								
to 2 ¾-in. diam. and to ¾-in. thick	1.70	25	5 600	4 000	1.8	27	100	7
3 to 11-in. diam. and 2 to 12-in. thick	1.70	24	5 600	2 700	1.7	30	95	7
20 to 24-in. diam. and 20 × 20 cross section	1.75	24	5 500	2 200	1.2	35	83	12
Graphite electrodes, anodes, cylinders, and plates								
to 2 ¾-in. diam. and to ¾-in. thick	1.58	30	4 000	2 000	1.5	33	88	6
6 to 12-in. diam. and 6 to 12-in. thick	1.57	30	2 900	1 300	1.5	33	88	7
14 to 35-in. diam. and 20 to 24-in. thick	1.58	32	2 000	1 000	0.8	33	88	5

Table courtesy of Carbon Products Division, Union Carbide Corporation.

Table 5.1–3 (continued)
TYPICAL ENGINEERING APPLICATIONS OF CARBON PRODUCTS

Typical application	*Type and form of carbon product*	*Desirable properties*[b]
Electrodes for arc furnaces, welding, and lighting arcs	Extruded or molded carbon or graphite	1, 2, 3, 4, 5, 6, 7, 8, 9, 10, 12
Elements for electric-resistance furnaces	Resistors fabricated from round or rectangular stock, usually graphite	1, 2, 3, 4, 5, 8, 10, 12
Metallurgical crucibles, boats, trays	Castings from carbon or graphite	1, 2, 4, 7, 8, 9, 10, 12
Foundry molds, chills, cores, risers, cupola linings	Castings, rods, bricks, shapes	1, 2, 4, 7, 8, 9, 10, 12
Brazing jigs; extrusion guides; dies	Blocks; machined shapes	1, 2, 7, 8, 9, 10, 12
High-temperature refractories and insulations	Solid or foamed bricks and shapes; tapes and laminates	1, 2, 3, 4, 7, 8, 9, 12
Rocket and missile nozzles, vanes, cones, shields	Castings; built-up shapes from graphite fabric, tape, and resins	1, 2, 3, 4, 5, 7, 8, 9, 10
Chemical reactor vessels; heat exchangers; pipes and fittings	Cast, machined, or extruded graphite or carbon, pure or impregnated with resins	1, 2, 3, 4, 7, 8, 9, 10, 12
Packing and sealing rings; bushings, joint packings	Graphite blocks, sheet, cloth, and yarn	1, 2, 3, 4, 7, 8, 9, 11, 12
Electrical contacts, brushes, resistors	Blocks, rods, machined parts	1, 3, 5, 6, 7, 10, 11
Anodes, electrodes, battery components	Extruded and molded rod, granular carbon	6, 12
Air purification, gas separations, odor, and vapor removal	Activated carbon granules	13
Nuclear reactor—moderator and reflector	Purified "nuclear graphite" bars and machined shapes	1, 2, 3, 4, 14, 15

[a]Protection from oxidizing atmosphere is necessary in all high-temperature uses of carbon or graphite, e.g., above 700° F.
[b]Numbers refer to items listed above.

From Bolz, R. E. and Tuve, G. L., Eds., *Handbook of Tables for Applied Engineering Science,* 2nd ed., CRC Press, Cleveland, 1973, 181.

Table 5.1–2
VARIATION OF PROPERTIES WITH TEMPERATURE

Property	500 deg C / 932 deg F	1 000 deg C / 1 832 deg F	1 500 deg C / 2 732 deg F	2 000 deg C / 3 632 deg F	2 500 deg C / 4 532 deg F
Thermal expansion[a] as percent elongation from room temperature					
Anthracite carbon	0.14	0.38	0.63	—	-
Graphite	0.12	0.32	0.55	—	—
Thermal conductivity as percent of room temperature value					
Fabricated anthracite carbon	100	103	123	—	—
Fabricated graphite	60	40	30	25	—
Instantaneous specific heat[b] as percent of room temperature value: graphite	225	262	282	294	302
Short-time breaking strength[c] as percent of room temperature value: graphite	107	120	135	153	181

[a]These are longitudinal expansion; transverse expansion is 10–60% greater.
[b]The specific heat at room temperature is about 0.17.
[c]Strength increases up to 2 500 deg C, then decreases rapidly; above 2 200 deg C appreciable creep will occur at high-stress levels.

From *Carbon Products Pocket Handbook*, Carbon Products Division, Union Carbide Corporation, New York, 1964.

Table 5.1–3
TYPICAL ENGINEERING APPLICATIONS OF CARBON PRODUCTS

Key to Advantageous Properties of Carbon Products:[a]

1. Stability and strength at high temperatures (to 4,500°F in nonoxidizing atmospheres)
2. High resistance to thermal shock
3. High thermal conductivity of solid; low conductivity of porous foam, cloth, and tape
4. Low coefficient of thermal expansion
5. High radiation emissivity
6. Good electrical conductivity
7. High compressive strength
8. Stiffness of solid; flexibility of filament, cloth, or tape
9. High resistance to erosion
10. Good machinability
11. Low friction; self-lubrication
12. High resistance to chemical attack and corrosion
13. High adsorption of gases and vapors
14. High moderating ratio, i.e., ratio of fast neutron slowing-down power to bulk neutron absorption coefficient
15. High ratio of thermal neutron scattering to absorption cross section

SPECIFIC PROPERTIES OF COMMERCIAL GRAPHITIC MATERIALS

The following five sections constitute a partial listing of commercially available graphites that have been produced in recent years, along with their representative properties. Approximate sizes have been given for a particular grade because properties tend to show significant variance depending on this factor. For a more complete listing of graphitic products consult individual producer data books and AFML TR 67-113, *Directory of Graphite Availability,* by J. Glasser and W. J. Glasser (August 1967). Carbon-based fibers are listed in Section 4.1 of Volume II of the *CRC Handbook of Materials Science.*

5.2 MOLDED GRAPHITES

Table 5.2–1
GRADE D-775

Description
Molded, fine grained; low coefficient of thermal expansion

Analysis
Average impurity content: Ni, 200 ppm; Ca, 200 ppm; Fe, 100 ppm; Si, 75 ppm; Al, 75 ppm; Co, 25 ppm; Na, 100 ppm; Ti, 10 ppm; Mo, 10 ppm, purified grade, 50 ppm total impurities

		With grain		Against grain		Typical high-temperature properties	
Properties	Units	Average value	Standard deviation, %	Average value	Standard deviation, %	1,300°F	4,000°F
Young's modulus	10^6 psi	1.3	15	1.1	15	1.4	1.9
Tensile strength	10^3 psi	2.1	20	2.0	20	2.1	5.9
Compressive strength	10^3 psi	8.4	20	8.0	20	8.8	12.8
Flexural strength	10^3 psi	4.2	20	4.0	20	4.3	7.3
Density	g/cm^3	1.75	5				
Coefficient of thermal expansion	$10^{-6}/°C$	3.4	5	3.3	5	4.1	
Thermal conductivity	$cal\text{-}cm/sec \cdot cm^2 \cdot K$	0.35	15	0.34	15		
Specific resistance	10^{-4} ohm-cm	13.0	1	13.5	1		

Source
Duramic Products

Size
18 × 7 × 4 in. maximum

Manufacturing data
Not available

Table 5.2–2
GRADE D-657

Description
Molded, fine grained; low cost

Analysis
Average impurity content: Ni, 400 ppm; Ca, 400 ppm; Fe, 200 ppm; Na, 200 ppm; Si, 150 ppm; Al, 150 ppm; Co, 50 ppm; Ti, 20 ppm; Mo, 20 ppm, purified grade, 50 ppm total impurities

		With grain		Against grain		Typical high-temperature properties	
Properties	Units	Average value	Standard deviation, %	Average value	Standard deviation, %	1,300°F	4,000°F
Young's modulus	10^6 psi	1.5	15	1.3	15	1.6	2.1
Tensile strength	10^3 psi	3.3	20	2.8	20	3.0	6.3
Compressive strength	10^3 psi	8.6	20	10.5	20	9.0	13.0
Flexural strength	10^3 psi	4.4	20	4.0	20	4.5	7.5
Density	g/cm^3	1.65	5				
Coefficient of thermal expansion	$10^{-6}/°C$	4.0	5	5.1	5	3.3	4.1
Thermal conductivity	$cal\text{-}cm/sec \cdot cm^2 \cdot K$	0.350	15	0.330	15		
Specific resistance	10^{-4} ohm-cm	12.7	1	14.0	1		

Source
Duramic Products

Size
24 × 20 × 9 in. maximum

Manufacturing data
Not available

Table 5.2–3
GRADE D-555

Description
Molded, fine grained; high purity

Analysis
Average impurity content: Ni, 3 ppm; Ca, 3 ppm; Fe, 2 ppm; Na, 2 ppm; Si, 1 ppm; Al 1 ppm

		With grain		Against grain		Typical high-temperature properties	
Properties	Units	Average value	Standard deviation, %	Average value	Standard deviation, %	1,300°F	4,000°F
Young's modulus	10^6 psi	1.4	15	1.4	15	1.4	2.0
Tensile strength	10^3 psi	1.8	20	1.79	20	1.8	4.9
Compressive strength	10^3 psi	8.0	20	7.9	20	8.6	13.0
Flexural strength	10^3 psi	4.0	20	3.9	20	4.1	7.6
Density	g/cm^3	1.55	5				
Coefficient of thermal expansion	$10^{-6}/°C$	5.7	5	5.6	5	6.6	
Thermal conductivity	$cal\text{-}cm/sec \cdot cm^2 \cdot K$	0.39	15	0.38	15		
Specific resistance	10^{-4} ohm-cm	36	1	37	1		

Source
Duramic Products

Size
15 × 6 × 3 in. maximum

Manufacturing data
Not available

Table 5.2–4
GRADE ME 14

Description
Molded, fine grained; low porosity

Analysis
Average impurity content: ash, 0.1–0.5%, Fe, <0.05%; V, < 0.005%; B, >1ppm

Properties	Units	With grain: Average value	With grain: Standard deviation, %	Against grain: Average value	Against grain: Standard deviation, %	Typical high-temperature properties: 1,300°F	Typical high-temperature properties: 4,000°F
Young's modulus	10^6 psi	2.5	<10				
Tensile strength	10^3 psi						
Compressive strength	10^3 psi						
Flexural strength	10^3 psi	5–10	5–10				
Density	g/cm^3	1.65–1.8					
Coefficient of thermal expansion	10^{-6}/°C	2–10					
Thermal conductivity	cal-cm/sec·cm^2·K						
Specific resistance	10^{-4} ohm-cm	10–50	5–10				

Source
General Electric

Size
Cylinder 1/8–45 in.; block 6 in. maximum; rod 1/16–1/8 in.

Manufacturing data
Raw materials may be combinations of resin, metal inorganic salt, calcined petroleum coke, lamp black, coal tar pitch, petroleum pitch, natural and artificial graphite; graphitized over 2,500°C; 100–2,000 lb batch size

Table 5.2–5
GRADE H205

Description

Molded, fine grained; high strength; high hardness

Analysis

Average impurity content: ash, 0.25%; Ni, 0.04%; Ca, 0.04%; Fe, 0.02%; Na, 0.02%; Si, 0.015%; Al, 0.015%

Properties	Units	With grain: Average value	With grain: Standard deviation, %	Against grain: Average value	Against grain: Standard deviation, %	Typical high-temperature properties: 1,300°F	Typical high-temperature properties: 4,000°F
Young's modulus	10^6 psi	1.6	10	1.4	10		
Tensile strength	10^3 psi	3.2	10	2.8	10		
Compressive strength	10^3 psi	8.5	10	10.0	10		
Flexural strength	10^3 psi	4.5	10	4.0	10		
Density	g/cm^3	1.75	2				
Coefficient of thermal expansion	$10^{-6}/°C$	3.9	5	5.0	5		
Thermal conductivity	$cal\text{-}cm/sec \cdot cm^2 \cdot K$	0.35	10	0.33	10		
Specific resistance	10^{-4} ohm-cm	12	10	14	10		
Hardness (Brinell) – 136 kg load-10 mm ball				15.0	10		
Permeability (D'Arcy)		0.2	10	0.004	10		

Source

Great Lakes Carbon

Size

Cylinder 10–22 in.; block 9 × 20 × 24 in.

Manufacturing data

Graphitized over 2,500°C; Acheson electric furnace; 1–20 ton batch size

Table 5.2–6
GRADE H205-85

Description
Molded, fine grained; high strength; high density; high hardness

Analysis
Average impurity content: ash, 0.25%; Ni, 0.04%; Ca, 0.04%; Fe, 0.02%; Na, 0.02%; Si, 0.015%; Al, 0.015%

Properties	Units	With grain		Against grain		Typical high-temperature properties	
		Average value	Standard deviation, %	Average value	Standard deviation, %	1,300° F	4,000° F
Young's modulus	10^6 psi	1.7	10	1.5	10		
Tensile strength	10^3 psi	3.5	10	3.0	10		
Compressive strength	10^3 psi	10.0	10	12.0	10		
Flexural strength	10^3 psi	4.7	10	4.3	10		
Density	g/cm^3	1.81	2				
Coefficient of thermal expansion	10^{-6}/°C	41	5	50	5		
Thermal conductivity	cal-cm/sec·cm^2·K	0.37	10	0.35	10		
Specific resistance	10^{-4} ohm-cm						
Hardness (Brinell) – 136 kg load-10 mm ball				18.0	10		
Permeability (D'Arcy)		0.006	10	0.001	10		

Source
Great Lakes Carbon

Size
Cylinder 10–22 in.; block 9 × 20 × 29 in.

Manufacturing data
Graphitized over 2,500°C; Acheson electric furnace; 1–20 ton batch size

Table 5.2–7
GRADE 2D9B

Description
Molded, fine grained

Analysis
Average impurity content: ash, 0.1–0.5%

Properties	Units	With grain: Average value	With grain: Standard deviation, %	Against grain: Average value	Against grain: Standard deviation, %	Typical high-temperature properties: 1,300°F	Typical high-temperature properties: 4,000°F
Young's modulus	10^6 psi						
Tensile strength	10^3 psi						
Compressive strength	10^3 psi	5–10					
Flexural strength	10^3 psi	5–10					
Density	g/cm^3	1.5–1.65					
Coefficient of thermal expansion	10^{-6}/°C						
Thermal conductivity	cal-cm/sec·cm^2·K						
Specific resistance	10^{-4} ohm-cm	10–50					
Hardness (scleroscope)		65					

Source
Ohio Carbon

Size
Cylinder 1/8–45 in.; block 1–6 in.; pipe <1/2–10 in.

Manufacturing data
Lamp black, calcined petroleum coke, and coal tar pitch; graphitized over 2,500°C in Acheson electric furnace; ground and machined; 100–2,000 lb batch size

Table 5.2–8
GRADE W97

Description
Molded, fine grained; high strength; high hardness

Analysis
Average impurity content: ash, >0.5%

Properties	Units	With grain: Average value	With grain: Standard deviation, %	Against grain: Average value	Against grain: Standard deviation, %	Typical high-temperature properties: 1,300° F	Typical high-temperature properties: 4,000° F
Young's modulus	10^6 psi						
Tensile strength	10^3 psi						
Compressive strength	10^3 psi	10–50					
Flexural strength	10^3 psi	5–10					
Density	g/cm^3	1.65–1.8					
Coefficient of thermal expansion	10^{-6}/°C						
Thermal conductivity	cal-cm/sec·cm^2·K						
Specific resistance	10^{-4} ohm-cm	10–50					
Hardness (scleroscope)		80					

Source
Ohio Carbon

Size
Cylinder 1/8–45 in.; block 1–6 in.; pipe <1/2–10 in.

Manufacturing data
Calcined petroleum coke, natural graphite, and coal tar pitch; processed below 2,500°C; ground and machined; 100–2,000 lb batch size

Table 5.2–9
GRADE AXF

Description

Molded, fine grained; high strength; high electrical resistance; high reproducibility; low porosity; chemical resistance; abrasion resistant; small sizes; isotropic

Analysis

Average impurity content: ash, 0.1%

Properties	Units	With grain: Average value	With grain: Standard deviation, %	Against grain: Average value	Against grain: Standard deviation, %	Typical high-temperature properties: 1,300°F	Typical high-temperature properties: 4,000°F
Young's modulus	10^6 psi	1.8	>20	1.6	>20		1.5
Tensile strength	10^3 psi	9.4					10.2
Compressive strength	10^3 psi	20.0	5–10				
Flexural strength	10^3 psi	10.0	10–20	10.0	10–20		
Density	g/cm^3	1.80–1.88					
Coefficient of thermal expansion	$10^{-6}/°C$	9.0					
Thermal conductivity	$cal\text{-}cm/sec \cdot cm^2 \cdot K$	0.1–0.5					
Specific resistance	10^{-4} ohm-cm	14–16					
Hardness (scleroscope)		78					

Source

Poco Graphite, Inc.

Size

Rod 1/8–5/8 in.; cylinder 8 in. maximum; block 4 × 8 × 18 in. maximum

Manufacturing data

Not available

Table 5.2–10
GRADE AXM

Description

Molded, fine grained; high strength; high electrical resistance; high reproducibility; small sizes

Analysis

Average impurity content: ash, 0.1%

		With grain		Against grain		Typical high-temperature properties	
Properties	Units	Average value	Standard deviation, %	Average value	Standard deviation, %	1,300° F	4,000° F
Young's modulus	10^6 psi	1.7		1.6			1.40
Tensile strength	10^3 psi	9.4					9.5
Compressive strength	10^3 psi	16.0					25.0
Flexural strength	10^3 psi	8.0					
Density	g/cm^3	1.70–1.79					
Coefficient of thermal expansion	10^{-6}/°C	8.0					
Thermal conductivity	cal-cm/sec·cm^2·K	0.1–0.5		0.1–0.5			
Specific resistance	10^{-4} ohm-cm	16–22					
Hardness (scleroscope)		70 typical					

Source

Poco Graphite, Inc.

Size

Rod 1/8–5/8 in.; cylinder 8 in. maximum; block 4 × 8 × 18 in. maximum

Manufacturing data

Not available

Table 5.2–11
GRADE AXZ

Description
Molded, fine grained; high electrical resistance; high reproducibility; small sizes

Analysis
Average impurity content: ash, <0.1%

		With grain		Against grain		Typical high-temperature properties	
Properties	Units	Average value	Standard deviation, %	Average value	Standard deviation, %	1,300°F	4,000°F
Young's modulus	10^6 psi	0.75		0.85		0.9	
Tensile strength	10^3 psi						
Compressive strength	10^3 psi	9.0				9.5	
Flexural strength	10^3 psi	6.0					
Density	g/cm^3	1.50–1.59					
Coefficient of thermal expansion	10^{-6}/°C	7.0					
Thermal conductivity	cal-cm/sec·cm^2·K	0.1–0.5		0.1–0.5			
Specific resistance	10^{-4} ohm-cm	28–34					
Hardness (scleroscope)		57 typical					

Source
Poco Graphite, Inc.

Size
Rod 1/8–5/8 in.; cylinder 8 in. maximum; block 4 × 8 × 18 in. maximum

Manufacturing data
Not available

Table 5.2–12
GRADE G-88-C

Description
Molded, fine grained; high strength; low coefficient of thermal expansion; good electrical and thermal conductivity; high reproducibility; low friction

Analysis
Average impurity content: ash, 0.5%

Properties	Units	With grain		Against grain		Typical high-temperature properties	
		Average value	Standard deviation, %	Average value	Standard deviation, %	1,300° F	4,000° F
Young's modulus	10^6 psi						
Tensile strength	10^3 psi						
Compressive strength	10^3 psi						
Flexural strength	10^3 psi	3	15				
Density	g/cm^3	1.55	1.5				
Coefficient of thermal expansion	$10^{-6}/°C$						
Thermal conductivity	$cal\text{-}cm/sec \cdot cm^2 \cdot K$						
Specific resistance	10^{-4} ohm-cm	75	75				
Hardness (scleroscope)		50	6				

Source
Pure Carbon

Size
Cylinder 1/8–6 in.; block 1–6 in.; rod 0.01–1/8 in.; pipe <1/2–6 in.

Manufacturing data
Lamp black, graphite, and pitch; graphitized over 2,500°C; Acheson electric furnace; finishing operations as required; 100–2,000 lb batch size

Table 5.2–13
GRADE L-55

Description
Molded, fine grained; high strength; low coefficient thermal expansion; good electrical and thermal conductivity; high reproducibility; low friction

Analysis
Average impurity content: ash, 0.3%

Properties	Units	With grain		Against grain		Typical high-temperature properties	
		Average value	Standard deviation, %	Average value	Standard deviation, %	1,300°F	4,000°F
Young's modulus	10^6 psi	<1					
Tensile strength	10^3 psi						
Compressive strength	10^3 psi						
Flexural strength	10^3 psi	3	15				
Density	g/cm^3	1.60	1.5				
Coefficient of thermal expansion	$10^{-6}/°C$						
Thermal conductivity	$cal\text{-}cm/sec \cdot cm^2 \cdot K$						
Specific resistance	10^{-4} ohm-cm	75	7.5				
Hardness (scleroscope)		52	<5				

Source
Pure Carbon

Size
Cylinder 1/8–6 in.; block 1–6 in.; rod 0.01–1/8 in.; pipe <1/2–6 in.

Manufacturing data
Lamp black, graphite, and pitch; graphitized over 2,500°C; Acheson electric furnace; impregnated in secondary processing; finishing operations as required; 100–2,000 lb batch size

Table 5.2–14
GRADE L-56

Description

Molded, fine grained; high strength; low coefficient of thermal expansion; good electrical and thermal conductivity; high reproducibility; low friction; high temperature oxidation resistance; abrasion resistant

Analysis

Average impurity content: ash, <0.1%; Fe, <0.05%; Si, 10 ppm

		With grain		Against grain		Typical high-temperature properties	
Properties	Units	Average value	Standard deviation, %	Average value	Standard deviation, %	1,300°F	4,000°F
Young's modulus	10^6 psi	1.5					
Tensile strength	10^3 psi	4					
Compressive strength	10^3 psi	30					
Flexural strength	10^3 psi	7.5	15				
Density	g/cm^3	1.60	1.5				
Coefficient of thermal expansion	10^{-6}/°C	6					
Thermal conductivity	cal-cm/sec·cm^2·K						
Specific resistance	10^{-4} ohm-cm	30	15				
Hardness (scleroscope)		64	<10				
Admittance	H^2/sec, He	10^{-2}					
Abrasion resistance	hr/mil	4					

Source

Pure Carbon

Size

Cylinder 1/8–12 in.; block 1–6 in.; rod 0.01–1/8 in.; pipe <1/2–10 in.

Manufacturing data

Lamp black, graphite, and pitch; graphitized over 2,500°C; Acheson electric furnace; finishing operations as required; 100–2,000 lb batch size

Table 5.2–15
GRADE P-9

Description

Molded, fine grained; carbon-graphite; high strength; high reproducibility; good electrical and thermal conductivity; low porosity; low friction; low coefficient of thermal expansion

		With grain		Against grain		Typical high-temperature properties	
Properties	Units	Average value	Standard deviation, %	Average value	Standard deviation, %	1,300° F	4,000° F
Young's modulus	10^6 psi	1.5					
Tensile strength	10^3 psi	7.5					
Compressive strength	10^3 psi	30					
Flexural strength	10^3 psi	7.5	15				
Density	g/cm^3	1.75	15				
Coefficient of thermal expansion	10^{-6}/°C	6					
Thermal conductivity	cal-cm/sec·cm^2·K						
Specific resistance	10^{-4} ohm-cm	75					
Hardness (scleroscope)		76	10				
Abrasion resistance	hr/mil	4	50				

Source

Pure Carbon

Size

Cylinder 1/8–19 in., block 1–6 in.; rod 0.01–1/8 in.; pipe <1/2–10 in.

Manufacturing data

Graphite, pitch; not graphitized; no secondary processing; finishing operations as required; 100–2,000 lb batch size

Table 5.2–16
GRADE P-3W

Description

Molded, fine grained; high strength; low coefficient of thermal expansion; good electrical and thermal conductivity; high reproducibility; low friction; low porosity; high temperature oxidation resistance; good mechanical properties

Properties	Units	With grain		Against grain		Typical high-temperature properties	
		Average value	Standard deviation, %	Average value	Standard deviation, %	1,300° F	4,000° F
Young's modulus	10^6 psi	<1					
Tensile strength	10^3 psi	3					
Compressive strength	10^3 psi	30					
Flexural strength	10^3 psi	7.5	15				
Density	g/cm^3	1.60	>2				
Coefficient of thermal expansion	10^{-6}/°C	6					
Thermal conductivity	cal-cm/sec·cm^2·K						
Specific resistance	10^{-4} ohm-cm	30	15				
Hardness (scleroscope)		46	10				
Abrasion resistance	hr/mil	8	50				

Source

Pure Carbon

Size

Cylinder 1/8–12 in.; block 1–6 in.; rod 0.01–1/8 in.; pipe <1/2–10 in.

Manufacturing data

Graphite and pitch; graphitized over 2,500°C; Acheson electric furnace; finishing operations as required 100–2,000 lb batch size

Table 5.2–17
GRADE P-03

Description

Molded, fine grained; high strength; low coefficient of thermal expansion; good electrical and thermal conductivity; high reproducibility; low friction; low porosity; high temperature oxidation resistance; good mechanical properties

Properties	Units	With grain		Against grain		Typical high-temperature properties	
		Average value	Standard deviation, %	Average value	Standard deviation, %	1,300°F	4,000°F
Young's modulus	10^6 psi	1.5					
Tensile strength	10^3 psi	5					
Compressive strength	10^3 psi	30					
Flexural strength	10^3 psi	10	15				
Density	g/cm^3	1.8	2				
Coefficient of thermal expansion	$10^{-6}/°C$	6					
Thermal conductivity	$cal\text{-}cm/sec \cdot cm^2 \cdot K$						
Specific resistance	10^{-4} ohm-cm	30					
Hardness (scleroscope)		78	10				
Abrasion resistance	hr/mil	30					
Oxidation rate in air	wt %/hr						

Source

Pure Carbon

Size

Cylinder 1/8–8 in.; block 1–6 in.; rod 0.01–1/8 in.; pipe <1/2–8 in.

Manufacturing data

Graphite and pitch; graphitized over 2,500°C; Acheson electric furnace; finishing operations as required; 100–2,000 lb batch size

Table 5.2–18
GRADE 9RL

Description

Molded, fine grained; high purity; high reproducibility; high temperature oxidation resistance

Analysis

Average impurity content: Fe, <10 ppm; V, 1 ppm; B, <1 ppm; Si, 10 ppm; Ca, <10 ppm; Al, 5 ppm; Mg, <1.0 ppm

		With grain		Against grain		Typical high-temperature properties	
Properties	Units	Average value	Standard deviation, %	Average value	Standard deviation, %	1,300°F	4,000°F
Young's modulus	10^6 psi	0.87		1.18			
Tensile strength	10^3 psi	1.8		1.6		2.2	3.8
Compressive strength	10^3 psi	6.8		7.2		7.0	9.8
Flexural strength	10^3 psi	3.6		3.1		4.2	7.0
Density	g/cm^3	1.68					
Coefficient of thermal expansion	10^{-6}/°C	3.3		4.4			
Thermal conductivity	cal-cm/sec·cm^2·K					0.15	
Specific resistance	10^{-4} ohm-cm	9.6				9.1	10.7
Hardness (scleroscope)		37					

Source

Speer Carbon

Size

Cylinder 10 in. maximum

Manufacturing data

Calcined petroleum coke and coal tar pitch; graphitized over 2,500°C; machined; 1–20 ton batch size

Table 5.2–19
GRADE 39RL

Description
Molded, fine grained; high strength; high purity; high reproducibility; high temperature oxidation resistance

Analysis
Average impurity content: ash, 50 ppm; Fe, <10 ppm; V, 1 ppm; B, <1 ppm; Si, 10 ppm; Ca, <10 ppm; Al, 5 ppm; Mg, <10 ppm

		With grain		Against grain		Typical high-temperature properties	
Properties	Units	Average value	Standard deviation, %	Average value	Standard deviation, %	1,300°F	4,000°F
Young's modulus	10^6 psi	1.5					
Tensile strength	10^3 psi	1.8		1.6		2.2	3.8
Compressive strength	10^3 psi	6.4		6.8		7.0	9.8
Flexural strength	10^3 psi	3.7		3.2		4.2	7.0
Density	g/cm^3	1.65					
Coefficient of thermal expansion	$10^{-6}/°C$	3.3		4.4			
Thermal conductivity	$cal\text{-}cm/sec \cdot cm^2 \cdot K$					0.12	
Specific resistance	10^{-4} ohm-cm	8.6				7.1	11.7

Source
Speer Carbon

Size
Block 12 × 12 × 2½ in.

Manufacturing data
Calcined petroleum coke and coal tar pitch; graphitized over 2,500°C; machined; 1–20 ton batch size

Table 5.2–20
GRADE 350

Description
Molded, fine grained; high strength; abrasion resistant; high hardness

Analysis
Average impurity content: ash, 3.5%

Properties	Units	With grain: Average value	With grain: Standard deviation, %	Against grain: Average value	Against grain: Standard deviation, %	Typical high-temperature properties: 1,300° F	Typical high-temperature properties: 4,000° F
Young's modulus	10^6 psi	2.0					
Tensile strength	10^3 psi	1.6		1.5			
Compressive strength	10^3 psi	10.0		11.0			
Flexural strength	10^3 psi	3.5		2.9			
Density	g/cm^3	1.63					
Coefficient of thermal expansion	$10^{-6}/°C$	4.3		5.4			
Thermal conductivity	$cal\text{-}cm/sec \cdot cm^2 \cdot K$	55.9					
Specific resistance	10^{-4} ohm-cm						
Hardness (Rockwell)		78					

Source
Speer Carbon

Size
Block 12 × 6 × 2 in.

Manufacturing data
Calcined petroleum coke and coal tar pitch; processed below 2,500°C; machined; 100–2,000 lb batch size

Table 5.2–21
GRADE 3499

Description
Molded, fine grained; high strength; high reproducibility; high production

Analysis
Average impurity content: ash, 0.03%

		With grain		Against grain		Typical high-temperature properties	
Properties	Units	Average value	Standard deviation, %	Average value	Standard deviation, %	1,300°F	4,000°F
Young's modulus	10^6 psi	1.5					
Tensile strength	10^3 psi	1.8		1.6		2.2	3.8
Compressive strength	10^3 psi	6.4		6.8		7.0	9.8
Flexural strength	10^3 psi	3.7		3.2		4.2	7.0
Density	g/cm^3	1.65					
Coefficient of thermal expansion	$10^{-6}/°C$	3.3		4.4			
Thermal conductivity	$cal\text{-}cm/sec \cdot cm^2 \cdot K$					0.12	
Specific resistance	10^{-4} ohm-cm	8.6				7.1	11.7
Hardness (scleroscope)		36					

Source
Speer Carbon

Size
Block 12 × 12 × 5 in.

Manufacturing data
Calcined petroleum coke and coal tar pitch; graphitized over 2,500°C; machined; 1–20 ton batch size

Table 5.2–22
GRADE 3499S

Description
Molded, fine grained; high strength; high reproducibility

Analysis
Average impurity content: ash, 0.03%

Properties	Units	With grain		Against grain		Typical high-temperature properties	
		Average value	Standard deviation, %	Average value	Standard deviation, %	1,300° F	4,000° F
Young's modulus	10^6 psi	0.87		1.18			
Tensile strength	10^3 psi	1.8		1.6		2.2	3.8
Compressive strength	10^3 psi	6.8		7.2		7.0	9.8
Flexural strength	10^3 psi	3.6		3.1		4.2	7.0
Density	g/cm^3	1.68					
Coefficient of thermal expansion	$10^{-6}/°C$	3.3		4.4			
Thermal conductivity	$cal\text{-}cm/sec \cdot cm^2 \cdot K$					0.15	
Specific resistance	10^{-4} ohm-cm	9.6				9.1	10.7
Hardness (scleroscope)		37					
Permeability	cm^2/sec^{-1}	6.0×10^{-1}		5.8×10^{-1}			

Source
Speer Carbon

Size
Cylinder 2–5/8–8 in. (also in cylinder 13 in. maximum)

Manufacturing data
Calcined petroleum coke and coal tar pitch; graphitized over 2,500°C; 1–20 ton batch size

Table 5.2–23
GRADE 4007

Description
Molded, fine grained; high reproducibility

Analysis
Average impurity content: ash, 0.03%

Properties	Units	With grain: Average value	With grain: Standard deviation, %	Against grain: Average value	Against grain: Standard deviation, %	Typical high-temperature properties: 1,300°F	Typical high-temperature properties: 4,000°F
Young's modulus	10^6 psi	1.3					
Tensile strength	10^3 psi	1.5					
Compressive strength	10^3 psi	6.3		6.5			
Flexural strength	10^3 psi	3.2		2.1			
Density	g/cm^3	1.68					
Coefficient of thermal expansion	10^{-6}/°C	2.7		4.1			
Thermal conductivity	cal-cm/sec·cm^2·K						
Specific resistance	10^{-4} ohm-cm	8.9					
Hardness (scleroscope)		40					

Source
Speer Carbon

Size
Block 10 × 4 × 2½ in.

Manufacturing data
Calcined petroleum coke and coal tar pitch; graphitized over 2,500°C; machined; 1–20 ton batch size

Table 5.2–24
GRADE 9135; 9139

Description
Molded, fine grained; high strength; high reproducibility; abrasion resistant

Analysis
Average impurity content: ash, 0.04%

Properties	Units	With grain		Against grain		Typical high-temperature properties	
		Average value	Standard deviation, %	Average value	Standard deviation, %	1,300°F	4,000°F
Young's modulus	10^6 psi	1.4		1.2			
Tensile strength	10^3 psi	2.4		2.1		4.0	5.7
Compressive strength	10^3 psi	9.2		9.6		10.5	14.8
Flexural strength	10^3 psi	4.4		4.1		6.0	11.3
Density	g/cm^3	1.79					
Coefficient of thermal expansion	$10^{-6}/°C$	3.0		4.7			
Thermal conductivity	$cal\text{-}cm/sec \cdot cm^2 \cdot K$					0.2	
Specific resistance	10^{-4} ohm-cm	11.4				8.4	7.6
Hardness (scleroscope)		48					

Source
Speer Carbon

Size
Finished shapes with <3-in. wall thickness

Manufacturing data
Calcined petroleum coke and coal tar pitch; graphitized over 2,500°C; 100–2,000 lb batch size

Table 5.2–25
GRADE 9134

Description
Molded, fine grained; high strength; high reproducibility; abrasion resistant

Analysis
Average impurity content: ash, 0.04%

Properties	Units	With grain		Against grain		Typical high-temperature properties	
		Average value	Standard deviation, %	Average value	Standard deviation, %	1,300°F	4,000°F
Young's modulus	10^6 psi			1.1			
Tensile strength	10^3 psi	2.2		2.1		3.2	4.8
Compressive strength	10^3 psi	8.1		8.4		9.7	13.2
Flexural strength	10^3 psi	4.2		3.6		5.8	9.3
Density	g/cm^3	1.70					
Coefficient of thermal expansion	$10^{-6}/°C$	3.2		4.5			
Thermal conductivity	$cal\text{-}cm/sec \cdot cm^2 \cdot K$					0.2	
Specific resistance	10^{-4} ohm-cm	9.1				7.6	10.2
Hardness (scleroscope)		44					

Source
Speer Carbon

Size
Machined shapes to 13-in. diameter maximum with <3-in. wall thickness

Manufacturing data
Calcined petroleum coke and coal tar pitch; graphitized over 2,500°C; 100–2,000 lb batch size

Table 5.2–26
GRADE 8882

Description
Molded, fine grained; high strength; high reproducibility

Analysis
Average impurity content: ash, 0.03%

Properties	Units	With grain: Average value	With grain: Standard deviation, %	Against grain: Average value	Against grain: Standard deviation, %	Typical high-temperature properties: 1,300° F	Typical high-temperature properties: 4,000° F
Young's modulus	10^6 psi			1.3			
Tensile strength	10^3 psi	2.0		1.8		3.2	5.0
Compressive strength	10^3 psi	8.2		9.0		10.0	15.0
Flexural strength	10^3 psi	4.0		3.8		6.2	10.00
Density	g/cm^3	1.73					
Coefficient of thermal expansion	$10^{-6}/°C$	3.4		4.5			
Thermal conductivity	$cal\text{-}cm/sec \cdot cm^2 \cdot K$					0.25	
Specific resistance	10^{-4} ohm-cm	9.9				7.6	13.5
Hardness (scleroscope)		45					
Permeability	cm^2/sec^{-1}	0.42		0.48			

Source
Speer Carbon

Size
Cylinder 10–13 in.

Manufacturing data
Calcined petroleum coke and coal tar pitch; graphitized over 2,500°C; machined; 1–20 ton batch size

Table 5.2–27
GRADE 9420

Description
Molded, fine grained; high strength; high electrical resistance; high reproducibility; abrasion resistant; high hardness

Analysis
Average impurity content: ash, 0.2%

		With grain		Against grain		Typical high-temperature properties	
Properties	Units	Average value	Standard deviation, %	Average value	Standard deviation, %	1,300°F	4,000°F
Young's modulus	10^6 psi	1.2					
Tensile strength	10^3 psi	2.5					
Compressive strength	10^3 psi	10.0		12.0			
Flexural strength	10^3 psi	5.0		4.6			
Density	g/cm^3	1.67					
Coefficient of thermal expansion	10^{-6}/°C	5.6		5.5			
Thermal conductivity	cal-cm/sec·cm^2·k						
Specific resistance	10^{-4} ohm-cm	30.5					
Hardness (scleroscope)		71					
Hardness (Rockwell) (M)		77					

Source
Speer Carbon

Size
Block 10 × 4 × 3 in.

Manufacturing data
Lamp black, graphitized over 2,500°C; machined; 1–20 ton batch size

Table 5.2–28
GRADE 9429

Description
Molded, fine grained; high strength; high electrical resistance; high reproducibility

Analysis
Average impurity content: ash, 0.12%

Properties	Units	With grain: Average value	With grain: Standard deviation, %	Against grain: Average value	Against grain: Standard deviation, %	Typical high-temperature properties: 1,300°F	Typical high-temperature properties: 4,000°F
Young's modulus	10^6 psi	1.1					
Tensile strength	10^3 psi	2.3					
Compressive strength	10^3 psi	10.2		10.5			
Flexural strength	10^3 psi	4.5		4.2			
Density	g/cm^3	1.58					
Coefficient of thermal expansion	$10^{-6}/°C$	5.7		5.6			
Thermal conductivity	$cal\text{-}cm/sec \cdot cm^2 \cdot K$						
Specific resistance	10^{-4} ohm-cm	35.6					
Hardness (scleroscope) (A)		60					
Hardness (Rockwell) (L)		85					

Source
Speer Carbon

Size
Block 12 × 6 × 2¼ in.

Manufacturing data
Lamp black; graphitized over 2,500°C; 1–20 ton batch size

Table 5.2–29
GRADE E-3

Description
Molded, fine grained; high reproducibility

Analysis
Average impurity content: ash, 0.05%

Properties	Units	With grain		Against grain		Typical high-temperature .properties	
		Average value	Standard deviation, %	Average value	Standard deviation, %	1,300° F	4,000° F
Young's modulus	10^6 psi						
Tensile strength	10^3 psi						
Compressive strength	10^3 psi						
Flexural strength	10^3 psi	3.5	1				
Density	g/cm^3	1.70	0.02				
Coefficient of thermal expansion	$10^{-6}/°C$						
Thermal conductivity	$cal\text{-}cm/sec \cdot cm^2 \cdot K$						
Specific resistance	10^{-4} ohm-cm	8.4	1				
Hardness (scleroscope)		35.5	5				
Hardness (Rockwell) (R)		73					

Source
Speer Carbon

Size
Block 12 X 12 X 2½ in. (fabricated brushes)

Manufacturing data
Calcined petroleum coke; graphitized over 2,500°C; machined; 1–20 ton batch size

Table 5.2–30
GRADE H

Description
Molded, fine grained; high strength; high electrical resistance; high reproducibility

Analysis
Average impurity content: ash, 0.25%

Properties	Units	With grain: Average value	With grain: Standard deviation, %	Against grain: Average value	Against grain: Standard deviation, %	Typical high-temperature properties: 1,300°F	Typical high-temperature properties: 4,000°F
Young's modulus	10^6 psi	1.7					
Tensile strength	10^3 psi	2.1					
Compressive strength	10^3 psi	10.5		11.0			
Flexural strength	10^3 psi	4.6		4.0			
Density	g/cm^3	1.66					
Coefficient of thermal expansion	$10^{-6}/°$	3.7		4.9			
Thermal conductivity	$cal\text{-}cm/sec \cdot cm^2 \cdot K$						
Specific resistance	10^{-4} ohm-cm	25.4					
Hardness (scleroscope)		52					
Hardness (Rockwell) (M)		50					

Source
Speer Carbon

Size
Block 12 × 12 × 2½ in.

Manufacturing data
Calcined petroleum coke and coal tar pitch; processed under 2,500°C; machined; 1–20 ton batch size

Table 5.2–31
GRADE KK-10

Description
Molded, fine grained; good electrical conductor; high reproducibility

Analysis
Average impurity content: ash, 0.03%

Properties	Units	With grain		Against grain		Typical high-temperature properties	
		Average value	Standard deviation, %	Average value	Standard deviation, %	1,300° F	4,000° F
Young's modulus	10^6 psi	2.0					
Tensile strength	10^3 psi	2.0					
Compressive strength	10^3 psi	9.6		8.8			
Flexural strength	10^3 psi	4.6					
Density	g/cm^3	1.79					
Coefficient of thermal expansion	10^{-6}/°C	3.4		4.5			
Thermal conductivity	cal-cm/sec·cm^2·K						
Specific resistance	10^{-4} ohm-cm	8.6					
Hardness (scleroscope)		45					

Source
Speer Carbon

Size
Block 12 X 12 X 2½ in.

Manufacturing data
Calcined petroleum coke; graphitized over 2,500°C; machined; 1–20 ton batch size

Table 5.2–32
GRADE L31

Description
Molded, fine grained; high reproducibility

Analysis
Average impurity content: ash, 0.07%

Properties	Units	With grain: Average value	With grain: Standard deviation, %	Against grain: Average value	Against grain: Standard deviation, %	Typical high-temperature properties: 1,300° F	Typical high-temperature properties: 4,000° F
Young's modulus	10^6 psi	1.3		1.1			
Tensile strength	10^3 psi						
Compressive strength	10^3 psi	14		15			
Flexural strength	10^3 psi	6		5			
Density	g/cm^3	1.65		1.65			
Coefficient of thermal expansion	$10^{-6}/°C$	6.1					
Thermal conductivity	$cal\text{-}cm/sec \cdot cm^2 \cdot K$						
Specific resistance	10^{-4} ohm-cm	32					

Source
Stackpole Carbon

Size
Block 12 × 12 × 12 in. maximum

Manufacturing data
Lamp black and coal tar pitch; graphitized over 2,500°C; machining and grinding as required; 100–2,000 lb batch size

Table 5.2–33
GRADE 331

Description
Molded, fine grained; high reproducibility

Analysis
Average impurity content: ash, 0.08%

Properties	Units	With grain: Average value	With grain: Standard deviation, %	Against grain: Average value	Against grain: Standard deviation, %	Typical high-temperature properties: 1,300°F	Typical high-temperature properties: 4,000°F
Young's modulus	10^6 psi	1.7		1.2			
Tensile strength	10^3 psi						
Compressive strength	10^3 psi	16		12			
Flexural strength	10^3 psi	5.7		5.4			
Density	g/cm^3 psi	1.74		1.74			
Coefficient of thermal expansion	10^{-6}/°C	3.2		4.9			
Thermal conductivity	cal-cm/sec·cm^2·K						
Specific resistance	10^{-4} ohm-cm	2.0					

Source
Stackpole Carbon

Size
Block 12 × 12 × 3 in.

Manufacturing data
Calcined petroleum coke and coal tar pitch; graphitized over 2,500°C; finishing as required; 100–2,000 lb batch size

Table 5.2–34
GRADE 2000

Description
Molded, fine grained; high reproducibility

Analysis
Average impurity content: ash, 0.15%

Properties	Units	With grain		Against grain		Typical high-temperature properties	
		Average value	Standard deviation, %	Average value	Standard deviation, %	1,300°F	4,000°F
Young's modulus	10^6 psi	2.1		1.2			
Tensile strength	10^3 psi						
Compressive strength	10^3 psi	19		17.5			
Flexural strength	10^3 psi	7.3		5.8			
Density	g/cm^3	1.82		1.82			
Coefficient of thermal expansion	10^{-6}/°C	4.1		7.2			
Thermal conductivity	cal-cm/sec·cm^2·K						
Specific resistance	10^{-4} ohm-cm	2.1					

Source
Stackpole Carbon

Size
Block 12 × 12 × 1¼ in.

Manufacturing data
Calcined petroleum coke and coal tar pitch; graphitized over 2,500°C; machining and grinding as required; 100–2,000 lb batch size

Table 5.2–35
GRADE 2020

Description
Molded, fine grained; high reproducibility; large sizes

Analysis
Average impurity content: ash, 0.15%

		With grain		Against grain		Typical high-temperature properties	
Properties	Units	Average value	Standard deviation, %	Average value	Standard deviation, %	1,300°F	4,000°F
Young's modulus	10^6 psi	1.3		1.4			
Tensile strength	10^3 psi						
Compressive strength	10^3 psi	12.5		12.0			
Flexural strength	10^3 psi	4.2		4.8			
Density	g/cm^3	1.72					
Coefficient of thermal expansion	$10^{-6}/°C$	4.3		3.4			
Thermal conductivity	$cal\text{-}cm/sec \cdot cm^2 \cdot K$						
Specific resistance	10^{-4} ohm-cm	21					

Source
Stackpole Carbon

Size
Block 13 × 13 × 72 in. maximum

Manufacturing data
Calcined petroleum coke and coal tar pitch; graphitized over 2,500°C; machining and grinding as required; 100–2,000 lb batch size

Table 5.2–36
GRADE ATJ

Description

Molded, fine grained; high strength; high reproducibility

Analysis

Average impurity content: ash, 0.15%

Properties	Units	With grain		Against grain		Typical high-temperature properties	
		Average value	Standard deviation, %	Average value	Standard deviation, %	1,300° F	4,000° F
Young's modulus	10^6 psi	1.4	11	1.0	8		
Tensile strength	10^3 psi	3.4	11	2.9	10		
Compressive strength	10^3 psi	8.3	12	8.6	13		
Flexural strength	10^3 psi	4.0	19	8.5	13		
Density	g/cm^3	1.7	2				
Coefficient of thermal expansion	$10^{-6}/°C$	2.2	10	3.4	6		
Thermal conductivity	$cal\text{-}cm/sec \cdot cm^2 \cdot K$	0.28		0.21			
Specific resistance	10^{-4} ohm-cm	11.0	15	14.5	10		

Source

Union Carbide

Size

Cylinder 13–17 in., block 9 × 20 × 24 in.

Manufacturing data

Calcined petroleum coke and coal tar pitch; graphitized over 2,500°C; Acheson electric furnace; impregnated in secondary processing; over 20-ton batch size

Table 5.2–37
GRADE ATJS

Description

Molded, fine grained; high strength; high reproducibility; high density; low porosity; high temperature oxidation resistance

Properties	Units	With grain: Average value	With grain: Standard deviation, %	Against grain: Average value	Against grain: Standard deviation, %	Typical high-temperature properties: 1,300°F	Typical high-temperature properties: 4,000°F
Young's modulus	10^6 psi	1.8		1.1			
Tensile strength	10^3 psi	4.4		3.3			
Compressive strength	10^3 psi	11.7		12.8			
Flexural strength	10^3 psi	5.7		4.3			
Density	g/cm^3	1.83					
Coefficient of thermal expansion	10^{-6}/°C	1.8		3.2			
Thermal conductivity	cal-cm/sec·cm^2·K						
Specific resistance	10^{-4} ohm-cm	8.0		11.5			

Source

Union Carbide

Size

Cylinder 13–17 in., block 9 × 20 × 24 in.

Manufacturing data

Calcined petroleum coke and coal tar pitch; graphitized over 2,500°C; Acheson electric furnace; impregnated in secondary processing; machined; 1–20 ton batch size

Table 5.2–38
GRADE CCT

Description

Molded, fine grained; high strength; high purity

Analysis

Average impurity content: ash, 15 ppm

Properties	Units	With grain: Average value	With grain: Standard deviation, %	Against grain: Average value	Against grain: Standard deviation, %	Typical high-temperature properties: 1,300°F	Typical high-temperature properties: 4,000°F
Young's modulus	10^6 psi	1.4		1.2			
Tensile strength	10^3 psi	3.5		2.9			
Compressive strength	10^3 psi	8.2		8.5			
Flexural strength	10^3 psi	4.0		3.6			
Density	g/cm^3	1.73					
Coefficient of thermal expansion	10^{-6}/°C	2.2		3.4			
Thermal conductivity	cal-cm/sec·cm^2·K	0.28		0.21			
Specific resistance	10^{-4} ohm-cm	11.0		1.4			

Source

Union Carbide

Size

6½ in. diameter × 24 in. long maximum

Manufacturing data

Calcined petroleum coke and coal tar pitch; graphitized over 2,500°C; electric resistance furnace; impregnated in secondary processing; 100–2,000 lb batch size

Table 5.2–39
GRADE CGW

Description
Molded, fine grained; high strength; high reproducibility; high density; low porosity

Analysis
Average impurity content: ash, 0.15%

Properties	Units	With grain		Against grain		Typical high-temperature properties	
		Average value	Standard deviation, %	Average value	Standard deviation, %	1,300°F	4,000°F
Young's modulus	10^6 psi	1.6		1.4			
Tensile strength	10^3 psi						
Compressive strength	10^3 psi	11		10.5			
Flexural strength	10^3 psi	4.4		4.1			
Density	g/cm^3	1.80					
Coefficient of thermal expansion	$10^{-6}/°C$	2.2		3.4			
Thermal conductivity	$cal\text{-}cm/sec \cdot cm^2 \cdot K$	0.28		0.21			
Specific resistance	10^{-4} ohm-cm	11.0		14.5			

Source
Union Carbide

Size
Cylinder 14–17 in., block 9 X 20 X 24 in.

Manufacturing data
Calcined petroleum coke and coal tar pitch; graphitized over 2,500°C; Acheson electric furnace; 1–20 ton batch size

Table 5.2–40
GRADE CDJ-83

Description
Molded, fine grained; high strength; abrasion resistant

Properties	Units	With grain: Average value	With grain: Standard deviation, %	Against grain: Average value	Against grain: Standard deviation, %	Typical high-temperature properties: 1,300° F	Typical high-temperature properties: 4,000° F
Young's modulus	10^6 psi	3.5	<10	3.5	<10		
Tensile strength	10^3 psi	7.5	7.5	7.5	7.5		
Compressive strength	10^3 psi	30	7.5	30	7.5	Maximum	
Flexural strength	10^3 psi	7.5	7.5	7.5	7.5	useful	
Density	g/cm^3	1.72	<1			temperature	
Coefficient of thermal expansion	10^{-6}/°C	<2	15	6	15	1,000° F	
Thermal conductivity	cal-cm/sec·cm^2·K	<0.1		0.3			
Specific resistance	10^{-4} ohm-cm						
Hardness (Rockwell) (E)		109					

Source
Union Carbide

Size
Cylinder 1/8–20 in.

Manufacturing data
Lamp black and natural graphite; processed below 2,500°C in a fuel-fired furnace; impregnated in secondary processing; machined and ground; 1–20 ton batch size

Table 5.2–41
GRADE RVD

Description
Molded, fine grained; high strength; high reproducibility; high density; large sizes

Analysis
Average impurity content: ash, 0.16%

Properties	Units	With grain		Against grain		Typical high-temperature properties	
		Average value	Standard deviation, %	Average value	Standard deviation, %	1,300°F	4,000°F
Young's modulus	10^6 psi	2.1	3	1.1	4		
Tensile strength	10^3 psi	4.1	11	2.7	8		
Compressive strength	10^3 psi	11.6	9	12.3	6		
Flexural strength	10^3 psi	4.7	8	3.1	7		
Density	g/cm^3	1.87	2				
Coefficient of thermal expansion	$10^{-6}/°C$	1.7		3.5			
Thermal conductivity	$cal\text{-}cm/sec \cdot cm^2 \cdot K$	0.27		0.20			
Specific resistance	10^{-4} ohm-cm	12.6	3	21.6	4		

Source
Union Carbide

Size
Cylinder 1/8–18 in.

Manufacturing data
Calcined petroleum coke and coal tar pitch; graphitized over 2,500°C; Acheson electric furnace; impregnated in secondary processing; 1–20 ton batch size

Table 5.2–42
GRADE 2

Description

Molded, fine grained; carbon-graphite; low coefficient of friction; will stand oxidizing atmosphere to 700°F; good electrical conductor; chemical resistant

Properties	Units	With grain		Against grain		Typical high-temperature properties	
		Average value	Standard deviation, %	Average value	Standard deviation, %	1,300°F	4,000°F
Young's modulus	10^6 psi	2.3					
Tensile strength	10^3 psi	4.5					
Compressive strength	10^3 psi	23.0					
Flexural strength	10^3 psi	5–10					
Density	g/cm^3	1.8					
Coefficient of thermal expansion	10^{-6}/°C						
Thermal conductivity	cal-cm/sec·cm^2·K						
Specific resistance	10^{-4} ohm-cm	10–50					
Hardness (scleroscope)		85					

Source

U.S. Graphite

Size

Cylinder 1/8–12 in., block 1–4 in., ring up to 13 3/4 × 10 × 1 3/4 in.

Manufacturing data

Carbon and graphite powders; compacted under high pressure; furnaced at temperatures up to 4,500°F; machined or ground to tolerance

Table 5.2–43
GRADE 1

Description

Molded, fine grained; low coefficient of thermal expansion; good electrical conductivity; good thermal insulator; high purity; good nuclear properties; high reproducibility; low friction; low porosity; chemical and abrasion resistance; large sizes

Analysis

Average impurity content: carbon, 99.99%

		With grain		Against grain		Typical high-temperature properties	
Properties	Units	Average value	Standard deviation, %	Average value	Standard deviation, %	1,300°F	4,000°F
Young's modulus	10^6 psi	3.5					
Tensile strength	10^3 psi						
Compressive strength	10^3 psi	100.0					
Flexural strength	10^3 psi	10–30					
Density	g/cm^3	1.3	1.5				
Coefficient of thermal expansion	10^{-6}/°C	2.0				3.2	
Thermal conductivity	cal-cm/sec·cm^2·K	0.02					
Specific resistance	10^{-4} ohm-cm	10–50					
Permeability to He	10^{-11} cm^2/sec	<0.25					
Hardness		820 Knoop (107 Shore)					
Specific heat	cal/g per °C	0.3					
Maximum usable temperature		3,000°C (inert atmosphere)					

Source

Vitreous Carbon

Size

Cylinder 30 in. diameter

Manufacturing data

Resin; processed below 2,500°C; 100–2,000 lb batch size

Table 5.2–44
GRADE MHLM

Description
Molded, medium grained; good electrical conductivity; high reproducibility; large sizes

Analysis
Average impurity content: ash, 0.40%; Fe, 0.10%; Ca, 0.06%; S, 0.06%; Si, 0.04%; Al, 0.02%; V, 0.01%

Properties	Units	With grain		Against grain		Typical high-temperature properties	
		Average value	Standard deviation, %	Average value	Standard deviation, %	1,300°F	4,000°F
Young's modulus	10^6 psi	1.2	10	1.0	10		
Tensile strength	10^3 psi	1.3	10	1.0	10		
Compressive strength	10^3 psi	5.2	10	5.2	10		
Flexural strength	10^3 psi	2.2	10	1.7	10		
Density	g/cm^3	1.75	2				
Coefficient of thermal expansion	10^{-6}/°C	2.6	5	2.4	5		
Thermal conductivity	cal-cm/sec·cm^2·K	0.09	10	0.1	10		
Specific resistance	10^{-4} ohm-cm	9	10	10	10		
Permeability (D'Arcy)		0.36	5				

Source
Great Lakes Carbon

Size
Cylinder 16–56 in.

Manufacturing data
Calcined petroleum coke and coal tar pitch; graphitized over 2,500°C; Acheson electric furnace; over 20-ton batch size

Table 5.2–45
GRADE MHLM-85

Description

Molded, medium grained; good electrical conductivity; high reproducibility

Analysis

Average impurity content: ash, 0.40%; Fe, 0.10%; Ca, 0.06%; S, 0.06%; Si, 0.04%; Al, 0.02%; V, 0.01%

Properties	Units	With grain		Against grain		Typical high-temperature properties	
		Average value	Standard deviation, %	Average value	Standard deviation, %	1,300°F	4,000°F
Young's modulus	10^6 psi	1.5	10	1.5	10		
Tensile strength	10^3 psi	1.8	10	1.5	10		
Compressive strength	10^3 psi	6.0	10	5.8	10		
Flexural strength	10^3 psi	3.1	10	2.4	10		
Density	g/cm^3	1.83	2				
Coefficient of thermal expansion	10^{-6}/°C	2.8	5	2.7	5		
Thermal conductivity	cal-cm/sec·cm^2·K	0.16	10	0.18	10		
Specific resistance	10^{-4} ohm-cm	8	10	7	10		
Permeability (D'Arcy)		0.009	5	0.005	5		

Source

Great Lakes Carbon

Size

Cylinder 16–56 in.

Manufacturing data

Calcined petroleum coke and coal tar pitch; graphitized over 2,500°C; Acheson electric furnace; over 20-ton batch size

Table 5.2–46
GRADE ATL

Description
Molded, medium grained; large sizes; high production

Analysis
Average impurity content: ash, 1.0%

		With grain		Against grain		Typical high-temperature properties	
Properties	Units	Average value	Standard deviation, %	Average value	Standard deviation, %	1,300° F	4,000° F
Young's modulus	10^6 psi						
Tensile strength	10^3 psi	1.1	23	1.1	24		
Compressive strength	10^3 psi	5.1	24	5.1	25		
Flexural strength	10^3 psi	2.2	15	2.4	19		
Density	g/cm^3	1.78	1				
Coefficient of thermal expansion	10^{-6}/°C	2.4	9	2.4	7		
Thermal conductivity	cal-cm/sec·cm^2·K	0.27		0.26			
Specific resistance	10^{-4} ohm-cm	11.3	8	11.8	10		

Source
Union Carbide

Size
Cylinder 30–50 in.; block 20 × 47 in.

Manufacturing data
Calcined petroleum coke and coal tar pitch; graphitized over 2,500°C; Acheson electric furnace; impregnated in secondary processing; machined; over 20-ton batch size

Table 5.2—47
GRADE CFW

Description
Molded, medium grained; high reproducibility; high density; low porosity; large sizes

Analysis
Average impurity content: ash, 0.34%

Properties	Units	With grain: Average value	With grain: Standard deviation, %	Against grain: Average value	Against grain: Standard deviation, %	Typical high-temperature properties: 1,300°F	Typical high-temperature properties: 4,000°F
Young's modulus	10^6 psi	1.4	8	1.3	7		
Tensile strength	10^3 psi	1.7	10	1.6	8		
Compressive strength	10^3 psi	8.2	14	8.2	13		
Flexural strength	10^3 psi	2.3	9	2.3	9		
Density	g/cm^3	1.87	2				
Coefficient of thermal expansion	10^{-6}/°C	2.0	9	2.3	6		
Thermal conductivity	cal-cm/sec·cm^2·K	0.36	5	0.34	5		
Specific resistance	10^{-4} ohm-cm	10.0	7	11.0	8		

Source
Union Carbide

Size
Cylinder 30–103 in.

Manufacturing data
Calcined petroleum coke and coal tar pitch; graphitized over 2,500°C; Acheson electric furnace; impregnated in secondary processing; over 20-ton batch size

Table 5.2–48
GRADE CFZ

Description

Molded, medium grained; high strength; high reproducibility; high density; low porosity; large sizes

Analysis

Average impurity content: ash, 0.25%

Properties	Units	With grain		Against grain		Typical high-temperature properties	
		Average value	Standard deviation, %	Average value	Standard deviation, %	1,300° F	4,000° F
Young's modulus	10^6 psi	1.9	4	1.5	3		
Tensile strength	10^3 psi	3.0	9	2.5	10		
Compressive strength	10^3 psi	10	14	12	6		
Flexural strength	10^3 psi	4.0	9	3.4	8		
Density	g/cm^3	1.91	1				
Coefficient of thermal expansion	10^{-6}/°C	1.9		2.64			
Thermal conductivity	cal-cm/sec·cm^2·K	0.32		0.25			
Specific resistance	10^{-4} ohm-cm	12.7		16.1			

Source

Union Carbide

Size

Cylinder 30-in. maximum

Manufacturing data

Calcined petroleum coke and coal tar pitch; graphitized over 2,500°C; Acheson electric furnace; impregnated in secondary processing; 1–20 ton batch size

Table 5.2–49
GRADE RVA

Description
Molded, medium grained; high strength; high reproducibility; high density; large sizes

Analysis
Average impurity content: ash, 0.30%

Properties	Units	With grain		Against grain		Typical high-temperature properties	
		Average value	Standard deviation, %	Average value	Standard deviation, %	1,300°F	4,000°F
Young's modulus	10^6 psi	1.7	9	1.3	9		
Tensile strength	10^3 psi	3.0	15	2.1	8		
Compressive strength	10^3 psi	8.4	13	8.1	15		
Flexural strength	10^3 psi	3.7	8	3.0	10		
Density	g/cm^3	1.84	2				
Coefficient of thermal expansion	$10^{-6}/°C$	1.8	5	2.7	3		
Thermal conductivity	$cal\text{-}cm/sec \cdot cm^2 \cdot K$	0.26		0.21			
Specific resistance	10^{-4} ohm-cm	12.2	3	15.7	6		

Source
Union Carbide

Size
Cylinder 30 in

Manufacturing data
Calcined petroleum coke and coal tar pitch; graphitized over 2,500°C; Acheson electric furnace; impregnated in secondary processing; 1–20 ton batch size

Table 5.2–50
GRADE RVC

Description
Molded, medium grained; high reproducibility; nearly isotropic high CTE

Analysis
Average impurity content: ash, 0.1–0.5%; Fe, 0.05–0.2%

Properties	Units	With grain: Average value	With grain: Standard deviation, %	Against grain: Average value	Against grain: Standard deviation, %	Typical high-temperature properties: 1,300°F	Typical high-temperature properties: 4,000°F
Young's modulus	10^6 psi	1.8		1.4			
Tensile strength	10^3 psi	2.7		1.3			
Compressive strength	10^3 psi	11		11			
Flexural strength	10^3 psi	3.2		2.0			
Density	g/cm^3	1.84					
Coefficient of thermal expansion	$10^{-6}/°C$	3.69		4.45			
Thermal conductivity	$cal\text{-}cm/sec \cdot cm^2 \cdot K$	0.27		0.24			
Specific resistance	10^{-4} ohm-cm	13.0		16.4			

Source
Union Carbide

Size
Cylinder 17 in. diameter × 14 in. long

Manufacturing data
Calcined petroleum coke and coal tar pitch; graphitized over 2,500°C; Acheson electric furnace; impregnated in secondary processing; machined

5.3 EXTRUDED GRAPHITES

Table 5.3–1
GRADE GSP

Description
Extruded, fine grained; maximum grain 0.008 in.; high purity

Analysis
Average impurity content: ash, 0.06% maximum

		With grain		Against grain		Typical high-temperature properties	
Properties	Units	Average value	Standard deviation, %	Average value	Standard deviation, %	1,300°F	4,000°F
Young's modulus	10^6 psi	1.1		0.8			
Tensile strength	10^3 psi	1.6		1.1			
Compressive strength	10^3 psi	4.7		4.7			
Flexural strength	10^3 psi						
Density	g/cm^3	1.55					
Coefficient of thermal expansion	$10^{-6}/°C$						
Thermal conductivity	$cal\text{-}cm/sec \cdot cm^2 \cdot K$						
Specific resistance	10^{-4} ohm-cm	11.4		16.0			

Source
Carborundum Company

Size
Cylinder 3/8–5 in.

Manufacturing data
Calcined petroleum coke and coal tar pitch; graphitized over 2,500°C; machining and grinding

Table 5.3–2
GRADE GSX

Description

Extruded, fine grained; good electrical and thermal conductor; maximum grain size 0.008 in.; high reproducibility; chemical resistant

Analysis

Average impurity content: ash, 0.2% maximum

Properties	Units	With grain: Average value	With grain: Standard deviation, %	Against grain: Average value	Against grain: Standard deviation, %	Typical high-temperature properties: 1,300°F	Typical high-temperature properties: 4,000°F
Young's modulus	10^6 psi	1.2		0.8			
Tensile strength	10^3 psi	2.1		1.5			
Compressive strength	10^3 psi	6.5		6.5			
Flexural strength	10^3 psi						
Density	g/cm^3	1.68					
Coefficient of thermal expansion	$10^{-6}/°C$						
Thermal conductivity	$cal\text{-}cm/sec \cdot cm^2 \cdot K$						
Specific resistance	10^{-4} ohm-cm	11.4		16.0			

Source

Carborundum Company

Size

Cylinder 3/8–2 in.; pipe 1 1/4–5 1/4 in.

Manufacturing data

Calcined petroleum coke, coal tar pitch; graphitized over 2,500°C; electric resistance furnace; machined and ground; 100–2,000 lb batch size

Table 5.3–3
GRADE GRAPH-I-TITE® "A"

Description

Extruded, fine grained; high strength; high density; low porosity; chemical resistant; good thermal conductor; high reproducibility

Analysis

Average impurity content: ash, 0.6–1.0% depending on size

Properties	Units	With grain		Against grain		Typical high-temperature properties	
		Average value	Standard deviation, %	Average value	Standard deviation, %	1,300°F	4,000°F
Young's modulus	10^6 psi	1.6		1.1			
Tensile strength	10^3 psi	3.6		2.5			
Compressive strength	10^3 psi	12.5		12.5			
Flexural strength	10^3 psi						
Density	g/cm^3	1.91		1.91			
Coefficient of thermal expansion	$10^{-6}/°C$						
Thermal conductivity	$cal\text{-}cm/sec \cdot cm^2 \cdot K$						
Specific resistance	10^{-4} ohm-cm	11.4		16			

Source

Carborundum Company

Size

Cylinder 3/8–30 in.; pipe 7/8–5 1/4 in.

Manufacturing data

Calcined petroleum coke, coal tar pitch; gaseous hydrocarbon, resin; graphitized under 2,500°C; electric resistance furnace; impregnated; machined and ground; 100–2,000 lb batch size

Table 5.3–4
GRADE H249

Description
Extruded, fine grained; high strength; low coefficient of thermal expansion; good electrical and thermal conductivity; high density

Properties	Units	With grain: Average value	With grain: Standard deviation, %	Against grain: Average value	Against grain: Standard deviation, %	Typical high-temperature properties: 1,300°F	Typical high-temperature properties: 4,000°F
Young's modulus	10^6 psi	2.0	10	1.3	10		
Tensile strength	10^3 psi	2.6	10	2.1	10		
Compressive strength	10^3 psi	7.9	10	7.9	10		
Flexural strength	10^3 psi	4.1	10	3.2	10		
Density	g/cm^3	1.88	2				
Coefficient of thermal expansion	$10^{-6}/°C$	2.0	5	3.3	5		
Thermal conductivity	$cal\text{-}cm/sec \cdot cm^2 \cdot K$	0.48	10	0.40	10		
Specific resistance	10^{-4} ohm-cm	9	10	8	10		
Permeability (D'Arcy)		0.04	10	0.03	10		

Source
Great Lakes Carbon

Size
Cylinder 3–24 in.

Manufacturing data
Graphitized over 2,500°C; Acheson electric furnace; 1–20 ton batch size

Table 5.3–5
GRADE 780GL

Description

Extruded, fine grained; high strength; low coefficient of thermal expansion; high purity; good nuclear properties; high temperature oxidation resistance

Analysis

Average impurity content: ash, 100 ppm maximum; Si, 10 ppm; Al, <10 ppm; Fe, 10 ppm; Ca, <10 ppm; Zn, <10 ppm; Na, <10 ppm; Mg, 2ppm

		With grain		Against grain		Typical high-temperature properties	
Properties	Units	Average value	Standard deviation, %	Average value	Standard deviation, %	1,300°F	4,000°F
Young's modulus	10^6 psi	2.0					
Tensile strength	10^3 psi	2.3		1.6		3.5	5.0
Compressive strength	10^3 psi	9.0		9.4		9.3	14.0
Flexural strength	10^3 psi	4.3		3.7		5.8	8.8
Density	g/cm^3	1.79					
Coefficient of thermal expansion	$10^{-6}/°C$	1.8		2.9			
Thermal conductivity	$cal\text{-}cm/sec \cdot cm^2 \cdot K$					0.06	
Specific resistance	10^{-4} ohm-cm	10.9				9.7	11.7

Source

Speer Carbon

Size

Cylinder 2 1/2–5 in.

Manufacturing data

Calcined petroleum coke and coal tar pitch; graphitized over 2,500°C; machined; 100–2,000 lb batch size

Table 5.3–6
GRADE 580

Description
Extruded, fine grained; high strength; low coefficient of thermal expansion

Analysis
Average impurity content: ash, 0.08%

Properties	Units	With grain: Average value	With grain: Standard deviation, %	Against grain: Average value	Against grain: Standard deviation, %	Typical high-temperature properties: 1,300° F	Typical high-temperature properties: 4,000° F
Young's modulus	10^6 psi	2.0					
Tensile strength	10^3 psi	2.3		1.6		3.5	5.0
Compressive strength	10^3 psi	9.0		9.4		9.3	14.0
Flexural strength	10^3 psi	4.3		3.7		5.8	8.8
Density	g/cm^3	1.79					
Coefficient of thermal expansion	10^{-6}/°C	1.8		2.9			
Thermal conductivity	cal-cm/sec·cm^2·K					0.06	
Specific resistance	10^{-4} ohm-cm	10.9				9.7	11.7
Hardness (scleroscope)		47					

Source
Speer Carbon

Size
Cylinder 2 1/2–5 in.

Manufacturing data
Calcined petroleum coke and coal tar pitch; graphitized over 2,500°C; 1–20 ton batch size

Table 5.3–7
GRADE 890RL

Description

Extruded, fine grained; high strength and purity; good nuclear properties; high-temperature oxidation resistant

Analysis

Average impurity content: ash, 100 ppm maximum; Al, 10 ppm; Ca, 10 ppm; Fe, 10 ppm; Mg, 1 ppm; Ni, 1 ppm; Si, 10 ppm; Ti, 1 ppm; V, 5 ppm

		With grain		Against grain		Typical high-temperature properties	
Properties	Units	Average value	Standard deviation, %	Average value	Standard deviation, %	1,300°F	4,000°F
Young's modulus	10^6 psi	1.9					
Tensile strength	10^3 psi	1.75		1.30		2.8	5.0
Compressive strength	10^3 psi	5.50		7.0		6.4	9.6
Flexural strength	10^3 psi	3.2		2.07		4.5	6.5
Density	g/cm^3	1.70					
Coefficient of thermal expansion	$10^{-6}/°C$	1.8		3.7			
Thermal conductivity	cal-cm/sec·cm^2·K					0.2	
Specific resistance	10^{-4} ohm-cm	6.1				6.6	10.2

Source

Speer Carbon

Size

Cylinder 2 1/2–9 in.

Manufacturing data

Calcined petroleum coke and coal tar pitch; graphitized over 2,500°C; 1–20 ton batch size

Table 5.3–8
GRADE 890S

Description
Extruded, fine grained; high strength; low coefficient of thermal expansion; good electrical and thermal conductivity

Analysis
Average impurity content: ash, 0.03%

		With grain		Against grain		Typical high-temperature properties	
Properties	Units	Average value	Standard deviation, %	Average value	Standard deviation, %	1,300°F	4,000°F
Young's modulus	10^6 psi						
Tensile strength	10^3 psi	1.87				2.3	5.0
Compressive strength	10^3 psi	1.75		1.30		6.4	9.6
Flexural strength	10^3 psi	3.20		2.07		4.5	6.5
Density	g/cm^3	1.70					
Coefficient of thermal expansion	$10^{-6}/°C$	1.8		3.7			
Thermal conductivity	$cal\text{-}cm/sec \cdot cm^2 \cdot K$					0.20	
Specific resistance	10^{-4} ohm-cm	10.2				6.6	10.2
Permeability	cm^2/sec	0.44–0.55		0.22–0.66			

Source
Speer Carbon

Size
Cylinder 2 1/2–9 in. diameter; block <40 $in.^2$

Manufacturing data
Calcined petroleum coke and coal tar pitch; graphitized over 2,500°C; machined; 1–20 ton batch size

Table 5.3–9
GRADE 890W

Description
Extruded, fine grained; high strength and purity; high-temperature oxidation resistant

Analysis
Average impurity content: ash, 0.03%

Properties	Units	With grain		Against grain		Typical high-temperature properties	
		Average value	Standard deviation, %	Average value	Standard deviation, %	1,300°F	4,000°F
Young's modulus	10^6 psi	1.9					
Tensile strength	10^3 psi	1.6				2.8	5.0
Compressive strength	10^3 psi	5.4		5.6		6.4	9.6
Flexural strength	10^3 psi	2.7		2.4		4.5	6.5
Density	g/cm^3	1.7					
Coefficient of thermal expansion	$10^{-6}/°C$	1.7		3.4			
Thermal conductivity	$cal\text{-}cm/sec \cdot cm^2 \cdot K$					0.2	
Specific resistance	10^{-4} ohm-cm	6.9				6.6	10.2

Source
Speer Carbon

Size
Cylinder 2 1/2–9 in.; block <40 $in.^2$

Manufacturing data
Calcined petroleum coke and coal tar pitch; graphitized over 2,500°C; machined; 1–20 ton batch size

Table 5.3–10
GRADE 900

Description
Extruded, fine grained; good electrical conductivity; low porosity

Analysis
Average impurity content: ash, 0.05%

Properties	Units	With grain: Average value	With grain: Standard deviation, %	Against grain: Average value	Against grain: Standard deviation, %	Typical high-temperature properties: 1,300° F	Typical high-temperature properties: 4,000° F
Young's modulus	10^6 psi	2.2		1.3			
Tensile strength	10^3 psi	2.0		1.4			
Compressive strength	10^3 psi	7.2		5.9			
Flexural strength	10^3 psi	4.0		3.0			
Density	g/cm^3	1.73					
Coefficient of thermal expansion	10^{-6}/°C	2.0		3.7			
Thermal conductivity	cal-cm/sec·cm^2·K						
Specific resistance	10^{-4} ohm-cm	7.5					
Hardness (scleroscope)		39					
Hardness (Rockwell) (R)		85					

Source
Speer Carbon

Size
Block 30 in.^{2} maximum

Manufacturing data
Calcined petroleum coke and coal tar pitch; graphitized over 2,500°C; over 20-ton batch size

Table 5.3–11
GRADE K1

Description

Extruded, fine-grained; good thermal conductivity; high reproducibility; low friction; small sizes

Analysis

Average impurity content: ash, 0.7%

		With grain		Against grain		Typical high-temperature properties	
Properties	Units	Average value	Standard deviation, %	Average value	Standard deviation, %	1,300°F	4,000°F
Young's modulus	10^6 psi	1.4		0.85			
Tensile strength	10^3 psi						
Compressive strength	10^3 psi	4.9		5.0			
Flexural strength	10^3 psi	5.0					
Density	g/cm^3	1.60		1.60			
Coefficient of thermal expansion	10^{-6}/°C	1.3					
Thermal conductivity	cal-cm/sec·cm^2·K						
Specific resistance	10^{-4} ohm-cm	7.6					

Source

Stackpole

Size

Cylinder 1 in. diameter × 60 in. maximum

Manufacturing data

Calcined petroleum coke and coal tar pitch; graphitized over 2,500°C; machining and grinding; 100–2,000 lb batch size

Table 5.3–12
GRADE AGSR

Description
Extruded, fine grained; low cost; large and small sizes

Analysis
Average impurity content: ash, 0.12%

Properties	Units	With grain		Against grain		Typical high-temperature properties	
		Average value	Standard deviation, %	Average value	Standard deviation, %	1,300° F	4,000° F
Young's modulus	10^6 psi	1.5	8				
Tensile strength	10^3 psi	1.1	15				
Compressive strength	10^3 psi						
Flexural strength	10^3 psi	2.6	15	0.980			
Density	g/cm^3	1.58	2				
Coefficient of thermal expansion	10^{-6}/°C	1.1	26				
Thermal conductivity	cal-cm/sec·cm^2·K	0.37		0.21			
Specific resistance	10^{-4} ohm-cm	8.4	7	15.0	8		

Source
Union Carbide

Size
Cylinder 1–2 1/2 in. diameter; block 3/4 × 5 in. cross section maximum

Manufacturing data
Calcined petroleum coke, graphitized over 2,500°C; Acheson electric furnace; machined; 1–20 ton batch size

Table 5.3–13
GRADE AUC

Description

Extruded, fine grained; high purity; small sizes; jigs and fixtures

Analysis

Average impurity content: ash, 0.08%. (Low gas evolution. Guaranteed maximum ash, 0.08%; average, 0.03%.)

		With grain		Against grain		Typical high-temperature properties	
Properties	Units	Average value	Standard deviation, %	Average value	Standard deviation, %	1,300° F	4,000° F
Young's modulus	10^6 psi	1.7	11	0.9	11		
Tensile strength	10^3 psi						
Compressive strength	10^3 psi						
Flexural strength	10^3 psi	2.3					
Density	g/cm^3	1.68	3				
Coefficient of thermal expansion	10^{-6}/°C	1.13	25	3.4	4		
Thermal conductivity	cal-cm/sec·cm^2·K	0.39		0.25			
Specific resistance	10^{-4} ohm-cm	8.2					

Source

Union Carbide

Size

Cylinder 1/4–1 1/8 in.

Manufacturing data

Calcined petroleum coke and coal tar pitch; graphitized over 2,500°C; Acheson electric furnace; impregnated in secondary processing; machined; 100–2,000 lb batch size

Table 5.3–14
GRADE AGSX

Description
Extruded, fine grained

Analysis
Average impurity content: ash, 0.13%

Properties	Units	With grain		Against grain		Typical high-temperature properties	
		Average value	Standard deviation, %	Average value	Standard deviation, %	1,300°F	4,000°F
Young's modulus	10^6 psi	1.8	7	0.8	6		
Tensile strength	10^3 psi	1.4	16				
Compressive strength	10^3 psi						
Flexural strength	10^3 psi	3.1	13	1.3	22		
Density	g/cm^3	1.67	2				
Coefficient of thermal expansion	10^{-6}/°C	0.1	22				
Thermal conductivity	cal-cm/sec·cm^2·K	0.39		0.23			
Specific resistance	10^{-4} ohm-cm	8.0	11	13.3	5		

Source
Union Carbide

Size
Cylinder 1–2 3/4 in.

Manufacturing data
Calcined petroleum coke and coal tar pitch; graphitized over 2,500°C; Acheson electric furnace; impregnated in secondary processing; machined; 1–20 ton batch size

Table 5.3–15
GRADE GRAPH-I-TITE® "G90"

Description
Extruded, medium grained; high strength; good electrical and thermal conductivity; high purity; good nuclear properties; high reproducibility; high density

Analysis
Average impurity content: ash, 0.06–0.08% depending on size

Properties	Units	With grain: Average value	With grain: Standard deviation, %	Against grain: Average value	Against grain: Standard deviation, %	Typical high-temperature properties: 1,300°F	Typical high-temperature properties: 4,000°F
Young's modulus	10^6 psi	1.2		0.9			
Tensile strength	10^3 psi	2.7		1.9			
Compressive strength	10^3 psi	7.7–9.7		7.7–9.7			
Flexural strength	10^3 psi						
Density	g/cm^3	1.91		1.91			
Coefficient of thermal expansion	$10^{-6}/°C$						
Thermal conductivity	$cal\text{-}cm/sec \cdot cm^2 \cdot K$						
Specific resistance	10^{-4} ohm-cm	7.9		11.0			

Source
Carborundum Company

Size
Cylinder 3–30 in.

Manufacturing data
Calcined petroleum coke, graphitized over 2,500°C; electric resistance furnace; impregnated in secondary processing; final product machined; 100–2,000 lb batch size

Table 5.3–16
GRADE HC

Description
Extruded, medium grained; good electrical conductivity

Analysis
Average impurity content: ash, 0.30%; S, 0.10%; Si, 0.04%; Fe, 0.04%; Ca, 0.03%; Al, 0.03%; V, 70 ppm

Properties	Units	With grain		Against grain		Typical high-temperature properties	
		Average value	Standard deviation, %	Average value	Standard deviation, %	1,300°F	4,000°F
Young's modulus	10^6 psi	1.2	10	1.0	10		
Tensile strength	10^3 psi	0.8	10	0.6	10		
Compressive strength	10^3 psi	3.5	10	3.5	10		
Flexural strength	10^3 psi	1.2	10	1.0	10		
Density	g/cm^3	1.55	2				
Coefficient of thermal expansion	$10^{-6}/°C$	1.2	5	2.4	5		
Thermal conductivity	cal-cm/sec·cm^2·K	0.33	10	0.30	10		
Specific resistance	10^{-4} ohm-cm	9	10	12	10		
Permeability (D'Arcy)		0.37	10	0.34	10		

Source
Great Lakes Carbon

Size
Cylinder 7–12 in.

Manufacturing data
Calcined patroleum coke and coal tar pitch; graphitized over 2,500°C; Acheson electric furnace; over 20-ton batch size

Table 5.3–17
GRADE HL

Description
Extruded, medium grained; good electrical conductivity; high purity; high reproducibility

Analysis
Average impurity content: ash, 0.20%; S, 0.03%; Si, 0.05%; Fe, 0.03%; Ca, 0.03%; Al, 0.03%; V, 60 ppm; Ti, 30 ppm; Na, 20 ppm

Properties	Units	With grain		Against grain		Typical high-temperature properties	
		Average value	Standard deviation, %	Average value	Standard deviation, %	1,300°F	4,000°F
Young's modulus	10^6 psi	1.5	10	1.2	10		
Tensile strength	10^3 psi	0.8	10	0.6	10		
Compressive strength	10^3 psi	3.5	10	3.5	10		
Flexural strength	10^3 psi	2.0	10	1.8	10		
Density	g/cm^3	1.6	2				
Coefficient of thermal expansion	10^{-6}/°C	1.8	5	2.2	5		
Thermal conductivity	cal-cm/sec·cm^2·K	0.33	10	0.30	10		
Specific resistance	10^{-4} ohm-cm	9	10	12	10		
Porosity (apparent)	%	30					

Source
Great Lakes Carbon

Size
Cylinder 3–6 in., maximum lengths to 100 in.; block 3/4–6 in. thick × 2–18 in. wide

Manufacturing data
Calcined petroleum coke and coal tar pitch; graphitized over 2,500°C; Acheson electric furnace; ground; over 20-ton batch size

Table 5.3–18
GRADE HL 8

Description
Extruded, medium grained; good electrical conductivity; high purity; high reproducibility

Analysis
Average impurity content: ash, 0.12%; S, 0.2%; Si, 0.04%; Fe, 0.03%; Ca, 0.02%; Al, 0.02%; V, 30 ppm; Ti, 10 ppm; Na, 10 ppm

		With grain		Against grain		Typical high-temperature properties	
Properties	Units	Average value	Standard deviation, %	Average value	Standard deviation, %	1,300°F	4,000°F
Young's modulus	10^6 psi	1.5	10	1.2	10		
Tensile strength	10^3 psi	0.8	10	0.6	10		
Compressive strength	10^3 psi	3.5	10	3.5	10		
Flexural strength	10^3 psi	2.0	10	1.8	10		
Density	g/cm^3	1.6	2				
Coefficient of thermal expansion	$10^{-6}/°C$	1.8	5	2.2	5		
Thermal conductivity	$cal\text{-}cm/sec \cdot cm^2 \cdot K$	0.33	10	0.30	10		
Specific resistance	10^{-4} ohm-cm	9	10	12	10		
Porosity (apparent)	%	30					

Source
Great Lakes Carbon

Size
Cylinder 3–6 in.; block 3/4–6 in. thick × 2–18 in. wide

Manufacturing data
Calcined petroleum coke and coal tar pitch; graphitized over 2,500°C; Acheson electric furnace;ground; over 20-ton batch size

Table 5.3–19
GRADE HLM

Description
Extruded, medium grained; good electrical conductivity

Analysis
Average impurity content: ash, 0.25%; Fe, 0.04%; Ca, 0.03%; S, 0.06%; Si, 0.02%; Al, 0.01%; V, 60 ppm

		With grain		Against grain		Typical high-temperature properties	
Properties	Units	Average value	Standard deviation, %	Average value	Standard deviation, %	1,300°F	4,000°F
Young's modulus	10^6 psi	1.8	10	1.5	10		
Tensile strength	10^3 psi	1.8	10	1.5	10		
Compressive strength	10^3 psi	6.5	10	6.0	10		
Flexural strength	10^3 psi	3.0	10	2.7	10		
Density	g/cm^3	1.75	2				
Coefficient of thermal expansion	$10^{-6}/°C$	1.8	5	3.3	5		
Thermal conductivity	cal-cm/sec·cm^2·K	0.39	10	0.36	10		
Specific resistance	10^{-4} ohm-cm	8	10	11	10		

Source
Great Lakes Carbon

Size
Cylinder 1–3 in.; block 1–2 in. thick × 2–6 in. wide

Manufacturing data
Calcined petroleum coke and coal tar pitch; graphitized over 2,500°C; Acheson electric furnace; over 20-ton batch size

Table 5.3–20
GRADE HLM-50

Description
Extruded, medium grained; high reproducibility

Analysis
Average impurity content: ash, 0.40%; Fe, 0.10%; Ca, 0.06%; S, 0.06%; Si, 0.04%; Al, 0.02%; V, 0.01%

		With grain		Against grain		Typical high-temperature properties	
Properties	Units	Average value	Standard deviation, %	Average value	Standard deviation, %	1,300°F	4,000°F
Young's modulus	10^6 psi	1.2	10	1.0	10		
Tensile strength	10^3 psi	0.85	10	0.7	10		
Compressive strength	10^3 psi	4.0	10	3.5	10		
Flexural strength	10^3 psi	1.5	10	1.2	10		
Density	g/cm^3	1.62	2				
Coefficient of thermal expansion	10^{-6}/°C	2.0	5	2.9	5		
Thermal conductivity	cal-cm/sec·cm^2·K	0.31	10	0.29	10		
Specific resistance	10^{-4} ohm-cm	10	10	14	10		
Permeability (D'Arcy)		0.30	10	0.34	10		

Source
Great Lakes Carbon

Size
Cylinder 1–55 in.; block 1–24 × 2–48 in.; block 24 × 30 in.

Manufacturing data
Calcined petroleum coke and coal tar pitch; graphitized over 2,500°C; Acheson electric furnace; over 20-ton batch size

Table 5.3–21
GRADE HLM-85

Description
Extruded, medium grained; good electrical conductivity; high reproducibility

Analysis
Average impurity content: ash, 0.25%; Fe, 0.04%; Ca, 0.03%; S, 0.06%; Si, 0.02%; Al, 0.01%; V, 60 ppm

		With grain		Against grain		Typical high-temperature properties	
Properties	Units	Average value	Standard deviation, %	Average value	Standard deviation, %	1,300° F	4,000° F
Young's modulus	10^6 psi	2.2	10	1.8	10		
Tensile strength	10^3 psi	2.4	10	2.0	10		
Compressive strength	10^3 psi	8.3	10	8.0	10		
Flexural strength	10^3 psi	4.1	10	3.5	10		
Density	g/cm^3	1.83	2				
Coefficient of thermal expansion	$10^{-6}/°C$	2.1	5	3.5	5		
Thermal conductivity	$cal\text{-}cm/sec \cdot cm^2 \cdot K$	0.39	10	0.36	10		
Specific resistance	10^{-4} ohm-cm	8	10	11	10		
Permeability (D'Arcy)		0.06	5	0.04	5		

Source
Great Lakes Carbon

Size
Cylinder 1–3 in.; block 1–2 × 2–6 in.

Manufacturing data
Calcined petroleum coke and coal tar pitch; graphitized over 2,500°C; Acheson electric furnace; over 20-ton batch size

Table 5.3–22
GRADE HLM-85

Description

Extruded, medium grained; good electrical conductivity; high reproducibility

Analysis

Average impurity content: ash, 0.25%; Fe, 0.04%; Ca, 0.03%; S, 0.06%; Si, 0.02%; Al, 0.01%; V, 60 ppm

		With grain		Against grain		Typical high-temperature properties	
Properties	Units	Average value	Standard deviation, %	Average value	Standard deviation, %	1,300°F	4,000°F
Young's modulus	10^6 psi	1.9	10	1.3	10		
Tensile strength	10^3 psi	2.1	10	1.5	10		
Compressive strength	10^3 psi	7.2	10	7.0	10		
Flexural strength	10^3 psi	3.4	10	2.2	10		
Density	g/cm^3	1.83	2				
Coefficient of thermal expansion	10^{-6}/°C	1.9	5	3.2	5		
Thermal conductivity	cal-cm/sec·cm^2·K	0.48	10	0.37	10		
Specific resistance	10^{-4} ohm-cm	6	10	8	10		
Permeability (D'Arcy)		0.02	5	0.01	5		

Source

Great Lakes Carbon

Size

Cylinder 4–14 in.; block 2–6 × 4–6 in.

Manufacturing data

Calcined petroleum coke and coal tar pitch; graphitized over 2,500°C; Acheson electric furnace; over 20-ton batch size

Table 5.3–23
GRADE HPC

Description
Extruded, medium grained; good electrical conductivity

Analysis
Average impurity content: ash, 0.03%; Si, 0.04%; S, 0.10%; Fe, 0.04%; Ca, 0.03%; Al, 0.03%; V, 70 ppm

Properties	Units	With grain		Against grain		Typical high-temperature properties	
		Average value	Standard deviation, %	Average value	Standard deviation, %	1,300° F	4,000° F
Young's modulus	10^6 psi	1.2	10	1.0	10		
Tensile strength	10^3 psi	1.5	10	1.3	10		
Compressive strength	10^3 psi	5.5	10	5.0	10		
Flexural strength	10^3 psi	2.5	10	2.2	10		
Density	g/cm^3	1.70	2				
Coefficient of thermal expansion	$10^{-6}/°C$	2.1	5	2.5	5		
Thermal conductivity	$cal\text{-}cm/sec \cdot cm^2 \cdot K$	0.39	10	0.36	10		
Specific resistance	10^{-4} ohm-cm	8	10	11	10		

Source
Great Lakes Carbon

Size
Cylinder 1–3 in.

Manufacturing data
Calcined petroleum coke and coal tar pitch; graphitized over 2,500°C: Acheson electric furnace; over 20-ton batch size

Table 5.3–24
GRADE HPL

Description

Extruded, medium grained; good electrical conductivity; high purity; high reproducibility; long experience; high production; used for electrolytic anodes

Analysis

Average impurity content: ash, 0.20%; S, 0.03%; Si, 0.05%; Fe, 0.03%; Ca, 0.03%; Al, 0.03%; V, 60 ppm; Ti, 30 ppm; Na, 20 ppm

		With grain		Against grain		Typical high-temperature properties	
Properties	Units	Average value	Standard deviation, %	Average value	Standard deviation, %	1,300° F	4,000° F
Young's modulus	10^6 psi	1.3	10	1.1	10		
Tensile strength	10^3 psi	1.3	10	1.1	10		
Compressive strength	10^3 psi	4.5	10	4.5	10		
Flexural strength	10^3 psi	2.5	10	2.3	10		
Density	g/cm^3	1.75	2				
Coefficient of thermal expansion	10^{-6}/°C	1.6	5	2.0	5		
Thermal conductivity	cal-cm/sec·cm^2·K	0.39	10	0.36	10		
Specific resistance	10^{-4} ohm-cm	8	10	12	10		
Porosity (apparent)	%	25					

Source

Great Lakes Carbon

Size

Cylinder 3–6 in.; block 3/4–6 × 2–8 in.

Manufacturing data

Calcined petroleum coke and coal tar pitch; graphitized over 2,500°C; Acheson electric furnace; over 20-ton batch size

Table 5.3–25
GRADE TL

Description

Extruded, medium grained; good electrical conductivity; high purity; high reproducibility

Analysis

Average impurity content: ash, 0.20%; S, 0.10%; Si, 0.05%; Fe, 0.03%; Co, 0.05%; Pb, 0.04%; Ca, 0.03% Al, 0.03%; Na, 20%; Mg, 20%

		With grain		Against grain		Typical high-temperature properties	
Properties	Units	Average value	Standard deviation, %	Average value	Standard deviation, %	1,300° F	4,000° F
Young's modulus	10^6 psi	1.5	10	1.2	10		
Tensile strength	10^3 psi	0.8	10	0.6	10		
Compressive strength	10^3 psi	3.5	10	3.5	10		
Flexural strength	10^3 psi	2.0	10	1.8	10		
Density	g/cm^3	1.7	2				
Coefficient of thermal expansion	$10^{-6}/°C$	1.8	5	2.2	5		
Thermal conductivity	$cal\text{-}cm/sec \cdot cm^2 \cdot K$	0.33	10	0.30	10		
Specific resistance	10^{-4} ohm-cm	9	10	12	10		
Porosity (apparent)	%	16					

Source

Great Lakes Carbon

Size

Cylinder 3–6 in.; block 3/4–6 × 2–18 in.

Manufacturing data

Calcined petroleum coke and coal tar pitch; graphitized over 2,500°C; Acheson electric furnace; over 20-ton batch size

Table 5.3–26
GRADE 700

Description
Extruded, medium grained; low coefficient of thermal expansion; good electrical and thermal conductivities

Analysis
Average impurity content: ash, <1.0%

Properties	Units	With grain: Average value	With grain: Standard deviation, %	Against grain: Average value	Against grain: Standard deviation, %	Typical high-temperature properties: 1,300°F	Typical high-temperature properties: 4,000°F
Young's modulus	10^6 psi	1.8		1.0		2.1	3.8
Tensile strength	10^3 psi	1.5		1.0		4.8	7.3
Compressive strength	10^3 psi	5.6		5.85		4.0	6.2
Flexural strength	10^3 psi	3.3		2.4			
Density	g/cm^3	1.77					
Coefficient of thermal expansion	10^{-6}/°C	2.5		4.5			
Thermal conductivity	cal-cm/sec·cm^2·K					0.23	
Specific resistance	10^{-4} ohm-cm	6.5					
Hardness (scleroscope)		35					

Source
Speer Carbon

Size
Cylinder 3–7 in.

Manufacturing data
Calcined petroleum coke and coal tar pitch; graphitized over 2,500°C; over 20-ton batch size

Table 5.3–27
GRADE 873 RL

Description

Extruded, medium grained; high strength; low coefficient of thermal expansion; high purity; good nuclear properties; high-temperature oxidation resistance

Analysis

Average impurity content: ash, 100 ppm maximum; Al, <10 ppm; B, <1 ppm; Ca, <1 ppm; Fe, 5 ppm; Mg, <1 ppm; Ni, <10 ppm; Si, 30 ppm; Ti, <10 ppm; V, 1 ppm

Properties	Units	With grain		Against grain		Typical high-temperature properties	
		Average value	Standard deviation, %	Average value	Standard deviation, %	1,300°F	4,000°F
Young's modulus	10^6 psi	1.8		1.1			
Tensile strength	10^3 psi	1.6		1.4		2.0	4.0
Compressive strength	10^3 psi	6.4		6.8		6.2	9.6
Flexural strength	10^3 psi	3.2		2.6		3.8	5.8
Density	g/cm^3	1.77					
Coefficient of thermal expansion	10^{-6}/°C	2.4		4.2			
Thermal conductivity	cal-cm/sec·cm^2·K					0.23	
Specific resistance	10^{-4} ohm-cm	6.35				6.25	10.3

Source

Speer Carbon

Size

Cylinder 14 in. maximum diameter

Manufacturing data

Calcined petroleum coke and coal tar pitch; graphitized over 2,500°C; over 20-ton batch size

Table 5.3–28
GRADE 873S

Description
Extruded, medium grained; high strength; low coefficient of thermal expansion; long experience

Analysis
Average impurity content: ash, 0.05%

Properties	Units	With grain: Average value	With grain: Standard deviation, %	Against grain: Average value	Against grain: Standard deviation, %	Typical high-temperature properties: 1,300°F	Typical high-temperature properties: 4,000°F
Young's modulus	10^6 psi	1.8		1.1			
Tensile strength	10^3 psi	1.6		1.4		2.0	4.0
Compressive strength	10^3 psi	6.4		6.8		6.2	9.6
Flexural strength	10^3 psi	3.2		2.6		3.8	5.8
Density	g/cm^3	1.77					
Coefficient of thermal expansion	10^{-6}/°C	2.4		4.2			
Thermal conductivity	cal-cm/sec·cm^2·K					0.23	
Specific resistance	10^{-4} ohm-cm	6.4				6.25	10.8
Hardness (scleroscope)		38					
Permeability	cm^2/sec	2.6–5.9		0.5–8.8			

Source
Speer Carbon

Size
Cylinder 12–16 in.

Manufacturing data
Calcined petroleum coke and coal tar pitch; graphitized over 2,500°C; 1–20 ton batch size

Table 5.3–29
GRADE HBX

Description

Extruded, medium grained; good electrical conductor; good thermal conductor; high reproducibility

Analysis

Average impurity content: ash, 0.08%

Properties	Units	With grain: Average value	With grain: Standard deviation, %	Against grain: Average value	Against grain: Standard deviation, %	Typical high-temperature properties: 1,300°F	Typical high-temperature properties: 4,000°F
Young's modulus	10^6 psi	1.5		0.9			
Tensile strength	10^3 psi						
Compressive strength	10^3 psi	3.7		3.6			
Flexural strength	10^3 psi	2.4		1.9			
Density	g/cm^3	1.61					
Coefficient of thermal expansion	10^{-6}/°C	2.1		4.1			
Thermal conductivity	cal-cm/sec·cm^2·K						
Specific resistance	10^{-4} ohm-cm	2.7		5.0			
Hardness (scleroscope)		35					

Source

Stackpole Carbon

Size

Cylinder 1/8–5 1/2 in. (up to 80 in. long); block 1–4 in.; rod 10 mil–1/8 in.

Manufacturing data

Calcined petroleum coke and coal tar pitch; graphitized over 2,500°C; machining and grinding as required; 100–2,000 lb batch size

Table 5.3–30
GRADE AGR

Description

Extruded, medium grained; long experience; large and small sizes; thermal shock resistance

Analysis

Average impurity content: ash, 0.30%

		With grain		Against grain		Typical high-temperature properties	
Properties	Units	Average value	Standard deviation, %	Average value	Standard deviation, %	1,300°F	4,000°F
Young's modulus	10^6 psi	1.4	10	0.8	14		
Tensile strength	10^3 psi	1.1	15	0.76	23		
Compressive strength	10^3 psi	4.4	19	4.0	15		
Flexural strength	10^3 psi	2.2	14	1.4	25		
Density	g/cm^3	1.58	2				
Coefficient of thermal expansion	$10^{-6}/°C$	1.38	15				
Thermal conductivity	$cal\text{-}cm/sec \cdot cm^2 \cdot K$	0.36		0.24			
Specific resistance	10^{-4} ohm-cm	8.6	8	12.8	8		

Source

Union Carbide

Size

Cylinder 3–5 3/4 in.

Manufacturing data

Calcined petroleum coke; graphitized over 2,500°C; Acheson electric furnace; machined; over 20-ton batch size

Table 5.3–31
GRADE AGX, AGLX

Description
Extruded, medium grained; long experience

Analysis
Average impurity content: ash, 0.42%

		With grain		Against grain		Typical high-temperature properties	
Properties	Units	Average value	Standard deviation, %	Average value	Standard deviation, %	1,300°F	4,000°F
Young's modulus	10^6 psi	1.6	12	0.9	8		
Tensile strength	10^3 psi	1.4	16	1.0	14		
Compressive strength	10^3 psi	5.6	24	5.3	22		
Flexural strength	10^3 psi	2.7	17	1.8	25		
Density	g/cm^3	1.69	1.5				
Coefficient of thermal expansion	$10^{-6}/°C$	1.6	12				
Thermal conductivity	$cal\text{-}cm/sec \cdot cm^2 \cdot K$	0.38		0.22			
Specific resistance	10^{-4} ohm-cm	8.2	12	13.9	10		

Source
Union Carbide

Size
Cylinder 3–5 3/4 in.; block 1–5 in.

Manufacturing data
Calcined petroleum coke and coal tar pitch; graphitized over 2,500°C; Acheson electric furnace; impregnated in secondary processing; machined; 1–20 ton batch size

Table 5.3–32
GRADE AUC

Description
Extruded, medium grained; high purity; large and small sizes

Analysis
Average impurity content: ash, 0.03%. (Low gas evolution. Guaranteed maximum ash, 0.08%; average, 0.03%.)

Properties	Units	With grain		Against grain		Typical high-temperature properties	
		Average value	Standard deviation, %	Average value	Standard deviation, %	1,300°F	4,000°F
Young's modulus	10^6 psi	1.7	11	0.9	11		
Tensile strength	10^3 psi	1.2	18	0.9	20		
Compressive strength	10^3 psi	4.6	17	4.6	13		
Flexural strength	10^3 psi	2.9	29	1.7	18		
Density	g/cm^3	1.68	3				
Coefficient of thermal expanison	$10^{-6}/°C$	1.1	25	3.4	4		
Thermal conductivity	$cal\text{-}cm/sec\cdot cm^2\cdot K$	0.39		0.25			
Specific resistance	10^{-4} ohm-cm	7.9	10	12.3	8		

Source
Union Carbide

Size
Cylinder 1 1/4–8 in.

Manufacturing data
Calcined petroleum coke and coal tar pitch; graphitized over 2,500°C; Acheson electric furnace; impregnated in secondary processing; machined; 1–20 ton batch size

Table 5.3–33
GRADE 700

Description
Extruded, coarse grained; low coefficient of thermal expansion; good electrical conductor; high temperature oxidation resistance

Analysis
Average impurity content: ash, 0.45%; Fe, 600 ppm; Si, 300 ppm

Properties	Units	With grain: Average value	With grain: Standard deviation, %	Against grain: Average value	Against grain: Standard deviation, %	Typical high-temperature properties: 1,300°F	Typical high-temperature properties: 4,000°F
Young's modulus	10^6 psi	0.6					
Tensile strength	10^3 psi	6.0		4.0		6.0	1.0
Compressive strength	10^3 psi	2.2		2.2		2.4	3.6
Flexural strength	10^3 psi	9.1		8.2		1.3	2.1
Density	g/cm^3	1.59					
Coefficient of thermal expansion	10^{-6}/°C	1.4		2.9			
Thermal conductivity	cal-cm/sec·cm^2·K						
Specific resistance	10^{-4} ohm-cm	8.6		15.2			

Source
Speer Carbon

Size
Cylinder 16–24 in. diameter × 60–96 in. long

Manufacturing data
Calcined petroleum coke and coal tar pitch; graphitized over 2,500°C; over 20-ton batch size

Table 5.3–34
GRADE HC

Description
Extruded, coarse grained; good electrical conductor

Analysis
Average impurity content: ash, 0.30%; S, 0.10%; Si, 0.04%; Fe, 0.04%; Ca, 0.03%; Al, 0.03%; V, 70 ppm

Properties	Units	With grain		Against grain		Typical high-temperature properties	
		Average value	Standard deviation, %	Average value	Standard deviation, %	1,300°F	4,000°F
Young's modulus	10^6 psi	1.2	10	1.0	10		
Tensile strength	10^3 psi	0.6	10	0.5	10		
Compressive strength	10^3 psi	2.0	10	2.0	10		
Flexural strength	10^3 psi	0.9	10	0.8	10		
Density	g/cm^3	1.55	2				
Coefficient of thermal expansion	$10^{-6}/°C$	1.1	5	2.1	5		
Thermal conductivity	$cal\text{-}cm/sec \cdot cm^2 \cdot K$	0.33	10	0.30	10		
Specific resistance	10^{-4} ohm-cm	9	10	12	10		
Permeability (D'Arcy)		0.65	10	0.47	10		

Source
Great Lakes Carbon

Size
Cylinder 14 in. diameter

Manufacturing data
Calcined petroleum coke and coal tar pitch; over 20-ton batch size

Table 5.3–35
GRADE HL

Description
Extruded, coarse grained; good electrical conductor; high purity; high reproducibility; large sizes

Analysis
Average impurity content: ash, 0.20%; S, 0.03%; Si, 0.05%; Fe, 0.03%; Ca, 0.03%, Al, 0.03%; V, 60 ppm; Ti, 30 ppm; Na, 20 ppm

		With grain		Against grain		Typical high-temperature properties	
Properties	Units	Average value	Standard deviation, %	Average value	Standard deviation, %	1,300°F	4,000°F
Young's modulus	10^6 psi	0.9	10	0.8	10		
Tensile strength	10^3 psi	0.5	10	0.5	10		
Compressive strength	10^3 psi	2.5	10	2.5	10		
Flexural strength	10^3 psi	1.0	10	0.9	10		
Density	g/cm^3	1.60	2.0				
Coefficient of thermal expansion	$10^{-6}/°C$	1.8	5	2.2	5		
Thermal conductivity	$cal\text{-}cm/sec \cdot cm^2 \cdot K$	0.31	10	0.28	10		
Specific resistance	10^{-4} ohm-cm	9	10	12	10		
Porosity (apparent)	%	28					

Source
Great Lakes Carbon

Size
Cylinders 16 in., 17 in., 19 in. diameter

Manufacturing data
Calcined petroleum coke and coal tar pitch; graphitized over 2,500°C; Acheson electric furnace; over 20-ton batch size

Table 5.3–36
GRADE AGR

Description
Extruded, coarse grained; long experience; large and small sizes; thermal shock resistant

Analysis
Average impurity content: ash, 0.96%

		With grain		Against grain		Typical high-temperature properties	
Properties	Units	Average value	Standard deviation, %	Average value	Standard deviation, %	1,300°F	4,000°F
Young's modulus	10^6 psi	0.5	13	0.5	21		
Tensile strength	10^3 psi	0.44	17	0.42	11		
Compressive strength	10^3 psi	1.9	22	2.0	18		
Flexural strength	10^3 psi	0.84	17	0.84	17		
Density	g/cm^3	1.54	2.5				
Coefficient of thermal expansion	10^{-6}/°C	1.2	31	1.9	16		
Thermal conductivity	cal-cm/sec·cm^2·K	0.32		0.27			
Specific resistance	10^{-4} ohm-cm	9.6	10	11.3			

Source
Union Carbide

Size
Cylinder 14–35 in. diameter; blocks 24 × 24 × 100 in. maximum

Manufacturing data
Calcined petroleum coke; graphitized over 2,500°C; Acheson electric furnace; machined; over 20-ton batch size

Table 5.3–37
GRADE HC

Description
Extruded, very coarse grained; good electrical conductivity

Analysis
Average impurity content: ash, 0.30%; S, 0.10%; Si, 0.04%; Fe, 0.04%; Ca, 0.30%; Al, 0.03%; V, 70 ppm

		With grain		Against grain		Typical high-temperature properties	
Properties	Units	Average value	Standard deviation, %	Average value	Standard deviation, %	1,300°F	4,000°F
Young's modulus	10^6 psi	0.9	10	0.8	10		
Tensile strength	10^3 psi	0.6	10	0.6	10		
Compressive strength	10^3 psi	3.0	10	3.0	10		
Flexural strength	10^3 psi	1.0	10	0.8	10		
Density	g/cm^3	1.60	2				
Coefficient of thermal expansion	$10^{-6}/°C$	2.4	5	2.8	5		
Thermal conductivity	$cal\text{-}cm/sec \cdot cm^2 \cdot K$	0.27	10	0.25	10		
Specific resistance	10^{-4} ohm-cm	7.5	10	10	10		

Source
Great Lakes Carbon

Size
Cylinder 30–56 in. diameter

Manufacturing data
Calcined petroleum coke and coal tar pitch; graphitized over 2,500°C; over 20-ton batch size

5.4 HOT-WORKED GRAPHITES

Table 5.4–1
GRADE ZTB

Description
Hot worked; very high density; high reproducibility; high strength; good thermal conductivity; highly oriented; low porosity

Analysis
Average impurity content: ash, 0.1%

Properties	Units	With grain: Average value	With grain: Standard deviation, %	Against grain: Average value	Against grain: Standard deviation, %	Typical high-temperature properties: 1,300°F	Typical high-temperature properties: 4,000°F
Young's modulus	10^6 psi	3.4		0.8			
Tensile strength	10^3 psi						
Compressive strength	10^3 psi	9.1		1.3			
Flexural strength	10^3 psi	6.2		2.5			
Density	g/cm^3	2.0					
Coefficient of thermal expansion	10^{-6}/°C	0.6		8.6			
Thermal conductivity	cal-cm/sec·cm^2·K	0.47		0.17			
Specific resistance	10^{-4} ohm-cm	6.7		19.7			

Source
Union Carbide

Size
Cylinder 8 1/2–14 in. diameter

Manufacturing data
Calcined petroleum coke and coal tar pitch; graphitized over 2,500°C and finally hot worked; 100–2,000 lb batch size

Table 5.4–2
GRADE ZTA

Description

Hot worked; high density fine grained; high reproducibility; high strength; good thermal conductivity; highly oriented; low porosity; grade is certified to be free of internal cracks, voids, or other structural defects as detected by radiographic inspection

Analysis

Average impurity content: ash, 0.1%

Properties	Units	With grain		Against grain		Typical high-temperature properties	
		Average value	Standard deviation, %	Average value	Standard deviation, %	1,300°F	4,000°F
Young's modulus	10^6 psi	2.6	9	0.8	5		
Tensile strength	10^3 psi	4.0	15	1.2	14		
Compressive strength	10^3 psi	7.2	18	1.2	13		
Flexural strength	10^3 psi	5.4	14	2.4	14		
Density	g/cm^3	1.95	1.5				
Coefficient of thermal expansion	10^{-6}/°C	0.7	0.35	8.2	4		
Thermal conductivity	cal-cm/sec·cm^2·K	0.52		0.20			
Specific resistance	10^{-4} ohm-cm	7.1	7	19.9	7		

Source

Union Carbide

Size

Cylinder 8 1/2–14 in. diameter

Manufacturing data

Calcined petroleum coke and coal tar pitch; graphitized over 2,500°C and hot worked; 100–2,000 lb batch size

Table 5.4–3
GRADE D-857

Description
Hot worked; high density; low porosity

Analysis
Average impurity content: Ni, 200 ppm; Ca, 200 ppm; Fe, 100 ppm; Si, 75 ppm; Al, 75 ppm; Co, 25, ppm; Mo, 10 ppm; Ti, 10 ppm; Na, 100 ppm. (Available in purified grade, 50 ppm total impurities.)

Properties	Units	With grain		Against grain		Typical high-temperature properties	
		Average value	Standard deviation, %	Average value	Standard deviation, %	1,300°F	4,000°F
Young's modulus	10^6 psi	2.1	15	1.9	15	2.2	2.8
Tensile strength	10^3 psi	5.0	20	4.9	20	5.0	9.0
Compressive strength	10^3 psi	20.0	20	19.0	20	20.5	25.0
Flexural strength	10^3 psi	10.0	20	9.5	20	10.1	14.0
Density	g/cm^3	1.85	5				
Coefficient of thermal expansion	10^{-6}/°C	4.2	5	4.1	5	5.4	
Thermal conductivity	cal-cm/sec·cm^2·K	0.30	15	0.29	15		
Specific resistance	10^{-4} ohm-cm	15.0	1	15.5	1		

Source
Duramic Products

Size
15 × 6 × 3 in. maximum

Manufacturing data
Not available

5.5 PYROLYTIC GRAPHITES

Table 5.5–1
GRADE pyro

Description

Pyrolytic graphite; high strength; low coefficient of thermal expansion; good electrical and thermal conductor; good thermal insulator, high purity; good nuclear properties; high reproducibility; low porosity; highly oriented; chemical resistant

Analysis

Average impurity content: ash, 0.0038 ± 0.0026%

		With grain c direction		Against grain ab direction		Typical high-temperature properties	
Properties	Units	Average value	Standard deviation, %	Average value	Standard deviation, %	1,300°F	4,000°F
Young's modulus	10^6 psi	1.7		4.4	6		
Tensile strength	10^3 psi	1.25	20	18.5	10		
Compressive strength	10^3 psi	68		14			
Flexural strength	10^3 psi	1.5		21.3	10		
Density	g/cm^3	2.212	1				
Coefficient of thermal expansion	10^{-6}/°C	19.44 at 500°F		0.36 at 500°F			
Thermal conductivity	cal-cm/sec·cm^2·K	0.004 at 500°F		0.826 at 500°F		0.003 (c) 0.371 (ab)	0.003 (c) 0.210 (ab)
Specific resistance	10^{-4} ohm-cm	4,840		4.29	6	2,540 (c) 2.03 (ab)	– (c) 3.82 (ab)

Source

General Electric

Size

Special shapes (up to 45 in.); pipe <1/2 – >10 in., flexibles >144 in.^{2}

Manufacturing data

Gaseous hydrocarbon; processed below 2,500°C; machined and ground; 100–2,000 lb batch size

Table 5.5–2
GRADE PYROLYTIC GRAPHITE

Description
Pyrolytic graphite; high strength; low coefficient of thermal expansion; good electrical and thermal conductivity; high electrical resistance; good thermal insulator; high purity; good nuclear properties; low porosity; highly oriented

Analysis
Average impurity content: C, 99.99%

		With grain		Against grain		Typical high-temperature properties	
Properties	Units	Average value	Standard deviation, %	Average value	Standard deviation, %	1,300° F	4,000° F
Young's modulus	10^6 psi	4.4				1.89	
Tensile strength	10^3 psi	18.7		0.5		22.500	25.000
Compressive strength	10^3 psi	66.1		14.5	7		
Flexural strength	10^3 psi	23.5		2.0	10	31.000	
Density	g/cm^3	2.20		2.20			
Coefficient of thermal expansion	$10^{-6}/°C$	1.30		23.7			
Thermal conductivity	$cal\text{-}cm/sec \cdot cm^2 \cdot K$	1.24		0.002		0.39	0.0015
Specific resistance	10^{-4} ohm-cm	4.79		8,000		2.50	
						1,690	

Source
Raytheon Company

Size
Various geometric shapes

Manufacturing data
Gaseous hydrocarbon; processed below 2,500°C; machined; less than 100-lb batch size

Table 5.5–3
GRADE 100, 101A

Description

Pyrolytic graphite; bulk or massive free standing; high strength, purity, density, and reproducibility; good electrical and thermal conductor; good thermal insulator; good nuclear properties; highly oriented; low porosity; chemical and abrasion resistant

Analysis

Average impurity content: ash, 0.01% maximum (Metallic impurities total less than 20 ppm.)

Properties	Units	With grain		Against grain		Typical high-temperature properties	
		Average value	Standard deviation, %	Average value	Standard deviation, %	1,300° F	4,000° F
Young's modulus	10^6 psi	2–5	10–20	1–4	10–20	Same	Decrease
Tensile strength	10^3 psi	10–30	5–10	0.5	10–20	Same	Increase
Compressive strength	10^3 psi	10–45	5–10	60	10–20	Same	Same
Flexural strength	10^3 psi	15–25	10–20	–	10–20	Same	Increase
Density	g/cm^3	2–2.2	5–10	–	–	Same	–
Coefficient of thermal expansion	$10^{-6}/°C$	2	5–10	20	5–10	Same	Same
Thermal conductivity	$cal\text{-}cm/sec \cdot cm^2 \cdot K$	1	5–10	0.1	5–10	Decrease	Decrease
Specific resistance	10^{-4} ohm-cm	1.0	10–20	2,000	10–20	–	–

Source

Space Age Materials

Size

Plate 1/16–1 1/4 in., 40 × 70 in. maximum; plate 1/16–1 1/4 in., 8 × 12 in. maximum

Manufacturing data

Gaseous hydrocarbon; processed to graphite below and above 2,500°C; no secondary processing; finishing operations including machining and grinding; less than 100-lb batch size

Table 5.5–4
GRADE PYROLYTIC GRAPHITE

Description

Pyrolytic graphite; high strength; low coefficient of thermal expansion; good electrical and thermal conductivity; high purity and resistance; good nuclear properties; high reproducibility; low friction; low porosity; chemical resistant; low hardness

Analysis

Average impurity content: Al, 0.01 ppm; B, <0.01 ppm; Ca, <0.007; Co, <0.100 ppm; Cu, 0.01 ppm; Fe, 0.40 ppm; Mg, <0.001 ppm; Nb, <0.07 ppm; Ti, 0.01 ppm; Zn, <0.1 ppm; Ta, <1.00 ppm

Properties	Units	With grain		Against grain		Typical high-temperature properties	
		Average value	Standard deviation, %	Average value	Standard deviation, %	1,300°F	4,000°F
Young's modulus	10^6 psi	2–5	10–20				
Tensile strength	10^3 psi	10–30	5–10	<1	5–10	16	24
Compressive strength	10^3 psi	10–50	>20	>50	<5		
Flexural strength	10^3 psi	>20	10–20			20	27
Density	g/cm^3	2–2.2	<1				
Coefficient of thermal expansion	$10^{-6}/°C$	<2	<2	10–20	<2	0.7	1.6
Thermal conductivity	$cal\text{-}cm/sec \cdot cm^2 \cdot K$	0.5–1		<0.1		0.6	0.2
Specific resistance	10^{-4} ohm-cm	<1	<5	>2,000	<5	3.7	5
Emissivity		0.8 at 2,000°F					
Thermal neutron abs cross section		3.4 mb					

Source

Super-temp

Size

Plate material 1/16–1 in. thick, 16 × 65 in. maximum; cylinder 1/4–20 in. diameter, 36 in. maximum length

Manufacturing data

Gaseous hydrocarbon; graphitized over 2,500°C; machining and grinding; 100–2,000 lb batch size

Table 5.5–5

GRADE REINFORCED PYROLYTIC GRAPHITE

Description

Pyrolytic reinforced graphite; high strength; high electrical resistance; good thermal insulator; high purity; good nuclear properties

Properties	Units	With grain		Against grain		Typical high-temperature properties	
		Average value	Standard deviation, %	Average value	Standard deviation, %	1,300°F	4,000°F
Young's modulus	10^6 psi						
Tensile strength	10^3 psi	0.4–2		0.2–1.5			
Compressive strength	10^3 psi	27.0		16.3			
Flexural strength	10^3 psi	0.5–17.0		3.5			
Density	g/cm^3	0.2–2.0					
Coefficient of thermal expansion[a]	10^{-6}/°C						
Thermal conductivity[a]	cal-cm/sec·cm^2·K						
Specific resistance	10^{-4} ohm-cm						

Source

Super-Temp

Size

Plate 1/4–1 in.; block 1–3 in.

Manufacturing data

Gaseous hydrocarbon; synthetic fiber, cellulose fiber; processed below 2,500°C; machining and grinding; less than 100-lb batch size. Individual fibers of a felt or clothlike material are infiltrated with pyrolytic carbon or graphite.

[a]Thermal expansion and thermal conductivity vary with density.

Table 5.5–6
GRADE PYROLYTIC

Description
Pyrolytic graphite; high strength; low coefficient of thermal expansion; good electrical and thermal conductor; good electrical resistance; good thermal insulator; high purity; good nuclear properties; high density; low porosity; highly oriented; chemical resistant

Analysis
Average impurity content: ash, <0.1%; Fe, <0.05%; V, <0.005%; S, <0.01%, B, <1 ppm

		With grain		Against grain		Typical high-temperature properties	
Properties	Units	Average value	Standard deviation, %	Average value	Standard deviation, %	1,300°F	4,000°F
Young's modulus	10^6 psi	4	15		3	3.5	2.8
Tensile strength	10^3 psi	15	<5	1	<20	15	30
Compressive strength	10^3 psi	13	<10	50	<10		
Flexural strength	10^3 psi	14	<10	19	<10	14	14
Density	g/cm^3	2.2	<5				
Coefficient of thermal expansion	mils/in. to 4,000°F	4		60			
Thermal conductivity	Btu/ft/hr/ft^2/°F	200	<15	1.4	<15	0.8	0.3
Specific resistance	10^{-4} ohm-cm	0.05	<15	5,000	<15	0.4	0.5

Source
Union Carbide

Size
Plate 1/16–1 in; pipe 1/2–10 in.

Manufacturing data
Gaseous hydrocarbon; processed below 2,500°C; electric resistance furnace; machined and ground; less than 100-lb batch size

5.6 FOAMED GRAPHITES

Table 5.6–1
GRADE POROUS GRAPHITE 25

Description
Graphite foam; good thermal insulator; high reproducibility; high porosity; chemical resistant; high permeability

Analysis
Average impurity content; ash, 0.1–0.5%

Properties	Units	With grain		Against grain		Typical high-temperature properties	
		Average value	Standard deviation, %	Average value	Standard deviation, %	1,300° F	4,000° F
Young's modulus	10^6 psi	0.2					
Tensile strength	10^3 psi	0.07					
Compressive strength	10^3 psi	0.4					
Flexural strength	10^3 psi	0.2					
Density	g/cm^3	1.03					
Coefficient of thermal expansion	$10^{-6}/°C$	2.0					
Thermal conductivity	cal-cm/sec·cm^2·K	16					
Specific resistance	10^{-4} ohm-cm	38.0					
Porosity	%	48					

Source
Union Carbide

Size
Cylinder 7 1/4 in.; block 9 × 14 × 14 in., pipe 1 3/4 in. outside diameter

Manufacturing data
Calcined petroleum coke; graphitized over 2,500°C; Acheson electric furnace; machined; 1–20 ton batch size

Note: Average permeability, water (70°F, 5 psi), 1-in. thick plate: 90 $gal/ft^2/min$.

Table 5.6–2
GRADE POROUS GRAPHITE 45

Description
Graphite foam; good thermal insulator; high reproducibility; high porosity; chemical resistant; high permeability

Analysis
Average impurity content: ash, 0.1–0.5%

		With grain		Against grain		Typical high-temperature properties	
Properties	Units	Average value	Standard deviation, %	Average value	Standard deviation, %	1,300°F	4,000°F
Young's modulus	10^6 psi	0.3					
Tensile strength	10^3 psi	0.07					
Compressive strength	10^3 psi	0.5					
Flexural strength	10^3 psi	0.3					
Density	g/cm^3	1.04					
Coefficient of thermal expansion	10^{-6}/°C	2.0					
Thermal conductivity	cal-cm/sec·cm^2·K	0.18					
Specific resistance	10^{-4} ohm-cm	33.0					
Porosity	%	48					

Source
Union Carbide

Size
Cylinder 7 1/4 in.; block 9 × 14 × 14 in.; pipe 1 3/4 in. outside diameter

Manufacturing data
Calcined petroleum coke; graphitized over 2,500°C; Acheson electric furnace; machined; 1–20 ton batch size

Note: Average permeability, water (70°F, 5 psi), 1-in. thick plate: 30 gal/ft^2/min.

Table 5.6–3
GRADE POROUS GRAPHITE 60

Description
Graphite foam; good thermal insulator; high reproducibility; high porosity; chemical resistant; high permeability

Analysis
Average impurity content; ash, 0.1–0.5%

Properties	Units	With grain		Against grain		Typical high-temperature properties	
		Average value	Standard deviation, %	Average value	Standard deviation, %	1,300°F	4,000°F
Young's modulus	10^6 psi	0.3					
Tensile strength	10^3 psi	0.2					
Compressive strength	10^3 psi	0.6					
Flexural strength	10^3 psi	0.4					
Density	g/cm^3	1.05					
Coefficient of thermal expansion	$10^{-6}/°C$						
Thermal conductivity	$cal\text{-}cm/sec \cdot cm^2 \cdot K$	0.21					
Specific resistance	10^{-4} ohm-cm	30.0					
Porosity	%	48					

Source
Union Carbide

Size
Cylinder 7 1/4 in.; block 9 × 14 × 14 in.; pipe 1 3/4 in. outside diameter

Manufacturing data
Calcined petroleum coke; graphitized over 2,500°C; Acheson electric furnace; machined; 1–20 ton batch size

Note: Average permeability, water (70°F, 5 psi), 1-in. thick plate: 10 $gal/ft^2/min$.

Section 6

Materials Information

A GUIDE TO SOURCES OF INFORMATION ON MATERIALS

R. S. Marvin
G. B. Sherwood
National Bureau of Standards

I. INTRODUCTION

A substantial fraction of all technical literature deals with some type of information about materials. The *Directory of Information Resources in the United States* (see Library of Congress listing in Part II, "General Reference Centers") identifies 2,891 sources of information, other than publications, in the physical sciences and engineering. A large fraction of these also deal with some aspect of information about materials.

The general card catalogs, lists of publications, and abstracts that cover the published literature are presumably known to readers of this *Handbook* and will not be discussed here. However, this should not be taken as a suggestion that a traditional literature search can safely be omitted when seeking information about an unfamiliar material. Some knowledge of the material is essential both to pose a question that will elicit a useful response and to judge which source is likely to contain the information needed.

Means for locating unpublished reports and other sources of information are not as widely known. In Part II we describe five general referral centers. Although only one is so named, all provide a service by which an inquirer is referred to a source of information he is seeking. In Part III we describe three publications which include some of this same type of information.

One important type of information about materials is numerical data on material properties. A major problem facing scientists and engineers today involves, first, finding, and, second, evaluating the reliability of the enormous volume of property data in the literature. A number of data centers have been established to collect and evaluate such data in defined areas. In Part IV we describe some major continuing data centers. Part V gives a selected list of tables and other data publications.

The criteria on which the selections for these listings were made are given at the start of each part.

II. GENERAL REFERRAL CENTERS

There are now so many sources of information about materials aside from formal publications that four major operations, each covering a different type of information, are concerned with merely listing and indexing such sources. The Library of Congress National Referral Center maintains a very general listing of information resources with a broad definition of both information and resources. The NBS Standards Information Service concentrates on voluntary engineering standards and federal and state government specifications. The National Technical Information Service covers reports of government agencies and their contractors. The Smithsonian Science Information exchange covers research projects in progress. All four are exhaustive rather than selective. They include varying amounts of information about the sources they list, and the information they have is not independently evaluated or assessed. The National Bureau of Standards Office of Standard Reference Data has the primary responsibility within the federal government for the National Standard Reference Data System program which concentrates on the critical evaluation of numerical data expressing physical and chemical properties of well-characterized materials. As an adjunct to this role, it attempts to guide inquirers to the best existing sources of such data. These five centers and their referral services are described below.

Many other more specialized centers, some designed to serve a particular group or to cover a specific type of information, exist in both the federal and private sectors. The federal centers are included in the Library of Congress listing. Their coverage of those in the private sector is by design rather incomplete. Some additional referral services are included in the *Encyclopedia of Information Systems and Services* described in Part III, "General Referral Publications."

A critical selection and evaluation of the information contained in the files of the National Referral Center and the National Technical Information service would greatly enhance their usefulness, but it is difficult to see how such general lists could be evaluated; this would require highly qualified specialists in each of the many fields covered.

A. National Referral Center
Science and Technology Division
Library of Congress
Washington, D.C. 20540
(202) 426–5670

The National Referral Center maintains an inventory of information resources in the United States covering the physical, biological, social, and engineering sciences and related technical areas. This includes formal data compilation and/or analysis centers, university departments, producers of materials, federal and state government agencies, societies, and associations. Coverage of sources in or supported by the federal government and of major societies and associations with information services and/or extensive publishing programs is essentially complete, that of university departments and industrial sources much less so. The inventory includes many other specialized listings of information sources, particularly those maintained by other agencies of the federal government. Information is supplied by each individual source and not independently evaluated to establish validity of coverage claimed. Technical books and journals, except those issued by sources listed, are not included. The Center directs those posing questions concerning specific subjects to appropriate sources. It does not provide answers to technical questions, nor does it offer bibliographic assistance. Service is available without charge by telephone, correspondence, or through personal visits. Listings in the inventory are published in *A Directory of Information Resources in the United States,* the volumes of which are

Water, September 1966 (LC1.31:D62/3) $2.95

Federal Government, June 1967, with a supplement including government-sponsored sources (LC1.31:D62/4-974) $4.25

General Toxicology, June 1969 (LC1.31:D62/5) $3.30

Physical Sciences, Engineering, 1971 (LC1.31:D62/6) $7.40

Biological Sciences, 1972 (LC1.31:D62/7) $5.00

Social Sciences, June 1973 (LC1.31:D62/2-973) $6.90

These may be ordered from the Superintendent of Documents, U.S. Government Printing Office, Washington, D.C. 20402, citing the numbers that follow titles.

B. Standards Information Service
National Bureau of Standards
Washington, D.C. 20234
(301) 921–2587

The Standards Information Service maintains a reference collection of over 25,000 voluntary engineering standards issued by 380 U.S. technical societies, professional organizations, and trade associations. It also includes standards and specifications of U.S. Government agencies, specifications of state purchasing offices, standards and specifications of the major foreign and international standardizing bodies, and related reference books, directories, encyclopedias, guides, manuals, periodicals, and newsletters. The collection may be consulted on site (at Gaithersburg, Md., about 20 miles northwest of Washington, D.C.). Lists of standards with organizations from which copies of the standards may be obtained will be sent in response to mail or telephone inquiries. Though most of these standards are concerned with allowable dimensions, materials permitted for a given application, recommended practices, and the like, there are many specifications that deal with material properties.

Publications[a]

Breden, L. H., Ed., *A World Index of Plastics Standards*, NBS Special Publication 352, November 1971, 454 pages, $7.65, SD Catalog No. C13.10:352.

Chumas, S. J., Ed., *Directory of U.S. Standardization Activities,* NBS Special Publication 417, expected June 1975, approximately 600 pages, SD Catalog No. C13.10:417.

Chumas, S. J., *Tabulation of Voluntary Standards and Certification Programs for Consumer Products,* NBS Technical Note 762,

[a]Order from Superintendent of Documents, U.S. Government Printing Office, Washington D.C. 20402; cite SD catalog numbers.

March 1973, 119 pages, \$1.55, SD Catalog No. C13.46:762.

Grossnickle, L. L., *An Index of State Specifications and Standards,* NBS Special Publication 375, September 1973, 377 pages, \$3.70, SD Catalog No. C13.10:375.

Keysar, B. C., *Specifications and Tolerances for Reference Standard and Field Standard Weights and Measures. 2. Specifications and Tolerances for Field Standard Measuring Flask,* NBS Handbook 105-2, January 1971, 6 pages, \$0.45, SD Catalog No. C13.11:105-2.

Keysar, B. C., *Specifications and Tolerances for Reference Standards and Field Standard Weights and Measures. 3. Specifications and Tolerances for Metal Volumetric Field Standards,* NBS Handbook 105-3, May 1971, 8 pages, \$0.45, SD Catalog No. C13.11:105-3.

Slattery, W. J., Ed., *An Index of U.S. Voluntary Engineering Standards, Covering those Standards, Specifications, Test Methods, and Recommended Practices Issued by National Standardization Organizations in the United States,* NBS Special Publication 329, March 1971, 1,000 pages, \$12.25, SD Catalog No. C13.10:329.

Slattery, W. J., Ed., *An Index of U.S. Voluntary Engineering Standards, Supplement 1,* NBS Special Publication 329, Suppl. 1, December 1972, 459 pages, \$8.25, SD Catalog No. C13. 10:329/Suppl. 1.

Slattery, W. J., Ed., *An Index of U.S. Voluntary Engineering Standards, Supplement 2,* NBS Special Publication 329, Suppl. 2, expected May 1975, approximately 500 pages, SD Catalog No. C13.10:329/Suppl. 2.

Stabler, T. M., *Specifications and Tolerances for Reference Standards Field Standard Weights and Measures. 1. Specifications and Tolerances for Field Standard Weights (NBS Class F),* NBS Handbook 105-1, July 1972 (revision of NBS Handbook 105-1, April 1969), 12 pages, \$0.45, SD Catalog No. C13.11:105-1.

The above indices are organized according to a permuted title scheme so a search can be conducted on any word appearing in the title of a standard.

C. National Technical Information Service

U.S. Department of Commerce
Springfield, VA 22151
(703) 321–8523

The National Technical Information Service is the central source for the public sale of government-sponsored research and development reports and other government analyses prepared by federal agencies or their contractors or grantees, for translations by the Joint Publications Research Service, and for many publications of the Atomic Energy Commission, National Aeronautics and Space Administration, and other agencies. It offers a computer-generated custom bibliographic search system of the more than 300,000 abstracts of federally sponsored documents published since 1964. Current abstracts are published in a series of journals available by subscription. This includes a *Weekly Government Abstracts–Materials Science* publication (\$22.50 per year). Other series are listed in a free brochure entitled "NTIS Information Services." Various microform and tape services are also described, and the complete text of most items is also available. Custom searches of the whole file can be made. Write NTISearch at the above address, or telephone the above number.

D. Smithsonian Science Information Exchange, Inc.

1730 M Street, N.W.
Washington, D.C. 20036
(202) 381–5511

The Smithsonian Science Information Exchange is a source of information on research programs in progress covering a wide range of disciplines and sources of funding, both federal and nonfederal. The disciplines covered include physics, chemistry, and engineering, the environmental, agricultural, electronic, materials, and earth sciences, and many others. The SSIE data base contains information on over 100,000 current research projects, and is updated annually; the services offered include custom searches, a program of pre-run searches in selected areas, and a recently established survey of science information resources available within the federal government. Information on these services and lists of currently available searches are contained in the *Smithsonian Science Information Exchange Newsletter* (SSIE Science Newsletter), which publishes ten issues per year (\$6 [foreign \$8] payable to the Smithsonian Institution; order from Columbia Publishers Management, Box 882, Columbia Station, Columbia, Md. 21044). Examples of pre-run searches available in the

March 1973 Newsletter include the following: superalloys-corrosion, phase studies, applications, powder metallurgy and other research on superalloys; metallography – use of metallography in the study of weld joints, corrosion, fatigue damage, diffusion, and other current research. Costs of the services vary from \$35.00 to \$65.00 for most types offered. The data provided on each research project include the supporting agency, title of the project, names of principal investigators, their associates, and department of specialty, location of the research, funding for the project, and period of the project.

E. Office of Standard Reference Data
National Bureau of Standards
Washington, D.C. 20234
(301) 921–2228

The Office of Standard Reference Data has the primary responsibility within the federal government for promoting and coordinating the critical evaluation of numerical data in the physical sciences. It sponsors a number of continuing data analysis centers, individual evaluation projects, and the publications of the National Standard Reference Data System (NSRDS). It maintains a file of all such publications and of many handbooks and other sources of numerical data. Its coverage of the areas of thermodynamic and transport properties, atomic and molecular data, chemical kinetics, solid state data, nuclear data, and colloid and surface properties is established well enough that it can usually locate existing data or sources of information on properties of pure and well-characterized materials in these areas. Its coverage of mechanical properties and commercial materials is not yet established. The Office will respond to inquiries using NSRDS publications or other sources in its files or by reference to an appropriate data center or individual specialist. References or (if the information sought is not too extensive) copies or extracts of data are provided. Inquiries may be made by mail or phone. There is no charge for the service provided by this Office, but an inquiry requiring appreciable time of an associated center to answer may require a reimbursement.

Data publications sponsored by OSRD appear primarily in NBS publications and in the quarterly *Journal of Physical and Chemical Reference Data,* edited by David R. Lide, Jr. These publications are listed in Part V, "Selected Data Publications." Information on current publications and related matters are covered in the monthly *NSRDS News,* available without charge from OSRD.

III. GENERAL REFERRAL PUBLICATIONS

The publications listed below, although duplicating much of the information given by the publications and referral centers listed in Part II, are listed here for the following reasons. The *Encyclopedia of Information Systems and Services* includes a number of industrial centers and information sources not covered in the Library of Congress *Directories.* The COSATI *Directory* covers a highly selected list of continuing data centers with more information than given in the Library of Congress *Directory.* The CODATA *Compendium* is the only compilation that attempts international coverage.

A. *Encyclopedia of Information Systems and Services,* Kruzas, A. T., Ed., Edwards Brothers, Ann Arbor, MI, 1971, 1109 pages, \$55.00. Second edition in press.

This is a listing similar in form to the National Referral Center *Directory.* There is a great deal of duplication, but the *Encyclopedia* has wider coverage of industrial sources, including many which restrict their services to members of the sponsoring organization or to interlibrary loans of documents. It includes the following indices: alphabetical, personal name, subject, abstracting and indexing services, computer applications and services, consulting and planning services, data collection and analysis centers, micrographic applications and services, networks and cooperative programs, research and research projects, selective dissemination of information services, and serial publication index. It should be noted that "data" as used here are not limited to materials property data, but include political, demographic, public opinion, and other types.

B. *Directory of Federally Supported Information Analysis Centers,* 3rd ed., 1974. Report No. COSATI-70-1, prepared by the National Referral Center for Science and Technology for the Committee on Scientific and Technical Information; available from National Technical Information Service, 5285 Port Royal Road, Springfield,

VA 22151 as ISBN 0-8444-0128-5; price $3.75 (hardcopy) or $1.45 (microfiche).

This includes a small list of centers (108) supported by the federal government and selected to include only those which emphasize analysis, interpretation, evaluation, and repackaging of information, generally by subject specialists. It does not include raw data files, conventional libraries, abstracting, indexing, and accession services, but does include name of director, year started, and size of staff.

C. *CODATA International Compendium of Numerical Data Projects,* A Survey and Analysis, by the Committee on Data for Science and Technology (CODATA) of the International Council of Scientific Unions, Springer-Verlag, New York, 1969, xxiii + 295 pages, $14.30.

This provides a thorough description of the operations and publications of continuing data analysis centers throughout the world, including descriptions of the national data programs in the U.S.A., U.K., and U.S.S.R., and a number of internationally known handbooks such as Landolt-Börnstein. Most of the information is of more interest to physicists and chemists than to engineers.

IV. SELECTED DATA EVALUATION CENTERS

This part lists continuing data evaluation centers. The first four subdivisions cover centers whose primary goal is the critical evaluation of numerical data within a defined field, leading to the selection of best values with associated limits of uncertainty within which the true value is expected to lie. The results obtained are disseminated primarily through published compilations, reviews, and bibliographies as indicated. Most centers will attempt to answer specific questions about data in their field to the extent that time and personnel permit. In some cases a fee for such services is charged. Prices and ordering information for these publications are given in Part V, "Selected Data Publications."

The variability of composition and structure found in many commercial materials and the influence of processing on many of their significant properties often preclude the assignment of "true" values to their properties. Rather, a range, an average, or a minimum value may be sought, depending on the intended use. Also, qualitative or semiquantitative information about behavior (e.g., machinability, corrosion resistance) is often as important as data on properties.

A number of organizations that can provide information about both properties and behavior of commercial materials are listed in the publications of the National Referral Center described in Part II.A. In Part IV.E we describe two of these centers which cover properties of metals, ceramics, and plastics.

A. Solid State Data Centers

1. *Title:* Crystal Data Center (H. M. Ondik)
   Location: Inorganic Materials Division, National Bureau of Standards, Washington, D.C. 20234
   Coverage: Materials coverage includes crystalline inorganic compounds, minerals, and intermetallic compounds. Data coverage includes cell constants, axial ratios, reduced cell parameters, and density. Other properties included are melting point, color, twinning, cleavage, crystal habit, indices of refraction, and optic axial angle.
   Form of output: Crystal Data Determinative Tables

2. *Title:* Phase Diagrams for Ceramists Data Center (E. M. Levin)
   Location: Inorganic Materials Division, National Bureau of Standards, Washington, D.C. 20234
   Coverage: Phase diagrams for one-, two-, three-, and multicomponent systems involving pressure, temperature, and composition as variables. Materials covered are those of special interest to ceramists: metal oxide systems (including Si), systems with oxygen-containing radicals, systems containing halides, sulfides, cyanides.
   Form of output: Publications of the American Ceramic Society

3. *Title:* Alloy Data Center (G. C. Carter)
   Location: Metallurgy Division, National Bureau of Standards, Washington, D.C. 20234
   Coverage: Materials coverage includes metals, semimetals, intermetallic compounds, and alloys. Physical properties coverage includes

electronic transport and magnetic and mechanical properties. Emphasis is on NMR (Knight or metallic shifts) and soft X-ray data.

Form of output: Metallic shifts in NMR – *Progress in Materials Science*; *NBS Technical Notes; Journal of Physical and Chemical Reference Data*

4. *Title:* Diffusion in Metals Data Center (J. R. Manning)
   Location: Metallurgy Division, National Bureau of Standards, Washington, D.C. 20234
   Coverage: Diffusion in metals and alloys. Primary emphasis is on diffusion coefficients and activation energies.
   Form of output: Journal of Physical and Chemical Reference Data

5. *Title:* Superconductive Materials Data Center (B. W. Roberts)
   Location: General Electric Research and Development Center, P.O. Box 8, Schenectady, N.Y. 12301
   Coverage: Properties of superconductive materials including critical temperatures, critical magnetic fields, crystallographic parameters, critical magnetic fields, crystallographic parameters, critical magnetic fields of high field superconductors, and some thermodynamic data.
   Form of output: NBS Technical Notes (Latest are NBS Technical Notes 724 and 825.)

6. *Title:* Research Materials Information Center (T. F. Connolly)
   Location: U.S. Atomic Energy Commission, Oak Ridge National Laboratory, P.O. Box X, Oak Ridge, TN 37830
   Coverage: Materials coverage includes semiconductors, metals, alloys, and single crystals. Physical properties coverage includes magnetic, superconductive, electronic, optical, and electrical properties. Information provided on preparation and properties of ultrapure inorganic research materials.
   Form of output: Solid State Literature Guides
7. *Title:* Cambridge Crystallographic Data Center (O. Kennard)
   Location: University Chemical Laboratory, Lensfield Road, Cambridge CB2 1EW, Eng.
   Coverage: Materials coverage includes crystalline organic and organometallic compounds. Data coverage includes cell constants, axial ratios, reduced cell parameters, and density. Other properties included are melting point, color, twinning, cleavage, crystal habit, indices of refraction, and optical axial angle.
   Form of output: Crystal Data Determination Tables; Molecular Structures and Dimensions (published by N. V. A. Oosthoek's Uitgevers Mij, Utrecht, for the Crystallographic Data Center and the International Union of Crystallography; three volumes to date)

8. *Title:* Electronic Properties Information Center (Y. S. Touloukian)
   Location: Purdue University, West Lafayette, IN 47906
   Coverage: Electronic, electrical, magnetic, and optical properties.
   Note: EPIC was operated by the Hughes Aircraft Co. from 1961–1972. It was combined with the Thermophysical Properties Research Center (see Part IV.B.7, below) in June 1973 and now operates as a component of the Center for Information and Numerical Data Analysis and Synthesis.

B. Thermodynamic and Transport Properties Data Centers

1. *Title:* Chemical Thermodynamics Data Group (D. D. Wagman, W. H. Evans)
   Location: Physical Chemistry Division, National Bureau of Standards, Washington, D.C. 20234
   Coverage: Heats, entropies, and free energies of formation of all inorganic substances in the standard state, heats of transition, heat capacities, and thermodynamic properties of selected organic compounds.
   Form of output: NBS *Technical Note 270* series, *Journal of Physical and Chemical Reference Data,* reports, *Bulletin of Thermodynamics and Thermochemistry*

2. *Title:* Thermodynamics Research Center (B. J. Zwolinski)
   Location: Texas A&M University, College Station, TX 77843
   Coverage: Thermodynamic properties, selected physical properties, and spectral behavior of

organic compounds (IR, NMR, Mass, Raman, UV).

Form of output: API 44 Tables, TRC (formerly MCA) Tables, *Journal of Physical and Chemical Reference Data,* other technical journals, *Bulletin of Thermodynamics and Thermochemistry*

3. *Title:* Cryogenic Data Center (N. A. Olien)
   Location: Cryogenics Division, National Bureau of Standards, Boulder, CO 80302
   Coverage: Low temperature physics and chemistry and cryogenic engineering including PVT properties of fluids, thermodynamic properties, and some physical and mechanical properties.
   Form of output: NBS publication series including NSRDS-NBS series, government, especially NASA, reports, *Journal of Physical and Chemical Reference Data,* other technical journals, three current awareness services

4. *Title:* Molten Salts Data Center (G. J. Janz)
   Location: Rensselaer Polytechnic Institute, Troy, NY 12181
   Coverage: Density, electrical conductivity, viscosity, surface tension, and thermodynamic and electrochemical properties of molten salt systems.
   Form of output: Handbooks, NSRDS-NBS series, *Journal of Physical and Chemical Reference Data*, other technical journals, current awareness service

5. *Title:* Aqueous Electrolyte Data Center (R. L. Nuttall)
   Location: Physical Chemistry Division, National Bureau of Standards, Washington, D.C. 20234
   Coverage: Thermodynamic properties of aqueous electrolyte systems.
   Form of output: NSRDS-NBS series, NASA reports, *Journal of Physical and Chemical Reference Data*

6. *Title:* High Pressure Data Center (H. T. Hall, L. Merrill)
   Location: Brigham Young University, Provo, UT 84601
   Coverage: High pressure properties of solids including measurement of pressure, structure, and other properties of high pressure phases.
   Form of output: Journal of Physical and Chemical Reference Data, other technical journals, NSRDS bibliographic series, current awareness service

7. *Title:* Thermophysical Properties Research Center (Y. S. Touloukian)
   Location: Purdue University, West Lafayette, IN 47906
   Coverage: Thermal conductivity, specific heat, thermal radiative properties, thermal diffusivity, viscosity, and thermal expansion at all temperatures.
   Form of output: Thermophysical Properties of Matter (Plenum Press, 13 volumes), *Thermophysical Properties of High Temperature Materials* (Macmillan, 6 volumes), *Thermophysical Properties Research Literature Retrieval Guide*, 2nd ed. (Plenum Press, 3 volumes), NSRDS-NBS series, government reports, *Journal of Physical and Chemical Reference Data,* and other technical journals.
   Note: This Center and EPIC (see Part IV.A.8. above) comprise the Thermophysical and Electronic Properties Information Analysis Center (TEPIAC); these two and the Underground Excavation and Rock Properties Information Center (UERPIC) constitute the Center for Information and Numerical Data Analysis and Synthesis (CINDAS).

C. Atomic and Molecular Properties Data Centers

1. *Title:* Atomic Energy Levels Data Center (W. C. Martin)
   Location: Optical Physics Division, National Bureau of Standards, Washington, D.C. 20234
   Coverage: Atomic spectra and atomic energy levels for both neutral and ionic species.
   Form of output: Journal of Physical and Chemical Reference Data, NBS publications

2. *Title:* Atomic Transition Probabilities Data Center (W. L. Wiese)
   Location: Optical Physics Division, National Bureau of Standards, Washington, D.C. 20234
   Coverage: Radiative transition probability of

atoms and atomic ions in the gas phase. All elements are covered.

Form of output: Journal of Physical and Chemical Reference Data, NBS publications

3. *Title:* Data Center on Atomic Line Shapes and Shifts (W. L. Wiese)

   Location: Optical Physics Division, National Bureau of Standards, Washington, D.C. 20234

   Coverage: Data on atomic spectral line shapes and shifts. All elements are covered.

   Form of output: Journal of Physical and Chemical Reference Data, NBS publications

4. *Title:* Diatomic Molecule Spectral Data Center (P. H. Krupenie)

   Location: Optical Physics Division, National Bureau of Standards, Washington, D.C. 20234

   Coverage: Spectroscopic data for diatomic molecules, with emphasis on the optical region, and molecular parameters derived from spectroscopic measurements.

   Form of output: Journal of Physical and Chemical Reference Data, NBS publications

5. *Title:* Microwave Spectral Data Center (F. J. Lovas)

   Location: Optical Physics Division, National Bureau of Standards, Washington, D.C. 20234

   Coverage: Microwave absorption spectra of gases, including transition frequencies and intensities as well as molecular constants derived from the spectral data.

   Form of output: Journal of Physical and Chemical Reference Data, NBS publications

6. *Title:* Atomic Collision Cross Section Information Center (L. J. Kieffer)

   Location: Joint Institute for Laboratory Astrophysics, University of Colorado, Boulder 80302

   Coverage: Cross sections (low energy range) for collision of electrons and photons with atoms, ions, and small molecules.

   Form of output: Critical reviews and data compilations, *Journal of Physical and Chemical Reference Data,* NBS publications, *Reviews of Modern Physics*, *Atomic Data*; bibliographies–JILA Information Center Reports and NSRDS OSRDB series available from NTIS.

7. *Title*: Atomic and Molecular Processes Information Center (C. F. Barnett)

   Location: Oak Ridge National Laboratory, P.O. Box Y, Oak Ridge, TN 37831

   Coverage: Atomic and small molecule collisions, specifically heavy particle-heavy particle atomic collision cross sections, particle interactions with quasistatic electric and magnetic fields, particle penetration into macroscopic matter, and energetic particle interactions with surfaces.

   Form of output: Critical monographs on various classes of heavy particle interactions published by John Wiley & Sons; bibliographies and a directory of workers in the field published as AMPIC publications.

8. *Title*: Data Center for Atomic and Molecular Ionization Processes (H. M. Rosenstock)

   Location: Physical Chemistry Division, National Bureau of Standards, Washington, D. C. 20234

   Coverage: Ionization and appearance potentials and the properties of excited ionic states – the energetics of gaseous ionization. All atomic and molecular species are included.

   Form of output: Journal of Physical and Chemical Reference Data, NBS publications

9. *Title*: X-ray Attenuation Coefficient Information Center (J. H. Hubbell)

   Location: Center for Radiation Research, National Bureau of Standards, Washington, D. C. 20234

   Coverage: Attenuation coefficients for high energy photon (X-ray, gamma ray) interaction with matter, including Compton and Rayleigh scattering, atomic photo-effect, and electron-positron pair production.

   Form of output: Journal of Physical and Chemical Reference Data, NBS publications

D. Chemical Kinetics Data Centers

1. *Title*: Chemical Kinetics Information Center (D. Garvin)

   Location: Physical Chemistry Division, National Bureau of Standards, Washington, D. C. 20234

Coverage: Rates of homogeneous chemical reactions in gaseous, liquid, and solid phases. Among properties of concern are rate constants, frequency factors, heats, energies, and entropies of activation.

Form of output: Journal of Physical and Chemical Reference Data, NBS publications, and bibliographies on kinetic data

2. *Title*: Radiation Chemistry Data Center (A. Ross)

Location: Radiation Laboratory, University of Notre Dame, Notre Dame, IN 46556

Coverage: Radiation yields and kinetic data on elementary processes in irradiated substances (organic and inorganic, aqueous and non-aqueous solutions, solids, and gases).

Form of output: Journal of Physical and Chemical Reference Data, NBS publications, and periodic indexes of literature covered.

E. Metals, Ceramics, and Plastics Data Centers

1. *Title*: Metals and Ceramics Information Center (H. Dana Moran; inquiries to Roy Endebrock)

Location: Battelle Columbus Laboratories, 505 King Avenue, Columbus, OH 43201

Coverage: Metals: titanium, aluminum, magnesium, beryllium, refractory metals, high-strength steels, superalloys, rhenium, vanadium; ceramics: borides, carbides, carbon/graphite, nitrides, oxides, sulfides and silicides, intermetallics, selected glasses and glass-ceramics; composites containing metallic or ceramic elements; mechanical and physical properties, plus other information including applications and test methods.

Form of output: Weekly and monthly reviews of current technology, special reports and handbooks available from National Technical Information Service (see Part II.C); monthly newsletter (no charge – request from MCIC, above address); technical advice, literature searches at fees quoted in advance.

2. *Title*: The Plastics Technical Evaluation Center (PLASTEC) (H. E. Pebly, Jr.)

Location: Picatinny Arsenal, Dover, NJ 07801

Coverage: Technological information and data on plastics, including composites and adhesives. Coverage includes mechanical and electrical properties, degradation, processing and uses.

Form of output: Bibliographies, patent surveys, guides to test methods, and occasional state-of-the-art reports available from National Technical Information Service (see Part II.C). Special literature searches and inquiry service at fees quoted in advance.

V. SELECTED DATA PUBLICATIONS

Our selection of a limited number of the many data compilations available is of necessity somewhat arbitrary. The examples listed present primarily data compiled directly from the original sources, either with an attempt at evaluation of reliability or a selection of recommended values by recognized authorities.

Part V.A lists three multivolume handbooks covering a broad range of both materials and properties, plus one single-volume handbook with broad coverage, selected because of its currency. Parts V.B through V.E list a number of more specialized compilations grouped according to the properties covered. The titles of the individual volumes (except for the International Critical Tables) and publications give a reasonable indication of their coverage. Explanatory notes have been appended where the titles are clearly inadequate. Within each section we list first any special periodicals or series, except for the *Journal of Physical and Chemical Reference Data* and the NSRDS-NBS series, since their coverage is so broad that it seems more useful to list their individual articles in the appropriate places. All spectral data have been included in Part V.B., "Nuclear, Atomic, and Molecular Data Compilations." Phase diagrams are included in Part V.C, "Thermodynamic and Transport Properties; Phase Diagrams; Surface Properties." These divisions list compilations primarily covering pure and well-characterized materials. Part V.F, "Handbooks Emphasizing Properties of Commercial Materials," lists a few compilations dealing primarily with commercial alloys, plastics, and ceramics. All of the publications listed in Parts V.B through F are reviewed periodically by the organizations that prepared or sponsored them with the object of ensuring revision when required.

Prices[a] and ordering information are included.

[a]Prices are given as of January 1975, and are subject to change without notice.

Publications in the NSRDS-NBS and NBS Technical Note series, NBS Miscellaneous Publications, and NBS Special Publications should be ordered from the Superintendent of Documents, U.S. Government Printing Office, Washington, D.C. 20402 at the prices shown, giving both the series and SD numbers. Publications marked with an asterisk (*) should be ordered from the National Technical Information Service, 5285 Port Royal Road, Springfield, VA 22151. Publications marked JPCRD appear in the quarterly, *Journal of Physical and Chemical Reference Data*; subscriptions are $25.00 per volume to members of the American Chemical Society and to members of societies affiliated with the American Institute of Physics, and $75.00 to others ($28.00 and $78.00 outside the U.S.A., Canada, and Mexico). Reprints of individual papers are available at the prices shown. Both subscriptions and orders for individual reprints should be sent to the American Chemical Society, 1155 16th Street, NW, Room 604, Washington, D.C. 20036.

A. General Handbooks

1. International Critical Tables of Numerical Data, Physics, Chemistry, and Technology

International Critical Tables of Numerical Data, Physics, Chemistry, and Technology, published for the National Academy of Sciences-National Research Council by McGraw-Hill, 1221 Avenue of the Americas, New York. Volumes I–VII plus index, 1926–33, $37.50 each, $265.00 for the set.

These Tables were initiated as a project of the International Union of Theoretical and Applied Chemistry. They were compiled by specialists throughout the world, chosen for their qualifications to select the "best" values in each category. Values selected are primarily from the literature published through 1923, and there is comprehensive coverage of physical and chemical properties of pure substances and of many commercial materials (alloys, clays, wood, etc.). Arrangement is not strictly logical, but the comprehensive index permits ready location of specific information. Although this source is still often quoted, subsequent improvements in purification and characterization of materials and in methods of measurements have made many of the tables obsolete. The original goal of compiling periodic revisions was never realized.

2. Landolt-Börnstein

"Landolt-Börnstein," Landolt, Hans Heinrich: *Zahlenwerte und Funktionen aus Physik, Chemie, Astronomie, Geophysik und Technik, Secheste Auflage,* Euken, A. and Hellwege, K. H., Eds., Springer-Verlag, Berlin (in the U.S.A., Springer-Verlag, New York, 175 Fifth Ave., New York 10010), 1950 –.

This is one of the oldest and probably the most extensive of the general compilations. The first edition (one volume) was published in 1883; so far 26 parts of the sixth edition have appeared, starting in 1950. It is in German with few exceptions. The original aim of "Landolt-Börnstein" was "to make exact results (that can be estimated numerically) of physical, chemical, and technical research as conveniently and completely (comprising all literature data) accessible as possible. . ." A consistent organization is followed with only a few exceptions.

The editors have concluded that such a systematic organization is no longer feasible. Hence, starting in 1961, the same editors and publisher initiated the "New Series": *Zahlenwerte und Funktionen aus Naturwissenschaften und Technik, Neue Serie – Numerical Data and Functional Relationships in Science and Technology, New Series,* Hellwege, K. H., Ed. These tables will be published in an order dictated by the development of individual fields of specialization. In effect, it is an open-ended series with no predetermined order of publication or termination. As suggested by the title, this New Series includes parallel text in German and English.

The various volumes in both series contain tables, graphs, and explanatory text (more text than the *International Critical Tables*). The authors of various sections are recognized authorities from throughout the world, but the sponsorship of both works is private. We give below a listing of the volumes of both series that have appeared to date.

Zahlenwerte und Funktionen aus Physik, Chemie, Astronomie, Geophysik und Technik, Sechste Auflage

Band I: *Atom und Molekularphysik*

1. Teil: *Atome und Ionen,* Euken, A., Ed., 1950, xii + 441 pages, $53.70.
2. Teil: *Molekeln I (Kerngerüst),* Euken, A. and Hellwege, K. H., Eds., 1951, viii + 571 pages, $71.50.

3. Teil: *Molekeln II (Elektronenhülle),* Euken, A. and Hellwege, K. H., Eds., 1951, xi + 724 pages, $92.90.
4. Teil: *Kristalle,* Hellwege, K. H., Ed., 1955, xi + 1,007 pages, $135.50.
5. Teil: *Atomkerne und Elementarteilchen,* Hellwege, K. H., Ed., 1952, viii + 470 pages, $62.90.

Band II: *Eigenschaften der Materie in Ihren Aggregatzuständen*

1. Teil: *Mechanische-thermische Zustandgrössen,* d'Ans, J. et al., 1971, xv + 944 pages, $207.20.
2. Teil: *Gleichgewichte ausser Schmelzgleichgewichten*
   - 2a. *Gleichgewichte Dampf-Kondensat und Osmotische Phänomene,* Schäfer, K. and Lax, E., Eds., 1960, xi + 974 pages, $190.60.
   - 2b. *Losungsgleichgewichte I,* Schäfer, K. and Lax, E., Eds., 1962, x + 98 pages, $217.20.
   - 2c. *Losungsgleichgewichte II,* Schäfer, K. and Lax, E., Eds., 1964, viii + 731 pages, $165.10.
3. Teil: *Schmelzgleichgewichte und Grenzflächenerscheinungen,* Schäfer, K. and Lax, E., Eds., 1956, xi + 535 pages, $84.40.
4. Teil: *Kalorische Zustandsgrössen,* Schäfer, K. and Lax, E., Eds., 1961, xii + 863 pages, $186.50.
5. Teil: *Transportphänomene*
   - 5a. *Transportphänomene I (Viskosität und Diffusion),* Schäfer, K., Ed., 1969, xi + 729 pages, $158.40.
   - 5b. *Transportphänomene II, Kinetik Homogene Gasgleichgewichte,* Schäfer, K., Ed., 1968, xi + 397 pages, $98.10.
6. Teil: *Elektrische Eigenschaften I,* Hellwege, K. H. and Hellwege, A. M., Eds., 1959, xv + 1,018 pages, $190.60.
7. Teil: *Elektrische Eigenschaften II (Elektrochemische Systeme),* Hellwege, K. H., Hellwege, A. M., Schäfer, K., and Lax, E., Eds., 1960, xii + 959 pages, $203.50.
8. Teil: *Optische Konstanten,* Hellwege, K. H. and Hellwege, A. M., Eds., 1962, xv + 901 pages, $202.40.
9. Teil: *Magnetische Eigneschaften I (Magnetic Properties I),* Hellwege, K. H. and Hellwege, A. M., Eds., 1962, xxv + 1,934 pages, $210.90.
10. Teil: *Magnetische Eigenschaften II (Magnetic Properties II),* Hellwege, K. H. and Hellwege, A. M., Eds., 1967, iv + 173 pages, $45.20.

Band III: *Astronomie und Geophysik,* Bartels, J. and ten Bruggencate, P., Eds., 1952, xviii + 795 pages, $105.50.

Band IV: *Technik*

1. Teil: *Stoffwerte und mechanisches Verhalten von Nichtmetallen,* Schmidt, E., Ed., 1955, xvi + 881 pages, $122.50.
2. Teil: *Stoffwerte und Verhalten von metallischen Werstoffen*
   - 2a. *Grundlagen, Prüfverfahren, Eisenwerkstoffe,* Borchers, H. and Schmidt, E., Eds., 1963, xii + 888 pages, $199.10.
   - 2b. *Sinterwerkstoffe, Schwermetalle (ohne Sonderwerkstoffe),* Borchers, H. and Schmidt, E., Eds., 1964, xx + 1,000 pages, $225.70.
   - 2c. *Leichtmetalle, Sonderwerkstoffe, Halbleiter, Korrosion,* Borchers, H. and Schmidt, E., Eds., 1965, xx + 976 pages, $220.60.
3. Teil: *Elektrotechnik, Lichttechnik, Röntgentechnik,* Schmidt, E., Ed., 1957, xv + 1,076 pages, $168.40.
4. Teil: *Wärmetechnik*
   - 4a. *Wärmetechnische Messverfahren Thermodynamische Eigenschaften homogenerstoffe,* Hausen, H., Ed., 1967, xii + 944 pages, $217.20.
   - 4b. *Thermodynamische Eigenschaften von Gemischen, Verbrennung; Wärmeübertragung,* Hausen, H., Ed., 1972, xviii + 771 pages, $177.60.

Zahlenwerte und Funktionen aus Naturwissenschaften und Technik, Neue Serie – Numerical Data and Functional Relationships in Science and Technology, New Series

Group I: *Kernphysik und Kerntechnik –*

Nuclear Physics and Technology

Volume 1. *Energy Levels of Nuclei: A = 5 to A = 257,* Ajzenberg-Selove, F. et al.; Hellwege, A. M. and Hellwege, K. H., Eds., 1961, xii + 813 pages, $90.30.

Volume 2. *Nuclear Radii,* Colland, H. R., Elton, L. R. B., and Hofstadter, R.; Schopper, H., Ed., 1967, vi + 54 pages, $16.30.

Volume 3. *Numerical Tables for Angular Correlation Computations in α-, β,- and γ-Spectroscopy: 3j-, 6j,- 9j- Symbols, F- and Γ- Coefficients,* Apple, H.; Schopper, H., Ed., 1968, vi + 1,202 pages, $139.50.

Volume 4. *Numerical Tables for Beta-Decay and Electron Capture,* Behrens, H. and Jänecke, J.; Schopper, H., Ed., 1969, vi + 316 pages, $43.70.

Volume 5. *Q-values and Excitation Functions of Nuclear Reactions,* Schopper, H., Ed.

Part a. *Q-values,* Keller, K. A., Lange, J., and Münzel, H., 1973, vii + 666 pages, $118.40.

Part b. *Excitation Functions for Charged-particle Induced Nuclear Reactions,* Keller, K. A., Lange, J., Münzel, H., and Pfennig, G., 1973, iv + 493 pages, $91.80.

Part c. *Systematics of Excitation Functions for Nuclear Reactions Induced by p, d, ^{3}He and α,* in preparation.

Volume 6. *Properties and Production Spectra of Elementary Particles,* Diddens, A. N., Pilkuhn, H., and Schlüpmann, K.; Schopper, H., Ed., 1972, xi + 164 pages, $43.70.

Group II: *Atom und Molekularphysik – Atomic and Molecular Physics*

Volume 1. *Magnetic Properties of Free Radicals,* Fischer, H.; Hellwege, K. H. and Hellwege, A. M., Eds., 1965, x + 154 pages, $28.90.

Volume 2. *Magnetic Properties of Coordination and Organo-Metallic Transition Metal Compounds,* König, E.; Hellwege, K. H. and Hellwege, A. M., Eds., 1966, xi + 578 pages, $98.80.

Volume 3. *Luminescence of Organic Substances,* Schmillen, A. and Legler, R.; Hellwege, K. H. and Hellwege, A. M., Eds., 1967, vii + 416 pages, $83.30.

Volume 4. *Molecular Constants from Microwave Spectroscopy,* Starck, B.; Hellwege, K. H. and Hellwege, A. M., Eds., 1967, ix + 225 pages, $47.00.

Volume 5. *Molecular Acoustics,* Schaaffs, W.; Hellwege, K. H. and Hellwege, A. M., Eds., 1967, xi + 286 pages, $66.30.

Volume 6. *Molecular Constants from Microwave-, Molecular Beam-, and ESR-Spectroscopy* (supplement and extension to *Sechste Auflage,* II/4), 1974, $182.90.

Group III: *Kristall- und Festkörperphysik – Crystal and Solid State Physics*

Volume 1. *Elastic, Piezoelectric, Piezooptic, and Electrooptic Constants of Crystals,* Bechmann, R. and Hearmon, R. F. S.; Hellwege, K. H. and Hellwege, A. M., Eds., 1966, ix + 160 pages, $28.90.

Volume 2. *Elastic, Piezoelectric, Piezooptic, and Electrooptic Constants, and Nonlinear Dielectric Susceptibilities of Crystals* (supplement and extension to Volume 1), Bechmann, R., Hearmon, R. F. S., and Kurtz, S. K.; Hellwege, K. H. and Hellwege, A. M., Eds., 1969, ix + 232 pages, $50.40.

Volume 3. *Ferro- and Antiferroelectric Substances,* Mitsui, T. et al.; Hellwege, K. H. and Hellwege, A. M., Eds., 1969, viii + 584 pages, $117.00.

Volume 4. *Magnetic and Other Properties of Oxides and Related Compounds,* Hellwege, K. H. and Hellwege, A. M., Eds.

Part a. Goodenough, J. B. et al., 1970, xv + 367 pages, $80.70.

Part b. Bonnenberg, D. et al., 1970, xvi + 666 pages, $158.40.

Volume 5. *Structure Data of Organic Crystals,* Schudt, E. and Weitz, G.; Hellwege, K. H. and Hellwege, A. M., Eds. (in two parts, not sold separately)

Part a. $C \ldots C_{13}$, 1971, xxi + 736 pages,

Part b. $C_{14} \ldots C_{120}$, 1971, vii + 737 pages, $355.20, for both parts.

Volume 6. *Structure Data of Elements and Intermetallic Phases,* Eckerlin, P. and Kandler, H.; Hellwege, K. H. and

Hellwege, A. M., Eds., 1971, xviii + 1,019 pages, $229.40.

Volume 7. *Structure Data of Inorganic Compounds,* Pies, W. and Weiss, A.; Hellwege, K. H. and Hellwege, A. M., Eds.; compounds that are not recorded in Volumes III/5 and III/6.

Part a. *Key Elements F, Cl, Br, I (VII Main Group), Halides and Complex Halides,* 1973, xxxii + 647 pages, $178.80.

Parts b through f. In preparation.

Part g. References for Group III, Volume 7, 1974, 457 pages, $90.20.

Part h. Index for Group III, Volume 7, in preparation.

Volume 8. *Epitaxy Data of Inorganic and Organic Crystals,* Gebhardt, M. and Neuhaus, A., 1973, vii + 186 pages, $43.70.

Group IV: *Macroscopic and Technical Properties of Matter Phosphoresyenz anorganischer Substanzen – Phosphorescence (Luminescence) of Inorganic Substances,* in preparation.

Group V: *Geophysics and Space Research Aeronomy; Cosmic Radiation,* in preparation.

Group VI: *Astronomie, Astrophysik und Weltraumforschung – Astronomy, Astrophysics and Space Research*

Volume 1. *Astronomy and Astrophysics,* Voigt, H. H., Ed., 1965, xxxix + 711 pages, $133.80.

3. Tables de Constantes Selectionnees

Tables de Constantes Selectionnees – Tables of Selected Constants, Pergamon Press, Maxwell House, Elmsford, N.Y.

Starting in 1910, a series of Annual Tables transcribing results from the original literature into a single series was prepared in Paris. The original coverage was broad and comprehensive rather than selective; the present series aims at selectivity, but it covers primarily nuclear, atomic, and molecular properties. Pergamon Press is the publisher of Volume 8 et sequnetes below, and also handles distribution of the earlier volumes and issues of the Annual Tables (up to 1936) still available.

Volume 1. *Longuers d'Onde d'Emissions et des Discontinuités d'Absorption X–Wavelengths of Emission and Discontinuities in Absorption of X-Rays,* Cauchois, Y. and Hulubei, H., 1947, 199 pages (out of print).

Volume 2. *Physique Nucléaire – Nuclear Physics,* Grégoire, R., Joliot-Curie, F., and Joliot-Curie, I., 1948, 131 pages (out of print).

Volume 3. *Pouvoir Rotatoire Magnétique (Effet Faraday) – Magnetic Rotatory Power (Faraday Effect),* de Mallemann, R.; *Effet Magnéto-Optique de Kerr – Magneto-Optic Effect (Kerr),* Suhner, F., 1951, 137 pages, $4.50.

Volume 4. *Données Spectroscopiques Concernant les Molécules Diatomiques – Spectroscopic Data for Diatomic Molecules,* Rosen, B., Barrow, R. F., Caunt, A. D., Downie, A. R., Herman, R., Huldt, E., MacKellar, A., Miescher, E., and Wieland, K., 1951, 361 pages, $15.00.

Volume 5. *Atlas des Longueurs d'Onde Caractéristiques des Bandes d'Emission et d'Absorption des Molécules Diatomiques – Atlas of Characteristic Wavelengths for Emission and Absorption Bands of Diatomic Molecules,* a continuation of Volume 4 by the same authors, 1952, 389 pages, $17.50.

Volume 6. *Pouvoir Rotatoire Naturel I-Steroides – Optical Rotatory Powers I-Steroids,* Mathieu, J.-P. and Petit, A., 1956, 507 pages (out of print).

Volume 7. *Diamagnétisme et Paramagnétisme – Diamagnetism and Paramagnetism,* Foëx, G.; *Relaxation Paramagnétique – Paramagnetic Relaxation,* Gorter, C. J. and Smits, L. J., 1957, 317 pages, $29.00.

Volume 8. *Potentiels d'Oxydo-Réduction – Oxidation-Reduction Potentials,* Charlot, G., Bézier, D., and Courtot, J., 1958, 41 pages, $6.00.

Volume 9. *Pouvoir Rotatoire Naturel II. Tritérpenoïdes – Optical Rotatory Power II. Triterpenoids,* Mathieu, J.-P. and Ourisson, G., 1958, 302 pages, $21.00.

Volume 10. *Pouvoir Rotatoire Naturel III. Amino-acides – Optical Rotatory Power III. Amino Acids,* Mathieu, J.-P., Roche, J., and

Desnuelle, P., 1959, 61 pages, $6.50.

Volume 11. *Pouvoir Rotatoire Naturel IV. Alcaloïdes – Optical Rotatory Power IV. Alkaloids,* Mathieu, J.-P. and Janot, M. M., 1959, 211 pages, $24.00.

Volume 12. *Semi-Conducteurs – Semi-Conductors,* Aigrain, P. and Balkanski, J., 1961, 78 pages, $10.50.

Volume 13. *Rendements Radiolytiques – Radiolytic Yields,* Haïssinsky, M. and Magat, M., 1963, 230 pages, $25.00.

Volume 14. *Pouvoir Rotatoire Naturel Ia. Stéroïdes – Optical Rotatory Power Ia. Steroids,* Jacques, J., Kagan, H., Ourisson, G., and Allard, S., 1965, 1,046 pages, $60.00.

Volume 15. *Données Relatives aux Sesquiterpenoïdes – Data Relative to Sesquiterpenoids,* Ourisson, G. and Munavalli, S., 1966, 70 pages, $18.00.

Volume 16. *Metaux, Données Thermiques et Mechaniques – Metals, Thermal and Mechanical Data,* Allard, S., 1969, 252 pages, $40.00.

Volume 17. *Données Spectroscopiques Relatives aux Molécules Diatomiques – Spectroscopic Data Relative to Diatomic Molecules,* Rosen, B., Ed., 1970, 55 pages, $70.00.

4. American Institute of Physics Handbook

American Institute of Physics Handbook, 3rd Ed., Gray, D. E., Coordinating Ed., McGraw-Hill, New York, vii + 2,442 pages, $49.50.

Most branches of science and engineering are served by one or more single-volume handbooks, including the popular *CRC Handbook of Chemistry and Physics* and several others from the publishers of this volume. The *AIP Handbook* is singled out for mention here because all sections of its current edition have been revised or completely rewritten by recognized authorities. Therefore, both the data and the extensive bibliographies are reasonably current. The texts are brief, but generally adequate to make it a self-contained reference. The major sections, all of which (except the first) include selected property data, are Mathematics Bibliography, SI Units; Mechanics; Acoustics; Heat; Electricity and Magnetism; Optics; Atomic and Molecular Physics; Nuclear Physics; and Solid State Physics. Each section includes tables and extensive references, and a complete index follows the last section.

B. Nuclear, Atomic, and Molecular Data Compilations

This part lists selected compilations of various spectra,[a] energy levels, cross sections, absorption coefficients, transition probabilities, dipole moments, and related quantities.

The following periodicals are edited by Katherine Way and published by Academic Press, 111 Fifth Avenue, New York, NY 10003. *Nuclear Data Tables* presents compilations and evaluations of experimental and theoretical data in various areas of nuclear-structure physics. Cumulative indexes appear at the end of each volume.

Nuclear Data Tables "A," Volumes 1–8, 1966–1971, $22.00 per volume

Nuclear Data Tables "A," Volumes 9–10, 1971, $24.00 per volume

Nuclear Data Tables "A," Volume 11, 1972, $ 36.00

Nuclear Data Sheets presents periodically revised collections of experimental information on nuclear structure organized by the mass number, A. A cumulative list of A-chains published and the references are included.

Nuclear Data Sheets, 10 parts, 1959–1965, $54.00 total

Nuclear Data Sheets "B," Volumes 1–13, 1966–1974, $20.00 per volume

Nuclear Data Sheets "B," Volumes 14–16, 1975, $24.00 per volume.

Atomic Data presents tables and compilations of experimental and theoretical data in various areas of atomic physics. Collections and evaluations of data on energy levels, wavefunctions, line-broadening parameters, collision processes, various interaction cross sections of atoms and simple molecules, transition probabilities, and penetration through matter of charged particles are included.

Atomic Data, Volumes 1–3, 1969–1971, $20.00 per volume

[a] A detailed survey of analytical spectral data sources is given by Gevantman, L. H., Survey of analytical spectral data sources and related data compilation activities, *Anal. Chem.*, 44(7), 30, 1972.

Atomic Data, Volumes 4–5, 1972–1973, $22.00 per volume

In 1973 the two journals, *Nuclear Data Tables* and *Atomic Data,* were merged into *Atomic Data and Nuclear Data Tables.* The coverage of this new journal combines that of its two predecessors.

Atomic Data and Nuclear Data Tables, Volumes 12–14, 1974, $30.00

Atomic Data and Nuclear Data Tables, Volumes 15 and 16, 1975, $34.50.

Nonperiodical publications are listed below, in an arrangement which groups related topics to the extent feasible. Ordering information for the *National Bureau of Standards Technical Note* and NSRDS-NBS series and for the *Journal of Physical and Chemical Reference Data* and its reprints is given in Part V, "Selected Data Publications." The *Journal of Physical and Chemical Reference Data* papers, though appearing in a quarterly journal, are listed here with nonperiodical publications since there is no particular arrangement of continuity of publications dealing with nuclear, atomic, or molecular properties or with the other classifications used in these lists.

NSRDS-NBS-3, Sec. 1, *Selected Tables of Atomic Spectra, Atomic Energy Levels and Multiplet Tables, Si II, Si III, Si IV,* Moore, C. E., 1965, $1.00, SD Catalog No. C13.48:3/Sec.1.

NSRDS-NBS-3, Sec. 2, *Selected Tables of Atomic Spectra, Atomic Energy Levels and Multiplet Tables, Si I,* Moore, C. E., 1967, $0.70, SD Catalog No. C13.48:3/Sec. 2.

NSRDS-NBS-3, Sec. 3, *Selected Tables of Atomic Spectra, Atomic Energy Levels and Multiplet Tables, C I, C II, C III, C IV, C V, C VI,* Moore, C. E., 1970, $1.70, SD Catalog No. C13.48:3/Sec. 3.

NSRDS-NBS-3, Sec. 4, *Selected Tables of Atomic Spectra, Atomic Energy Levels and Multiplet Tables, N IV, N V, N VI, N VII,* Moore, C. E., 1971, $1.15, SD Catalog No. C13.48:3/Sec. 4.

NSRDS-NBS-3, Sec. 6, *Selected Tables of Atomic Spectra, Atomic Energy Levels and Multiplet Tables, H I, D, T,* Moore, C.E., 1972, $0.40, SD Catalog No. 13.48:3/Sec. 6.

NSRDS-NBS-14, *X-Ray Wavelengths and X-Ray Atomic Energy Levels,* Bearden, J. A., 1967, $1.14, SD Catalog No. C13.48:14.

NSRDS-NBS-29, *Photon Cross Sections, Attenuation Coefficients, and Energy Absorption Coefficients from 10 keV to 100 GeV,* Hubbell, J. H., 1969, $1.25, SD Catalog No. C13.48:29.

NSRDS-NBS-4, *Atomic Transition Probabilities, Vol. I, Hydrogen Through Neon,* Wiese, W. L., Smith, M. W., and Glennon, B. M., 1966, $2.50, SD Catalog No. C13.48:4/Vol. I.

NSRDS-NBS-22, *Atomic Transition Probabilities, Vol. II, Sodium Through Calcium, A Critical Data Compilation,* Wiese, W. L., Smith, M. W., and Miles, B. M., 1969, $8.60, SD Catalog No. C13.48:22/Vol. II.

Reprint No. 20. Smith, M. W. and Wiese, W. L., Atomic transition probabilities for forbidden lines of the iron group elements, *J. Phys. Chem. Ref. Data,* 2(1), 85, 1973, $4.50.

NSRDS-NBS-35, *Atomic Energy Levels as Derived from the Analyses of Optical Spectra, Vol. I,* ^{1}H *to* ^{23}V*; Vol. II,* ^{24}Cr *to* ^{41}Nb*; Vol. III,* ^{42}Mo *to* ^{57}La*,* ^{72}Hf *to* ^{89}Ac, Moore, C. E., 1971, Vol. I, $9.25; Vol. II, $7.95; Vol. III, $8.30, SD Catalog No. C13.48:35/Vols. I, II, and III.

Reprint No. 26. Martin, W. C., Energy levels of neutral helium (^{4}HeI), *J. Phys. Chem. Ref. Data,* 2(2), 257, 1973, $3.00.

NSRDS-NBS-6, *Tables of Molecular Vibrational Frequencies, Part 1,* Shimanouchi, T., 1967. Superseded by NSRDS-NBS-39.

NSRDS-NBS-11, *Tables of Molecular Vibrational Frequencies, Part 2,* Shimanouchi, T., 1967. Superseded by NSRDS-NBS-39.

NSRDS-NBS-17, *Tables of Molecular Vibrational Frequencies, Part 3,* Shimanouchi, T., 1968. Superseded by NSRDS-NBS-39.

NSRDS-NBS-39, *Tables of Molecular Vibrational Frequencies, Consolidated Volume I,* Shimanouchi, T., 1972, $5.10, SD Catalog No. C13.48:39.

Reprint No. 5. Shimanouchi, T., Tables of molecular vibrational frequencies, Part 5, *J. Phys. Chem. Ref. Data,* 1(1), 189, 1972, $4.00.

Reprint No. 21. Shimanouchi, T., Tables of molecular vibrational frequencies, Part 6. *J. Phys. Chem. Ref. Data,* 2(1), 121, 1973, $4.00.

Reprint No. 25. Shimanouchi, T., Tables of molecular vibrational frequencies, Part 7, *J. Phys. Chem. Ref. Data,* 2(2), 225, 1973, $4.00.

NSRDS-NBS-10, *Selected Values of Electric*

Dipole Moments for Molecules in the Gas Phase, Nelson, R. D., Jr., Lide, D. R., Jr., and Maryott, A. A., 1967, $0.95, SD Catalog No. C13.48:10.

Reprint No. 28. Young, K. F. and Frederikse, H. P. R., Compilation of the Static Dielectric Constant of Inorganic Solids, *J. Phys. Chem. Ref. Data,* 2(2), 313, 1973, $6.00.

NSRDS-NBS-40, *A Multiplet Table of Astrophysical Interest, Part I – Tables of Multiplets; Part II – Finding List of All Lines in the Table of Multiplets* (reprint of 1945 edition), Moore, C. E., 1972, $3.65, SD Catalog No. C13.48:40.

NSRDS-NBS-34, *Ionization Potentials and Ionization Limits Derived from the Analyses of Optical Spectra,* Moore, C. E., 1970, $0.75, SD Catalog No. C13.48:34.

NSRDS-NBS-26, *Ionization Potentials, Appearance Potentials, and Heats of Formation of Gaseous Positive Ions,* Franklin, J. L., Dillard, J. G., Rosenstock, H. M., Herron, J. T., Draxl, K., and Field, F. H., 1969, $6.20, SD Catalog No. C13.48:26.

NSRDS-NBS-25, *Electron Impact Excitation of Atoms,* Moiseiwitsch, B. L. and Smith, S. J., 1968, $2, SD Catalog No. C13.48:25.[a]

Reprint No. 14. Johnson, D. R., Lovas, F. J., and Kirchhoff, W. H., Microwave spectra of molecules of astrophysical interest. I. Formaldehyde, formamide, and thioformaldehyde, *J. Phys. Chem. Ref. Data,* 1(4), 1011, 1972, $4.50.

Reprint No. 17. Kirchhoff, W. H., Johnson, D. R., and Lovas, F. J., Microwave spectra of molecules of astrophysical interest. II. Methylenimine, *J. Phys. Chem. Ref. Data,* 2(1), 1, 1973, $3.00.

Reprint No. 23. Lees, R. M., Lovas, F. J., Kirchhoff, W. H., and Johnson, D. R., Microwave spectra of molecules of astrophysical interest. III. Methanol, *J. Phys. Chem. Ref. Data,* 2(2), 205, 1973, $3.00.

Reprint No. 24. Helminger, P., DeLucia, F. C., and Kirchhoff, W. H., Microwave spectra of molecules of astrophysical interest. IV. Hydrogen sulfide, *J. Phys. Chem. Ref. Data,* 2(2), 215, 1973, $3.00.

NSRDS-NBS-38, *Critical Review of Ultraviolet Photoabsorption Cross Sections for Molecules of Astrophysical and Aeronomic Interest,* Hudson, R. D., 1971, $1, SD Catalog No. C13.48:38.

Reprint No. 29. McAlister, A. J., Dobbyn, R. C., Cuthill, J. R., and Williams, M. L., Soft X-ray emission spectra of metallic solids: a critical review of selected systems, *J. Phys. Chem. Ref. Data,* 2(2), 411, 1973, $3.00.

Reprint No 8. Krupenie, P. H., The spectrum of molecular oxygen, *J. Phys. Chem. Ref. Data,* 1(2), 423, 1972, $6.50.

Reprint No. 4. Tilford, S. G. and Simmons, J. D., Atlas of the observed absorption spectrum of carbon monoxide between 1060 and 1900 Å, *J. Phys. Chem. Ref. Data,* 1(1), 147, 1972, $4.50.

NSRDS-NBS-5, *The Band Spectrum of Carbon Monoxide,* Krupenie, P. H., 1966, $2.05, SD Catalog No. C13.48:5.

NSRDS-NBS-47, *Tables of Collision Integrals and Second Virial Coefficients for the (m, 6, 8) Intermolecular Potential Function,* Klein, M., Hanley, H. J. M., Smith, F. J., and Holland, P., $1.90, 1974, SD Catalog No. C13.48:47.

NSRDS-NBS-12, *Tables for the Rigid Asymmetric Rotor: Transformation Coefficients from Symmetric to Asymmetric Bases and Expectation Values of* P_z^2, P_z^4, *and* P_z^6, Schwendeman, R. H., 1968, $1.45, SD Catalog No. C13.48:12.

C. Thermodynamic and Transport Properties; Phase Diagrams; Surface Properties

Selected Values of the Thermodynamic Properties of the Elements, Hultgren, R., Desai, P. D., Hawkins, D. T., Gleiser, M., Kelley, K. K., and Wagman, D. D., American Society for Metals, Metals Park, Oh., 1973, $20.00. $20.00.

NBS Tech. Note 270-3, *Selected Values of Chemical Thermodynamic Properties, Tables for the First Thirty-Four Elements in the Standard Order of Arrangement,* Wagman, D. D., Evans, W. H., Parker, V. B., Halow, I., Bailey, S. M., and Schumm, R. H., 1968, $2.75, SD Catalog No. C13.46:270-3.

NBS Tech. Note 270-4, *Selected Values of Chemical Thermodynamic Properties, Tables for Elements 35 through 53 in the Standard Order of Arrangement,* Wagman, D. D., Evans, W. H., Parker, V. B., Halow, I., Bailey, S. M., and Schumm, R. H., 1969, $2.10, SD Catalog No. C13.46:270-4.

NBS Tech. Note 270-5, *Selected Values of Chemical Thermodynamic Properties, Tables for*

[a]Available from the National Technical Information Service, U.S. Department of Commerce, Springfield, VA 22151.

Elements 54 through 61 in the Standard Order of Arrangement, Wagman, D. D., Evans, W. H., et al., 1971, $0.95, SD Catalog No. C13.46:270-5.

NBS Tech. Note 270-6, *Selected Values of Chemical Thermodynamic Properties, Tables for the Alkaline Earth Elements (Elements 92 through 97 in the Standard Order of Arrangement),* Parker, V. B., Wagman, D. D., and Evans, W. H., 1971, $1.55, SD Catalog No. C13.46:270-6.

NBS Tech. Note 270-7, *Selected Values for Chemical Thermodynamic Properties, Tables for the Lanthanide (Rare Earth) Elements (Elements 62 through 76 in the Standard Order of Arrangement),* Schumm, R. H., Wagman, D. D., Bailey, S., Evans, W. H., and Parker, V. B., 1973, $1.25, SD Catalog No. C13.48:270-7.

Selected Values of the Thermodynamic Properties of Binary Alloys, Hultgren, R., Desai, P. D., Hawkins, D. T., Gleiser, M., and Kelly, K. K.. American Society for Metals, Metals Park, Oh., 1973, $30.00.

NSRDS-NBS-37, *JANAF Thermochemical Tables, 2nd ed.,* Stull, D. R., Prophet, H., et al., 1971, $13.40, SD Catalog No. C13.48:37. Basically a related and consistent set of enthalpies and Gibbs energies of formation of 13 elements, e^-, and their simple compounds.

Thermophysical Properties of Matter, A Comprehensive Compilation of Data, Touloukian, Y. S. and Ho, C. Y., Series Eds., Thermophysical Properties Research Center, Purdue University, IFI/Plenum, New York, 13-volume set, price $575.00.

Volume 1. *Thermal Conductivity – Metallic Elements and Alloys,* Touloukian, Y. S., Powell, R. W., Ho, C. Y., and Klemens, P. G., 1970, xxvii + 53a + 1,469 + A46 pages, individual volume price $95.00.

Volume 2. *Thermal Conductivity – Nonmetallic Elements,* Touloukian, Y. S., Powell, R. W., Ho, C. Y., and Klemens, P. G., 1970, xxviii + 55a + 1,172 + A46 pages, individual volume price $85.00.

Volume 3. *Thermal Conductivity – Nonmetallic Liquids and Gases,* Touloukian, Y. S., Liley, P. E., and Saxena, S. C., 1970, xxii + 108a + 531 + A46 pages, individual volume price $55.00.

The three-volume *Thermal Conductivity* subset price is $185.00.

Volume 4. *Specific Heat – Metallic Elements and Alloys,* Touloukian, Y. S. and Buyco, E. H., 1970, xxii + 32a + 750 + A26 pages, individual volume price $65.00.

Volume 5. *Specific Heat – Nonmetallic Solids,* Touloukian, Y. S. and Buyco, E. H., 1970, xxx + 32a + 1,649 + A26 pages, individual volume price $100.00.

Volume 6. *Specific Heat – Nonmetallic Liquids and Gases,* Touloukian, Y. S. and Makita, T., 1970, xviii + 27a + 312 + A26 pages, individual volume price $40.00.

The three-volume *Specific Heat* subset price is $155.00.

Volume 7. *Thermal Radiative Properties – Metallic Elements and Alloys,* Touloukian, Y. S. and DeWitt, D. P., 1970, xxxix + 61a + 1,540 + A14 pages, individual volume price $100.00.

Volume 8. *Thermal Radiative Properties – Nonmetallic Solids,* Touloukian, Y. S. and DeWitt, D. P., 1972, xxxiv + 52a + 1,763 + A41 pages, individual volume price $70.00.

Volume 9. *Thermal Radiative Properties – Coatings,* Touloukian, Y. S., DeWitt, D. P., and Hernicz, R. S., 1972, xxxiv + 62a + 1,378 + A95 pages, individual volume price $100.00.

The three-volume *Thermal Radiative Properties* subset price is $220.00.

Volume 10. *Thermal Diffusivity,* Touloukian, Y. S., Powell, R. W., Ho, C. Y., and Nicolaou, M. C., 1973, xxvi + 64a + 661 + A9 pages, individual volume price $50.00.

Volume 11. *Viscosity,* in press, 1975, approximately 800 pages, individual volume price $50.00.

Volume 12. *Thermal Expansion – Metallic Elements and Alloys,* in press, 1975, approximately 1,300 pages, individual volume price $80.00.

Volume 13. *Thermal Expansion – Nonmetallic Solids,* in press, 1975, approximately 1,300 pages, individual volume price $80.00.

The two-volume *Thermal Expansion* subset price is $120.00.

Selected Values of Properties of Hydrocarbons and Related Compounds, American Petroleum Institute Research Project 44, Zwolinski, B. J. et al., Thermodynamics Research Center,[a] 2,814 loose-leaf data sheets available as of June 1, 1973, 30 centers per data sheet.

Selected Values of Properties of Chemical Compounds, Thermodynamics Research Center Data Project, Zwolinski, B. J. et al., Thermodynamics Research Center,[a] 1,209 loose-leaf data sheets available as of June 1, 1973.

Comprehensive Index of API44-TRC Selected Data on Thermodynamics and Spectroscopy, 2nd ed., Zwolinski, B. J. and Wilhoit, R. C., Thermodynamics Research Center,[a] Publ. 100, 1974, 600 pages, $14.00.

Handbook of Vapor Pressures and Heats of Vaporization of Hydrocarbons and Related Compounds, Wilhoit, R. C. and Zwolinski, B. J., Thermodynamics Research Center,[a] Publ. 101, 1971, 329 pages, $10.00.

NSRDS-NBS-27, *Thermodynamic Properties of Argon from the Triple Point to 300 K at Pressures to 1000 Atmospheres,* Gosman, A. L., McCarty, R. D., and Hust, J. G., 1969, $1.80, SD Catalog No. C13.48:27.

NSRDS-NBS-19 *Thermodynamic Properties of Ammonia as an Ideal Gas,* Haar, L., 1968, $0.20, SD Catalog No. C13.48:19.

Reprint No. 30. Chao, J., Wilhoit, R. C., and Zwolinski, B. J., Ideal gas thermodynamic properties of ethane and propane, *J. Phys. Chem. Ref. Data,* 2(2), 427, 1973, $3.00.

NBS Tech. Note 384, *Thermophysical Properties of Oxygen from the Freezing Liquid Line to 600 R for Pressures to 5000 Psia,* McCarty, R. D. and Weber, L. A., 1971, $1.50, SD Catalog No. C13.46:384.

NSRDS-NBS-18, *Critical Analysis of the Heat-Capacity Data of the Literature and Evaluation of Thermodynamic Properties of Copper, Silver, and Gold from 0 to 300 K,* Furukawa, G. T., Saba, W. G., and Reilly, M. L., 1968, $0.40, SD Catalog No. C13.48:18.

Thermodynamic Properties of Copper and Its Inorganic Compounds, Monograph II, King, E. G., Mah, A. D., and Pankratz, L. B., International Copper Research Association, New York, 1973, xii + 257 pages, $10.00.

NSRDS-NBS-15, *Molten Salts: Vol. 1, Electrical Conductance, Density, and Viscosity Data,* Janz, G. J., Dampier, F. W., Lakshminarayanan, G. R., Lorenz, P. K., and Tomkins, R. P. T., 1968, $3.00, SD Catalog No. C13.48:15/Vol. 1.

NSRDS-NBS-28, *Molten Salts: Vol. 2, Section 1. Electrochemistry of Molten Salts: Gibbs Free Energies and Excess Free Energies from Equilibrium-Type Cells,* Janz, G. J. and Dijkhuis, C. G. M.; *Section 2. Surface Tension Data,* Janz, G. J., Lakshminarayanan, G. R., Tomkins, R. P. T., and Wong, J., 1969, $4.70, SD Catalog No. C13.48:28/Vol. 2.

Janz, G. J., Krebs, U., Siegenthaler, H. F., and Tomkins, R. P. T., Reprint No. 10. Molten salts: Volume 3, nitrates and mixtures, electrical conductance, density, viscosity, and surface tension data, *J. Phys. Chem. Ref. Data,* 1(3), 581, 1972, $8.50.

NSRDS-NBS-7, *High Temperature Properties and Decomposition of Inorganic Salts, Part 1. Sulfates,* Stern, K. H. and Wiese, E. L., 1966, $0.85, SD Catalog No. C13.48:7/Pt. 1.

NSRDS-NBS-30, *High Temperature Properties and Decomposition of Inorganic Salts, Part 2. Carbonates,* Stern, K. H. and Wiese, E. L., 1969, $0.75, SD Catalog No. C13.48:30/Pt. 2.

Reprint No 11. Stern, K. H., High temperature properties and decomposition of inorganic salts – Part 3, nitrates and nitrites, *J. Phys. Chem. Ref. Data,* 1(3), 747, 1972, $4.00.

Reprint No. 18. Cook, F. J., Analysis of specific heat data in the critical region of magnetic solids, *J. Phys. Chem. Ref. Data,* 2(1), 11, 1973, $3.00.

Reprint No. 6. Domalski, E. S., Selected values of heats of combustion and heats of formation of organic compounds containing the elements C, H, N, O, P, and S, *J. Phys. Chem. Ref. Data,* 1(2), 221, 1972, $5.00.

Reprint No. 16. Hanley, H. J. M. and Prydz, R., The viscosity and thermal conductivity coefficients of gaseous and liquid fluorine, *J. Phys. Chem. Ref. Data,* 1(4), 1101, 1972, $3.00.

Reprint No. 7. Ho, C. Y., Powell, R. W., and Liley, P. E., Thermal conductivity of the elements, *J. Phys. Chem. Ref. Data,* 1(2), 279, 1972, $7.50.

NSRDS-NBS-8, *Thermal Conductivity of*

[a]Order from Thermodynamics Research Center, Texas A & M University, College Station, TX 77843.

Selected Materials, Powell, R. W., Ho, C. Y., and Liley, P. E., 1966, PB189698.[a]

NSRDS-NBS-16, *Thermal Conductivity of Selected Materials, Part 2,* Ho, C. Y., Powell, R. W., and Liley, P. E., 1968.[a]

Reprint No. 1. Marrero, T. R. and Mason, E. A., Gaseous diffusion coefficients, *J. Phys. Chem. Ref. Data,* 1(1), 1, 1972, $7.00.

Reprint No. 3. Pound, G. M., Selected values of evaporation and condensation coefficients of simple substances, *J. Phys. Chem. Ref. Data,* 1(1), 119, 1972, $3.00.

NSRDS-NBS-24, *Theoretical Mean Activity Coefficients of Strong Electrolytes in Aqueous Solutions from 0 to 100°C,* Hamer, W. J., 1968, $6.10, SD Catalog No. C13.48:24.

Reprint No. 15. Hamer, W. J. and Wu, Y. C., Osmotic coefficients and mean activity coefficients of uni-univalent electrolytes in water at 25°C, *J. Phys. Chem. Ref. Data,* 1(4), 1047, 1972, $5.00.

NSRDS-NBS-2, *Thermal Properties of Aqueous Uni-univalent Electrolytes,* Parker, V. B., 1965, $1.10, SD Catalog No. C13.48:2.

NSRDS-NBS-33, *Electrolytic Conductance and the Conductances of the Halogen Acids in Water,* Hamer, W. J. and DeWane, H. J., 1970, $0.85, SD Catalog No. C13.48:33.

NSRDS-NBS-32, *Phase Behavior in Binary and Multicomponent Systems at Elevated Pressures: n-Pentane and Methane-n-Pentane,* Berry, V. M. and Sage, B. H., 1970, $1.15, SD Catalog No. C13.48:32.

Constitution of Binary Alloys, 2nd ed., Hansen, M. and Anderko, K., McGraw-Hill, 1958, xix + 1,305 pages, $42.75.

Constitution of Binary Alloys, first suppl., Elliott, R. P., McGraw-Hill, 1965, xxxii + 877 pages, $39.50.

Constitution of Binary Alloys, second suppl., Shunk, F. A., McGraw-Hill, 1969, xi + 720 pages, $39.50.

Phase Diagrams for Ceramists, 1969 suppl., Levin, E. M., Robbins, G. R., and McMurdie, H. F.; Reser, M. K., Ed., American Ceramic Society, Columbus, Oh., 1969, 625 pages, $30.00.

Phase Diagrams for Ceramists, 1964 ed., Levin, F. M., Robbins, C. R., and McMurdie, H. F.; Reser, M. K., Ed., American Ceramic Society, Columbus, Oh., 1964, 601 pages, $18.00.

Selected Thermodynamic Values and Phase Diagrams for Copper and Some of Its Binary Alloys, Monograph I, Hultgren, R. and Desai, P. D., International Copper Research Association, New York, 1971, ix + 204 pages, $10.00.

NSRDS-NBS-36, *Critical Micelle Concentrations of Aqueous Surfactant Systems,* Mukerjee, P. and Mysels, K. J., 1971, $5.70, SD Catalog No. C13.48:36.

Reprint No. 13. Jasper, J. L., The surface tension of pure liquid compounds, *J. Phys. Chem. Ref. Data,* 1(4), 841, 1972, $8.50.

D. Solid State

Data on crystal structures:

Crystal Data, Determinative Tables, 3rd ed., Volume 1: Organic Compounds, 1972, ix + 854 pages, $30.00; Volume 2: Inorganic Compounds, 1973, ix + 1,757 pages, $50.00; available from the Joint Committee on Powder Diffraction Standards, 1601 Park Lane, Swarthmore, PA 19081.

Crystal Data, Systematic Tables, 2nd ed., ACA Monograph No. 6, Nowacki, W., Edenharter, A., and Matsumoto, T., American Crystallographic Association, Washington, D.C., 1967, v + 195 pages, $7.50; available from Polycrystal Book Service, P.O. Box 11567, Pittsburgh, PA 15238.

Crystal Data, Determinative Tables is a comprehensive listing of cell dimensions and angles covering the literature through 1966. *Crystal Data, Systematic Tables* lists the compounds covered in the second edition of the *Determinative Tables* according to their space groups and gives statistics on the distribution of substances among space groups, crystal classes, etc. It is not clear whether or not a third edition of the *Systematic Tables* will be prepared.

Standard X-ray Diffraction Powder Patterns (NBS Circular 539, Volumes 1–10, and NBS Monograph 25, Sections 1–10), Swanson, H. E. et al. All volumes of Circular 539, and Sections 1–3 of Monograph 25 are available from the National Technical Information Service, U.S. Department of Commerce, Springfield, VA 22151. Order by the PB or COM number and/or the series number:

Circular 539, Volume 1, PB 178902

[a]Available from the National Technical Information Service, U.S. Department of Commerce, Springfield, VA 22151.

Circular 539, Volume 2, PB 178903
Circular 539, Volume 3, PB 178904
Circular 539, Volume 4, PB 178905
Circular 539, Volume 5, PB 178906
Circular 539, Volume 6, PB 178907
Circular 539, Volume 7, PB 178908
Circular 539, Volume 8, PB 178909
Circular 539, Volume 9, PB 178910
Circular 539, Volume 10, PB 178911
Monograph 25, Section 1, PB 178429
Monograph 25, Section 2, PB 178430
Monograph 25, Section 3, PB 178431

Monograph 25, Sections 4–10 are available from the Superintendent of Documents, U.S. Government Printing Office. Order by SD catalog number shown (add 25% more for foreign orders):

Monograph 25, Section 4, use series number
Monograph 25, Section 5, use series number
Monograph 25, Section 6, use series number
Monograph 25, Section 7, use series number
Monograph 25, Section 8, PB194874
Monograph 25, Section 9, COM 72-50002
Monograph 25, Section 10, COM 72-51079
Monograph 25, Section 11[a]
Monograph 25, Section 12, COM 75-50162/AS

Structure Reports, Volume 8, 1950 (covering the literature for 1940–1941) through Volume 29, 1972 (covering the literature for 1964), is available from the Polycrystal Book Service, P.O. Box 11567, Pittsburgh, PA 15238 at prices ranging from $16.50 to $70.00 with discounts to members of recognized scientific societies and for sets.

Structure Reports gives detailed structural information appearing in the literature for the period covered. Standard X-ray Diffraction Powder Patterns gives experimental results obtained on specially prepared, high-purity samples, plus some patterns calculated from published structure data. A cumulative index appears in each volume or section. These results are included in the Powder Diffraction File which is a compilation from many sources produced by the Joint Committee on Powder Diffraction Standards in cooperation with the American Society for Testing and Materials, the American Crystallographic Association, the Institute of Physics, and the National Association of Corrosion Engineers. The data in these sources are used for identification of unknown crystalline materials by matching spacings and intensities.

The three following publications present other data on solids as indicated by their titles.

NSRDS-NBS-41, *Crystal Structure Transformations in Binary Halides,* Rao, C. N. R. and Natarajan, M., 1972, $10.95, SD Catalog No. C13.48:41.

Reprint No. 22. Strehlow, W. H. and Cook, E. L., Compilation of energy band gaps in elemental and binary compound semiconductors and insulators, *J. Phys. Chem. Ref. Data,* 2(1), 163, 1973, $4.50.

NBS Tech. Note 724, *Properties of Selected Superconductive Materials,* Roberts, B. W., 1972, $1.40, SD Catalog No. C13.46:724.

E. Chemical Kinetics

NSRDS-NBS-20, *Gas Phase Reaction Kinetics of Neutral Oxygen Species,* Johnston, H. S., 1968, $0.95, SD Catalog No. C13.48:20.

Reprint No. 9. Wilson, W. E., Jr., A critical review of the gas-phase reaction kinetics of the hydroxyl radical, *J. Phys. Chem. Ref. Data,* 1(2), 535, 1972, $4.50.

NSRDS-NBS-21, *Kinetic Data on Gas Phase Unimolecular Reactions,* Benson, S. W. and O'Neal, H. E., 1970, $9.30, SD Catalog No. C13.48:21.

NSRDS-NBS-9, *Tables of Bimolecular Gas Reactions,* Trotman-Dickenson, A. F. and Milne, G. S., 1967.[b]

Reprint No. 19. Schofield, K., Evaluated chemical kinetic rate constants for various gas phase reactions, *J. Phys. Chem. Ref. Data,* 2(1), 25, 1973, $5.00.

NSRDS-NBS-42, *Selected Specific Rates of Reactions of the Solvated Electron in Alcohols,* Watson, E., Jr. and Roy, S., 1972, $0.30, SD Catalog No. C13.48:42.

[a]Available in microfiche only from the National Technical Information Service; use series number. For paper copy, order from the Superintendent of Documents, U.S. Government Printing Office, Washington, D.C. 20402, as C13.44:25/Sec. 11, for $1.55.

[b]Available from the National Technical Information Service, U.S. Department of Commerce, Springield, VA 22151.

NSRDS-NBS-44, *The Radiation Chemistry of Gaseous Ammonia,* Peterson, D. B., 1974, $0.70, SD Catalog No. C13.48:44.

NSRDS-NBS-43, *Selected Specific Rates of Reactions of Transients from Water in Aqueous Solution. I. Hydrated Electron,* Anbar, M., Bambenek, M., and Ross, A. B., 1972, $0.90, SD Catalog No. C13.48:43.

NSRDS-NBS-45, *Radiation Chemistry of Nitrous Oxide Gas. Primary Processes, Elementary Reactions and Yields,* Johnson, G. R. A., 1973, $0.60, SD Catalog No. C13.48:45.

NSRDS-NBS-46, *Reactivity of the Hydroxyl Radical in Aqueous Solutions,* Dorfman, L. M. and Adams, G. E., 1973, $0.90, SD Catalog No. C13.48:46.

NSRDS-NBS-48, *Critical Data Review on the Radiation Chemistry of Ethanol: Yields, Reaction Rate Parameters and Spectral Properties of Transients,* Freeman, G. R., 1974, $0.80, SD Catalog No. C13.48:48.

NSRDS-NBS-13, *Hydrogenation of Ethylene on Metallic Catalysts,* Horiuti, J. and Miyahara, K., 1968, $3.00, SD Catalog No. C13.48:13.

Reprint No. 27. Hampson, R. F., Ed., Survey of photochemical and rate data for twenty-eight reactions of interest in atmospheric chemistry, *J. Phys. Chem. Ref. Data,* 2(2), 267, 1973, $4.50.

F. Handbooks Emphasizing Properties of Commercial Materials

Handbook of High-temperature Materials, No. 2 – Properties Index, Samsonov, G. V., Plenum Press, New York, 1964, 418 pages, $35.00.

This is a translation from Russian of *Refractory Compounds – Handbook of Properties and Applications,* 1963, revised to include material published in 1962 and 1963. It covers refractory compounds, mainly binary compounds between a transition metal and boron, carbon, silicon, nitrogen, sulfur, or phosphorus, and gives data on the physical, technical, mechanical, chemical, and refractory properties selected by the author as most reliable. The introduction discusses reliability of data.

Metals Reference Book, 4th ed., Smithells, C. J., Ed., Plenum Press, New York; Butterworths, London, 1967, $92.50. Three volumes, total of 1,250 pages plus contents and index, repreated in each volume.

The source contains data pertaining to metallurgy and metal physics, selected by a number of contributors as the most reliable, with bibliographies of the more important original sources. The data are presented in tables or diagrams with short monographs on particular subjects where required. Coverage includes crystal chemistry, thermochemical data, metallography, phase diagrams, physical and mechanical properties, and other properties important in production and fabrication of pure metals and alloys.

Metals Handbook, 8th ed., American Society for Metals, Metals Park, OH 44073.

Volume 1. Properties and Selection, 1961, 1,300 pages, $45.00.

Volume 2. Heat Treating, Cleaning and Finishing, 1964, 672 pages, $42.50.

Volume 3. Machining, 1967, 550 pages, $42.50.

Volume 4. Forming, 1969, 527 pages, $42.50.

Volume 5. Forging and Casting, 1970, 300 pages, $37.50.

Volume 6. Welding and Brazing, 1971, 734 pages, $45.00.

Volume 7. Atlas of Microstructures of Industrial Alloys, 1972, 366 pages, $42.50.

Volume 8. Metallography, Structures, and Phase Diagrams, 1973, 466 pages, $42.50.

Volume 9. Fractography, 1974, 499 pages, $45.00.

Volume 10. Failure Analysis and Prevention, publication expected Autumn 1975.

Two additional volumes are planned, but no preparatory work has yet been done:

Volume 11. Nondestructive Inspection
Volume 12. Mechanical Testing

A number of authorities have contributed to this reference (1,335 to Volume 1 alone). Most of the properties data are contained in Volume 1, whose contents are definitions and reference tables; carbon and low-alloy steels; cast irons; stainless steels and heat-resisting alloys; tool materials; magnetic, electrical, and other special purpose materials; nonferrous metals; and properties of pure metals. Tables and graphs showing the properties of major commercial alloys that are important in their principal uses are interspersed throughout sections describing their production, uses, formability, and cost; descriptions of the various standard test methods and samples employed are also included. Where appropriate, ranges of values, influence of temperature and processing, and effects of other significant variables and conditions are shown and discussed.

Polymer Handbook, Brandrup, J. and Immergut, E. H., Eds., Interscience, New York, 1966, 1,249 pages, $27.95

Only fundamental constants and parameters were compiled for this handbook. This was interpreted to include data that are either physical or chemical constants of the polymer molecules within reasonable or predictable limits or constants of existing physical laws describing the properties and behavior of polymers. Constants that depend on the particular processing conditions or sample history were not compiled. No critical evaluation of published values was attempted. All data found in the literature were listed except those that were obviously erroneous. A bibliography is included at the end of each section. The sections included in this handbook are

I. Nomenclature Rules
II. Polymerization
III. Solid State Properties
IV. Solution Properites
V. Miscellaneous Properties
VI. Physical Constants of Some Important Polymers
VII. Physical Data of Oligomers
VIII. Physical Properties of Monomers and Solvents
IX. Contemporary Thermoplastic Materials, Property and Price Chart
X. Subject Index.

The *Aerospace Structural Metals Handbook,* published by the Mechanical Properties Data Center, 13919 West Bay Shore Drive, Traverse City, MI 49684 (1972, $125, including revisions and additions for one year), is a four-volume, 2,800-page loose-leaf handbook updated quarterly. It contains information on physical, chemical, and mechanical properties of over 200 metals (properties covered are those of interest to designers and materials engineers), and alloys. Also included are data source references, a general discussion of properties, a glossary of terms, a discussion of fracture toughness, and a cross index of the alloys covered. The contents of the four volumes are

Volume 1. Introduction and General Discussion of Alloys and Their Properties, Carbon and Low Alloy Steels (FeC), Ultra High Strength Steels (FeUH), Austenitic Stainless Steels (FeA), and Martensitic Stainless Steels (FeH).

Volume 2. Introduction and General Discussion of Alloys and Their Properties, Age Hardening Steels (FeAH), Nickel Chromium Steels (FeNC), Aluminum Alloys–Cast (AlC), Aluminum Alloys–Wrought, Heat Treatable (AlWT), and Aluminum Alloys–Wrought, Not Heat Treatable (AlWN).

Volume 3. Introduction and General Discussion of Alloys and Their Properties, Magnesium Alloys–Cast (MgWT), Magnesium Alloys–Wrought, Heat Treatable (MgWT), Magnesium Alloys–Wrought, Not Heat Treatable (MgWN), Titanium Alloys (Ti), and Titanium Alloys–Cast (TiC).

Volume 4. Introduction and General Discussion of Alloys and Their Properties, Nickel Base Alloys (<5% Co)(Ni), Nickel Base Alloys (>5% Co)(NiCo), Cobalt Base Alloys (Co), Beryllium Alloys (Be), Columbium (Niobium) Alloys (Cb), Molybdenum Alloys (Mo), Tantalum Alloys (Ta), Tungsten Alloys (W), Vanadium Alloys (V), and Zirconium Alloys (Zr).

The *Structural Alloys Handbook,* also published by the Mechanical Properties Data Center (1st ed., 1972, plus first year supplements, $75.00), is a companion volume covering the more common metals and alloys of importance in the construction, machine tool, heavy equipment, automotive, and general manufacturing industries. Fewer than 300 alloys are covered in this initial edition, but substantially expanded coverage, based on the extensive data files of the MPDC, is expected in the supplements. There is extensive coverage of room temperature mechanical properties plus data on corrosion, hydrogen embrittlement, heat treatment, and other technologically important properties, but less extensive coverage of elevated and low temperature properties.

Damage Tolerant Design Handbook, first edition of *Fracture and Crack Growth Data for High-Strength Alloys,* Campbell, J. E., Berry, W. E., and Fedderson, C. E., MCIC-HB-01, Metals and Ceramics Information Center, Battelle, Columbus Laboratories, 505 King Avenue, Columbus, OH

43201, December 1972, is available from the National Technical Information Service as AD753774 ($37.50 for hard copy or microfiche [400 pages]); the first supplement (700 pages) is also available ($25.00). This is the first edition of a compilation of fracture mechanics data on high-strength aluminum alloys, alloy steels, stainless steels, and titanium alloys. The data represent state-of-the-art information on critical plane-strain stress intensity factors (K_{Ic}), plane stress and transitional stress intensity factors (K_c), threshold stress intensity factors in corrosive media (K_{Issc}), sustained-load crack-growth rates in corrosive media (da/dt vs. K_I), and fatigue crack-growth rates (da/dN vs. ΔK). Data related to these parameters were collected from many sources which cover the period 1962 to 1972. This handbook is intended to be the primary source of fracture mechanics data for Air Force contractors for design of damage-tolerant structures according to requirements of MIL-STD-1530 Aircraft Structural Integrity Program and the supporting MIL-A-8866 service specifications. Annual supplements will be issued incorporating latest fracture mechanics data and additional alloys.

Metallic Materials and Elements for Aerospace Vehicle Structures, MIL-HDBK-5B, September 1, 1971, is a loose-leaf, two-volume set which may be ordered from the Superintendent of Documents, U.S. Government Printing Office, Washington, D.C. 20401 (D7.6/2:5B/VI & II, $9.60 for two volumes). The handbook was prepared by the Department of Defense and the Federal Aviation Administration as a source of strength properties of metals and elements (primarily fasteners) that have received general acceptance for use in design of flight vehicle structures. Individual standards are updated periodically. The primary strength properties (ultimate tensile and bearing stress, tensile and compressive yield stress at which permanent strain equals 0.0002, bearing yield stress, and ultimate stress in pure shear) and the elongation and reduction in area values given are the minimum values at room temperature above which at least 99% (A basis) or 90% (B basis) of values are expected to fall with a confidence of 95%. If the test information is not available to establish these statistical limits, the specified minimum value of the appropriate government or SAE Aerospace Material Specification for the material is given. Values of density, specific heat, thermal conductivity, and mean coefficient of thermal expansion are generally average or typical values.

Plastics for Aerospace Vehicles, Part 1. Reinforced Plastics, MIL-HDBK-17A, January 1971 (superseding MIL-HDBK-17, November 1959; looseleaf) was prepared by the Plastics Technical Evaluation Center (PLASTEC), Picatinny Arsenal, and may be ordered from the Superintendent of Documents, U.S. Government Printing Office, Washington, D.C. 20401 (D7.6/2:17A/Part 1; $2.50). This handbook provides (1) a standardized basis for the design and construction of fiber-reinforced plastic composites primarily for use in aerospace vehicles, (2) appropriate data on the mechanical properties of such composites, and (3) design techniques and analytical procedures which account for anistropy of these composites and permit computation of directional properties. The mechanical properties contained herein have been developed for laminates made from commonly used material systems and fabricated by processes representative of current industrial practice. The bulk of the data pertains to fiberglass composites, principally of E glass. Additional data are provided for S glass composites and for boron-reinforced composites. Chapters on design and analysis of composites, and on the fastening and joining of composites to each other and to metallic parts are also provided. Property data are provided over the normal temperature range from −60 to 160°F and beyond the maximum intended use temperature for high service temperature materials. Effects of exposure to 100% relative humidity atmospheres are also shown.

Engineering Properties of Selected Ceramic Materials, Lynch, J. F., Ruderer, C. G., and Duckworth, W. H., Eds., Defense Ceramic Information Center, Battelle/Columbus Laboratories, published and distributed by The American Ceramics Society, Columbus, Oh., 1966 (783 loose-leaf pages; $12.00 to members of the American Ceramic Society, $16.00 to nonmembers). This source covers the physical, thermal, and mechanical properties, and gives information on resistance to thermal stress, oxidation, and corrosion of refractory ceramics, selected ceramic composites, refractory intermetallic compounds, and selected metalloid elements; it does not cover glasses, carbons, graphites, and ceramics and intermetallics with melting points less than 2,000°F. An appendix defines the properties covered and discusses measurement methods and the influence of structure and composition on the values. All data available to the compilers up to 1965 are included in the source. A loose-leaf format permits insertion of supplements which were sent periodically to initial purchasers from Battelle/Columbus by the compilers. A completely revised and updated edition is currently in preparation under the auspices of the Metals and Ceramics Information Center, Battelle/Columbus Laboratories, 505 King Avenue, Columbus, OH 43206.

Critical Surveys of Data Sources: Mechanical Properties of Metals, Gavert, R. B., Moore, R. L., and Westbrook, J. H., NBS Special Publication 396-1, 1974, vii + 81 pages. Order from Superintendent of Documents, U.S. Government Printing Office, Washington, D.C. 20402, citing SD Catalog No. C13.10:396-1. The price is $1.25.

This is the first in a series designed to describe the major sources of data covering selected types of materials, with emphasis on commercial materials, in enough detail to enable a user to select the source most suitable for his particular needs. Coverage of materials and properties is thorough, and the source(s) and selectivity of the data, as well as comments on their usefulness, are given. Additional surveys covering electrical and magnetic properties of metals, properties of ceramics, and corrosion data are in preparation. This first survey on mechanical properties of metals describes 21 publications and 9 information centers, plus data activities of 12 technical societies and trade associations. The emphasis is on U.S. alloys and sources, but a few foreign publications and centers are included.

G. Addendum

This section contains an unclassified listing of publications appearing in the *Journal of Physical and Chemical Reference Data* and other selected NSRDS data publications which have appeared between the preparation of this manuscript in 1973 and January 1975.

The following reprints from the *Journal of Physical and Chemical Reference Data* may be ordered from the Subscription Service Department, American Chemical Society, 1155 Sixteenth Street, N. W., Washington, D. C. 20036, at the (1975) prices shown.

Reprints from the *Journal of Physical and Chemical Reference Data*
(reprint number precedes each)

31. Stein, A. and Allen, G. F., An analysis of coexistence curve data for several binary liquid mixtures near their critical points, 2(3), 1973, $4.00.

32. Herron, J. T. and Huie, R. E., Rate constants for the reactions of atomic oxygen (O_3P) with organic compounds in the gas phase, 2(3), 1973, $5.00.

33. Humphreys, C. J., First spectra of neon, argon, and xenon 136 in the 1.2–4.0 μm region, 2(3), 1973, $3.00.

34. Ledbetter, H. M. and Reed, R. P., Elastic properties of metals and alloys, I. Iron, nickel, and iron-nickel alloys, 2(3), 1973, $6.50.

35. Hanley, H. J. M., The viscosity and thermal conductivity coefficients of dilute argon, krypton, and xenon, 2(3), 1973, $4.00.

36. Butrymowicz, D. B., Manning, J. R., and Read, M. E., Diffusion in copper and copper alloys, Part I. Volume and surface self-diffusion in copper, 2(3), 1973, $3.00.

37. Cohen, E. R. and Taylor, B. N., The 1973 least-squares adjustment of the fundamental constants, 2(4), 1973, $5.50.

38. Hanley, H. J. M. and Ely, J. F., The viscosity and thermal conductivity coefficients of dilute nitrogen and oxygen, 2(4), 1973, $4.00.

39. Jacobsen, R. T. and Stewart, R. B., Thermodynamic properties of nitrogen including liquid and vapor phases from 63 K to 2000 K with pressures to 10,000 bar, 2(4), 1973, $8.50.

40. McCarty, R. D., Thermodynamic properties of helium 4 from 2 to 1500 K at pressures to 10^8 Pa, 2(4), 1973, $7.00.

41. Janz, G. J., Gardner, G. L., Krebs, U., and Tomkins, R. P. T., Molten salts: Volume 4. Part 1, Fluorides and mixtures, electrical conductance. density, viscosity, and surface tension data, 3(1), 1974, $7.00.

42. Rodgers, A. S., Chao, J., Wilhoit, R. C., and Zwolinski, B. J., Ideal gas thermodynamic properties of eight chloro- and fluoromethanes, 3(1), 1974, $4.00.

43. Chao, J., Rodgers, A. S., Wilhoit, R. C., and Zwolinski, B. J., Ideal gas thermodynamic properties of six chloroethanes, 3(1), 1974, $4.00.

44. Furukawa, G. T., Reilly, M. L., and Gallagher, J. S., Critical analysis of heat-capacity data and evaluation of thermodynamic properties of ruthenium, rhodium, palladium, iridium, and platinum from 0 to 300 K. A survey of the literatute data on osmium, 3(1), 1974, $4.50.

45. DeLucia, F. C., Helminger, P., and

Kirchhoff, W. H., Microwave spectra of molecules of astrophysical interest, V. Water vapor, 3(1), 1974, $3.00.

46. Maki, A. G., Microwave spectra of molecules of astrophysical interest, VI. Carbonyl sulfide and hydrogen cyanide, 3(1), 1974, $4.00.

47. Lovas, F. J. and Krupenie, P. H., Microwave spectra of molecules of astrophysical interest, VII. Carbon monoxide, carbon monosulfide, and silicon monoxide, 3(1), 1974, $3.00.

48. Tiemann, E., Microwave spectra of molecules of astrophysical interest, VIII. Sulphur monoxide, 3(1), 1974, $3.00.

49. Schimanouchi, T., Tables of molecular vibrational frequencies, Part 8, 3(1), 1974, $4.50.

50. Chase, M. W., Curnutt, J. L., Hu, A. T., Prophet, H., Syverud, A. N., and Walker, L. C., JANAF thermochemical tables, 1974 supplement, 3(2), 1974, $8.50.

51. Stern, K. H., High temperature properties and decomposition of inorganic salts – Part 4. Oxy-salts of the halogens, 3(2), 1974, $4.50.

52. Butrymowicz, D. B., Manning, J. R., and Read, M. E., Diffusion in copper and copper alloys, Part II. Copper-silver and copper-gold systems, 3(2), 1974, $5.50.

53. Lovas, F. J. and Tiemann, E., Microwave spectral tables, I. Diatomic molecules, 3(3), 1974, $8.50.

54. Martin, W. C., Hagan, L., Reader, J., and Sugar, J., Ground levels and ionization potentials for lanthanide and actinide atoms and ions, 3(3), 1974, $3.00.

55. Cannon, J. F., Behavior of the elements at high pressure, 3(3), 1974, $4.50.

56. Kaufman, V. and Edlén, B., Reference wavelengths from atomic spectra in the range 15Å to 25000Å, 3(4), 1974, $5.50.

57. Ledbetter, H. M. and Naimon, E. R., Elastic properties of metals and alloys, II. Copper, 3(4), 1974, $4.50.

58. Hendry, D. G., Mills, T., Howard, J. A., Piszkiewicz, L., and Eigenmann, H. K., Critical review of hydrogen atom transfer reactions in the liquid phase, 3(4), 1974, $4.50.

59. Hanley, H. J. M., McCarty, R. D., and Haynes, W. N., The viscosity and thermal conductivity coefficients for dense gas and liquid argon, krypton, zenon, nitrogen, and oxygen, 3(4), 1974, $4.50.

Supplements to the *Journal of Physical and Chemical Reference Data*[a]

Wilhoit, R. C. and Zwolinski, B. J., *Physical and Thermodynamic Properties of Aliphatic Alcohols,* Vol. 2, Suppl. 1, 1973, 420 pages; hardcover, $33.00; softcover, members $10.00, nonmembers $30.00.

Ho, C. Y., Powell, R. W., and Liley, P. E., *Thermal Conductivity of the Elements: A Comprehensive Review,* Vol. 3, Suppl. 1, 1974, 796 pages; hardcover, $60.00; softcover, members $25.00, nonmembers $55.00.

The following publications may be ordered from the Superintendent of Documents, U.S. Government Printing Office, Washington, D.C., 20402, citing the SD Catalog numbers given. Prices are as of January 1975.

NSRDS-NBS-49, *Transition Metal Oxides, Crystal Chemistry, Phase Transition, and Related Aspects,* Rao, C. N. R. and Subba Rao, G. V., 1974, SD Catalog No. C13.48:49, $1.70.

NSRDS-NBS-50, *Resonances in Electron Impact on Atoms and Diatomic Molecules,* Schulz, G. J., 1973, SD Catalog No. C13.48:50, $1.35.

NSRDS-NBS-51, *Selected Specific Rates of Reactions of Transients from Water in Aqueous Solution, II. Hydrogen Atom,* Anbar, M., Farhataziz, and Ross, A. B., SD Catalog No. C13.48:51, in press.

NSRDS-NBS-52, *Electronic Absorption and Internal and External Vibrational Data of Atomic and Molecular Ions Doped in Alkali Halide Crystals,* Jain, S. C., Warrier, A. V. R., and Agarwal, S. K., 1974, SD Catalog No. C13.48:52, $0.95.

NSRDS-NBS-53, *Crystal Structure Transformations in Inorganic Nitrites, Nitrates and Carbonates,* Rao, C. N. R., Prakash, B., and Natarajan, M., SD Catalog No. C13.48:53, in press.

NSRDS-NBS-54, *Radiolysis of Methanol: Product Yields, Rate Constants and Spectroscopic Parameters of Intermediates,* Baxendale, J. H. and Wardman, P., SD Catalog No. C13.48:54, in press.

NBS Tech. Note 825, *Properties of Selected Superconductive Materials – 1974 Supplement,* Roberts, B. W., 1974, SD Catalog No. C13.46:825, $1.25.

[a]Available from the JPCRD Reprint Service, American Chemical Society, 1155 18th Street, N.W. Room 604, Washington, D.C. 20036.

Index

INDEX

A

B

C

D

E

F

G

H

I

K

L

M

N

O

P

S

T

U

V

W